ASIMOV'S
CHRONOLOGY
OF SCIENCE AND
DISCOVERY

Over 4,000,000 Years of Science at a Glance

Scientific Inventions and Discoveries from the Dawn of Time to the Present

Begin reading time line at lower left.

4,000,000 B.C.	Bipedal species emerged
2,000,000 B.C.	Stone tools used
500,000 B.C.	Fire tamed
200,000 B.C.	*Homo sapiens neanderthalensis* emerged
30,000 B.C.	*Homo sapiens sapiens* became dominant
20,000 B.C.	Oil lamp invented Bow and arrow used
12,000 B.C.	Animals domesticated
8000 B.C.	Agriculture developed
7000 B.C.	Pottery used
6000 B.C.	Linen cords used for nets Rafts invented Sickles invented
5000 B.C.	Irrigation used Scales for measurement developed
4000 B.C.	Copper obtained from ore Sundial invented
3600 B.C.	Bronze discovered
3500 B.C.	Wheeled carts invented River boats used Writing developed
3100 B.C.	First nation united in Egypt
3000 B.C.	Candles used
2800 B.C.	Calendar used
2650 B.C.	Stone monuments built
2500 B.C.	Glass used
2340 B.C.	First empire established
2000 B.C.	Horses tamed
1800 B.C.	Number system based on 60 developed Seven-day week devised Five planets, twelve constellations of zodiac named Uses of fermentation discovered
1775 B.C.	First surviving law code established
1550 B.C.	First surviving account of medical remedies compiled
1500 B.C.	Alphabet developed
1375 B.C.	Monotheism espoused
1200 B.C.	Dyes resistant to sun and water developed
1100 B.C.	Phoenicians began exploring by sea
1000 B.C.	Steel developed; Iron Age began
750 B.C.	Arches used in building
700 B.C.	Aqueducts constructed First zoo established Sundial improved

640 B.C.	First library established; Coins used
585 B.C.	Solar eclipse predicted
580 B.C.	Water identified as element
520 B.C.	Irrational numbers defined
510 B.C.	First realistic map drawn
500 B.C.	First forays into Atlantic Ocean made; Human cadaver first dissected; Abacus used; Venus named
480 B.C.	Rational view of dreams proposed
440 B.C.	Notion of atoms maintained
420 B.C.	Physical cure sought for epilepsy
400 B.C.	Catapult devised
387 B.C.	Advanced schools founded
350 B.C.	Heliocentrism suggested; System of logic described; Arguments for spherical Earth summarized; Five elements suggested; Animals classified; First star map drawn
320 B.C.	First systematic book on botany written
312 B.C.	Paved roads built
300 B.C.	Geometry developed; Tides observed; Arteries and veins distinguished
280 B.C.	Parts of brain described; Size of Moon and Sun estimated; First major lighthouse built
270 B.C.	Waterclock developed
260 B.C.	Theory of lever action described
240 B.C.	Circumference of Earth estimated; Standard system of numbering years established
214 B.C.	Great Wall in China begun
170 B.C.	Parchment developed
150 B.C.	Distance of Moon calculated
134 B.C.	Star map improved
100 B.C.	Glass-blowing developed
85 B.C.	Waterwheel invented
46 B.C.	Julian calendar, with leap year, devised
25	Earth's climatic zones named
50	First important work on pharmacology written; Steam power used
105	Paper made
140	Geocentric universe described
180	Function of spinal cord studied
250	Algebra developed
300	Alchemy summarized; Metal stirrups made
400	Wheelbarrow invented
537	Windowed dome built
552	Silk production in West began
600	Moldboard plow invented
673	"Greek fire" invented
700	Porcelain made
750	Acetic acid studied
770	Horseshoes used
810	Zero added to numerical system
850	Coffee discovered
870	Arctic Circle crossed by sea
874	First permanent colony in Iceland founded
900	Horsecollar used
982	Greenland settled
1000	Labrador and Newfoundland discovered
1025	Science of optics founded
1050	Crossbow invented
1054	New star appeared
1066	Bright comet appeared
1071	Forks used
1137	Flying buttresses designed
1180	Windmills built in Europe; Magnetic compass invented
1194	Spitzbergen discovered
1202	Arabic numerals introduced to Europe
1228	Coal-mining begun in England
1241	Rudders used on European ships
1249	Eyeglasses invented; Gunpowder used in Europe
1252	Alfonsine planetary tables prepared
1269	Magnetic poles described
1291	Mirrors invented
1298	Far East explored; Spinning wheels introduced to Europe; Longbow invented
1300	Sulfuric acid discovered; Distilled liquor made
1304	Comet first painted realistically
1312	Canary Islands visited
1316	First book devoted to anatomy published
1335	Mechanical clocks invented
1346	Cannon invented
1403	First quarantine imposed
1405	Chinese explored Indian Ocean

1418	Madeira discovered
1427	Azores discovered
1436	Perspective used in art
1439	Artillery improved
1450	Harquebus invented
1451	Concave lenses developed
1454	Gutenberg Bibles printed
1472	Path of comet traced
1487	Cape of Good Hope discovered
1492	America discovered Magnetic declination noted
1495	First outbreak of syphilis occurred
1497	Portuguese reached India
1502	America recognized as new world
1504	First watch made
1513	"South Sea" (Pacific Ocean) discovered Florida discovered
1519	Mexico taken over by Spain
1523	Earth circumnavigated
1531	Peru claimed by Spain
1535	Cubic equations solved
1538	Position of comets' tails noted
1541	Mississippi River seen by Europeans
1542	Amazon River explored
1543	Mathematics of heliocentric system published Illustrated book of human anatomy published Scientific Revolution began
1545	Negative numbers described Techniques of surgery improved
1551	Trigonometric tables produced Prussian planetary tables prepared
1552	Eustachian tube described
1553	Northeast Passage attempted
1555	Homologies (similarities) among vertebrate skeletons described
1556	Science of mineralogy founded Tobacco introduced to Europe
1560	First scientific association founded
1565	Muskets used
1568	World maps based on Mercator projection found modern geography
1572	Supernova observed
1576	Greenland discovered by Europeans
1577	First astronomical observatory established
1578	Drake Strait discovered
1581	Pendulum movement studied Siberia explored
1582	Gregorian calendar adopted
1583	Science of hydrostatics founded
1586	Decimal fractions used
1589	Experiments on falling bodies done Cryptanalysis used in war Knitting machine invented
1590	Microscope invented
1591	Algebraic symbols used
1592	Thermometer invented Ruins of Pompeii discovered
1596	Dutch trade expanded to East Indies Value of pi to twenty decimal places obtained
1597	First chemical textbook written
1600	Magnetism of Earth proposed
1603	Vein valves studied
1607	Jamestown founded
1608	Telescope invented
1609	Planetary orbits described as ellipses Milky Way seen by telescope Moon seen by telescope
1610	Jupiter's moons discovered Venus seen by telescope Sunspots discovered
1612	Andromeda nebula discovered
1614	Table of logarithms published Human metabolism studied
1616	Baffin Bay discovered Tierra del Fuego circumnavigated
1620	Stagecoaches used Scientific method described
1621	Mathematics of refraction studied
1622	Slide rule invented
1624	Gases named and studied
1627	Rudolphine planetary tables published Last auroch died
1628	Principles of blood circulation published, founding science of physiology
1633	Galileo forced by Inquisition to renounce heliocentrism
1635	Observations of magnetic declination published
1637	Algebra and geometry combined to produce analytic geometry Fermat's Last Theorem proposed
1640	Coke produced from coal
1641	Cross hairs used in telescopes
1642	Knowledge of quinine reached Europe Adding machine invented South Pacific islands discovered
1643	Torricellian vacuum (first barometer) invented
1645	Air pump invented
1648	Air pressure related to altitude Pascal's principle of fluid pressure discovered

1650 — First double star detected; Age of Earth based on Bible proposed

1651 — Moon's craters named

1653 — Lymphatics discovered

1654 — Study of probability begun; Power of air pressure demonstrated

1656 — Saturn's ring and satellite discovered; Pendulum clock invented

1657 — Bodies falling in vacuum studied

1658 — Red blood corpuscles discovered

1659 — Syrtis Major discovered on Mars

1660 — Capillaries discovered; Static electricity demonstrated

1661 — Chemical element defined; Acid-base balance as basis of health proposed

1662 — Gas volume and pressure related by Boyle's law; Royal Society (scientific association) founded

1664 — Jupiter's Great Red Spot observed

1665 — Cells discovered; Study of light diffraction published; Planetary rotations measured

1666 — Light spectrum discovered

1668 — Law of conservation of momentum proposed; Spontaneous generation disproved; First reflecting telescope built

1669 — Calculus invented; Phosphorus discovered; View of fossils as remains of living things proposed; Double refraction described; Reaction of blood with air noted

1670 — Symptoms of diabetes recognized

1671 — Second satellite of Saturn discovered

1672 — Distance of Mars determined

1675 — Speed of light calculated; Separate rings around Saturn noted

1676 — Microorganisms seen with microscope

1678 — Catalog of southern stars published; Wave theory of light proposed

1679 — Pressure cooker invented

1680 — Bone-muscle mechanics described

1681 — Last dodo died

1682 — Sexual reproduction of plants described

1683 — Bacteria discovered

1684 — Size of Earth measured more precisely

1685 — Imaginary numbers defined

1686 — Study of winds, with meteorological map, published; Modern plant classification published

1687 — Three laws of motion published in Newton's *Principia*; Law of universal gravitation proposed; Earth as oblate spheroid with equatorial bulge proposed

1688 — Plate glass manufactured

1691 — Modern animal classification begun

1693 — Calculating machine invented; First mortality tables prepared

1698 — Steam-powered pump (Miner's Friend) invented; First ocean voyage for scientific investigation begun

1699 — Gas volume related to temperature

1700 — Uses of binary system proposed

1705 — Study of comets' orbits refined; return of Halley's comet predicted; Air's role in plant nourishment suggested

1706 — Carriage springs used; Static-electricity generator improved

1707 — Pulse watch invented

1709 — Coke used in iron-smelting

1710 — Pennsylvania rifle designed

1712 — Newcomen steam engine invented

1713 — News of smallpox inoculation spread

1714 — Mercury thermometer invented

1715 — Solar eclipse predicted and widely studied

1718 — Stellar motion determined

1722 — Easter Island discovered

1728 — Ship's chronometer invented; Aberration of light proposed; speed of light calculated more precisely; Bering Strait discovered; First book on dentistry published

1729 — Electrical conductors and insulators defined and studied

1733 — Achromatic lens invented; Blood pressure measured; Similarities between electricity and magnetism noted; Flying shuttle invented

1735 — Earth's equatorial bulge measured; Plant classification improved; modern system of taxonomy proposed; Description of winds improved

1736 — First book devoted to mechanics published

1737 — Cobalt discovered

1738 — Kinetic theory of gases proposed

1739 — Rocky Mountains seen by Europeans

1740 — Fresh-water hydra discovered

1742 — Celsius scale devised; Goldbach's conjecture proposed; Franklin stove invented

1744 Transcendental numbers defined

1745 Leyden jar invented
Iron found in blood

1747 Cure for scurvy found in diet

1748 Osmosis studied
Characteristics of platinum described

1749 Origin of Earth from collision of Sun with comet proposed
Speculations about biological evolution published

1751 Nickel discovered
Publication of first modern encyclopedia begun

1752 Lightning rod invented
Chemical action of digestion demonstrated
Heat as factor in creating Earth's surface proposed

1754 Quantitative analysis applied to study of carbon dioxide

1755 Concept of galaxies proposed

1756 Land bridges between continents suggested

1758 Halley's comet returned
Flame tests used in chemistry

1759 Study of tissues founded science of embryology

1760 Study of earthquakes founded science of seismology
Scientific study of heat begun
Science of pathology founded

1761 Transit of Venus studied
Percussion as means of medical diagnosis proposed

1762 Latent heat defined

1763 Pollination studied

1764 Steam engine improved

1765 Plutonism (volcanic action as factor in creating Earth's surface) proposed

1766 Hydrogen studied
Study of nerves founded science of neurology

1768 Spontaneous generation of microorganisms disproved
Australia discovered
Carbonated water invented

1769 Quantitative chemistry advanced
Spinning frame invented

1770 Source of Blue Nile discovered
Ocean currents studied
Water-soluble gases studied

1771 List of Messier objects published
Plants' use of carbon dioxide studied

1772 Quantitative studies of combustion made
Relationship between diamond and coal noted
Nitrogen discovered

1773 Antarctic Circle crossed
Bacilli and spirilla identified

1774 Oxygen discovered
Chlorine isolated
Psychosomatic illness treated by "Mesmerism"

1775 Action of digitalis described

1776 Classification of races made

1777 Torsion balance used for measurement

1778 Molybdenum discovered
Hawaiian Islands discovered

1779 Fertilization in humans studied
Process of photosynthesis further described

1780 Electrical stimulation of muscles studied

1781 Uranus discovered
Binary stars discovered
Science of crystallography founded
Axial inclination of Mars determined
Steam engine improved; key of Industrial Revolution

1782 Eclipsing variable stars discovered

1783 Movement of Sun in relation to stars studied
Respiration seen as form of combustion
First flight by hot-air balloon made
Tungsten discovered

1784 Effect of volcanic eruption on weather suggested
Bifocal lenses invented
Hydrogen and oxygen discovered to form water
Martian ice caps observed
Alaska settled by Europeans
Tellurium discovered

1785 Globular clusters observed
Number and density of stars in Galaxy estimated
Theory of uniformitarianism proposed

1786 First ascent of Mont Blanc, popularized mountain-climbing

1787 Relationship of volume to temperature of gas rediscovered
Standard system of chemical nomenclature proposed
First workable steamboat invented

1788 Algebra and calculus applied to mechanical problems
"Affinities" of chemicals tabulated

1789 Two more satellites of Saturn discovered
Acids found that did not contain oxygen
Law of conservation of mass proposed
Uranium discovered

1790 Industrial Revolution reached United States
Metric system devised

1791 Titanium discovered
Columbia River discovered

1793 Cotton gin invented
Humane treatment of insane introduced
Vancouver Island circumnavigated

1794 Rational explanation of meteorites offered
"Rare earth" discovered

1795 Development of food canning process begun

1796 Smallpox vaccination founded science of immunology
Nebular hypothesis proposed
Heptadecagon constructed

1797 Chromium discovered
Parachute used by human being

1798
- Mass of Earth calculated
- Science of comparative anatomy founded
- Malthusian analysis of population pressure published
- Ammonia liquefied
- Interchangeable parts for muskets produced
- Beryllium discovered

1799
- Law of definite proportions proposed
- Strata of rock studied
- Perturbations of Solar System studied
- Rosetta Stone discovered

1800
- Electric battery invented
- Water decomposed by electrolysis
- Infrared radiation discovered
- Gas lighting invented
- Nitrous oxide discovered
- Study of tissues published
- Malleable platinum invented

1801
- Jacquard loom invented
- First asteroid, Ceres, discovered
- Invertebrates classified, founding science of invertebrate zoology
- Ultraviolet radiation discovered
- Existence of light waves demonstrated
- Niobium discovered

1802
- More asteroids discovered
- Tantalum discovered

1803
- Concept of atomic weight proposed
- Existence of meteorites investigated and accepted
- Cerium, osmium, and iridium discovered

1804
- Ballooning used for scientific research
- Steam locomotive on railroad demonstrated
- Missouri River traced to source

1805
- Morphine discovered

1806
- First amino acid, asparagine, discovered

1807
- Sodium and potassium discovered
- Steamboat achieved commercial success

1808
- Polarized light discovered

1809
- Inheritance of acquired characteristics proposed
- Science of aerodynamics founded

1810
- Treatise on nervous system published
- Chlorine identified as element

1811
- Avogadro's hypothesis proposed
- Iodine discovered

1812
- Catalysis used to transform substances
- Theory of catastrophism proposed; science of paleontology founded
- Mechanistic universe proposed

1813
- System of chemical symbols proposed
- Encyclopedia of plant life begun

1814
- Spectral lines discovered

1815
- Plane of polarized light studied
- Organic radicals studied
- Prout's hypothesis proposed
- Road construction improved

1816
- Stethoscope invented

1817
- Chlorophyll isolated
- Cadmium, lithium and selenium discovered

1818
- Treatise on transverse light published
- Orbit of Encke's comet established
- Table of atomic weights published

1819
- Specific heat related to atomic weight
- First steamship crossed Atlantic

1820
- Phenomenon of electromagnetism established
- Amino acid glycine discovered
- Antarctic land sighted
- Diffraction gratings used to produce light spectra

1821
- Motion produced by electrical forces
- Seebeck effect demonstrated
- More extensive glacial past proposed

1822
- Heat flow studied
- Modern computer conceived
- Projective geometry developed
- Iguanodon (dinosaur) remains found
- Hieroglyphics deciphered

1823
- Hydrochloric acid discovered in stomach secretions
- Platinum used as catalyst
- Isomers discovered
- Chlorine liquefied by cold and pressure
- Electromagnet invented

1824
- Portland cement invented
- Steam engine efficiency studied; science of thermodynamics founded
- Distance of Sun measured more precisely
- Algebraic solution of quintic equations proved impossible
- Silicon discovered

1825
- Steam locomotive achieved commercial success
- Aluminum discovered
- Gastric digestion observed
- Candles made of fatty acids
- Lenses designed to correct astigmatism

1826
- System of non-Euclidean geometry published
- Bromine discovered

1827
- Ohm's law proposed
- First turbine built
- Screw propeller invented
- Mammalian ova studied
- Foods classified into carbohydrates, fats, and proteins
- Brownian motion described

1828
- Urea synthesized
- Presence of notochord noted in all vertebrate embryos
- Thorium discovered

1829
- Nicol prism invented

1830
- Achromatic microscope invented
- Group theory invented in mathematics
- Uniformitarianism accepted

1831
- Electric generator devised
- Electric motor invented
- Friction matches invented
- North magnetic pole discovered
- Cell nucleus recognized
- Rate of diffusion of gases related to molecular weight
- Chloroform discovered
- Theory of direction of cyclonic storms published

1832
Laws of electrolysis proposed

1833
Diastase discovered

1834
Mechanical reaper invented
Cellulose discovered

1835
Dry ice formed
Coriolis effect described
Revolver patented

1836
Pepsin discovered
Daniell cell invented

1837
Ice Age postulated
Photosynthesis related to chlorophyll
Trisecting an angle and doubling a cube proved impossible

1838
Distance of stars calculated
Cell theory proposed
Yeast cells studied
Protein named
Morse code invented

1839
Photography invented
Moon photographed
Vulcanized rubber invented
Fuel cell devised
Antarctica discovered
Ross Ice Shelf discovered
First bicycle designed
Lanthanum discovered

1840
Science of thermochemistry founded
Ozone discovered

1841
Hypnotism studied
Photographic negative invented
Needle gun invented
Standardization of screw threads suggested

1842
Chemical fertilizer invented
Doppler effect noted
Cranial index to classify humans proposed

1843
Mechanical equivalent of heat measured
Sunspot cycle discovered
Algebra of quaternions devised
Higher analytic geometry developed
Wheatstone bridge made popular
First transatlantic liner launched

1844
Message sent by telegraph
Dark companions of Sirius and Procyon hypothesized

1845
More asteroids discovered
Spiral nebulas discovered
Six gases remained "permanent" (nonliquefiable)

1846
Ether used as anesthetic
Neptune discovered
Planet Vulcan hypothesized
Effect of crystal asymmetry on polarized light studied
Protoplasm named
Cuneiform writings deciphered
Sewing machine patented

1847
Law of conservation of energy proposed
Sterilization of doctors' hands proposed
Anesthesia used in childbirth
Nitroglycerine invented
Study of symbolic logic founded
Silver fillings introduced

1848
Absolute scale of temperature proposed
Crab nebula discovered
Spectral line shift noted

1849
Speed of light calculated by earthbound experiment
Roche limit explained Saturn's rings
Nerve fibers shown to be outgrowths of cells

1850
Second law of thermodynamics established
Infrared radiation studied

1851
Earth's rotation demonstrated
More satellites of Saturn and Uranus discovered

1852
Joule-Thomson effect noted
Theory of valence published
Use of gyroscope as compass demonstrated
Effect of sunspots on Earth's magnetic field described
Elevator invented

1853
Age of Sun calculated
First glider flight made
Kerosene produced for use in lamps

1854
Cholera related to water supply
Telegraph Plateau on ocean floor discovered
New form of non-Euclidean geometry developed

1855
Lines of force given mathematical form
Geissler tube invented
Seismograph invented
Synthetic plastic invented

1856
Glycogen discovered
Blast furnace invented
Synthetic dye invented
Remains of Neanderthal man discovered
Process of pasteurization developed

1857
Composition of Saturn's rings proposed

1858
Theory of evolution by natural selection published
Organic molecular structure deciphered

1858
Science of cellular pathology founded
First successful refrigerator designed
Electricity in vacuum studied

1859
First oil well drilled
Storage battery invented
Distinct spectral lines found in each element
Solar flare observed
Mathematics of molecular movement in gases determined in detailed mathematical form

1860
Spontaneous generation finally disproved
Numerous organic molecules synthesized
Internal-combustion engine developed
Solar prominences discovered from photograph
Avogadro's hypothesis accepted
Theory of black bodies suggested

1861
Fossil of archeopteryx found
Function of Broca's convolution shown
Thallium discovered

1862
Germ theory of disease published
Dim companion of Sirius observed
Hydrogen discovered in Sun
Chloroplasts discovered
Source of White Nile found
Ironclad warships used
Machine gun invented
Hemoglobin named

1863
Characteristics of greenhouse effect described
Law of octaves suggested for classifying elements

1863
- Constitution of stars determined
- Barbituric acid discovered
- Indium discovered

1864
- Nature of Orion nebula determined

1865
- Mendelian laws of genetics published, founding science of genetics
- Benzene ring conceived
- Avogadro's number calculated
- Antiseptic surgery instituted
- Maxwell's equations begun
- Möbius strip invented
- Cylinder lock patented

1866
- Dynamite invented
- Clinical thermometer invented
- Kirkwood gaps discovered
- Nickel-iron core suggested for Earth
- Spectrum of nova studied

1867
- Dry cell invented
- Typewriter invented
- Law of mass action proposed

1868
- Air brake invented
- Helium discovered
- Skeletons of Cro-Magnon man found
- Deep-sea life discovered

1869
- Periodic table of elements published
- Nucleic acid discovered
- Critical temperature of gases suggested
- Science of biogeography founded
- Islets of Langerhans discovered in pancreas
- Celluloid patented
- Stock ticker invented

1870
- Ruins of Troy found

1871
- Theory of human evolution published
- Dry plates used in photography
- New plant strains developed

1872
- Epic of Gilgamesh uncovered
- Science of bacteriology founded
- Star's spectrum photographed
- Science of experimental psychology founded

1873
- More precise gas laws introduced
- Cause of leprosy identified
- Transcendental numbers defined
- Platelets discovered

1874
- Gallium discovered
- Tetrahedral carbon atom suggested
- Transfinite numbers defined
- Anomaly of electric current in crystals noted

1875
- Fertilization in sea urchin observed
- Radiometer invented

1876
- Telephone patented
- Four-stroke engine invented
- Principles of chemical thermodynamics published
- Bacteria cultured
- Cathode rays named

1877
- Molecular weights of proteins determined
- Oxygen liquefied
- Phonograph invented
- Martian "canals" detected
- Martian satellites observed

1878
- Enzymes named
- Varves (layers) of glacial sediment studied
- Ytterbium discovered

1879
- Electric light invented
- Origin of Moon suggested
- Saccharin synthesized
- Scandium discovered
- Thulium, holmium, and samarium discovered
- Temperature of a body related to radiation

1880
- Cause of malaria discovered
- Orion nebula photographed
- First modern seismograph devised
- Gadolinium discovered
- Causes of ailments found in unconscious mind
- Electromechanical calculator invented
- Cathode rays shown to be charged particles
- Materials studied under high pressures
- Piezoelectricity detected

1881
- Interferometer devised
- Anthrax inoculation made
- Pneumococcus isolated
- Venn diagram proposed

1882
- Chromatin studied in cell division
- Speed of light measured more accurately
- Diffraction gratings improved
- Cause of tuberculosis discovered
- Pi proved transcendental

1883
- First alloy steel patented
- Alternating current used
- Edison effect described
- Rayon invented
- Meiosis studied
- Function of phagocytes determined

1883
- Cause of diphtheria discovered
- Maxim gun invented
- Eugenics proposed

1884
- Science of statistical mechanics founded
- Theory of ionic dissociation proposed
- Sugar structure studied
- Cocaine used as local anesthetic
- Bacterial staining developed
- First successful steam turbine constructed
- Linotype machine patented
- Fountain pen invented

1885
- Rabies inoculation made
- Purines and pyrimidines isolated
- Praseodymium and neodymium separated
- Welsbach mantle invented
- Transformer invented
- Gasoline-powered automobile built
- Individuality of fingerprints established

1886
- Cheap way of making aluminum devised
- Germanium discovered
- Fluorine isolated
- Dysprosium discovered
- Canal rays discovered
- Raoult's law proposed
- Nitrogen-fixing bacteria discovered in plants

1887
- Michelson-Morley experiment failed
- Photoelectric effect observed
- Air resistence to high-speed objects studied
- Rubber tire invented

1888
- Radio waves detected
- Le Châtelier's principle stated

1895

X rays discovered

Cathode rays proved to be charged particles

1894

Argon discovered

1893

Theory of psychoanalysis proposed

Cause of cattle fever discovered

Wavelength related to absolute temperature

Mathematics of alternating current circuitry described

Arctic Ocean explored

1892

Fifth satellite of Jupiter discovered

Pressure of light measured

FitzGerald contraction suggested

Dewar flask invented

1891

Asteroids discovered by photography

Gravitational and inertial mass measured

Electron as fundamental unit of electricity suggested

Carborundum produced

Glider improved

1890

Tetanus antitoxin produced

Remains of Java man found

Spectroheliograph invented

Surgical gloves used

1889

Neuron theory proposed

Cause of tetanus isolated

Symbolic logic applied to mathematics

Energy of activation studied

Spectroscopic binaries discovered

Mercury's rotation studied

Cordite invented

Motion pictures invented

1888

Chromosomes named

Greenland ice cap crossed

Kodak camera invented

1895

Velocity related to mass

Helium discovered on Earth

Effect of heat on magnetism shown

Radio antennas invented

1896

Radiation from uranium discovered

Dietary deficiency disease identified

Zeeman effect of magnetism on light demonstrated

"Ferments" and enzymes proved identical

Science of architectural acoustics founded

1897

First subatomic particle (electron) discovered

Radiation traced to uranium atom

Alpha and beta rays identified

Nickel used as catalyst

Malaria related to mosquitoes

Oscilloscope invented

Largest refracting telescope built

Diesel engine invented

1898

Polonium and radium discovered

Neon, krypton, and xenon discovered

Hydrogen liquefied

Ninth satellite of Saturn discovered

First "Earth-grazing" asteroid discovered

Filterable virus discovered

Mitochondria identified

Epinephrine studied

First modern submarine launched

1899

Actinium discovered

Logic applied to geometry

Hydrogen solidified

1900

Quantum theory proposed

1903

First airplane flight made

Possibility of spaceflight proposed

Electrocardiogram invented

1902

Chromosomes related to genetic factors

First hormone, secretin, discovered

Laws of genetics applied to animals

Anaphylactic shock studied

Sutures developed

Radioactive series identified

Photoelectric effect on electron emission studied

Kennelly-Heaviside layer identified

Stratosphere and troposphere identified

Application of symbolic logic to mathematics extended without success

Ultramicroscope invented

1901

Radioactive energy measured

Radio invented

Europium discovered

Grignard reagents used

1900

Increase in mass with velocity measured

Beta particles determined to be electrons

Gamma rays discovered

Radon discovered

Radioactivity as change of one atom to another suggested

Electron emission from heated metals noted

Concept of mutations suggested

Blood types O, A, B, and AB distinguished

Cause of yellow fever discovered

Interpretation of Dreams published

Tryptophan isolated

Free radicals studied

Dirigible invented

Ruins of Knossos found

1904

Electronic rectifier (first radio tube) invented

Atomic structure proposed

Coenzymes discovered

Organic tracers used

Novocaine discovered

Star streams proposed

Jupiter's outer satellites discovered

1905

Theory of special relativity proposed

Law of conservation of mass-energy demonstrated

Photoelectric effect related to quantum theory

Brownian motion described in terms of molecular bombardment; size of atoms determined

Stellar luminosity related to absolute magnitude

Planetesimal hypothesis suggested

Metabolic intermediate isolated

Term *hormone* suggested

Linkage of characteristics on a single chromosome suggested

High-pressure apparatus invented

IQ test devised

1906

Radio waves converted to sound

Triode invented

Trojan asteroids discovered

Alpha particles related to helium

Characteristic X rays produced

Third law of thermodynamics proposed

Vitamin concept suggested

Place of magnesium in chlorophyll molecule determined

Technique of chromatography developed

Radioactivity in Earth's crust related to volcanic action

1907

Theory of space-time published

1907

- Radioactive dating suggested
- Lutetium discovered
- Structure of protein molecule revealed by synthetic peptides
- Chemotherapy developed
- Fruit flies used to study genetic inheritence
- Tissues cultured
- Conditioned response studied

1908

- Approximate size of atoms determined from observation
- Helium liquefied
- Geiger counter invented
- Sunspots shown to be subject to electromagnetic field
- Rickettsia discovered
- Assembly line invented
- Haber process developed

1909

- Compound effective against syphilis produced
- Carrier of typhus discovered
- Ribose identified
- Term *gene* suggested
- Tungsten filaments developed
- Bakelite invented
- Mohorovičić discontinuity noted
- North Pole reached

1910

- Neon light produced
- Sex-linked characteristics studied
- Attempt to relate mathematics to logic extended

1911

- Theory of nuclear atom proposed
- Cloud chamber perfected
- Electron charge calculated
- Cosmic rays detected
- Superconductivity observed

1911

- Chromosome map presented
- Tumor virus identified
- Earthquakes related to faults
- First practical seaplane built
- South Pole reached
- Electric starter developed for automobile

1912

- Luminosity of Cepheid variables related to period
- Velocity of Andromeda nebula measured
- Theory of continental drift proposed
- X-ray diffraction pattern studied
- Two varieties of neon atoms observed
- Concept of dipole moments proposed
- Term *vitamins* suggested
- Coal hydrogenation used to form gasoline

1913

- Concept of isotopes proposed
- Variations in atomic weight of lead measured
- Quantum theory applied to atom
- Coolidge tube invented
- Nitrogen used in light bulbs
- Stark effect demonstrated
- Distance of Magellanic Clouds determined
- Ozonosphere detected
- Vitamins A and B studied
- Michaelis-Menten equation derived
- Glycolysis explained

1914

- Atomic numbers assigned to elements
- X-ray wavelengths calculated
- Ionic structure of crystals proposed
- Beta particle energies studied

1914

- Proton studied and named
- Main Sequence described
- White dwarf discovered
- Jupiter's ninth satellite discovered
- Acetylcholine isolated
- Earth's mantle and core distinguished
- Behaviorism proposed

1915

- Pellagra shown to be diet-related
- Thyroxine isolated
- Bacteriophages discovered
- Elliptical electron orbits suggested
- Hydrogen-helium conversion hypothesized

1916

- General theory of relativity proposed
- Concept of black hole suggested
- Role of electrons in chemical bonding proposed
- Superheterodyne receiver invented

1917

- Expanding universe suggested
- Microcrystalline diffraction demonstrated
- 100-inch telescope first used
- Protactinium discovered
- Technique of sonar developed

1918

- Center of Galaxy estimated
- Spectra of stars classified
- Radioactive tracers used
- Development of embryo studied

1919

- Mass spectrometer invented
- Nuclear reaction produced by particle bombardment
- Deflection of starlight measured to test general relativity
- Bee communication studied

1920

- Stellar diameter measured
- Distance of Andromeda nebula debated
- Asteroid Hidalgo discovered
- Dendrochronology (dating by tree rings) proposed
- Climatic cycles suggested
- Liver proved to correct anemia
- Air masses detected in atmosphere

1921

- Insulin isolated
- Vagusstoffe (acetylcholine) proved chemical involved in nerve impulse
- Vitamin D shown to inhibit rickets
- Glutathione isolated
- Magnetron developed
- Tetraethyl lead added to gasoline
- Concepts of introvert and extrovert popularized
- Rorschach test introduced

1922

- Sumerian remains excavated
- Tutankhamen's tomb entered
- Vitamin E discovered
- Growth hormone discovered
- Lysozyme isolated
- Origin of life investigated
- Velocity of nerve impulses measured
- Theory of expanding Universe extended

1923

- Particle aspects of electromagnetic waves called photons
- Particles' associated matter-waves proposed
- Debye-Huckel equations developed
- Acid-base pairs suggested
- Coenzyme structure determined
- Cepheids observed in Andromeda
- Hafnium discovered
- Ultracentrifuge developed

1924

- Australopithecine skull found
- Bose-Einstein statistics developed
- Ionosphere detected
- Cytochromes discovered
- Effects of irradiation determined

1925

- Concept of packing fraction proposed
- Exclusion principle suggested
- Fourth quantum number interpreted as particle spin
- Matrix mechanics developed
- Magnetism as means of reaching absolute zero suggested
- Gravitational red shift detected
- Rhenium discovered
- Morphine synthesized
- Parathormone isolated
- Cytochromes found to contain iron

1926

- Wave mechanics developed
- Wave packets suggested
- Fermi-Dirac statistics developed
- Galactic rotation proposed
- First liquid-fuel rocket fired
- Enzyme crystallized
- Pernicious anemia proved dietary deficiency disease

1927

- Uncertainty principle proposed
- Electron diffraction demonstrated
- Speed of light measured more accurately
- Theory of cosmic egg proposed
- Quantum mechanics applied to electron bonds
- Molar of Peking man discovered
- Fruit-fly mutations produced by X rays
- M and N blood groups discovered
- Talking pictures invented

1928

- Penicillin discovered
- Diels-Alder reaction produced
- Raman spectra described
- Game theory devised
- Hexuronic acid isolated

1929

- Receding galaxies proposed
- Composition of sun determined
- Source of solar energy suggested
- Coincidence counter devised
- Particle accelerator invented
- Isotopes of oxygen discovered
- Deoxyribose identified
- Structure of heme identified
- Estrone isolated
- Intrinsic factor hypothesized
- Electroencephalography developed
- Cardiac catheter invented
- Four stages of children's mental growth described

1930

- Pluto discovered
- Moon's surface temperature measured
- Coronagraph invented
- Schmidt camera invented
- Presence of interstellar matter proposed
- Antimatter hypothesized
- Cyclotron developed
- Computer capable of solving differential equations produced
- Pepsin crystallized
- Structure of vitamin A elucidated
- Chlorofluorocarbons synthesized

1931

- Gödel's proof advanced
- Existence of neutrino proposed
- Deuterium discovered
- Concept of resonance proposed
- Androsterone isolated
- Neoprene invented
- Nylon invented
- Size of viruses determined
- Viruses cultured
- Stratospheric balloons used

1932

- Neutron discovered
- Proton-neutron nucleus proposed
- Positron discovered
- Nuclear reaction produced with particle accelerator
- Radio waves from space detected
- Electron miscroscope invented
- Prontosil found to cure streptococcus infections
- Ascorbic acid isolated
- Urea cycle identified
- Polaroid invented
- Quinacrine developed to combat malaria

1933

- Vitamin C synthesized
- Molecular beams studied
- Absolute zero approached more closely

1934

- Uranium bombarded with neutrons
- Weak interaction proposed
- Artificial radioactivity created
- Cherenkov radiation observed
- Existence of supernovas suggested
- Existence of neutron star proposed
- Progesterone isolated
- Bathysphere invented

1935

- Uranium-235 discovered
- Isotopic tracers used
- Virus crystallized
- Strong interaction proposed
- Sulfanilamide isolated
- Riboflavin synthesized
- Cortisone isolated
- Prostaglandin isolated
- Radar developed
- Imprinting described
- Richter scale established

1936

- Neutron absorption described
- Thiamine synthesized
- Perfusion pump used

1937

- Technetium created
- Muon discovered
- Electrophoresis developed
- Electron microscope outperformed optical microscope
- Field-emission microscope invented
- Radio telescope built
- Ribonucleic acid found in virus
- Citric acid cycle discovered
- Pellagra cured by niacin
- Vaccine against yellow fever developed
- Mutation related to evolution

1938

- Mechanism of solar energy source described
- Technique of magnetic resonance developed
- Vitamin E synthesized
- Phase-contrast microscope invented
- First practical TV camera (iconoscope) built
- Xerography invented
- Ballpoint pen invented
- Coelacanth found

1939

- Nuclear fission discovered
- Nuclear bomb proposed
- Francium discovered
- Possibilities of neutron star analyzed
- Magnetic moment of neutron calculated
- Vitamin K synthesized
- Rh factor discovered

1939

- Antibacterial agent in penicillin isolated
- Tyrothricin isolated
- First essential mineral, zinc, identified
- DDT introduced as pesticide
- First flight by helicopter made
- Frequency modulation devised

1940

- Neptunium and plutonium discovered
- Uranium hexafluoride used to prepare enriched uranium
- Astatine discovered
- Betatron devised
- Streptomycin discovered
- System of color television devised

1941

- High-energy phosphate identified
- Polarimetry perfected
- Cardiac catheterization introduced
- Distance of Sun measured more accurately
- First jet plane flown
- Function of gene deduced

1942

- First nuclear reactor demonstrated
- Biotin synthesized
- Structure of bacteriophage seen in photograph

1943

- Adrenocorticotrophic hormone isolated
- Lysergic acid diethylamide produced
- Seyfert galaxy discovered
- Aqualung invented

1944

- DNA identified as genetic material
- Paper chromatography invented
- Teflon produced for commercial use

1944

- Quinine synthesized
- 2,4-D introduced as herbicide
- New nebular hypothesis proposed
- Theory of radio wave emission from hydrogen proposed
- Americium and curium created
- V-2 missiles fired

1945

- Nuclear fission ("atomic") bomb detonated
- Synchrocyclotron devised
- Promethium discovered
- Viral mutations demonstrated
- Jet streams studied
- Artificial kidney designed

1946

- First practical electronic digital computer (ENIAC) devised
- Distance of Moon determined by microwave reflection
- Technique of nuclear magnetic resonance developed
- Noradrenaline identified as transmitting chemical of sympathetic nerves
- Sexual reproduction in bacteria identified
- Complex reproduction of viruses identified
- Cloud-seeding demonstrated

1947

- Pion discovered
- Carbon-14 dating developed
- Crab nebula found to be radio source
- Martian atmosphere analyzed
- Coenzyme A isolated
- Chloramphenicol isolated
- Theory of holography proposed
- Land camera invented
- First supersonic flight made
- Television introduced to homes

1948

- Transistor invented
- Long-playing record invented
- Theory of cybernetics published
- Nuclear shell numbers proposed
- Theory of quantum electro-dynamics proposed
- "Big bang" theory described
- Fifth satellite of Uranus discovered
- Structure of nucleic acid molecules studied
- Cyanocobalamine discovered
- Cortisone found to relieve arthritis
- First tetracycline used
- Tissue transplantation studied
- Virus culture improved
- Starch chromatography introduced
- Bathyscaphe invented

1949

- Asteroid, Icarus, discovered
- Second satellite of Neptune discovered
- Atomic clock invented
- Berkelium and californium produced
- First Soviet fission bomb exploded
- Cause of sickle-cell anemia determined
- Embryonic immunological tolerance demonstrated
- Essential amino acids identified
- "Dirty snowball" theory of cometary structure suggested

1950

- Existence of cometary cloud suggested
- Pluto's diameter measured
- Turing machine invented
- Chess-playing computer suggested
- Endoplasmic reticulum discovered
- Carbon-14 used as tracer

1951

- Breeder reactor developed
- Stellarator constructed
- Hydrogen radiation from outer space detected
- Spiral arms of Milky Way Galaxy deduced
- Twelfth satellite of Jupiter discovered
- Theory of superconductivity proposed
- UNIVAC designed
- Steroids synthesized
- Acetylcoenzyme A isolated
- Fluoridation of water begun

1952

- Nuclear fusion ("hydrogen") bomb exploded
- Einsteinium and fermium detected
- Kaons and hyperons discovered
- Experiment on origin of life attempted
- X-ray diffraction studies of DNA made
- Structure of insulin identified
- Nucleic acid of viruses studied
- Nerve growth factor identified
- Radioimmune assay devised
- REM sleep identified
- First tranquilizer, reserpine, studied
- Gas chromatography developed
- Zone refining introduced

1953

- Double helix suggested as DNA structure
- Isotactic polymers developed
- Theory of plate tectonics proposed
- Bubble chamber invented
- Strange particles studied
- Maser developed
- Heart-lung machine used
- Transistorized hearing aids introduced
- Spray cans produced more cheaply

1954

Salk vaccine introduced
First successful kidney transplant performed
First controlled fission reactor built for electric power
Oxytocin synthesized
Chloroplasts isolated
Strychnine synthesized
Multinucleotide genetic code suggested
Photovoltaic cells developed
Robotic device patented
Bevatron built
Oral contraceptive proved effective
Plastic contact lenses produced

1955

Active galaxies suggested
Birth of stars witnessed
Radio waves from Jupiter detected
Pluto's rotation determined
Antiprotons detected
Medelevium discovered
Diamonds synthesized
Field ion microscope devised
Nucleic acids formed with help from catalyzing enzymes
Structure of cyanocobalamin determined

1956

Antineutrinos detected
Law of conservation of parity proposed
Antineutron hypothesized
Continuous maser devised
High temperature of Venus deduced
Ribosomes found to be site of protein manufacture
Transfer RNA discovered
Structure of pituitary hormones determined

1957

Sputnik sent into orbit
Jodrell Bank radio telescope built

1957

Details of photosynthesis described
Giggerellins studied in West
Interferon discovered
Sabin vaccine used against polio
Compact pacemaker devised
Esaki tunnel diodes invented
Borazon formed

1958

Mössbauer effect discovered
Solar X rays detected
Magnetosphere detected
Nobelium formed
Photocopying perfected

1959

Moon probes launched
Shape of Earth determined by satellite
Existence of solar wind verified
Structure of hemoglobin molecule determined
Remains of *Homo habilis* found
Spark chamber constructed
New theory of color vision proposed

1960

First laser constructed
General relativity supported by experiment
Standard meter defined
Integrated circuits constructed
Resonance particles detected
Sea-floor spreading suggested
Weather satellites launched
Structure of cyclic-AMP determined
Chlorophyll synthesized

1961

First human being sent into space
Microwave reflections received from Venus

1961

Heliosphere detected
Existence of quarks postulated
Lawrencium discovered
Genetic code further deciphered
Gene regulators suggested
Electronic watches introduced

1962

First American placed in orbit
Communications satellite launched
Venus probe launched
Rotation of Venus determined
Noble gas compounds formed
Absolute zero approached more closely
First practical light-emitting diode produced
Effect of pesticides on environment publicized

1963

Quasars studied
Largest single radio telescope first used
Rockets used to detect X rays from space
Hydroxyl groups detected in space
First woman placed in orbit
Magnetic reversals observed in sea-floor sediments

1964

Background microwave radiation detected
Omega-minus particle discovered
CPT symmetry suggested
Transfer RNA structure determined
First multiperson spaceflight launched
Rutherfordium formed

1965

Martian craters photographed

1965

Rotation of Mercury determined
Spacewalks begun
First space rendezvous performed
First communications satellite for commercial use launched
Venus probe reached Venus
First holograms produced
Microfossils identified
Proteins synthesized

1966

Moon's surface mapped
First space-docking performed

1967

Pulsars discovered
Atmosphere of Venus analyzed
First space casualties suffered
Successful heart transplant performed
Clone of vertebrate produced
Hahnium formed

1968

Electroweak interaction proposed
Solar neutrinos detected
Water and ammonia molecules detected in space, founding science of astrochemistry
Pulsars identified as rotating neutron stars
Moon circumnavigated

1969

Human beings landed on the Moon
Cosmonauts transferred between spaceships in flight
Optical pulsar detected
Meteorites found on Antarctic ice cap
Protein structure further elucidated

1969

Artificial heart implanted

Technique of coronary bypass developed

1970

Black hole evaporation proposed

Meteoritic amino acids studied

Genelike molecule synthesized

Technique of recombinant DNA developed

Reverse transcriptase discovered

Megavitamin therapy proposed

Fiber optics developed

Scanning electron microscope built

Instruments soft-landed on Venus

Astronauts escaped from Apollo XIII

Supersonic transport (SST) introduced

1971

Surface of Mars mapped

Moon rocks brought back to Earth

First space station placed in orbit

Black hole indirectly observed

Mini-black holes suggested

Pocket calculator introduced

1972

Vitamin B-12 synthesized

Punctuated evolution suggested

Speed of light measured more precisely

Earth resources satellite launched

Science of quantum chromodynamics founded

CAT scan introduced

Laser disks introduced

1973

Jupiter probe reached Jupiter

Skylab launched

Origin of Universe as quantum fluctuation proposed

Genetic engineering begun

Proton decay suggested

1974

Surface of Mercury mapped

New theory of Moon's origin proposed

Thirteenth satellite of Jupiter discovered

Potential of chlorofluorocarbons to destroy ozone layer proposed

Tauon produced

Charmed quarks produced

1975

Microchips produced

Surface of Venus photographed

Endorphins discovered

1976

Experiments made to detect life on Mars

Pluto's surface deduced from light spectrum

Synthetic gene placed in living cell

String theory proposed

1977

Rings of Uranus detected

Farthest asteroid, Chiron, discovered

Inflationary universe hypothesized

Pulsar discovered in Vela nebula

Deep-sea life not dependent on photosynthesis discovered

Remains of bipedal hominid (Lucy) discovered

Nonbacterial DNA studied

Last case of smallpox recorded; first case of AIDS discovered

Fiber optics used in experimental telephone setups

Balloon angioplasty developed

1978

Surface of Venus mapped by radar

Satellite of Pluto discovered

Oncogenes produced

Virus genome determined

Test-tube baby born

1979

Jupiter's satellites observed by probes

Extinction of dinosaurs by comet proposed

Evidence of gluons observed

1980

Satellites and rings of Saturn observed by probes

Experiments suggest mass of neutrino

1981

Space shuttle launched

Unsymmetrical rings of Neptune suggested

1982

Millisecond pulsar discovered

Magnetic monopole detected

Jarvik heart implanted

Laser printers introduced

1983

Electroweak theory confirmed

Extrasolar planets suggested

Nuclear winter described

1984

DNA analysis applied to human evolution

Brown dwarf reported

1985

Hole in ozone layer detected

Surfaces of Pluto and Charon determined

1986

Uranus explored by probe

Halley's comet returned

1987

Supernova in Large Magellanic Cloud studied

Higher-temperature superconductivity obtained in ceramic substances

1988

Distant galaxies detected

Shroud of Turin proved fake by carbon-14 dating

Greenhouse effect appeared to intensify

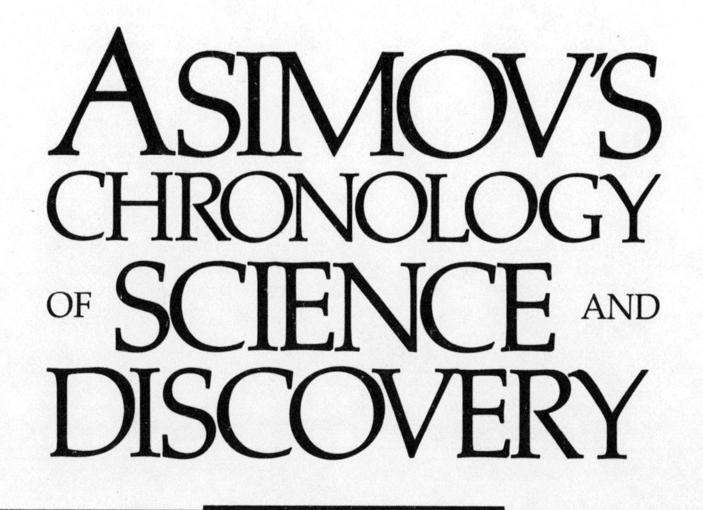
ASIMOV'S
CHRONOLOGY
OF SCIENCE AND
DISCOVERY

ALSO BY ISAAC ASIMOV

Asimov's New Guide to Science

Asimov's Biographical Encyclopedia of Science and Technology

The History of Physics

Asimov's Guide to the Bible (2 volumes)

Asimov's Guide to Shakespeare (2 volumes)

Isaac Asimov's Treasury of Humor

Asimov's Annotated Don Juan

Asimov's Annotated Paradise Lost

The Annotated Gulliver's Travels

Asimov's Annotated Gilbert and Sullivan

ISAAC ASIMOV

HARPER & ROW, PUBLISHERS, New York

Grand Rapids, Philadelphia, St. Louis, San Francisco

London, Singapore, Sydney, Tokyo, Toronto

 For information address Harper & Row, Publishers, Inc., 10 East 53rd Street, New York, NY 10022.

FIRST EDITION

Designed by Alma Orenstein

Library of Congress Cataloging-in-Publication Data

Asimov, Isaac, 1920–
[Chronology of science and discovery]
Asimov's chronology of science and discovery / Isaac Asimov. — 1st ed.
p. cm.
Includes index.ry.
ISBN 0-06-015612-0
1. Science—History. 2. Science—Social aspects. 3. Inventions—History. I. Title. II. Title: Chronology of science and discovery.
Q125.A765 1989
509—dc20 89-45024

---

91 92 93 DT/RRD 10 9 8 7 6 5 4 3 2

To Robert Esposito

Reader Extraordinaire

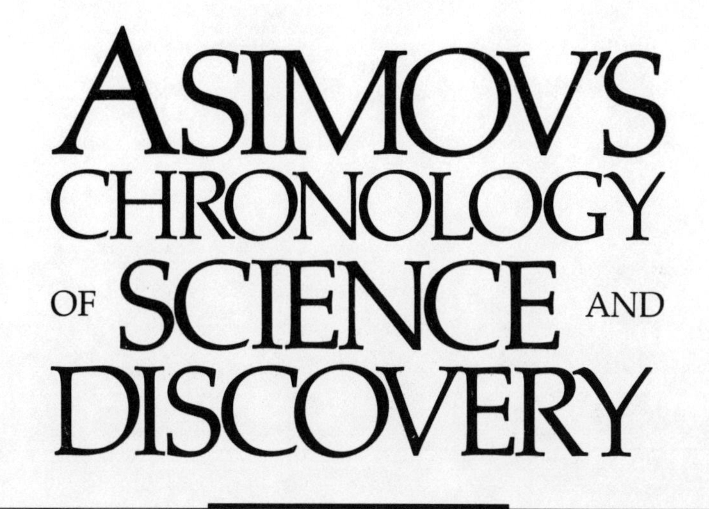
ASIMOV'S
CHRONOLOGY
OF SCIENCE AND
DISCOVERY

4,000,000 B.C.

BIPEDALITY

The first human advance was biological. It consisted of becoming human.

We might ask what makes a human being human. What part is sufficiently *human* so that we can at once point to it and say, "This is human. Without it, this organism would be something else."

The human being of today has, of course, evolved many characteristics that are now considered human, so many that it is difficult to point to any one of them and consider it the key. What we must do, then, is go back in time, watching humanity become more and more primitive, less human, more ape-like.

Yet we must stop at some point where our ancestors are still more nearly human than they are apish. Any organism that is more human than ape is called a *hominid* (from a Latin word for "man"). Any organism that is more ape than human is called a *pongid* (from a Congolese word for "ape").

Therefore, we might recast the first sentence of this book to read "The first hominid advance was biological. It consisted of becoming hominid."

As we go back in time, studying the bones and teeth (all that remain) of earlier forms of hominid life, we come to an organism that was perhaps the size of a modern chimpanzee or even a little smaller and with a brain that was no larger. Yet in one crucial respect, it was much closer to the human than to the ape. This human characteristic is so obvious that, if we were to see the organism in real life, we would at once say, "This is no ape."

It was the first hominid, and what made it so was its bipedality. It walked on two legs, as we can tell from the shape of its spinal column, its pelvic girdle, and its thighbones.

The fact that human beings walk on two legs strikes us as characteristically human. We are *bipeds* (from Latin words for "two legs"), while other mammals are *quadrupeds* (four legs).

Of course birds walk, run, or hop on two legs, and the Greek philosopher Plato (*ca.* 427–*ca.* 347 B.C.) therefore defined a human being as a "featherless biped." That is insufficient, however, for there are bipeds with fur (kangaroos and jerboas) and bipeds with scales (various dinosaurs), which Plato knew nothing about.

Let's consider bipedality, then, to see what makes human bipedality different from other types.

Animals that are bipedal are often restricted to two legs because two others have been devoted to some other (and preferred) form of locomotion. Most birds are designed to be flyers, and the forelegs have become wings for that purpose. Penguins are swimmers, and the forelegs have become flippers. In either case, walking, running, or hopping is secondary.

There are, of course, nonflying birds like ostriches, for which walking or running on two legs is the only means of locomotion. In such cases, the body is designed for it and is essentially horizontal, so that there are roughly equal amounts of it at front and rear. With the two legs centered, bipedality is a mechanically easy thing to maintain. This is also true of bipedal reptiles and mammals like the tyrannosaur and the kangaroo. Long tails provide balance, and the body remains essentially horizontal.

Suppose, though, a quadruped's body

ends at the hips and there is no tail to act as a balance. In that case, the only way the body's center of gravity can be brought above the hind legs is to tip the entire body into a vertical position.

Some tailless animals actually do this. Bears and chimpanzees can stand upright on their hind legs and even walk about in this fashion, but they are clearly uncomfortable doing so and prefer to let the forelegs share the work. Penguins, too, hold their bodies upright, but they are swimmers and clumsy on land. Though they can walk long distances if they must, they prefer to belly-whop on ice when they can.

The human being, then, is a tailless, habitual, and *comfortable* biped. But what is it that makes the human a comfortable walker on two legs?

It is that the spinal column, just above the pelvis, bends backward in human beings, assuming a shallow S-shape and adding a spring to the human walk that makes it comfortable. No other organism has that backward bend of the spine in the small of the back. (Bipedality also produces problems. There are slipped disks, inflamed sinuses, and accidental falls, for instance. None of these would be likely were it not that human beings are still not entirely adapted to walking upright.)

The earliest hominids were first identified by an Australian-born South African anthropologist, Raymond Arthur Dart (1893–1988), to whom a skull, rather human-looking except for its extraordinarily small size, was brought from a South African limestone quarry in 1924. Dart, in 1925, named the type of organism to which the skull belonged *Australopithecus* (from Greek words meaning "southern ape"). Further finds made it clear that it was not an ape but a hominid, and at least four different species have now been identified, which are lumped together as *australopithecines*.

In 1974 an American anthropologist, Donald Johanson, unearthed an unprecedentedly complete and ancient skeleton of an australopithecine female, who was given the name Lucy. (You can tell a female from a male by the shape of its pelvis.) It was possibly as much as four million years old judging from the age of the rocks in which it was found.

Lucy is an example of *Australopithecus afarensis,* because Afars is the name of the region in east-central Africa where her remains were found. Australopithecines existed only in eastern and southern Africa, so east-central Africa may have been the cradle of humanity.

Lucy was the size of a chimpanzee and slighter in build. Her australopithecine relatives seem to have ranged between 3 and 4 feet in height and to have weighed perhaps 65 pounds. Their brains were no larger than a chimpanzee's and about a quarter the size of our own.

The early australopithecines probably lived much as chimpanzees do, may have spent their time partly in trees, must have been very largely vegetarian, and undoubtedly could not speak. However, they were as bipedal as we are and walked as easily and as comfortably on their hind legs as we do.

Why did the australopithecines develop the backward bend in the spine? Why did the processes of evolution invent the hominid, in other words?

Four million years ago the Earth had been warm for quite a while, and large tropical animals, such as elephants, rhinoceroses, and hippopotamuses, tended to lose their hair because such insulation kept them too warm. For some reason, the hominids also lost their hair though they were much smaller than other hairless mammals. We don't know at what stage hair was lost.

The Earth of the australopithecines was turning cooler, however. The forests shrank and were replaced by grasslands. Those organisms whose habitat was the forest and who did not give up the trees naturally retreated with the forest.

Some arboreal prehominids managed to adapt to the grasslands in east-central Africa and to spend more and more time out of the trees. It must have been a difficult transition. As they spent more time on the ground, they had an increasing tendency to rise to their hind legs and peer about over the tall grass, searching for food or watching out for predators. Those who could stand upright more easily and could do so for longer intervals were better able to survive.

Even a slight crook in the spine, which made standing upright slightly easier, gave those who had it a better chance of surviving, and of having children to inherit that crook. What we call *natural selection* would thus drive the prehominids toward bipedality and a true hominid character.

Bipedality had side effects that were also beneficial and that reinforced the natural-selection drive. The forelimbs were freed for duty as something other than support. The freed hands could more easily manipulate portions of the environment, feel them, and bring them close to the eyes, ears, and nose, so that the brain was continually flooded with sensations.

Any change that made the brain a bit larger or more complex made it possible for the brain to handle the flood of sensation more efficiently, and this led to an improved chance of survival. Thus, natural selection introduced a drive for bigger and better brains.

The early australopithecines, with a brain as large as that of a chimpanzee but with a body that was slighter, already had a larger brain-to-body ratio than any pongid then or since. Since the brain-to-body ratio is a crucial factor in what we call intelligence (provided the brain is reasonably large), the australopithecines were probably the most intelligent land animals existing in their time.

2,000,000 B.C.

STONE TOOLS

Sometimes, we think of the human being as a tool-using animal. Tool-using is not really unique to human beings, however. For example, sea otters routinely smash shellfish against a rock they carry for the purpose on their abdomens as they float belly-up. A long litany of other examples could also be given.

If we change that to tool-*making* animal, we are somewhat better off, but not entirely. Chimpanzees have been seen to strip leaves from twigs and then use the bare twigs as devices for capturing termites, which to them are tasty snacks.

Undoubtedly, australopithecines could do anything that chimpanzees could do. We have no evidence, but we can be fairly certain that they used tree branches and long bones as clubs. They could surely throw rocks or use them as sea otters do.

The australopithecines may have existed on Earth for three million years, not becoming finally extinct till as late as 1,000,000 B.C. For the final third of their existence, though, they were no longer the only hominids. Some australopithecines had evolved to the point where they had become sufficiently human to be placed in the same genus with ourselves.

About two million years ago, in other words, genus *Homo* came into existence. For a time, it coexisted with the australopithecines, but there was bound to be conflict between them, and in this, the larger and brainier hominids won out, contributing (very largely, perhaps) to the extinction of the australopithecines.

In the 1960s, the English anthropologist Louis Seymour Bazett Leakey (1903–1972), his wife, Mary, and their son, Jonathan, located remains of the oldest of the genus *Homo* in the Olduvai Gorge, in what is now Tanzania. The hominids thus uncovered were named *Homo habilis* (Latin for "able man," because with them were discovered objects that seemed to indicate that they made simple stone tools).

Homo habilis was smaller than some of the larger species of australopithecines. When, in the summer of 1986, fossil remains of *Homo habilis* were discovered that were some 1.8 million years old (the first time that both skull fragments and limb bones of the same individual of that species had been located), they seemed to represent a small, light adult about 3½ feet tall and with arms that were surprisingly long.

Though members of *Homo habilis* may have been small, they had more rounded heads than any of the australopithecines and larger brains, nearly half as large as those of modern human beings. They had thinner skull bones, and based on the brain configuration, if they could not talk, they could at least make a greater variety of sounds. Their hands were more like our modern hands, and their feet were completely modern. Their jaws were less massive, so that their faces looked less apelike.

These creatures apparently used stone tools to chip pieces of flint and so create a sharp edge. This meant that for the first time, hominids had sharp edges in quantity and did not have to depend on the chance finding of one. Moreover the edges could be made truly sharp, and the sharpness could be renewed when the rock was blunted.

These stone knives increased the food supply. *Homo habilis* could not tear away at the tough hides of animals, as fanged predators such as the various cats, dogs, and bears could. Without knives, hominids had to make do with carcasses that had already been mangled by other predators and to make off with what they could scavenge.

With knives, however, *Homo habilis* had artificial fangs that could slit hides and scrape the meat off hides and bones. What's more, *Homo habilis* no longer had to scavenge. Hominids could now kill animals on their own, even fairly large animals. And once they caught on to the trick of tying stone axes to wooden branches and creating the first crude spears, they could stab animals at a distance. If the spears were thrown, the distance could be made long enough to obviate immediate retaliation.

Hominids became hunters and killed off the competing australopithecines, no doubt, so that for the last million years, all hominids, without exception, have been part of genus *Homo*.

500,000 B.C.

FIRE

By 1,600,000 B.C., *Homo habilis* was gone. For one thing, it had evolved into a new species, *Homo erectus,* with members about as large and as heavy as modern human beings. If any examples of *Homo habilis* existed after the new species had established itself, they didn't last long.

Between 1,000,000 B.C. and 300,000 B.C., *Homo erectus* was the only hominid in existence. It was the first hominid that could, in some cases, attain a height of as much as 6 feet and weigh over 150 pounds. The brain was larger, too, sometimes three-fourths as heavy as that of modern human beings.

Homo erectus made much better stone tools than had been made before. As hunters, its members were capable of taking on the biggest animals they could find. They were the first hominids who could hunt mammoths successfully.

Homo erectus made two particularly enormous advances.

All the hominids, for as much as three and a half million years, had been confined to the southeastern half of Africa. *Homo erectus* was the first hominid to expand that range significantly. By 500,000 B.C. *Homo erectus* had moved into the rest of Africa, into Europe and Asia, and even into the Indonesian islands.

In fact, the first discoveries of the remains of *Homo erectus* were made on the Indonesian island of Java, where the Dutch anthropologist Marie Eugene Dubois (1858–1940) discovered a skullcap, a femur, and two teeth in 1894. No hominid with so small a brain had yet been discovered, and Dubois named it *Pithecanthropus erectus* (from Greek words meaning "erect ape-man").

Similar finds were made near Peking, beginning in 1927, by a Canadian anthropologist, Davidson Black (1884–1934). He named his hominid *Sinanthropus pekinensis* (Greek for "Chinaman from Peking").

Eventually it was recognized that both sets of remains, along with others, were of the same species and deserved to belong to genus *Homo.* Dubois's term *erectus* was kept, even though hominids had been walking erect for at least two and a half million years before *Homo erectus* had evolved. This was not known in Dubois's time, however.

By the time *Homo erectus* had evolved, the Earth was in a glacial period. When the glaciers were at their peak, they withdrew so much water from the sea that the sea level dropped as much as 300 feet, exposing the continental shelves under the shallow sections. It was this that made it possible for *Homo erectus* to wander from the Asian continent to the Indonesian islands.

The cold weather enforced new habits. *Homo erectus* traveled in bands, as earlier hominids must have, but made shelters by building up stones to break the wind, or by suspending hides from a central pole. These were the first crude homes. Where caves existed, *Homo erectus* found shelter there. The first traces of *Homo erectus* in Asia (Black's findings near Peking) were in a filled-up cave.

The cave near Peking had traces of campfires. This meant that fire had been "discovered" some five hundred thousand years ago. Here is a characteristic that marks off human beings from all other organisms.

Every human society in existence, however primitive, has understood and made use of fire. No living creature other than human beings uses fire in even the most primitive fashion.

I have put *discovered* in quotation marks, for fire was not discovered in the usual sense. Lightning could start a fire ever since Earth's atmosphere gained enough oxygen to sustain one and Earth's land surface possessed a forest cover that could burn, and that means for some four hundred million years. From that fire, then as now, any animal capable of fleeing would flee.

By the discovery of fire, then, we really mean its taming. At some time, *Homo erectus* learned how to retrieve some burning object from the edge of a naturally caused fire, keep the fire alive by feeding it judicious amounts of fuel whenever it showed signs of dying down, and make good use of it.

How this happened we don't know. My own guess is that it began when children grew fascinated by leaping flames. With their overactive curiosity, and with the lack of bitter experience of what happens when one is burnt, they might be more likely to play with it than adults would be. Presumably, the nearest adult would snatch the child away from the fire and stamp it out. On the other hand, the time might come when an adult, more venturesome than most, might see the advantage of continuing the play in a more purposeful manner.

The use of fire changed human life completely. For one thing, it gave light in the dark and warmth at any time. This made it possible to extend activity into nighttime and winter, which would be particularly important in a glacial period for it meant *Homo erectus* could extend its range into the cooler regions.

Of course with fire alone one is condemned in cold weather to hug the hearth, but a hunting society could easily get the notion of scraping an animal hide clean and wrapping oneself in it. In this way, animal fur would replace the hair human beings had lost.

Fire was also useful as a means of protection from other animals, even the fiercest. A fire in a cave or within a circle of stones would keep the predators away. They might snarl and slink about the outskirts, but if they were not intelligent enough to stay away from fire, one experience of its nearness was all they would need. In fact, *Homo erectus* could now carry burning branches to scare game and send it stampeding into traps and off cliffs.

Then, too, fire made it possible to cook food. This is more important than might appear. Meat is tenderer and tastier if roasted. What's more, roasting kills parasites and bacteria so that the meat is safer to eat. Eventually fire made plant food, otherwise inedible, supremely edible. Try eating rice fresh from the stalk, or uncooked grain of any kind, and you'll understand what a little application of heat from a fire can do.

Finally, fire made possible various chemical changes in inanimate matter, such as the smelting of metals. In short, fire introduced humanity's first age of comparatively "high-tech."

To begin with, of course, fire could be obtained only after it had been started by natural means. Once one had a fire, it had to be kept burning continuously, and if it ever died out, the search for another fire had to be instituted at once. If there was not a nearby tribe from which fire could be obtained (assuming they were friendly enough to grant it —and they probably would be, for it might be their turn next) then it would be necessary to wait for a natural fire again and hope that conditions would be such that some could be taken safely.

The time came, however, when techniques were developed for starting a fire where none had been before. This could be done by friction: by turning a pointed stick in a depression in another stick, a depression that contained very dry shreds of wood, leaves, or fungus (tinder). The heat of friction might eventually ignite the tinder. We don't know when such methods were first developed, but the technique for *starting* a fire represents another enormous step forward.

200,000 B.C.

RELIGION

By 200,000 B.C., the last individuals who might be considered members of *Homo erectus* were dead and the species was extinct. By then, though, some had evolved into hominids whose brains were every bit as large as ours, although they were somewhat differently proportioned, less massive in front and more massive in the rear. They had first appeared not long before this time and were probably instrumental in bringing about the end of the older species.

The first traces of such hominids were discovered in 1856, in the valley of the Neander River in western Germany. Neander Valley is *Neanderthal* in German, and the skeletal remains were referred to as *Neanderthal man* at first, or simply *Neanderthals*.

These were the first hominids to be discovered. They were clearly different from modern human beings. Their skulls were distinctly less human than our own. They had pronounced eyebrow ridges, large teeth, protruding jaws, and receding foreheads and chins.

Because they were the first hominids to be discovered, and because the Western world still believed strongly that the Earth was only a few thousand years old (as the Bible seemed to imply), there was a certain reluctance to accept the Neanderthal bones as remains of an *early* form of *Homo sapiens*. Some preferred to consider them remains of ordinary members of *Homo sapiens* who were suffering from some sort of bone disease or other abnormality.

When more examples of Neanderthal skeletons were found and all had the same kind of skulls, however, the notion of abnormality could not be maintained. The French anthropologist Paul Broca (1824–1880) marshalled the arguments in favor of Neanderthals being a more primitive form of life than we ourselves, and that turned the tide.

The formal name for the Neanderthals was at first *Homo neanderthalensis,* but they were so like us in all but a few details of the skull that they were finally recognized as being of our species. And why not? There is evidence that they may have interbred with human beings of the present type. They are now referred to as *Homo sapiens neanderthalensis* and are considered one of the two known subspecies of *Homo sapiens*. We moderns are the other.

The Neanderthals lived from 200,000 B.C. to 30,000 B.C. in Africa and Eurasia. They lived during glacial times and hunted the mammoth, the woolly rhinoceros, and the giant cave bear. Their stone tools were greater in variety and more delicate and precise than had ever been seen before. They definitely knew how to start fires.

They were the first hominids to bury their dead. Earlier hominids, like animals generally, simply left their dead lying where they had fallen so that they were scavenged by predators and what was left over rotted. The fact that Neanderthals buried their dead, thus preserving them from scavengers, if not from decay bacteria, tends to show that they valued life somehow, felt affection, and cared for individuals. Sometimes the dead were old and crippled and could only have lived as long as they did with the loving help of others of the tribe.

What's more, food and flowers were often buried with the corpse, and this seems to indicate that Neanderthals felt life continued on an individual basis after death. If they felt that there was life after death, then this might indicate the first stirrings of what we can call religion—a feeling that there is more to the Universe than is apparent to the senses.

20,000 B.C.

ART

Sometime after 50,000 B.C., a variety of Neanderthal existed with less pronounced eyebrow ridges, even in male adults, with a high forehead and a distinct chin, and with smaller teeth. In short, this was the kind of hominid, quite exactly, that *we* are. We are *Homo sapiens sapiens,* and we are sometimes referred to as *modern man,* though *modern human beings* is more appropriate terminology to make it clear that men, women, and children are meant and not only men.

Between 50,000 B.C. and 30,000 B.C., the two varieties of *Homo sapiens* coexisted, but by the latter date, some interbreeding and probably a great deal of slaughter had put an end to the Neanderthals, so that for the last thirty thousand years or so, all living hominids have been of the modern type.

Modern human beings were extremely successful. For the first time, they extended the range beyond where *Homo erectus* had left it. Between 40,000 B.C. and 30,000 B.C., human beings took advantage of the existence of land bridges that the fall in sea level had produced. They entered Australia from southeastern Asia, and North America from northeastern Asia. No hominids had existed in either continent before. They also found their way to the Japanese islands.

The new lands were steadily overrun, and by 10,000 B.C. human beings had reached the southernmost part of South America, and even Tierra del Fuego, the island to the south of that continent. All the continental areas except Antarctica and the glaciated areas in the north were settled.

Human beings were hunters, of course, and they had developed rituals to improve their success. One, apparently, was to draw pictures of animals being successfully hunted, in the conviction, perhaps, that life would imitate art, or that the spirits that animated animals would be mollified in this way and would cooperate.

In 1879, a Spanish archaeologist, Marcellino de Sautuola (d. 1888), was excavating Altamira Cave in northern Spain, when his twelve-year-old daughter, who was with him, spied paintings on the ceiling and cried out "Bulls! Bulls!" There were paintings of bison, deer, and other animals in red and black, drawn perhaps as long ago as 20,000 B.C.

The drawings spoke highly of the artists' skill. If anything were needed to show that early human beings were our intellectual equals, this would do. In the last twenty millennia, we have gained enormous knowledge and experience, but we are not one whit more human than those ancient cave artists.

So excellent was the art, in fact, that many people refused to believe it was truly ancient. Many felt it to be a fraud of some sort, a modern hoax. It was only with the finding of other caves and other cave paintings that the art was finally accepted as ancient.

The cave paintings were found in remote places and were invisible except by artificial light, so that it is believed they were drawn for religious and ritualistic purposes rather than for display. Nevertheless, they are clearly the result of infinite pains, and it is hard to believe that the artists did not derive joy from their work.

BOWS AND ARROWS

In some of the early art, there are clear depictions of bows and arrows being used. How old the bow and arrow are is uncertain, but they were in use by 20,000 B.C. at least.

The bow and arrow is an important device because it is the first one invented by human beings in which energy is slowly stored and then released all at once. It made possible attack from a greater distance than with a thrown spear and so was the first truly long-distance weapon. The value of attacking an infuriated animal much larger than yourself from as great a distance as possible is clear.

Bows and arrows were eventually used by humans against other humans (as has been true of any object capable of inflicting damage, no matter what the purpose for which it might originally have been designed). The bow remained a prime weapon in warfare right down to the beginning of the fifteenth century.

OIL LAMPS

A campfire of burning wood or brush gives light, but it is not portable. It cannot give light precisely where that light might be needed. Wood, however, is not the only possible fuel. People when roasting meat might well have noticed that the fat dripped and flamed.

A smaller fire, then, in more concentrated form, might be obtained by dipping some porous wood in oil and setting it ablaze, thus forming a torch. Still more convenient would be some oil in a container (a hollowed-out stone, for instance) with a wick of some plant fiber inserted into it. The oil would soak into the wick and would burn at its end. It could be carried from place to place to suit one's needs.

There are indications that primitive lamps of this sort may have been in use as long ago as 20,000 B.C.

12,000 B.C.

ANIMAL DOMESTICATION

In the 1950s, fossil remains of dogs were found along with remains of humans in caves near Kirkuk in what is now northern Iraq. They dated from about 12,000 B.C.

How dogs came to be domesticated is not known, of course. My own guess is that, again, children were responsible. A child could form a close bond with a puppy that had been found abandoned, or that was left over when the mother was killed either in self-defense or for food. Once the bond was formed, the child would object strenuously to the use of the puppy as food, and parents might oblige.

It would quickly have turned out that dogs, being hunters and pack animals, will accept a human master as the pack leader. Dogs would go hunting with their masters, help in killing the game, wait for the human beings to take what they wanted, and be satisfied to be thrown a minor share.

In this way human beings, for the first time, obtained the services of another species of animal.

By 10,000 B.C., another step had been taken, with the domestication of goats in the Middle East. The goats would be cared for, fed, and encouraged to reproduce. They would supply milk, butter, and cheese and, by dint of judicious culling, meat. What's more, since goats eat grass and other substances humans find inedible, the food supply was increased at no cost. (Dogs have to be given food that would otherwise fill human stomachs.)

Until then, people had found their food by hunting and gathering, with all the insecurities attendant on that. Herding provided a much more secure food supply.

IN ADDITION

At this time the glaciers were beginning to recede.

8000 B.C.

AGRICULTURE

Human beings led a nomadic life. As long as hunting was a major source of food, they had to be prepared to follow migrating herds of animals. Even if they lived on plants and on nonmigrating animals, a tribe lingering in one place too long would consume the available food and have to move elsewhere for fresh foraging.

Even when human beings became herders, they remained nomads. The herds had to be taken to fresh pastures now and then, either because of seasonal changes or because of overgrazing.

About 8000 B.C., however, in the same region where animals were first domesticated, something new arrived, heralding a change greater than any since the first use of fire.

What it amounted to was that plants were domesticated. Somehow it occurred to human beings to plant seeds deliberately, to wait for them to grow, to water them, and to wait for them to ripen while destroying competing plants. Then the plants could be harvested and would serve as food.

It was tedious and back-breaking work, but the net result was that a great deal of food could be obtained, far more than by hunting and gathering, or even herding, since plant life is more copious than animal life.

The coming of herding and agriculture, particularly agriculture, meant that a given area of land could support a larger population than before. There was less starvation, more children survived, and the population increased.

Agriculture began in northern Iraq where wheat and barley grew wild, and it was these that were domesticated. The kernels of grain could be ground into flour that could be stored for months on end without spoiling and could be baked into a tasty and nutritious bread.

Despite the increase in the food supply, farmers must have been very conscious of their labor, a kind of slavery foisted upon them, which even the use of animals didn't much mitigate. The tale of the Garden of Eden in the Bible may be the product of farmers who remembered with nostalgia a kind of "golden age" when human beings hunted and gathered in freedom and comparative idleness and wondered what had happened to evict them from this Elysium and force them to earn their bread by the sweat of their brow.

Then too, Adam's first two sons were depicted as Abel, a herder, and Cain, a farmer. The farmers increased in population faster than the herders did, and we can well imagine that areas devoted to farming spread and steadily took up space that had earlier been freely used by herders. (The same thing happened in the American West, when the farmers settled the land and fenced in their plots to the discomfiture of the nomadic cowboys.) No wonder the Bible pictures Cain as killing Abel.

Agriculture, for the first time, condemned human beings to a sedentary existence. Once a farm was established, there could be no further wandering. The farmers had to remain with the farm, which was fixed in one place.

A sedentary life had its dangers. As long as human beings hunted and gathered, or even herded, danger could be avoided. If a hungry, marauding tribe approached, intent on taking what food they could find, the tribe already present, if they decided fighting would be too dangerous, could always flee.

Farmers could not run, at least not without abandoning their farms and seeing their lifework ruined and themselves faced with starvation. Once the population had increased, thanks to agriculture, they could not possibly find enough food to maintain themselves except by continuing with agriculture—they had a tiger by the tail.

Farmers therefore had to be prepared to fight at whatever cost, and they gathered together for mutual self-protection. They would find a site on an elevation (so that they could throw missiles downward, whereas an enemy would have to throw them upward, with lesser effect) and with a secure water supply (you can go without food for a period of time but not without water). There they would build houses and surround them with a protective wall. The result would be a city, and the inhabitants would be city-dwellers, or *citizens*.

In northern Iraq, for instance, near the site where herding and agriculture developed, are the remains of a very ancient city, founded perhaps in 8000 B.C., at a site called Jarmo. It is a low mound into which, beginning in 1948, the American archaeologist Robert J. Braidwood dug carefully. He found the remains of houses built with thin walls of packed mud and divided into small rooms. The city may have held no more than a hundred to three hundred people, but cities grew rapidly larger.

Agriculture made it possible for farmers to produce more food than their own families needed. This meant that people could do things other than farming—for example, engage in artisanry or art—and trade their products for some of a farmer's excess food. For the first time, human beings could find time to think of something other than the next meal. In addition, living in close quarters in a city, they could interact easily, and the innovations and ideas of one could be rapidly transmitted to the others.

As a result, the coming of agriculture and of cities meant the coming also of a new and more complex way of life, which we call *civilization* (from a Latin word for "city-dweller"). The civilized area was small at first, but it spread outward steadily until it now occupies virtually the entire world.

IN ADDITION

By now, the glaciers had receded and the Earth's climate had become more or less what it is today. The Arctic shores became open to human habitation, and Inuit (Eskimos), Lapps, and Siberians began to occupy them. The sea level rose to present heights, cutting off the Americas and Australia from Asia, and they remained in isolation for nearly ten thousand years.

The human population of Earth may have been no more than 3 million altogether in 10,000 B.C. (and it may have been only 2 million in Neanderthal times). With the coming of herding, it increased, and was 5 million in 8000 B.C. With agriculture it continued to increase.

7000 B.C.

POTTERY

It has always been important for human beings to carry things, and the obvious way to carry them is in the hands, or in the crook of the arm. There is a limit to how much can be carried in this fashion, however. What we needed were artificial hands, so to speak, that were considerably larger than our natural hands.

Objects could be carried in hides, but hides are an inconvenient shape and heavy. Gourds might do, but they have to be taken as they come. Eventually human beings learned to weave twigs or other fibers into baskets. These were light and could be made in any shape.

Baskets, however, were only useful in carrying solid, dry objects made up of particles considerably larger than the interstices of the weave. Baskets could not be used to carry flour or olive oil, for instance, or most important, water.

It might seem natural to daub baskets with clay, which upon drying would cake the holes and make the basket solid. The dried mud would tend to fall away, however, especially if the basket was shaken or struck. But if the basket was placed in the sun and allowed to bake in direct sunlight, the mud would harden further, and the basket might then become fairly serviceable for carrying powders and fluids.

But then why use the basket? Why not simply begin with clay, mold a container out of it, and let it dry in the sun? You would then have a pot made of crude earthenware, and some of these may have been formed as early as 9000 B.C. Such pots are soft, however, and don't last long.

Some stronger heat was required. When earthenware was placed in the fire, it became hard *pottery,* and such pottery can be traced back to perhaps 7000 B.C. This may have been the first use of fire for something other than light, heat, and cooking.

Pottery not only made it possible to carry liquids, it also introduced a new form of cooking. Until then, food had usually been roasted, exposed directly to the flames or to dry heat. Once a pot existed that could hold water and withstand the heat of flames, food could be heated in the water—it could be boiled. In this way stews and casseroles came into existence.

And of course pottery could be decorated and well shaped. Cleverly decorated examples would be in special demand. Artisans could exchange them for other material they found needful. And since pottery lasts indefinitely, if well cared for, it can change hands often, and one group of people can use it in trading with another group.

In early pottery, the clay was pressed and pounded into the shape of a pot and the result was something quite lumpy and asymmetric but serviceable.

If the pot could be turned, however, a relatively light pressure from the hand would produce a symmetrical cylindrical shape, and by appropriate increases in pressure or by downward pushes, complicated modifications of the basic cylinder could be made while retaining symmetry. This could be done if the clay was placed on a horizontal, circular slab of wood or stone (a *potter's wheel),* which had a central spike underneath that could be balanced in a depression and the whole turned rapidly.

The potter's wheel was one of the first examples of the use of a wheel and one of the first uses of rotary motion. We don't know when it was first used, but it may have led to the idea of *wheel* generally, and to wheeled transportation.

IN ADDITION

Jericho, in what is now Israel, may have been the largest city in the world at this time, with a population of 2,500.

6000 B.C.

LINEN

Flax produces a fiber that can be interwoven much as twigs or bark might be interwoven to form a basket. To make a strong thread, a number of flax fibers would be twisted together. We call the result *linen.* (Like the word *linen,* the word *line* comes from the word for "flax.")

The first use of linen, perhaps as early as 6000 B.C., was to produce linen cords that could be used in fishing. Interweaving these cords produced nets.

Eventually very fine nets were made—in other words, *cloth,* or *textiles* (from a Latin word for "weaving"). The formation of cloth from linen and eventually from other plant and animal fibers, such as cotton and wool, revolutionized clothing. Until then, furred hides had been worn. These were all right in cold temperatures but too hot at other times. They weren't porous, they were heavy, and they smelled.

Textiles, on the other hand, were light, flexible, porous, and could be easily cleaned. They have remained the preferred material for clothing ever since.

RAFTS

Human beings could not avoid bodies of water, especially since fresh water was needed for drinking. They would cluster near rivers and lakes for that purpose.

Water would offer them a source of food, too, and they would venture out into it to catch fish. People would learn to swim. In addition, they could not help but notice that wood floats. By 6000 B.C., they must have learned how to lash logs together to form rafts that would keep them out on the surface of quiet bodies of water for a time. By paddling with their hands, if nothing else, they could even cross small stretches of water.

SICKLES

Human beings had to invent devices that would be of help in dealing with the plants they were trying to exploit. The ripe stalks of grain would have to be cut down, and as early as 6000 B.C., *sickles* (from a Latin word meaning "to cut") were used for that purpose. They were essentially knives (originally of sharpened stone) at the end of sticks, which could be used to slash at the base of the stalks.

Once the stalks were cut down, the grain could be rubbed between two stones to eliminate the chaff and reduce the starch to powder. One stone would have a depression into which the grain could be poured, and the other would be rounded and could be used for grinding, impelled by muscle power. Such a hand-mill is called a *quern.*

IN ADDITION

About this time, the powerful wild ox (the "unicorn" of the Bible) was tamed and gave rise to the domestic cattle of today.

5000 B.C.

IRRIGATION

Agriculture requires a steady supply of water to keep the plants alive, so it naturally started where rainfall could be counted upon to a reasonable extent. Rainfall, however, even at best, is a chancy thing, and droughts are all too common.

One reliable source of fresh water (the salt water of the sea won't work) is a sizable river. Farms therefore began to develop along riverbanks.

Whereas rain falls directly on the crops, however, river water generally remains within its banks. In order to correct that situation, it is necessary to dig ditches so that water will flow outward from the river and soak the ground in which the plants are growing. These ditches have to be kept in order and not allowed either to silt up or to overflow.

What's more, when the river's water level falls during a drought, the ditches have to be dug deeper. And since the river occasionally rises in periods of heavier-than-usual rains (not necessarily at the site of the farms, but many miles away, nearer the river's source), levees have to be built to confine the rising water within the banks. These have to be tended continually to prevent leaks and breakdowns.

All this *irrigation* (from Latin words meaning "to water inward") more or less guarantees a good harvest and an ample food supply, at the cost of unremitting labor.

This labor cannot be done alone, nor can it be done by various people, each in his or her own way and at his and her own time. Irrigation requires cooperation. Many farms depend upon it, and the labor of many has to be supervised to make a coherent whole, so that the levees are in good shape everywhere.

As a result, the farms depend upon control by capable leaders who can supervise the work and allot the tasks, encourage the industrious and capable, and punish the idle or incompetent. In short, irrigation leads to the formation of what we call *government,* so that a cluster of farms surrounding a defensible city becomes a *city-state,* with a ruler and established rules of behavior.

The first such city-states formed along the lower courses of the Euphrates and Tigris rivers in what is now southern Iraq (but was then known as Sumeria) about 5000 B.C. Other city-states developed at nearly the same time along the Nile River in Egypt. It almost never rains in Egypt, but the Nile remains a reliable water supply and overflows regularly once a year when the rainy season takes place far to the south, nearer its source. The Nile flood deposits fertile mud over the farms on its banks.

SCALES

Trade is bound to lead to measurement—so much of this for so much of that. You can heft things by hand, but that is subjective and buyer and seller will never agree. The easiest way to be objective is to hang two pans from opposite ends of a rod that is held up in the middle. The thing being weighed is placed in one pan, and standard weights are placed in the other until the two pans are in balance. The principle is so simple and the device itself so easy to make that it may have been used as early as 5000 B.C. in Egypt and have been reasonably accurate.

4000 B.C.

COPPER

From the early days of *Homo habilis* to 4000 B.C. or thereabouts, a period of two million years, tools and weapons were made of stone, wood, or bone. Stone is the most durable of these and the most likely to remain as evidence of long-past human activity. As a result, that long period is known as the *Stone Age,* a term first used by the Roman poet Titus Lucretius Carus (95–55 B.C.) and reintroduced by a Danish archaeologist, Christian Jürgensen Thomsen (1788–1865), in 1834.

The Stone Age is divided into the *Paleolithic,* the *Mesolithic,* and the *Neolithic* (from Latin words meaning "Old Stone," "Middle Stone," and "New Stone," respectively) based on advancing techniques of handling the stone.

Occasionally, though, pebbles that were not like other pebbles must have been found by Stone Age people. For one thing, these occasional odd pebbles were shiny and were heavier than ordinary pebbles of that size. What's more, if these shiny pebbles were struck with a stone hammer, they did not split or shatter as ordinary pebbles did, but they distorted.

These occasional pebbles were *metals.* Dozens of different metals are known, but most of them remain in combination with nonmetallic substances, and rocky substances are the result. Only those metals that are inert and tend not to combine with other substances are likely to be found in a free state. The three inert metals most likely to be found free are rather rare even for metals. These are copper, silver, and gold. Their rareness is shown by the fact that the very word *metal* is from a Greek term meaning "to search for."

Metal nuggets exist that were dealt with by human beings as long ago as 5000 B.C. or even earlier. Because of their metallic luster and the fact that they can be beaten into interesting shapes, they were used at first almost exclusively as ornaments. Of these, gold was the most desired, for it was the most beautiful in color (a gleaming yellow), the heaviest, and the most inert. It simply doesn't change with time. Silver, a very light yellow, tends to darken with time, and copper, which is reddish, may even turn green. (*Copper* is from the word for Cyprus, the island that served as an early source of the metal.)

It was only when human beings discovered that metals could be obtained from special rocks called *ores* that metals became common enough to use for other purposes. Of these, the first to be discovered was copper. Copper is combined with oxygen, carbon, or both in certain rocky ores, and the discovery that copper could be obtained from them in pure form took place about 4000 B.C.

Undoubtedly it was accidental at first. A fierce wood fire might be built on such copper ore. Under the heat of the fire, the carbon in the wood and in the ore would combine with the oxygen in the ore to form the gas carbon dioxide, which would escape, leaving metallic copper behind. Some observant person might notice the reddish globules among the ashes of the fire, and eventually the circumstances would be understood and the ores searched for and deliberately heated. In this way fire made possible *metallurgy*—the obtaining of metals from their ores.

Copper ornaments became more common thereafter, but copper could not be used as a tool, though one might think it could. After all, a sharp-edged piece of rock, if chipped, loses its edge, and this cannot be restored without laborious chipping. If a sharp-edged piece of metal is blunted, it can simply and easily be beaten sharp again. However, copper was too easily blunted. It couldn't very well be beaten after every minor use.

SUNDIALS

From very early times, people could count the days as a measure of time, but they often desired to measure parts of days as well. One way of doing so was to note the progress of the Sun from east to west. The motion seems to take place at a steady speed.

Of course, one can't stare at the Sun, but it is the easiest thing in the world to thrust a stick into the ground and watch the shadow. At sunrise, the shadow extends far to the west. Then, as the day progresses, it grows shorter and shorter, swinging somewhat to the north. It reaches its shortest extent (northward) at midday, lengthening toward the east as sunset approaches.

This tool is most likely to have been developed in Egypt, where the sunshine is constant, and Egyptians may have divided the day into twelve equal *hours* as early as 4000 B.C.

IN ADDITION

City-states had spread to a third river, the Indus, in what is now Pakistan. This *Indus civilization* remained unknown to moderns till excavations at the site of Mohenjo-Daro began in 1922.

3600 B.C.

BRONZE

Copper obtained from some ores is harder than from others. The reason is that copper ore is not necessarily pure; it may be mixed with other substances that, on being heated, combine with copper to form an *alloy.*

One such mixture consists of copper and arsenic, but arsenic is poisonous, and people who worked with it must have fallen sick. Such mixed ores were therefore abandoned (perhaps the first known case in which worker safety was a factor in technology).

Fortunately, another type of ore mixture was discovered that also resulted in the smelting of a hard form of copper. This was a tin ore, and the hard copper was actually a copper-tin alloy. The alloy was called *bronze* (possibly from a Persian word for "copper").

Bronze was hard enough to compete with rock. It could hold an edge better and could, of course, be beaten back into shape if necessary, though that was not often required.

Increasingly bronze came to be used for tools and for weapons and armor, too. By 3000 B.C., the Middle East was in the *Bronze Age,* and this spread outward slowly in all directions as the methods of copper-smelting and bronze formation diffused.

The great cultural product of the Bronze

Age was Homer's *Iliad*, the tale of the Trojan War (fought about 1200 B.C.), in which both Greek and Trojan heroes fought in bronze armor, carried bronze shields, and struck out with bronze swords and bronze-tipped spears.

3500 B.C.

WHEELED CARTS

Where objects are too heavy to be carried by hand, transportation over land becomes a problem. Even where land is fairly smooth, there is considerable friction, whether the ground is sandy, pebbly, or grassy.

Heavy objects had to be dragged on sledges by sheer force at first, and even when animals stronger than humans were used (oxen, for instance), it was slow going.

It could be made easier if crude rollers in the form of wooden logs were placed under sledges. The rollers turned rather than dragged, and that cut down on friction considerably. It meant less work, but might actually take more time, as rollers had to be picked up from the back and put down again in front. What is needed is an *axle and wheels*.

We don't know how someone came to think of attaching two rollers to the sledge back and front in such a way that they rolled inside the straps that held them and remained with the sledge at all times. At the end of each roller, solid wooden wheels were then placed to lift the sledge off the ground, and the wheels turned freely.

A wheeled cart moves more quickly and with far less effort than a sledge, even a sledge on rollers, so that such carts marked a revolution in land transportation. They made trade easier, for one thing.

Such carts had showed up in Sumeria by 3500 B.C.

RIVER BOATS

It is certainly easier to carry heavy loads over water than over land. Water offers much less friction than land does, and there are no permanent unevennesses in it: no rocks, no ridges, no uphill stretches.

In this connection, the Nile was ideal. Not only was it a source of water for rainless Egypt—not only did its annual flood periodically fertilize the soil—but its current was gentle and there were no storms. The Nile did not damage and overturn boats as the unruly Tigris in Sumeria did. (The very name of that river is the Greek word for "tiger.")

What's more, the Nile flows almost due north, while the wind is almost always from the north. Therefore a boat can move smoothly down river and, when the time comes to return, a sail can be hoisted to catch the wind and the ship will be blown upstream.

Egypt is not a forest nation, but it had luxuriant stands of reeds (called *papyrus*) along the river in those days, and the reeds could be used in bundles to build boats. The boats were built to form hollows, so that they would displace more water and carry more weight without sinking. These reed boats were not particularly sturdy, but the gentle Nile did not require sturdy boats. (When Moses was set afloat in the Nile River, he was placed in a little boat—or *ark*—of bulrushes; that is, papyrus.)

Egypt thus had a most convenient form of

communication that perhaps accounted for its slowness to adopt the Sumerian wheeled cart. Egypt had scarcely any need for land transportation.

By 3500 B.C. Egyptian boats had begun to ply the Nile, and by 3000 B.C. they were venturing out of the Nile into the Mediterranean, hugging the shore, and making their way past Sinai and Canaan to Lebanon. There they obtained the tree-trunks they lacked in Egypt and brought them back for construction purposes.

WRITING

Because Sumeria was the most advanced and complex civilization in the world, life was more complex too. People had to keep track of the grain they produced, how much they traded, what they bought and sold, and what they contributed to the common fund (in what we call *taxes*).

It became more and more difficult to keep it all in mind. We can say this without casting aspersions on the extraordinary powers of a trained memory. It was necessary to keep score.

Almost anyone can think of making some sort of mark in the ground to stand for a basket full of fruit and then eventually counting the marks in order to know how many baskets have been delivered. It's just that if society is simple, there is no need to take the trouble to do this.

However, as the memory becomes strained, such marks are made. In order to simplify matters, different marks would be made for one, five, and ten, perhaps so that there would not be too many unit-marks to count. This would lead on to a mark that meant *fruit* and another that meant *grain* and still another that meant *man* and so on. Eventually there would be marks, perhaps crude pictures at first, for every different object. If different people agree on the marks and learn how to make them and how to interpret them, you have *writing*.

The Sumerians seem to have been the first to work out a writing system, using a stylus that produced wedge-shaped marks in soft clay. This was later called *cuneiform*, from Greek words meaning "wedge-shaped." As time went on, the symbols became more stylized and somewhat simpler and lost their function as pictorial representations. Nevertheless, each symbol stood for a word, more or less, so that anyone who wanted to read and write had to memorize hundreds and even thousands of different symbols.

This meant that the literates (those who could read and write) were always a small minority of the population, but that was all that was needed in those days to run the society.

Egypt picked up the notion of writing and invented a totally different set of signs, as complicated in its way as the cuneiform. The Egyptian writing is called *hieroglyphic* (from Greek words meaning "priestly writing," because the Greeks came into contact with it mostly in Egyptian temples). It was brushed onto thin flexible sheets of papyrus pith.

Writing makes an enormous difference. It is a kind of frozen speech. Thoughts and records remain much more permanent than the spoken word and, if carefully copied now and then, persist indefinitely and remain more precise than the memory of the spoken word usually is.

This means that each generation can learn, more precisely and quickly, the accumulated experience and wisdom of the previous generation, and advances quicken as a result.

Furthermore, the records kept in writing give us a reasonably exact version of events that took place in the past, complete with names, places, and details. In order to understand what went on in a society without writing, we must try to interpret matters from the artifacts they left behind—from their pottery, their art, and even their garbage.

A society that possesses writing is therefore *historic.* One that does not is *prehistoric.* In other words, history began with Sumeria about 3500 B.C.

PLOUGHS

In the early days of agriculture, seed used to be scattered over the ground, where it grew in anarchic fashion. Eventually it was discovered that if the grain was planted in separated rows, it was then easier to water, to weed, and to harvest.

In its simplest form, the *plough* was a forked stick that was dragged through the soil making a furrow in which the seeds were planted. This greatly hastened the rate of sowing. A plough was first used in Sumeria about 3500 B.C.

3100 B.C.

NATIONS

As city-states enlarged their boundaries and grew more populous, their territories were bound to run into each other and their interdependence to grow.

Just as it was necessary to organize matters within a city-state when irrigation came into use, it became necessary to organize matters still further throughout all the city-states of a particular river. It did no good for one city-state to keep its irrigation ditches and levees in tip-top shape if another one upstream allowed its own to go to pieces and produce an untimely flood, or cut the flow of water downstream.

There was therefore pressure for union, and this was carried out first in Egypt. The ease of communication along the Nile tended to even out differences, and all the city-states for 500 miles along the river had the same culture and the same language.

About 3100 B.C., the city-states of the Nile delta (lower Egypt) were united with those south of the delta (upper Egypt) under the rule of Menes, the first king of the First Dynasty. (An Egyptian priest, Manetho, about 300 B.C., wrote a history of Egypt and divided its rulers into dynasties, each one representing a family whose members ruled over Egypt for a period of time.)

Since the Egyptian city-states shared a common language and culture, Egypt can be viewed as what we would today call a *nation.* It was the first nation the world had seen.

3000 B.C.

CANDLES

Oil lamps had been used for thousands of years and would continue to be used for thousands more, but the oil could be spilled, which could spread fire dangerously. If some solid fat was melted and then allowed to solidify again about a wick, the solid would be illuminant and container at once. Such a *candle* could be carried about without danger of spilling.

The earliest candles are shown in Egyptian paintings dating back to about 3000 B.C. and they have been in use ever since, though at the present time they are used more for decoration than illumination.

IN ADDITION

A civilization was developing in Crete, an island in the Mediterranean lying between Egypt and Greece. It was the first civilization to form on what might be considered European territory rather than on the continents of Asia or Africa.

2800 B.C.

CALENDARS

Use of the day (or the alternation of day and night) and parts of the day as determined by a sundial is insufficient as a means for keeping track of time. Certain phenomena, such as the changing of the seasons, have periods that are several hundred days long. It is tedious to count those days, and the chances of mistakes are great.

There is a cycle of intermediate length, however, that of the phases of the Moon. It takes 29 or 30 days for the Moon to go through its cycle of phases, and it takes 12 or 13 of these cycles (*months,* from the word *Moon*) to make up the cycle of the seasons.

We don't know when people first began to attach importance to the months. There are indications that even prehistoric people counted them, but it was the people of the Tigris-Euphrates region who first systematized the matter. They worked out a cycle of 19 years, in which certain years had 12 lunar months and others had 13 lunar months. Such a cycle kept the years even with the seasons. This *lunar calendar* was adopted by the Greeks and the Jews and is still used as the Jewish liturgical calendar.

The Egyptians, however, did not make use of the Moon primarily. To them, the important feature of the year was the periodic flooding of the Nile. The priests in charge of irrigation carefully studied the height of the river from day to day and eventually discovered that, on the average, the flood came every 365 days. That was also the time it took the Sun to make an apparent circuit of the sky relative to the stars. (In modern times, we view this as the time it takes the Earth to go around the Sun.) This is the *solar year*, and a calendar based on it is a *solar calendar*.

The Egyptians were aware that there were 12 new Moons to the year, so they had 12 months, but they made each month 30 days long, paying no attention to the actual phase of the Moon. That made 360 days, to which they added 5 more days at the end.

This calendar was much simpler and handier than any other calendar invented in ancient times. Historians are uncertain of the date when it was first adopted, but priests may have been using it for their private computations (it obviously made them more powerful if they alone knew when the Nile would flood) as early as 2800 B.C.

Nothing better than the Egyptian calendar was devised for nearly three thousand years, and even then what was produced was a mere modification of it—and not all the changes were for the better. Our present calendar is still based on the Egyptian calendar, with changes that are, in some cases, also not for the better. This makes our calendar, in essence, nearly five thousand years old.

2650 B.C.

STONE MONUMENTS

Thanks to the Nile, the Egyptians were able to grow a surplus of food so that many could devote themselves, for at least part of the year, to other tasks. That meant that the Egyptian rulers could put the Egyptian people to work on public projects designed to show the greatness of their rulers and, through them, of the nation and the people. The projects would also serve as memorials to that greatness to future generations.

Thus the Egyptian rulers built elaborate houses (or *palaces*, as we now call them). Indeed, the ruler was referred to as *pharaoh*, which is the Greek version of an Egyptian word meaning "big house." (This is similar to our present habit of saying "the White House" when we mean the president.)

It was customary for the important citizens of the nation to build themselves elaborate burial tombs, since Egyptian religion dealt in detail with life after death, and it was felt that, to insure immortality, the body had to be preserved. The tombs were oblong objects called *mastabas*. (Nowadays, to insure the immortality of our presidents, we build colossal presidential libraries.)

In about 2686 B.C., when Djoser, the second king of the Third Dynasty came to power, he decided to build a particularly elaborate tomb as a memorial to his greatness. He had a counselor named Imhotep who supervised the building of six mastabas of stone, one on top of the other, each smaller than the one below. The result was basically pyramidal in shape, but it was set back periodically as modern skyscrapers sometimes are. Because these setbacks were like steps a giant would use in climbing to the top, the structure is called the Step Pyramid. The base is an oblong about 400 feet by 350 feet, and the top is almost 200 feet high.

The Step Pyramid was the first large stone structure ever built and is the oldest structure built by humans that remains substantially intact today.

The Step Pyramid set a fashion, and for a couple of centuries afterward the pharaohs kept the people busy in their spare time building more and more elaborate pyramids. Larger and larger stones were used, and the climax came when the Pharaoh Khufu (Cheops to the Greeks) supervised construction of the Great Pyramid, the largest of all, in about 2530 B.C.

When that pyramid was finished, its square base was 755 feet on each side, so that it covered an area of 13 acres. The four sides sloped upward evenly (for the notion of steps had been abandoned) to a point 481 feet high. It was solidly composed of slabs of rock—2,300,000 of them, it is estimated, with an average weight of 2½ tons apiece. Each had to be brought some 600 miles, by water, of course, from quarries far up the Nile.

In among these rocky slabs were passages leading to a chamber near the center of the huge pile, which was to contain the king's coffin, his mummy, and his treasures after his death.

The fad for such large, vainglorious structures did not persist for long. They took too much time and too much work even for Egypt. The urge to build big objects, some useful, some symbolic, some vainglorious, has never left humanity, however. Some of the medieval cathedrals finally surpassed the pyramids in height (after 3,500 years or so) and today, of course, we have our skyscrapers, our huge bridges and dams, and so on.

2500 B.C.

LITERATURE

Telling stories is probably as old as speech, and gifted storytellers were probably as much in demand fifty thousand years ago as they are today. Eventually, quite elaborate stories were memorized and told, or chanted, to audiences. Homer's *Iliad* and *Odyssey* were probably recited many times before being reduced to permanent form.

Once writing was invented, it could only be a matter of time before it was used to record the more famous oral tales and sagas. As long as a tale is available only in oral form, a person can only hear it when a bard is available and is willing to recite it. It represents a rare theatrical performance. Once the same tale is in written form, it can be read at will and at any time. A written story is a permanently available bard.

The Sumerians, who invented writing, were very likely the first to put tales into writing. One of these tales was discovered in the remains of the library of Ashurbanipal, an Assyrian king who ruled from 668 to 626 B.C., nearly two thousand years after the Sumerians had established written literature.

The discovery was made in 1872 by an English archaeologist, George Smith (1840–1876). He found twelve clay tablets inscribed with cuneiform writing, telling the tale of a Sumerian king named Gilgamesh and his search for immortality.

The story may have first appeared in written form as early as 2500 B.C. It contains, as a subplot, the tale of a great flood that had ravaged the Tigris-Euphrates valley some centuries earlier. This story was borrowed by the biblical writers, who described it as Noah's Flood and imagined it to have been worldwide.

The story of Gilgamesh is the oldest written tale that has survived to this day almost intact and may be considered to represent the founding of written literature.

GLASS

Glass is made out of sand rather than out of clay as pottery is. Glass is not really a solid at all but a liquid that is so stiff it does not flow perceptibly and *seems* to be solid. It is much more fragile and easily broken than pottery is, and it would not be considered a reasonable competitor for a moment except for its beauty. Glass has a certain transparency, and the presence of impurities (sometimes deliberately added) can give it deep and lovely colors.

The earliest glass objects known were found in Egyptian tombs dating back to 2500 B.C., but they were simply ornaments. Not for another thousand years was glass used for vessels.

IN ADDITION

By this time, agriculture had been developed independently in the valley of the Huang (Yellow) River in what is now northern China, and as a result civilization was established there as well. Agriculture was also beginning, again independently, in Central America.

2340 B.C.

EMPIRES

Conflict of some sort is as old as life. With human beings, conflict reached new pitches of danger, thanks to intelligence. Human beings can remember past wrongs, brood upon them, and plot revenge. They can also, after victory, realize that the defeated parties may plot revenge, and therefore take further action to wipe them out completely. What's more, the advance of technology put weapons in human hands that made conflict ever bloodier.

Sumeria was not as fortunately placed as Egypt was. The Tigris and Euphrates did not offer quite the same gentle watery highway that the Nile did. Communications were not as effective and there seemed less community of interests among the people of the region. The Sumerian city-states, with their wheeled carts and bronze weapons, fought each other more assiduously than the Egyptian city-states did.

What's more, whereas Egypt was protected on both sides of the Nile by desert, Sumeria was more open to invasion. The result was that non-Sumerian people settled on the upper reaches of the Tigris and Euphrates.

Just to the north of Sumeria, where the Euphrates and Tigris come closest together, cities were founded by Akkadians, who spoke a non-Sumerian language. The Akkadian language was of a type that came to be called *Semitic.* (The most important Semitic language extant today is Arabic.) The Sumerian language was not Semitic and, in fact, is not related to any other language we know of. There was even less community of interest between the Sumerian city-states and their Akkadian neighbors than between the Sumerian city-states themselves.

Thus it came about that, even seven centuries after Egypt had formed a nation, the city-states of the Tigris and Euphrates were unable to unite in peace. It was clear that the region would prosper best under unified rule, but there was no agreement as to which city and which ruler should take over the leadership, and the matter had to be settled by force.

About 2350 B.C., a man named Sargon (*ca.* 2334–2279 B.C.) took over the rule of Agade, one of the Akkadian cities. He was successful in war and established his rule over all of both Akkad and Sumeria. He also sent his armies northward and eastward and established his control over the upper reaches of the Tigris-Euphrates valley, a region that eventually came to be known as Assyria, and over the territory to the east of the Tigris, which was known as Elam.

While the Egyptian unification involved city-states of similar language and culture, Sargon ruled over peoples with different languages and cultures, his Akkadians being dominant over the others.

If one cultural group dominates others politically and militarily, the result is commonly referred to as an *empire.* Sargon established the first empire we know of. It was not, of course, the last.

IN ADDITION

Crete was now becoming a naval power, the first the world had seen. Since it was an island, it had to trade by means of ships, and the ships served also to guard Crete against invasion. Thanks to its navy, Crete dominated the islands of the Aegean Sea and Greek coastline. It also secured for itself a thousand years of peaceful civilization.

2000 B.C.

HORSES

Until now, the animals that had been used for pulling carts and ploughs were oxen and donkeys. The ox was strong, but it was lumbering, stupid, and slow. The donkey was more intelligent, but it was smaller and weaker than the ox. Neither could pull the heavy, solid-wheeled carts rapidly.

Animal transport could not be used in warfare with any great success, therefore. Armies consisted of masses of foot soldiers, who slogged into each other, wielding spears and swords and cowering behind their shields, until one side or the other broke and ran. The carts could serve only as a ceremonial means of keeping the ruler and other military leaders from walking, or to carry arms and supplies.

But then, about 2000 B.C., a fleet beast was tamed—the wild horse—and not by any of the civilizations but by nomadic dwellers in the steppes of what we now call Iran. The horse was larger and stronger than the donkey, and faster and more intelligent than the ox. At first, though, it seemed useless for transport, for it was difficult to harness. A harness that was suitable for an ox placed pressure on the horse's windpipe and cut its speed.

Then, some time before 1800 B.C., someone devised a method of using the horse for specialized light traction. A cart was made as light as possible, becoming scarcely more than a small platform between two large wheels, a platform just large enough to hold a human being. Even the wheels were lightened, without loss of strength, by being made spoked rather than solid, and they were so fixed to the axle that they could turn individually. The result was a *chariot* (a word not too different from *cart*).

A horse, or horses, pulling so light a load could run fleetly, much faster than a foot soldier could. With only two wheels, the chariot was almost as maneuverable as the horse itself and could skid into a new direction with little trouble.

It did not take long for the nomads to discover that a body of charioteers, driving in fiercely, could not be stopped by the foot soldiers of the day. Indeed, foot soldiers broke and fled in horror at the first sight of the animals thundering toward them.

This is the first clear case we have of a new war weapon catching those without it by surprise and bestowing a kind of universal victory on those who had it. The raiding nomads ripped into the Tigris-Euphrates valley, which came under "barbarian rule" for a period of time. The nomads established the kingdom of *Mitanni* in what is now Syria and northern Iraq, and the *Hittite* kingdom in what is now eastern Turkey. In 1700 B.C., the horsemen drove down into Canaan and then even into Egypt, which fell to alien invaders for the first time, and also into India.

Such invasions spread devastation in the settled areas, but they did tend to stir things up. They helped change ways of life that had perhaps become somewhat decadent, and encouraged the flow of new ideas from one settled place to another.

IN ADDITION

The Phoenician city-states along the eastern shore of the Mediterranean were beginning to grow prominent. The Phoenicians built ships and learned seafaring skills, although for some centuries the Cretans remained dominant in this respect.

1800 B.C.

MATHEMATICS AND ASTRONOMY

The concept of mathematics is as old as human beings. Even some animals have a primitive number sense.

It is certainly difficult to believe that the pyramids, for instance, could have been built without substantial ability in geometry.

The Sumerians, and the Babylonians who succeeded them, were the first to make significant advances in mathematics and in astronomy. By 1800 B.C., they had developed a number system based on 60 that we still follow in some ways, since we still have 60 seconds to the minute and 60 minutes to the hour. Why 60? Because it can be evenly divided by 2, 3, 4, 5, 6, 10, 12, 15, and 30, thus eliminating the frequent need of fractions—with which the ancients had trouble.

In addition, there are 360 degrees (6 × 60) in the circle. Again, it is a number easily divided. In addition, the Sun takes 365 days to move completely around the sky, so that it travels, relative to the stars, just about 1 degree per day. That too may have influenced the choice of 360.

The sky-watchers of the Tigris-Euphrates valley eventually discovered that, in addition to the Sun and Moon, five bright stars changed position against the remaining "fixed" stars. These moving stars, which we call *planets* (from a Greek word for "wandering") were given the names of gods and goddesses, and we still do that today. We call the five bright stars Mercury, Venus, Mars, Jupiter, and Saturn. The presence of seven such planets (if we include the Sun and the Moon) gave rise to the seven-day *week,* with one of the planets in charge of each day. The week as a unit of time was thus instituted by the Babylonians and was adopted first by the Jews, then by the Christians, and through them, by virtually all the modern world.

The seven planets had paths around the sky, paths that passed through certain conglomerations of stars that were termed, in total, the *zodiac* by the later Greeks. This was divided into twelve *constellations,* so that the Sun remained in each constellation for about a month. Eventually the Sumerians and Babylonians worked out the pathways in some detail and could predict, in at least a rough manner, where the planets would be at future times. That represented the beginning of mathematical astronomy.

Since the Sun affected the Earth profoundly, making the difference between the day and night, and the Moon's phases marked the length of the month, it seemed natural to suppose that the other planets also had significance to human beings. Imaginative suggestions were made about the influence of each, as it varied in position with respect to the stars and other planets, and an intricate system of foretelling the future from planetary positions was evolved. This is *astrology.*

Astrology is utter nonsense, but people strongly want the security of knowing the future, so that even today astrology is accepted by many people who are uneducated, unsophisticated, or simply silly.

FERMENTATION

Fruit juices that are left standing will sometimes *ferment;* that is, undergo changes that alter the taste. The same is true of moistened grain. Human beings, driven by thirst or hunger, might consume such fermented materials and then find that they liked the taste and the aftereffects. They were, of course, consuming alcohol, formed from sugars and starches by yeast, and they would grow elated as a result, because somewhat intoxicated. (This is not an exclusively human trait. Birds and animals will sometimes greedily feed on fermented fruits and become obviously intoxicated too.)

This may have happened in prehistoric times, but by 1800 B.C. the use of fermented materials was so common that laws had to be passed directing what was to be done in the case of misdeeds committed under the influence of too much beer.

From the beginning of agriculture, grain was converted into flour, which was moistened and made into flat, hard, but nourishing bread. Every once in a while, though, the moistened dough fermented and released gases (carbon dioxide) that caused the bread to rise and grow spongy. The result was *leavened bread* (from a Latin word meaning "to rise"), which was just as nutritious as flat bread but softer and much more pleasant to eat.

The Egyptians discovered this not long after 1800 B.C. and eventually learned that the process could be controlled. If some of the fermenting bread was saved before it was baked and added to dough that had not yet begun to ferment, the fresh dough would ferment in its turn. One would not have to depend on chance.

IN ADDITION

At this time, both the Sumerian civilization and the Indus civilization were decaying rapidly. The Sumerian culture was disappearing under the strains of foreign invasion. The Indus people, through overirrigation, had increased the salt content of their fields to such an extent that crops would no longer grow well enough to support them.

1775 B.C.

LAW

Human beings must always have had customs that they followed assiduously, even when enforcement was not an issue.

In a simple society, custom is, in fact, enough. Everyone knows what behavior is expected, and conforms almost automatically. If not, there is social ostracism, and this is sufficiently undesirable to enforce the rule of custom.

As a society grows more complex, however, there are more varieties of behavior that must be controlled and regulated, more perplexing conditions, more complicated questions, more puzzling interactions. It becomes difficult to remember all the rules, and the suspicion is sure to arise that powerful people make up or alter rules to suit themselves. The demand, then, is for the rules of society to be put in writing, so that all can see for themselves what they are, and so that they cannot be unfairly or arbitrarily twisted or modified.

We don't know when the first laws were written down, but the first relatively complete law code that we still have was established by Hammurabi, king of Babylon (reigned 1792–1750 B.C.), who founded a rather short-lived Babylonian Empire in the Tigris-Euphrates valley, one that succeeded the Akkadian Empire. (After this time, the people of the valley were referred to as Babylonians for nearly two millennia.)

Perhaps about 1775 B.C., Hammurabi had his law code inscribed on an 8-foot-high stone pillar of hard diorite. It was clearly intended to be permanent, and it was, for we still have it.

The stele is topped by a relief that shows Hammurabi standing before the Sun-god, Shamash. (It was usual in ancient times to suppose that a law code was received by a king from a god. That tended to lend the law authority. Thus, Moses received the Jewish law code from God on Mount Sinai, according to the Bible.)

Down along the face of the stele are twenty-one columns of finely written cuneiform, outlining nearly three hundred laws that were to govern people's actions and guide the king and his officials in dispensing justice.

The stele originally stood in the town of Sippar, some 30 miles upstream from Babylon, but an invading Elamite force plundered the city and carried away the stele as spoil. It remained in Elam's capital, Susa, thereafter and was still there, in Susa's ruins, in 1901, when a French archaeologist, Jacques Jean Marie de Morgan (1857–1924), found it and brought it back to Europe.

1550 B.C.

MEDICINE

People are sure to get sick at times, or have accidents and be hurt. The problem of getting well, or being made well, naturally concerns everyone.

To aid the coming of wellness, people might attempt to conciliate various gods by proper incantations or rituals, make use of various forms of ritualized behavior, or use parts of plants or animals thought to have curative value.

The first written compilation of such cures that we know of is an Egyptian papyrus dated about 1550 B.C. It was discovered in 1873 by a German archaeologist, Georg Moritz Ebers (1837–1898), and is called the Ebers Papyrus in consequence. It contains about seven hundred magical remedies and descriptions of folk medicine for the treatment of various ailments.

IN ADDITION

The Egyptians, making their capital at Thebes, evicted the charioteers from the north in 1570 B.C. and went on to conquer parts of the east Mediterranean coast. They set up an Egyptian Empire and for three centuries enjoyed the most powerful period of their history.

The Greeks were at this time establishing a civilization on the Greek mainland. Their most powerful city was Mycenae and they were referred to as the Mycenaeans. Because Greece is a mountainous land of separate valleys and no centralizing river, the Greeks established city-states and in fourteen centuries of history never managed to unite.

1500 B.C.

ALPHABET

By 1500 B.C., Egyptian hieroglyphic writing and Babylonian cuneiform writing (inherited from the Sumerians), together with Chinese writing in the Far East, were the most important written languages in the world. All remained terribly complicated, and there is no reason they should not have remained complicated until the present time. The Chinese language has.

Between the Egyptians and the Babylonians were the Canaanites, inhabiting the eastern shore of the Mediterranean Sea. (The Greeks called them Phoenicians.) They were traders who acted, among other things, as intermediaries between the Egyptians and Babylonians. It was necessary for such traders to know both the Egyptian and Babylonian languages, and that was a hard chore indeed.

It occurred to some nameless Canaanite to simplify writing by adopting a kind of shorthand. Why not give a separate symbol to each of the common sounds made by human beings in speaking a language? You could then build up any words of any language by using those sound-symbols. Sound-symbols had, in fact, been used by the Egyptians, but they also preserved symbols for syllables and for whole words. The Canaanite inventor had the notion that the sound-symbols should be used *exclusively* and that words should be built up out of them.

The first two symbols of this collection were *aleph* (the usual symbol for an ox) and *beth* (the usual symbol for a house). To the Greeks, who eventually adopted the system, these became *alpha* and *beta,* and we still call the system of symbols the *alphabet.*

The *Phoenician alphabet,* which first came into use about 1500 B.C., revolutionized writing, making it far easier to write and to read, so that the chances of literacy were increased. This is one invention that seems to have occurred only once in human history. The alphabet was not invented independently by any other society. All alphabets in use today (including the one in which this book is written and printed) are descended from that first Phoenician one.

IN ADDITION

The Chinese were beginning to push forward technologically at this time. They had developed horse-drawn vehicles, tamed the water buffalo, and were beginning to make use of silk, an animal fiber from the cocoons of certain caterpillars.

1375 B.C.

MONOTHEISM

It is customary for people to believe in a multitude of separate supernatural influences. Every object—the Sun, the Moon, trees, animals, even abstractions such as tribes and nations—must have its supernatural accompaniment, cause, or protector.

The first person we know of to suppose that there was only one divine influence that controlled everything was an Egyptian pharaoh named Amenhotep IV, who reigned over Egypt from 1379 to 1362 B.C. He accepted the Sun-god as the one and only god, and this makes some sense, since the Sun is an overriding influence on Earth and on humanity. The Sun-god was, to him, Aton, and he called himself Akhenaton, meaning "Aton is satisfied." His seventeen-year reign was a failure, however, because his views were not accepted by the priesthood or by the Egyptian people, who clung to their old ways.

It is possible, however, that the Akhenatonic tradition lingered among a few and influenced Moses, the legendary leader who brought the Israelite slaves out of Egypt, according to the Bible, about a century and a half after Akhenaton's time. (The later Jews attributed monotheism to the legendary

Abraham, about four or five centuries before Akhenaton, but there is no record of Abraham outside the Bible.)

Monotheism was a clear advance over polytheism, since it reduced the chaos of the supernatural and made possible a more orderly theology.

IN ADDITION

In 1470 B.C. a terrific volcanic explosion destroyed the island of Thera, north of Crete. The ashes from the explosion blanketed Crete, and the *tsunami* (tidal wave) that resulted buffeted its shoreline. Crete appears to have been so devastated by the event that its fifteen-century civilization sank to its close.

This gave the Mycenaean Greeks of the mainland a chance to take over Crete and maintain supremacy in the Aegean for a couple of centuries. It also gave the Phoenicians their chance to become the supreme seafarers of the ancient world, and they remained so for a thousand years.

1200 B.C.

DYES

The human urge to beautify is irresistable, and since we can see colors, we usually find them, singly and in combination, to be more attractive than black and white. The Old Stone Age artists used colored earths to make their paintings.

As early as 3000 B.C., dyes were used to color otherwise white or yellowish cloth, both in Egypt and in China. *Indigo,* a blue color extracted from a plant, was used and *madder*, a red color extracted from the root of another plant. By 1400 B.C., cloth could be dyed in virtually all colors.

The chief trouble with most early dyes was that they tended to bleach in the sun and to wash out in water, so that dyed fabrics quickly grew dim and blurred.

One dye that was very resistant both to sun and water was obtained from a snail in the eastern Mediterranean. To obtain the dye was a tedious task, but the resultant red-purple color was brilliant and *stayed* brilliant. The city of Tyre in Phoenicia had developed this dye industry by 1200 B.C. so that the color was called *Tyrian purple.* Between the difficulty of amassing a quantity of it, and its great desirability, its price went sky-high, but it still sold, albeit only to the rich and powerful. Tyre grew wealthy out of its dye trade and was able to support its merchant fleet and undertake trading ventures that made it richer still.

Some believe that the Greek name *Phoenicia,* applied to the land in which Tyre was a city, came from a Greek word meaning "red-purple" in reference to the dye.

IN ADDITION

The confusion in the eastern Mediterranean set off by the explosion at Thera was still convulsing the area. Sea-raiders (including Cretans fleeing their doomed civilization) invaded Canaan and set up the Philistine cities. They also attacked Egypt, which managed to fight them off but at a tremendous cost, and the Egyptian nation began a long decline from which it was never to recover.

Meanwhile, the Mycenaeans reached the peak of their power when they destroyed the city of Troy in northwestern Asia Minor in 1184 B.C. Troy had controlled the straits between the Aegean Sea and the Black Sea, and now the Mycenaeans could trade freely past that bottleneck.

1100 B.C.

SEA NAVIGATION

Boats had been in existence for over two thousand years, but they had been confined to rivers. When they did venture out to sea, they generally hugged the coast. Even the Cretans, the boldest seagoers up to this time, had confined themselves to the eastern Mediterranean and felt safest in the Aegean Sea, where numerous islands made possible short sails from land to land.

Greek legends treated distant portions of the sea as places of myth and mystery. The tale of Jason and the Argonauts reflects early ventures into the large and island-free Black Sea. Homer's *Odyssey* describes the adventures of Odysseus in the even larger western Mediterranean.

The first to cast out boldly into the open sea were the Phoenicians. They noted that the seven stars that made up the familiar grouping of the *Big Dipper* were always located to the north and were visible all night long at every season of the year (barring the presence of clouds). This must surely have been known for a long time, but the Phoenicians seem to have been the first who were willing to risk their ships and lives on the fact. By observing the Big Dipper, they knew always which direction was north, and from that they knew all other directions. This obviated the fear of being lost once out of sight of land and "landmarks." There were always the "skymarks," so to speak.

In addition, the Phoenicians refused to depend on sails and wind alone. The wind was erratic, after all, so the Phoenicians used oars, which the Egyptians had been using on the Nile for twenty centuries. Oared ships (galleys) now ruled the Mediterranean for twenty-six centuries.

Using oars, and keeping the Big Dipper on the right, the Phoenician ship captains could sail boldly westward, knowing they could return by remembering to keep the Big Dipper on the left. Beginning about 1100 B.C. then, the Phoenicians explored the coasts of northern Africa west of Egypt and southern Europe west of Greece, trading and in some cases settling.

IN ADDITION

In western Asia at about this time, the Israelites were subject to the Philistines. The Assyrians, on the other hand, who lived on the upper courses of the Euphrates and Tigris, were making their mark as conquerors for the first time. Under their king, Tiglath-pileser I, who reigned from about 1115 to 1077 B.C., they reached the Mediterranean.

1000 B.C.

IRON

Iron is the second most common metal in the Earth's crust (only aluminum is more common), but it always occurs in combination with other substances. It is not found in free metallic form except in some meteorites, which are not of Earth but fall from the sky.

Such meteorites were occasionally found by the ancients and used even in the earliest days of civilization. Compared to gold, silver, and copper, iron is an ugly metal, but the meteoric iron that was found revealed itself to be harder and tougher even than bronze. Since it held its edge far better than bronze did, it was much in demand for edged portions of tools.

The result is that no iron meteorite from the past is ever found where the earliest civilizations flourished. The ancients scavenged them all.

Yet ores did not yield iron. Gold, silver, copper, lead, tin, and eventually mercury were obtained with ease by use of wood fires, but such fires never yielded iron. Iron held on to other substances more tightly than the other metals did, and a higher temperature was needed.

Eventually, though, charcoal was obtained by burning wood with an insufficient supply of air so that more-or-less pure carbon was left behind when other substances burned away. Charcoal burns flamelessly but reaches higher temperatures than wood does.

About 1500 B.C. the Hittites of Asia Minor found they could obtain iron from certain ores by heating those ores with burning charcoal. Iron was at first a disappointment. In pure form, it is tough but not as hard as the best bronze. (Meteoric iron is *not* pure iron but is a 9-to-1 mixture of iron and nickel, something the ancients couldn't duplicate since they knew nothing of nickel.)

By 1200 B.C., undoubtedly through hit-and-miss, it had been discovered that iron, properly smelted, could appear in hard form. This came about when some of the carbon in charcoal mixed with iron to form an iron-carbon alloy we call *steel*.

By 1000 B.C., such carbonized forms of iron could be formed in quantity and the *Iron Age* began, the period when iron was the chief metal used in arms and tools.

IN ADDITION

The coming of iron tipped the scales in warfare. The bronze-armed Mycenaeans faced an invasion of semibarbarous Greeks (Dorians) from the north. The Dorians had the advantage of iron arms, while the Mycenaeans were still using bronze. The Dorians swept over Greece, destroying the Mycenaean civilization. A "dark age" followed in Greece that lasted a couple of centuries.

In Canaan, the Israelites had also obtained iron arms. The result was that they defeated the Philistines and, under their new king, David, were at this time establishing an empire of their own, occupying the entire eastern coast of the Mediterranean.

750 B.C.

ARCHES

The easiest way to build an opening is to set up two vertical pieces of wood, stone, or other material and then balance a horizontal piece above the two.

The horizontal piece, unsupported in the middle, can break with relative ease, and the weakness increases as it grows longer. If instead one uses relatively small pieces arranged in a vertical semicircle so that each piece helps support the piece above, and if one uses mortar to make the pieces adhere to each other, one has an arch.

An arch can span a much wider distance and carry a much heavier load than a horizontal piece can.

Small, primitive arches were used as early as Sumerian times, but a true arch, properly built for maximum strength, showed up for the first time among the Etruscans in 750 B.C.

IN ADDITION

The Etruscans had arrived on the west coast of Italy to the north of Rome about 900 B.C. and were now the strongest power in Italy. The city of Rome was founded, according to legend, in 753 B.C., and during its early centuries it was dominated by the Etruscans.

The city of Carthage, later to be Rome's rival, was, also according to legend, founded in 814 B.C., in what is now Tunisia, by the Phoenicians.

The Israelite Empire of David proved to be short-lived. It broke up into the two nations of Israel and Judah in 933 B.C. and both lived under the growing power of Assyria, which now dominated western Asia.

In Greece, the dark age was finally lifting. Homer composed his epics of the Trojan War about 850 B.C., and the first Olympian games were celebrated in 776 B.C. (The Homeric epics and the Olympian games, along with the Greek language, united the Greeks culturally, though politically, they continued to fight each other.)

700 B.C.

AQUEDUCTS

As cities grew, supplying what was needed to so many people so densely packed together became a problem. Air itself, the most immediate and pressing necessity, was available everywhere more or less (though the use of fires in every house could fill that air with unpleasant smoke).

Water was more of a problem. Cities are usually built where there is a water supply, but as they grow, the water supply may become insufficient. Wells within the city limits or just outside may not supply enough. It may then become necessary to fetch water from some distance, either through canals or tunnels or along artificial structures of masonry.

These last are called *aqueducts* (from Latin words meaning "a drawing off of water") and by 700 B.C., Sennacherib, who was king of Assyria from 704 to 681 B.C., had an aqueduct constructed that would bring water into his capital, Nineveh. At about the same time, Hezekiah, king of Judah from about 715 to about 686 B.C., built an aqueduct to supply Jerusalem with water.

ZOOS

From the beginning of genus *Homo,* hominids and modern human beings had been hunting animals for food. No thought was given to conserving the animals. Either people took it for granted that every species of animal would always exist, or they gave it no thought at all. The Assyrian monarchs were great hunters themselves (for sport, rather than for food), and Assyrian art is full of representations of the kings slaying lions and other animals.

Nevertheless, there was also sometimes the impulse to conserve. If an animal was rare, there might be a certain prestige in keeping it for one's pleasure, as one might keep a rare work of art. An early example of this could be found on the palace grounds of the Assyrian king Sennacherib (see above), where there was a zoo and also a botanical garden.

SUNDIALS

The original sundial was a stick pounded into the ground so that its shadow might be studied (see 4000 B.C.). Such a stick was called a *gnomon* (a Greek word meaning "indicator") since it gave a rough indication of the time.

Eventually people learned how to prepare a circular bowl with the hours marked off on the rim and with a gnomon in the center that was tilted due north. The shadow then stayed the same length as it traveled from west to east around the rim of the bowl. This greatly increased the convenience and use of sundials.

Sundials of this sort appeared in Egypt at least as early as 700 B.C. (We still find them in modern gardens, by the way, not to tell the hours but as ornaments.)

IN ADDITION

Assyria under Sennacherib (see above) controlled all the older civilizations of western Asia at this time. It destroyed the nation of Israel in 722 B.C. and in 701 B.C. laid siege to Jerusalem. Judah survived but only at the cost of paying a heavy tribute. The Phoenician cities were also tributaries.

640 B.C.

LIBRARIES

Books, whether clay bricks covered with cuneiform or papyrus covered with hieroglyphics and rolled up (the word *volume* is from the Latin word for "to roll up") were hard to come by in ancient times. To get an additional copy of a book, it had to be copied over, stroke by stroke, by a meticulous and thoroughly literate scribe. Such copying took a long time and was hard work, so that books were both rare and expensive.

Only a few people could own books, and a *library* (from a Latin word for "book") consisting of several books must have been the mark of a rich man or the painfully accumulated store of a scholar. Only monarchs with the resources of kingdoms behind them could accumulate large libraries in the modern sense.

The first such monarch that we know of was Ashurbanipal (see 2500 B.C.). He arranged to have every book in his kingdom copied and the copy placed in his library at Nineveh. He ended up with thousands of books, all carefully cataloged.

COINS

Trading originally consisted of barter: you give me this and I will give you that. If two people both had something they didn't need that the other badly wanted, the trade was easy. Usually, however, both sides wanted to make sure they didn't give up something more valuable for something less valuable. Since comparative values are hard to judge, there must have been many times when both traders felt cheated.

Eventually the custom arose of using metals, especially gold, as a medium of exchange. Gold was beautiful and much to be desired as ornamentation. It didn't rust or corrode, and it was rare, so that a little bit went a long way. Once everything was valued as so many unit weights of gold, a person could buy an object for that amount or exchange it for an object worth that same amount.

In all transactions, it was then necessary to have a scale (see 5000 B.C.) that could be used to weigh little odds and ends of gold, with the usual fears on all sides that the scale, or the weights, might be crooked.

In western Asia Minor, the kingdom of Lydia was founded about 680 B.C. by Gyges, who ruled till about 648 B.C. Under his son, Ardys, who ruled from about 648 to about 613 B.C., the Lydian government issued pieces of gold of standard weight, with the weight marked on them and a portrait of the monarch included as a governmental guarantee. In any transaction, it was now only necessary for a number of coins to be passed over; no weighing was necessary. (*Coin* is from a word meaning "stamp" because of the weight and portrait figure that are stamped upon it.)

The development of coins greatly accelerated trade, and the idea was so transparently good that it was adopted by other governments as well.

IN ADDITION

Assyria was still growing in might. In 675 B.C., the Assyrian king Esarhaddon, who ruled from 680 to 669 B.C., attacked and conquered Egypt.

Meanwhile the nation of Japan, according to legend, came under the rule of its first emperor, Jimmu Tennō, in 660 B.C.

585 B.C.

ECLIPSES

In studying the motion of the planets along the path of the zodiac, the Babylonian astronomers could not help but note that sometimes the motions would bring two of them fairly close together. This would be most spectacular in the case of the Sun and the

Moon. Every once in a while the Moon would pass in front of the Sun and obscure part or sometimes all of it. At times, too, the Sun would be on one side of the Earth, and the Moon would be directly on the other. The Earth's shadow would then fall on the Moon, obscuring *it*. Thus, there could be either a *solar eclipse* or a *lunar eclipse*. (*Eclipse* is from Greek words meaning "to leave out," since when one happened, the Sun or the Moon would seem to be left out of the sky.)

An eclipse is a frightening spectacle. Those who become aware of it may actually think that the Sun or the Moon is dying, with incalculable consequences. Even if it is understood that the Sun or Moon is only obscured temporarily, there is a feeling that it is an omen of evil sent by the gods as a warning.

However, by studying the movements of the Sun and Moon, early astronomers learned to predict when eclipses would take place. Since that made the eclipses appear to be automatic and unavoidable phenomena, it removed their unexpected and ominous connotations. (There is some feeling that even prehistoric watchers of the sky learned to tell when lunar eclipses would occur, and that the stones at Stonehenge in southwestern England were arranged as a kind of observatory that allowed prediction of such phenomena.)

The Greek philosopher Thales (624–546 B.C.) seems to have learned the Babylonian methods and predicted an eclipse of the Sun, one that we now know (by calculating backward) took place on May 28, 585 B.C. This added greatly to Thales' prestige and also helped make eclipses less frightening, since they were demonstrated to be predictable.

IN ADDITION

For all of Assyria's apparent might, the career of conquest and the difficulty of keeping subject nations subdued sapped its strength. Once Ashurbanipal died in 626 B.C., Assyria crumbled rapidly under his incapable successors and by 609 B.C. Assyria no longer existed. Now the Chaldeans ruled over the Tigris-Euphrates valley and the eastern shores of the Mediterranean. To the north of the Chaldean Empire was the Median Empire.

In Greece, the city of Sparta in the south was developing a militaristic society and making itself into the most powerful of the Greek city-states. Athens was moving toward a democracy.

580 B.C.

ELEMENTS

Thales (see 585 B.C.) was the first to ask himself what the Universe was made of and to seek an answer that did not depend on gods or the supernatural. He thus represents the birth of rationalism.

His answer, which he may have reached about 580 B.C., was that all matter was fundamentally water, and that everything that did not seem to be water originated in water or was modified water. Water was thus, in his opinion, the primary *element* (from a Latin word of uncertain meaning), or fundamental substance, of the Earth.

IN ADDITION

At this time, the city of Tyre (see 1200 B.C.) was under siege by Nebuchadrezzar II (*ca.* 630–562 B.C.), the ruler of the Chaldean Empire. Tyre fell, after a thirteen-year siege, in 573 B.C. While it continued to be an important city for two more centuries, its great days were over, and Carthage became the most important Phoenician city in the world. Meanwhile, under Nebuchadrezzar II, the city of Babylon was at its height. It was the richest and most populous city in the world.

520 B.C.

IRRATIONAL NUMBERS

The Greek philosopher Pythagoras (*ca.* 580–*ca.* 500 B.C.) believed that whole numbers, including fractions, since they are ratios of whole numbers, were the basis of the Universe. Thus, $3/4$ is the ratio of 3 to 4. If you begin with 3 pies and divide them equally among 4 people, each person gets $3/4$ of a pie. Whole numbers and fractions together make up the *rational numbers* (those that can be expressed as ratios), and it is easy to suppose that rational numbers are all that exist.

However, suppose you have a right triangle with each side equal to 1 unit. What is the length of the hypotenuse? The answer can be obtained by remembering that the square of the hypotenuse is equal to the sum of the squares of the sides. This was long known, but Pythagoras worked out a good proof and it is called *the Pythagorean theorem* as a result.

The square of each side is 1, so the square of the hypotenuse is 2, and the length of the hypotenuse is the square root of 2, or that number which, when multiplied by itself, equals 2. The number $7/5$ is nearly right, since $7/5 \times 7/5 = 2.04$. The number $707/500$ is even closer, since $707/500 \times 707/500$ is a little over 1.999.

It can be shown quite easily, just the same, that there is no fraction, no fraction *at all*, however complicated, that when multiplied by itself gives exactly 2. The square root of 2 is therefore not a rational number. It is an *irrational number*, and as it turns out, there are an infinite number of such irrationals.

IN ADDITION

In India, Siddhārtha Gautama (the Buddha) (*ca.* 563–*ca.* 483 B.C.) established Buddhism. At about the same time, in Persia, Zoroaster (*ca.* 628–*ca.* 551 B.C.) founded Zoroastrianism. In China, Lao-tzu (6th century B.C.) founded Taoism.

Neither the Chaldean nor the Median empires were long-lived. In Persia (a province of the Median Empire), a local ruler named Cyrus II (*ca.* 585–*ca.* 529 B.C.) overthrew the Median king and founded the *Persian Empire.* He conquered Lydia and the Chaldean Empire, and his son, Cambyses II (reigned 529–522 B.C.) conquered Egypt. The Persian Empire was now the largest the Western world had yet seen and may have had a population of 15 million. China, however, may have had a population of 20 million at this time.

510 B.C.

MAPS

Egyptians and Babylonians alike attempted to draw maps of the world they knew. In early times, however, traveling was difficult, and most people either knew only their immediate neighborhood or, if they traveled, had difficulty keeping directions and distances in mind.

The first map that we can recognize as having some vague relation to reality was drawn by a Greek traveler named Hecataeus (6th–5th century B.C.). He had the advantage of living after the Persian Empire had been firmly established, so that it was possible for him to travel for thousands of miles without encountering war or disorder.

Hecataeus drew a map about 510 B.C. in which the land area of the world was shown as a circle, with the sea around it. An arm of the sea cuts halfway into the circle from the west. It is the Mediterranean. Europe lies to the north, Africa to the south, and Asia to the east.

IN ADDITION

The city of Rome, which had existed under a monarchy for two and a half centuries, evicted its king in 509 B.C. and established the Roman Republic, which was to exist for nearly five centuries.

In a curiously parallel phenomenon, the city of Athens, which had been under a dictatorship, established a democracy in 510 B.C.

500 B.C.

ATLANTIC OCEAN

The Phoenician navigators, who had made themselves at home the length and breadth of the Mediterranean over the past six centuries, even ventured through what we now call the Strait of Gibraltar and into the Atlantic Ocean.

One of the driving forces behind them was the depletion of eastern Mediterranean tin mines, since tin is a rather rare metal. (This was the first time that human beings had to contend with the loss of a necessary resource.) Since tin was an essential component of bronze, it had to be obtained somewhere; if not in the Mediterranean lands, then elsewhere.

The Phoenicians found *Tin Islands* somewhere in the Atlantic. They kept the location secret in order to retain a monopoly of the tin ore, but it is thought that they reached Cornwall, at the southwestern tip of England, where tin ore was produced right into modern times.

By 500 B.C., the Phoenicians are even reported to have circumnavigated Africa, a voyage that took them three years. The Greek historian Herodotus (*ca.* 484–between

430 and 420 B.C.), writing half a century later, described the voyage but doubted the whole thing because the Phoenicians reported that in the far south the noonday Sun was in the northern half of the sky. Herodotus felt this to be impossible.

We moderns, however, know that the Sun *is* always in the northern half of the sky when seen from the South Temperate Zone. The Phoenicians would not have made up such an apparently ridiculous story if they had not actually witnessed it, so the very item that caused Herodotus to doubt the story convinces us that it must be true.

DISSECTION

The interior of a human body cannot ordinarily be seen. Animals, however, have been butchered since dim prehistory, so that much was known of animal organs. There were also suggestions that one could foretell the future by studying animal livers, for instance, which meant that studies of animal *anatomy* (from Greek words meaning "to cut up") were more detailed and careful than one would expect from the process of butchering.

However, one cannot treat a dead human being as one would treat a dead animal. There is a feeling that human beings should be treated with respect, even if dead. Some human beings might be hurt and cut open in the course of wars, private fights, or hunting, but the studies this made possible could only be limited and unsystematic.

A Greek physician, Alcmaeon (6th century B.C.) was the first to take the chance of deliberately and carefully dissecting human cadavers, possibly about 500 B.C. In this way, he could see the difference between arteries and veins and tell that the sense organs were connected to the brain by nerves.

ABACUS

No one knows when the abacus first came into use, but it was probably known in Egypt at least as long ago as 500 B.C.

It consists essentially of rows of beads, sometimes strung on wires. In the simplest form there are ten beads on each wire, the first row being units, the second row being tens, the third row being hundreds and so on.

The beads can be manipulated much as we manipulate the fingers on a hand in simple adding and subtracting. The advantage is that you may have nine or ten "hands" present as so many rows of beads, and the movements you make are easier and quicker than manipulating your fingers would be.

A skilled operator can use the abacus flashingly to multiply, to divide, and to perform many complicated arithmetical manipulations. It was the first really important computing device worked out by human beings.

VENUS

The Greeks were not, at first, as advanced in astronomy as the Babylonians were. They knew the evening star, a bright planet that appeared in the western sky after sunset, and they called it *Hesperos* (the Greek word for "evening"). There was also a morning star, a bright planet that appeared in the eastern sky before sunrise, and they called it *Phosphoros* (the Greek word for "light-bringer," because once it appeared, the Sun was not far behind).

Pythagoras (see 520 B.C.) was the first Greek to realize the two were the same object, since when the evening star was in the sky there was no morning star and vice versa. (He is supposed to have traveled in Babylonia, and he may have learned this there.) About 500 B.C., he named this single planet, which swung from one side of the Sun to the other and back again, Aphrodite, after the Greek goddess of love and beauty. The Romans (and we) called it by their equivalent, *Venus*.

IN ADDITION

The Greek cities on the Asia Minor coast revolted against their Persian overlords in 499 B.C. The city of Athens sent twenty ships to help the rebels, which infuriated the Persian monarch, Darius I (reigned 522–486 B.C.). He had crushed the revolt by 494 B.C. and then had time to turn his attention to Athens and to Greece generally.

480 B.C.

DREAMS

Dreams have always seemed to human beings to be a doorway into some strange and different world. Dreams in which people who were dead appeared and seemed to live and speak might have given rise to notions of ghosts and a spirit world, and reinforced belief in a life hereafter. Dreams that made little sense might seem like obscure messages from divine beings. Dreams are described as messages from Zeus in Homer, and as messages from God in both the Old and New Testaments.

The Greek philosophers, however, were wedded to rationalism. They felt that the Universe ran according to laws of nature that could be understood by observation and reasoning and did not require any supernatural force—that is, any force outside of or superior to the laws of nature.

Thus, about 480 B.C., the Greek philosopher Heracleitus (*ca.* 540–*ca.* 480 B.C.) maintained that dreams had no meaning outside a person's own thoughts.

IN ADDITION

In 492 B.C. the Persians seized control of Thrace and Macedonia to the north of Greece. In 490 B.C., a Persian force landed in the territory of Athens itself but was defeated at the Battle of Marathon. This kept Greece from falling under Persian domination. When Darius I died, his son and successor, Xerxes I (reigned 486–465 B.C.), had to deal with an Egyptian revolt.

In China, the philosopher Kung Fu-tzu (551–479 B.C.; usually known by the Latinized version of his name as Confucius) was coming to the end of his life. He did not found a religion but he did promulgate a philosophy of morality that has had great influence in China ever since.

440 B.C.

ATOMS

The Greek philosopher Leucippus (5th century B.C.) was the first to state categorically that every event has a natural cause. This rules out all intervention by the supernatural and represents the scientific view held today.

Leucippus's student Democritus (*ca.* 460–*ca.* 370 B.C.) adopted and extended Leucippus's notions. He maintained, from about 440 B.C., as Leucippus had earlier, that all matter was composed of tiny particles so small that nothing smaller was conceivable. Hence they were indivisible, and he called them *atoms* from a Greek word meaning "indivisible."

Of course Leucippus and Democritus had no evidence for their atomistic views. They were only speculations, and most other philosophers of the time vehemently rejected them. It was to be two thousand years before atomistic views began to gain ascendancy.

IN ADDITION

In 480 B.C. Xerxes sent a large army into northern Greece that forced its way southward to Athens and burned it. The Athenian population had escaped to the island of Aegina, however, where they were protected by the Athenian fleet. After the Battle of Salamis on September 23, 480 B.C., and the Battle of Plataea the next year, the Persians were driven out of Greece, the Greek cities on the Persian coast were freed, and the Athenians established a naval empire in the Aegean Sea. By 460 B.C., the Athenians had entered a Golden Age, in which art, drama, philosophy, and history were all represented by a sudden flourishing of genius, under the leadership of the statesman Pericles (*ca.* 495–429 B.C.). The Athenian city-state had a population of about 250,000 at this peak in its history, but a third of its people were slaves.

At about this time China entered the Iron Age, some five hundred years after the Middle East.

420 B.C.

EPILEPSY

Following the rationalist view, the Greek physician Hippocrates (*ca.* 460–*ca.* 377 B.C.) maintained that all diseases had natural causes and were not to be viewed as divine visitations or punishments.

In particular, about 420 B.C., he applied this rule to epilepsy, a disease whose sufferers were likely to fall to the ground suddenly and behave as though they were no longer in control of their moaning, twitching bodies. It was called the "sacred disease," and epileptics were thought to be in the grip of gods or demons. Hippocrates, on the other hand, sought a physical cure or amelioration.

Hippocrates believed that health depended on the proper balance of the four *hu-*

mors (or *fluids*) of the body: blood, phlegm, bile, and black bile. In this he was wrong, but at least he sought the cause of disease among natural phenomena, and in that he was not wrong.

IN ADDITION

The *Peloponnesian War* between Sparta and Athens (each with its allies) began in 432 B.C. and involved all of Greece. A plague in 429 B.C. killed thousands, and thereafter the war settled down to a slogging match that slowly ruined the country.

400 B.C.

CATAPULTS

The Greeks of this period were good at war. They had developed *hoplites* (from a Greek word for "heavy shield"), or heavily armed foot soldiers. The hoplites' helmets, breastplates, and leg armor were made of good steel. They carried a shield on one arm (instead of around the neck) and a sword in the other. They also had long spears to thrust with, rather than to hurl. They were trained to fight in close formation as a unit—it was not the individual champion but the weight of the entire formation that counted. A line of hoplites could wipe out the lightly armed disorderly mob that made up most non-Greek infantry, and it was for that reason that the Greeks managed to defeat the enormous Persian Empire.

The most important Greek city in the West was Syracuse, on the eastern coast of Sicily, which reached its period of greatest power under Dionysius (reigned 405–367 B.C.). He encouraged work on new weapons, and about 400 B.C. his workers devised the *catapult* (from Greek words meaning "to hurl down"). In its first form it was like a giant bow that was immobile and took many men to cock. When it was released, however, it hurled down upon a city's walls, not a little arrow but a huge rock—or hurled it over the wall and into the city.

It was the first long-range weapon that could hurl heavy objects, or the first piece of *artillery* (from a French word relating to a bow, which was the first long-range weapon).

The one big disadvantage of the catapult was its slowness. The enemy could see the cocking going on and had plenty of time to prepare for the blow or avoid it. Nevertheless it was a premonitory example of things to come.

IN ADDITION

A short period of peace during the Peloponnesian War was broken when the Athenian general Alcibiades (*ca.* 450–404 B.C.) persuaded Athens to launch a huge naval attack against Syracuse in 415 B.C. When Alcibiades' enemies framed him on a blasphemy charge and secured his recall, he fled to Sparta, and the attack on Syracuse ended in utter disaster for the Athenians. Sparta won a complete victory over Athens in 404 B.C.

387 B.C.

ADVANCED SCHOOLS

The Greek philosopher Plato (*ca.* 428–*ca.* 348 or 347 B.C.) founded a school in the western suburbs of Athens in 387 B.C. Intended for advanced study, it might be called the world's first university. Because it was on the grounds that had once belonged to a legendary Greek named Academus, it came to be called the *Academy*.

Plato's pupil Aristotle (384–322 B.C.) founded a school of his own in Athens in 335 B.C. It was called the *Lyceum*, because the building it occupied had been dedicated to Apollo Lyceus, god of shepherds. Aristotle's lectures at the school were collected into nearly a hundred and fifty volumes, representing a one-man encyclopedia of the knowledge of the times. Much of it represented the original thought and observations of Aristotle himself.

Some fifty of these volumes have survived through a fortunate chance. They were found in a pit in Asia Minor about 80 B.C. by soldiers of the Roman general Lucius Cornelius Sulla (138–78 B.C.). They were then taken to Rome and copied.

IN ADDITION

For a time the Athenians suffered under a right-wing tyranny, but they threw this off. In 399 B.C., they condemned to death the most eminent philosopher in history, Socrates (*ca.* 470–399 B.C.). This is usually considered a great blot on the Athenian democracy, but Socrates was himself a right-winger and several of his followers had been antidemocratic. Plato was one of his students and wrote up his teachings in a book that has lived ever since, presenting Socrates in a better light, perhaps, than he deserved.

In Italy, Rome was still a small city of not much account, fighting endless wars with its neighbors. In 390 B.C., tribes of Celts called Gauls invaded Italy from the north and sacked Rome. The Gauls then settled permanently in the Po Valley, leaving a shattered city behind them. No one at that time would have expected to hear anything further from Rome.

350 B.C.

OTHER CENTERS OF THE UNIVERSE

At this time it seemed obvious to almost everybody that the Earth was solid, immovable, and the center of the Universe, with everything in the sky moving about it. It certainly looked that way, and why should the evidence of one's eyes be denied?

Nevertheless, the Greek philosopher Philolaus (5th century B.C.), a student of Pythagoras (see 520 B.C.), felt that the Earth, with all the visible planets including the Sun, rotated about a central fire, which could not be seen. He was the first person we know of to suggest that the Earth moved and was not at the center of the Universe, but his suggestion was more mystical than rational and it won little credence.

The Greek astronomer Heracleides Ponticus (*ca.* 390–after 322 B.C.) did not go so far. He felt the Earth was the immovable center of the Universe, but he pointed out, about 350 B.C., that Mercury and Venus were never very far from the Sun. This could be accounted for by schemes the Greeks worked out that had each planet circling independently about the Earth, but to do so was difficult. Heracleides maintained that it was much simpler to suppose that Mercury and Venus circled the Sun, and that the Sun, with these two subsidiary bodies in attendance, circled the Earth. He was the first to suggest that there was at least some heliocentrism to the Universe, that at least some bodies revolved about the Sun and only secondarily about the Earth.

LOGIC

Everyone reasons after a fashion. It is impossible not to. Primordial hunters would reason from footprints that an animal had passed that way and would identify it from the nature of the markings. Everything you do, if you are in a normal state of mind, has some reason behind it. Unfortunately, however, there are innumerable ways of reasoning in a faulty manner, and reasoning in general can be swayed by emotions, by self-interest, and so on. The result is that people frequently, and under some circumstances almost always, behave in an irrational manner.

Aristotle (see 387 B.C.) was the first thinker we know of who undertook to work out a legitimate system of reasoning (*logic,* from the Greek for "word"). His book *Organon* developed the study of logic in great and satisfying detail, describing the art of reasoning from premise to necessary conclusion and thereby demonstrating how to establish the validity of a line of thought.

SPHERICAL EARTH

Anyone looking at the Earth can see that the land surface is bumpy and uneven but on the whole flat. This is particularly true if we look out over the surface of a lake.

The first person we know of who seems to have suggested that the Earth was *not* flat but spherical was Pythagoras (see 520 B.C.). It was Aristotle (see 387 B.C.), however, who summarized the reasons, possibly about 350 B.C., reasons that still hold today.

As one moves north, stars rise above the northern horizon and sink below the southern horizon, while as one moves south, the reverse happens. The shadow of the Earth on the Moon during a lunar eclipse is always a circular arc. When ships sail away from you at sea, the hull always seems to disappear before the superstructure, and this happens in the same way in any direction. All these indicate the Earth to be spherical.

These arguments were accepted by educated people even at low points in intellectual history, yet there are people today, with an education of sorts, who cling to something equivalent to a Flat-Earth Society. This, of all antiscience movements, seems the most indefensible, and one suspects that they are either joking or just a little mad.

FIVE ELEMENTS

Aristotle (see 387 B.C.) also summarized earlier thoughts on the elements of which the world was composed. Thales (see 580 B.C.) had suggested that it was composed of water, while later philosophers had named other candidates.

Aristotle suggested that it was made of four elements—earth, water, air, and fire—

and that it was built in consecutive shells. At the center was a ball of earth, surrounded by a ball of water (through which some earth protruded in places). This was in turn surrounded by a ball of air, and then, still farther out, by a ball of fire (sometimes visible as lightning).

The heavens themselves, however, were fundamentally different from the world, in Aristotle's view, and were made up of a fifth element, which he called *aether* (from a Greek word meaning "blazing"). After all, the heavenly bodies were all luminous, while the world was dark, except when it reflected light. The heavenly bodies moved in endless circles, while on Earth bodies fell or rose. The heavenly bodies were unchanging and incorruptible, while on Earth everything changed and deteriorated.

This view of Aristotle's turned out to be wrong, but to this day we still speak of "fighting the elements" when we face wind and rain, and when we want to say that something is the pure and abstract representation of anything we call it a *quintessence,* from Latin words meaning "fifth element."

ANIMAL CLASSIFICATION

Aristotle (see 387 B.C.) was a careful and meticulous observer who was fascinated by the task of classifying animal species and arranging them into hierarchies. He dealt with over five hundred animal species in this way and dissected nearly fifty of them. His mode of classification was reasonable and in some ways strikingly modern.

He was particularly interested in sea life. He observed that the dolphin brought forth its young alive, nourishing its young by a special organ called a placenta before birth and by milk after it was born. No fish did this, but all mammals did, so Aristotle classified the dolphin with the beasts of the field rather than with the fish of the sea. It took biologists generally about two thousand years to catch up with Aristotle in this regard.

Classification is important in itself, for it helps to organize a field of study. In the case of biology it was particularly important, for it led eventually to thoughts of biological evolution.

STAR MAPS

The Greek mathematician Eudoxus (*ca.* 400–*ca.* 350 B.C.), perhaps about 350 B.C., drew a better map of the Earth than Hecataeus had managed (see 510 B.C.), and was the first to attempt a map of the sky.

The sky was more difficult to map than the Earth was. On Earth there were physical landmarks: coastlines, rivers, mountain ranges, and so on. In the sky there were only stars.

The reasonable thing to do was to *create* landmarks, so Eudoxus drew imaginary lines diverging out from the pole star, and other imaginary lines meeting the first set at right angles. The diverging lines are what we now call *longitude,* and the ones at right angles are *latitude.* In this way Eudoxus could locate stars unmistakably in the otherwise featureless sky.

IN ADDITION

The disaster of the Peloponnesian War taught no lesson to the Greeks. The fighting among the cities became even worse. For example, Thebes, which was under Spartan domination, struggled to gain its own freedom.

Epaminondas (*ca.* 410–362 B.C.), who led the rebellion against Sparta, organized the Theban army in such a way that one wing had forty-eight rows of hoplites that could strike like a battering ram. It was called a *phalanx* (from the Greek word that originally meant "battering ram"). Epaminondas further arranged that it strike first, with the rest of the line coming up afterward. The Theban phalanx smashed the Spartan line at the Battle of Leuctra in 371 B.C., and with that one military defeat, the Spartans lost their leadership forever.

Thebes could not maintain its own leadership, however, for the other cities immediately turned against it.

In Macedon, in northern Greece, a remarkable man named Philip II (382–336 B.C.) became king in 359 B.C. He organized the Macedonian army into a magnificent fighting force and began a clever program designed to make himself ruler over all of Greece. One Athenian politician, Demosthenes (384–322 B.C.) realized the danger and did his best to rouse the Athenians, but because of the damage they had suffered over the previous century, they could not rise to the occasion.

Meanwhile, far to the west, Rome, though damaged by the Gallic invasion, recovered faster than the towns around it and slowly asserted its power over them, assuming leadership of the *Latin League.*

320 B.C.

BOTANY

The Greek scholar Theophrastus (*ca.* 372–*ca.* 287 B.C.) was a student of Aristotle (see 387 B.C.) and headed the Lyceum after the latter's retirement. He was interested in the plant world and wrote a book about 320 B.C. that included descriptions of 550 plant species. It was the first systematic book on botany and included some species from as far away as India.

IN ADDITION

When Philip II (see 350 B.C.) was assassinated, his young son succeeded as Alexander III (356–323 B.C.). The new king proved even more capable than his father. He defeated the Greeks all around in lightning strokes, crossed over into Persia in 334 B.C., and in ten years of fighting conquered all the Persian Empire, destroying vast armies and never losing a battle. He has been called Alexander the Great ever since. He died in Babylon after a drunken carouse.

It was because the Persian Empire was now entirely in Greek hands that Theophrastus was able to study plant species from as far away as India.

In Italy, the only power that rivaled Rome in the center of the peninsula were the Samnite tribes to the east of the Latin League. They were a formidable foe, and all the time that Alexander was fighting in Persia, the Romans were fighting the Samnites in central Italy. By 320 B.C., the war was far from decided, and it was still possible for the Romans to be crushed.

312 B.C.

ROADS

Once carts were developed, roads were needed. Vehicles could not progress rapidly over rocks and underbrush, and if they tried, the wheels would be quickly ruined. That meant that everywhere roads, reasonably wide, reasonably straight, and reasonably smooth, had to be built. This Rome understood.

In the years after its humiliation by the Gauls, Rome had perfected the *legion,* which was a much more flexible formation than the phalanx. The phalanx could *only* fight in close order, and any unevenness in the ground would upset that order. The legion, on the other hand, could disperse on uneven ground without falling into disarray and could come together again when conditions allowed.

In 312 B.C. Appius Claudius (4th–3rd century B.C.), a high Roman official, initiated the building of the *Appian Way,* the best road the world had yet seen. It extended from Rome to Capua, a distance of 132 miles. At first it was covered by gravel, but eventually it was paved with blocks of stone and extended to the heel of Italy.

The road allowed for the rapid movement of troops and gave Rome an enormous advantage in bringing up reinforcements or achieving surprise. Eventually Rome would build 50,000 miles of roads throughout its dominions. Some were over 30 feet wide. This meant that Roman armies could be hurried from one border to another at quick step, and relatively small forces could protect its boundaries.

IN ADDITION

Alexander the Great died without leaving any obvious heir. His generals, all of them capable men and tested war-leaders, fell out at once and tore the empire apart. By 312 B.C., eleven years after Alexander's death, it was still not certain what the outcome would be.

300 B.C.

GEOMETRY

Geometry, as a practical study, may have begun in Egypt, where the building of pyramids and the necessity of reestablishing boundaries after the Nile flood made it essential. The Greeks made it theoretical, however, working with ideal points, lines, curves, planes, and solids. They attempted to prove things by reason alone and without actual measurement. (Reason was the mark of the philosopher, measurement was only for the artisan, and the Greek scholars were snobs. Snobbery turned out to be useful in mathematics, though not in experimental science, where the Greeks fell badly short.)

A number of Greek mathematicians contributed to the development of geometry, particularly Eudoxus (see 350 B.C., Star Maps). It was Euclid (fl. *ca.* 300 B.C.) who brought geometry to maturity, however. He worked in Alexandria, Egypt, and thereby hangs a tale.

The city of Alexandria, on the seacoast at the westernmost branch of the Nile delta, had been founded by Alexander III (see 320 B.C.) and named after himself. It was a largely Greek city, though it contained many Egyptians and Jews as well. It quickly became the largest and most cosmopolitan city of the Greek world. Ptolemy I (*ca.* 366–*ca.* 283 B.C.), who ruled over Egypt after Alexander's death, established his capital at Alexandria.

Ptolemy I saw himself as a patron of the arts and sciences and founded the *Museum,* so-called because it was an institution devoted to the Muses, the patron goddesses of learning. He and his son, Ptolemy II (308–246 B.C.), made the Museum the largest and most important of all the ancient universities. Associated with it was the largest of all the ancient libraries.

The Ptolemies encouraged scientists and thinkers generally to come to Alexandria and subsidized them well. Euclid, a Greek mathematician who may have studied at the Academy, moved from Athens to Alexandria as a result, symbolizing the "brain drain" that followed, as Greeks poured out of Greece proper into the new dominions of the *Hellenistic kingdoms* that succeeded Alexandria.

Perhaps about 300 B.C., Euclid began compiling all the geometrical findings of earlier mathematicians into a textbook eventually called *Elements.* He added comparatively little himself, but what he did was beyond price.

He began with a minimum number of statements that were so self-evident that they required no proof. From these *axioms,* he proceeded systematically to prove theorem after theorem, each proof depending only on the axioms and on previous proofs, so that geometry was given a firm foundation and structure.

It was the most successful textbook ever written and has been used, in more or less modified form, to this day.

TIDES

The Greeks were not quite the sea-voyagers the Phoenicians had been, though by this time they could follow the Phoenicians' path and sail all over the Mediterranean. Only one Greek, Pytheas (fl. 300 B.C.), followed the Phoenicians out of the Mediterranean and into the Atlantic, however.

He too sailed north to the British Isles and even farther north to *Thule,* which may have been either Norway or Iceland. He also sailed past Denmark and into the Baltic. His tales of what he had seen, although moderns find them authentic-sounding, met with disbelief on the part of his contemporaries, and nothing came of his explorations.

From the scientific standpoint, the most important observations he made were of the tides. The Mediterranean is a virtually tideless sea, because by the time some of the high tidal water flows through the narrow Strait of Gibraltar and raises the sea level a couple of inches, it is time for it to flow out again.

In the Atlantic Ocean, however, Pytheas observed the existence of true tides and described them—and was disbelieved.

ARTERIES

The Greek physician Praxagoras (4th century B.C.) distinguished between the two kinds of vessels we know as arteries and veins. However, he thought arteries carried air (they are usually empty in corpses). That turned out to be wrong, but the idea remains in the name, which is from Greek words meaning "air-carrier." He also noticed that the brain and the spinal cord were connected.

IN ADDITION

In 301 B.C. the final battle among the generals of Alexander the Great was fought at Ipsus. Twenty-one years had passed since the death of Alexander, and all the fighting merely confirmed the fact that the empire was forever broken up. The generals and their descendants called themselves kings, and the Hellenistic kingdoms continued to fight each other as the Greek city-states had done and with the same result. All grew weaker.

280 B.C.

BRAIN

The Museum at Alexandria was the site of important early work on anatomy by Herophilus (*ca.* 355–*ca.* 280 B.C.) and his successor, Erasistratus (fl. 250 B.C.). Both were particularly interested in the brain and the nerves. About 280 B.C., Herophilus divided nerves into sensory (those that received sense impressions) and motor (those that stimulated motion). He described the liver and spleen as well, described and named the retina of the eye, and named the first section of the small intestine the duodenum. He noticed that the arteries pulsed, and thought that they carried blood, not air.

Erasistratus distinguished between the cerebrum (the main section of the brain) and the cerebellum (the smaller section behind it). He was struck by the fact that the convolutions of the human brain were more numerous than those in other animals and suggested that this was related to superior human intelligence. Unlike Herophilus, he did not think the arteries carried blood.

This promising beginning came to a sudden end. The Egyptian population believed it was necessary to keep the body intact if a decent status in the afterlife was to be achieved, and public opinion forced an end to all dissection at the Museum. The study of the human body ceased for over fifteen centuries as a result.

SIZE OF MOON AND SUN

It is easy to suppose that the heavenly bodies are insignificant in size compared to the enormous Earth. After all, the stars are just specks of light on the vault of a sky that looks as though it barely clears Earth's high mountains. The Sun and Moon have visible orbs, but they seem small too. To suggest otherwise would certainly have been considered foolish, or worse.

Thus, when the Greek philosopher Anaxagoras (*ca.* 500–*ca.* 428 B.C.) suggested that the Sun was a rock the size of southern Greece, he horrified the Athenian conservatives, who accused him of irreligion, brought him to trial, and forced him into exile.

Two centuries had passed since then, and the Greek world had expanded enormously. With its horizons broadened, daring thoughts were better tolerated, and the Greek astronomer Aristarchus (fl. *ca.* 270

B.C.) was the first to try to determine the size of the heavenly bodies.

About 280 B.C., he noted the size of the shadow cast by Earth on the Moon and, following a correct line of mathematical argument, estimated that the Moon was a body with a diameter one-third that of the Earth. His result was a little high because he lacked the instruments with which to measure the shadow accurately.

Aristarchus also tried to determine the relative size of the Moon and the Sun by trigonometry. He noted that at the time the Moon was in its half-phase, the Moon, Sun, and Earth were at the apexes of a right triangle. Thus, if the angles were measured, the lengths of the sides could be calculated. Aristarchus's mathematics was correct but again he was victimized by the lack of instruments with which to make accurate measurements. He ended by deciding that the Sun was twenty times as far from the Earth as the Moon was and that the Sun had therefore seven times the diameter of the Earth. This turned out to be a gross underestimate, but Aristarchus was nevertheless the first to show, by scientific reasoning, that the heavenly bodies were objects comparable to the Earth in size.

A consideration of the Sun's huge size possibly led Aristarchus to maintain that the Sun, and not the Earth, was the center of the Universe, and that the various planets, including the Earth, revolved about the Sun. He had no evidence for this, and the notion did not convince anyone. Even if the Sun were a huge body, it was considered an insubstantial ball of light, and the thought of the solid, heavy Earth revolving around it seemed ridiculous.

LIGHTHOUSE

The Hellenistic realms did not hesitate to show their advanced technologies in the form of large architectural undertakings. The island of Rhodes, for instance, celebrated its successful resistance to a siege by a Macedonian general in 305–304 B.C. by having a large statue of the Sun-god constructed overlooking its harbor. That statue was 105 feet high and was completed in 280 B.C. It was called *the Colossus of Rhodes.* It stood for sixty years before being destroyed by an earthquake, and after its destruction its size was greatly exaggerated.

In Alexandria a much more useful and even larger structure was built, the first major lighthouse. It was called the *Pharos* from the name of the spit of land on which it was built. It was at least 280 feet high, rested on a bulky square base, and had stairs up which loads of resinous wood had to be carried. (No elevators, of course.) The light of the burning wood could be seen 35 miles out to sea. It, too, was completed in 280 B.C. and it stood for sixteen centuries before being destroyed by an earthquake.

Both the Colossus of Rhodes and the Pharos were listed by the ancients as among the *Seven Wonders of the World.*

IN ADDITION

Rome now dominated all of central Italy between the region of the Po, where the Gauls had settled, and the southern regions, which were dominated by Greek city-states. Of these city-states the foremost was Tarentum. Fearing Rome, it called for help from Pyrrhus (319–272 B.C.), king of Epirus, who was the most skilled at using the phalanx. For the first time the Romans were going to have to face the Greeks on the field of battle.

270 B.C.

WATER CLOCK

The sundial gave people an idea of the passage of hours, but it did so only by day when the sun was out, and it wasn't portable.

Other ways of measuring time were also used, since any process that remained steady over a period of time would serve the purpose at least crudely. There was the hourglass, in which fine, dry sand dripped from an upper chamber to a lower one in a known time. There were candles that burned down a given length in a known time, which could be marked off to indicate the passage of hours. In Egypt and in China, water dripping from an upper chamber to a lower one had long been used as a way of telling time.

About 270 B.C., a Greek inventor, Ctesibius (2d century B.C.), devised a model of the water clock that gained great popularity. It included a float in the water, which rose as the water in the lower chamber accumulated. This float was attached to a notched rod that turned a gear as it pushed upward, which in turn twisted a pointer that marked off numbers from 1 to 12. The Greeks called the device a *clepsydra* (water-stealer), since the water quietly leaked out of the upper chamber.

Water clocks could tell time well enough to govern the length of time a person might talk in court or before an assembly, but they were nevertheless crude timepieces at best.

IN ADDITION

Pyrrhus defeated the Romans at Heraclea in southern Italy, since the Romans had not previously encountered the phalanx or the elephants that Pyrrhus had brought with him. But in 275 B.C., the Romans defeated Pyrrhus outright at the Battle of Beneventum and he was forced to return to Greece. The Romans annexed the Greek city-states and controlled all of Italy south of the Po Valley.

Carthage, just across a narrow section of the Mediterranean from Italy, was at the height of its prosperity at this time, and it viewed the rising power of Rome with concern. Between itself and Italy was the island of Sicily, of which Carthage controlled the western portion. The eastern portion was controlled by the Greek city-state of Syracuse, which was in alliance with Rome.

260 B.C.

LEVER

Levers were used in prehistoric times. It is no great trick for a lively mind to try to pry up a rock with a long stick and to find that it works better if another, smaller rock is placed under the stick to give the stick something to push against. It would then quickly be discovered that the closer the small rock is to the big rock being pried up, the easier the prying gets.

Nevertheless, the precise mathematics of

lever action was not worked out until the Greek scientist Archimedes (*ca.* 287–212 B.C.) did it about 260 B.C.

You might say, "What difference does it make that scholars worked out fancy theories and mathematics for levers, when practical people had been actually *using* such devices for thousands of years?"

The point is that use without theory is largely hit and miss. Advances are made, yes, but slowly. Once a useful theory is worked out, however, it is like removing a blindfold. It becomes obvious how a device might be improved, or what new observations need to be made. With a theory, advances speed up enormously.

Therefore we give Archimedes credit for the principle of the lever, regardless of how long the lever had been in use before his time.

Archimedes also worked out the principle of buoyancy, the manner in which any object immersed in a fluid displaces a volume of fluid equal to its own volume. This provided a way of measuring volume, a way of explaining why some things float and some don't, and so on. Archimedes grasped the principle quite suddenly when he lowered himself into a public bath and noticed the water overflow.

The story is that he sprang out of the bath and raced home nude, shouting "Eureka! Eureka!" ("I have it! I have it!"). He had been given the problem of checking whether a golden crown was adulterated with a less dense metal or not, without damaging the crown. For that he had to know the volume, and the buoyancy effect would give it to him. (The ancient Greeks, by the way, did not mind nudity, so Archimedes' action was not as bizarre as might be thought.)

IN ADDITION

War between Rome and Carthage was inevitable, and it started in 264 B.C. in a quarrel over Sicily. This was *First Punic War* (*Punic* was the Roman way of saying *Phoenician*). At first Rome was helpless, since it had no ships and Carthage had the best navy in the world. However, a Carthaginian ship was wrecked on the Italian coast, and a Greek from southern Italy used it to design similar ships for Rome. The Romans placed beaks on their ships so they could drive them into the Carthaginian vessels and fix them together. The Romans could then fight little land battles on the ships' decks. In 260 B.C., the Romans won a naval battle in this manner, and the war began to swing to the Roman side.

240 B.C.

SIZE OF EARTH

Even after the Earth was known to be spherical, there was the question of how large that sphere might be. It was bound to be huge, since no traveler had ever been totally around it. There was always more Earth, unknown and as yet unvisited, up ahead.

But then a Greek scholar, Eratosthenes (*ca.* 276–*ca.* 194 B.C.), working at Alexandria, found a way to measure the Earth's circumference without leaving Egypt. At the summer solstice, he had been told, the Sun at noon cast no shadow at the city of Syene (modern Aswan), which was far south of Alexandria. That meant that the Sun was then directly overhead at Syene. At the same time, though, the Sun was 7 degrees from the zenith in Alexandria. This difference had to result from the fact that the Earth curved between Syene and Alexandria. Knowing the north-south distance from Syene to Alexandria, Eratosthenes could use mathematics to calculate how far it would take the curvature of the Earth, worked out between Syene and Alexandria, to take him all the way around the Earth.

He came out with a length of 25,000 miles for Earth's circumference. He was correct.

The ancients, however, thought his figure was too high and preferred to accept a lower one.

CHRONOLOGY

No two political groupings among the ancients counted the years in the same way. Usually it was by strictly local methods—a given year was "the year when so-and-so ruled the city," or "the seventh year of King so-and-so." Not only was it difficult to match the chronologies of one political unit with another, but even within a given political unit things were hazy if you forgot the order of the rulers or how long each had ruled.

Eratosthenes (see above) was the first person to try to make sense out of chronology and to match one system with another. He did his best to stretch dating back to the time of the Trojan War.

Meanwhile, Alexander the Great's general Seleucus I (*ca.* 358–281 B.C.) had marched into Babylon in 312 B.C. and counted that as year 1 of the *Seleucid Era*. The years were counted upwards indefinitely after that without regard to the succession of monarchs.

The actual year in which events took place in ancient times is still a bit hazy, and the farther back one goes the hazier it gets. However, thanks to the establishment of the Seleucid Era and to the work of Eratosthenes, there is less problem with dates than one might expect, especially for those after 312 B.C.

IN ADDITION

By 241 B.C., the First Punic War was over, and the Romans had won. They took over western Sicily as their first province. The Carthaginians, humiliated, planned revenge.

Ptolemaic Egypt, which was shrewd enough to make an alliance with Rome, reached the peak of its power under Ptolemy III (reigned 246–221 B.C.).

India was under the rule of Aśoka from about 273 to 232 B.C. He controlled almost all the peninsula and would have taken the rest but for his pacifist Buddhist principles. His reign was unusually enlightened.

214 B.C.

GREAT WALL

Although China had by now been civilized for some two thousand years at least, and though its technology and science were remarkable—in some ways they continued to be in advance of the West until the beginning of modern times—I have little to say about it. There are two reasons for this.

First, in 221 B.C. a new dynasty began to rule over China, the Ch'in, the first emperor of that dynasty being Shih Huang Ti (259–210 B.C.). Shih Huang Ti was a reforming emperor who wanted to make a new start and therefore had all books burned (except for those in the practical arts) to free the land from the stultifying hand of the past. And indeed, the land took its name, China, from this dynasty. History before Shih Huang Ti is most uncertain because of the destruction.

Secondly, we can only judge scientific advance by its effect on the world of today. Some discovery that was made a long time ago but came to nothing must be more or less ignored. Discoveries and inventions can only count when they affect society. As it happens, the modern world was carved out by Europe when it engaged in the Age of Exploration in the fifteenth and sixteenth centuries, in the Scientific Revolution in the sixteenth and seventeenth centuries, and in the Industrial Revolution in the eighteenth and nineteenth centuries. History naturally must concern itself with all people and all cultures in all aspects, but in this book, I concern myself with scientific advances that affect our life today, and for that reason consider whatever happened prior to the contemporary era only insofar as it affected Europe. This is not provincialism on my part but a decision based on the way the world has developed.

Nevertheless, sometimes a non-European event must be discussed even if it did not directly affect Europe at the time, and here we have a case.

From its earliest history, China found itself threatened by the nomads of central Asia, who were only too ready to raid the hard-working Chinese peasantry and carry off their crops for food and them for slaves.

It struck Shih Huang Ti that the best thing to do would be to build a wall across the countryside, a large wall, tall and wide, not so much to keep the nomads out, as to keep their horses out. The nomads were helpless without their horses, and though human beings may climb even a difficult wall, a horse cannot.

This wall was begun in 214 B.C. Initially it was of earth, but later it was made of brick. It eventually extended for 1,500 miles, from the Pacific Ocean to a point deep in central Asia. It had periodic watchtowers, and on the whole it did its job. China was not made invulnerable by the Great Wall but it was surely made less vulnerable.

The Great Wall is the largest construction project ever carried out, the only object made by human beings that outdoes the pyramids (which were, to be sure, built twenty-five centuries earlier).

IN ADDITION

Meanwhile in Europe, Carthage tried to make up for what it had lost in Sicily by building a new empire in Spain. When Rome tried to interfere with Carthaginian efforts in Spain, the Carthaginian general Hannibal (247–183 B.C.), one of the greatest generals the world has ever seen, had his chance for revenge. In 218 B.C. he crossed the Alps and came down into the north Italian plain, catching the Romans entirely by surprise.

He then defeated a Roman army at Trebbia, a still larger Roman army at Lake Trasimene, and a larger Roman army still at Cannae in 215 B.C. No general had defeated the Romans so disgracefully for centuries, and no general was to do it again until centuries later. In 215 B.C. it looked as though Rome would be crushed.

170 B.C.

PARCHMENT

Throughout ancient times, papyrus had been the material on which people had written, but Egypt was the only source, and the papyrus reed could not be grown quickly enough to supply the demand. Besides, the Egyptian rulers were not anxious to have other nations build up libraries. So when the small Hellenistic kingdom of Pergamum in western Asia Minor, under Eumenes II, who ruled from 197 to somewhere around 160 B.C., wished to build a library that would rival the one at Alexandria, naturally the Ptolemies would not cooperate by shipping the necessary papyrus.

The scholars working under Eumenes II therefore invented a new way of treating hides, about 170 B.C. Hides had often been used as writing material, but the Pergamese learned to stretch them, scrape them, and clean them, so that they ended with a thin white sheet that could be covered with writing on both sides. In later times, it came to be called *parchment,* which may be a mispronunciation of *Pergamum.*

Parchment is much stronger than papyrus and lasts practically forever. It can be scraped clean and used again (not always a good thing), whereas papyrus cannot be. Its greatest flaw is that it is much more expensive than papyrus. Then too, parchment cannot be formed into long sheets that can be rolled into a volume. Instead, separate sheets have to be glued together into a *codex,* which is the form in which modern books appear.

IN ADDITION

After the disaster at Cannae, Rome fought a very cautious war against Hannibal, refusing battle and hoping to wear him down. This worked largely because the conservative Carthaginian government did not wish Hannibal to grow too powerful and therefore refused him support.

Eventually the Romans sent an army into Africa to threaten Carthage itself. The faithful Hannibal hastened back to defend the city, and in 202 B.C. he was finally defeated. The vengeful Romans took over all Carthaginian dominions, including Spain, leaving only the city itself. Rome was now supreme in the western Mediterranean.

The Romans then turned on Philip V, who ruled Macedon from 238 to 179 B.C., and who had helped Carthage. They defeated him in Thessaly in 197 B.C., drove him out of Greece, and forced him to pay a large tribute.

Meanwhile, Antiochus III, who ruled over the Seleucid Empire from 223 to 187 B.C., turned to face the Romans, whom he imagined he could easily defeat. Instead, the Romans defeated him in 190 B.C. and again in 189 B.C., and after that Hellenistic kingdoms rarely dared challenge Rome.

150 B.C.

DISTANCE OF THE MOON

In astronomy it is necessary to work with angles. You can't measure the distance between two heavenly bodies by holding a yardstick against the sky. You can only measure the angle you must turn your head in looking first at one, then at the other.

If you make the angle part of a right triangle, then the sides have fixed ratios to one another. The ratios have names like *sine, cosine,* and *tangent.* They are examples of *trigonometric functions.*

The Greek astronomer Hipparchus (fl. 146–127 B.C.), usually considered the greatest of ancient astronomers, was the first to make up careful tables relating angles to side ratios, so that if you knew the angle, you could look up the ratios, and vice versa. For this reason, Hipparchus is usually considered the founder of trigonometry.

Hipparchus used trigonometry to calculate the distance from the Earth to the Moon. First he noted the Moon's position against the stars from different positions on Earth, because when your point of view changes, a relatively near object seems to change position in comparison to a relatively distant one; this is called *parallax.* The smaller the shift with a fixed change of position, the greater the distance to the nearer object. Once he had measured the parallax of the Moon, he could determine by trigonometry the distance of the Moon, at least in terms of the Earth's size. In this way Hipparchus calculated that the Moon was at a distance equal to 30 times the Earth's diameter.

If, as Eratosthenes had determined, Earth had a circumference of 25,000 miles, it had to have a diameter of 8,000 miles. That meant that the Moon was 30 × 8,000 miles, or 240,000 miles away (which is very nearly correct). That's nearly a quarter of a million miles away, and yet the Moon was well known to be the *nearest* of the heavenly bodies.

This was the first indication that the Universe was far larger than had been thought, but it was also a dead end. Parallaxes grow smaller as objects grow more distant, and the Moon is the only object close enough to give a measurable parallax to the unaided eye.

Hipparchus also worked out in detail the mathematics of an Earth-centered planetary system.

IN ADDITION

As a result of further wars, Macedon was made a Roman province by 148 B.C., and the land of Philip and Alexander the Great lost its independence forever.

134 B.C.

STAR MAP

In 134 B.C. Hipparchus (see 150 B.C.) observed a star in the constellation Scorpio of which he could find no record in previous observations. This was a serious matter, for the heavens were supposed to be eternal and changeless. Was it really a new star, or had Hipparchus simply overlooked it earlier?

Hipparchus determined to prepare a star map so that from then on any astronomer sighting a star that seemed new could check it against the map. In order to make his map, he plotted the position of each star according to its latitude and longitude, as Eudoxus had done (see 350 B.C., Star Maps). Hipparchus included over a thousand stars, and both in quantity and in accuracy it was a far better map than anything attempted before. What's more, it occurred to Hipparchus to transfer the system of latitude and longitude to maps of the Earth. That has been done ever since.

In the course of preparing his map, Hipparchus compared the locations of stars with those recorded by earlier astronomers and discovered a uniform shift of all the stars from west to east, at a rate that would bring them completely around the sky in 26,700 years. Because it meant that the vernal equinox came a bit ahead of its previous mark each year, this motion was called the *precession of the equinoxes*.

Perhaps it was about this time, too, that Hipparchus divided the stars into classes, later called *magnitudes*. The brightest twenty were of the *first magnitude;* those somewhat dimmer were *second magnitude,* and so on, until *sixth magnitude* stars were just barely visible.

IN ADDITION

Rome, unable to forgive Carthage, launched an unprovoked attack on it and, after a three-year war, destroyed the city completely in 146 B.C. Its nearly seven-century existence was over.

In western Asia, an even older city vanished. Babylon had been declining for some time, and now the city that four centuries earlier was the largest in the world was gone forever.

As Rome loomed larger and larger over the Mediterranean world, China under its new Han dynasty became more united and stronger than ever before. Trade between the two great nations was beginning at about this time, though because of the distance between them it was only a trickle.

100 B.C.

GLASS-BLOWING

Glass-making was, for many centuries, a slow and tedious affair, so that glass was rare and was used only for ceremonial purposes. The turning point came about 100 B.C., apparently in Syria, when someone discovered that molten glass could be blown out of a pipe like a soap bubble, so that a round hollow shape could be made at once, and that shape could easily be changed into enchanting curves, with other bits of glass fitted onto it. The whole glass vessel could then be allowed to cool and be broken off the blowpipe. In this way, artistic vases and cups and vessels of all kinds could be made. Glass at once grew cheaper and more common and was popular throughout the Mediterranean world. It remained colored, however. The art of producing colorless glass was still not known.

IN ADDITION

Rome had no longer anything to fear from the military might of any nation in the Mediterranean world. All were either conquered or had been converted into puppets. But that didn't mean there was nothing to fear at all.

Even if organized governments could no longer resist Rome, there were still barbarians from outside the Mediterranean world who could invade, puppet kingdoms that could corrupt the Roman government, and slaves in Italy who could rebel. Yet Rome managed to handle all these things.

85 B.C.

WATERWHEELS

Human beings began by using their own muscles for power and eventually added the muscles of tamed animals. But animals, like human beings, have to be fed and cared for. Would it be possible to use inanimate forces for power, forces always present and requiring no care?

One such power was the wind, which could fill sails and push a ship across a water surface against the force of a current. Was there a force that could turn a millstone and grind grain? That must be done every day all day if the world was to be fed.

Someone at some time thought of using a river current to do so. The water would turn a wheel by flowing past blades emerging from it, and the wheel would turn the millstone to which it was attached by appropriate gears. Waterwheels were also used to power other kinds of devices.

This lifted some of the burden of work from humans and animals. The first mention of it is in a poem written in 85 B.C., though it probably came into use earlier than that.

IN ADDITION

There were still Hellenistic kingdoms in eastern Asia Minor that dreamed of throwing out the Romans. Before 85 B.C. they experienced certain successes, but the Roman general Lucius Cornelius Sulla (138–78 B.C.) put an end to these.

46 B.C.

LEAP YEAR

The Romans, whatever their success in war, politics, and law, were poor in science. There was not one really important Roman scientist. The Romans left science to the Greeks, and as Roman fortunes grew and Greek fortunes declined, science too declined into what was eventually a dark age.

It is not surprising then that the Romans used a calendar far worse than those used by any of the nations to its east. And since that calendar was occasionally manipulated by political priests for partisan advantage, it grew worse with time rather than better.

The Roman statesman Gaius Julius Caesar (100–44 B.C.) admired the Egyptian solar calendar, however, and brought in a Greek astronomer, Sosigenes (1st century B.C.), to work out a version of that calendar for use at Rome. Thus originated the *Julian calendar* (named in Caesar's honor) of 365 days, with some months of 30 days and some of 31, plus an added day every four years *(Leap Year).* The additional day was added because the year is actually 365¼ days long, and in this respect the Julian calendar was an improvement over the Egyptian. (The Julian calendar, with a small correction made sixteen centuries later, is still in use today.)

IN ADDITION

In 48 B.C. Caesar became dictator of the Roman dominions. He was assassinated on March 15 (the famous *Ides of March*), 44 B.C.

By that time, the only Hellenistic monarchy that still existed was Ptolemaic Egypt, which was ruled by Queen Cleopatra VII, from 51 to 30 B.C.

25

EARTH'S ZONES

Anyone who traveled was bound to find that the climate was different in different places. In the north European forest, it was colder than it was in Greece, with longer, snowier winters. In Egypt it was warmer than in Greece, and cold weather was rare.

The idea was formalized by the Roman geographer Pomponius Mela (1st century A.D.) about A.D. 25. (From now on, any dates not marked B.C. will be A.D.) Pomponius Mela, accepting a spherical Earth, suggested that it be divided into a North Frigid Zone and a South Frigid Zone in the neighborhood of the poles, a Torrid Zone in the neighborhood of the Equator, and a North Temperate Zone and a South Temperate Zone in between. This notion still holds today, even though variations in climate are far more complicated than would be thought from a consideration of zones alone.

IN ADDITION

After Julius Caesar's assassination in 44 B.C., his grand-nephew, Gaius Octavius (63 B.C.–14), kept the formal trappings of Roman government but made himself supreme. In 27 B.C. he adopted the name of Augustus Caesar, and that is usually taken as the moment when the Roman Republic ceased to be and the Roman Empire took its place. By that time Egypt had become a Roman province.

In 4 B.C., or perhaps a little earlier, Jesus was born. In the year 29 he was crucified.

50

PHARMACOLOGY

The Greek physician Pedanius Dioscorides (*ca.* 40–*ca.* 90) served in the Roman armies and had an opportunity to study the plant life in large parts of the Mediterranean world. He was interested chiefly in the medical applications of plants, and in his book *De Materia Medica,* he described about six hundred plants and nearly a thousand drugs. This was the first important work on *pharmacology* (from Greek words meaning "the study of drugs").

STEAM POWER

Although Alexandria was now Roman and well past its great days, the Museum and the Library still existed, and a Greek engineer, Hero (1st century), worked there at about this time. Hero constructed a hollow sphere to which two bent tubes were attached, the openings pointing in opposite directions. When water was boiled in the sphere, the steam escaped through the tubes and, as a result of what we now call the law of action and reaction, caused the sphere to rotate rapidly. (The modern lawn sprinkler works in precisely this fashion, using the force of flowing water rather than steam.)

Hero had produced a steam engine. This does not represent the invention of the device, since it did not affect society. It is mentioned only as a curiosity and because it makes one wonder what might have happened if Greek science had continued unabated, uncrushed by the weight of Roman lack of interest.

IN ADDITION

When Augustus died in 14, his stepson, Tiberius (42 B.C.–37), succeeded to the throne. The emperorship continued in Augustus's family till 68, but in fact no reasonable system of succession was ever established, so that the throne was often up for grabs, which helped destabilize the empire.

105

PAPER

About 105, a Chinese eunuch, Tsai Lun (50?–?118), invented a method for making a thin, smooth writing surface so like papyrus that in Europe the name was retained. (In English, the new surface was called *paper*, a word obviously derived from *papyrus*.) The advantage of paper over papyrus was that, instead of being made from an increasingly rare reed, as papyrus was, it could be made from bark, hemp, rags, even low-quality wood—almost any form of useless cellulose. Since cellulose is the most common organic compound, there has never been any long-term shortage of paper.

It took a thousand years for knowledge of paper to reach Europe.

IN ADDITION

In 79, Vesuvius, a mountain near the city of Naples, which had not erupted in the memory of human beings, *did* erupt and buried the towns of Pompeii and Herculaneum.

Under Marcus Ulpius Traianus (Trajan) (reigned 98–117), the Roman Empire attained its maximum extent, for he conquered Dacia (modern Romania), Armenia, and Mesopotamia. The population of the empire at this time may have been as much as 40 million.

China, under the Han dynasty, had also reached a peak, with a population of approximately 50 million. The two empires, taken together, contained at that time about a third of the Earth's population.

140

GEOCENTRIC UNIVERSE

Claudius Ptolemaeus (2d century), better known as Ptolemy, was the last important astronomer of the ancient world. He wrote a summary of ancient astronomy, known later to the Arabs as *Almagest* (the greatest). He drew largely from Hipparchus.

In this synthesis, he described the Earth as the center of the Universe and all the planets as going around it in combinations of circular motions. In order to account for the visible motions of the planets across the skies, those combinations of circular motions had to be complicated indeed, but Ptolemy worked out mathematical methods for predicting planetary motions that seemed adequate to his contemporaries and to future generations for fourteen centuries. (His chief instrument was an astrolabe, a device for determining the latitude of the heavenly bodies. It had been invented a couple of centuries before and is considered the oldest of scientific instruments.)

IN ADDITION

In 135, the Roman Emperor Hadrian (reigned 117–138) crushed the Jews remaining in Judea and evicted them from the land. From then on, united only by their religion, they remained a people without a land, scattered over the world, for eighteen centuries. Hadrian abandoned the outlying provinces conquered by Trajan. There were to be no further Roman conquests.

180

SPINAL CORD

A Greek physician, Galen (129–*ca.* 199), worked at a gladiatorial school at Pergamum, the city of his birth, and there was able to gain some rough-and-ready hints on human anatomy. In Rome, from 161 on, he could dissect only animals, which misled him now and then as far as human anatomy was concerned.

Nevertheless, he did good work on muscles, identifying many for the first time, and showed that they worked in groups. He also showed the importance of the spinal cord by cutting it at various levels in animals and noting the extent of the resulting paralysis.

IN ADDITION

In 165, a plague—it may have been smallpox—struck the Roman Empire. In 167 came the first full-scale attack by barbarian tribes from the north. After the death of Emperor Marcus Aurelius in 180, the story of the Roman Empire is of a long decline and fall.

The city of Rome at this time held a million people, possibly a million and a half, and was the largest city in the world.

250

ALGEBRA

Through most of Greek history, mathematicians concentrated on geometry, although Euclid (see 300 B.C.) considered the theory of numbers. The Greek mathematician Diophantus (3rd century), however, presented problems that had to be solved by what we would today call algebra. His book is thus the first algebra text.

He is best known for problems that had to be solved with whole numbers, and such problems involve what are still called *Diophantine equations.* He also showed that fractions could be treated as numbers, thus reducing much of the discomfort they usually caused.

IN ADDITION

By about 250, the Chinese had invented gunpowder, but they never really used it for much besides fireworks and as a psychological weapon with which to frighten the enemy. They also began to use tea, which gave boiled water a pleasant taste and cut down on the infections that would follow if they drank water that had not been boiled.

300

ALCHEMY

Creating chemical change has been part of human life from the start. Cooking involved chemical changes and so did fermentation. The production of pottery out of clay, metals out of ore, charcoal out of wood, and glass out of sand all involved chemical changes.

It was not until the years after Alexander the Great (see 320 B.C.), however, that scholars began to study chemical change systematically. This may have been the result of a fusion of Greek and Egyptian thinking, and it flourished first in Ptolemaic Egypt.

As Euclid, in Egypt, summarized ancient geometry, and Ptolemy, again in Egypt, summarized ancient astronomy, so Zosimus (240–?), also in Egypt, about the year 300, summarized ancient alchemy.

The early work in alchemy was highly mystical and not very useful and was sidetracked eventually in a vain effort to find some way of changing "base metals," such as lead or iron, to gold. Nevertheless, inquiring minds, even when misled, could not help making some discoveries in the long run, and this the alchemists did.

STIRRUPS

The Greeks and Romans relied on infantry. Stolid, well-trained foot soldiers, whether organized into a phalanx or a legion, could withstand horsemen, so that the cavalry was reduced to auxiliary importance in Greek and Roman times. Cavalry could charge in an attempt to panic the enemy, and could chase the enemy once a panic had set it, and could fight each other, but cavalry rarely decided a battle.

The horse and chariot had faded in importance as a larger horse was bred capable of running with the full weight of an armed soldier on its back. Saddles made it easier to sit on a horse's bony spine, but riding was still precarious, and thrusting with a spear was dangerous, for if the thrust was parried the rider might easily be pushed off the horse. It was safer to shoot arrows from a distance.

In India, about 100 B.C., the notion arose of having a leather loop suspended from the saddle, into which the big toes could be thrust in order to steady the rider. The Chinese, living in a colder climate and wearing shoes, needed a wider loop into which the whole shoe might be thrust. By 300, such *stirrups* (from an old Teutonic word meaning "climbing rope," because they could be used to hoist oneself up on a tall horse) were made of metal and were wide enough to allow the foot to withdraw quickly in case of need.

The stirrup made it possible for a rider to

sit firmly in the saddle and strike at an enemy with spear or sword. From the Chinese, the notion of the metal stirrup spread to the central-Asian nomads and from there westward.

IN ADDITION

The Roman Empire was continuing to fade. Since 180, it had been suffering under the periodic blows of invading Germanic tribes from the north. Every once in a while, a capable emperor would throw them back, as did Claudius II (who reigned from 268 to 270) and Aurelian (who reigned from 270 to 275). Then, every once in a while, an emperor would reorganize the Empire in an attempt to make it stronger, as did Diocletian (who reigned from 284 to 305). All this merely postponed the end, however. The empire on the whole grew weaker and the invaders grew stronger.

400

WHEELBARROWS

A wheelbarrow is essentially a one-wheeled cart, with the wheel far in front so that it acts as the fulcrum and, by lever action, allows considerable weights of material to be lifted clear of the ground—more than could be carried by a human being alone. It does not require the use of an animal, and it can be maneuvered easily along narrow lanes and crowded streets.

It was invented in China about 400 (perhaps considerably earlier) but did not reach Europe for centuries. It is one of those inventions that, in hindsight, seems so obvious that one wonders why it wasn't invented the instant the wheel was—but hindsight is cheap.

IN ADDITION

The Roman Emperor Constantine I (who reigned from 306 to 337) accepted Christianity in 313. He founded a new city on the site of older Byzantium and named it Constantinople after himself. As the weight of the empire shifted eastward, Constantinople began to replace Rome as the imperial capital.

The last strong emperor to rule over the whole of the Roman Empire was Theodosius I, who reigned from 379 to 395. On his death, the eastern half of the empire was given to his older son, Arcadius, who reigned in Constantinople till 408; the western half was given to his younger son, Honorius, who reigned in Ravenna, Italy, till 423. The Roman Empire was never united again.

As the use of stirrups continued to spread, cavalry became unstoppable, and for a thousand years war became aristocratic again, since only the ruling classes could afford horses. The middle class and peasantry could only rarely stand up against their rulers.

537

DOMES

A dome is a semispherical structure on top of a building, which looks impressive in itself and gives an opportunity for vertical windows that will introduce light inside. (Skylights on a flat roof are less impressive and are a source of weakness.)

The Romans first introduced domes, placing one on the Pantheon, a building begun in 27 B.C. This was the largest dome built before modern times. It is heavy, however, and is built on a circular building, with no openings except one at the very top, so its esthetic appeal is limited.

About 480 architects in the East Roman Empire perfected a system of placing a hemispherical dome upon a square support in such a way that the bottom of the dome could be pierced by many windows without sacrificing its strength.

This discovery was given its chance when the East Roman Emperor Justinian (who reigned from 527 to 565) decided to rebuild the church of Hagia Sophia after its destruction during a period of rioting. The ruins were cleared, a larger area was marked off, and for six years, ten thousand laborers worked on it. The large dome was so cleverly designed, so skillfully pierced with windows, that the whole interior of the church —108 feet across and 180 feet high—was bathed in sunlight. The enormous dome, as seen from below, seemed to have no support at all but to be suspended from heaven.

IN ADDITION

After 400 the Roman Empire could no longer withstand the tribal invaders from the north. In 476 the last West Roman Emperor was forced into retirement, and for this reason 476 is considered the date of *the fall of the Roman Empire,* but the East Roman Empire remained intact.

The Huns, who entered the Roman Empire about 410, were the most feared of the invaders. Under their king, Attila (406?–453), they eventually reached central Gaul—the farthest west any central-Asian invaders were ever to penetrate. They were defeated at the Battle of Chalons in 451, however. Two years later, Attila died and the Hunnish Empire melted away.

At about this time the Polynesians, who had been sailing over the vast Pacific without compasses by following the stars and the currents, were settling island after island in the greatest overall feat of ocean navigation of all time. About 450 they reached Hawaii.

And at the same time, the Mayan civilization in what is now Central America was building what was to be their chief city, Chichen Itza.

552

SILK

Chinese legends have it that silk culture was introduced in China in 2640 B.C., but this may be taken with a certain skepticism.

During the time of the Roman Empire, silk reached the Western world over the *Silk Road,* which ran the width of Asia. With every tribe en route charging to let the silk through, it arrived in Rome worth its weight in gold. Since the Roman aristocracy insisted on silk and on other exotic luxuries from the East, however, the trade balance was heavily against Rome, and this undoubtedly contributed to its decay.

Then a neo-Persian Empire was formed, which was so hostile to the Romans that silk was not allowed to pass through at all.

Justinian (see 537) therefore arranged to have two Persian monks who had lived in China go there and bring back silkworm eggs secreted in hollow bamboo canes. In 552 Constantinople began silk production itself, and from that time to this the West has had its own silk.

IN ADDITION

Pagan learning had just about come to an end. The Library at Alexandria had suffered much depredation at the hands of Christian fanatics, and Justinian in 529 had put an end to the Academy, which had been founded by Plato nine centuries before.

600

MOLDBOARD PLOWS

The Slavs of eastern Europe were hard-working peasants who were exposed in their vast plains to invasion from the north and the east. Both the Goths and the Huns dominated them, and later on, so would other invaders. (The word *slave* may have come from *Slav* because they seemed so easily enslaved.) Nevertheless, they endured and multiplied and contributed an important advance.

About 600, they are supposed to have invented the *moldboard plow.* This had a knife blade, or *coulter,* that cut deep into the ground, a plowshare that cut grass and stubble at ground level, and a shaped *moldboard* above the plowshare that lifted and turned the soil. This was a useful device for damp, moist ground. It wasn't needed in the light Mediterranean soil, but as it slowly spread through eastern and northern Europe, food production took a jump and population grew.

IN ADDITION

In 611, Khosrau II (king of the Sassanid Empire of Persia from 590 to 628) began an invasion of the East Roman Empire. He had amazing success, taking all of the Roman dominions in Asia, together with Egypt, by 619. At the same time, Asian invaders called Avars swept through the Balkans. Nothing seemed left but the city of Constantinople itself and the North African province. However, the head of that province, Heraclius (575–641) became Roman Emperor in 610. In 622, he took an army into Asia and, like a new Alexander, totally defeated the Persians. By 630, the East Roman dominions had been entirely restored.

Meanwhile in Arabia, a young man named Muhammad (570–632) began preaching a new religion called *Islam* (submission to the will of God). On September 20, 622, he was forced to flee his home city of Mecca for Medina. This was the *Hegira* (from an Arabic word for "flight"), and from that day, his followers, the Muslims, count their years.

673

GREEK FIRE

The Arabs, emerging from their peninsula in 632 B.C., began an amazing series of conquests that in the space of fifty years seemed to bring to life again the old Persian Empire with Arabia and North Africa added. Nothing more seemed needed but the city of Constantinople itself before all the European dominions of the old Roman Empire would fall. In 673 the Arab army was just across the Hellespont from Constantinople, and their fleet was offshore. It seemed that nothing could possibly save the city.

There was in the city, however, an alchemist named Callinicus (7th century), of Egyptian or Syrian birth, who had arrived in Constantinople as a refugee.

He had invented a mixture containing naphtha, plus potassium nitrate and calcium oxide, perhaps (we don't know the exact recipe), which not only burned but would continue to burn on water even more fiercely. This *Greek fire* was spurted out of pipes into the paths of the wooden ships of the Arabs. The fear of being set on fire, and the horror of watching fire burn on water, forced the Arab fleet to retreat and Constantinople was saved.

IN ADDITION

Even as the Roman Empire seemed to be coming to an end, the Han Empire, its alter ego in China, *did* come to an end in 618 with the murder of the last Han emperor. China, however, did not fall apart. A new dynasty, the T'ang, was instituted, which was to be even more successful.

700

PORCELAIN

About 700, the Chinese learned to make *porcelain,* a kind of pottery that was shiny, almost glassy, very hard, and very white. What's more, it rang like a bell when struck. Porcelain eventually reached Europe, where it was known as *china* and where it became the material of choice for dishes (when it could be afforded), replacing wood, ordinary pottery, and metal.

About this time, other products of the Orient were making their way toward Europe—notably cotton and sugar from India.

IN ADDITION

At the time when the Huns were terrorizing western Europe, refugees fled to certain offshore islands near the northern end of the Adriatic. There they lived by fishing and by manufacturing salt from seawater. Gradually, the islands grew into a city named Venice, and in 687, it elected its first *doge* (duke, or leader) and prepared for a thousand years as a Mediterranean power.

750

ACETIC ACID

Once the Arabs had conquered the territories of the old Greek Hellenistic kingdoms, they were exposed to the old Greek books on science, and they loved them. Whereas Greek learning had been almost forgotten in western Europe, the Arabs preserved it and translated the great books of Euclid, Aristotle, Ptolemy, and others into Arabic. For several centuries the Arabs were the leading scientists of the Western world, excelling in astronomy, medicine, and alchemy.

The greatest of the Arabian alchemists was Jābir ibn Hayyān (*ca.* 721–*ca.* 815), known to Europeans later as *Geber.* He sought for methods of forming gold, and also for some mysterious dry powder (*elixir,* from Arabic words meaning "the dry one") that would do the job. It was thought that such a magical substance could also cure all disease, and it was known as the *elixir of life* or the *panacea* (from Greek words meaning "all-healing"). Centuries of effort went into a useless search for this substance.

Geber, however, in his researches, also managed to make important discoveries. Up to his time, the strongest acid known was vinegar, a dilute solution of acetic acid. By distilling vinegar, Geber obtained purer samples of acetic acid, which were, naturally, stronger than vinegar. This was important, for until then the one agent known to induce chemical change was heat. Acids, if strong enough, are another important agent for change, and they bring about changes other than those brought about by heat.

IN ADDITION

The Muslims took to the sea well and became traders. By 701, they were visiting the Indonesian islands and bringing back spices. These lent taste to food and helped to mask high odors and tastes in food that was slightly spoiled (a common matter when there was no such thing as refrigeration). The spices eventually reached Europe and were a powerful driving force in setting off the Age of Exploration.

The Muslims continued to expand. They were finally stopped at the Battle of Tours by the Frankish leader Charles Martel (*ca.* 688–741), who had developed a heavy artillery—large horses in armor, carrying knights in armor. They were like living tanks.

A second Arabic attempt to take Constantinople, in 718, met determined opposition, which drove the Muslims off. However, the shrunken remnant of the East Roman Empire (largely Asia Minor and the Balkan Peninsula) are referred to after the Muslim conquest as the *Byzantine Empire.*

In Central America, the Mayan civilization was now at its height.

770

HORSESHOES

Horses are by far the most useful animals. Strong and fleet, they were indispensable in war and could be indispensable on the farm if they could be used properly. A moldboard plow, cutting deep into heavy damp soil (see 600) needs a strong pull.

One step forward was to take care of the horse's tender hooves, which could easily be hurt by rocks and pebbles. About 770, iron horseshoes were coming into common use, and that took care of that. What was still needed, however, was some method of harnessing that would allow a horse to pull hard without closing down its own windpipe.

IN ADDITION

In 751 Charles Martel's son Pépin III (714?–768) persuaded Pope Stephen II (who held the seat from 752 to 757) to declare him king of the Frankish realm. In return, Pépin protected Stephen against the Lombards, a Germanic tribe that now controlled most of Italy, and in 755 gave the Pope a strip of land in central Italy to rule for himself. This was the *Donation of Pépin,* and it created the *Papal States,* which would endure for eleven centuries. King Pépin was the founder of the *Carolingian line.*

In 754 the Muslim Empire came under a new dynasty, the Abbasids, who were to rule in a new capital, Baghdad, to which they moved in 762. Under the Abassids, the Muslim Empire reached its peak.

810

ZERO

Human beings had been manipulating numbers ever since they had begun writing, about twenty-three centuries earlier. In general, there would be a tendency to make score marks for units, so that 4 would be / / / /. Different marks would be introduced for fives, tens, and fifties, to avoid having to make too many score marks. Or else, as in the case of the Jews and Greeks, letters of the alphabet would be used (which introduced nonsignificant connections between words and numbers and introduced the superstitious folly of numerology).

It might have occurred to someone to use the same numbers for units, tens, hundreds, and so on, merely placing the numbers in different positions for each level, as on an abacus (see 500 B.C.). No one tried this positional notation, however, because no one thought of using a symbol for an abacus level in which no beads had been moved.

For example, if you want to indicate 507 on an abacus, you move 5 beads at the hundreds level and 7 at the units level. You can record the 5 and the 7, but how do you indicate that the tens level hasn't been touched?

About the year 500, some Indian mathematician suggested that such an untouched abacus level be given a special symbol. (Our symbol is 0 and we call it *zero*.) This meant that you could no longer confuse 507 with 57 or 570. The Arabs may have picked it up from the Hindus about the year 700.

The first important mathematician to make use of this *positional notation* was an Arab, Muhammad ibn Al-Khwarizmi (780–850), who wrote a book featuring it about 810. (In the book he coined a term that in English became *algebra*.)

The new system slowly penetrated Europe, which took centuries to give up its clumsy Roman numerals and take up the new *Arabic numerals* (although, of course, the numerals were Indian to start with). It took centuries to overcome the habit of sticking to something inconvenient but customary rather than adopting something good but new. Still, it was done in the end, and the transition democratized arithmetical computation, bringing it within reach of everyone.

IN ADDITION

Pépin III, king of the Frankish realm, died in 768 and was succeeded by his sons, the elder of whom was Charles (742–814). He was successful in all he did and was called Charles the Great (Charlemagne in French—and it is as Charlemagne that he is best known in English, too). He destroyed the Lombard kingdom, drove the Muslims back in Spain, and Christianized still-pagan Germans at the point of the sword. On Christmas Day in the year 800, he was crowned Emperor of the West by Pope Leo III (who held the seat from 795 to 816). Although after Charlemagne the western empire proved to be nothing much (it was called the *Holy Roman Empire,* because it was formed and existed under the Pope's auspices) and only had power under an unusually strong ruler, it continued to exist for a thousand years.

The people of Scandinavia at this time began to make their mark on history. They were the Viking sea-rovers, who struck at England in 787 and at Ireland in 795. That was only the beginning.

850

COFFEE

In many regions of the world, it is necessary to treat water before it is safe to drink. Alcohol will kill germs, so that many people drank beer or wine rather than water. (They didn't know about germs, but beer or wine tasted better anyway.) Others boiled water and added tea leaves to get rid of the flat taste.

The Muslims were not allowed to drink wine and didn't know about tea. Naturally, they would be on the watch for something.

Coffee may have originally grown wild in Ethiopia in the province of Kaffe and been brought to southern Arabia. There, in 850, according to tradition, a goatherd noticed that his goats acted frisky after eating berries of the plant. He tried it himself, liked the sensation, and told others. In time, people learned how to roast the beans obtained from the berries, steep them in boiling water, and produce coffee. Centuries later, coffee was introduced to western Europe.

IN ADDITION

The three grandsons of Charlemagne fought each other as though they were Greek city-states, and in 843, they signed the Treaty of Verdun, which split up Charlemagne's empire forever. The western half developed into France, the eastern half into Germany.

Viking raiders attacked all the coasts, even entering the Mediterranean. Central power broke down, and various landowners took over the defense of their own lands. Feudalism was established.

Vikings from Sweden entered Russia and established their capital at Kiev. Thus, Russia entered history.

The Abbasid Empire reached the height of its power as Arabs took the island of Crete in 826 and invaded Sicily in 827, making themselves supreme in the Mediterranean. Thereafter, however, the Abbasids declined rapidly.

870

ARCTIC CIRCLE

The Vikings were sea-raiders and the terrors of the European coasts in the ninth and tenth centuries, but they were great sea-explorers as well. Among Europeans, they were the greatest since the Phoenicians, thirteen centuries earlier.

A Viking named Ottar sailed northward in 870 out of nothing more than sheer curiosity, apparently. He said he wanted to see how far north land existed, and whether it was populated. He succeeded in rounding the northern end of the Scandinavian peninsula (North Cape), and sailing on eastward, he eventually entered the White Sea.

When passing the North Cape, Ottar was 125 miles north of the Arctic Circle. He was the first human being, as far as we know, ever to cross the Arctic Circle by sea.

IN ADDITION

Two missionaries, Cyril (*ca.* 827–869), and his brother, Methodius (*ca.* 825–884), spread Christianity among the Slavs in what is now Czechoslovakia. They are supposed to have invented the *Cyrillic alphabet,* based on the Greek, which is still used by the Russians, Bulgarians, and Serbs.

874

ICELAND

Between 500 and 800, Ireland had experienced a kind of Golden Age of learning, which was ended by the coming of the Vikings.

An Irishman, Brendan (*ca.* 484–578), is supposed to have sailed northward about 550 and explored the islands off the Scottish coast, the Hebrides to the west and the Shetlands to the north, and he may have probed further. There are stories that the Irish reached Iceland and settled there, but if so, they did not stay for long or they died out.

In 874 a Viking chieftain named Ingolfur Arnarson sailed westward and landed in Iceland, which is 650 miles west of Norway. By that time the Irish settlers if any were gone, or if any remained, they were killed by the Vikings. In either case, it was the Norse who founded the first *permanent* colony in Iceland.

This was Europe's first expansion to new lands overseas, but it took place in a vacuum. Outside of the people actually involved, the Vikings, Europe knew nothing of the event.

IN ADDITION

In 871 Alfred (849–899) came to the throne in England. He was the most capable of all the Anglo-Saxon monarchs.

900

HORSE COLLARS

With moldboard plow (see 600) and horseshoes (see 770), all that was needed to convert the horse into a farm animal was some good way of harnessing it.

In 900, or even some time before, the horse collar came into use. This allowed the horse to pull with the shoulder instead of the windpipe and increased the horse's available force fivefold.

Now, finally, the food supply and therefore the population began to increase in northern Europe. For the first time, power began to shift from the Mediterranean area, where it had been from the beginning of civilization, to the north—a process that was to continue for nine centuries.

IN ADDITION

Alfred of England defeated the Danes in 878 and forced them to adopt Christianity. The Danes were not totally crushed, however, and new invasions were bound to come.

982

GREENLAND

Once Iceland was settled by the Vikings (see 874), tales arose of another island to the west, and in fact there was an enormous one only 200 miles away.

In 982, an Icelander, Erik Thorvaldson (10th century) (usually called *Erik the Red* from the color of his hair) was sent into a three-year exile for some reason and used it to explore westward. He discovered the island, and by 985 he was back in Iceland searching for settlers who would volunteer to go there. He called the place *Greenland*, in a shameless effort to make it sound attractive.

In 986 the first settlers arrived and established a colony on the southwestern shores of the island. Despite the horrible climate, the Vikings clung to that patch of Greenland for over four centuries. Again, though, the rest of Europe knew nothing about the venture.

IN ADDITION

The T'ang dynasty of China came to an end in 907.

Russian ships ventured into the Black Sea to attack Constantinople. They, too, were driven off by Greek fire—the last time it was used.

The last major Viking raid against France was that of Rollo (860–931) in 911. He was bought off by Charles III, who was known as Charles the Simple and who reigned over France from 893 to 923. Rollo and his men were given a section of the channel coast for their own use, and thus *Normandy* (named for the Northmen) was established.

The Magyars, an Asian tribe, invaded Germany but were defeated at the Battle of the Lech in 955 by Otto I, who reigned from 936 to 973. He revived Charlemagne's Holy Roman Empire and was crowned emperor in 962. The Magyars settled down in what is now Hungary (named for the Huns, with whom the Magyars were confused).

1000

VINLAND

A Viking named Bjarne Herjulfson told, in 1000, of being caught in a storm and driven past Greenland to land still farther west. The son of Erik the Red (see 982), Leif Eriksson (fl. 1000), went exploring westward to check the matter.

Eriksson found the land we now call Labrador and Newfoundland, and he called it *Vinland* (land of vines), perhaps in an attempt to dress up the new discovery in brighter colors. In 1002 a settlement was made in Vinland but it didn't last long. Internal quarrels and resistance from the Native Americans brought it quickly to an end.

Europeans had now for the first time landed on the soil of North America—but of this, Europeans, other than the Vikings, were unaware.

IN ADDITION

Under Basil II, who ruled over the Byzantine Empire from 976 to 1025, that realm became a strong military power for the last time.

1025

OPTICS

The Arabic physicist known to later Europeans as Alhazen (965–1039) was the first to maintain that vision was made possible by rays of light falling on the eye and was *not* the result of the eyes giving out rays of light, as physicists had thought till then.

Alhazen also worked with lenses and attributed their magnifying effect to the curvature of their surfaces and not to any inherent property of the substances making them up. His work represented the beginning of the scientific study of *optics*.

IN ADDITION

Sweyn I of Denmark, who ruled from about 987 to 1014, conquered England in 1013, died soon afterward, and was succeeded by his son Canute, who reigned till 1035. Danish overlordship was mild, and Canute was on the whole popular with his subjects. In Ireland, however, the Vikings were finally driven out in 1014 by Brian Boru, who reigned over Ireland from 1002 to 1014.

In central America, the Mayan civilization began a rapid decline. Historians are not sure of the reason for this.

1050

CROSSBOWS

The greater the force with which a bow must be bent, the greater the force with which the arrow will be sent forth when the tension is released. The greater the force of the shot, the greater the range and the penetrating power. Naturally, the larger the bow, or the stiffer, the better—except that eventually human muscle doesn't suffice for pulling back the string of the bow.

In France, sometime about 1050, machinery was brought into play: the bow was drawn back by a two-handed crank or the equivalent. Eventually the bows were made of steel, and a short bolt was shot out that had a range of about 1,000 feet and could penetrate chain mail.

This was the first mechanized hand weapon, and the bolt it shot forth seemed so terrifying that the weapon seemed too horrible to use. At least a Church council of 1139 tried to ban its use except against non-Christians. (The ban didn't work.)

The chief disadvantage of the crossbow was its slowness. It took a long time to crank it up and make ready to shoot and once it was fired, the enemy might easily swoop down before it could be cocked again. (Hence the expression "to have shot their bolt," meaning to have taken action and to be helpless thereafter.)

IN ADDITION

England was under the rule of a native monarch again when Edward the Confessor (1003?–1066) became king in 1042. He was a mild king but greatly under the influence of Normans. Normandy had been under the rule of Duke William (*ca.* 1028–1087) since 1035, who turned out to be the most capable ruler of his time.

1054

NEW STAR

Hipparchus was supposed to have seen a new star some twelve centuries before (see 134 B.C.), but no European had seen one since, though Chinese astronomers had reported a number of new stars in that interval.

On July 4, 1054, a bright new star blazed forth in the constellation of Taurus. For three weeks it shone so brightly it could be seen in daylight. At its peak, it was two or three times as bright as Venus at its brightest and could cast a dim shadow. It remained visible for nearly two years before finally fading away into invisibility.

This new star was recorded by the Chinese astronomers, but in Europe it went unnoted (or at least no reference to it has survived). This is an indication of the low state of astronomy, and of science generally, in Europe, which was just then emerging from a five-century-long *Dark Age*.

IN ADDITION

A group of Normans under Robert Guiscard (*ca.* 1015–1085) established a kingdom in southern Italy in 1053, which for two centuries rose to its highest pitch of power since Greek days, thirteen centuries before.

The western Christian church, which acknowledged the leadership of the Pope at Rome, and the eastern Christian church, which acknowledged the leadership of the Patriarch at Constantinople, frequently quarreled over points of doctrine, and these quarrels were poisoned by the rival ambitions of the two religious leaders. In 1054, Pope Leo IX (who held the seat from 1048 to 1054) excommunicated the Patriarch, creating a permanent split between the Roman Catholic and the Greek Orthodox Church.

1066

COMET

Comets made periodic appearances in the sky. They were frightening because they came unheralded and followed an unpredictable path. Furthermore, their shapes were irregular, rather like a woman's head with long streaming hair as though in mourning. (*Comet* is from the Greek word for "hair.")

The unheralded coming made them appear like special warnings from heaven, and the streaming "hair" made it seem certain that the warning was of disaster. Sure enough, disaster always came when a comet blazed in the sky. (Disaster always came when no comet blazed in the sky, too, but people paid no heed to that.)

In 1066, there was a bright comet in the sky that attracted much attention, especially because of events that were than taking place in Normandy and in England.

IN ADDITION

Edward the Confessor of England had died in 1066, and William of Normandy wanted the throne. He launched an invasion force while the comet of 1066 blazed in the sky and while Harold II (*ca.* 1022–1066), the Anglo-Saxon contender for the throne, was in the north fighting off a Norse invasion. William made capital out of the comet, claiming it presaged disaster for Harold, and so it did. At the Battle of Hastings on October 14, 1066, Harold made several tactical mistakes, and William won. He took over England, becoming *William the Conqueror,* and reigning from 1066 to 1087 as William I of England. All later English monarchs, right down to Elizabeth II, the present queen, are descended from him.

1071

FORKS

Knives and spoons are of prehistoric origin, but forks are relatively new. Byzantine aristocrats used them at a time when everyone in western Europe, high and low, ate with their fingers. A Byzantine princess who married a doge of Venice brought her forks with her. The Venetian aristocracy picked up this obviously cleanly habit and the fashion spread.

Nevertheless, it did seem a bit persnickety to many people, who felt it was an example of false gentility and prissiness. To this day we sometimes hear the phrase "fingers were made before forks." So they were, and so were dirty fingers.

IN ADDITION

A Turkish tribe, the *Seljuk Turks* (so-called after an early tribal leader), became powerful in 1037. Their second *sultan* (from an Arabic word meaning "ruler") was Alp Arslan (*ca.* 1030–1072 or 1073). In 1071 he brought an army against the Byzantine Emperor Romanus IV Diogenes (d. 1071) at Manzikert in eastern Asia Minor. The Turks won overwhelmingly, and most of Asia Minor fell to them. The Byzantine Empire was permanently weakened as a result, and although it continued to exist for nearly four centuries, it depended on western help for survival from then on.

1137

FLYING BUTTRESSES

Roman architects had found it impossible to build tall structures without thick walls. When stone came to be used for roofs, the weight was all the heavier and the walls had to become enormously thick. Nor could there be more than a few narrow windows if fatal weaknesses were not to be introduced. The result was that the *Romanesque* churches of earlier centuries bore a predominant atmosphere of squat gloom.

In the twelfth century, however, the notion arose of designing large structures in such a way as to concentrate the weight of the roof in certain areas where outside buttresses of masonry could be built. For further strength, buttresses that stood well away from the building could be connected to the key points needing support by diagonal structures, called *flying buttresses.*

Since the buttresses carried the weight, those sections of the walls not directly involved in support could be left thin and pierced with numerous windows. When the windows were filled with stained glass, the interior of the structure would be drenched with colored light in a beautiful and impressive manner. What's more, now cathedrals hundreds of feet high could be built. For the first time, Egypt's Great Pyramid was surpassed in height.

The first important example of the new style was the Abbey of St. Denis, just north of Paris, completed in 1137 under the direction of the French statesman Suger (1081–1151).

Those who clung to the old tradition called the new architecture *Gothic* (that is, *barbarian*) in derision. The name stuck, but the insult vanished. Gothic architecture became one of the artistic glories of the twelfth and thirteenth centuries.

IN ADDITION

The Byzantine Empire, now under Alexius I Comnenus, who reigned from 1081 to 1118, was under attack not only by the Turks from the east but by the Normans from the west and appealed to the Western powers for help.

Pope Urban II, who was Pope from 1088 to 1099, was willing to help. First, he was anxious to free the Holy Land from the Turks. Second, because the population of western Europe had grown due to agricultural advances, the aristocracy had increased and there wasn't enough land to go around, so that fighting was endless. In 1095, therefore, Urban preached a Crusade, and landless knights flocked eastward along with crowds of others, out of religious fervor and out of a longing for loot, too.

What was really important about the Crusades was not the question of who would own the Holy Land but the fact that the European Crusaders came into contact with a more advanced civilization.

1180

WINDMILLS

The chief source of inanimate power continued to be the waterwheel. Unfortunately, the waterwheel only worked where a rapidly flowing stream existed, or where a stream could be dammed to produce a waterfall that would turn the wheel. What was needed was a source of inanimate energy not so constrained.

Moving air can turn a vaned wheel as well as moving water can, and its force was already well understood in connection with sailing vessels. What's more, moving air exists everywhere. The earliest windmills were developed in Persia about 700, and returning Crusaders brought back word of these devices.

The first windmill was built in France in 1180 and windmills quickly spread over western Europe. The wheels in the Middle East had usually been horizontal, but those in Europe were made vertical. The wheels could be turned so that they would catch the wind from whichever direction it blew, and eventually they were so designed that the force of the wind itself forced the mill to turn to face it. Windmills became a major way of securing the power to grind grain and pump water.

MAGNETIC COMPASSES

In the sixth century B.C. it was discovered (by a shepherd, according to legend) that a certain kind of ore attracted iron. Since the ore was found near the Asia Minor city of Magnesia, it came to be called the *Magnesian stone,* or in English, a *magnet,* and the phenomenon was *magnetism.* The phenomenon was first studied by Thales (see 585 B.C.). It was eventually found that stroking with the magnetic ore could turn a sliver of iron or steel into a magnet.

Somehow it was discovered that if a magnetic sliver was allowed to turn freely, it would come to rest pointing in a north-south direction. We don't know how this fact was discovered, but the Chinese were the first to be aware of it. It is referred to in Chinese books dating as far back as the second century.

The Chinese never used the magnet for direction-finding in navigation, because by and large they were not great navigators. The Arabs may have learned of it from them, however, and perhaps some Crusaders learned of it from the Arabs.

In 1180 the English scholar Alexander Neckam (1157–1217) was the first European to make reference to this directional ability of magnetism. As soon as the Europeans heard of it, they began to try to put it to use as a navigation aid, and they began to improve it. Eventually a magnetic needle was put on a card marked with various directions, and because the needle was free to move all around the card, it was referred to as a magnetic *compass* (from a French word meaning "to go around").

If a single point in time can be picked as the moment when Europe first took the road that was to lead to world dominion, it was the moment when Europeans heard of the magnetic compass and put it to use. With the compass, Europeans could eventually go wherever they wished over the wide oceans, so that they eventually dominated the whole world as no other relatively small group of people had ever done before or (in all likelihood) will ever do again.

IN ADDITION

A Second Crusade was preached in 1147 by the powerful French ecclesiastic Bernard of Clairvaux (1090–1153). It was led by Louis VII of France, who reigned from 1137 to 1180, and by Conrad III of Germany, who reigned from 1138 to 1152. It was a complete disaster.

1194

SPITZBERGEN

Icelandic Vikings apparently discovered islands they called *Svalbard* (cold coast), which are perhaps better known to non-Scandinavians as *Spitzbergen* (mountain peaks). These are 900 miles north of Iceland and 500 miles north of the northern tip of Norway.

They represent the farthest point north reached by the Viking explorers, or by any ships without compasses. The discovery, like all Viking discoveries, remained unknown to Europe generally.

IN ADDITION

After the failure of the Second Crusade, a capable Muslim leader arose in the person of Salāh al-Din Yūsuf ibn Ayyūb (1137 or 1138–1193), better known to Europeans as *Saladin*. He united large sections of the Muslims and drove back the Crusaders. In 1187 he recaptured Jerusalem, which had been in Christian hands for not quite ninety years.

A Third Crusade was organized in 1189, under the leadership of Richard I (the Lion-Hearted) of England, who reigned from 1189 to 1199, Philip II (Augustus) of France, who reigned from 1179 to 1223, and Holy Roman Emperor Frederick I (Barbarossa, or Redbeard), who reigned from 1152 to 1190. Frederick died en route, and Philip and Richard spent their time quarreling. It is not surprising, then, that the Crusade was a failure and that Jerusalem remained Muslim.

1202

ARABIC NUMERALS

An Italian mathematician, Leonardo Fibonacci (*ca.* 1170–after 1240), had occasion to travel widely in North Africa, since his father was a merchant. There he learned of Arabic numerals and positional notation, which had been advocated by Al-Khwarizmi (see 810).

Fibonacci wrote a book on the subject in 1202, *Liber Abaci (Book of the Abacus)*. This served to introduce Arabic numerals to Europe, but Roman numerals held their own for three more centuries before succumbing.

IN ADDITION

The Italian port cities of Venice, Genoa, and Pisa, particularly Venice, were at the peak of their commercial prosperity. They traded with the Byzantine Empire (what was left of it) and with the Muslims. This enabled Italian scholars like Fibonacci to be at the forefront of intellectual advance.

1228

COAL

The first fuel used for fire, and still a very common one, is wood. Wood grows constantly, so that ideally it should last as long as Earth does in approximately its present form. However, it is possible to use wood faster than it replaces itself and, in fact, this becomes inevitable as population grows and the uses for fire increase.

Coal (from an old word meaning a "burning ember") is actually the remains of very ancient wood. When coal was found, it was burned. When quantities were found partly buried, the ground was sometimes dug into to find additional quantities. There are records of coal having been burned in China about 1000 B.C., in ancient Greece, among pre-Columbian Native Americans, and so on.

For a long time, these were merely opportunistic operations, but surface coal was getting hard to find, and people started digging in earnest—first in China.

In England, coal-mining became a serious operation in the early thirteenth century, and by 1228 London was receiving shipments of coal by sea from Newcastle. (Londoners called it *sea-coal,* for that reason.)

Coal continued to be burned as a substitute for wood, all the more so as England was gradually being deforested.

IN ADDITION

A Fourth Crusade was preached, and one of its leaders was Enrico Dandolo (1107?–1205), Doge of Venice. He had been blinded when he was on a diplomatic mission to Constantinople nineteen years earlier, and he did not forget. Though ninety-two years old at the time of the Crusade, he managed to divert it against Constantinople, which was in the midst of a civil war. Constantinople was taken and mercilessly sacked in 1204. By then, it was the last place in the world where the full corpus of Greek literature existed—and when it was destroyed, only scraps were left to us.

In England, King John, who reigned from 1199 to 1216, had to face a rebellion by his nobles and, in 1215, was forced to sign the *Magna Carta,* which defined his powers and the rights of the nobles. The common people were not involved, but insofar as it limited the powers of the king, it was rightly considered a step away from autocracy.

The central Asian nomads were coming under the influence of someone who may have been the greatest military genius of all time, a Mongol known as Genghis Khan (*ca.* 1162–1227). By the time he died, he had conquered northern China, Afghanistan, and Iran and plundered northern India.

1241

RUDDERS

To steer ships, it was customary for someone to hold a broad oar out back of the ship and by turning it this way and that cause the ship to curve in its path as desired. Eventually it occurred to someone to make the steering mechanism part of the ship itself and to control it from within the ship. This is a *rudder,* from an old word related to the verb *to row.*

The Arabs had it first, and it may have been another item brought back by the Crusaders. Ships of the Hanseatic League (a trading combine of north European seaports that was increasingly important at this time) were using it about 1241.

IN ADDITION

After the death of Genghis Khan, his son Ögödei (1185–1241) succeeded to the Mongol throne, and the career of Mongol conquest did not falter. In 1237, they returned to Europe and in the space of three years conquered Russia, Poland, and Hungary. In 1241 they were approaching Vienna and Venice when the news of the death of Ögödei reached them. The armies had to turn back to help elect a successor, and they never returned to western Europe, though they retained their hold on Russia for a century and a half. While the Mongol Empire lasted, communications between China and Europe were better than ever before, once again to the benefit of European technological advances.

1249

EYEGLASSES

The English scholar Roger Bacon (*ca.* 1220–1292) mentioned lenses for use in improving vision about 1249. Both China and Europe developed eyeglasses at roughly the same time, and it might well be that the news traveled from one place to the other through the Mongol Empire. The first eyeglasses (or spectacles) had convex lenses for the aged who had become far-sighted. Lenses for the near-sighted came much later.

GUNPOWDER

Roger Bacon also wrote of gunpowder in 1249, but there is no question about its place of origin: China had it first by many centuries, and the Mongols may just possibly have brought it west with them.

There are Chinese books that still survive, dating back to 1044, that give the proportions of saltpeter, charcoal, and sulfur for the making of gunpowder. The Chinese exploded the gunpowder in bamboo tubes and rockets and used it against the Mongols. These were not powerful weapons. They may have served only to frighten horses and in any case did not stop the Mongols.

As in the case of the magnetic compass (see 1180), the Europeans, once they learned of gunpowder, moved quickly to make it a serious weapon.

IN ADDITION

Although the Russians were helpless before the Mongol juggernaut, they could hold their own against lesser enemies. The prince of Novgorod was Alexander (*ca.* 1220–1263). He made sure that Novgorod paid tribute to the Mongols and did not in any way offend them and in that manner kept the dreaded horsemen from taking over direct control of Novgorod. In the meanwhile, he defeated the Swedes in 1240 on the Neva River near the site of what is now Leningrad. He was called *Alexander Nevsky* after that. He then went on to defeat the Teutonic Knights (a group of German soldiers dedicated to extending German control over the eastern Slavs) on the ice of Lake Peipus in 1242.

1252

PLANETARY TABLES

Nothing better than Ptolemy's tables of planetary motion (see 140) had been prepared in eleven centuries. Now a new set was prepared under the sponsorship of Alfonso X of Castile, who reigned from 1252 to 1284. He was very interested in astronomy and was no fool at it and was called *Alfonso the Wise* in consequence. He is famous for a remark he made in reference to the calculations necessary to the preparation of the tables: "If

God had asked my advice, I would have suggested a simpler design for the Universe."

He was right. The Universe is certainly enormously more complex than Ptolemy thought it was, but with respect to the information necessary for preparing planetary tables, it *was* simpler in design than Ptolemy had thought. Nevertheless, the *Alfonsine tables* were an improvement on what had gone before.

IN ADDITION

Louis IX of France launched a Seventh Crusade in 1248 and struck at Egypt, assuming that if he could take it, the Holy Land would fall to him almost automatically. By 1250 it proved to be another Crusading disaster. Louis IX was taken prisoner and had to be ransomed.

1269

MAGNETIC POLES

In 1269 a French scholar, Pèlerin de Maricourt (13th century), also known by his Latin name of Petrus Peregrinus de Maricourt, was taking part in the slow and dull siege of an Italian city. To pass the time, he wrote a letter to a friend describing his researches on magnets. He described the existence of *magnetic poles,* regions on a magnet where the magnetic force was most intense, and showed how to determine the north and south poles of a magnet, since like poles repelled each other and unlike poles attracted each other. He also explained that one could not isolate a pole, for if a magnet were broken into smaller pieces, each piece would have both a north and south pole.

This is about the first piece of good scientific experimentation in the modern sense, although it would be more than three centuries before experimental science was well established.

In the same letter, Peregrinus explained that a compass would work better if the magnetic sliver or needle were put on a pivot rather than allowed to float on a piece of cork, and that a graduated circular scale should be placed under it to allow directions to be read more accurately. This was another major step toward making navigation of the open sea a practical task.

IN ADDITION

By 1260 the Mongols controlled virtually all of the Muslim dominions in Asia. They had even taken Baghdad in 1258 and destroyed the canal system that had existed for five thousand years or more, something from which the Tigris-Euphrates valley never completely recovered.

1291

MIRRORS

Until now, glass had almost always been colored (see 100 B.C.). It was in Venice that the technique for adding decolorizing material to glass was developed and where glass was first formed that was reasonably clear and transparent. Though uncolored glass might seem boring, it turned out not to be so. Clear glass struck people as beautiful, and to have cups and other objects made of it proved very desirable.

In 1291 Venice removed its glass-manufacturing establishment to a guarded island and set up stiff penalties for anyone revealing any manufacturing secrets. It did its best to maintain a strict monopoly of the valuable material, and *Venetian glass* continued to be considered the height of luxury.

One thing that clear glass made possible was the modern mirror. In ancient times, people could see themselves in still water or in the polished surface of a metal such as bronze. Water rarely remained still for long, however, and polished metal was expensive. The result was that very few people knew what they looked like or could do something as simple as arrange their own hair.

A sheet of clear glass, however, could be backed with a film of metal, and the result was a luminously clear mirror in which it was possible to study one's own face to one's heart's content. (It is not for nothing that a mirror is also called a *looking-glass.*)

IN ADDITION

In 1259 Kublai Khan (1215–1294), after having conquered southern China, became ruler of the Mongol Empire. During his thirty-five-year reign, the Mongol Empire, stretching from the Pacific Ocean to the Baltic Sea, was at its peak.

Edward I (1239–1307) became king of England in 1272 and ten years later conquered Wales. From then on, the oldest son of the king and heir to the throne would be known as the *Prince of Wales.*

Rudolf I (1218–1291) became Holy Roman Emperor in 1273. He was the first of the House of Hapsburg, which was to remain prominent in European history for over six centuries.

In 1290 a Turkish chieftain named Osman I (1258–*ca.* 1326) organized a warrior band that was to be known as the Ottoman Turks.

In 1291 three Alpine cantons, Uri, Schwyz, and Unterwalden, formed a union that represented the beginning of the nation of Switzerland.

1298

THE FAR EAST

The existence of the Mongol Empire made it easier for travelers to make their way from Europe to China than ever before. In 1260 two brothers, Nicolo and Maffeo Polo, well-to-do Venetian merchants, made their first trading trip eastward. In 1275 they returned to northern China, where Kublai Khan maintained his capital, and this time they brought Nicolo's son, Marco Polo (1254–1324). Marco stayed in China for twenty years, was in high favor most of the time, and had an opportunity to study the land, the people, and the customs. He found a nation far in advance of Europe in population, in wealth, in technology, and in all the civilized amenities.

He didn't return to Venice till 1295. Then, caught up in a war between Venice and Genoa, he was taken prisoner and, while in prison, began to dictate his reminiscences of China. The book was published in 1298 and was immensely popular, but it was largely disbelieved.

It did create an impression of the "gorgeous East," however, so that the Far East became a wonder-goal for European dreamers, with important consequences for the future.

SPINNING WHEELS

For thousands of years, fibers had been spun into thread by the action of the hand: the fiber was held on a *distaff* and was twisted into yarn by turning a spindle. It was a slow, painstaking process that thoroughly occupied much of the time of the females of a household. (It was "women's work" par excellence, and the females of a family are still referred to jocularly as "the distaff side.")

In India the process had eventually been mechanized: a foot-powered treadle turned a large wheel that twisted the spindle. Such a *spinning wheel* greatly hastened the work of spinning thread out of fiber. About 1298 the spinning wheel made its way into Europe.

The wheel and the spindle were connected by a belt, the first example of a *belt drive* in machinery. The spinning wheel was also one of the first important mechanical means of lightening women's work.

LONGBOWS

The longbow was invented in the thirteenth century by the Welsh. It was more than 6 feet long and fired arrows that were 3 feet long. A skilled longbow archer could shoot an arrow accurately for 250 yards and reach an extreme range of 350 yards. This was twice the range of the average crossbow (see 1050), and much more important, while the crossbow was being cranked up once, the longbow could be fired five or six times. If equal numbers of archers with longbows and crossbows were to encounter each other, the crossbow archers would be riddled.

The disadvantage of the longbow, however, was that the archer had to exert a force of 90 to 100 pounds to draw the bow and to maintain that pull evenly until the feather of the arrow was aligned with the archer's ear. That required strength and a great deal of training.

Edward I of England (see 1291) recognized the value of the weapon and set about training a corps of English longbow archers. He put the longbow to the test against the Scots at the Battle of Falkirk on July 22, 1298.

The Scottish infantry had pikes, but the English archers with their longbows shot them down from a distance, and when enough had been destroyed to reduce the rest to a disorderly mob, the English cavalry came in to finish the job.

The English went on to use the longbows in other battles, and no other nation ever adopted this obvious weapon. As a result, the English were a great military power for the next century and a half.

1300

SULFURIC ACID

The greatest of all the alchemical discoveries was made by someone whose name is not known. This alchemist wrote about 1300 using the name of Geber in order to gain credence by association with that great Arabian alchemist (see 750).

The result is that we know nothing about this alchemist, who can only be referred to as the *False Geber*. This is a pity, for the False Geber was the first to describe sulfuric acid, the most important single industrial chemical used today (barring such "always-known" substances as air, water, and salt).

Sulfuric acid is much stronger than acetic acid and made possible many chemical changes that would not have been possible before.

DISTILLED LIQUOR

Natural fermentation has its limits. As fruit or other materials ferment, alcohol accumulates and eventually grows sufficiently concentrated to kill the yeasts or other microorganisms that were producing the fermentation.

The alchemists had learned how to distill: how to heat a substance and drive off the volatile materials, which could condense into liquid elsewhere, leaving behind dissolved matter. Thus, if sea water is heated, the vapors consist only of water, and if cooled, this is drinkable. The salt is left behind and has its own uses.

Eventually, alcoholic beverages were distilled. Since alcohol boils at a lower temperature than water does, the initial vapors of the beverage are higher in alcohol than the original liquid. If the vapors are then cooled and condensed, the result is a stronger liquor with a good deal more of a "kick" than the original has.

In 1300 the Spanish alchemist Arnau de Villanova (*ca.* 1235–1312) distilled wine and obtained reasonably pure alcohol for the first time. In the process, of course, he prepared brandy, which is distilled wine with a much higher alcohol content than ordinary wine. Not only brandy, but whiskey (made by distilling fermenting grain) became available in sizable quantities.

1304

GIOTTO'S COMET

A bright comet was visible in Europe's sky in 1301. It created the usual panicky stir, but the Italian artist Giotto di Bondone (*ca.* 1267–1337), usually known by his first name, observed it with an artist's eye.

Till then, and for a considerable time afterward, those who drew comets let their panic be their guide and presented the silliest pictures imaginable. In 1304, however, Giotto painted *The Adoration of the Magi,* in which he pictured the star of Bethlehem as a comet, and seems to have let the comet of 1301 guide his brush. Giotto's is the first *realistic* depiction of a comet.

IN ADDITION

In 1302, at the Battle of Courtrai, a determined band of Flemish townsmen equipped with pikes slaughtered the disorderly mass of French horsemen who tried to ride them down. This is the first premonitory rumble of the end of cavalry domination of the battlefield, which had begun at the Battle of Adrianople a thousand years before. The French were not to learn the lesson for over a century, however.

Philip IV (the Fair) of France, who reigned from 1285 to 1314, was angered by Pope Boniface VIII's claims of papal sovereignty and sent agents to arrest him. The Pope was arrested on September 8, 1303, and the result was a blow to papal prestige from which the papacy never recovered. Thereafter, the Pope might still have influence, but only in a moral sense.

1312

CANARY ISLANDS

The Canary Islands are off the shore of Morocco, which is in northwestern Africa. The King of Mauritania (located where Morocco is now) had sent an expedition there in 40 B.C. and found them already inhabited. In 999 Arabs landed but did not stay.

The early visits to the Canary Islands did not remain in European consciousness. However, in 1312, a Genoese vessel reached the Canary Islands, and though they didn't stay either, this visit was known to Europe and remembered. It was the first small and abortive step toward European expansion overseas.

IN ADDITION

The Middle Ages were concerned chiefly with theology; that is, the human being's relationship with God. After the thirteenth century, a movement began toward a primary interest in human beings themselves, a kind of *humanism* reminiscent of that which had existed in the Greek heyday. This movement came to be called the *Renaissance* (French for "rebirth"), and it was the people of this Renaissance who first called the past centuries the *Middle Ages*—intermediate between ancient humanism and modern humanism.

In 1309 Philip IV of France forced the Pope to leave Rome for the first time and established the papacy in the city of Avignon, deep in French territory, in order that it might be more nearly under his control. This too damaged the prestige of the papacy.

1316

DISSECTION

The stirrings of humanism allowed scholars to grow more interested in science. In the medical schools of Italy, it even became possible to dissect cadavers once again. The greatest of the new group of anatomists was the Italian Mondino de Luzzi (*ca.* 1275–1326), who taught at the medical school of Bologna.

In 1316 he published the first book in history to be devoted entirely to anatomy. He remained under the influence of the Greek and Arabic writers and clung to them sometimes in preference to the evidence of his own eyes. Nevertheless, his book remained the best there was for two and a half centuries.

IN ADDITION

Edward I (see 1291) had tried to conquer Scotland and make it part of his English dominion, as he had done to Wales. He nearly succeeded, too, but died before he could put down Scottish rebellions against his rule. He was succeeded by his son, Edward II, who reigned from 1307 to 1327. Edward II was not a capable monarch and, mishandling his longbow brigade, led his army into defeat at the Battle of Bannockburn on June 24, 1314. This victory insured that Scotland would remain independent for three more centuries.

1335

MECHANICAL CLOCKS

The first advance over the water clock (see 270 B.C.) came in the fourteenth century. Instead of being driven by a rise in water level, the dial on the clock face was driven by the downward pull of gravity on weights.

The resulting mechanical clocks did not tell time more accurately than water clocks did, but they were more convenient and required less care. They could be mounted in a tower (either of the city hall or of the town church) for all to see. One was erected in Milan, Italy, in 1335, for instance. It struck the hour, and for the first time citizens could learn the time (to the nearest hour, at any rate) by listening to the number of times the bell rang. (The very word *clock* is from the French word for "bell.")

IN ADDITION

Paper finally reached Europe in 1320. Paper money had been instituted in China in 1236, and Marco Polo had noted it in wonder. Under the Mongols, however, overprinting of it had led to inflation, especially after Kublai Khan's death. Paper money was discontinued in 1311, but the experience had greatly weakened the Mongol government.

In Mexico, in 1325, the Aztec Empire was beginning its rise, and Tenochtitlán (what is now Mexico City) was founded.

In 1328 Charles IV of France, who reigned from 1322 to 1328, died, leaving no sons or brothers. He had a first cousin, Philip of Valois, who was descended from the French king Philip III by way of his father, and a nephew, Edward III of England (1312–1377), who was descended from the French king Philip IV by way of his mother. Edward was the closer relative, but the French preferred a French nobleman to an English king and invalidated Edward's claim by refusing to accept inheritance through a woman. Since Edward III refused to accept the invalidation, there was clearly trouble ahead.

In 1333 a serious disease made itself evident in the Far East. It was spread by fleas, which bit rats and human beings, transferring infected blood to those not yet infected. It was the *bubonic plague*.

1346

CANNON

Once the Europeans got their hands on gunpowder, it didn't take them long to place it in a strong metal tube from which its explosive force could hurl out a ball of rock or metal much more forcefully than any catapult could manage. We don't know who first attempted to build these tubes, or *cannon* (from the Italian word for "tube"). Some claim that primitive cannons were used at a siege of the city of Metz in 1324.

There is no doubt, however, that they were in use by 1346. Edward III of England, intent on claiming the throne of France, went to war in 1337 over the matter, thus beginning what was eventually to be called the *Hundred Years War*.

The first great land battle of the war was at Crécy in north-central France on August 26, 1346. The French outnumbered the English, particularly in mounted knights. The French also had crossbow archers from Genoa. The English, however, had longbow archers, and it was no contest. The English archers had it all their own way, and the French were massacred. Edward III also had cannon at Crécy. They were primitive things and accomplished nothing—but they were a portent of the future.

IN ADDITION

The bubonic plague (or the *Black Death* as it was called) marched inexorably westward. It had reached the Crimea by 1343 and Genoese traders there were infected. The Genoese managed to bring their ship back to Italy, with everyone on board either dead or dying, and the plague began to spread over western Europe.

The disease killed quickly (sometimes within twenty-four hours of the first onset of symptoms). Since people knew nothing of the causes of disease and were ignorant of even the simplest hygienic principles, the only recourse was to flee and spread the disease. The disaster became unimaginable. By some estimates it killed one-third of humanity worldwide. (The wonder is, perhaps, that two-thirds of the population managed to survive.)

No disaster that we know of, either before or since, came so close to wiping out humanity.

1403

QUARANTINE

Despite the fact that nothing was known about how disease came to be (except for the usual theories of punishment by God or infestation by demons), people did tend to avoid those who were sick with some particularly fatal or loathsome disease.

Thus, leprosy (undoubtedly along with less drastic skin diseases) was treated as something that required isolation. Lepers were driven out of society.

When the Black Death struck, people instinctively fled from those affected (sometimes leaving the dying to die and the dead to remain unburied).

In 1403 the city of Venice, always rationally ruled, decided that recurrences of the Black Death could best be averted by not allowing strangers to enter the city until a certain waiting period had passed. If by then they had not developed the disease and died, they could be considered not to have it and would be allowed to enter.

The waiting time was eventually standardized at forty days (perhaps because forty-day periods play an important role in the Bible). For that reason, the waiting period was called *quarantine,* from the French word for "forty."

In a society that knew no other way of fighting disease, quarantine was better than nothing. It was the first measure of public hygiene deliberately taken to fight disease.

IN ADDITION

In China the Mongols were finally driven out, a century and a half after Genghis Khan had led them in, and in 1368 a new native dynasty, the Ming, took control of the land.

This did not mean, however, that the Mongols were quite reduced to nothing. A descendant of Genghis Khan, Timur Lenk (Timur the Lame), usually known in English as Tamerlane (1336–1405), began a career of conquest that made it seem almost as though Genghis Khan had been reborn, for he never lost a battle. He conquered central Asia, making his capital at Samarkand, then went on to take Persia. In 1391, Tamerlane smashed the Golden Horde in Russia and made it possible for Russia to emerge from the Mongol yoke. He went on to invade India and Asia Minor and defeated the Ottoman Turks in 1402 (granting Constantinople another half-century of life). He died as he was on his way to China to restore Mongol control there.

1405

INDIAN OCEAN

For a brief period under the Emperor Yung-lo of China (who reigned from 1402 to 1424), it seemed as though China might become a sea power. A Muslim eunuch, Cheng Ho (*ca.* 1371–*ca.* 1433), led a series of expeditions southward and westward through the Indian Ocean. His first expedition sailed forth in 1405, with twenty-seven thousand men and three hundred ships, and forced the potentates of the Indonesian islands to accept Chinese overlordship (at least until the ships left).

In a second voyage in 1409, Cheng Ho visited India and Ceylon. When the Ceylonese attacked the ships, he defeated them and brought the ruler of Ceylon a prisoner to China. In later voyages (there were seven in all), he went still farther west, reaching the Red Sea and visiting Mecca and Egypt.

After Emperor Yung-lo's death, however, succeeding emperors decided that China had no need to venture far afield in order to visit and deal with the inferior barbarians. China was a world in itself and it was enough for the Chinese.

And so, in effect, China abdicated its chance at world influence and left it to other much smaller, much weaker, and much less advanced nations.

1418

MADEIRA

Little Portugal, at the opposite end of the vast Eurasian continent, had an attitude quite different from that of China. China was aware of its immense self-sufficiency; Portugal was aware of its own insufficiencies. China needed nothing from abroad; Portugal wanted silk, spices, and all sorts of other things. What's more, Portugal was at the very end of the line, farthest from the source, so that it received least and paid most. And now that the Mongols were gone and the hostile Ottoman Turks were in control of the Middle East (despite the fierce temporary shock of Tamerlane's career), trade between China and western Europe had dwindled further.

This was the situation that faced Prince Henry of Portugal (1394–1460). It occurred to him that there was no use in trying to trade with the gorgeous East that Marco Polo had so fetchingly described—at least not by land. Why not instead bypass the Turks altogether and sail around Africa to get there?

The trouble was that no one knew how far south Africa extended, whether the voyage was practical, the seas navigable, the Tropic Zone passable, and so on. (No one paid any attention to Herodotus's wild tale of the Phoenicians having circumnavigated Africa two thousand years before.)

Prince Henry therefore established an observatory and school for navigation at Sagres on Cape St. Vincent in 1418. This was in southernmost Portugal, the southwestern tip of Europe. Year after year, he outfitted and sent out ships that inched their way farther and farther down the African coast. With Prince Henry, now universally known as Henry the Navigator, the great European *Age of Exploration* began, and the role that China was giving up at just about this time, Portugal (and later other European powers) were accepting.

The result was that Chinese vessels never came to Portugal, but Portuguese vessels (and those of other European powers) eventually came to China, and China was to pay dearly for its period of self-satisfaction.

The first fruit of Prince Henry's efforts came in 1418, when Portuguese navigators discovered *Madeira*. It was a heavily wooded island (the name is from the Portuguese word for "wood") and was uninhabited. Prince Henry ordered the island settled. The woods were burned down and the land was given over to cultivation, that of sugarcane in particular.

IN ADDITION

Henry V (1387–1422) succeeded to the English throne in 1413 and felt ready to take up the quarrel with France once again. He therefore led an invasion force and, at Agincourt, battled the French on October 25, 1415. Again the English longbow archers won a complete and one-sided victory over a much larger French force. Henry V then went on to capture Normandy. He took its capital, Rouen, in 1418, and France was suddenly in greater danger than it had been eighty years before at Edward's hands.

1427

AZORES

The Azores are a group of islands in the Atlantic, of which the easternmost is about 750 miles west of Portugal. They were discovered in 1427 by the Portuguese navigator Diogo de Sevilha. Like Madeira, they were uninhabited, and like Madeira, they are still part of Portugal. Since they are one-third of the way across the Atlantic in the direction of the American continents, it is clear that already, thanks to the compass, the Portuguese were ranging far.

IN ADDITION

Henry V of England died young in 1422 and was succeeded by his nine-month-old son, who reigned as Henry VI from 1422 to 1461. The war in France was carried on by Henry V's capable younger brother, John, Duke of Bedford (1389–1435).

The French king, Charles VI, had also died in 1422, and he was succeeded by his son Charles VII, who reigned from 1422 to 1461. Charles VII could not be crowned, however, for Reims (where the French kings were traditionally crowned) was in the hands of his enemies. Philip, Duke of Burgundy, who controlled much of eastern France, was fighting on the English side. The situation looked desperate for France.

1436

PERSPECTIVE

The Renaissance was a great age of realism in art, and the Italian painters wanted their canvases to appear to have three dimensions. In order to do that, they had to have proper *perspective:* lines had to come together as they seem to do in real life. The Italian artist Leon Battista Alberti (1404–1472) published a book in 1436 in which he described the proper method of achieving perspective, handling the matter in a careful mathematical manner. This proved to be a forerunner of *projective geometry,* which was developed four centuries later.

IN ADDITION

In 1428 English forces advanced to lay siege to French-held Orleans at the bend of the Loire river. The only thing that kept the French from breaking the siege was their reluctance to fight the apparently invincible English.

And then in 1429, a peasant girl arrived, named Jeanne Darc (*ca.* 1412–1431) (incorrectly translated as Joan of Arc). She claimed to be sent by God, and that was all that was needed to hearten the French and dishearten the English. The French broke the siege, and Joan of Arc led them to Reims against virtually no resistance, where the French dauphin was crowned as Charles VII.

Joan then attempted to take Paris, but the French generals felt that you could press luck (and English fears of a witch) only so far. Joan foolishly tried to press on anyway and was taken prisoner. She was convicted of witchcraft and burned to death in Rouen on May 30, 1431.

1439

ARTILLERY

Charles VII, now truly king, bestirred himself to reform his army, hiring two brothers, Jean and Gaspard Bureau, to reorganize the artillery. They improved the design of cannon and the quality of gunpowder, oversaw the production of cannon in greater numbers, and placed them under the control of specialists.

Charles VII's armies became the first to make an able and systematic use of artillery. This marked the end of the medieval fashion of making war and it completed the job of restoring the cavalry to its one-time position as a mere auxiliary arm.

City walls, which, unlike personal armor, were impervious to longbows, began to fall before the new artillery. Just as the French could not understand why the longbow won, so the English could not understand why it stopped winning, and they proceeded to lose the Hundred Years War.

IN ADDITION

In South America, a new dynasty began its rule over the Inca Empire, which had most ingeniously adapted itself to life in the Andes mountains, even though it lacked writing, and was approaching its height.

1450

HARQUEBUSES

One problem with artillery, of course, was that cannon were heavy and had to be dragged, with much effort, from place to place. There would obviously be a benefit in having a cannon small enough to be carried by one person.

About 1450 the first gun small enough to be fired by one person was invented in Spain. It was called a *harquebus* (or *arquebus*) from a Dutch word meaning "hook-gun." Perhaps it got this name because the early harquebuses were used in association with pikes, pikes being "hook-spears."

The harquebus was not easily portable. The first ones were so heavy they needed supports. The powder behind the bullet had to be lit before the gun would fire, and it was as hard to reload as a crossbow, so that pikewielders were needed to protect the harquebusiers while they were reloading.

Nevertheless, the harquebus marked the beginning of *small arms.* It was improved and made lighter so that it could be fired from the shoulder. It continued to be used for a century before it was superseded.

IN ADDITION

All cultures had made use of slaves, and it seemed to the Portuguese, as they explored the African coast, quite natural to enslave people for the crime of being unable to withstand better-armed and better-disciplined soldiers.

In 1441 black slaves were for sale in Lisbon, the capital of Portugal. That was the beginning of a slave trade that saw an estimated twenty million blacks removed from Africa by force, resulting in incalculable suffering, and not for blacks alone.

1451

CONCAVE LENSES

Until now, only convex lenses had been used in eyeglasses (see 1249). Convex lenses are thicker in the center than at the edges, and they curve the light inward so that on passing through the lens of the eye it reaches a focus sooner than it otherwise would. This is useful for eyes that are too short and are ordinarily far-sighted (usually among the aged).

In 1451, however, the German scholar Nicholas of Cusa (1401–1464) suggested the use of concave lenses, thinner in the center than at the edges, to bend light outward and bring it to a focus later than would otherwise take place. This is useful for eyes that are too deep and that are otherwise near-sighted. This made eyeglasses a boon to young people (who are often near-sighted) as well as old.

IN ADDITION

In 1451 a new sultan ascended the Turkish throne. The Ottoman Empire had been badly shaken as a result of its defeat by Tamerlane, but it had recovered by now. The new sultan, Mohammed II (1430–1481), was determined to restore the Ottoman expansion to the full. For the purpose, he intended once and for all to make Constantinople Turkish.

Constantinople had a new emperor, too, Constantine XI Palaeologus (1404–1453), who had come to the throne in 1449. After a long line of incompetent emperors, Constantinople finally had a vigorous and capable one, but all he had to rule over was the city itself and a small portion of southern Greece.

1454

PRINTING

There is no way to overestimate the importance of writing (see 3500 B.C.). Nevertheless, it can't be denied that writing is a tedious process, and there have always been efforts to hasten it. The Egyptians worked out faster ways of scrawling their complicated symbols, and the Romans worked out systems of shorthand.

The ancient Sumerians, however, developed little cylinders of hard stone with designs incised into them that could be rolled onto soft clay so that the designs were impressed, and these could be made permanent by baking. The cylinders could be used over and over again and served as a signature for the owner.

Why could not the same system be used for pressing symbols onto a sheet of paper? If a block with a raised reversed symbol on it is smeared with ink, it can press an inked symbol (nonreversed) onto paper. The Chinese started doing this about the year 350, and by 800 they were carving entire pages on wooden blocks. Such a page could then be printed any number of times, all the impressions being exactly alike. But then, it took a long time to carve the wooden block into a bas-relief with all the symbols perfectly formed.

Later the Chinese got the idea of using a different block for each symbol, so that the blocks could be arranged into any desired combination to give any desired page. By 1450 they had movable wooden characters of this type, and by 1500 they were making use of metal characters.

By then, though, the Europeans had outstripped them (though it is possible that news concerning movable type had reached Europe from China and given the Europeans a head start).

The German inventor Johannes Gutenberg (*ca.* 1390–1468) had been working out the matter of movable type since 1435. He had paper to work with (it had long ago reached Europe from China) and he experimented with different inks. He also designed a printing press, a device that would press the paper down on all those little metal characters evenly.

By 1454 Gutenberg had worked out all the bugs in his process and was ready for the big task: he began to put out a Bible, in double columns, with forty-two Latin lines to the column. He produced three hundred copies of each of 1282 pages and that produced the three hundred *Gutenberg Bibles.* It was the first printed book, and many people consider it the most beautiful ever produced—so that the art was born at its height. The Gutenberg Bibles that remain today are the most valued books in the world.

IN ADDITION

Mohammed II began his assault on Constantinople in 1452. Constantine XI bravely set up a determined resistance. On May 30, 1453, Constantinople fell. It became a Turkish city and has remained one ever since. Mohammed II made it the capital of the Ottoman Empire. Constantine XI was the last of the Byzantine emperors.

There was another ending in 1453 in western Europe, though it was not quite as colossal as the one in eastern Europe. The war between England and France had been continuing since 1337 (fully 116 years, off and on, though it is called the Hundred Years War). It had been a losing war for England since the time of Joan of Arc (see 1436). But in 1453 the English made one last effort and sent John Talbot (1388–1453) to Bordeaux to reestablish English sovereignty there. Talbot could do nothing against the French artillery. He was killed in battle and Bordeaux became French forever on October 19, 1453.

Some historians place the end of the Middle Ages and the beginning of modern times in 1453, with the end of the Byzantine Empire and the Hundred Years War. Others place it a half-century later, more or less, with the discovery of America or the beginning of the Protestant Reformation.

These are military, exploratory, and religious ways of looking at it. To those interested in scientific history, though, surely a far sharper dividing point is 1454, when Gutenberg began to prepare the first printed book.

1472

POSITION OF COMETS

Comets had always been viewed with such terror that almost no one had been able to observe them rationally. Then in 1472, when a bright comet appeared in the sky, a German astronomer, Johann Müller (1436–1476), refused to allow himself to be governed by fear. (He is better known by his self-chosen name of Regiomontanus, which means King's Mountain, as does the German name of his birthplace, Königsberg.)

Regiomontanus observed the comet from night to night and noted its position against the stars. In this way, for the first time, the exact path of a comet across the sky was plotted. It marked the beginning of rationalism with respect to those bodies.

IN ADDITION

In 1455, partly due to frustration over the loss of France and partly because Henry VI of England was showing signs of insanity, the English nobles began a civil war over the succession. It was eventually called the *War of the Roses* and lasted thirty years, on and off.

France, too, was enduring a sort of civil war, between Burgundy, which carried on an independent foreign policy that was often anti-French, and Louis XI (1423–1483), the new king of France.

Spain had been divided into a western portion called Castile and an eastern portion called Aragon for over four centuries. In 1469, however, Isabella (1451–1504), heiress to the throne of Castile, married Ferdinand (1452–1516), heir to the throne of Aragon. Isabella succeeded to her throne in 1474 and Ferdinand to his in 1479. They ruled jointly over a united Spain and the nation has remained united ever since.

1487

CAPE OF GOOD HOPE

In February 1487 the Portuguese navigator Bartholomeu Diaz (1450–1500) set forth to search for the southernmost part of the African continent. In a way, he didn't find it, for a storm drove him past it into the open sea.

He turned north again and reached a portion of the African coast that was running eastward. He followed it eastward until it began turning northward. By that time, his rebellious crew forced him to turn back. He retraced his steps, located the southernmost point of the continent, and reported it to King John II of Portugal (1455–1495) as the *Cape of Storms,* for obvious reasons. King John II, however, realizing that one more push would now bring his ships to the Far East, renamed it the *Cape of Good Hope,* the name it bears to this day.

IN ADDITION

Moscow gradually united the Russian principalities under its own leadership, and Ivan III (1440–1505) made himself the first national Russian sovereign.

In 1477 the French portions of Burgundy were absorbed by Louis XI of France, but the Holy Roman Emperor, who was now routinely chosen from the Austrian House of Hapsburg, took over the Low Countries (known today as the Netherlands and Belgium). This was a bad break for France, which was to have the unfriendly House of Hapsburg on its northeastern border for three centuries.

In 1485 the War of the Roses in England came to an end with the Battle of Bosworth and the victory of Henry VII (1457–1509).

In 1487 Tomás de Torquemada (1420–1498) was made Grand Inquisitor of Spain by Innocent VIII (1432–1492), who had become Pope in 1484. Torquemada made the *Spanish Inquisition* a byword for cruelty and horror. He was not to be outdone in this till the mid-twentieth century.

1492

NEW WORLD

While the Portuguese were working their way around Africa, there were those who felt that the same result could be achieved another way. Since it was understood that the Earth was spherical, it was bound to occur to people that it might be circumnavigated and that the Far East could be reached by sailing west.

The concept was a simple one and, in fact, it had been suggested by Roger Bacon (see

1249) two centuries before. What stopped people from making the effort was the thought that between the western coast of Europe and the eastern coast of Asia might be a vast stretch of ocean that the ships of the day could not be expected to manage.

If Eratosthenes was correct and the Earth was 25,000 miles in circumference (see 240 B.C.), then between Europe and Asia were some 12,000 miles of unbroken sea. Yet other authorities, such as Ptolemy (see 140), had thought the Earth was smaller than that, and Marco Polo had thought that Asia extended farther east than it really did.

Combining a smaller Earth with a more eastward Asia, an Italian navigator, Christopher Columbus (1451–1506), was convinced that a westward trip from Europe to Asia was a matter of only 3,000 miles. He thought this could be managed, and he shopped about the various nations of western Europe for financial support so that he could outfit an expedition.

Portugal was a natural target, of course, but the Portuguese experts thought the Earth was larger than Columbus's figure (the Portuguese were right) and were convinced it would not be long before they circumnavigated Africa and reached their goal.

Columbus tried elsewhere without luck and was almost on the point of giving up when things turned out well for him in Spain.

With Ferdinand and Isabella ruling jointly over a united Spain, the nation could assault the last scrap of Muslim rule. This was the nation of Granada in the far south of Spain. The joint monarchs prosecuted a vigorous war against Granada and on January 2, 1492, it fell. What's more, Torquemada engineered the expulsion of the Jews from Spain in 1492. (This was not a new phenomenon, for the Jews had earlier been expelled from England and France. They found a refuge in Poland, which needed a merchant class, and in the Muslim world, which was more tolerant than the Christians, because for one thing, they were at the time more civilized.)

The Spanish monarchs, feeling Spain to be united and strong, decided to give Columbus a minimum of financial backing. With three old ships and a crew of prisoners released from jail for the purpose, he set forth on August 3, 1492. For seven weeks he sailed westward, encountering no land but also encountering no storms. Finally on October 12, he sighted land—an island, as it turned out, in the Bahamas.

He sailed southward and encountered islands of the West Indies. (To his dying day, Columbus was convinced he had reached the *Indies;* that is, the eastern coast of Asia. The name of the West Indies and the habit of calling Native Americans *Indians* are the result of that delusion.)

It was not Asia, of course, that he had come upon, but the American continents, a *New World,* and the *Old World* would never be the same again.

Columbus was not, of course, the first human being to set foot on these continents. Siberian natives had done so at least thirty thousand years earlier (see 20,000 B.C.). He was not even the first European to do so, for Leif Eriksson had done it five centuries before (see 1000). Columbus's feat, however, led almost at once to the beginning of permanent European settlements on the new continents, and that marked their entry into the common current of world history. It is for this reason that Columbus is generally given credit for the "discovery."

Incidentally, the fact that new continents existed that were wholly unknown to the ancients helped eliminate the notion that the ancient thinkers had known everything and had solved all problems. Europeans gained the heady feeling that they now moved beyond the ancients, and that helped make possible the Scientific Revolution that was to start in half a century.

MAGNETIC DECLINATION

The magnetic compass, which made a voyage like Columbus's possible, naturally focused considerable attention on the phenomenon of magnetism. No one knew at this time why the needle pointed north, but people took it for granted that whatever the reason, the needle pointed north steadily and without change.

The person who found out differently was Columbus himself. As he sailed westward, he couldn't help but notice that the magnetic needle changed somewhat in direction. When he started, the compass needle pointed a bit west of true north as determined by the stars. As he traveled westward, the magnetic needle veered eastward, pointed to true north at a certain point, and finally pointed a bit east of north.

Columbus, being scientifically minded, noted this in his journal, but he kept it a secret from his crew. Had they known that the compass wasn't to be relied on, they would surely have panicked, killed Columbus, and headed back east in a desperate effort to regain land before they were lost forever. Without Columbus's firm hand, they would probably not have made it, and had the mission not returned, it would have been a long time before any European monarch risked money on a second such venture.

IN ADDITION

The Renaissance reached its height in Florence under the rule of the Medici family, especially under Lorenzo the Magnificent (1449–1492) who became ruler in 1469. The Medicis were munificent patrons of arts and letters and welcomed refugee Byzantine scholars into Florence after Constantinople fell in 1453.

1495

SYPHILIS

In 1495 there was an outbreak of a new disease in the Italian city of Naples, which was under siege by a French army. This disease spread quickly, carried by soldiers from place to place. Nearly half a century later, an Italian astronomer, Girolamo Fracastoro (*ca.* 1478–1553), wrote a poem about a shepherd who caught this new disease, which the Italians called the French disease and the French called the Neapolitan disease. Since the shepherd's name was given by Fracastoro as Syphilis, Europeans generally, and then the world, used that as the name of the disease.

Syphilis may not have been a completely new disease, since what the ancient and medieval people called leprosy might, in some cases, have been a form of syphilis. To the people of the time, however, it seemed new.

Since it made its appearance so soon after the discovery of America, and since the story spread that some of Columbus's sailors were in the armies of Naples at the time, the feeling arose that it represented a native American disease that had been carried into Europe. Later it was said that the Europeans carried the disease to the Americas. Which is actually so, however, we can't say.

IN ADDITION

In a second voyage to the New World, in 1493, Columbus discovered the island of Hispaniola (Little Spain), on which the modern nations of Haiti and the Dominican Republic are to be found.

Charles VIII (1470–1498), who became king of France in 1483, invaded Italy, planning to make the Kingdom of Naples his own (it was at this time that syphilis made its appearance). This began a cycle of wars, with France on the one hand and Spain and the Holy Roman Empire on the other, that devastated Italy, which was then the most civilized portion of Europe, and left it in a kind of dark age that lasted for three and a half centuries.

1497

INDIA

On July 8, 1497, the Portuguese navigator Vasco da Gama (1460–1524) set sail from Lisbon with four vessels. He rounded the Cape of Good Hope on November 22, passed the farthest point reached by Diaz (see 1487), sailed up the eastern coast of Africa, and finally reached Calicut, India, on May 20, 1498.

The deed was done. What Prince Henry the Navigator had begun (see 1418) was complete, nearly four decades after his death. The Portuguese had bypassed the Ottoman Empire and the Italian trading cities, such as Venice. From this point on, the Mediterranean region declined in power and wealth, and the Atlantic powers took the lead.

Da Gama's trip was the first one long enough to induce scurvy in his men. It was a debilitating disease that eventually killed those who suffered from it. Da Gama lost three-fifths of his crew to it.

IN ADDITION

Financed by the English, an Italian navigator, Giovanni Caboto (*ca.* 1450–*ca.* 1498), better known in English as John Cabot, set sail in 1497 and discovered Newfoundland and Nova Scotia. He was the first European after the Vikings to reach the continental lands of the New World. (Up to that point, Columbus had only reached various islands.)

1502

AMERICA

The Italian navigator Amerigo Vespucci (1454–1512) is better known by the Latinized version of his name, Americus Vespucius. Beginning in 1497 he was involved in expeditions that explored the coast of South America.

The voyages were not of the first importance in themselves, but Vespucius derived something important from them. It struck him that none of the lands he had seen, or those reported by other explorers of the New World, in the least resembled the Asia described by Marco Polo, and he was the first to conclude that they were *not* Asia. In 1502 he published his assertion that they were a new continent, and that Asia lay far to the west beyond a second ocean.

If Columbus had been the first to land in the New World, Vespucius was the first to *recognize* that it was a new world. A German geographer, Martin Waldseemüller (*ca.* 1470–*ca.* 1518), was impressed by this and, in 1507, when he published a map showing the new lands as a separate continent and *not* as part of Asia, he proposed that the new continent be named America after Americus Vespucius.

The name caught on. Since we now know there are two continents, connected by a narrow strip of land, we speak of *North America* and *South America,* while the connecting region is *Central America*.

IN ADDITION

In 1498 Columbus made a third voyage to the New World, and this time he landed at the mouth of the Orinoco River in what is now Venezuela. This was the first time he reached the American continents themselves.

The Portuguese explorer Pedro Álvars Cabral (1467 or 1468–1520) landed on what is now the coast of Brazil in March 1500. He claimed it for Portugal, and the result is that the modern nation of Brazil speaks Portuguese, while the rest of the American continents south of the United States speak Spanish.

In 1501 Spanish settlers in the West Indies brought in black slaves from Africa. This continued, and blacks now form an important fraction of the population of both American continents.

In a fourth and last voyage, in 1502, Columbus touched the shores of Central America.

1504

WATCHES

A mechanical clock had to remain vertical if it was powered by weights being pulled downward by gravity. Nor could it be small, for the weights could not be reduced in size very much without ceasing to work. An alternative to weights came into being, however, when the mainspring was invented about 1470. This was a spiral spring that could be wound tightly, so that its tendency to unwind would then power the clock.

A German locksmith, Peter Henlein, realized that a mainspring would work just as well when it was small as when it was large, and since it didn't depend on gravity, it could work in any position. About 1504, therefore, he placed a small mainspring in a clock small enough to be carried about in a pocket.

Such a small clock was called a *watch,* because sailors or others who had to maintain a watch at sea or elsewhere for a fixed period of time would find it useful. The original watches had hour hands only and were not particularly accurate, but better things were to come.

1513

SOUTH SEA

Columbus had reached the coast of what we now call the Isthmus of Panama in his final voyage, and Spanish settlers arrived soon after. One of the leading settlers was Vasco Nuñez de Balboa (1475–1519).

At that time, those Spaniards who thought they were in Asia kept looking for the vast wealth that Marco Polo had talked about, and gold was a particularly concentrated form of wealth, one they particularly lusted for. After all, since the people who lived in these continents (whether Asians or not) were not Christian, any gold they possessed would rightly belong to any Christians who cared to help themselves to it.

Balboa heard rumors of gold to the west and he organized an expedition to find it. Panama, however, is a narrow isthmus and you can't wander far in it without striking the other coast. Balboa's expedition left on September 1, 1513, and on September 25, he climbed a hill and found himself staring at the limitless expanse of what seemed to be an ocean. Since the coasts of Panama run east-west and the Atlantic Ocean is on the north shore, Balboa named the new ocean on the south shore the *South Sea.*

Balboa probably did not realize that this was the second ocean Vespucius had referred to, the one that lay between the American continents and Asia—but it was.

FLORIDA

Puerto Rico (rich port) had been discovered in 1492 in the course of Columbus's first voyage, and he had left some men behind when he returned to Spain. When he reached Puerto Rico on his second voyage, the settlers were gone, but others soon followed. By 1513 the Spaniards were well established in Puerto Rico, and one of the settlers was Juan Ponce de León (1460–1521).

Ponce de León dealt in slaves and on March 3, 1513, he sailed northwestward in search of more. He reached the North American continent during the Easter season and called the land *Florida* (flowery) because of its appearance. It became the first portion of what is now the United States to be settled by Europeans.

IN ADDITION

By 1503 the Portuguese had reached the Indonesian islands and brought back shiploads of spices that broke the Venetian monopoly. Prices were greatly reduced.

Basil III (1479–1533) became czar of Russia in 1505 on the death of his father, Ivan III. He incorporated Pskov, the last independent realm in Russia, into his dominions in 1510, so that the Russian principalities were now completely united. Its territory stretched over the northwestern third of modern European Russia.

In 1512 a Portuguese ship landed in Canton, China. This was the consequence of China having retreated from the sea while Portugal reached out to embrace it.

1519

MEXICO

After a quarter-century of sailing about the Caribbean, Spanish explorers had still not encountered the civilizations that existed in the American continents. In 1517, Francisco Fernández de Córdoba (1475?–1525 or 1526), had sailed westward from Cuba and discovered the peninsula of Yucatan. There he was the first to encounter traces of the Mayan civilization, but that was a civilization in ruins.

Westward, however, across the Gulf of Mexico, lay the still flourishing Aztec Empire, which controlled all of central and southern Mexico and had a population of about 5 million.

In 1519 some six hundred Spaniards under Hernán Cortés (1485–1547) landed with seventeen horses and ten cannon. That this small force sufficed to destroy the Aztec Empire is not as surprising as it may sound and does not imply that Europeans were innately superior to the Aztecs. For one thing, the Aztecs had nothing to counter either the horses or the cannon. For another, the subject peoples of the Aztecs were restive and ready to fight on the side of Cortés. Finally, the Aztecs and their king, Montezuma II (1466–1520), had the superstitious feeling that the Spaniards were gods whose coming had been predicted, and they did not resist strongly until it was too late.

So the Aztec Empire was destroyed and Mexico was taken over by Spain. No effort was made to save the Aztec culture or retain information concerning it. It was, after all, non-Christian.

1523

CIRCUMNAVIGATION OF THE EARTH

Ferdinand Magellan (*ca.* 1480–1521) is the English name of a Portuguese navigator, financed by Spain, who sailed west with five ships on September 20, 1519, in search of the Far East. When he reached the eastern bulge of South America, he began looking for a southern end to that continent, which he found on October 21. For over five weeks, he felt his way through what is now known as the *Strait of Magellan,* amid storms, and on November 28, it opened into an ocean and the storms ceased. As Magellan sailed on and on through good weather, he called this new ocean the *Pacific.*

However, the Pacific Ocean was far larger than anyone would have expected, and was sadly free of land. For ninety-nine days, the ships sailed through unbroken sea, and the men underwent tortures of hunger and thirst. Finally they reached the island of Guam, then they sailed westward to the Philippine Islands. There, on April 17, 1521, Magellan died in a skirmish with the inhabitants.

The expedition continued westward, however, and a single ship with eighteen men aboard, under the leadership of Juan Sebastián de Elcano (*ca.* 1476–1526), finally arrived back in Spain on September 7, 1522. This first circumnavigation of the globe had taken three years, and if the loss of life can be set aside, the single returning ship carried enough spices to make the voyage a complete financial success.

The voyage showed beyond doubt, at last, that the Earth was 25,000 miles in circumference, as Eratosthenes had calculated (see 240 B.C.). It showed also that the Earth possessed a worldwide ocean in which the continents existed like huge islands.

IN ADDITION

Selim I, called the Grim (1467–1520), became Ottoman sultan in 1512. He conquered Syria in 1516 and Egypt in 1517, thus converting the Ottoman Empire into the largest and most powerful Muslim realm since the Abbasid Empire was at its height seven centuries before.

On October 31, 1517, a German monk, Martin Luther (1483–1546), nailed a sheet of paper to the door of a church in Wittenberg, Saxony. On it were listed ninety-five theses he was prepared to dispute with others. Luther's ideas spread rapidly throughout western Europe.

That Luther could succeed where previous reformers had failed might be credited to the existence of the printing press. Luther fought his battle by means of pamphlets that flooded Germany and surrounding countries in immense numbers; the Church could not counteract those pamphlets. The nailing of the theses to the church door is usually hailed as the beginning of the *Protestant Reformation,* as it is called by the Protestants, or the *Protestant Revolt,* as it is called by Catholics.

1531

PERU

There remained the Inca Empire, a realm stretching along the Andes mountain range, with its center in what is now Peru and with a population of about 7 million. It was ruled by Atahualpa (*ca.* 1502–1533), who had become head of the empire in 1530.

In 1531 Francisco Pizarro (*ca.* 1475–1541) sailed to Peru with 180 men, twenty-seven cannon, and two horses. In the course of the next three years, what had happened in Mexico happened again in Peru. The too-trusting Incas were overcome by a combination of force and treachery.

After this, the Spaniards claimed and settled all of the American continents up to what is now the southern United States (except for Brazil, which was claimed and settled by the Portuguese) and held it for three centuries.

IN ADDITION

In 1520 the Ottoman Sultan Selim I died, and was succeeded by his son, Süleyman I (1494 or 1495–1566), who reigned as Süleyman the Magnificent and brought the Ottoman Empire to the peak of its power.

Sweden, which for a time had been under the

rule of Denmark, managed to regain its independence under Gustav I Vasa (1496?–1560), who established himself as king in 1523.

In 1524 an Italian navigator in the pay of France, Giovanni da Verrazano (1485?–1528), explored the east coast of North America and was the first to sail into what is now called New York Bay.

In Asia, Babur (1483–1530), who maintained he was a descendant of Tamerlane, took Delhi and Agra in India, in 1526, and established the *Mogul Empire* (*Mogul* is a form of *Mongol*). This empire endured for over three centuries.

In 1530 the potato was discovered by the Spanish explorer Gonzalo Jiménez de Quesada (*ca.* 1495–1579), who conquered what is now Colombia and founded Bogota. The potato, along with maize and tobacco, was one of the most important botanical contributions of the American continents to Europe.

1535

CUBIC EQUATIONS

At this time, mathematicians found it easy to solve equations of the first degree (*linear equations*, involving x) and of the second degree (*quadratic equations*, involving x^2). Equations of the third degree (*cubic equations*, involving x^3) defeated them.

In 1535, however, the Italian mathematician Niccolò Tartaglia (1499–1557) found a general method for solving cubic equations. In those days, mathematicians would often keep their discoveries secret and parade their ability to solve problems others could not. This raised their reputations and gave them a feeling of power. However, another Italian mathematician, Geronimo Cardano (1501–1576), wheedled the cubic equation solution out of Tartaglia and then published it, so that Cardano is often given credit for the discovery.

Tartaglia objected loudly, but the event set an important precedent. Scientific discoveries do not belong to the discoverer but to the world. If discoverers all clutched their discoveries close to their chests in order to garner personal glory, the progress of science would creak to a halt. It has therefore become a rule that credit for a discovery goes not necessarily to the original discoverer, but to the first who publishes the discovery.

This encourages publication and allows all scientists to learn of discoveries as soon as possible. Science as we know it would not exist without the "first publication" rule, and Cardano, by his dishonorable action, pushed science in that direction, thus doing the world more good than he did Tartaglia harm.

IN ADDITION

Henry VIII of England (1491–1547), who became king in 1509, abandoned his wife, Catherine of Aragon (1485–1536) in 1531 because the Pope would not grant him a divorce, married Anne Boleyn (1507?–1536) in 1533, and in 1534 began the process that established Anglicanism, which is essentially like the Catholic Church except that the English king and not the Pope is the supreme head.

In 1534 a French navigator, Jacques Cartier (1491–1557), thought he had found the Northwest Passage—that is, a navigable passage through North America from the Atlantic to the Pacific and to Asia—when he came upon an opening between

Labrador and Newfoundland, which we now call the Strait of Belle Isle. He went through it into what seemed a wide inlet of the ocean. Since he entered it on August 10, a day dedicated to St. Lawrence, it is now called the Gulf of St. Lawrence. It turned out to be the outlet of the St. Lawrence River and not a strait leading to the Pacific. Nevertheless, France based its claim to Canada on Cartier's voyage and eventually held Canada for over two centuries as a result.

1538

COMETS' TAILS

In the 1530s, no fewer than six comets appeared in the sky. Fired by the example of Regiomontanus (see 1472), astronomers viewed them calmly. One of these was Girolamo Fracastoro, who had coined the word *syphilis* (see 1495). In 1538 he published a book in which he recorded his observations and mentioned that a comet's tail always pointed away from the Sun.

A German astronomer, Peter Bennewitz (in Latin Petrus Apianus; 1501–1552), also studying these comets, published a book in 1540 in which he came to the same conclusion independently. He published the first scientific drawing of a comet, in which he indicated the position of the tail with reference to the Sun.

IN ADDITION

Henry VIII, having broken with the Catholic Church so that he might marry Anne Boleyn, found that she produced only a daughter and had her beheaded in 1536 on trumped-up charges of adultery. He then married Jane Seymour (1509?–1537), who died in childbirth after giving him the son he so ardently desired.

1541

MISSISSIPPI RIVER

The interior of the American continents was being increasingly explored. Cortés explored what is now northern Mexico and in 1536 discovered the peninsula of Baja California.

In 1539 the Spanish explorer Hernando de Soto (*ca.* 1500–1542) was commissioned by Charles V to land in Florida and add it to the Spanish dominions. De Soto did more than that. Over the space of three years, he explored what is now the southeastern portion of the United States. In 1541 he and his men were the first Europeans to see the Mississippi River. He crossed it but returned to it in the spring of 1542 and died on its banks.

IN ADDITION

The French theologian John Calvin (1509–1564) preached a more extreme form of Protestantism than Luther had. It came to be called *Calvinism* and gave rise to Presbyterianism.

1542

AMAZON RIVER

One of Pizarro's aides during his conquest of Peru was Francisco de Orellana (*ca.* 1490–*ca.* 1546), who was exploring eastward past the Andes Mountains when he came upon the headwaters of a river. He felt it would be easier to see where the river led him than to make his way back across the formidable mountain barrier.

From April 1541 to August 1542, he progressed down the river, which as it happened, was by far the greatest in the world in terms of water volume delivered to the ocean and area drained. His report mentioned tribes that appeared to be led by women. This reminded people of the Amazons, the women warriors of Greek legend, and the river was named the Amazon in consequence.

Orellana was the first European to cross South America from ocean to ocean.

IN ADDITION

Henry VIII, having married and quickly divorced a fourth wife, Anne of Cleves (1515–1557), in 1540, at once married a fifth wife, Catherine Howard (1520?–1542), and soon had her executed for adultery.

1543

HELIOCENTRIC SYSTEM

The speculations of Aristarchus (see 280 B.C.), about a heliocentric system in which the Sun was at the center of the Universe, with the planets, including the Earth, revolving about it, had been disregarded, and the geocentric system of Hipparchus (see 150 B.C.) and Ptolemy (see 140) had been accepted without question.

However, the mathematics needed to work out the planetary motions on a geocentric basis was very difficult. While the Sun and Moon moved steadily west-to-east against the stars, the other planets occasionally reversed direction *(retrograde motion)* and grew markedly brighter and dimmer as they progressed across the sky.

The Polish astronomer Nicolaus Copernicus (1473–1543) thought, as early as 1507, that if one went back to Aristarchus's view and supposed that all the planets, including

Earth, were moving about the Sun, it would become easy to explain retrograde motion. It would also be easy to explain why Venus and Mercury always remained near the Sun and why planets grew dimmer and brighter. In addition to all this, the mathematics for working out planetary motions and positions would be simplified.

Copernicus did not, in his suggestions, abandon all Greek ideas. He clung to the notion that planets had to move in orbits that were circles and combinations of circles, and in this way he retained much unnecessary complexity.

The difference between Aristarchus and Copernicus was that Aristarchus merely presented his notion as a logical way of looking at the planets. Since others thought it wasn't logical, that ended it. Copernicus, however, used the Aristarchean idea to work out the actual mathematics of the planetary motions and show the reduction in complexity. This meant that even if people denied that the heliocentric system could be true, they would still be apt to use it as a simplified device for calculations.

Nevertheless, Copernicus hesitated to publish his theory and his computations, because he knew that the geocentric theory was felt by the Church to be in accordance with the Bible. To advance a heliocentric theory that would seem to be going against the Bible would surely create a storm. He therefore quietly circulated his book only in manuscript form.

Finally he let himself be persuaded by enthusiastic friends to have the book printed. It was entitled *De Revolutionibus Orbium Coelestium (Concerning the Revolution of Heavenly Bodies).* He dedicated it to Pope Paul III (1468–1549) as a placatory gesture, and then died. The story is that Copernicus was given the very first copy of his book on the day of his death.

As Copernicus had forseen, the book created a loud and violent storm. The Catholic Church put the book on its Index, forbidding the faithful to read it, and did not lift the ban till 1835. The Lutherans were equally hostile. The book could not be suppressed, however. With the coming of printing, far too many copies flooded the libraries of the scholars.

Copernicus's book totally overturned Greek astronomy, and though it was fifty years before astronomers generally could bear to turn their backs on Ptolemy and accept the fact that the vast Earth flew through space on an annual journey about the Sun, the book marked the birth of what came to be called the *Scientific Revolution.* With it came final proof that the ancients did *not* know it all and that moderns might strike out on their own in new directions and reach new heights—and they certainly did.

It could be argued that just as printing made the Protestant Reformation possible, it also made the Scientific Revolution possible.

NEW ANATOMY

Just as Copernicus was overturning Greek notions of astronomy, a Flemish anatomist, Andreas Vesalius (1514–1564), was upsetting Greek notions of anatomy. Unlike other anatomists of the period, Vesalius was willing to trust his own eyes when they disagreed with the statements of the Greeks.

He put together the results of his researches in a book entitled *De Corporis Humani Fabrica (Concerning the Structure of the Human Body)* in which he corrected over two hundred errors of Galen (see 180). What's more, the book took advantage of the technique of printing to reproduce careful illustrations of anatomical facts, drawn by the Flemish artist Jan Stephan van Calcar (*ca.* 1499–after 1545), who had been a student of Titian.

Vesalius's book was published in 1543, the same year in which Copernicus's book appeared, and this double appearance strengthens the feeling that the year marks the beginning of the Scientific Revolution.

IN ADDITION

Henry VIII married his sixth and last wife, Catherine Parr (1512–1548), in 1543. In that year also, Europeans reached Japan for the first time. They introduced the musket, and the Japanese adopted it at once.

1545

NEGATIVE NUMBERS

Until this time, mathematicians had assumed that all numbers, whether integers, fractions, or irrationals, had to be greater than zero. It might seem, after all, that one could not possibly have less than nothing.

On the other hand, mathematicians knew there were such things as debts. To have no money and to owe a sum to someone else means having less than no money. This might seem merely like practical business, nothing to do with ethereal numbers, but Cardano (see 1535) showed, in 1545, that debts and similar phenomena could be treated as negative numbers, which would follow rules of mathematics very similar to those that ordinary numbers did. You could have negative integers, negative fractions, and negative irrationals.

In that same year, Cardano worked out a general solution for equations of the fourth degree, involving x^4.

SURGERY

In ancient and medieval times, surgery was looked down upon as an inferior branch of medicine, since it involved working with one's hands and was uncomfortably akin to butchery and vivisection. Physicians therefore left the matter of cutting the flesh to those who also cut hair, so that the profession of the barber-surgeon was recognized.

One of the French barber-surgeons was Ambroise Paré (1510–1590), who was skillful enough to serve as surgeon to Henry II of France (1519–1559) and to his three sons. Paré is best known for his improvements in battle surgery. Most surgeons of the time practiced searing, disinfecting gunshot wounds with boiling oil and stopping bleeding by cauterizing arteries (without anesthetics, of course). A torture chamber could scarcely have been worse than this treatment.

Paré practiced cleanliness instead. He used soothing oils instead of boiling oils. He also tied off the arteries rather than burning them shut. He managed to bring about far more cures, with an infinitesimal fraction of the pain, and he is therefore considered the father of rational surgery.

He wrote a report of his findings in this area in 1545. In those days, and for a century and a half afterward, learned books were routinely written in Latin. Paré did not have a classical education and was forced to write in French. For this he was scorned by the learned ignoramuses of the day.

IN ADDITION

In 1545 the Catholic Church opened a Council at the city of Trent in northeastern Italy. The *Council of Trent* sat for eighteen years and introduced many needed reforms into the Church, to the point where one could speak of a *Counter-Reformation*. Until then, Protestantism had spread rapidly, fueled by the notorious corruption of the Catholic higher clergy. Afterward, the Protestants could win no further easy victories and an impasse developed, which led both sides into resorting to war.

1551

TRIGONOMETRIC TABLES

The German mathematician Georg Joachim Iserin von Lauchen (1514–1576), better known as Rhäticus from the region of his birth, was a student of Copernicus and had been instrumental in persuading him to publish his book (see 1543). To help out with the mathematics required to determine planetary movements, Rhäticus prepared trigonometric tables; that is, ratios of the length of the sides of triangles to each other for different angles. Such tables had been prepared in the time of the Greeks, but Rhäticus prepared the best yet and, for the first time, related the ratios to the size of the angle (as is now invariably done) rather than to arcs of circles.

These trigonomeric tables, combined with Copernicus's heliocentric view, made it possible for computational astronomy to take a big leap forward.

PLANETARY TABLES

Copernicus had pointed the way to better determinations of planetary motions, but he had not made a major effort himself to produce the necessary tables.

This was done in 1551 by the German mathematician Erasmus Reinhold (1511–1553), with the backing of Albert (1490–1568) who was the last grand master of the Teutonic Knights. (This was a Catholic order, but Albert led it into Lutheranism in 1544 and made himself duke of Prussia, the easternmost German province.)

Reinhold went over Copernicus's mathematics, sharpening and improving it, then prepared what he called *Tabulae Prutenicae (Prussian Tables)* in honor of his patron. They were better than the Alfonsine Tables of three centuries earlier, which had been based on Ptolemy's mathematics—but not much better. The planetary system had to be improved beyond Copernicus if tables were to be seriously improved, and that wasn't to happen for another half-century.

IN ADDITION

In 1546 Holy Roman Emperor Charles V (1500–1558) took up arms against Lutheran principalities in order to force their allegiance to emperor and Church. This marked the beginning of the *Wars of Religion*, which were to devastate western Europe for a century.

1552

EUSTACHIAN TUBE

Vesalius's new anatomy (see 1543) spurred on the field generally.

The Italian anatomist Bartolommeo Eustachio (1520–1574) prepared a book on anatomy in 1552, nine years after Vesalius's. In some respects it was more accurate than the earlier book, but the illustrations were not as beautiful. Eustachio described a narrow tube connecting the ear and the throat, which has been known as the *Eustachian tube* ever since, though it may have been discovered by a Greek physician, Alcmaeon (see 500 B.C., Dissection), two thousand years before.

Eustachio was also the first to describe the adrenal glands.

IN ADDITION

In Russia, Ivan IV Vasilyevich (called Ivan the Terrible) (1530–1584) had become ruler in 1533 and in 1547 had himself crowned *czar.* He was the first Russian ruler to adopt the name formally. In 1552 he began a successful campaign against the Tatars, who had ruled eastern Russia since the Mongol conquest. By 1555 he ruled two-thirds of what is European Russia today.

In 1552 the French astrologer Michel de Notredame (1503–1566), better known by the Latinized version of his name as *Nostradamus,* began to put out gibberish verses that purported to foretell the future. These have been in vogue among the simple-minded ever since.

1553

NORTHEASTERN PASSAGE

Portugal had reached the Far East by going around the southern end of Africa (the *Southeast Passage*) in 1497, and Spain had reached the Far East by going around the southern end of South America (the *Southwest Passage*) in 1521. Both passages remained closed to other European nations as long as Portugal and Spain remained strong at sea.

The *Northwest Passage,* a possible route to Asia around the northern shores of North America, had been searched for by France, through Verrazano (see 1531) and Cartier (see 1535), without success. In 1553 the British tried the *Northeast Passage,* a possible route to Asia around the northern shores of Asia.

This proved totally impractical, but one English ship, under Richard Chancellor (d. 1556), managed to make its way into the White Sea, as Ottar the Viking had done (see 870). Chancellor landed at the Russian port of Arkhangelsk and was taken overland to greet Ivan IV of Russia. Thereafter, trade between England and Russia flourished.

IN ADDITION

Protestantism in England had flourished under Edward VI (1537–1553), who succeeded his father, Henry VIII, but he was succeeded in 1553 by his older sister, Mary I (1516–1558), the daughter of Catherine of Aragon. Mary was a devout Catholic, and during her reign, she made strenuous efforts to return England to the Church.

The Ottoman Empire was busily engaged at this time in spreading its control over the Mediterranean shores of northern Africa.

1555

HOMOLOGIES

Anyone can see that living organisms can be grouped. Dogs and wolves resemble each other more than either resemble rabbits, for instance. Cats, lions, and tigers resemble each other. Sheep and goats resemble each other. Insects have a common resemblance that separates them from noninsects, and so on.

This might have given rise to evolutionary notions, that some basic doglike animal had given rise to both dogs and wolves, as an example. However, the Bible described all animals as being created simultaneously and separately, and it is possible to argue that God may just have decided to create animals in groups for his own purposes.

It would be more impressive if similarities could be found in more widely diverse organisms, similarities that were not too obvious to the eyes. This was done by a French naturalist, Pierre Belon (1517–1564).

Francis I of France (1494–1547) was engaged in a prolonged struggle with Charles I of Spain, who was also the Emperor Charles V. Francis was desperate enough to ally himself with the Ottoman Empire, and in 1546 he sent Belon on a diplomatic mission to the Ottomans.

This gave Belon a chance to study plant and animal life in the eastern Mediterranean and to compare it with life in France. He was the first to describe, in writings he published in 1555, basic similarities *(homologies)* in the skeletons of all vertebrates, from fish to humans. Such details as the number of bones in the limbs were remarkably constant from animal to animal, regardless of the difference in outer appearance.

This could not help but encourage evolutionary thought, although it was to be three centuries before it came to fruition.

IN ADDITION

The Treaty of Augsburg, signed in Germany in 1555, allowed each German ruler to choose either Lutheranism or Catholicism for his or her people. No provision was made for Calvinism, and religious hostility continued unabated so that no true toleration was possible in most places.

1556

MINERALOGY

Mining had been of interest to human beings since the development of metallurgy some forty-five centuries before. At this time it also became of interest to medicine, as physicians began to investigate mineral remedies. An example was Theophrastus Bombastus von Hohenheim (1493–1541), better known as *Paracelsus,* a Swiss physician who pioneered the use of opium extracts but also used compounds of mercury and antimony, even after they had proved toxic.

Another physician who grew interested in mining was Georg Bauer (1494–1555), better known by his Latinized name as Georgius Agricola. (*Agricola* in Latin and *bauer* in German both mean "farmer.")

Agricola studied mining processes carefully and wrote a book, *De Re Metallica (Concerning Metallic Things),* which was published posthumously in 1556. In it he summarized all the practical knowledge he had gained from German miners. It was clearly written and had excellent illustrations of mining machinery. It was the first important book on minerals ever written and is usually considered to have founded the science of *mineralogy.*

TOBACCO

Native Americans were quite willing to teach European arrivals the mysteries of tobacco—how to prepare the leaves, burn them, and inhale the smoke. They probably didn't intend this as revenge for being enslaved and killed, but it worked out that way. Tobacco addiction settled down on Europe and eventually on the rest of the world. There is no estimate of how much discomfort it has caused in the way of stench, how much natural and property damage it has caused in the way of forest fires and building fires, and how many people (smokers and nonsmokers alike) it has killed in those fires or through lung cancer and heart disease.

The first tobacco seeds reached Spain in 1556.

A French diplomat, Jean Nicot (*ca.* 1530–1600), served as ambassador to Portugal between 1559 and 1561 and sent tobacco seeds to France. His name is immortalized in *nicotine,* the name of the highly poisonous substance that is the active ingredient in tobacco. An English naval commander, John Hawkins (1532–1595), introduced it to England in 1565.

IN ADDITION

An earthquake struck in the Chinese province of Shansi on January 24, 1556, and the death toll is supposed to have been some eight hundred thousand. If this is so, it was the deadliest earthquake in recorded history.

The Holy Roman Emperor Charles V abdicated in 1556. He left his German dominions and the imperial title to his younger brother, Ferdinand I (1503–1564), and left Spain, the Low Countries, various Italian regions and all his dominions overseas to his son, Philip II (1527–1598).

1560

SCIENTIFIC SOCIETIES

Throughout history, scientists have usually worked alone because of the difficulty of communication. Sometimes they gathered in some particular center of learning, as in Athens, Alexandria, and Baghdad, but even then their companionship was a haphazard thing.

The coming of printing made it easier to record and publish advances, however, and the affair of Tartaglia and Cardano (see 1535) made it important to publish if one wanted credit. There would be value in exchanging information, then, for it would benefit all scientists in their search for reputation.

In 1560 an Italian physicist, Giambattista della Porta (1535?–1615), founded the first scientific association designed particularly for this interchange of ideas, the *Academia Secretorum Naturae (Academy of the Mysteries of Nature)*. It was closed down by the Inquisition, which was oversensitive to any gatherings in those harsh days of religious conflict, but the idea was too good to let go, and other scientific societies were formed in time, and persisted.

These societies helped give rise to a *scientific community* that was as superior to an individual scientist as a phalanx or legion was to an individual soldier.

IN ADDITION

In 1557 the Portuguese opened a trading station at Macao, near the south-Chinese city of Canton. This was the beginning of the establishment of European footholds in China. To this day, Macao is still Portuguese.

In 1558 Mary I of England who died and was succeeded by her younger half-sister, Elizabeth I (1533–1603), daughter of the ill-fated Anne Boleyn. Elizabeth was Protestant, and her throne was disputed by her cousin (once-removed) Mary (1542–1587; usually called *Mary, Queen of Scots*), who was Catholic.

1565

MUSKET

By this time, the harquebus had been refined to a *musket*. The word is from the Latin word for "fly" and had earlier been applied to the bolts of crossbows, perhaps because the crossbow bolt or the musket ball whizzed past one's ear with the buzz of a fly.

Muskets could fire balls that penetrated armor, so armor began to disappear. There was no use in bearing the weight if it offered no protection.

For two centuries, muskets remained the principle small arms used by soldiers, but they were still difficult to handle and reload and musketeers needed the protection of pikebearers.

IN ADDITION

In 1562 civil war began in France between French Catholics and French Protestants (called *Huguenots*). It was to continue, on and off, for over a quarter of a century.

In 1565 the Spaniards under Pedro Menendez de Avilés (1519–1574) established a settlement at St. Augustine, on the northeastern shore of Florida. It was the first permanent European settlement on what is now the territory of the United States. That same year, another Spanish soldier, Miguel López de Legazpi (*ca.* 1510–1572), conquered the island group on which Magellan had died nearly half a century before and named them the Philippine Islands, after Philip II of Spain.

1568

WORLD MAPS

Once the Age of Exploration began, it became important to try to map the world in general and to do so accurately, so that navigators could reach their destinations more easily.

The difficulty was that one cannot map a spherical surface on a plane surface without distortion. Since distortion is unavoidable, therefore, one must find a distortion that will remain useful.

The answer was supplied by a Flemish geographer, Gerhard Kremer (1512–1594), better known by his Latinized name, Gerardus Mercator. In 1568, Mercator perfected his *cylindrical projection*.

Imagine a hollow cylinder encircling the Earth and touching it all around the equator. A light at the Earth's center can then be imagined as casting the shadow of the surface features on the cylinder, and the cylinder, when unwrapped, will carry a map of the world, a *Mercator map*.

In this map, the meridians of longitude are vertical and parallel. Since, on the sphere, the meridians of longitude approach each other and meet at the poles, this means that in a *Mercator projection*, the east-west distances are increasingly exaggerated as one travels north or south from the equator. The parallels of latitude run horizontal and parallel, as they do on the sphere, but as one goes north and south from the equator, they spread out more widely, too.

On such a map, Greenland looks larger than Africa, although in actual fact, Africa is thirteen times as large as Greenland.

Nevertheless, the Mercator projection is useful for navigators, because a ship that travels in any constant compass direction moves in a straight line on the Mercator projection but in a curved line on any other type of projection.

The cover of Mercator's book of maps showed Atlas, the mythological Greek Titan, holding the world on his shoulders. As a result, that book and all later books of maps have been called *atlases*.

With Mercator, Greek geography ended and modern geography began.

IN ADDITION

When Süleyman the Magnificent died, the Ottoman Empire, which had reached its peak under a succession of remarkably capable sultans, began to decline.

In 1568 the Netherlands, largely Protestant, rebelled against its strongly Catholic overlord, Philip II of Spain. The intense struggle that resulted was to last for eighty years.

1572

SUPERNOVA

A supernova, like the one that had appeared in 1054, blazed out in the constellation of Cassiopeia, high in the northern sky, in November 1572. The supernova of 1054 had gone unobserved in Europe, but times had changed.

A young Danish astronomer, Tycho Brahe (1546–1601), usually known by his first name, watched the *new star* carefully from night to night. When he first saw it, it was brighter than Venus, but it gradually faded until in March 1574, it could no longer be seen at all. Tycho had watched it for 485 days.

The Greeks had assumed that the heavens (unlike Earth) were perfect and unchanging. Anything in the sky that seemed to change (or to move in anything but a regular and predictable path) could not be part of the sky, they thought, but must be part of the atmosphere of the imperfect Earth. This included clouds, shooting stars, and comets.

The new star, being a temporary phenomenon, ought to be part of the atmosphere, too, but although Tycho tried to determine its parallax (see 150 B.C.), he could detect none. The new star must be beyond the Moon and therefore be part of the heavens, possibly a very distant part. The notion of heavenly perfection and immutability was destroyed.

In 1573 Tycho published a small book detailing all his observations of the star, with a title usually given in short form as *De Nova Stella (Concerning the New Star)*. Because of that title, stars that suddenly appear in the sky are now known as *novae* (the Latin plural) or *novas* (the English plural).

Tycho was suddenly the most famous astronomer in Europe.

IN ADDITION

In 1569 Poland and Lithuania were combined into a single nation that was larger than any other nation west of Russia. It was poorly organized, however, and had a most turbulent and unmanageable aristocracy.

In 1570 the Ottoman Empire declared war on Venice and proceeded to attack the island of Cyprus, which was then Venetian. Pope Pius V (1504–1572) organized an alliance against the Turks, and 208 Catholic galleys (oared ships) clashed with 273 Ottoman galleys in a three-hour fight at Lepanto. The Catholics won a complete victory. It was the first great defeat for the Ottoman Empire. Its reputation for invincibility was forever gone and its decline continued slowly but inexorably. Lepanto was the last important battle to be fought by galleys. From now on, improvements in sails and rudders made sailing ships more reliable.

In 1572 one of the darkest moments in the Wars of Religion came when, during a period of supposed peace between the Catholics and Huguenots in France, the Catholics suddenly attacked on August 23 (St. Bartholomew's Day) and killed fifty thousand unarmed and defenseless Huguenots throughout France. This *Bartholomew Day's Massacre* remained a blot the Catholics could never wipe out.

1576

NORTHWEST PASSAGE

The English, having failed at the Northeast Passage (see 1553), tried their hand at the *Northwest Passage* around the northern shores of North America.

In 1576 an English navigator, Martin Frobisher (*ca.* 1535–1594), sailed to North America with three ships and thirty-five men. He sailed northward from the Labrador area and discovered what we now call Baffin Island, a large island west of Greenland.

In a second voyage, in 1578, Frobisher sighted Greenland itself (see 982). By the time Frobisher arrived, the Viking settlers had died or departed and its shores were occupied only by Inuit (Eskimos). From this time on, Greenland remained part of world geography.

Frobisher did not discover any practical Northwest Passage, however.

IN ADDITION

The revolt of the Netherlands was in full flame. Leading the Netherlanders was William I of Nassau (1533–1584), known as William the Silent. He was the founder of the Dutch Republic. The Spanish army was the best in Europe at the time and the Netherlanders could not withstand it in the field. They resolutely underwent sieges of their cities, however, opened the dikes when necessary, and used their ships to bring in supplies. The Spaniards won battles, but they could not win the war.

1577

DISTANCE OF COMETS

Tycho Brahe (see 1572), with the help of the Danish king, set up the first real astronomical observatory, on the island of Hven in the strait between Denmark and Sweden, outfitting it with the best instruments he could make.

In 1577 a bright comet appeared in the sky and Tycho observed it carefully. By Greek notions, it was an atmospheric phenomenon, and therefore it should have a large parallax. However, Tycho could find none, and was certain that it was at a distance far beyond that of the Moon. This was another serious blow against Greek astronomy.

1578

DRAKE STRAIT

The English navigator Francis Drake (1540 or 1543–1596) had made a career out of raiding the Spanish possessions in the Americas in the course of an undeclared war between England and Spain. It occurred to him that the Spanish settlements on the Pacific coast of the Americas were entirely undefended because none of Spain's enemies had made their way into the Pacific, so in 1572 he had landed at Panama, crossed the Isthmus, and became the first Englishman to see the Pacific.

In 1577 he set sail on an expedition that he hoped would carry him through the Strait of Magellan, which so far only Spanish ships had passed through. No one knew the exact extent of Tierra del Fuego, the land south of the strait, and some thought it was part of a vast Antarctic continent.

Drake passed through the Strait of Magellan in 1578 and was then struck by a storm in the Pacific and driven far enough south to see that open water lay to the south of Tierra del Fuego, which turned out to be nothing more than a moderately sized island. The water to the south of that island has been known as *Drake Passage,* or *Drake Strait,* ever since.

Drake sailed up the Pacific coast of the Americas as far as what we now call San Francisco Bay. He found no water route that would connect with the Atlantic Ocean, so he decided to sail westward across the Pacific. He reached England in 1580, the first person to circumnavigate the globe since Magellan's ship had done so six decades earlier (see 1523).

1581

PENDULUM

In measuring time intervals of less than a day, the point is to find some physical action that proceeds at a constant rate. The sifting of sand or the dripping of water through a small hole, the burning of a candle, or the progress of the Sun across the sky are all fairly constant motions, but might there not be some convenient action that was even more steadily constant?

The first hint of a constant action that was unknown to the ancients came in 1581, when a seventeen-year-old Italian boy, Galileo Galilei (1564–1642), who is usually known by his first name, was attending services at the Cathedral of Pisa.

His attention was caught by a chandelier that swayed as air currents caught it. Sometimes it swung through a small arc, sometimes through a larger one, but to Galileo's inquiring mind, there seemed an anomaly: long or short, the time it took the chandelier to complete its swing back and forth seemed the same. He timed it by the beating of his pulse. Upon returning home, he set up two pendulums of equal length and swung one in larger, one in smaller sweeps. They kept together and he found he was correct.

Nevertheless, when in later life Galileo conducted experiments in which it was necessary to know the time elapsed, he had to continue to use his pulse or dripping water. Not for another seven decades would the steady beat of the pendulum be put to use as a device to help measure time.

SIBERIA

Despite Russia's vast extent over eastern Europe, its long sleep under the Mongols had left it technologically backward. On Russia's western boundaries were the Swedes, the Poles, and the Germans, with none of whom Russia could compete militarily.

Toward the east, however, were vast spaces that at the moment contained no formidable enemy. This was also a cold region that would ordinarily not seem very enticing except that, as in Russia's European north, there were animals living there whose pelts, adapted to the Arctic cold, were thick and valuable.

In 1581, as Ivan IV's reign approached its end, the Stroganovs, a Russian family that had made a fortune in the fur trade, employed a Cossack named Yermak Timofievich (?–1584) to explore eastward and expand the Stroganov fur trade. Yermak conquered a Mongol kingdom east of the Urals, named Sibir. The name (*Siberia* in English) came to be applied to the entire northern third of Asia. This was the beginning of a process that would eventually lead the Russians to the Pacific Ocean and put an end forever to the raids of central Asian nomads against the settled regions to its south and west.

IN ADDITION

In 1578 Sebastian (1554–1578), who had become king of Portugal in 1557, was defeated and slain in Morocco. He was succeeded by a great-uncle, Henry (1512–1580), who died without heirs. Philip II of Spain, who was married to Sebastian's aunt, sent an army into Portugal in 1580, defeated the Portuguese, and made himself king of Portugal as well as of Spain. The Iberian peninsula was united for the first time since the Muslim invasion over eight and a half centuries before. The Portuguese overseas empire passed under Spanish control, and Spain reached the peak and pinnacle of its power.

1582

GREGORIAN CALENDAR

The Julian calendar, adopted by Julius Caesar (see 46 B.C.), was not quite correct. It assumed a year that was 365.25 days long, but the year was more nearly 365.2422 days long.

If the year were exactly 365.25 days long, that extra quarter-day could be made up for by adding an extra day every fourth year. Then every fourth year would be 366 days long (leap year), and in the space of 400 years there would be 100 leap years.

A year that is 365.2422 days long is about 365$^{97}/_{400}$ days long. This means that there should be only 97 leap years every 400 years, not 100. By the Julian calendar, 3 days too many were added every 400 years, and the vernal equinox fell earlier and earlier. If the vernal equinox was on March 21 when the Julian calendar was established, then by 1582 it would fall on March 11, 10 days too early.

The Church was very involved in this because the holy days depended on the calendar, and if the drift continued, Easter would eventually come in winter and Christmas in the autumn. However, earlier attempts to reform the calendar had failed, for people are conservative about such things.

By 1582, however, the situation had come to seem intolerable to the Church. The Bavarian astronomer Christoph Clavius (1537–1612) worked out a scheme for a more correct calendar and Pope Gregory XIII (1502–1585) adopted it.

On October 4, 1582, 10 days were dropped and the next day was October 15. Thereafter, any year that ended in 00 but was not divisible by 400 was *not* a leap year. Thus 1600 was a leap year, but 1700, 1800, and 1900 were *not* leap years. However, 2000 will be a leap year. In this way, there are only 97 leap years every 400 years.

Catholic Europe adopted the *Gregorian calendar* (named in the Pope's honor) almost at once. The new Protestant states were more reluctant, preferring to disagree with the Sun rather than agree with the Pope. Great Britain didn't adopt the new calendar for two centuries; Russia, for three and a half centuries.

IN ADDITION

In Japan, Hideyoshi Toyotomi (1537–1598) rose from poverty to become military dictator of Japan in 1582. He completed the unification of Japan, and the nation has remained unified ever since.

1583

HYDROSTATICS

The Dutch mathematician Simon Stevin (1548–1620) showed that the pressure of a liquid upon a given surface depends on the height of the liquid above the surface and upon the area of the surface, but does *not* depend on the shape of the vessel containing the liquid. This finding is considered to have founded the modern science of *hydrostatics.*

IN ADDITION

In 1583 the English navigator Humphrey Gilbert (*ca.* 1539–1583) succeeded in establishing a settlement at what is now St. John's in Newfoundland, an island that had first been discovered by John Cabot (see 1497). This was the first English overseas colony.

1586

DECIMAL FRACTIONS

Mathematicians had found fractions difficult to handle ever since the days of the Sumerians. Special rules had to be worked out to deal with them. In 1586, however, Stevin (see 1583) showed that they could be made part of ordinary positional notation. To the right of the units column (on the other side of a *decimal point*) could be the tenths column, then the hundredths column, and so on. Thus 2¼ would become 2.25; 2⅛ would become 2.125; 2⅞ would become 2.875; and so on.

The disadvantage of such *decimal fractions* is that some never end. Thus 2⅓ is 2.3333333 . . . forever; 2⅚ is 2.8333333 . . . forever; and so on. Despite this, decimal fractions greatly simplified computations involving fractions.

IN ADDITION

Walter Raleigh (1554–1618) also tried to establish settlements in North America. He named the east coast of the continent north of Florida *Virginia,* in honor of Elizabeth I, who was known as the *Virgin Queen.* He attempted settlements at what is now Roanoke Island in North Carolina in 1585, but they were not successful.

William the Silent of the Netherlands was assassinated on July 10, 1584, at the instigation of Philip II, who had offered a large reward for the deed. However, the Netherlands continued in rebellion under William's son, Maurice of Nassau (1567–1625), who was a better military commander than his father had been.

1589

FALLING BODIES

Aristotle had stated that the heavier an object was, the more rapidly it would fall. It seemed reasonable to say so. Why shouldn't a heavier body fall more rapidly? It is clearly being attracted to Earth more strongly, or it wouldn't be heavier. Furthermore, anyone who watches a feather, a leaf, and a stone falling will see at once that the stone falls faster than the leaf, which in turn falls faster than the feather.

The trouble is that light objects are impeded by air resistance, and in order to avoid that one should consider only relatively heavy objects. Thus, if one observes the falling of a rock that weighs 1 pound and another that weighs 10 pounds, air resistance is insignificant in either case. Do we then see that the 10-pound rock nevertheless falls faster than the 1-pound rock?

In 1586, Simon Stevin (see 1583) is supposed to have dropped two rocks at the same time, one considerably heavier than the other, and showed that they struck the ground at the same time. Later accounts said it was Galileo who demonstrated this by dropping different weights simultaneously from the Leaning Tower of Pisa. Both stories may or may not be true.

What is certain, though, is that in 1589 Galileo started a series of meticulous tests with falling bodies. Such bodies fall too rapidly to make it easy to measure the rate of fall accurately, especially since no accurate way of measuring short periods of time had yet been worked out.

Galileo therefore allowed balls to roll down inclined planes. The more nearly level the plane, the more slowly the balls moved under the pull of gravity and the more easily their rate of fall could be measured by primitive methods such as water dripping out of a small hole. In this way Galileo found it quite easy to show that as long as the balls were heavy enough to be relatively uninfluenced by air resistance, they rolled down an inclined plane at the same rate.

He was also able to show that the balls rolled down the inclined plane with a constant acceleration—that is, they gained speed with time at a constant rate, under the constant pull of gravity.

This settled another important point. Aristotle had held that in order to keep a body moving, a force had to be continually applied. This again seemed to fit observation. If an object were sent sliding across a floor, it would quickly slow to a stop. To keep it moving, you would have to keep pushing.

For this reason, it was felt that the planets in their eternal movement about the Earth had to be continually pushed by angels.

Galileo's observations showed that a constant push was not necessary to keep an object moving, if friction was removed. If a constant pull *was* exerted by gravity, for instance, an object moved with a constantly *increasing* speed. Consequently, no angels were required to keep the planets moving.

Galileo's experiments on moving bodies were so impressive that, even though he was not the first to conduct experiments—Peter Peregrinus had done so more than three centuries before (see 1269)—he is usually given credit for having founded *experimental science.*

CRYPTANALYSIS

Simple codes are almost as old as writing itself. After all, by substituting or rearranging words or letters according to some prearranged scheme, something that remains plain to the people involved may be made totally obscure to others. In such cases, you have a *code* or *cryptogram.*

Codes that can be made can be broken, and as the years went by, the use of ever more subtle methods of encoding messages were countered by ever more subtle methods of decoding them.

An early example of this came in 1589, when France was in the last stages of a civil war. Henry III (1551–1589) had no direct heirs, and his second cousin, Henry of Navarre (1553–1610), was the logical successor. Henry of Navarre was a Huguenot, however, and was bitterly opposed, not only by French Catholics but by Philip II of Spain.

Philip II was using a code that the French mathematician François Viète (1540–1603; in Latin Franciscus Vieta), working for Henry of Navarre, was able to decode in 1589. Philip II, unable to account for the fact that his messages were being read, complained to Pope Sixtus V (1521–1590) that the French were using sorcery and ought to suffer ecclesiastical punishment as a result.

KNITTING MACHINES

It is possible to build a device that will imitate the motions of the hands or feet, where such motions are endlessly repeated and do not call for the guide of continuous intelligence.

In 1589 an English clergyman, William Lee (1550?–1610), invented a device called a stocking frame that would knit faster than hand-knitters could produce their products. This great advantage of the stocking frame was also its great disadvantage, for if it were put into operation on a large scale, it would throw hand-knitters out of work. Elizabeth I of England refused to grant Lee a patent for that very reason.

Lee therefore took his machine to France where he received the necessary support.

Lee's experiences in England are an early example of how the threat of *technological unemployment* can slow technological advance. To be sure, technological advance creates more new jobs than it destroys, but there is always a painful transition period, and a humane government must help those who suffer, not so much out idealism (though what is wrong with that?) as to keep society stable and to make it easier to extend the benefits of technological advance to society in general.

IN ADDITION

Mary, Queen of Scots, was driven out of Scotland and into England by her rebellious nobles in 1568. She was Catholic and they were largely Protestant. Elizabeth I of England held Mary prisoner for the rest of her life. Because Mary was the center of continuing plots to make her queen of England in Elizabeth's place, on February 8, 1857, Elizabeth had her beheaded.

Philip II of Spain, furious at this, sent a fleet of 132 vessels (the *Invincible Armada*) to gain control of the English Channel and make it possible for the Spanish army in the Netherlands to invade England.

The English ships were smaller and fewer but more maneuverable and more skillfully handled by Francis Drake (see 1578) and John Hawkins (1532–1595). What's more, there were violent storms in the channel that wreaked more havoc on the large and clumsy Spanish ships than on the English ships that had friendly ports into which to escape. By August 8, 1588, the Armada had been defeated, and Spanish control of the sea was destroyed. From this point on, England could scour the seas at will and indeed exerted increasing control of the sea for three and a half centuries.

In 1588 Abbās I (1571–1629) became shah of Persia, introducing Persia's greatest period of power since the time of the Sassanids, a thousand years before.

1590

MICROSCOPES

It must have dawned on people rather early that there were ways of making small objects appear larger in size. Dewdrops on a leaf or on a blade of grass will make the surface they rest on look larger. Spheres of glass will do the same.

The people who would most notice this sort of thing and be most concerned with it were the spectacle-makers, since the convex lenses used to correct far-sightedness acted to magnify objects.

The spectacle-making industry was most advanced in the Netherlands at this time. It occurred to a Dutch spectacle-maker, Zacharias Janssen (1580–*ca.* 1638), that if one lens magnified somewhat, two would magnify to a greater extent. He placed a convex lens at each end of a tube and found that magnification was indeed improved. The improvement wasn't great, but Janssen's tube can be viewed as the first microscope, and its descendants were to revolutionize biology.

1591

ALGEBRAIC SYMBOLS

Until now, mathematicians had described quantities, relationships, and problems in words (it seemed the only way), and what they described was often hard to envisage.

Vieta (see 1589, Cryptanalysis) began to symbolize constants and unknowns by letters of the alphabet, the now familiar *x*'s and *y*'s of algebra. In 1591, he wrote a book about algebra that was the first a present-day high-school student would recognize at a glance to be dealing with that subject.

The progression from words to symbols was to mathematics something like the progression from ideographs to letters in ordinary writing, or from Roman numerals to Arabic numerals in arithmetical computation.

1592

THERMOMETER

The concept of hot and cold must be as old as humanity. We can all tell when an object is hot or cold by placing the hand near it (not even necessarily on it). We can also tell if one subject is substantially warmer than another. Such subjective sensations are useless, however, where small differences in temperature are concerned and are, in any case, untrustworthy. A humid day feels hotter than a dry day at the same temperature; a windy day colder than a calm day at the same temperature.

What is needed is some physical phenomenon that changes regularly in a measurable way with changes in temperature, and the first person to attempt to find such a phenomenon was Galileo (see 1581).

He warmed an empty bulb with a long tube extending from it, then placed the open end of the long tube into a container of water. As the warm air within the bulb cooled, it contracted, and water was drawn up into the tube. As temperature changed and the air within the bulb cooled or warmed, the water level rose or fell accordingly, and from the position of the level, one could judge the temperature.

This was a very crude device, because for one thing the water level was also affected by the air pressure upon the water reservoir. Nevertheless, it was the first *thermometer* (from Greek words meaning "to measure heat").

ARCHAEOLOGY

The cities of Pompeii and Herculaneum, near the base of Mt. Vesuvius in southern Italy, had been buried under lava and ash when the volcano unexpectedly erupted on August 24, 79.

For fifteen centuries they remained hidden, until an Italian engineer, Domenico Fontana (1543–1607), began tunneling under a hill in order to establish an aqueduct. The ruins were discovered in the process.

This brought the realization that part of the past was actually preserved for investigation in the present. Excavation for the deliberate purpose of studying the past did not start for another century, but the subject matter of the study was known to be waiting, so this discovery might be viewed as the beginning of modern *archaeology*.

1596

EAST INDIES

After the defeat of the Spanish Armada, the Netherlanders were heartened and fought more fiercely than ever. This was especially true among the northern Protestant half of the country, which eventually came to be known as the *Dutch Republic.* The southern half remained Catholic and under Spanish dominion and was called the *Spanish Netherlands*.

The Dutch, growing stronger at sea and becoming wealthy through sea trade despite the Spanish soldiers who harassed them, looked for expansion overseas. They had no hesitation in invading lands that were supposedly reserved for the Spanish and Portuguese (both ruled by Philip II). After all, they were anti-Catholic, anti-Spanish, and promoney.

In 1596, therefore, the Dutch set up a factory at Palembang in the island of Sumatra, in what came to be called the *East Indies*. That was the start of a Dutch overseas empire.

PI

The ancient Greeks had their favorite problems and one of them was *squaring the circle;* that is, given a circle of a particular size, to construct a square with the same area. The rules were that you could only use a straightedge (something that would draw straight lines) and a compass (something that would draw circles) and you had to do it in a finite number of steps. Unfortunately, they could never solve that problem.

In working with it, though, they got involved with the ratio of the circumference of a circle to its diameter, a ratio that we now refer to as *pi* (one of the Greek letters). Anyone can measure the diameter and then run a piece of string around the circumference, straighten it, and measure that, too. It turns out that the circumference (for any circle at all) is just a little over 3 times as long as the diameter. But what is the ratio *exactly?*

There are geometric ways of trying to get the exact ratio, and Archimedes (see 260 B.C.) had gotten a figure of about 3.142. Better values were obtained in later centuries until, in 1596, the Dutch mathematician Ludolf van Ceulen (1540–1610) obtained a value of pi to twenty decimal places. (Later in life, he got it to thirty-five decimal places.)

Even that didn't give an *exact* value, but it was so nearly exact that no reasonable computation involving pi would need a more exact value. (In Germany, pi is still sometimes called *Ludolf's number*.) Since then, pi has been obtained to a vastly greater number of decimal places, but even so there is no exact figure.

IN ADDITION

The Dutch began searching for the Northeast Passage. A Dutch navigator, Willem Barents (*ca.* 1550–1597), left Amsterdam in 1594 and explored the stretch of ocean lying to the north of Scandinavia and western Russia, which is now called *Barents Sea* in his honor. In 1596 he sighted the two large islands of Novaya Zemlya, which no European had ever seen before, and which are now part of the Soviet Union. His ship was forced to winter in Novaya Zemlya in 1596–1597. Barents and a cabin boy didn't make it, but fifteen crew members did, the first European explorers to survive an Arctic winter.

1597

MEDIEVAL ALCHEMY

The medieval alchemists didn't achieve much of what they would have liked to achieve. They didn't manufacture gold out of lead, and they didn't find the elixir of life. However, they weren't totally useless, either.

In 1597 a German alchemist, Andreas Libau (*ca.* 1540–1616), better known by the Latinized form of his name as Libavius, wrote a book entitled *Alchemia* in which he summarized medieval achievements in alchemy. It was the first chemical textbook worthy of the name. Libavius wrote clearly rather than mystically. He was the first to describe the preparation of hydrochloric acid, and he gave clear directions for preparing other strong acids such as sulfuric acid and *aqua regia* ("royal water," a mixture of sulfuric and nitric acids that was so powerful it would even dissolve the royal metal, gold).

With Libavius's book, the stage was set for the birth of real chemistry two-thirds of a century later.

IN ADDITION

India was under the rule of Akbar (1542–1605) at this time. He had come to the throne in 1556 as the third of the Mogul dynasty and managed to unite under his rule virtually all of India.

England had maintained a foothold in eastern Ireland for four centuries but had never managed to conquer the entire nation. In 1597 another of the numerous Irish rebellions broke out, this one under Hugh O'Neill (1540?–1616). Elizabeth I sent her inept favorite, Robert Devereux, Earl of Essex (1566–1601), to put it down, and of course, he failed.

1600

EARTH AS MAGNET

Although the compass had been known for nearly five centuries, no one knew why it pointed north. The English physician and physicist William Gilbert (1544–1603) put it to the test and published a book, *De Magnete (Concerning Magnets),* in 1600 in which he described his experiments.

For instance, he tested the general opinion that garlic would destroy magnetism and that diamonds would produce it. He rubbed magnets with garlic and the magnetism did *not* disappear. He rubbed unmagnetized iron with diamonds and magnetism did *not* appear. He took the precaution of doing this before witnesses.

The most important thing he did, however, was to take a large piece of loadstone and fashion a globe out of it. He located its magnetic poles and showed that a compass needle would point *north* if placed near the surface of this spherical magnet.

What's more, if he arranged for the compass needle to swivel vertically, it showed what was called *magnetic dip,* for it pointed straight through the body of the object. In fact, if the compass needle was held above the magnetic pole, it pointed straight down. (Magnetic dip was first noted on Earth's surface by an English navigator, Robert Norman [1560–?], in 1576.)

Gilbert concluded, then, that compass needles acted the way they did because Earth itself was a huge magnet.

IN ADDITION

Henry IV of France, who had earlier been Henry of Navarre, put through the *Edict of Nantes* in 1598, in which Huguenots were given freedom of religion within certain cities and towns.

In 1598 in Japan, Hideyoshi (see 1582) died. His attempt to conquer Korea toward the end of his career had failed. In 1600 Ieyasu (1543–1616) established himself as *shogun* (chief military commander). He was of the Tokugawa family, and the shogunate remained in that family for over two and half centuries. He moved his capital from Kyoto to Edo (which is now called Tokyo).

The Italian philosopher Giordano Bruno (1548–1600) wrote and spoke of a multiplicity of worlds, of an infinity of space, of a moving Earth, and of atoms. In every respect he was right, but he irritated the conservatives of the day by his loud and tactless scorn of them, and he refused to recant when threatened with death. He was burned at the stake on February 17, 1600, and his death had a chilling effect on scientific advance, especially in the Catholic nations.

1603

VEIN VALVES

That blood moves is certain, for when an artery is cut, blood emerges in spurts. The spurts are in time with the beating of the heart, so that the heart clearly pumps the blood.

The classical belief, inherited from Galen (see 180), was that the blood was manufactured in the liver and carried to the heart, from which it was pumped outward through arteries and veins alike and consumed in the tissues.

Of course, the heart is actually two pumps, with a thick muscular wall between, but there seemed no reason for two pumps. It was Galen's notion that there were small holes in the wall, too small to be visible, and that through these the blood passed, converting the heart into a single pump.

In 1603 an Italian physician, Girolamo Fabrici (1537–1619), better known by his Latinized name of Fabricius ab Aquapendente, studied the veins of the legs and noted that they had little valves along their length.

It was clear that the valves prevented the blood from flowing downward. Muscular action during walking squeezed the leg veins and forced the blood upward because that was the only direction in which it could go. This meant that the blood in the leg veins could move only *toward* the heart.

Fabricius, however, did not dare to go against Galen's doctrine and he refused to draw this conclusion.

IN ADDITION

Elizabeth's favorite, the Earl of Essex, was executed in 1601 after a remarkably inept (even for him) attempt to foment an insurrection. Elizabeth herself died in 1603 after a reign of forty-five years that is widely considered the most successful in English history. She was succeeded by her cousin (twice removed), the son of Mary, Queen of Scots, James VI of Scotland (1566–1625). He reigned as James I of England, establishing the Stuart line of English monarchs.

In 1602 the English navigator Bartholomew Gosnold (d. 1607) explored the North American coast of what is now New England.

1607

JAMESTOWN

Since the defeat of the Spanish Armada (see 1589) the English had been trying to make settlements overseas. They had succeeded in Newfoundland but had failed at Roanoke.

Finally, on May 24, 1607, a party of English settlers, with John Smith (*ca.* 1580–1631) the most prominent member, landed in what is now the state of Virginia. They moved up the James River (named for James I of England) and founded Jamestown.

This was the first permanent English settlement in the territory of what is now the United States.

IN ADDITION

In 1604 Russia became immersed in dynastic problems and no czar could be found with undisputed right to the throne. There followed a few anarchic years called the *Time of Troubles* in which Sweden and Poland advanced from the west nearly to Moscow, and Russia seemed on the point of falling apart.

1608

TELESCOPE

After the invention of the microscope (see 1590), it wouldn't have taken much to work out lens combinations that would magnify distant objects, or make them appear closer.

The discovery seems to have come about in 1608 through accident. Hans Lippershey (*ca.* 1570–1619), a Dutch spectacle-maker, had an apprentice who was playing with lenses during an idle moment and found that if he held two lenses in front of his eyes, one at a distance from the other, and looked through both, he saw a distant church steeple that seemed considerably closer than it was, and upside down, too.

Startled, he told his master, who grasped the importance of the discovery at once. Lippershey mounted the lenses in a tube that placed them at the proper distance from each other and had the first primitive *telescope* (from Greek words meaning "to see far").

The Netherlands was still in rebellion against Spain, and Lippershey realized that a telescope would be an important war weapon, making it possible to see the approach of enemy ships or troops before they could be made out by the eye alone. He explained this to Maurice of Nassau (see 1586), who saw the point and made an attempt to keep the nature of the device secret. The effort failed, however. Rumors spread, and the device was too simple not to be reconstructed at once.

QUEBEC

The French explorer Samuel de Champlain (*ca.* 1567–1635) was commissioned by Henry IV of France to explore that portion of the North American coast reached by Cartier (see 1535). From 1603 on, he explored the St. Lawrence River together with the coast from Nova Scotia to Cape Cod.

In 1608 he established a settlement on the St. Lawrence River that was called Quebec. It was the first permanent French foothold in Canada. The next year he explored southward and discovered the body of water that is now known as *Lake Champlain* in his honor.

The French in Quebec and the English in Jamestown were 600 miles apart. However, settlements have a way of spreading, and the scene was set for a rivalry between the powers that was not to be settled for a century and a half.

IN ADDITION

For the first time, in 1608, an English ship *(Hector)* reached India, over a century after Portugal had blazed the way. India was at this time ruled by Jahangir (1569–1627), who had become the Mogul ruler in 1605 at the death of his father, Akbar. He granted trade concessions to the English, the first small step toward what was to become British domination of the country in the course of the next two and a half centuries.

1609

PLANETARY ORBITS

For nearly two thousand years, since Plato (see 387 B.C.), it had been taken for granted that planetary orbits were circles, if only because the circle was the simplest curve and therefore the most elegant and esthetic. Surely the heavens would not deal with anything else.

Planetary movements, however, did not match the notion of simple circular orbits, and the Greeks had to assume combinations of circles that grew more and more complicated as observations of actual planetary motions across the sky grew more precise.

Copernicus placed the Sun, rather than the Earth, at the center of the Universe, but kept circular orbits, which meant that there still had to be complicated combinations, although not quite as complicated as those required by the Greek system.

Tycho Brahe (see 1572) had carefully observed the position of Mars in the sky from night to night, making better measurements than anyone before him had done. His assistant during the last few years was a German astronomer, Johannes Kepler (1571–1630), and after Tycho's death in 1601, Kepler tried to work out an orbit that would best fit the data Tycho had gathered.

Kepler tried a number of things that didn't work and was finally driven to the all-but-unthinkable alternative of considering orbits that were *not* circles. Finally, he had his answer and, in 1609, published it in a book entitled *Astronomia Nova (New Astronomy)*. In it, he maintained that planets moved about the Sun in ellipses (flattened circles whose geometric properties had first been explained by the Greek mathematician Apollonius in the first century B.C.). The Sun was located at one of the two foci of the ellipse, and with such orbits, no curve combinations were required. Our present picture of the Solar System (that is, the Sun plus its retinue of planets and other bodies) remains essentially that worked out by Kepler. No substantial change is expected in the future.

The elliptical orbit is Kepler's First Law of planetary motion. He also advanced a Second Law in his book, which described how the planetary speed altered with distance from the Sun. With the Sun at one focus of an ellipse, the planet was closer to the Sun (and moved faster) in one half of the orbit than in the other.

THE MILKY WAY

The Milky Way is a dim band of foggy light that encircles the sky. Many were the speculations as to what it was. It might be a spurt of milk from a goddess's breast, or it might be a bridge used by the gods to travel from

Earth to Heaven or vice versa. Democritus (see 440 B.C.) suggested that the Milky Way was a conglomeration of vast numbers of stars that were individually too dim to be seen. That, however, was just a speculation.

In 1609, however, Galileo heard rumors of the construction of a telescope in the Netherlands the year before. From what he heard, he had little trouble devising a telescope for himself, and for the first time, he turned one on the sky.

When he looked at the Milky Way through the telescope, he found that it was indeed composed of a myriad of faint stars. Democritus had been absolutely correct.

In fact, wherever Galileo looked he saw additional stars that, without the telescope, were too dim to be seen. The sky was full of them.

MOON

Galileo also looked at the Moon through his telescope. He found that it contained craters, mountains, and dark areas that he thought might be seas. These dark areas are still called *maria,* which is the Latin word for "seas." It was quite obvious that the Moon was not a heavenly globe of light, but was a world that in some ways resembled Earth. Aristotle's notion that heavenly bodies were of a structure different from that of Earth thus received a blow.

IN ADDITION

The English navigator Henry Hudson (d. 1611) began his search for a Northwest Passage under Dutch auspices, in his ship *Half Moon.* In 1609 he sailed into New York Bay, as Verrazano had done (see 1531). Hudson entered the river that flowed into the bay (which is now called *Hudson River* in his honor) and sailed up it to the site of present-day Albany before he could be quite sure it wasn't a strait leading to the Pacific. It was because of these explorations that the Dutch Republic (which that year signed a truce with Spain) laid claim to the region later on.

Spain, in order to remove any possibility of religious conflict, drove 275,000 Moriscos (descendants of the Muslims who had dominated Spain for so long) out of the country. In this way, Spain lost a valuable portion of its population and hastened its own continuing decline.

1610

JUPITER

Other than the Sun and the Moon, the planets known to the ancients were seen merely as points of light. When Galileo turned his telescope on them, however, he found that they expanded into little orbs. Clearly, they were extended bodies but were too small, or too distant, or both, to show as orbs to the unaided eye. (The stars, however, remained points of light even when viewed by telescope.)

Jupiter was not only an orb but, in January 1610, Galileo observed four dimmer objects in its immediate vicinity. As he watched from night to night, he saw that they were circling Jupiter, as the Moon circles the Earth. They were, in short, four *moons* of Jupiter. Kepler (see 1609) later referred to them as *satellites,* a Latin word referring to those who remain close to someone rich or powerful in the hope of picking up scraps and favors.

Jupiter's four satellites were the first objects ever seen in the sky that clearly circled some object other than Earth, which was a strong point against Ptolemy's geocentrism. For that reason, the discovery displeased the rigidly religious. Some refused to look through the telescope in order to avoid seeing the satellites. One pointed out that since the satellites were not mentioned by Aristotle, they clearly did not exist.

Seeking support from Cosimo II (1590–1621) of the Medici family, who in 1609 had become grand duke of Tuscany (an Italian state with its capital at Florence), Galileo called the satellites "the Medicean stars." Fortunately, the name didn't stick. The German astronomer Simon Mayr, known by his Latin name of Simon Marius (1570–1624), saw the satellites soon after Galileo. He named them, in order of increasing distance from Jupiter, Io, Europa, Ganymede, and Callisto, after individuals closely associated with Jupiter (Zeus) in the Greek myths.

Galileo also noted that Jupiter and Saturn both had orbs that did not appear perfectly circular, as did the orbs of the Sun and Moon, but were somewhat elliptical.

VENUS

Galileo began observing Venus in 1610. According to geocentric notions, Venus should always be in a crescent phase. According to the heliocentric view, however, Venus should show the full range of phases that the Moon does. As he watched Venus from night to night, Galileo was satisfied, eventually, that the full range of phases *did* appear, and this was a particularly powerful piece of evidence in favor of heliocentrism.

SUNSPOTS

Galileo (like others at about the same time) also found that the disk of the Sun had dark spots on it. These are now called *sunspots.*

This was particularly annoying to the religious conservatives, since they had accepted the Sun as a symbol of God and felt that it, of all objects, ought to be perfect.

IN ADDITION

Henry IV of France was assassinated by a Catholic zealot after a reign of twenty-one years. He was succeeded by his son, who reigned as Louis XIII (1601–1643).

Henry Hudson, now under English sponsorship, continued his search for the Northwest Passage and was the first to enter the large northern stretch of sea now known as *Hudson Bay* in his honor. Hudson never emerged, however. He penetrated it to its southernmost extension, called *James Bay* in honor of James I of England, and there his crew mutinied in 1611 and left him to die.

The Jamestown settlement was on the point of being abandoned in 1610, when new men and supplies arrived with Thomas West, Baron De La Warre (1577–1618). He had been appointed governor, and his name was eventually given to *Delaware Bay* and the *Delaware River*.

1612

ANDROMEDA NEBULA

In 1612 Simon Marius (see 1610) noted a fuzzy spot in the constellation of Andromeda. It did not have the sharp, pointlike quality of a star, but seemed a tiny luminous cloud. Indeed, it came to be called the *Andromeda nebula,* from the Latin word for "cloud."

The discovery of the Andromeda nebula did not seem very important at the time, but three centuries later it would initiate a discussion that would lead to a fundamentally new understanding of the Universe.

IN ADDITION

The Jamestown settlement finally achieved economic stability and strength through the production and export of tobacco. The growth and prosperity of the colony of Virginia was now assured.

1614

LOGARITHMS

Numbers can be written in exponential form. Thus, 2^4 is four twos multiplied together, or 16; while 2^5 is five twos multiplied together, or 32. Nine twos multiplied together, or 2^9, is 512. Since $16 \times 32 = 512$, we can say that $2^4 \times 2^5 = 2^9$. Instead of multiplying numbers, we add exponents. This turns out to be a general rule. In the same way, instead of dividing numbers, we can subtract exponents.

If 16 is 2^4 and 32 is 2^5, then 22 must be 2 to some exponent that lies between 4 and 5. If we had the exponents for all numbers listed in a convenient table, multiplication would be reduced to addition, and division to subtraction, with a great saving in time and trouble.

The Scottish mathematician John Napier (1550–1617) spent years working out formulas that would give him appropriate exponents for a great many numbers, and he called them *logarithms* (from Greek words meaning "proportionate numbers"). In 1614, Napier published his table of logarithms and they at once became useful in all sorts of complicated computations that scientists were forced to make. Nothing better was to come along for over three centuries.

METABOLISM

In 1614 an Italian physician, Santorio Santorio (1561–1636), better known by his Latinized name as Sanctorius, reported on experiments he had conducted on himself. He had constructed an elaborate weighing machine and sat in it while he ate and drank and eliminated wastes. He found that he lost more weight than the wastes alone would account for and he attributed this to "insensible perspiration"; that is, perspiration that evaporated as fast as it was produced so that it was not seen.

Sanctorius's experimentation was the very beginning of the study of *metabolism,* which deals with the chemical changes that go on within living tissue.

IN ADDITION

The Russians, having made peace with the Poles and Swedes, crowned Michael Romanov (1596–1645) as czar. That put an end to the Time of Troubles and founded the Romanov dynasty, which would rule Russia for three centuries. Although the Russians had trouble holding on to their western provinces (they lost Novgorod to the Swedes in 1614, for instance), their explorers in the east had passed the Yenisei River in their penetration of Siberia. By the time Michael became czar, the Russians had reached a point 2,000 miles east of Moscow.

1616

BAFFIN BAY

The search for the Northwest Passage continued. The English explorer William Baffin (*ca.* 1584–1622) in 1615 sailed his ship up the western coast of Greenland into a stretch of water now known as *Baffin Bay.* By 1616 he had made his way northward to nearly 78 degrees North Latitude. That placed him within 800 miles of the North Pole, and no one would get closer to that for two and a half centuries. Baffin concluded, correctly, that there was no practical Northwest Passage.

TIERRA DEL FUEGO

Drake had seen open water south of Tierra del Fuego (see 1578). In January 1616 the Dutch navigator Jakob Le Maire (1585–1616) sailed around the southern end of Tierra del Fuego, studying its coasts in detail, and showed it to be an island with an area of about 18,000 square miles. It was the most southerly piece of land to have been reached by prehistorical humanity. Its southernmost point is named *Cape Horn,* after the Dutch city of Hoorn, where the ship's captain, Willem Corneliszoon Schouten (*ca.* 1580–1625), was born.

IN ADDITION

In 1616 tribes in eastern Manchuria were organized militarily and began an aggressive campaign that would place them in control of China in a little over a quarter of a century.

1620

STAGECOACHES

A stagecoach is any horse and carriage that travels between set places *(stages)* according to a fixed timetable and accepts passengers for pay. Such things came into use at about this time.

Stagecoaches allowed people without money enough to own a horse to be carried from one place to another more quickly than in any other way. There were disadvantages, of course. You had to ride with strangers, at a time and to a place that suited the coach-owners and not necessarily yourself. Nevertheless, it was much more convenient than having to walk or try to hitch a ride in a farmer's cart.

The stagecoach remained the fastest means of overland travel (for the relatively impecunious) for over two centuries.

SCIENTIFIC METHOD

The English philosopher Francis Bacon (1561–1626) published *Novum Organum (New Organon)* in 1620. The reference is to Aristotle's *Organon,* in which the rules of logic were drawn up (see 350 B.C., Logic).

Bacon argued strenuously that deduction might do for mathematics but it would not do for science. The laws of science had to be induced; that is, established as generalizations drawn out of a vast mass of specific observation. Such experimental science had already been put into practice, but Bacon supplied the theoretical backing for it, describing what is today called the *scientific method*.

IN ADDITION

The Wars of Religion reached their peak in 1618. The Protestants of Bohemia rebelled against Catholic governors that were put over them and threw a couple of them out a window. Foreign powers intervened, and the fighting, which continued for thirty years (it was called the *Thirty Years War*), devastated Germany.

The first black slaves arrived in Virginia in 1619 and a race problem began that plagues the United States to this day.

A hundred *Separatists* (Protestants who wanted to separate from the Church of England altogether, and some of whom had fled to the Dutch Republic to avoid English persecution) traveled to North America in the *Mayflower*. They arrived at Plymouth in what is now the state of Massachusetts in December 1620 and set up the first permanent English settlement in New England.

1621

REFRACTION

The action of lenses had been known from ancient times. Archimedes (see 260 B.C.), according to a doubtful story, used large lenses to focus sunlight and set the Roman ships afire when they were besieging Syracuse. Obviously, light had to be bent in passing through them.

The first to make a mathematical study of this was a Dutch mathematician, Willebrord Snel (1580–1626).

It was known that when a beam of light passed from air to a denser medium, such as water or glass, and struck the surface of the denser medium at an oblique angle, it was bent toward the vertical. Ptolemy (see 140) maintained that the angle to the vertical made by the light hitting the surface bore a fixed relationship to the angle to the vertical made by the light after passing through the surface into the medium beyond.

Snel showed that the constant relationship was not between the angles themselves but between the sines of the angles. Ptolemy had been deluded because, with small angles, the sines are almost proportional to the angles themselves.

IN ADDITION

Robert Burton (1577–1640) published, in 1621, *The Anatomy of Melancholy,* a medical treatise on the causes and cures of melancholy that also drifts into many other areas.

1622

SLIDE RULES

It was not long after the discovery of logarithms by Napier (see 1614) that the matter became mechanized. An English mathematician, William Oughtred (1574–1660), prepared two rulers along which logarithmic scales were laid off. By manipulating the rulers one against the other, calculations could be performed mechanically by means of logarithms.

The instrument, modified and improved, became the slide rule that scientists and engineers carried about with them constantly until it was replaced three and a half centuries later by pocket computers.

IN ADDITION

The population of Virginia stood at 1,500, but the death rate through disease and Indian attacks was high.

1624

GAS

Air was considered by the Greeks to be one of the four *elements* that made up the Earth. To them, and to those who followed them, any vapor was a kind of air.

The Flemish physician Jan Baptista van Helmont (1579–1644) worked with vapors, however, and it seemed clear to him that some vapors had properties so different from others, and from ordinary air, that they represented different substances. Just as there were different liquids and solids, so there were different *airs.*

In 1624 Helmont felt he needed a word to use for such airs generally. It seemed to him that since airs had no specific volume but filled any container, they were examples of matter in complete chaos. He called them *chaos,* spelling the word according to its pronunciation in Flemish so that it came out *gas.* The term did not become popular at once, but eventually it gained equality with *liquid* and *solid* to represent the three ordinary states of matter.

In particular, Helmont studied the gas produced by burning wood. He called this *gas sylvestre* (wood gas), but we know it as *carbon dioxide.*

Helmont also conducted the first important quantitative measurement involving a biological problem since Sanctorius (see 1614). He grew a willow tree in a weighed quantity of soil and showed that, after five years, during which time he added only water, the tree had gained 164 pounds, while the soil had lost only 2 ounces. From this he deduced that water was converted by the tree into its own substance. Unfortunately for Helmont, he did not take into consideration the fact that air was in continuous contact with the tree. While water was important to the nourishment of the tree, equally important was the carbon dioxide of the air, the very gas that Helmont had studied.

IN ADDITION

The Dutch Republic grew more aggressive in its overseas policy. In 1623 the Dutch killed a number of Englishmen on Amboyna, a small Indonesian island, and drove the English out of the East Indies. In that same year, Dutch settlers established themselves on Manhattan Island and began to move up the Hudson and Connecticut rivers. The region was called *New Netherlands,* and since it lay between the English settlements in New England and Virginia, the germ of future conflict existed.

1627

PLANETARY TABLES

If Kepler's elliptical planetary orbits (see 1609) really presented an advance over the circular orbits of both Ptolemy and Copernicus, then using them ought to result in improved planetary tables.

Kepler spent some years working out new

tables on the basis of his elliptical orbits, using Napier's logarithms (see 1614) in his calculations—the first important scientific use of the new technique. In 1627 they were published as the *Rudolphine Tables,* in memory of Rudolf II (1552–1612), who had become Holy Roman Emperor in 1576 and who had supported Kepler.

They were indeed the best tables of planetary motions yet prepared. The publication included tables of logarithms and a star map based on the work of Tycho and expanded by Kepler to include over a thousand stars.

AUROCHS

The cattle that populate the world in their hundreds of millions and supply us with beef, milk, butter, cream, cheese, and leather are descended, it is supposed, from the *aurochs,* large, black animals standing 6 feet at the shoulder and with long, forward-curving horns. As cattle multiplied, the aurochs declined, until the only ones left in the world were a herd in Poland. The herd continued to dwindle, and the last one died in 1627. This was an example of how easy it is for large and magnificent animals to vanish. They had not been killed in anger. There was merely no room left for them. The room was needed for their domesticated descendants.

IN ADDITION

In 1626 the Dutch official Peter Minuit (1580–1638) bought Manhattan Island from the Indians who inhabited it for trinkets that were, according to tradition, worth twenty-four dollars (but perhaps several thousand of today's dollars). In that same year, the French established a settlement on the island of Madagascar off the southeast coast of Africa.

In 1627 the Mogul ruler of India, Jahangir, died and was succeeded by his son, Shāh Jahān (1592–1666). Under him, the magnificence of the court (though not necessarily of the people) reached its peak, as he ordered the construction of the Peacock Throne, which was to be riddled with precious stones. It took seven years to build it.

1628

BLOOD CIRCULATION

The idea of Galen (see 180) that the heart was a single pump and that there were pores in the thick muscular wall separating the right ventricle from the left ventricle was not universally accepted.

In 1242 an Arabic scholar, Ibn an-Nafīs (d. 1288), wrote a book in which he suggested that the right and left ventricles were totally separate. Blood was pumped out of the right ventricle into arteries that led it to the lung. There, in the lungs, the arteries divided into smaller and smaller vessels, within which the blood picked up air from the lungs. The vessels were then collected into larger and larger vessels until they were brought back to the left ventricle from which the blood was pumped out to the body generally.

In this way, the double pump was explained. One pump was needed for the lungs and aeration; the other for the rest of the body. An-Nafīs had grasped the *lesser circulation*. However, his book was not known to the West until 1924, and it had no effect on later developments.

In 1553 a Spanish physician, Miguel Serveto, known as Michael Servetus (1511–1553), published a book in which he too described the lesser circulation. The major part of the book, however, dealt with the Servetus's theological views, which were Unitarian. Servetus, having ventured into Geneva, which was ruled by his deadly enemy, John Calvin (see 1541), was taken into custody and burned at the stake.

Calvin then attempted to destroy all copies of Servetus's book, and it wasn't till 1694 that some unburned copies were found.

In 1559 an Italian anatomist, Realdo Colombo (1516?–1559), became the third person to understand, independently, the lesser circulation. His work was the first to reach the medical profession, and it was much more detailed and careful than those of his two predecessors, so it is Colombo who gets credit for the discovery.

Then came the English physician William Harvey (1578–1657). He studied the heart carefully and noticed that each side had valves that allowed blood to enter each of the two ventricles but not to leave except by way of arteries.

He also knew about the valves in the veins, since he had studied under Fabrici (see 1603), who had discovered them. He experimented with animals, tying off a vein or an artery and noting that the blood piled up in a vein on the side away from the heart, but in an artery on the side toward the heart. It was clear to him that blood flowed away from the heart in arteries and back to the heart in veins.

By 1628, he had all the evidence he needed and he published a seventy-two-page book in the Netherlands with the title *De Motu Cordis et Sanguinis (Concerning the Motions of the Heart and Blood)*. In it, he advanced his findings concerning the circulation of the blood: It leaves the right ventricle, goes to the lungs, and returns to the left ventricle. It then leaves the left ventricle, goes to the body generally, and returns to the right ventricle to begin all over.

The book was received sourly by the medical profession, but Harvey lived long enough to see it accepted. His book represents the beginning of modern physiology.

IN ADDITION

France was now under the strong control of Armand-Jean du Plessis, cardinal and duc de Richelieu (1585–1642), who became Louis XIII's chief minister in 1624. It was Richelieu's policy to unify France by doing away with those towns and cities that the Edict of Nantes (see 1600) had placed under Huguenot control. In 1628 the only remaining Huguenot city was La Rochelle and it was under attack by France. England under Charles I's minister, George Villiers, Duke of Buckingham (1592–1628), tried vainly to help, but Buckingham was assassinated in 1628 and La Rochelle was taken. The Huguenots were no longer a military force in France, but they were still tolerated after a fashion.

The English were still moving to New England. Salem, Massachusetts, was founded in 1628.

1633

SCIENCE AND RELIGION

Galileo had long accepted the Copernican idea of a heliocentric planetary system, but he was reluctant to be open in this view, for the papacy was strong in Italy and geocentrism was the only allowable astronomical view in Catholic doctrine of the time.

Urban VIII (1568–1644) had become Pope in 1623, and Galileo thought him to be a friend. In 1632, therefore, he took the chance of publishing a book with the title, in English, *Dialogue on the Two Chief World Systems.* The dialogue has three actors: a Ptolemy-supporter, a Copernicus-supporter, and a neutral person seeking information.

The book created a stir. In the first place, it was written in Italian rather than in Latin, so that it was not confined to scholars but could reach the general public. In the second place, Galileo was a brilliant writer, much given to sarcasm, and he certainly gave the Copernican the best of it. What's more, it was easy to persuade the Pope that the Ptolemy-supporter was meant to be a satire on him personally.

Galileo was therefore brought before the Inquisition in the most famous confrontation between science and religion prior to the evolution controversy in the present century.

On June 22, 1633, under the threat (but not the use) of torture, he was forced to renounce any of his views that were at variance with geocentrism. Sometimes Galileo is blamed for giving in, but he was seventy at the time, and he had the example of Bruno (see 1600), a generation before, to keep in mind.

However, the victory of the Church was an empty one. The heliocentric theory continued to gain an ever-firmer hold on the minds of scientists and ordinary people everywhere.

IN ADDITION

Not only had the Protestant Huguenots been defeated in France, but the war in Germany, which had been raging for fifteen years, was also resulting in a resounding Protestant defeat. But in 1630 Gustavus II Adolph (1594–1632), a Lutheran who had been king of Sweden since 1611, landed with an army in Germany and turned the tide of the war. He was killed in the course of his third victory, but the Swedish army remained, the Protestant cause was saved, and the Thirty Years War dragged on, bloodier than ever, for another fifteen years.

A large group of English Puritans arrived in New England in 1630. The town of Boston was founded, and the first settlements were made in what is now New Hampshire.

1635

MAGNETIC DECLINATION

Gilbert's demonstration that the Earth was a magnet (see 1600) could be used to explain the fact that the compass needle did not necessarily point to the truth north. If the Earth's magnetic north pole were not located at the geographic north pole, and if the compass needle pointed to the magnetic north pole, then naturally it would not necessarily point true north. What's more, if the magnetic pole were on the Atlantic side of the geographic north pole, then as one crossed the Atlantic from east to west, the compass needle would begin by pointing west of north and end by pointing east of north, as Columbus found it did (see 1492, Magnetic Declination).

At any one place, however, Gilbert maintained, the compass needle would always point in the same direction.

The English astronomer Henry Gellibrand (1597–1636) showed, however, that this was not so. He kept track of the pointing of the compass needle in London, both by his own observations and by the recorded observations of others, and in 1635 published his findings: the compass needle had shifted direction some 7 degrees in the past half-century. This was an indication not only that the magnetic poles might exist away from the geographic poles but that they might be shifting position as well.

IN ADDITION

Cecilius Calvert, Lord Baltimore (1605–1675), received permission from Charles I of England to establish a colony to the north of the settlements in Virginia. The colony, which was established in 1634, was named *Maryland* in honor of Henrietta Maria (1609–1669), Charles I's queen. Meanwhile, Puritans from Massachusetts were establishing permanent settlements in what is now Connecticut, which became a separate colony in 1635.

In Canada, explorers were pushing westward. Jean Nicolet (1598–1642) was the first European to reach Lake Michigan.

1637

ANALYTIC GEOMETRY

The French mathematician René Descartes (1596–1650) published, in 1637, his *Discours de la méthode (Discussions on the Method)*—of finding scientific truth by good reasoning, that is.

In a hundred-page appendix to this book, Descartes combined algebra and geometry. He pointed out that if one drew two perpendicular straight lines, marked the intersection 0, and laid off units on each line, positive numbers to the right and up, negative numbers to the left and down, then every point in the plane could be represented by two numbers, one for its position along the horizontal axis and the second for its position along the vertical axis. (One could add a third axis, in and out, and locate every point in the Universe by three numbers.)

Straight lines and curves could then be expressed by algebraic equations, which would locate every point on the line or curve with reference to the two axes. This combination of disciplines, producing *analytic geometry,* strengthened both. Geometric problems could be solved algebraically, and algebraic equations could be illustrated geometrically.

It also laid the foundation for the development of the calculus, which is essentially the application of algebra to smoothly changing phenomena that can be represented geometrically by curves of various sorts.

FERMAT'S LAST THEOREM

The French mathematician Pierre de Fermat (1601–1665) had a bad habit of not publishing but of scribbling hasty notes in margins of books or of writing casually about his discoveries in letters to his friends. The result is that even though he understood analytic geometry before Descartes, he loses the credit.

Here is something that involved him.

It is possible to have two squares that add up to a third square: $3^2 + 4^2 = 5^2$, or $9 + 16 = 25$. There are an infinite number of such cases. Are there, however, two cubes that add up to a third cube, or two fourth powers that add up to a fourth power, and so on?

Fermat wrote in the margin of a book that there aren't—it works *only* for squares. He had a perfectly marvelous proof for it, he said, but there was no room in the margin to write it down.

Now Fermat often said he had a proof for some particular proposition, and in every case, proofs have been discovered for those cases even when Fermat didn't give his. In every case but one, that is. The proposition I have just mentioned is the last Fermat proposition that has yet to be proved. That is why it is called *Fermat's Last Theorem*.

If it were anyone but Fermat, we would conclude that he happened to be wrong in this one instance. Fermat was so good a mathematician, however, that it goes against the grain to say he made a mistake. Yet no mathematician has ever found the proof. Fermat's Last Theorem is the most famous unsolved problem in mathematics.

IN ADDITION

The English clergyman Roger Williams (1603?–1683) found the Massachusetts colonists too intolerant for him, so he traveled southward, bought land from the Indians, and founded the settlement of Providence in 1636. This formed the nucleus of Rhode Island, in which Williams was the first to establish absolute religious freedom.

In 1637 Russian fur traders caught their first glimpse of the Pacific Ocean at the far eastern end of Siberia.

1640

COKE

England's forests were diminishing, and for that reason the English were turning to coal for fuel, even though its fumes and smoke were neither appetizing nor healthy.

Wood was still needed for charcoal, however (see 1000 B.C.), which was used in iron-smelting. Coal could substitute for wood in heating homes but could not substitute for charcoal in industry.

If coal is subjected to the same kind of incomplete burning as wood, however, coal like wood will burn off the noncarbon material leaving virtually pure carbon behind. The pure-carbon residue of coal is called *coke* (a word of unknown derivation). It may have been first formed in 1603, but by 1640 it was certainly known.

Coke is very like charcoal and bore the promise of being useful in iron-smelting, once the proper technique had been developed.

IN ADDITION

Christina of Sweden (1626–1689), who became queen in 1632, authorized an expedition to the American shores. Under the Dutch navigator Peter Minuit (see 1627), a party of Swedes set up the colony of *New Sweden,* in what is now the state of Delaware.

In 1638 a clergyman named John Harvard (1607–1638) left his library and half his estate to a college founded two years earlier in what is now Cambridge, Massachusetts. The college was named Harvard in consequence, and it is the oldest in the United States.

The Scots had grown increasingly restless because of religious differences with England. Charles I of England was anxious to bring them to book by force, but for that he would need money. As a result, he convened Parliament for the first time in eleven years, but its members had scores to settle, and a deep and dangerous controversy began.

1641

CROSS HAIRS

Although telescopes had been in use for a generation by now, they still left the task of judging distances between stars to the naked eye. They showed more stars and they widened the distance between them, but the precision of measuring that distance remained low.

The English astronomer William Gascoigne (1612?–1644) invented a simple solution. In 1641 he placed fine cross hairs where the image reached a focus. A particular spot could be located and accurately centered at the crossing point of the hairs and the telescope then shifted to a neighboring star, making use of a device to measure how great an angle the telescope had turned through.

This began the conversion of the telescope from a mere viewing toy to an instrument of precision. However, Gascoigne was killed in battle before he could properly develop his discovery. It was twenty years before cross hairs were rediscovered and put to general use in telescopes.

IN ADDITION

Troubles were multiplying for Charles I. His chief minister, Thomas Wentworth, Earl of Strafford (1593–1641), pushed for war against Scotland and urged Charles I to take various despotic actions that angered the people. In 1641 Wentworth was tried for treason by parliamentarians, and

Charles I was forced to assent to his execution. The Archbishop of Canterbury, William Laud (1573–1645), who also backed Charles I in his attempts to rule autocratically, was imprisoned in the Tower of London in 1641 and eventually executed.

1642

QUININE

The Incas had used the bark of the cinchona tree as a treatment for malaria. The active ingredient in it eventually came to be known as *quinine.* The first knowledge of quinine reached Europe in 1642, and for three centuries it remained the only treatment for this common and debilitating disease. Without quinine, it is doubtful if Europeans could have long remained in tropical climates.

ADDING MACHINE

In 1642 the French mathematician Blaise Pascal (1623–1662) invented a calculating machine that could add and subtract. It had wheels that each had 1 to 10 marked off along its circumference. When the wheel at the right, representing units, made one complete circle, it engaged the wheel to its left, representing tens, and moved it forward one notch.

With such a machine, as long as the correct numbers were entered into the device, there was no possibility of a mistake.

He patented the final version in 1649, but it was a commercial failure. It was too expensive, and most people continued to add and subtract on their fingers, on an abacus, or on a sheet of paper.

SOUTH PACIFIC

The ancient Greeks had thought there might be a large continent in the southern hemisphere to balance the land they knew in the northern hemisphere. This was pure speculation, but the later Europeans took Greek speculation seriously.

Both South America and Africa extended into the southern hemisphere, but they didn't seem to fulfill the requirements. There ought to be a continent that was entirely in the southern hemisphere. The Pacific Ocean was clearly vast, covering nearly half the area of the globe. Since Magellan had first entered it (see 1523), no significant land area had been found, but little of it had as yet been explored.

The Indonesian islands, which straddled the Equator, seemed a good place to start. In 1606 a Spanish navigator, Luis Vaez de Torres (fl. 1606), sailed all around the island of New Guinea and showed that it was not part of a continental landmass. It proved to be the second-largest island in the world. The narrow stretch of water south of New Guinea is called *Torres Strait* in the explorer's honor.

When the Dutch became dominant in the Indonesian islands, the Dutch governor-general Anthony van Diemen (1593–1645) sent out an exploring expedition under Abel Janszoon Tasman (1603?–1659).

On August 14, 1641, Tasman left the island of Java and for ten months sailed the Pacific. He discovered an island he named *Van Diemen's Land,* in honor of his chief, but it is now known as *Tasmania* in his own honor. He also discovered the southern island of what is now known as New Zealand.

The astonishing thing about Torres's and Tasman's voyages is that neither managed to spot Australia, a piece of land as large as the United States. New Guinea is only 100 miles from northeastern Australia, and Tasmania is only 230 miles from southeastern Australia, but both somehow managed to miss it.

In 1644 Tasman did finally spot a portion of the Australian coast and called it *New Holland,* but he didn't explore it further.

IN ADDITION

In 1642 the crisis in England degenerated into outright civil war, with Charles I sending his army against the parliamentarians. The north and west stood for the king, the south and east for Parliament. The deciding factor was Oliver Cromwell (1599–1658), who fought for Parliament and unexpectedly turned out to be an outstanding general.

In Canada, the city of Montreal was founded. Though the French were rapidly increasing the area of North America they controlled, it contained but a thin French population. On the other hand, the English were flooding into New England, which at this time held about 16,000 colonists.

1643

BAROMETER

It had long troubled mining engineers and others that pumps could not lift water more than 33 feet above its natural level. The usual pump produced a partial vacuum, which the water rushed upward to fill, but apparently the rush had its limits. The Italian physicist Evangelista Torricelli (1608–1647) worked for Galileo in that scientist's last years, and Galileo suggested that his assistant investigate this pumping problem.

It occurred to Torricelli that the water was lifted not because it was pulled up by the vacuum, but because it was pushed up by the normal pressure of air. After all, the vacuum in the pump produced a low air pressure and the normal air outside the pump pushed harder.

In 1643, to check this theory, Torricelli made use of mercury. Since mercury's density is 13.5 times that of water, air should be able to lift it only $\frac{1}{13.5}$ times as high as water, or 30 inches. Torricelli filled a 6-foot length of glass tubing with mercury, stoppered the open end, upended it in a dish of mercury, unstoppered it, and found the mercury pouring out of the tube, but not altogether: 30 inches of mercury remained in the tube, as expected.

Above the mercury in the upended tube was a vacuum (except for a small quantity of mercury vapor). It was the first one ever artificially created—a *Torricellian vacuum.*

Torricelli noticed that the height of the mercury column varied slightly from day to day and surmised correctly that the atmosphere possessed a slightly different pressure at different times. He had invented the first barometer.

IN ADDITION

Louis XIII died in 1643 and was succeeded by his five-year-old son, who reigned as Louis XIV (1638–1715).

1645

AIR PUMPS

Once Torricelli had produced a vacuum by allowing mercury to pour out of a tube (see 1643), it seemed to some that vacuums could be produced in more direct ways. Perhaps air could simply be pumped out of any vessel, and larger volumes of vacuum could be formed than Torricelli had managed.

A German physicist, Otto von Guericke (1602–1686), produced the first practical air pump in 1645. It worked like a water pump but with parts sufficiently well fitted to be reasonably airtight. It was run by muscle power and was slow, but it worked.

Guericke produced a large enough vacuum to make useful experiments possible. He was able to show that a ringing bell within a vacuum could not be heard, thus bearing out Aristotle's contention that sound would not travel through a vacuum. Guericke also showed that candles could not burn in a vacuum and that animals could not live.

He also weighed a metal sphere before he evacuated it and then again afterward. The small loss in weight was obviously the weight of air that had been inside. From that, and the volume of the air, he was able to get the first measurement of air's density.

IN ADDITION

The Ming dynasty in China came to an end in 1644, and the Manchus were now in full control. They established the Ch'ing dynasty, which was to rule China for two and a half centuries.

Oliver Cromwell won his first great victory in the English Civil War at Marston Moor on July 2, 1644. His next victory, at the Battle of Naseby, on June 14, 1645, virtually doomed the king's cause.

In 1645 the Ottoman Empire began a long war with Venice over the island of Crete. Both were shadows of the past, actually, and this war resulted in the further decline of both.

1648

AIR PRESSURE AND ALTITUDE

If the mercury column of Torricelli's barometer (see 1643) were held up by air pressure, then if one went up high in the air, there should be less air above, and the air pressure should decrease. Therefore, so should the height of the mercury column.

To test this, Pascal (see 1642) sent his brother-in-law up a neighboring mountain with a couple of barometers. His brother-in-law climbed about a mile and found that the mercury columns had dropped from 30 to 27 inches.

This showed clearly that the atmosphere could only have a finite height. In fact, if it were as dense throughout as it was at sea level, it would be only 5 miles in height.

Even when it became apparent that air became less dense with height, so that quantities of it extended far higher than 5 miles, there had to be limits. At a height of 100 miles or so, air would be so thin that it might as well be vacuum, and so for all the rest of the distance to the Moon and to other heavenly bodies.

Experiments like those of Torricelli and Pascal's brother-in-law amounted to the discovery of *outer space.*

FLUID PRESSURE

About 1648, Pascal also studied fluid pressure by exerting pressure on water and noting how that pressure was transmitted against the walls of a closed vessel. He concluded that when pressure is exerted on a fluid in a closed vessel, it is transmitted undiminished throughout the fluid and that it pushes at right angles to all surfaces that it touches. This is called *Pascal's principle* and is the basis of the hydraulic press.

IN ADDITION

The Thirty Years War ended on October 24, 1648. The Holy Roman Empire had been enormously weakened, and Germany had lost a large proportion of its population. The Dutch Republic had won its independence from Spain at last. France was now the strongest military power in Europe and would remain so for over two centuries. At the moment, however, France was weakened by an insurrection of its nobles, who took advantage of the childhood of the new king and of the unpopularity of the chief minister, Giulio Mazarini (1602–1661), better known by the French version of his name as Jules Mazarin.

The English Civil War was renewed in rather chaotic fashion, but Cromwell grew steadily stronger, purging the Parliament of those he did not want in power and holding Charles I himself as prisoner.

In 1647 the religious reformer George Fox (1624–1691) formed a *Society of Friends of Truth,* which within a few years became far better known as the *Quakers.*

1650

DOUBLE STARS

In 1650 the Italian astronomer Giambattista Riccioli (1598–1671) observed, telescopically, that Mizar, the middle star of the handle of the Big Dipper, was actually two stars so close together that they could not be seen as separate with the unaided eye. This was the first *double star* to be detected.

AGE OF THE EARTH

Of the written materials available to Europeans at this time, the Bible was the only one that claimed to give the history of Earth from the creation, and it was generally accepted by all scientists as the authoritative word of God at this time and for two more centuries. (Many people accept it as such to this day.)

The Bible does not use any acceptable chronology for its early history, but by following back from the reign of King Saul and making use of hints here and there in the earlier historical sections, it is possible to decide what the biblical date of the creation might have been.

In 1650 James Ussher (1581–1656), an Anglican bishop, worked out the date of the creation in this way and decided it had taken place in 4004 B.C. Four years later an English theologian, John Lightfoot (1602–1675), sharpened the date and made it 9 A.M., October 26, 4004 B.C.

Such dating of creation has no valid basis whatever, but it has strongly influenced popular opinion to the present day.

IN ADDITION

On January 30, 1649, Charles I was beheaded. Oliver Cromwell crushed an Irish rebellion and brought the entire island under English subjection. He was now in full control of the British Isles.

The world had a population of about 500 million; England about 5 million; and London about 350,000.

1651

NAMES ON THE MOON

In 1651 Riccioli (see 1650, Double Stars) published *Almagestum Novum (New Almagest).* The reference to Ptolemy's old book (see 140) was not accidental, for Riccioli rejected heliocentricity and insisted on an Earth-centered astronomy a full century after Copernicus's book had been published.

The book, however, included a map of the Moon, with names given to the various craters. Riccioli began the habit of naming surface features on other worlds after astronomical notables, and many of his names are still used. Of course, he gave the most conspicuous crater on the Moon the name of *Tycho,* whom he admired extravagantly. Another large one was *Ptolemaeus. Copernicus* is a pretty notable crater, and *Kepler* is not bad either.

IN ADDITION

Charles I's son, who called himself Charles II (1630–1685), invaded England, hoping to win the crown, but was defeated by Cromwell on September 3, 1651, at the Battle of Worcester, and had to flee into exile again.

1653

LYMPHATICS

Veins and arteries had been known since Greek times, but in 1653 a third set of vessels was discovered. The Swedish naturalist Olof Rudbeck (1630–1702) demonstrated their existence in a dog. These new vessels resembled veins but had thinner walls and carried the clear watery fluid portion of the blood *(lymph)*, so that the vessels were called *lymphatics*.

The lymph is forced out of the smallest blood vessels and into the spaces around the cells, forming the *interstitial fluid*. This drains into the lymphatics and is carried back into the blood vessels in various parts of the body.

IN ADDITION

England and the Dutch Republic were now leading the world in overseas trading and it was only natural that they compete intensely. In 1652 the first of a series of Anglo-Dutch naval wars erupted.

The Dutch continued their vigorous program of overseas settlement, establishing Capetown on the southernmost portion of Africa in 1652.

The desultory insurrection in France (called the *Fronde*) was finally put down in 1653, after five years, and Mazarin (see 1648) was in full control. In England, Cromwell was declared "Lord Protector," and he was in full control also.

1654

PROBABILITY

People much given to gambling usually manage to work out rough-and-ready ways of measuring the likelihood of certain situations so as to know which way to bet their money, and how much. If they did not do this, they would quickly lose all their money to those who did.

A certain French gambler, Chevalier de Mere (1610–1685), was puzzled at the fact that he kept losing money in a certain game of dice, which he felt he ought to win. In 1654 he consulted Pascal (see 1642) on the matter, who in turn consulted Fermat (see 1637). Pascal and Fermat worked out mathematical techniques for judging the likelihood of certain combinations appearing in the fall of (honest) dice. In doing so, they laid the foundations for the theory of *probability*.

The chief function of probability was to deal with large numbers of events, which singly were random in nature but in total were predictable. As time went on, considerations of probability proved to be almost inconceivably important in the development of science.

AIR PRESSURE

Guericke, having invented the air pump (see 1645), used it to demonstrate the power of air pressure, beginning in 1654.

For instance, he affixed a rope to a piston and had fifty men pull on the rope while he slowly created a vacuum on the other side of the piston within the cylinder. Air pressure inexorably pushed the piston into the cylinder despite the struggles of the fifty men to prevent it.

Then he prepared two metal hemispheres that fitted together along a greased flange. (They were called *Magdeburg hemispheres* because Guericke was mayor of Magdeburg.) When the hemispheres were put together and the air within them evacuated, air pressure held them together even though teams of horses were attached to the separate hemispheres, straining to their utmost in opposite directions. When air was allowed to enter the joined hemispheres, they fell apart of themselves.

This demonstration took place before the eyes of Ferdinand III (1608–1657), who became Holy Roman Emperor in 1637. He was so impressed, he ordered Guericke's work to be written up and published.

1656

SATURN'S RING

Galileo had observed Saturn through his telescope in 1612 and noted something odd about it. There seemed to be projections on either side. He could not quite make them out, and after a while they disappeared. It annoyed Galileo. After all, he had been attacked by religionists who said that his telescope produced optical illusions, and here was one case where perhaps it did. He refused to look at Saturn again.

In 1655, however, the Dutch astronomer Christiaan Huygens (1629–1695), with the help of a Dutch philosopher and optician, Benedict Spinoza (1632–1677), worked out a new and better method for grinding lenses. He installed his improved lenses in a telescope that was 23 feet long, and with that he studied Saturn in 1656.

He could see what it was that had puzzled Galileo. Saturn was surrounded by a thin, broad ring that did not touch the planet at any point. No other object in the sky had so peculiar a structure, and Saturn is widely considered the most beautiful object in the sky because of it.

In addition, he discovered that Saturn had a satellite, which he named *Titan* (because Saturn, or Cronos as the Greeks called him, was the leader of a group of gods called Titans).

That same year, he discovered that the middle star of Orion's sword was not a star but a cloud of luminous gas. It is now known as the *Orion nebula*.

PENDULUM CLOCK

Up to this time, the best timepieces were still the mechanical clocks of medieval days that could not keep time to better than a large fraction of an hour.

Galileo's discovery of the principle of the pendulum did not help immediately. An ordinary pendulum swings in the arc of a circle, and when it does that it doesn't have a constant period. The period becomes a little longer through a wide arc than through a narrow one.

If the pendulum could swing through a *cycloid* (very like a circle over a small arc), then the period would remain truly constant. Huygens (see above) showed how to arrange matters so that the pendulum would swing through a cycloid. He then hooked it up to falling weights in such a way that the pendulum controlled the rate at which the weights fell and made it truly constant.

In 1656 Huygens had the first *pendulum clock* (or as it is sometimes called, *grandfather's clock*). It was the first timepiece that could tell time to a minute or better and was the first accurate enough to be useful to scientists.

IN ADDITION

The Dutch continued to expand overseas. In 1655 in New Netherlands, Peter Stuyvesant (*ca.* 1610–1672), who had been governor since 1647, sent his forces to take over New Sweden, which in seventeen years of existence had received very few settlers. New Netherlands controlled the coast from Connecticut to Delaware and was now at the peak of its strength and prosperity.

In the Indian Ocean, the Dutch took Colombo, Ceylon, from the Portuguese.

1657

FALLING BODIES

Galileo was quite satisfied that all bodies fell at equal rates, provided that air resistance didn't complicate matters. If objects fell through a vacuum, then there would be no resistance, and if this could be observed, the question could be settled directly instead of by inference.

The English physicist Robert Hooke (1635–1701) designed an air pump that worked better and more quickly than Guericke's (see 1645). When Hooke had a good vacuum in a large jar, he arranged for a feather and a coin to be released from the top of the jar at the same time, and behold, they dropped at the same speed.

IN ADDITION

In 1657 Cromwell allowed the Jews to return to England. They had been expelled by Edward I three and a half centuries before.

1658

RED BLOOD CORPUSCLES

Microscopes of sorts had been in existence for half a century, but they had not been very good. They magnified but slightly and were usually not perfectly in focus. It was not till the 1650s that microscopes improved in quality to the point where they were useful in studying the minutiae of living things.

The Dutch naturalist Jan Swammerdam (1637–1680) studied insects under the microscope and collected some three thousand species of them, so that he is considered the father of modern *entomology*.

Swammerdam's most famous discovery, however, made in 1658, was the red blood corpuscle. Present in the bloodstream in the billions, red blood corpuscles carry the chemical that absorbs oxygen from the air in the lungs, though this was not known till much later.

IN ADDITION

Oliver Cromwell died on September 3, 1658.

1659

SYRTIS MAJOR

The telescope, in converting planets from points of light to small orbs, made it possible to see markings on some of them.

Venus, which at times approaches Earth more closely than any other planet, remained featureless, because it seemed to have a heavy cloud layer.

Mars, next closest, was different. In 1659, Huygens (see 1656, Saturn's Ring) made out a dark, triangular marking he called *Syrtis Major* (large bog). It wasn't a bog any more than the Moon's "seas" were seas, but it was one marking that future astronomers continued to see.

IN ADDITION

The English Commonwealth was quickly reduced to chaos after Cromwell's death, and it became clear that the restoration of the monarchy was only a matter of time.

France and Spain had continued fighting even after the conclusion of the Thirty Years War, but in 1659 they made peace at last, entirely to France's advantage, and Spain was never again to play the part of a great power.

1660

CAPILLARIES

Harvey's discovery of the circulation of the blood (see 1628) had an important flaw. According to Harvey, the blood traveled from the heart through the arteries, into the veins, and back to the heart. But how did the blood get from the arteries to the veins? There was no visible connection, and Harvey was forced to maintain that the connection consisted of blood vessels that were too small to see.

The microscope was now an important tool, however, and a pioneer in this field was the Italian physiologist Marcello Malpighi (1628–1694). The thin wing-membranes of the bat contained what was virtually a two-dimensional network of blood vessels. Malpighi studied it under the microscope in 1660 and saw the tiniest arteries and veins connected by vessels too small to see without the microscope. He called the tiny vessels *capillaries,* from a Latin word meaning "hair-like." Harvey's theory was complete, but Harvey hadn't lived to see it. He had died three years before.

STATIC ELECTRICITY

When Thales (see 585 B.C.) studied the magnetic properties of loadstone, he is also supposed to have studied amber, which, when rubbed, attracts light objects. Whereas magnets attract only iron, rubbed amber attracts many things.

William Gilbert, who showed that Earth was a magnet (see 1600), found that rock crystal and a variety of gems showed the same attractive force when rubbed that amber did. Since the Greek word for amber was *elektron,* Gilbert called these substances *electrics,* and the phenomenon came to be called *electricity.* Because the electricity in electrics seemed to stay put if left undisturbed, it was referred to eventually as *static electricity,* from a Greek word meaning "to stand."

The first to demonstrate static electricity on a large scale was Guericke, who had invented the air pump (see 1645). The electrical phenomenon was produced by rubbing, and Guericke in 1660 fashioned a globe of sulfur that could be rotated on a crank-turned shaft. When it was stroked with the hand as it rotated, it accumulated quite a lot of static electricity. It could be discharged and recharged indefinitely, and Guericke could produce sparks from his electrified globe.

IN ADDITION

On May 8, 1660, Charles II was restored to the throne, after an eleven-year gap in which no king ruled in England. It was the only monarchless gap in fifteen centuries of English history.

1661

CHEMICAL ELEMENTS

It had been just about two thousand years now since Aristotle listed the four elements (earth, water, air, and fire) that made up the Earth and a fifth (aether) that made up the heavenly bodies. That still remained the dominant theory, although some alchemists had considered mercury, sulfur, and salt to be particularly important.

But the day of alchemy was over. The Irish-born physicist and chemist Robert Boyle (1627–1691) published a book in 1661 that was called *The Skeptical Chymist,* and as a result, the very name *alchemist* was changed to *chemist.* The dropping of the prefix *al-*, which is the Arabic *the,* seemed somehow to symbolize turning the back on medievalism. In this book, Boyle also divorced chemistry from medicine and made it a separate science.

Boyle's most important feat was to push chemistry in the direction of becoming an experimental science. He wanted chemical elements to be established by experimentation rather than by deduction.

He pointed out that an element was one of the simple components of the Earth, one that could not be converted into anything simpler. Therefore, anything that couldn't be converted into anything simpler was an element, while anything that could be so converted was not. The only way you could differentiate an element from a nonelement, then, was to try hard to make a substance simpler.

ACID-BASE BALANCE

Hippocrates (see 420 B.C.) had maintained that health was preserved by the proper balance of the four humors (blood, phlegm, bile, and black bile), and this idea had also lasted two thousand years, like the four elements of Aristotle.

The Dutch physician Franz Deleboe (1614–1672), better known by the Latinized name Franciscus Sylvius, denounced the four-humor theory in 1661 and suggested instead that health depended on the balance of acids and bases in the body. This was certainly an improvement on the older view.

Sylvius also studied digestive juices such as saliva and suggested that digestion was a chemical process (fermentation) rather than a mechanical one (grinding). In this, he was quite right.

IN ADDITION

Under Charles II, all clergymen, college fellows, and schoolmasters had to accept the Anglican Book of Common Prayer. Those Protestants who refused were known as Nonconformists.

In France, Mazarin died and Louis XIV, now twenty-three years old, took over personal control of the government.

1662

BOYLE'S LAW

As Boyle experimented with vacuums, it was he who employed Robert Hooke to build an improved air pump (see 1657).

The air pump got Boyle interested in gases, and in 1662 he discovered that air could be compressed. He did this by trapping some air in the short closed end of a J-shaped 17-foot-long glass tube into which he poured mercury to close off the bottom curve.

If he next added more mercury to the open end, the weight of the additional mercury squeezed the trapped air in the closed end more closely together, and its volume decreased. Indeed, Boyle found that the volume of the gas varied inversely with the pressure upon it. That is, if he doubled the weight of mercury upon it, the volume shrank to one-half the original; if he tripled the weight of mercury, the volume shrank to one-third, and so on. This is called *Boyle's Law*.

The most important conclusion of this experiment was that air, and presumably other gases, were atomic in nature and that the atoms were widely spread apart. With pressure, the atoms were forced more closely together and the volume shrank.

Liquids and solids could not be compressed with the same ease that gases could be, but that did not necessarily mean they weren't composed of atoms. In their cases, the atoms might already be in contact.

The idea of atomism had not entirely died out since the time of Democritus (see 440 B.C.). There had always been a few from time to time who accepted atoms. Boyle's experiments, however, were the strongest evidence yet, and Boyle himself became a convinced atomist. It was to be a century and a half, however, before atomism won out entirely.

ROYAL SOCIETY

There had been informal meetings of scientists in London in the mid-1600s, and the habit grew more ingrained after the *Restoration* (of Charles II, that is).

Like many monarchs of his time, Charles II patronized science as a way of gaining prestige for the nation and possibly material benefit as well. He therefore gave a legal charter to the *Royal Society* in 1662. For years, it represented the most brilliant assemblage of scientists since the great days of Alexandria.

The Royal Society received communications from members, both at home and abroad, and held meetings where members could inform one another of their work. It also published a journal called *Philosophical Transactions,* in which experimental work and findings could be published. (By *philosophical* is meant what we today would call *scientific.* The words *science* and *scientist* had not yet been invented.)

Other nations established scientific societies as well following the success of the Royal Society.

IN ADDITION

The colony of Connecticut received a charter from Charles II that gave it a virtually independent democratic government. This gave its settlers a tradition of going their own way that would give rise to rebellious notions when England tried to assert its authority a century later.

1664

GREAT RED SPOT

In 1664 Hooke (see 1657) noted a large oval marking on Jupiter that came to be called the *Great Red Spot*. The name was accurate, for it did seem to be red in color and it was great in the sense of being very large. The entire Earth could be dropped into it without touching its sides.

IN ADDITION

Until this time, it was difficult for the English settlements to move south of what is now Virginia because Spanish claims raised complications. The general decline of Spain, however, made it impossible for that nation to control any coastal territory north of Florida itself. In 1663, therefore, Charles II of England granted the right to colonize the coast south of Virginia to eight courtiers who had aided his restoration. From this grant, the colonies of North Carolina and South Carolina were established.

New hostilities were brewing between England and the Dutch Republic, and the chief early result of this was that on August 27, 1664, an English naval squadron forced the Dutch to surrender New Netherlands, which they had controlled for half a century. New Netherlands became New York, and the city of New Amsterdam became New York City. The English now possessed the North American coast from the Carolinas to Maine, plus Newfoundland. There still remained the Spaniards in the south and the French in the north, however.

1665

CELL

The use of the microscope was now spreading rapidly and one of the best of the early microscopists was Hooke (see 1657). In 1665 he published *Micrographia,* which outlined his work in this field. It had some of the most beautiful drawings of microscopic observations ever made.

His most important discovery (though it did not seem so at the time, no doubt) involved the structure of cork. Under the microscope, he found a thin sliver of cork to be composed of a finely serried pattern of tiny rectangular holes. These he called *cells,* from a Latin word for "small chamber"—especially one of the type that exists in rows, as monastery cells or prison cells do.

The cells that Hooke observed were empty only because they were found in dead tissue. In living tissue they are filled with fluid, and that disqualifies them from being called cells, strictly speaking. The name, however, clung.

LIGHT DIFFRACTION

It was about this time that the question of waves versus particles began to be raised, one that was to be argued for a long time.

Water waves could be seen, and the fact that they tended to curve around obstacles could be considered a characteristic of all waves. Particles, on the other hand, if moving in a straight line, did *not* curve around obstacles but either struck and bounced back (being reflected) or missed and continued moving in a straight line.

Since sound curved around obstacles, it was taken to be a wave phenomenon. Since light cast sharp shadows, it was natural to think of it as composed of tiny particles.

An Italian physicist, Francesco Maria Grimaldi (1618–1663), made an observation that was published posthumously in 1665. He had let a beam of light pass through two narrow apertures, one behind the other, and then fall on a blank surface. He found that the band of light on that surface was a trifle wider than the apertures. He believed, therefore, that the beam had been bent slightly outward at the edges of the aperture, a phenomenon he called *diffraction.*

This would make it seem that light was a wave phenomenon.

But a wave's bending about an obstacle depends on the relative size of the wave and the obstacle. Any wave will be reflected by a barrier considerably larger than itself: water waves will be reflected by a breakwater and sound waves will be reflected by a cliff wall. Since light waves were reflected from small objects, and diffraction was only very slight, it must follow that if light were composed of waves, they were very small waves indeed.

However, Grimaldi's work was largely neglected and the controversy went on for a century and a half without reference to him.

PLANETARY ROTATIONS

The fact that features could be made out on the surface of some of the planets meant that those features could be observed from night to night as they passed around the planet to the other side and across the face again. By watching long enough and dividing the time elapsed by the number of turns, an accurate measure of the rotation rate of the planet could be obtained.

In this way, in 1665, the Italian-born French astronomer Gian Domenico Cassini (1625–1712) determined that Mars rotated in 24 hours 40 minutes, while Jupiter rotated in 9 hours 56 minutes.

Since these planets rotated on their axis as Earth did, it made them seem more like our planet. As astronomical discoveries continued, Earth seemed less and less a special case —except insofar as we are here on Earth and not anywhere else.

IN ADDITION

London was struck by the plague and the result was tragic. Of those who could not flee the city, half died. But plague does not seem to stop war, and open hostilities began between the English and the Dutch that year.

In Spain, Philip IV (1605–1665) died. He was succeeded by his son, Charles II (1661–1700), who was only four years old and so sickly he was not expected to survive. Since there were no further immediate heirs, the question of who was to rule the vast Spanish dominions became of feverish interest to Europe and continued even when Charles II managed to remain alive for thirty-five years after becoming king. Since he was sick every day of those years, and never had heirs, tension continued for all that time.

In North America, the colony of New Jersey was founded.

1666

LIGHT SPECTRUM

The English scientist Isaac Newton (1642–1727), curious about light, conducted a series of crucial experiments in 1665 and 1666. He allowed a beam of light to pass through a prism (a triangular piece of glass) and then fall on a white wall.

What emerged was a band of colors, the least bent portion of the light being red, and then following, in order, orange, yellow, green, blue, and violet, each merging gradually into the next.

Were the colors produced by the glass? No, for when Newton passed the light that had emerged from the prism through a second prism oriented in the opposite direction, all he got was white light. The colors had recombined.

This meant that light had to be looked upon in a totally new way. It had always been assumed that white light was "pure" light and that color was introduced as an impurity through the effect of material substances on the light.

Newton's work made it clear that color was an inherent property of light and that white light was a mixture of different colors. Matter affected color only by absorbing some kinds of light and transmitting or reflecting others.

Exactly what it was that made light assume different colors was not yet clear, however.

IN ADDITION

London, having gone through the devastation of the plague, was now subjected to a great fire, which began on September 2, 1666, and raged for four days and nights, virtually destroying the older sections of the city.

1668

CONSERVATION OF MOMENTUM

In studying motion, it became clear that motion wasn't created out of nothing. If some moving object struck another object at rest, it might well impart motion to the second object. (Anyone playing billiards can see this.) Yet a great deal of motion in a light object would impart only a small motion to a heavy object. (Anyone kicking a cannonball can see this.)

Perhaps if one multiplied the mass by the velocity, it would be that product that stayed the same. The product of mass and velocity is called *momentum* (from a Latin word for "movement"). The English mathematician John Wallis (1616–1703) was the first to suggest, in 1668, that the total momentum of a closed system (one into which no momentum from the outside entered, and from which no momentum leaked into the outside) remained always unchanged. This is called the *law of conservation of momentum*.

Momentum can be shifted from one part of a system to another but can neither be created nor destroyed. Momentum can be in either of two directions—let us say, plus or minus. If one starts off with no momentum at all in a closed system, one part may start moving in one direction (the plus direction) *if* another part starts moving in the opposite (minus) direction. If the two *momenta* (movements) are equal and opposite, they cancel, and the total momentum remains zero. Similarly, if two bodies come together with equal and opposite momenta (total, zero), they can bounce off each other, having switched the plus and minus to some extent, or they can stick together and both be motionless (the total always remaining zero).

Such a conservation law explains a great many things about motion that would otherwise be puzzling. The law of conservation of momentum was the first of the conservation laws to be understood, but others would, in time, follow, and all are fundamental to our understanding of the structure and functioning of the Universe.

SPONTANEOUS GENERATION

It has always been commonly supposed that some forms of life can arise spontaneously from nonlife. This has been especially so in connection with life forms that human beings do not want—weeds and vermin. Useful forms of life require the utmost care, but useless or harmful forms seem to flourish despite human effort to eradicate them. It is tempting to believe that these pests arise out of nowhere. Besides, there is the evidence of the eyes. One can see how maggots, for instance, arise out of decaying meat. The meat is dead but gives rise to living maggots. *Spontaneous generation* seems unquestionable.

In 1668, however, the Italian physician Francesco Redi (1626–1697) decided to test the matter experimentally.

He prepared eight flasks with a variety of meats in them. Four he sealed and four he left open to the air. Flies could land only on the meat in the open vessels, and only those bred maggots. The meat in the closed vessels turned just as putrid—but no maggots.

To test whether it was the absence of fresh air that did it, Redi repeated the experiments without sealing any flasks but covering four with gauze instead. The air was not excluded from the gauze-covered flasks, but flies were, and there were no maggots in those flasks. (This was the first clear-cut case of the use of proper *controls* in a biological experiment.)

Redi concluded that maggots did not arise by spontaneous generation but hatched from eggs laid by flies, eggs too small to be easily visible. This did not settle the matter generally, but it was difficult thereafter to believe that life forms large enough to be visible to the unaided eye arose by spontaneous generation.

REFLECTING TELESCOPES

During the first sixty years of the telescope's use, its lenses curved light by refraction and focused it. In this way, the eye saw the image as brightened and expanded. Such telescopes were *refracting telescopes.*

Unfortunately, the lenses refracted different colors of light differently and formed a spectrum, so that the images in such telescopes were always blurred by colored rings, red or blue *(chromatic aberration).* This was minimized by using only the center of the lens and having the light come together only gradually and reach a focus at a considerable distance from the lens, but this meant that telescopes capable of considerable brightening and enlargement had to be long and unwieldy.

Newton, through his experiments with light (see 1666), supposed that one couldn't possibly have a lens without blurring the image with color. He therefore thought of an alternative. Why not use curved mirrors instead of curved lenses, and focus the light by reflection rather than refraction? Reflection did not produce a spectrum.

In 1668, therefore, he built the first *reflecting telescope,* and thereafter two varieties of telescope were available to astronomers.

IN ADDITION

A thirteen-year-long war between Russia and Poland ended with Russia taking Smolensk and Kiev. From this point, Poland, after half a century of power, began its decline.

1669

CALCULUS

In the years 1665–1666, Newton (see 1666) was staying at his mother's farm in order to escape the London plague. One evening he saw an apple fall from a tree at a time when the Moon was shining peacefully overhead and began to wonder why the Moon didn't fall. He then thought that perhaps it did, but that it was also moving horizontally and fell at each moment just enough to make up for the curvature of the Earth. Thus, the Moon fell forever but only succeeded in moving around the Earth in one of Kepler's ellipses.

Newton spent considerable time trying to work out how Earth might pull at the Moon as it pulled at the apple, and at what rate the Moon might be falling, but he was not satisfied with his calculations and put them to one side. Some say it was because there was no good determination of the exact size of the Earth at the time; others say it was because he didn't know how to allow for the fact that every bit of Earth was pulling at the Moon from slightly different distances and angles. He needed a mathematical tool that would help him solve that problem.

In 1669 he began working on such a tool, a mathematical technique that came to be called *calculus*. This was a more versatile technique than anything invented earlier, and science could not do without it these days. Calculus is the beginning of *higher mathematics*.

Working on the calculus at roughly the same time as Newton was a German mathematician, Gottfried Wilhelm Leibniz (1646–1716). Both he and Newton worked out the technique—Newton, perhaps, a little sooner—but Leibniz ended with a better symbolism.

This is not unusual. Frequently two scientists working independently come up with the same answers in response to the same problems. Often the solution is to give them joint credit. Sometimes, however, there are arguments about which scientist was really first, and this may even degenerate into accusations of plagiarism.

That is what happened this time. Exacerbated by national prejudices, English versus German, there was a Homeric battle that poisoned the scientific community for years without ever settling anything. Today, Newton and Leibniz are given joint credit for the calculus.

PHOSPHORUS

Of the substances chemists now consider to be elements, nine were known to the ancients. These included seven metals: gold, silver, copper, tin, iron, lead, and mercury; and two nonmetals: carbon and sulfur. Four more elements were probably known and were unmistakably described by the medieval alchemists: arsenic, antimony, bismuth, and zinc. We do not know who first discovered any of these elements, or when.

The situation changed when the German chemist Hennig Brand (d. *ca.* 1692) began searching for something that would enable him to create gold. For some reason, he thought he would find it in urine. He did not succeed in making gold, but perhaps as early as 1669, he obtained a white waxy substance that glowed faintly in the air and which he therefore called *phosphorus* (from Greek words meaning "light-bearer"). The faint glow was due to the spontaneous combustion of phosphorus in air.

All the ninety or so elements discovered after 1669 can be attributed to a specific person and a specific time. Brand's discovery of phosphorus is the first of which this can be said.

FOSSILS

The word *fossil* is from a Latin word meaning "to dig." At first anything that could be dug out of the earth was called a fossil. The word came to be applied, however, to those particular objects that were dug up and that, although made of rock, seemed remarkably like remnants of living things—bones and teeth particularly. Agricola (see 1556) had commented on these over a century earlier.

There were numerous theories about these fossils. Some thought them practice attempts by God to create living things. Some thought them failing attempts of Satan to imitate God. Some thought them the remains of animals that had drowned in Noah's Flood.

In 1669, however, the Danish geologist Nicolaus Steno (1636–1686) maintained that they were the remains of creatures that had lived long ago and whose remains had slowly petrified; that is, been converted into stone. This view gradually prevailed, and fossils were to be the most spectacular (though far from being the only) evidence in favor of biological evolution.

DOUBLE REFRACTION

Sometimes a discovery is so puzzling it must simply be put aside until science advances to the point where an explanation is possible.

Thus, in 1669 a Danish physician, Erasmus Bartholin (1625–1698), obtained a transparent crystal of a type now called *Iceland spar*. Bartholin noted that objects viewed through the crystal appeared double. It was as though some light were refracted by the crystal through one angle, with the rest refracted through a slightly different angle. The phenomenon was therefore called *double refraction*.

Bartholin further noticed that when he rotated the crystal, one of the images remained fixed and the other revolved about it.

Bartholin could not explain these observations, and neither could anyone else. The observations had to remain in suspended animation for a century and a half until enough was known about light for an explanation to become possible.

BLOOD COLOR

It was clear that blood went to the lungs to pick up air, and it was suspected that this must involve a chemical change in the blood. The first to notice evidence to that effect was an English physician, Richard Lower (1631–1691). In 1669 he noted that dark blood drawn from veins turned bright red on contact with air. The necessary details concerning the change had to wait another century, however.

IN ADDITION

The Ottoman Empire had won its long war with Venice and had occupied Crete. Venice had come to the end of the road as a significant power, but the Ottoman Empire had paid too much for too little.

Ruling in India at this time was Ālamgīr (1618–1707), also known as Aurangzeb, who had become the sixth ruler of the Mogul dynasty in 1658. He was the last powerful non-European ruler of India.

1670

DIABETES

Many communicable diseases had been described and differentiated, but there are also diseases that are not communicable. These are the result of some inborn deficiency of the body, which may make itself evident from birth or only later in life.

Of these, the most important is diabetes, a disorder in which the patient is unable to process sugar in the normal way. Sugar in a diabetic tends to accumulate in the blood and spill over into the urine.

Some of the ancient physicians may have been aware that the urine of diabetics was sweet, where normal urine was not. It may be that flies collecting about diabetic urine gave the first hint.

It was the English physician Thomas Willis (1621–1675) who in 1670 was the first modern to point out the sweetness. Naturally, recognizing and understanding symptoms is a step toward learning how to treat a disease, but where diabetes was concerned the treatment didn't come for a century and a half.

1671

SATURN'S SATELLITES

By now, six satellites were known: four circling Jupiter (Io, Europa, Ganymede, and Callisto), one circling Saturn (Titan), and, of course, one circling Earth (Moon). Huygens (see 1656), in an uncharacteristic moment of mysticism, thought that the six satellites matched the six planets in number (Mercury, Venus, Earth, Mars, Jupiter, and Saturn), which balanced things, so that no further satellite discoveries were to be expected.

Cassini (see 1665) blew that view to smithereens in 1671 when he discovered a second satellite of Saturn, which he named Iapetus. Over the next thirteen years he discovered three more: Rhea, Dione, and Tethys. These names are those of Titans: Iapetus was a brother of Saturn (Cronos), while the other three were among his sisters.

IN ADDITION

Stenka Razin (d. 1671), a peasant leader, had organized a vast rebellion against the Russian aristocracy and for a while in 1670 controlled the regions about the Volga River. Alexis I Mikhaylovich (1629–1676), the second Romanov czar, who had gained the throne in 1645, sent his western-trained army (sharpened in successful wars against Poland and Sweden) against the rebels. Razin was defeated, taken to Moscow, and executed on June 16, 1671.

1672

DISTANCE OF MARS

Nineteen centuries before, Hipparchus had determined the distance to the Moon (see 150 B.C.). Since then, no further heavenly distance had been determined accurately. The parallaxes of all other heavenly bodies were far too small to measure with the unaided eye, and the telescopes weren't quite good enough to do the job.

However, Kepler's elliptical orbits and his three laws of planetary motion had made it possible to build a model of the Solar System in the proper proportions. If any planetary distance could be obtained, then all the other distances would be known too.

Cassini (see 1665) tackled the job. Thinking his telescope might be good enough if he viewed Mars from two places sufficiently far apart, he sent another French astronomer, Jean Richer (1630–1696), to Cayenne in French Guiana on the northern shore of South America.

In 1672 Cassini determined the position of Mars against the stars as seen from Paris and, using the position of Mars as given in reports from Cayenne, worked out the parallax of Mars. That gave him the distance between Mars and Earth at that time, and from that he could calculate the other distances of the Solar System.

From his figures, he determined that the

Sun was 87,000,000 miles from the Earth, as compared with the 5,000,000-mile distance that Aristarchus (see 280 B.C.) had estimated. Cassini's estimate was 7 percent too low, but for a first try it was amazingly close.

For the first time, an appropriate idea of the size of the Solar System was obtained. Even allowing for the slightly low figure Cassini had, it was clear that the orbit of Saturn, which was then the farthest known planet, had to be over 1,600,000,000 miles across.

The stars must be farther still. No one yet knew how much farther the stars must be, but Cassini gave human beings the first shocking realization of how small they and their world were compared to the Universe. There were other and greater shocks yet to come.

IN ADDITION

Louis XIV was determined to make France militarily dominant over Europe. With the largest and best army anyone had seen since Roman times (and with artillery, which the Romans hadn't had), he invaded the Dutch Republic in 1672. The Dutch leader at the time was Johan de Witt (1625–1672), who with his brother, Cornelis de Witt (1623–1672), had been running the republic since 1653. In terror before the advancing French, a Dutch mob killed the de Witt brothers and called in their stead William III, Prince of Orange (1650–1702), a great-grandson of William the Silent.

The land we now call the Ukraine, north of the Black Sea, had been a kind of no-man's land. It was not clear whether the Cossacks who inhabited it would rule it independently or whether it would fall to Poland, Russia, or the Ottoman Empire. Fighting was continuous, and in 1672 Poland and the Ottoman Empire went to war over the region.

1675

SPEED OF LIGHT

At this time, nobody knew how quickly light traveled. Galileo (see 1581) had tried to measure it by standing on top of one hill and having a friend stand on top of another, each with dark lanterns. Galileo would reveal the flame in his lantern and at once his friend would reveal the flame in his. The time it took between Galileo's sending the light and seeing the return light would be the time it took light to make the round trip. However, the time was always the same, no matter how far apart the hills were, and Galileo realized that he was just measuring his friend's reaction time and gave up. Light obviously moved too swiftly for its speed to be measured in that way. (There were those who thought it moved with infinite speed.)

In 1675, however, a Danish astronomer, Olaus Roemer (1644–1710), was carefully observing the motions of Jupiter's satellites from the Paris Observatory, including the time when they passed behind Jupiter and were eclipsed. Cassini (see 1665) had timed those motions carefully and Roemer was checking them. To Roemer's surprise, he found that the eclipses came progressively earlier at those times of the year when Earth was approaching Jupiter in its orbit and progressively later when Earth was receding from Jupiter.

Roemer supposed this might be because light did *not* travel at infinite speeds but took longer to go from Jupiter to Earth when Earth was on the opposite side of the Sun from Jupiter, and less time when Earth was on the same side. From these time differences, Roemer calculated that light must travel at a speed of about 141,000 miles per second. This is only about three-fourths of its actual speed, but for a first determination it's not bad.

SATURN'S RINGS

Cassini (see 1665), studying Saturn's ring in 1675, noted that a dark line seemed to separate the ring into an outer one that was narrow and bright and an inner one that was broad and a little less bright. Some astronomers thought the ring was a single object with a dark line running around it, but the majority opinion favored the possibility that there were two separate rings with a true division. The majority proved to be correct. The dark line is *Cassini's division* to this day, and people speak of Saturn's *rings* in the plural.

IN ADDITION

William of Orange saved the Dutch by opening the sluice gates and flooding the land in 1673.

In 1674 John III Sobieski (1629–1696) was elected king of Poland. He was the last strong king Poland would have, but even he could not stem the disorder and decline of the land.

In 1675 the New England Native Americans, driven to desperation by the unending encroachment of the European settlers on their lands, attacked under the leadership of Metacomet, known to the Europeans as King Philip (1639?–1676). The resulting *King Philip's War* took the course of all such wars. After a series of surprise attacks by the Native Americans, along with the killing of settlers, the remaining settlers counterattacked and, after defeating the Native Americans, systematically destroyed them regardless of age or sex.

1676

MICROORGANISMS

Microscopists had been looking at minute sections of ordinary living organisms for twenty years and more, but a Dutch microscopist, Antoni van Leeuwenhoek (1632–1723), now surpassed them all.

Whereas other microscopists used combinations of lenses, Leeuwenhoek used small single lenses, but he ground them to perfection, so that they could magnify up to 200 times. He ground a total of 419 lenses in his long lifetime, even though he was already over forty when he took up this hobby.

In 1676, studying pond water, he found it contained living organisms so small they could not be seen with the unaided eye. He called them *animalcules,* but we call them *microorganisms.* By whatever name, Leeuwenhoek had opened up an entirely new microscopic zoo to the astonished eyes of humanity. (In 1677 he detected spermatazoa in semen.)

IN ADDITION

An astronomical observatory was established in Greenwich, a suburb of London, in 1676. The English astronomer John Flamsteed (1646–1719) was placed in charge as the first *astronomer royal.*

1678

SOUTHERN STARS

Throughout history, from the Sumerians onward, detailed astronomical observation has been confined to the northern hemisphere. The north celestial pole is high in the sky as seen from Europe and the Middle East, and the stars near it, in their daily circle about it, never dip below the horizon.

On the other hand, the south celestial pole, along with the stars near it, are never above the horizon as seen from Europe.

As a result, the southernmost sky remained unknown to European astronomers till the Age of Exploration began. During Magellan's voyage, when they were off the shores of southern South America, the sailors saw two hazy clouds that looked like detached portions of the Milky Way. They are called the *Magellanic Clouds* to this day. A bright star configuration, the *Southern Cross,* was also reported.

No systematic astronomical observations of the far southern skies were made, however, until the English astronomer Edmond Halley (1656–1742) traveled to the island of St. Helena in the south Atlantic Ocean. There he spent two years, and despite bad weather that severely limited astronomical observations, he returned in 1678 to publish a catalog of 341 southern stars that astronomers had, until then, known nothing about.

LIGHT WAVES

The particle-versus-waves controversy grew sharper.

Newton (see 1666) felt that light consisted of particles, partly because there was a vacuum between Earth and Sun and he didn't see how a wave could cross a gap where there was nothing to wave.

Huygens (see 1656), on the other hand, insisted that light consisted of waves of the same type as sound (so-called *longitudinal waves,* which wave in and out in the same direction the waves are traveling). As to what was waving, Huygens supposed that there was a very subtle fluid in the vacuum of space (this came to be referred to by Aristotle's term *aether*), which could not be detected in any ordinary way.

Neither Newton's particles nor Huygen's longitudinal waves could explain Bartholin's observation of double refraction nine years before, but that was ignored. The controversy continued to rage.

IN ADDITION

The Protestants in England were so jumpy that when, in 1678, an Englishman named Titus Oates (1649–1705) invented the tale of a "Popish plot" to massacre Protestants, he was widely believed and England went through a seven-year period remarkably like the McCarthy "witch-hunt" in the United States (see 1950 and 1954).

1679

PRESSURE COOKERS

Sixteen centuries after Hero had created a device that whirled because of steam pressure (see 50), the first useful application of steam was advanced.

In 1679 the French physicist Denis Papin (1647–?1712) developed the *pressure cooker,* a vessel with a tightly fitted lid in which water was boiled and the accumulated steam created a pressure that raised water's boiling point. At this higher temperature, bones softened and meat cooked faster than usual. A safety valve was included in case the steam pressure got too high. Papin cooked a meal for the Royal Society in his device and prepared one for Charles II as well.

IN ADDITION

In North America, the Great Lakes region continues to be explored by the French.

1680

MUSCLES AND BONES

There has been a continuing struggle between those who feel that life is something fundamentally different from nonlife and obeys different laws of nature *(vitalism),* and those who believe that one set of natural laws governs everything, life and nonlife alike. Generally, over the last three centuries, victory has come to the latter group.

Thus, in 1680 the Italian physiologist Giovanni Alfonso Borelli (1608–1679) posthumously published a book, *De Motu Animalium (Concerning Animal Motion),* in which he successfully explained muscular action on a mechanical basis, describing the actions of bones and muscles in terms of a system of levers. The laws that applied to inanimate levers also applied to our bone-muscle systems. (Of course, the bone-muscle action is one of the simplest aspects of life. Things got much more difficult as scientists tried to deal with the more complex aspects of life.)

1681

DODO

Mauritius is an island about half again as large as the state of Rhode Island, located in the Indian Ocean 500 miles east of Madagascar. In 1598 the Dutch settled it and named it after Maurice of Nassau (see 1586). The Dutch stayed, on and off, till 1710.

Mauritius had forms of life that had evolved in isolation and that were different from forms elsewhere. For example, there was the *dodo*, a flightless pigeon, larger than a turkey and with a huge hooked bill. It was harmless and unafraid (hence its name, perhaps) since there was nothing on Mauritius to threaten it.

Once settlers arrived, however, they and their domestic animals began to kill this inoffensive bird. This continued until the very last dodo died about 1680. Similar birds on nearby islands were also killed off. Now we have the expression "dead as a dodo."

It is hard to believe, nowadays, that such an unusual and interesting form of life should have been slaughtered so casually, with no attempt made to save a few, but it has happened over and over again. It is one of the brighter aspects of recent history that human beings are now desperately trying to save various endangered species, though considering the inexorable increase in the number of human beings and the space they must occupy, this is often a losing battle.

IN ADDITION

Charles II granted William Penn (1644–1718) the right to found a colony in North America. This was the beginning of the colony of Pennsylvania. Penn was a Quaker and campaigned for such things as religious freedom, frequent elections, and uncontrolled parliaments, which of course caused him to be viewed as a dangerous radical.

1682

PLANT SEXUALITY

Before modern times, plants were usually thought to be not truly alive in the sense that animals were. In the biblical story of creation, plants grow as soon as dry land appears. They seem to be part of the land and meant only for food. In contrast, on the fifth and sixth days, God is described as specifically creating animal life and instructing it to multiply. (Even today, vegetarians announce that their love of living things keeps them from eating animal food, although the plant life they eat is every bit as alive.)

Some of this disdain for plant life was broken down, in 1682, by the English botanist Nehemiah Grew (1641–1712), who showed that plants reproduced sexually, that they had sexual organs, and that individual grains of pollen were the equivalent of the sperm cells of the animal's world.

IN ADDITION

The French explorer René-Robert Cavelier de La Salle (1643–1687) sailed down the Mississippi River from its upper regions to its mouth, reaching the Gulf of Mexico on April 9, 1682. He was the first European to do so and, for all we know, the first human being of any kind to have done so. La Salle claimed the entire drainage basin of the Mississippi for France and named it *Louisiana* for Louis XIV.

In Russia, Peter I (1672–1725) became czar. He was only nine years old, but the time would come when he would be universally known as *Peter the Great.*

1683

BACTERIA

In 1683 Leeuwenhoek (see 1676) made his most remarkable discovery. Through one of his little lenses he saw what he thought were living things far smaller than his animalcules. They were just barely at the limit of what he could see in his lenses, and it is a tribute to the excellence of his grinding that he saw what no one else could see for another century. In hindsight, we know that what Leeuwenhoek had discovered were bacteria.

IN ADDITION

In this year, an almost forgotten nightmare suddenly swirled out of the East.

In 1676, an energetic grand vizier, Merzifonlu Kara Mustafa Pasa (1634–1683) had come to power in the Ottoman Empire under the weak Mohammed IV (1641–1691), who had become sultan in 1648. In 1683 Kara Mustafa dashed the Turkish army against Vienna and on July 17 laid siege to it. Vienna held firm, however, and a Polish army under John III Sobieski (see 1675) came thundering down on September 12 to break the siege. It was the last Ottoman spark. From here on, for two and a half centuries, there was only recession.

Incidentally, the Turks left coffee beans behind as they departed, and the coffee craze entered the Western world, and never left.

1684

SIZE OF THE EARTH

Eratosthenes' figure for the circumference of the Earth (see 240 B.C.) had not been bettered in a thousand years.

In 1684, however, certain observations of the French astronomer Jean Picard (1620–1682) were published posthumously. Instead of noting the distance of the Sun from the zenith (the point in the sky that is directly overhead) at different places on the Earth by measuring its shadow, as Eratosthenes had done, Picard measured the distance of a star from the zenith at different places. With telescopes, that was the more accurate method, and Picard calculated the Earth's circumference as 24,876 miles and its diameter as 7,900 miles, figures that are very close to the best we have today.

IN ADDITION

In 1684 the Boston-born preacher Cotton Mather (1663–1728) referred to the European settlers in the English colonies as *Americans*. This may have been the first use of the word in print.

1685

IMAGINARY NUMBERS

Mathematicians knew that the multiplication of two negative numbers yields a positive product. Thus, not only does $+1 \times +1 = +1$, but $-1 \times -1 = +1$. What number, then, multiplied by itself yields -1? To put it another way, what is the square root of -1?

Mathematicians can invent the necessary number, call it an *imaginary number*, and symbolize it as *i* for imaginary. You can then say that $+i \times +i = -1$. What's more, $-i \times -i = -1$.

Wallis (see 1668) succeeded in making sense out of such imaginary numbers in 1685.

Imagine a horizontal line. Mark off a point as zero and then imagine the positive numbers marked off to the right and the negative numbers marked off to the left, with all the fractions and irrational numbers appropriately marked off between the whole numbers. That is the *real number axis*.

Next, draw a vertical line passing through the zero point. Mark all the *i* numbers (*i, 2i, 3i,* and so on) upward, and all the $-i$ numbers downward, with all the imaginary fractions and irrational numbers marked off, too. That is the *imaginary number axis*.

Every point in the plane can then be marked off just as Descartes did in his analytical geometry (see 1637). Every point (a) on the real number axis becomes a + 0*i*; every point (b) on the imaginary number axis becomes 0 + b*i*; and every number on neither axis (the *complex numbers*) becomes a + b*i*.

Such a scheme proved enormously useful to mathematicians, scientists, and engineers.

IN ADDITION

On February 6, 1685, Charles II of England died, and in the absence of legitimate sons, his younger brother, a Catholic, succeeded to the throne as James II (1633–1701).

On October 18, 1685, Louis XIV revoked the Edict of Nantes (see 1600), which his grandfather had granted the Huguenots. Many Huguenots fled the country, moving to England, Prussia, or America, and not only depriving France of their brains and industry but taking with them wherever they went an ineradicable hatred of France. The revocation did Louis XIV and France far more harm than it did the Huguenots.

1686

METEOROLOGICAL MAP

The wind often seems to be totally unruly and unpredictable, but even the ancient Romans knew that it blew from Africa to India for six months and then in the other direction for six months. These *seasonal winds* were called *monsoons,* from an Arabic word meaning "season." There were also the steady trade winds that blew southwestward north of the equator and northwestward south of the equator.

The first person to try to make world sense out of the winds was Halley (see 1678), who in 1686 wrote a book on the subject. It included a world map that showed the prevailing winds over the tropic oceans—the monsoons and trade winds. Halley did not have quite the proper explanation for the winds, however. He knew they involved the rising of sun-heated air but did not understand the reason for the westward flow of tropical air.

PLANT CLASSIFICATION

There is a natural tendency for people interested in natural history to classify animals and plants. Aristotle (see 350 B.C., Animal Classification) classified the former, and Theophrastus (see 320 B.C.) classified the latter. The ancients, however, had limited access to living things because so much of the world was beyond their ken.

The first to do a modern job of classification was an English naturalist, John Ray (1627–1705). Starting in 1686, he published what eventually came to be a painstaking three-volume classification of 18,600 different plant species. While such work might seem to be mere listing, the *basis* for the classification requires a good deal of ingenuity, and on the whole, Ray made good decisions in these matters. Classifications such as his made the matter of biological evolution seem an overwhelming likelihood.

IN ADDITION

The Ottoman recession became obvious as Austrian troops, counterattacking after the siege of Vienna, marched into Budapest in 1686, a city that had belonged to the Ottoman Empire for a century and a half.

Western Europe, on the other hand, was expanding. France annexed Madagascar in 1686, and in India, the English East India Company was increasing its power and established itself on an island in the mouth of the Ganges around which the city of Calcutta would grow.

1687

LAWS OF MOTION

In the nearly eighty years since Kepler had come up with the elliptical orbits of the planets (see 1609), scientists had been trying to work out what it was that kept the planets in their orbits and made the orbits ellipses. It was clear that the Sun had to attract the planets somehow, but what was the attraction and how did it work?

A number of scientists got pretty close to what turned out to be the truth, notably Hooke (see 1657), who was a great enemy of Newton (see 1666). When Hooke boasted to Halley (see 1678), who was a great friend of Newton, that he (Hooke) had the answer. Halley went to Newton to check the matter with him. Newton said that he had worked out the answer in 1666 (see 1669, Calculus) but had never published it. Halley, in great excitement, urged publication.

Newton could do this now with much greater confidence than he could have done it twenty years earlier. For one thing, he now had calculus, which made some calculations easy that would have been difficult before. For another, he had Picard's figures on the Earth's size (see 1684), and accuracy in that respect was necessary to his calculations.

Newton took eighteen months to write the book and, in 1687, published *Philosophiae Naturalis Principia Mathematica (Mathematical Principles of Natural Philosophy)*, often known simply as the *Principia*. It was written in Latin and did not appear in English until 1729. It is generally considered the greatest science book ever written.

Despite the greatness of the book, Newton had trouble publishing it. Hooke was unalterably opposed, and the Royal Society hesitated to become involved in the controversy. Fortunately, Halley had inherited a fortune in 1684, when his father was murdered by unknown assailants. He saw to the proofreading of the book and had it published at his own expense.

In the book, Newton codified Galileo's findings concerning falling bodies (see 1589) into the *three laws of motion*.

The first enunciated the principle of *inertia:* A body at rest remains at rest and a body in motion remains in motion at a constant velocity (that is, constant speed in a constant direction) as long as outside forces are not involved.

The second law of motion defines force as the product of mass and acceleration. This was the first clear distinction between the mass of a body (representing its resistance to acceleration) and its weight (representing the extent to which it is acted on by a gravitational force).

The third law of motion states that for every action there is an equal and opposite reaction.

These laws of motion are equivalent to the axioms and postulates with which Euclid began his treatment of geometry. From the axioms and postulates, an incredible number of theorems can be derived, each one building on theorems that went before. In the same way, from the laws of motion, an enormous number of mechanical effects can be deduced.

UNIVERSAL GRAVITATION

From the laws of motion, Newton was able to deduce the manner in which the gravitational force of attraction between the Earth and the Moon could be calculated. He showed that it was directly proportional to the product of the masses of the two bodies and inversely proportional to the square of the distance between their centers. The proportionality could be made an equality by the introduction of a constant. In other words:

$$F = gmm'/d^2$$

where g is the gravitational constant, m and m′ are the masses of the Earth and the Moon, d is the distance between their centers, and F is the force of gravitational attraction between them.

Most important of all, Newton postulated that this law of gravitational attraction held not only between the Earth and the Moon, but between any two bodies at all throughout the Universe. It was not merely gravitation he was speaking of, but *universal gravitation.* This was another claim that the laws of nature were the same everywhere and the mightiest blow yet against the view that the heavenly bodies worked by some set of natural laws other than those that prevailed on Earth.

From this rather simple law of universal gravitation, all of Kepler's laws of planetary motion could be derived. It accounted for all the irregularities of planetary motion known in Newton's time, for while the Sun was the predominant attractive body, the planets had minor attractions for each other that resulted in slight alterations *(perturbations)* of their orbits from what they would be if the Sun alone were involved.

What Newton had done was to describe the machinery of the Universe effectively and to show that it was essentially simple. Despite the fact that forces other than gravitation are known today, and that Newton's description of gravitation has been refined since, it still remains true that gravitation is the overriding force that controls the great sweeps of the Universe, and that Newton's formulation is an excellent one if distances and velocities are not too great.

SHAPE OF THE EARTH

In the *Principia,* Newton referred to Richer's expedition to Cayenne, in French Guiana, which helped determine the parallax of Mars (see 1672). While there, Richer had found that a pendulum beat more slowly than it did in Paris, so that a clock that would have been correct in Paris lost two and a half minutes a day in Cayenne.

Newton pointed out that this could be so if the force of gravity were slightly weaker in Cayenne than in Paris or if Cayenne were farther from the Earth's center than Paris was. Since Cayenne, like Paris, was at sea level, the conclusion was that sea level must itself be higher at Cayenne than at Paris.

Newton showed that if a body rotated, a centrifugal effect would act to counter gravitation to a certain extent and that this countering force would be zero at the poles and grow stronger and stronger as one moved away from the poles until it reached a maximum at the Equator. The centrifugal effect would, in other words, give rise to an *equatorial bulge,* which would make the diameter across the Equator somewhat longer than the diameter from pole to pole.

Jupiter and Saturn, which are much larger than Earth, spin much faster and are, in addition, made of lighter material. In their cases, the equatorial bulges are so great that their orbs look clearly elliptical rather than circular. Earth's outline drawn from pole to pole and back would also be slightly elliptical (though not noticeably so to the unaided eye). Instead of being almost perfectly spherical as the Sun and Moon are, Earth would be an *oblate spheroid* according to Newton's reasoning. (Of course, the matter would have to be checked by actual measurement eventually.)

IN ADDITION

The Ottoman Turks and the Venetians were again at war, and the Venetians actually managed to take southern Greece and Athens temporarily. The Ottoman Turks were criminal enough to store gunpowder in the Parthenon, a temple to Athena that dated back to the time of Pericles and Greece's Golden Age. The Venetians, for their part, were criminal enough to bombard the Athenian Acropolis. The gunpowder was struck, blew up, and left the Parthenon in ruins.

1688

PLATE GLASS

For centuries clear glass had been a luxury item. Little by little, though, the art of pressing or casting glass (that is, the making of sheets by some means other than blowing) was developed. At first the sheets were quite small, but by 1688 in France, large sheets were being made for mirrors and for coach windows.

This meant that glass became increasingly common and cheap, so that its use became almost universal. Glass now allowed light to enter a room while keeping wind and rain out.

IN ADDITION

On June 10, 1688, Mary of Modena (1658–1718), queen of England, gave birth to a son. England had more or less impatiently borne with James II since his accession three years before, waiting for him to be succeeded by his Protestant daughter, Mary (1662–1694). Now his Catholic son was heir, and the English would not endure a Catholic dynasty. An appeal went out to William of Orange, Mary's husband, and on November 5 William landed in England. James II, finding himself deserted, fled to France on December 23. Since the coup had been carried through without significant bloodshed, it was called "the Glorious Revolution" in contrast to the civil war that had put an end to the reign of James II's father, Charles I.

1691

ANIMAL CLASSIFICATION

Ray, who had previously classified thousands of species of plants (see 1686), now began to work on a similar classification of animals. He classified animals logically on the basis of hooves, toes, and teeth, a system that, in some respects, has persisted to this day. His sober, matter-of-fact descriptions finally superseded the fanciful prose of the Roman writer Pliny.

IN ADDITION

In 1689 William of Orange established himself as William III of England, ruling jointly with his wife, Mary II. His chief aim in life was to beat Louis XIV, and he had no hesitation in using English resources for this purpose. As for Louis XIV, he championed James II and was perfectly willing to fight England. The result was that in 1689, England and France began fighting a series of wars that was to continue for a century and a quarter. (In a way, it was a second Hundred Years War between them.)

The war between England and France spilled over into North America, where the English colonists called it *King William's War*. It led to seven decades of warfare between the English and French (and their respective colonists and Indian allies) through the forests of North America. King William's War was desultory and ended where it began.

In Russia, Peter I began to rule in his own name in 1689. He was already interested in shipbuilding and in western technology, and he grew increasingly determined to modernize Russia. Meanwhile, in the Far East, Russian explorers made contact with Chinese forces at the Amur River, which forms part of the northern boundary of Manchuria. At the Treaty of Nerchinsk, Russia was forced to leave the lower Amur in Chinese hands.

In India, the city of Calcutta was founded by the East India Company in 1690.

1693

CALCULATING MACHINES

Leibniz (see 1669, Calculus) devised a calculating machine in 1693 that went beyond Pascal's (see 1642). Whereas Pascal's could only add and subtract, Leibniz's could multiply by automatically repeating addition, and divide by automatically repeating subtraction. Leibniz also invented a mechanical aid to the calculation of trigonometric and astronomical tables.

This showed much more clearly than Pascal's device did that arithmetical manipulations followed simple rules and repetitions and by no means required the creative imagination or reasoning power of the human brain.

MORTALITY TABLES

Death was death and human beings had to learn to accept it stoically. It did not occur to people to subject so universal and somber a fact to statistical evaluation until 1693, when Halley (see 1678) prepared the first mortality tables, relating the death rate to age. It may seem perfectly obvious that the older you get, the more likely you are to die, but it is always useful to know something by observation rather than by assumption (however reasonable). Besides, careful mortality tables might show aspects of death that were not the result of age.

IN ADDITION

In Scotland, the famous *Massacre of Glencoe* took place in 1692. The MacDonald clan had declared allegiance to William III, but the news was deliberately repressed, and a troop of soldiers accepted

MacDonald hospitality and then surprised them when they were asleep and slew thirty-six of them.

A preacher in Salem, Massachusetts, Samuel Parris (1653–1720), succumbed to the witch mania that had been sweeping Europe for some time. In 1692 he accused some harmless people of being witches, the hysteria caught on, and within the next two years nineteen were hanged and one was pressed to death. (None were burned.) As atrocities go, it wasn't much, but it has been a source of shame to Americans ever since.

1698

MINER'S FRIEND

England, short of wood, turned increasingly to coal, but coal mines in England's damp climate tended to fill with water, and the water had to be pumped out at the cost of much labor, involving both human and animal muscle.

An English engineer, Thomas Savery (1650?–1715), knew of the power of the vacuum as demonstrated by Guericke (see 1645). It occurred to him that one could create a vacuum by filling a vessel with hot steam and then cooling it so that the steam turned to a few drops of water. If the vessel were connected to a tube reaching down into the water at the bottom of the mine, the water could be sucked up as much as 33 feet (see 1643). It could then be blown out by more steam, which would fill the vessel—and be cooled again to suck up another batch of water.

Savery built such pumps in 1698 and called them the *Miner's Friend*. A few even came to be used. However, they made use of high-pressure steam, and the technology of the day was not able to handle such steam safely. Besides, it took a great deal of fuel to heat water to form the necessary steam, and most of the coal obtained with the help of the Miner's Friend had to be used in operating the device.

Nevertheless, Savery's pump got people to thinking about using steam power, and much was to come of that.

SCIENTIFIC VOYAGES

Voyages undertaken for exploration, settlement, or trade might also produce scientific knowledge as a side effect, as Columbus's first voyage did when he observed magnetic declination (see 1492), or Magellan's when the Magellanic Clouds were observed (see 1678). In 1698, however, an ocean voyage was undertaken, for the first time, that was for the specific purpose of scientific investigation and nothing more.

The ship was the *Paramour Pink* and it was under the command of Halley (see 1678). He remained at sea for two years, measured magnetic declinations all over the world, and prepared the first map of the world showing the wiggling lines of equal declination. He also did his best to determine accurate latitudes and longitudes for the various ports he stopped at.

IN ADDITION

The Bank of England was chartered on July 27, 1694. This made it possible for England to borrow money systematically and to handle a national debt rather than trying to force the population to pay expenses as they came up. This in turn meant that Britain could subsidize other nations and fight wars without economic harm, while France, a larger and richer nation, was constantly on the brink of bankruptcy.

In 1696 Russia defeated the Ottoman Turks and reached the Sea of Azov. In the Far East in that same year, Russia occupied the peninsula of Kamchatka, north of Japan. In 1697 Peter I of Russia toured western Europe incognito (although everyone knew who he was) learning western technology firsthand.

In that same year, Charles XI of Sweden (1655–1697) died and was succeeded by his fourteen-year-old son, Charles XII (1682–1718), whose career was to be intertwined with that of Peter I.

After an eight-year war, Louis XIV was forced to recognize William III as king of England.

1699

GAS VOLUME AND TEMPERATURE

The French physicist Guillaume Amontons (1663–1705) devised an air thermometer that was different from Galileo's (see 1592), for it measured temperature by the change in gas pressure rather than by the change in gas volume. He used the thermometer to show that a liquid such as water always boiled at the same temperature. This made it possible to use the temperature at which water boiled as a standard reference.

With his new thermometer, Amontons tested the volume of a fixed quantity of gas at different temperatures and, in 1699, showed that the volume increased at a steady rate as the temperature went up and decreased at the same steady rate as the temperature went down. Much more important, he showed that for each gas he studied, the volume change with temperature was the same. It seemed a property of all gases.

IN ADDITION

The Austrian-Ottoman war, which started with the siege of Vienna, ended in a total Turkish defeat. The Ottoman Empire had to give up all of Hungary, which now came under the rule of Austria. Never again would the Turks threaten western Europe.

In Virginia, Jamestown was destroyed by fire after having existed for nine decades, and the capital was moved to Williamsburg, 6 miles to the north. Jamestown was never rebuilt.

Russia, Poland, Denmark, and Saxony reached a secret agreement to divide up Swedish territory among themselves now that the land was ruled by a boy. Unfortunately for them, they had no idea what kind of a boy Charles XII of Sweden was. As it happened, he was the nearest approach to Alexander the Great that the modern world had to offer.

1700

BINARY SYSTEM

Our system of positional notation for numbers is based on ten, undoubtedly because we have ten fingers on our two hands. There is, however, nothing magic about the figure 10. Instead of units, tens, hundreds (10 × 10s), thousands (10 × 10 × 10s), and so on, we could have units, eights, sixty-fours (8 × 8s), five-hundred-twelves (8 × 8 × 8s), and so on, or ones, seventeens, two-hundred-eighty-nines (17 × 17s), four-thousand-nine-hundred-thirteens (17 × 17 × 17s), and so on—or any number.

This was pointed out by Leibniz (see 1669, Calculus) about 1700. Some bases for positional notation are, of course, more convenient than others. Using the base 12 or 8 each has some advantages over 10. Leibniz also showed that the *binary system* based on 2 had its uses. Its positions were units, twos, fours, eights, sixteens, and so on. The only symbols it needed were 1 and 0. The binary system is particularly useful for modern computers.

IN ADDITION

The nations around Sweden attacked it in what came to be called the *Great Northern War,* but eighteen-year-old Charles XII showed what he was made of. As soon as the war started, he struck rapidly at Denmark and knocked it out of the war with a blow. He then turned and completely destroyed a Russian army eight times the size of his own. But then Charles overestimated the value of his early victory over Russia, and spent years winning victories in Poland while the furious Peter I drove his Russians mercilessly to organize them into a nation that could fight Sweden on equal terms.

Meanwhile in Spain, Charles II died, leaving no heirs. However, Louis XIV of France had been married to Charles II's half-sister, Marie Thérèse (1638–1683). Their grandson might well be considered Charles II's closest relative who was not in direct line for the throne of France. Before he died, Charles II was persuaded to leave the kingdom to the young man, Philip V (1683–1746). Louis XIV promised that at no time would the monarchies of France and Spain be united under the same king, but for a while it seemed he was at the peak of his power, with Spain for his puppet, together with all its dominions overseas. However, France was on the rocks financially, and neither Austria nor William III of England (who had been ruling alone since the death of Mary II in 1694) would consent to any increase in Louis XIV's power. As a result, France faced war.

London, having survived plague and fire, had a population of about 550,000 in 1700 and was the largest city in Europe. The English colonies in North America had a total population of about 262,000. Their largest cities were Boston and Philadelphia with a population of 12,000 each.

1705

COMETS' ORBITS

For a century or more, astronomers had been trying to puzzle out the orbits of comets. It was clear that comets' orbits were nothing at all like those of the planets. Some astronomers thought comets passed through the Solar System in a straight line. Others thought they passed through in parabolas—coming in from far-off space, going around the Sun and out forever.

Once Newton's *Principia* was published, however (see 1687), it seemed to many that comets had to be bound by gravitation as the planets were.

Halley (see 1678), in an attempt to prove that this was so, began collecting data on comets. Eventually, when he had listed the movements of two dozen comets, he was struck by the similarity of the path across the sky of the 1682 comet (which he himself had observed) with the paths of the comets that had appeared in 1607, 1531, and 1456. These four had come at intervals of seventy-five or seventy-six years. It seemed to Halley that it must be the same comet, returning regularly.

If that were so, a comet would have an orbit that was an ellipse, just as Earth's orbit was, but the cometary ellipse would be extremely elongated. At one end, the comet would approach close to the Sun; at the other, it would recede far beyond Saturn, the farthest known planet.

Halley predicted, in a book written in 1705, that this same comet would return about 1758 and that it would cross the sky in the same path as in 1682. He was aware, he said, that the gravitational influence of the planets might alter the orbit somewhat and change the time of appearance a little.

Halley's claim was not taken seriously at the time, but it did rouse additional interest in comets.

PLANT NOURISHMENT

The English physiologist Stephen Hales (1677–1761) began experiments on plants in 1705. His most important suggestion was that air contributed to the nourishment of plants, thus correcting Helmont's misconception of a century before to the effect that only water counted (see 1624).

Hales was the first person to collect gases by bubbling them through water and trapping them in an upside-down vessel.

IN ADDITION

England, the Dutch Republic, and Austria formed an alliance and declared war on France in order to force the French prince, Philip V, from the Spanish throne. The war that resulted was called the *War of the Spanish Succession.* William III died on March 8, 1702, but he was succeeded by his sister-in-law, Anne (1665–1714), who was also Protestant, and the war continued. (On the North American Continent, this meant war between England and France also, and there it was called *Queen Anne's War.*) Leading the Allied forces against Francis was John Churchill, Duke of Marlborough. A first-class general, Marlborough defeated the hitherto invincible French at the Battle of Blenheim on August 13, 1704.

In the Great Northern War, Charles XII of Sweden continued to win all his battles and was rapidly reducing Poland to helplessness. As for Peter I of Russia, he had no intention of giving in. In 1702 he drove his people remorselessly to build the city of St. Petersburg on territory that had been Swedish. It was to be his "window to the west." He abandoned Moscow and made the new city his capital, and it remained the Russian capital for nearly two and a quarter centuries.

1706

CARRIAGE SPRINGS

All conveyances, from sedan chairs to carriages, were subject to the unevennesses of the road, jolting passengers as every projection or rut was encountered. It was not till 1706 that springs were used in carriages to absorb some of the shock. To be sure, this induced swaying, which has its unpleasantness, too, but is undoubtedly preferable to the lurchings and bangings that existed before springs. The devising of ever-more-efficient springs and, just as important, smoother roads gave land transportation less the feeling of being a devil's playground.

STATIC ELECTRICITY

Guericke's ball of sulfur (see 1660) was not an extremely efficient static-electricity generator. In 1706, however, the English physicist Francis Hauksbee (*ca.* 1666–1713) constructed a glass sphere turned by a crank, which, through friction, could build up a more intense electric charge than sulfur could. This greatly stimulated further experimentation with static electricity.

1707

PULSE WATCH

With the development of the pendulum clock by Huygens (see 1656) and the hairspring by Hooke, timepieces could tell time to the nearest minute. By the late 1600s, minute hands were added to clocks and watches, but second hands had not yet come in.

Then in 1707, the English physician John Floyer (1649–1734) devised a *pulse watch*, which after winding would run for a minute exactly. One could thus count the number of heartbeats in a minute. The pulse watch was the first precision instrument that could be used by physicians.

IN ADDITION

England and Scotland had had the same monarch for a century, but they were still, in theory, separate nations with separate parliaments. On May 1, 1707, however, the two were combined into the

United Kingdom of Great Britain. From now on, one spoke of Great Britain rather than of England and Scotland, and of the British rather than of the English and Scottish.

The Mogul Emperor Ālamgīr died on March 3, 1707. With his death, India promptly fell apart into hostile fragments, leaving the way clear for the British and other European powers to extend their hold over the land's territory and wealth.

1709

COKE AND IRON

Iron ore had required the carbon and the high temperatures of burning charcoal ever since iron-smelting had begun over three thousand years before (see 1000 B.C.). In England, however, the price of charcoal was climbing out of sight as the forests dwindled. Coke had been produced for half a century or more but its use in iron-smelting had not been worked out.

The British ironworks master Abraham Darby (1678–1717) made the first successful use of coke in iron-smelting in 1709. In fact, he found that lumps of coke were stronger than lumps of charcoal and could support a larger charge of iron ore, so that iron could actually be produced at a faster rate. A larger furnace meant more draft and a hotter fire so that iron production was improved still further.

In short, Great Britain was now producing the best and most iron in the world, and since iron could be used for building machinery of all kinds, thanks to its combination of strength and cheapness, Great Britain was set for what would be called the *Industrial Revolution*.

IN ADDITION

Charles XII invaded Russia in force in 1708, but it was now too late. Peter I was battle-hardened and he had a good army. He won a smashing victory at Poltava on July 8, 1709, and gained the sobriquet of Peter the Great. Charles XII had to flee south into the Ottoman Empire with a few men, his image as a new Alexander forever destroyed. Now Russia assumed the role of a great power, which it has held ever since.

In the War of the Spanish Succession, Marlborough continued his winning ways, beating the French at Oudenarde on July 11, 1708, and at Malplaquet on September 11, 1709. France was on the point of defeat, but the battles were increasingly bloody and Marlborough was detested by many in Great Britain as a butcher.

1710

RIFLE

If the muzzle of a gun is rifled—that is, outfitted with spiral grooves—the bullet is set to spinning, and a spinning bullet can be aimed with greater precision. Rifling was tried from the early days of gunnery, but rifling requires a greater force to push the bullet through the muzzle, which means that guns have to be better constructed and are harder to reload. On the whole, the smooth-bore muzzles of muskets seemed better.

About 1710 or soon thereafter, however, the *Pennsylvania rifle* was designed by the Pennsylvania Dutch (not Dutch, really, but German immigrants).

The Pennsylvania rifle took twice as long to reload as the musket, but it had two to three times the range and much greater accuracy. As long as soldiers fought with muskets, they had to maintain a straight line, and all had to shoot at once in the general direction of the enemy, hoping that by sheer luck some bullets would strike home. If they fought soldiers with rifles, the rifles could pick them off before they could even get in musket range.

IN ADDITION

In Great Britain, the Whigs, who had supported the War of Spanish Succession and Marlborough, were voted out, and the Tories, who opposed both, were voted in. This was the first peaceful and orderly change of government in Great Britain, and governments have continued to move in and out peacefully in that nation ever since.

In North America, Queen Anne's War brought about a territorial change when the British conquered Acadia in 1710 and changed its name to Nova Scotia, which it has remained. The French settlers were left in peace for the time being.

1712

NEWCOMEN STEAM ENGINE

The use of coke for iron smelting (see 1709) increased the demand for coal out of which to make coke, and that meant there was a greater need for some device to pump water out of the coal mines. Savery's Miner's Friend was too inefficient and too dangerous.

In 1712 an English engineer, Thomas Newcomen (1663–1729), devised a new kind of steam engine. It didn't depend on using steam to form a vacuum that would suck water upward and then using high-pressure steam to blow it out. Newcomen's engine used ordinary low-pressure steam to push a piston. That meant the pistons didn't have to fit as tightly as they must when high-pressure steam is used, and it meant the machine was less dangerous.

Newcomen engines proved quite popular,

but they were still terribly inefficient. Most of the heat of the burning fuel went into heating the chamber till water boiled into steam and pushed the piston. Then the chamber was cooled to let the piston move back, after which the chamber had to be refilled with water and heated all over again for another forward thrust.

IN ADDITION

In 1711 Marlborough was relieved of his post by the hostile Tory government of Great Britain, and the British army at once began to lose battles. Louis XIV was saved and the war wound toward its close.

Peter I of Russia, rendered overconfident by his victory at Poltava, attacked the Ottoman Turks but allowed himself to be surrounded and was forced to sign a disadvantageous peace on July 21, 1711.

Afghanistan established itself as an independent nation in 1711.

1713

SMALLPOX INOCULATION

Smallpox was the dread disease of this time. The plague, which had been the enormous danger of the last half of the 1300s (see 1346), still struck, but only in isolated places and briefly, and was never again quite the universal danger it had been in the earlier period, perhaps because human beings had gained greater immunity. Smallpox, however, was increasing, and the great dread it inspired was not merely because it frequently killed, but because even when it did not, it disfigured. The large puckered pockmarks it would form left faces visions of ugliness.

If one survived smallpox, however, one was safe from a second attack. A patient would only get it once. What's more, some smallpox cases were rather mild, didn't kill, and only slightly disfigured, but the mild cases led to immunity against a second attack just as efficiently as severe cases did. It followed that a mild case of smallpox was actually better than none at all.

The reasoning, then, was that if you knew someone with a mild case of smallpox, you might get close to him or her in the hope that you would catch a case just as mild.

In 1713, in fact, the English poet Lady Mary Wortley Montagu (1689–1762), whose husband was a British ambassador to Turkey and who accompanied him there, brought back the news that in Turkey they were actually inoculating people with pus from the pustules of people with mild cases, to make sure that the inoculated person would get it. The only trouble with that was that you couldn't always tell—what was a mild case in one person might prove to be a severe one, even a deadly one, in the person being inoculated. It was very much like playing Russian roulette.

Nevertheless, such was the fear of smallpox that, for over three-fourths of a century, many people submitted to such inoculation (also called *variolation* from *variola,* the medical term for smallpox).

IN ADDITION

Brandenburg was a province of northeastern Germany of no great prominence in early modern times. John Sigismund (1572–1619), who became Elector of Brandenburg in 1608, took over the duchy of Prussia, which lay outside the Holy Roman Empire and which had been part of Poland. In 1701 the Elector of Brandenburg was Frederick III (1657–1713), who called himself Frederick I, King of Prussia. Frederick was succeeded by his son Frederick William I (1688–1740), who became king of Prussia in 1713 and militarized it, with disastrous consequences for Europe two centuries later.

The War of Spanish Succession ended with the Treaty of Utrecht on April 11, 1713. Philip V remained on the throne, but France had been so battered that there was felt to be little chance that it would unite with Spain to threaten Europe. Britain ended much stronger economically, while France and Spain were weakened.

1714

MERCURY THERMOMETER

As long as thermometers were open to the air, like those of Galileo (see 1592) and Amontons (see 1699), they were affected by air pressure in one way or another and were always limited in accuracy. The first person to devise a closed thermometer was Ferdinand II de' Medici (1610–1670), who did so in 1654.

At first, sealed thermometers used water or alcohol or mixtures of the two, but these fluids produced vapors that introduced pressure effects. What's more, water didn't expand and contract as evenly with temperature as was needed for good precision, while alcohol boiled at too low a temperature.

The German physicist Daniel Gabriel Fahrenheit (1686–1736) worked with alcohol thermometers at first, but in 1714 he made the key advance of using mercury. Mercury stays liquid between quite low temperatures and quite high temperatures, produces very little vapor, and expands and contracts quite evenly with temperature change. It is an ideal fluid for thermometers and is still commonly used for the purpose today.

Fahrenheit made another advance toward the setting of good standard temperatures. He noted the height of the mercury column in a mixture of ice, water, and ammonium chloride, which resulted in the lowest temperature he could get, and he called that 0. A mixture of ice and water he set at 32, and the temperature of boiling water was then 212. This is the *Fahrenheit scale,* and it is still commonly used in the United States to measure temperature.

The Fahrenheit thermometer was the first that could measure temperature with sufficient accuracy to be useful to scientists.

IN ADDITION

Queen Anne of Great Britain died on August 1, 1714. She was the last monarch of the House of Stuart. Anne's nearest Protestant relative was George, (1660–1727), Elector of Hanover (a west-German province). He landed in England on September 18, 1714, and became king as George I of Great Britain, founding the *Hanover dynasty.*

Not everyone was pleased with that. There was still a son of James II and Mary of Modena living. He was James Francis Edward Stuart (1688–1766).

He called himself *James III* after his father died in 1701, and he was commonly known in Great Britain as the Old Pretender. The name arose because he eventually had a son, Charles Edward Stuart (1720–1788), who was called the Young Pretender. Both Stuarts had their supporters in Great Britain.

George I was a hard person to feel enthusiastic about. He was a dull man who was interested only in Hanover. He did not speak English, did not try to learn, and did not bother to govern. He was perfectly willing to leave all matters of government to the prime minister. It was his behavior that led to Great Britain's present form of government, in which the monarch is merely a figurehead.

As a further consequence of the Treaty of Utrecht, Spain agreed to let Austria take over the Spanish lands in Italy and the Netherlands. What had been the Spanish Netherlands for a century and a half was now the Austrian Netherlands.

1715

SOLAR ECLIPSE

On April 22, 1715, a solar eclipse was to take place and the path of totality was to cross Great Britain and parts of Europe. It had been twenty-three centuries since Thales had predicted an eclipse (see 585 B.C.) and it was perfectly understood by astronomers that eclipses were phenomena that were natural, harmless, and splendid. Nevertheless, superstition is immortal, and in order to prevent as much panic as possible, Halley (see 1678) carefully plotted out the path the eclipse would take and prepared maps of it well in advance so that everyone knew exactly when he or she would see the Sun lose its light.

Halley also organized a large number of observers throughout Europe to watch and time the eclipse. This was the first eclipse for which astronomers turned out en masse. From this point on, every eclipse would bring its crowd of observing astronomers.

IN ADDITION

On September 1, 1715, Louis XIV died, after having reigned seventy-two years, the longest reign in modern history. Like Philip II, he had lived during his country's most "glorious" period, but by attempting too much, had left it weaker than he found it. He was succeeded by his five-year-old great-grandson, who reigned as Louis XV (1710–1774).

The death of Louis XIV upset France's plans for supporting a rebellion in England designed to place James III on the throne, and the attempt failed. (Those who supported James—Jacobus, in Latin—were called *Jacobites*.)

1718

STELLAR MOTION

Ever since the motion of the Sun, Moon, and planets relative to the stars generally had been discovered in Sumerian times, the chief quality of the thousands of remaining stars in the sky had seemed to be that they did *not* move relative to each other. They were designated as the *fixed stars*—fixed to the solid firmament of the sky, that is—while the various planets circled Earth below the firmament.

In 1718, however, Halley (see 1678) determined the position of the bright stars Sirius, Procyon, and Arcturus and observed that they had changed their position markedly since Greek times. They had even changed position perceptibly since the time of Tycho Brahe (see 1572).

It didn't seem possible that the Greeks had made mistakes that large in recording their data, and it was even less possible that Tycho had made such mistakes. Halley therefore concluded that the stars were *not* fixed but moving. It was because they were so distant that their "proper motions" were too small to observe until they had accumulated over extended periods of time.

With Halley, then, the firmament—the solid vault of the sky that everyone had accepted till then, and that even had the authority of the Bible—did not exist. The stars were, rather, like a swarm of very distant and widely spread-out bees, each moving at its own speed and in its own direction.

IN ADDITION

On December 11, 1718, Charles XII was shot through the head, possibly by his own troops, while fighting in Norway. With his death, the Great Northern War was finally brought to a close.

1722

EASTER ISLAND

Dots of land scattered over the Pacific remained to be discovered by Europeans even two centuries after Magellan had made the first crossing (see 1523). In 1722 the Dutch navigator Jacob Roggeveen (1659–1729) came across a small island, 45 square miles in area, that was one of the most isolated bits of land in the world. It was 1,200 miles from the nearest land, which was another small island like itself. Since the island was sighted on Easter Sunday, it was named Passeisland in Dutch, or Easter Island in English.

Easter Island was probably the farthest point reached by the Polynesians in their settlement of the Pacific Islands (see 1642). The island is best known for six hundred stone statues of a type found nowhere else. This has lent the island an air of mystery that it probably doesn't deserve.

Later in the same voyage, Roggeveen discovered the Samoan Islands.

IN ADDITION

Investments in glorious ventures across the seas could be oversold and there could be sudden collapses. In 1720 the *South Sea Bubble* collapsed in Great Britain, while in France, the *Mississippi Bubble* did the same. In both cases, a large number of greedy get-rich-quick investors were ruined.

In 1721 Russia finally made peace with Sweden, taking over a large part of the Baltic coast that it would keep for nearly two centuries.

Robert Walpole (1676–1745) became prime minister of England in 1721 and remained in that post for twenty-one years. Because George I willingly left all business to him, he was Britain's first prime minister in the modern sense. He cultivated peace, encouraged trade, and left the American colonies to themselves, where they grew accustomed to running their own affairs—to the discomfort of Great Britain a half century later.

1728

SHIP'S CHRONOMETER

If a ship wants to locate itself on the ocean, it can measure its latitude (the distance north or south of the equator) by taking the position of the Sun at maximum height or the position of the North Star and calculating the distance of either from the zenith. To measure a ship's longitude (the distance east or west of its home port), however, can be done accurately only if the exact time is known, which presented a problem at this time. A pendulum clock would obviously fail to work on the swaying deck of a ship, and the watches of the time were insufficiently accurate.

In 1714 the British government offered a prize of twenty thousand pounds, an incredible fortune in those days, to anyone who would devise a method of determining a ship's longitude. The sum was clearly worth it, to be sure, considering the improvement it would make in a ship's ability to navigate accurately and the profits that would result from accelerated trade.

Beginning in 1728, an English instrument-maker, John Harrison (1693–1776), built a series of five clocks, each one better than the one before. Each clock was so mounted that it could take the sway of a ship without being adversely affected. He designed a pendulum of different metals so that it would stay the same length and give the same beat even as temperature changed. He also inserted a mechanism that would keep the clock going while it was being wound. Any one of Harrison's clocks met the demands of the prize conditions. In fact, they were more accurate at sea than any other clock then known was on land. One of Harrison's chronometers was off by less than a minute after five months at sea.

The British parliament, however, put on an extraordinary display of miserliness in this connection. It put off the payment to Harrison for years, and he did not get the full amount until 1773.

ABERRATION OF LIGHT

Since Copernicus's book (see 1543) nearly two centuries before, the question of the parallax of the stars had bothered astronomers. If the Earth did indeed move about the Sun, that should induce a parallactic displacement among the nearer stars as compared with the farther ones. Viewing a nearby star from one side of the Sun and then viewing it again from the other side, 186,000,000 miles away, should produce a displacement—but it didn't.

Those who accepted Copernicus and Kepler felt this was because even the nearest stars were so far away that their parallax was too small to measure. Still, telescopes were constantly improving, so astronomers kept trying.

One of those who tried was a British astronomer, James Bradley (1693–1762). Bradley, using a telescope 212 feet long, tried to measure the small displacement of stars in the course of the year and actually detected such a displacement. What he found, however, could not be parallax, because the displacement did not coincide with what would be expected if it were the result of Earth's changing position in its orbit.

Bradley looked for an alternate explanation and one occurred to him in 1728: The displacement arose because the telescope had to be tipped slightly to catch the light as the Earth moved (this is called adjusting to the *aberration of light*), just as an umbrella must be tipped when you're walking briskly through a rainstorm in which the drops are falling vertically. The amount by which the telescope must be tipped depends on the ratio of the speed of the Earth in its orbit to the speed of light.

This meant that although Bradley had not detected parallax, he had discovered a new way of calculating the speed of light, since the speed of the Earth in its orbit was known and the amount of the tipping of the telescope was known, too. This was the first determination of the speed of light since Roemer's a half-century before (see 1675) and it was a more accurate measurement. Bradley's figure was 176,000 miles per second, only 5 percent less than the true value.

What's more, the existence of light aberration was just as strong a proof that the Earth was moving as the existence of stellar parallax would have been.

BERING STRAIT

As Peter I's reign drew toward its close, the Russian occupation of Siberia was complete. The question remained, though, of whether there was a land connection between Siberia and North America. Peter commissioned a Danish navigator, Vitus Jonassen Bering (1681–1741), to look into it.

In 1725 Bering crossed Siberia overland and reached Kamchatka, which he was the first to map. From Kamchatka, he sailed north in 1728 and reached the ice of the Arctic Ocean without sighting land. He had sailed through what is now known as the *Bering Strait*, which separates Siberia from Alaska. The sea to the south is the Bering Sea.

Finally, two and a quarter centuries after Columbus (see 1492), it was shown definitely that North America was not part of Asia.

DENTISTRY

The first book to be devoted entirely to dentistry appeared in 1728. It was *Le chirurgien dentiste (The Dental Surgeon)*, and it was written by a French dentist, Pierre Fauchard (1678–1761). He discussed artificial dentures and crowns and described how to treat caries by cleaning out the decay and making use of metal fillings. Because of this, Fauchard is considered the *father of dentistry*.

IN ADDITION

Peter I of Russia died on January 28, 1725. His wife (of peasant origin) succeeded as Catherine I (1684–1727).

George I of England died on June 10, 1727, and was succeeded by his son, who reigned as George II (1683–1760). He was another dull German who cared nothing for Great Britain.

1729

ELECTRICAL CONDUCTANCE

The interest in static electricity produced by Hauksbee's friction machine (see 1706) began to bring about results. An English experimenter, Stephen Gray (1666–1736), found that when he electrified a long glass tube, the corks at the end were also electrified even though they had not been touched. The electricity, whatever it was, had clearly traveled from the glass into the corks. Gray thought, therefore, that electricity was a fluid.

He experimented further, causing the electrical fluid to travel through long stretches of twine (as long as 800 feet). He found that the fluid flowed more easily through some substances than through others. This led to the division of substances into *conductors* and *nonconductors.* Nonconductors might also be called *insulators,* from the Latin word for "island," since a nonconductor could pen up the electric fluid and keep it confined, as the sea confines an island.

IN ADDITION

Methodism got its start in 1729, when John Wesley (1703–1791), a student at Oxford, gathered like-minded students to meet on Sundays for prayer. Methodism was so-called because of Wesley's methodical way of studying, and it led to a religious revival in Great Britain.

1733

ACHROMATIC LENS

Newton had been convinced that a lens was bound to produce colors at the focus that would blur the image (see 1666). However, an English mathematician, Chester Moor Hall (1703–1771), noticed something Newton had missed: different kinds of glass produced spectra of different widths. Flint glass (containing lead) produced a rather wider spectrum than crown glass (ordinary window glass) did.

Hall therefore decided to make a convex lens out of crown glass and a concave lens out of flint glass, designed in such a way that the two would fit together to form a single biconvex lens. The crown glass would spread out the colors, while the flint glass would bring them together again without neutralizing all the magnifying value of the crown glass. The end result was an *achromatic* (from Greek words meaning "no color") *lens,* which would magnify an object without producing colors.

Hall did not publicize his lens properly, and John Dollond (1706–1761), who prepared an achromatic lens in 1757, often gets the credit. In any case, the achromatic lens made it possible to use the entire lens and to have a short focal length, so that telescopes could be made shorter and more convenient —and better, too.

BLOOD PRESSURE

Hales (see 1705) had studied the flow of sap in plants and went on to study the flow of blood in animals, measuring the rate of flow in different portions of the circulatory system. Most important of all, he was the first person to measure blood pressure, albeit in a crude way. He described his work in this field in a book entitled *Hemostaticks,* published in 1733.

TWO ELECTRICAL FLUIDS

The French physicist Charles-François de Cisternay du Fay (1698–1739) was experimenting with static electricity as so many scientists were doing at this time. In 1733 he found that two bits of cork electrified by the same means repelled each other.

He found, however, that if one cork ball was electrified by means of an electrified glass rod, and another by means of an electrified resin rod, the two cork balls attracted each other.

Du Fay decided there must be *two* electrical fluids and he called them *vitreous* (from a Latin word for "glass") *electricity* and *resinous electricity.* One form of electricity attracted the other, but each repelled itself. This was similar to the well-known magnetic property that like poles repelled each other but unlike poles attracted each other. Thus began the process of detecting similarities and interconnections between electricity and magnetism, which was to become particularly important a century later.

FLYING SHUTTLE

In weaving, a shuttle, with the yarn (the *woof*) attached, must be led across a long series of other yarns (the *warp*), in and out, to form a strong textile material.

In 1733 a British machinist, John Kay (1704–1764), invented the *flying shuttle,* a device whereby the weaver, by pulling a cord, could activate a driver that would send the shuttle flying across the loom. Pulling the cord in the opposite direction would send it on its return journey. The vertical (warp) yarns were automatically separated so that the flying shuttle could pass quickly in and out through them.

This was one of the early examples of a factory process being mechanized so that one person could do more work much more quickly with far less effort. It pointed the way to the coming Industrial Revolution.

IN ADDITION

The American colonies produced their first "world class" personality in Benjamin Franklin (1706–1790), who began to make himself famous (and well-to-do) when he began publishing *Poor Richard's Almanac* in 1732.

In 1732 a British philanthropist, James Edward Oglethorpe (1696–1785), who was interested in prison reform, established a colony in North America where men freed from debtor's prisons might make a new start. This colony was founded to the south of the Carolinas (Spain could not object by this time) and was named Georgia, after George II of Great Britain.

Parliament passed a *Molasses Act* in 1733, which placed duties on molasses, sugar, and rum imported into the North American Colonies from non-British sources. This raised the price of rum, which the colonies drank in large quantities, and they responded by smuggling it in massively without paying duty. This set up a behavior pattern that continued for nearly half a century. Parliament proposed and the colonists disposed.

1735

SHAPE OF THE EARTH

Newton had suggested that, on the basis of his gravitational theory, the Earth ought to be an oblate spheroid and have an equatorial bulge, because it was rotating (see 1687, Universal Gravitation). Plans were now being made to check this prediction by actual measurement.

If the polar regions were slightly flattened and the equatorial regions were slightly bulging, a degree of latitude near the poles should be slightly greater in mileage than a degree of latitude near the Equator. To see if that were so, two expeditions were sent out by the French in 1735. One, under the French geographer Charles-Marie de La Condamine (1701–1774), was sent to Peru, quite near the Equator. The other, under the French mathematician Pierre-Louis Moreau de Maupertuis (1698–1759), went to Lapland, which was about as close to the pole as Europeans could venture in those days.

The results, when finally achieved, completely supported Newton. The degree of latitude was 1 percent longer near the poles than near the Equator. Sea level at the Equator, we now know, is 13 miles farther from the Earth's center than sea level at the poles.

Before returning to Europe, by the way, La Condamine explored the Amazon River region. This was the first time it had been explored in depth by a European since Orellana (see 1542), and La Condamine brought back the first rubber and curare to Europe.

TAXONOMY

Carl von Linné (1707–1778), born in Sweden, is better known by the Latinized version of his name, Carolus Linnaeus. Always interested in plants, he traveled 4,600 miles through northern Scandinavia, where he discovered a hundred new species of plants. He also traveled through Great Britain and western Europe.

In 1735 he published *Systema Naturae (Systems of Nature)*, in which he classified numerous plants, extending the classification in later editions to animals.

Linnaeus was remarkable for the methodical way he went about his classifications. Similar species were grouped into *genera* (singular, *genus*); similar genera into *classes;* similar classes into *orders*. He described each species concisely and gave it a double name, genus and species *(binomial nomenclature)*. Thus, he was the first to call human beings members of the species *Homo sapiens*.

Linnaeus's systematic process made him the father of modern *taxonomy* (from Greek words meaning "naming in order"). What's more, his system of groups, groups of groups, groups of groups of groups, and so on gave his description of living things the appearance of a tree in which large branches divided into smaller branches, those into smaller still, until the final twigs were the species. This made the notion of biological evolution seem more natural than ever, but Linnaeus himself was strongly antievolution and clung to the tale in Genesis.

TRADE WINDS

Half a century before, Halley had tried to explain the trade winds and monsoon but had missed one essential point (see 1686). The British physicist George Hadley (1685–1768) now supplied it. He pointed out that air near the Equator had a faster west-to-east motion than air farther from the Equator. Therefore, winds moving away from the Equator tended to drift eastward, while winds moving toward the Equator tended to drift westward. This view proved satisfactory.

IN ADDITION

A German-born printer in New York, John Peter Zenger (1697–1746), had printed reports that William Cosby, the governor of New York, had tried to rig an election. Cosby declared this a libel and sued. A Scottish-born American lawyer, Andrew Hamilton (1676–1741), defended Zenger. He admitted that Zenger had printed the report but pointed out that the report was true and insisted that truth could not be a libel. The jury agreed with this, and the case was considered crucial to the establishment of a free press in the colonies.

1736

MECHANICS

Even Newton clung to ancient conventions when he could. He wrote his great book (see 1687) in Latin rather than in English, and although he obtained his results by the use of calculus, he managed to put the proofs into geometric forms in his book.

In 1736, however, the Swiss mathematician Leonhard Euler (1707–1783), the most prolific of all time, wrote the first book—*Mechanics*—to be entirely devoted to that subject. In place of Newton's geometry, he made use of algebra and calculus whenever he could.

IN ADDITION

The Russians and the Turks were at war again in 1736, and the Russians regained the foothold on the Sea of Azov that they had lost in Peter the Great's losing campaign.

1737

COBALT

It puzzled miners that a blue mineral resembling copper ore did not yield copper when smelted. The miners assumed it was copper ore that had been bewitched by kobolds—earth spirits who were thought to be malevolent at times.

In 1737 a Swedish chemist, Georg Brandt (1694–1768), investigated the blue ore and managed to obtain a metal from it but one that was definitely not copper. Brandt gave it the name of the earth spirit, spelling it *cobalt*, and that is still the name of the element.

Cobalt was the first new element discovered since Brand's discovery of phosphorus three-quarters of a century earlier (see 1669). Since phosphorus is not a metal, cobalt was also the first metal to be discovered that was not known to the ancients or to the medieval alchemists.

Brandt was perhaps the first important chemist to be completely free of any alchemical taint, and after him the discovery of new elements continued till contemporary times.

IN ADDITION

In Philadelphia, Benjamin Franklin set up the first police force anywhere to be paid by its city and would soon set up the first city-paid fire department.

1738

KINETIC THEORY OF GASES

Boyle had supposed gases to consist of widely spaced atoms, since that would account for the fact that gases could be compressed (see 1662).

This notion was carried further by a Swiss mathematician, Daniel Bernoulli (1700–1782). He assumed that the atoms making up gases were always in rapid, random movement, colliding with each other and with the walls of the containing vessel. (This is called the *kinetic theory of gases; kinetic* is from the Greek word for "motion.")

If the temperature rises, the atoms move faster and collide harder and thus move a bit farther away from each other. For that reason, the volume increases as the temperature rises and decreases as the temperature falls, if pressure remains the same. If the volume is kept from changing, then the pressure (the force with which the atoms hit the walls) increases as the temperature goes up and decreases as the temperature goes down.

This description turned out to be correct, but an adequate mathematical treatment of the subject did not arrive for another century and a quarter.

IN ADDITION

Active excavation began at the sites of Pompeii and Herculaneum (see 1592).

1739

ROCKY MOUNTAINS

The French were continuing their exploration of the interior of North America. The explorer Pierre Gaultier de Varennes de La Vérendrye (1685–1749) had been pushing westward from the Great Lakes since 1731 and by the end of the decade had discovered Lake Winnipeg and the Black Hills of South Dakota.

Two French brothers, Pierre and Paul Mallet, had reached Colorado in 1739 and were the first Europeans to get a glimpse of the Rocky Mountains.

IN ADDITION

The Mogul Empire of India, already broken into warring fragments, was dealt a shattering blow by the invading forces of Nāder Shāh (1688–1747), who had usurped the Persian throne in 1736. He captured and sacked Delhi in 1739, carrying off the Peacock Throne and the Kohinoor Diamond, and went on to make other conquests in central Asia. He left India all the more helpless in the face of European encroachments.

1740

HYDRA

The Swiss naturalist Abraham Trembley (1700–1784), working in the Dutch Republic, discovered the freshwater *hydra* in 1740. This was a small, very primitive organism, which resembled a plant somewhat in appearance, but which Trembley showed to be an animal. It had tentacles, and seemed a tiny and harmless replica of the fearsome Hydra destroyed by Hercules in the Greek myths.

The resemblance was all the closer in that the mythical Hydra grew back heads that were lopped off and the tiny freshwater hydra could regenerate parts that were cut off. In fact, Trembley showed that if a hydra was cut in two, each half could grow into a complete organism, and if two hydra were grafted together they would form a single individual. This showed that such properties, thought to be exclusive to plants, clearly existed in animals, too, if the animals were sufficiently primitive.

IN ADDITION

Frederick William I of Prussia died on May 31, 1740. He had kept the peace, but he had built up a large and superbly trained army, which his son and successor, Frederick II (1712–1786) could now use if he wished. When the Holy Roman Emperor, Charles VI, died on October 20, 1740, Frederick II sent his army into Silesia, the Austrian province immediately to the southeast of Prussia. This began the *War of the Austrian Succession* in which France, Spain, Bavaria, and Saxony joined Prussia in the hunt for Austrian spoils.

1742

CELSIUS SCALE

For nearly thirty years, the Fahrenheit scale (see 1714) had been commonly used for temperature measurements. It had some disadvantages, however. For instance, the freezing point of water was set at 32 degrees, a curiously uneven number.

It makes a huge difference, both to scientists and to human beings generally, whether water is liquid or solid, whether a pond is frozen over or not, whether it snows or rains. In 1742, therefore, the Swedish astronomer Anders Celsius (1701–1744) suggested that the freezing point of water be set at 0 degrees, so that a positive reading meant water and a negative reading meant ice. The boiling point of water would then be set at 100 degrees, rather than 212.

This new scale was at first called the *Centigrade scale* (from Latin words meaning "a hundred steps"—that is, from the freezing point to the boiling point of water), but this was converted to *Celsius scale* by international agreement in 1948. The entire world has now adopted the Celsius scale with only one significant exception—the United States.

GOLDBACH'S CONJECTURE

A mathematician who thinks some statement seems true but can't *prove* it's true may then advance it as a *conjecture*. Fermat's Last Theorem (see 1637) is not a conjecture; Fermat stated flatly that he had a proof—though he may have been mistaken, to be sure.

The most famous actual conjecture is one made by a German mathematician who worked in Russia, Christian Goldbach (1690–1764).

To explain it, we begin by saying that a prime number is any number greater than 1 that can only be divided by itself and 1. There are an infinite number of primes. The first few are 2, 3, 5, 7, 11, 13, 17, 19, and 23.

It seemed to Goldbach that any even number greater than 2 could be expressed as the sum of two primes (sometimes in more than one way). Thus, $4 = 2 + 2$; $6 = 3 + 3$; $8 = 5 + 3$; $10 = 5 + 5$; $12 = 7 + 5$; $14 = 7 + 7$; $16 = 11 + 5$; $18 = 13 + 5$; $20 = 13 + 7$; $22 = 11 + 11$; $24 = 13 + 11$; $26 = 13 + 13$; $28 = 23 + 5$; $30 = 23 + 7$; $32 = 19 + 13$; $34 = 17 + 17$; $36 = 23 + 13$; $38 = 19 + 19$; $40 = 23 + 17$; $42 = 23 + 19$; and so on.

No mathematician has ever found an even number higher than 2 that could not be expressed as the sum of two prime numbers. Every mathematician is convinced that no such number exists and that Goldbach's conjecture is true. However, no mathematician has ever been able to *prove* the conjecture.

But that sort of thing is the excitement of mathematics and of science generally. We shall never run out of problems, and some will always remain incredibly tantalizing.

FRANKLIN STOVE

Originally, campfires were built in the open, or inside a cave. Fires in confined areas presented a problem with smoke, so chimneys had to be invented. Fireplaces and chimneys are wasteful, however. Hot air from the fire goes straight up the chimney and does not warm much of the room. Indeed, the rising hot air creates a draft that brings cold air in from the outside.

It occurred to Benjamin Franklin (see 1733) that what was needed was an iron stove within a room. Inside that a fire could be built that would create no draft but would heat up the metal. That in turn would heat the air, and the warm air would stay inside the room instead of vanishing up a chimney. As for the smoke, that could pass through a stovepipe into the chimney.

Stoves of this sort grew instantly popular, and indeed, the modern home furnace in the basement is a kind of Franklin stove.

IN ADDITION

Frederick II demonstrated the worth of the Prussian army, and of his own leadership, by handily defeating the Austrians in Silesia in 1741 and taking over the province. Austria found itself menaced from all sides, but the young Queen Maria Theresa (1717–1780) proved herself the most capable reigning queen since Elizabeth I of England and resisted the various invaders grimly.

Also in 1741, Elizabeth Petrovna (1709–1762), the daughter of Peter the Great, overthrew the government of her half-cousin once removed, Anna Leopoldovna (1718–1746), and became czarina of Russia. She proved a deadly enemy of Frederick II of Prussia.

1744

TRANSCENDENTAL NUMBERS

One would think that mathematicians had dealt with all the different kinds of numbers there were—integers, fractions, negative numbers, irrational numbers, imaginary numbers, what else could there be?

In 1744 Leonhard Euler (see 1736) pointed out that polynomial algebraic equations, consisting of powers of x, can have all sorts of solutions—whole numbers, fractions, irrationals, negatives, imaginaries, complex numbers, and so on. These are all *algebraic numbers.*

Nevertheless, Euler pointed out, there are other numbers that can never be a solution for any algebraic equation that can possibly be written. Such numbers are *transcendental numbers,* from Latin words meaning "to climb beyond," because they climb beyond the equations, so to speak. We now know that there are an infinite number of transcendental numbers.

IN ADDITION

The British joined in the War of the Austrian Succession because Hanover fought on the Austrian side and George II was Elector of Hanover as well as King of Great Britain. George II led the British to victory over the French at the Battle of Dettingen on June 27, 1743. This was the last time a British monarch was to appear on the field of battle. By this time the war had spilled over to the North American colonies, where it was known as *King George's War*.

1745

LEYDEN JAR

Hauksbee's glass sphere (see 1706), first produced forty years earlier, was surpassed as an electricity-storing device by the work of a Dutch physicist, Pieter van Musschenbroek (1692–1761).

In 1745 he placed water in a metal container suspended by insulating silk cords, and led a brass wire through a cork into the water. He built up an electric charge in the water but did not realize how much had accumulated in the device until an assistant happened to pick it up and touch the brass wire outside the cork. The container promptly discharged all of the electric charge it had accumulated and gave the poor assistant a fearful shock. It was the first good-sized artifical electric shock anyone had ever received. (The lightning stroke is worse, of course, but it is a natural electric shock.)

A German physicist, Ewald Georg von Kleist (1700–1748), independently produced the same device at about the same time. He discovered the strength of the charge by accidentally discharging it into his own body. He said he wouldn't take another such shock to be the king of France and worked with the device no longer.

Because Musschenbroek popularized the device and because he worked at the University of Leyden in the Netherlands, the electricity-storing device came to be called a *Leyden jar*. It was at once made use of by other experimenters.

BLOOD AND IRON

Chemists had little or no idea of the composition of living tissue—even what elements it contained, since so little was known about the elements.

Of course, iron was known, and an Italian physician, Vincenzo Menghini (1704–1759), fed iron preparations to dogs in 1745 in order to see what would happen to the iron and whether any of it would be incorporated in the tissues. To make sure that no iron was present in dogs that didn't eat the iron preparation, he obtained blood from normal dogs and burned it, expecting to find no iron in the ash. However, he *did* find iron, to his surprise. What's more, he was able to ascertain that the iron was present, specifically, in the red blood corpuscles.

This represented the first discovery in living tissue of a trace element; that is, one present in only small quantities but nevertheless essential to life (though the fact that iron was essential to life was not yet understood).

IN ADDITION

The Jacobites took advantage of British preoccupation with the War of the Austrian Succession to stage an invasion. Charles Stuart (Bonnie Prince Charlie), the Young Pretender, invaded Scotland on July 25, 1745. The Jacobites won battles in Scotland and marched south, reaching Derby in England, only 120 miles north of London, on December 4. The British were forced to recall their army from Europe. There the French, on May 11, 1745, defeated an allied army under William Augustus, the Duke of Cumberland (1721–1765), a son of George II.

In North America, the British colonists under William Pepperell of Maine (1696–1759) took the strong French fortress of Louisburg at the northeast corner of Nova Scotia on June 16, 1745.

1747

SCURVY

Ever since Vasco da Gama's voyage (see 1497), scurvy had been an increasing menace. More men were disabled by scurvy on long voyages than by any other cause. This was particularly serious in Great Britain, which depended on its navy for protection and on its merchant ships for its prosperity.

A British physician, James Lind (1716–1794), had served in the navy and knew that the diet on board ship was monotonous in the extreme, consisting of hardtack, salt pork, and other food items whose only virtue was that they didn't spoil in an age without refrigeration or canning. He also knew that scurvy appeared in prisons, in besieged towns, on overland exploring expeditions—always where diet was limited and monotonous.

Lind therefore studied the effect of adding food items that didn't last well, especially fruits and vegetables, to the diet of people affected with scurvy. In 1747 he found that citrus fruits worked amazingly well in effecting relief. It was nearly half a century, however, before the British navy could bestir itself to make use of that fact and put an end to the menace of scurvy.

IN ADDITION

The Duke of Cumberland was brought back from the continent by the British to take charge of the defense against the Jacobite invaders. He drove them back and on April 16, 1746, smashed them at the Battle of Culloden Moor in northwestern Scotland. (This was the last land battle fought on British territory.) After the victory, Cumberland (called "the butcher" in consequence) had the wounded Jacobites killed and carried out other savage reprisals that broke the Jacobite power forever. Prince Charles escaped alive but was never a threat to the Hanover dynasty again.

1748

OSMOSIS

It is a common observation that liquids will soak through some materials but not others. But if a liquid soaks through a particular material, it seems reasonable to suppose that it will soak through in either direction.

Nevertheless, when a French experimenter, Jean-Antoine Nollet (1700–1770), covered an alcoholic solution with a section of pig's bladder and placed it in a tub of water, the bladder began to bulge over a period of time as though more water was entering the solution than was leaving it. Eventually, the bladder membrane burst.

In this way, Nollet discovered what is now called a *semipermeable membrane*—that is, one permeable to some fluids but not to others. If there are different fluids on the two sides of such a membrane (pure water on one side, for instance, and an alcoholic solution on the other), the fluid flow in one direction may be greater than in the other. This is called *osmosis* (from a Greek word meaning "push," since the fluid ends by pushing through in one direction).

The explanation for osmosis had to wait half a century till the notion of molecules of different sizes was understood.

PLATINUM

Gold, silver, and copper are not the only rare metals sometimes found in free metallic form. Another is *platinum*. A metallic casket found among the relics of seventh century B.C. Egypt is reported to be of platinum. In general, however, the existence of this metal remained unremarked and unknown. For one thing, platinum is as rare as gold, but it lacks gold's ostentatious beauty. Platinum has a dull, leaden color when unpolished that does not attract the eye.

In 1748, however, a Spanish scientist, Antonio de Ulloa (1716–1795), published a report on his travels in South America. In it, he reported on *platina* (from a Spanish word for "silver," chiefly because it was a free metal that lacked the pronounced color of gold or copper). He remarked on its peculiar properties, for when properly examined, it turned out to be denser than gold, higher-melting, and even less reactive. Eventually, it proved extremely useful to scientists for just these properties.

IN ADDITION

The War of the Austrian Succession came to an end in October 1748 with the signing of the Treaty of Aix-la-Chapelle. Frederick II of Prussia kept Silesia, but Austria retained its remaining territory intact and Maria Theresa was recognized as monarch. As part of the treaty, Great Britain handed Louisburg back to France, which infuriated the New Englanders who had captured it. They remembered this as an indication of British indifference to colonial interests.

1749

BIOLOGICAL EVOLUTION

Until now, natural historians who had been busily engaged in classifying life forms had refrained, out of religious conviction or out of prudence, from drawing the logical conclusion that biological evolution had taken place.

The first important scientist to speculate openly on evolution was the French naturalist Georges-Louis Leclerc de Buffon (1707–1788). In 1749, he began publishing volumes of his book *Natural History*, which was eventually to consist of forty-four volumes.

Buffon treated evolution as a matter of degeneration. It is, after all, a common observation that many things deteriorate with time. Why should not evolution exemplify this? Buffon maintained that apes were degenerated humans, donkeys degenerated horses, jackals degenerated wolves, and so on.

This view is quite wrong, but it implied that species did change with time, which was crucially important. It also sufficed to get Buffon into a certain amount of trouble, which he managed to smooth over with diplomatic recantations.

FORMATION OF THE EARTH

Buffon (see above) was also daring enough to think that there might be some natural cause, one not involving God, for Earth's coming into existence. In the first volume of his *Natural History*, he suggested that Earth (and presumably the other planets) had been formed by the collision of the Sun with another massive body (which he called a comet).

Buffon then tried to decide how old the Earth was by calculating how many years it would take an object the size of the Earth to cool from the temperature of the Sun to the temperature of the Earth today. This, he eventually announced, could be as much as 75,000 years; Earth had grown cool enough to support life about 40,000 years before, and it would exist for another 90,000 years before becoming too cold to support life any longer.

To be sure, Buffon's estimates of these values were much smaller than the time periods scientists were to accept later on, but he was the first important scientist to suggest that the Earth might be much older than the 6,000 years allotted it by Ussher (see 1650).

IN ADDITION

In North America, the collision of French and British colonials came closer as both moved into the region north of the Ohio river (the *Ohio Territory*) with opposing claims. In Nova Scotia, the British founded Halifax.

1751

NICKEL

Although Brandt had isolated cobalt fourteen years before (see 1737), the problem of copper ore that did not yield copper persisted. Some of this sort of ore didn't yield cobalt, either, and the miners called it *kupfernickel* (*Old Nick's copper,* with reference to the devil).

In 1751 a Swedish mineralogist, Axel Fredrick Cronstedt (1722–1765), who had studied under Brandt, tackled kupfernickel and isolated from it a white metal that was neither copper nor cobalt. Cronstedt called it *nickel,* taking the second half of the miners' name.

Cronstedt discovered that nickel, like iron but less strongly, was attracted by a magnet. It was the first time that anything but iron had been found to be attracted by a magnet. As a matter of fact, magnets were eventually found to attract cobalt as well.

Indeed, iron, cobalt, and nickel are very similar metals. This was the first hint that elements might be grouped into families and that some sort of order might be imposed on them—but that did not take place for another century.

ENCYCLOPEDIAS

The growing accumulation of knowledge and the growing self-confidence of the scientists of this period (which called itself the *Age of Reason*) made it natural to think of preparing a multivolume summary of all knowledge for the general public, under headings arranged in alphabetical order. This would be an *encyclopedia,* from Greek words meaning "general education."

A rather modest suggestion to this effect was made by a bookseller to a prolific French free-lance writer, Denis Diderot (1713–1784). Diderot caught fire and made a huge project out of it. He used collaborators but in the end did most of the work alone. The first volume appeared in 1751.

Diderot's product was the first modern encyclopedia. It took a completely rational view of the world and clearly opposed the shibboleths of state and church at that time. It impressed intellectuals as the supreme product of the Age of Reason.

IN ADDITION

In 1750 the Cumberland Gap was discovered through the Appalachian Mountains in what is now northeastern Tennessee, which facilitated movement westward. The discoverer, a land speculator and explorer named Thomas Walker (1715–1794), named it for the Duke of Cumberland who had recently put down the Jacobite rebellion so savagely.

1752

LIGHTNING ROD

The Leyden jar (see 1745) had become a favorite plaything of many scientists. One of them was Benjamin Franklin.

In 1747 Franklin rejected du Fay's notion of two electrical fluids (see 1733). He thought there was only one, which could exist in an excess (above normal) or in a deficiency (below normal). Excess repelled excess, since neither could accept the other's. Similarly, deficiency repelled deficiency, since neither could offer anything to the other. Excess, however, attracted deficiency, and the electrical fluid poured from the excess to the deficiency, neutralizing both and leaving each *uncharged*.

Franklin suggested that the excess be called *positive electricity* and the deficiency *negative electricity*. There was no telling which variety of electricity, vitreous or resinous, was positive and which negative. Franklin guessed arbitrarily and happened to guess wrong. That makes no difference, however. The names can be used and the literal meanings forgotten.

Franklin noted the manner of discharge of the Leyden jar. When the electrical charge was drawn off, it emitted a spark of light and a crackle of sound. Franklin was struck by the similarity to a tiny lightning stroke and an equally tiny crack of thunder. He at once reversed the thought. During a thunderstorm, did Earth and sky set up a gigantic Leyden jar, and was the lightning and thunder an equally gigantic discharge?

He decided to experiment. In 1751, he flew a kite in a thunderstorm. The kite carried a metal point to which a long silk thread was attached. At the bottom of the thread, near Franklin (who held onto the silk by way of a second thread that remained dry), was a metal key. As the thunderclouds gathered and the silk thread began to show signs of electrical charge (the separate fibers repelled each other), Franklin put his knuckle near the key and it sparked and crackled just like a Leyden jar. Moreover, Franklin charged a Leyden jar from the key just as easily as if it were a friction machine. The Leyden jar charged by heavenly electricity behaved precisely as though it had been charged by earthly electricity. The two electricities were identical.

Franklin was able to put his discovery to practical use at once. Lightning, he decided, hit a particular building when that building gathered charge during a thunderstorm. His experience with Leyden jars showed that they discharged much more easily if a sharp needle was attached to them. Indeed, the charge leaked out so easily through the needle that they couldn't be charged in the first place. Why not, then, attach a sharp metal rod to the top of a building, and ground it properly, so that any charge that gathered might leak away rapidly and silently and no charge would accumulate to the point where a disastrous discharge would be forced.

Franklin published his thoughts on the matter in 1752 in *Poor Richard's Almanac,* and the *lightning rods,* as they were called, began to go up at once, first in America and then in Europe. They proved efficacious, and for the first time in history, a natural catastrophe was averted not by prayer or by magical incantations of one sort or another, which never really worked, but by reliance on an understanding of natural laws, which *did* work. Once lightning rods appeared on church steeples (which, being the highest point in town, were particularly vulnerable), the point was made for all to see.

DIGESTION

Was digestion a physical action (the result of the grinding of the stomach) or a chemical one (the result of fermentation)? The controversy had been going on for a century.

In 1752, a French physicist, René-Antoine Ferchault de Réaumur (1683–1757), induced a hawk to swallow metal cylinders that were open at both ends, with those ends covered by wire gauze. Inside the cylinder was meat.

Ordinarily, a hawk swallows its food in large pieces, digests what it can, and regurgitates the indigestible remainder. Réaumur waited for the hawk to regurgitate the cylinder and found the meat partially dissolved. Clearly, the meat could not have been affected by grinding or by any mechanical action, since the metal cylinders protected it from that. Therefore, the stomach juices must have had a chemical effect on the meat.

He checked this further by persuading the hawk to swallow a small sponge. When that was regurgitated, it was soaked with stomach juice, which Réaumur squeezed out and collected. He found that this stomach juice slowly dissolved meat placed in it. He experimented with dogs, too, and obtained the same results. Muscles and bones might behave like mechanical systems, but the body was a chemical device as well, and as scientists increasingly discovered, the chemistry is even more important than the mechanics.

EARTH AND HEAT

There is ample evidence that the Earth's surface has undergone enormous changes in its history, and there must be titanic forces behind those changes. Up to this time, most Europeans took it for granted that the causative factor was water and, in particular, the action of Noah's Flood, which was considered a God-caused cataclysm far more forcible than any natural flood could have been. Those who believed this were called *Neptunists.*

In 1752, however, a French geologist, Jean-Étienne Guettard (1715–1786), was convinced by his observations that the rocks he saw in central France had experienced great heat at some time in the past. This began the belief in heat as the causative factor.

IN ADDITION

Finally, Great Britain and its colonies accepted the Gregorian calendar. By now it was necessary for Great Britain to drop eleven days, so that September 3 to September 13, 1752, simply vanished, to the disturbance of many unsophisticated people who thought they were losing eleven days of their lives. (To be sure, landlords charged rent for those eleven days.)

1754

CARBON DIOXIDE

Carbon dioxide had been studied by Helmont (see 1624), but until this time it had only been detected as a product of combustion or fermentation.

In 1754, however, a Scottish chemist, Joseph Black (1728–1799), in his thesis for a medical degree (which he published two years later), described how he had formed a gas by strongly heating limestone (calcium carbonate), leaving lime (calcium oxide) behind. The gas he produced would combine with calcium oxide to form calcium carbonate once again. He called the gas *fixed air*, because it could be *fixed* (made to take on a solid form) by combination with calcium oxide. Fixed air turned out to be carbon dioxide.

It was clear, then, that gases could be formed from ordinary solids and that they could take part in chemical reactions. That went far toward converting gases from mysteries to chemicals.

In fact, since calcium oxide slowly turned to calcium carbonate if simply allowed to remain exposed to air, it was clear that there must be some small fraction of the air that was carbon dioxide. That was the first indication that air was not a simple substance (an element) but a mixture of different gases.

In studying the action of heat on calcium carbonate, by the way, Black measured the loss of weight involved in its conversion to calcium oxide. He also measured the amount of acid that a given quantity of calcium carbonate would neutralize. This was the first application of quantitative analysis to chemical reactions, a chemical wave of the future that could come to fruition in a quarter of a century.

IN ADDITION

The clash between French and British forces in the Ohio Valley had become inevitable, as the French spread their fortifications even into what is now western Pennsylvana. In 1754 Virginia, the oldest of the colonies and the one with the most grandiose claims to all the western lands, decided it was time for action. The governor of Virginia, Robert Dinwiddie (1693–1770), sent a young Virginian surveyor, George Washington (1732–1799), to visit the French in western Pennsylvania and demand that they withdraw. The French paid no attention to this demand, and Dinwiddie sent Washington back with a small force of men. Washington was defeated by the French, and this was the opening battle of what came to be called the *French and Indian War*.

1755

GALAXY

Do the stars in the sky spread out evenly and indefinitely in all directions or are they contained in a finite volume with a particular shape. To the eye it might seem that the first alternative is correct—except for the Milky Way. Once Galileo showed that the Milky

Way consisted of very many, very dim stars (see 1609), it became clear that there are far more stars in the direction of the Milky Way than in other directions. In 1750, an English astronomer, Thomas Wright (1711–1786), maintained that the stars formed a flattened finite system, but his writings were so mystical that it was hard to take him seriously.

In 1755, however, the German philosopher Immanuel Kant (1724–1804) made a similar suggestion. He said the Sun was one of a large number of stars that existed in a lens-shaped conglomerate, and that the Milky Way was the result of looking into the sky along the long axis of the lens. This conglomerate came to be called the *Galaxy*, from a Greek word for the Milky Way. Kant also suggested that certain nebulas, such as the one in Andromeda, were other galaxies or, as he called them in a dramatic phrase, "island universes."

In this, Kant was quite correct, but it was to be over half a century before the Galaxy's existence could be clearly demonstrated and over a century and a half before the existence of other galaxies could be demonstrated.

IN ADDITION

Great Britain sent a substantial army to North America under Edward Braddock (1695–1755), which landed in Virginia on February 20, 1755. He led it into western Pennsylvania and, on July 9, 1755, tried to fight the French and their Indian allies in European fashion, with his men lined up in a row. The French and Indians, from behind trees, destroyed the British. Only Washington, leading a Virginian detachment that fought Indian-fashion themselves, managed to get a remnant off.

On November 1, 1755, a terrific earthquake utterly destroyed the city of Lisbon and shook large parts of western Europe and northern Africa. The earthquake, together with the floods and fires it gave rise to, may have killed as many as sixty thousand people. It shook the confidence of the Age-of-Reason generation of Europeans, who were far more impressed by the earthquake than by the tales of fighting in distant America.

1756

LAND BRIDGES

In the book of Genesis, God is described as separating land and sea on the third day of creation. It does not say that God fixed the boundaries eternally, but this was assumed. It was taken for granted by Europeans that the shape of the continents was fixed and eternal, barring certain trivial changes produced by storms.

In 1756, however, a French geologist, Nicolas Desmarest (1725–1815), pointed out the similarities of the channel shores of England and France and suggested that there had been a land bridge between them once, a bridge now covered by the sea. (In time, he was proved correct.) This was the first indication that continents might shift and change shape. Desmarest thought such changes were brought about by earthquakes, which was not a bad idea, but it was wrong.

IN ADDITION

Since the end of the War of the Austrian Succession, Maria Theresa had been scheming to get Silesia back. She formed secret alliances with France, Russia, and Sweden against Prussia, but Frederick II of Prussia attacked in 1756 before the alliance could entirely prepare itself. This began the *Seven Years War.* Since Prussia was fighting France, Great Britain supported Prussia. This meant that Great Britain was fighting France in North America, in Europe, and in India, so that the Seven Years War was a kind of world war, the first of its kind.

1758

HALLEY'S COMET

A little over half a century before, Halley had predicted that the comet of 1682 would return in 1758 (see 1705). The amateur astronomer Johann Georg Palitzsch (1723–1788) set up his telescope and trained it on the part of the sky where the comet was expected to appear, if it did return. On December 25, 1758, he spotted it, and once the news broke, professional astronomers zeroed their instruments in upon it. The comet has ever since been known as *Halley's comet,* or in line with the conventions of the present day, *Comet Halley.* Calculating backward, Halley's comet turned out to be the one that appeared at the time of the invasion of England by William of Normandy. It was also the comet that Giotto painted (see 1304).

The return of Halley's comet suddenly made comets the headliners of astronomy, and for several decades it seemed that the greatest feat any astronomer could achieve was to discover comets.

FLAME TESTS

It was always a difficult task for chemists to distinguish one substance from another when some clearly visible property difference (color, softness, and so on) didn't exist. Chemists then had to test for more subtle differences.

A new test that produced something clearly visible to the eye was discovered by a German chemist, Andreas Sigismund Marggraf (1709–1782), in 1758. Marggraf found that sodium compounds turned a flame yellow, while otherwise very similar potassium compounds turned it violet. (Of course, while the compounds were known, the elements sodium and potassium were not to be isolated for half a century.)

This introduced the flame test into chemistry. Later, Cronstedt (see 1751) introduced the *blowpipe,* which directed a thin jet of air into a flame, making it hotter and more effectively heating minerals to produce delicate gradations of color. For many decades, chemists had to be skillful at blowpipe analysis if they expected to be successful in research.

IN ADDITION

In 1758 the British took Calcutta and drove the French out of Bengal (a region almost as large as Great Britain). This was the beginning of British rule over India, a rule that was to endure nearly two centuries.

In Europe, Frederick II of Prussia won two great victories in 1757, one at Rossbach over the French on November 5 and one at Leuthen over the Austrians on December 5. On August 25, 1758, he beat the Russians at the Battle of Zorndorff.

The French were losing to Frederick II in Europe and to Great Britain in India, and they were losing to Great Britain on the African coast and in North America as well. On July 26, 1758, the British took Louisburg, this time permanently, destroying the fortifications. The British also drove the French out of western Pennsylvania.

1759

EMBRYOLOGY

It was customary at this time to think that seeds and eggs (or pollen and sperm) had miniature organisms in them that simply grew. Some even thought that an organism within an egg might have eggs of its own with ultraminiature organisms within and still smaller eggs—and so on.

In 1759, however, the German physiologist Kaspar Friedrich Wolff (1734–1794) showed that specialized organs developed out of unspecialized tissue. Thus the tip of a growing shoot consists of undifferentiated and generalized tissue. As it grows, however, specialization develops, and some bits of tissue develop into flowers, while other bits, originally undistinguishable, develop into leaves.

The same principle could, and did, apply to animals, so that the idea of miniature organisms in eggs disappeared. Wolff is considered another of the founders of modern *embryology*, in consequence.

IN ADDITION

Although Frederick II had won important victories and was clearly the greatest commander in the field, the weight of numbers was beginning to tell on him. He and his army were worn out and the French, Austrians, and Russians, no matter how often beaten, always returned to the fray.

In North America, the climactic battle between the French and the British was fought. The city of Quebec was under attack by a British force under the competent James Wolfe (1727–1759) and being defended by the equally competent Louis-Joseph de Montcalm-Gozon (1712–1759). On the morning of September 13, 1759, the French were surprised to find a British army of five thousand marching to attack them. The British won the battle and took Quebec, but both Wolfe and Montcalm were killed.

1760

EARTHQUAKES

Human beings have always known about earthquakes from sad and frightening experience. What they didn't know was the cause. Early theories attributed them to the restlessness of gods or demons imprisoned underground. The ancient Greek philosophers, in search of a more rational cause, supposed that there might be pent-up air underground, which occasionally, in its attempt to escape, shook the earth.

The Lisbon earthquake of 1755 forced serious thinking on the matter. In 1760, the English physicist John Michell (1724–1793) noted that earthquakes frequently appeared in volcanic areas. He felt that underground water might be boiled by volcanic heat at times, and that it was the imprisoned steam that caused earthquakes.

Furthermore, he said, the earthquakes set up waves that traveled through the Earth at some measurable speed. If the time at which waves reached different points were recorded, the point of origin (or *epicenter*) of the earthquake could be determined. Epicenters, he suggested, could exist in the rocks beneath the sea, and it was one of these that had destroyed Lisbon.

All of Michell's ideas were pretty sound, so that he is known as the founder of *seismology*.

HEAT CAPACITY

Until this time it was assumed that as more heat poured into a substance, its temperature rose steadily, and that given quantities of different materials all rose in temperature at the same rate. Why not? Heat was considered a very subtle fluid, and it made sense to suppose that it filled all substances equally, so that all given weights of substances had the same capacity for containing heat.

In 1760, however, Black (see 1754) showed that this notion, however sensible it seemed, was quite wrong. When he heated equal weights of mercury and water over the same flame, the mercury's temperature went up twice as fast as the water's temperature did. To Black, this indicated that the heat capacity of mercury was less than the heat capacity of water, so that mercury was more rapidly filled with heat.

It followed that if equal quantities of mercury and water were mixed, where mercury was at the higher temperature, the final temperature would not be precisely midway between but would be below the midpoint, because the quantity of heat present in the mercury and transferred to the water would not fill the water as much as it had the mercury.

This started the scientific study of heat, as distinct from temperature.

PATHOLOGY

In 1760 the Italian anatomist Giovanni Battista Morgagni (1682–1771) published a book in which he described the 640 postmortems he had conducted in his long life. He carefully detailed the lives of his patients, the presence and development of disease, and the manner of dying and tried to interpret everything from the anatomical point of view. As a result, he is usually considered the founder of modern *pathology*.

IN ADDITION

George II of Great Britain died on October 25, 1760, and his grandson succeeded as George III (1738–1820). While George I and George II spoke only German, George III was completely English, and while George I and George II were content to let the prime minister run the country, George III (with the advice of his mother) wanted "to be a king" in the French style. It was too late for that, though, and his attempt to assert himself was to bring disaster.

Frederick II was having more and more trouble as he raced his army from one problem area to another. On October 9, 1760, a Russian army took Berlin and burned it, then retreated when they heard that Frederick II was approaching.

1761

VENUS'S ATMOSPHERE

Unlike the other planets, Venus had no markings. It always presented a featureless white orb. It remained interesting, however, because periodically (being closer to the Sun than Earth is) it passed exactly between Earth and Sun. On such occasions, it appeared as a small black sphere moving across the face of the Sun. This is called a *transit of Venus.*

In 1761 there was a transit of Venus coming and astronomers organized expeditions to Newfoundland and to St. Helena to observe it from widely separated spots. If the exact time at which Venus's orb first touched the Sun's rim and the exact time at which it finally left on the other side were determined from those widely separated spots, Venus's parallax could be determined, and its distance, together with that of the Sun, might be determined to a higher degree of accuracy than was produced by Cassini's determination of the parallax of Mars.

The expedition failed. The time when Venus entered and left the solar sphere could not be determined accurately, to the frustration of all.

One observer, the Russian scientist Mikhail Vasilievich Lomonosov (1711–1765), pointed out that this could be due to Venus's possession of an atmosphere. The atmosphere would fuzz Venus's outline, so to speak, and make the points of contact difficult to time. Furthermore, if the atmosphere had a permanent cloud layer, that would help explain both Venus's brilliance (since the cloud layer would reflect most of the sunlight striking it) and its featurelessness.

PERCUSSION

Methods for medical diagnoses were not many in these days. In 1761, however, an Austrian physician, Leopold Auenbrugger von Auenbrugg (1722–1809), published a book entitled *Inventum novum (A New Invention)* in which he pointed out that tapping the body, especially the chest area, and listening to the sound produced would give some indication of certain disorders of the internal organs. (He checked this by comparing the sounds he elicited with the state of the organs as revealed by postmortems.)

It took forty years, however, before this diagnostic method became common in medicine.

1762

LATENT HEAT

Black (see 1754) found in 1762 that if a mixture of ice and water was heated, the heat was absorbed but the temperature did *not* change. All the heat went toward melting the ice into water, the water being at the same temperature as ice but containing more heat. The same happened to an even greater extent when water was boiled into vapor.

Black called this *latent heat* (from a Latin word for "hidden"), since the heat was there but did not make itself apparent in the form of temperature. The latent heat was not lost, of course, for when water vapor was condensed to water, or water was frozen to ice, the latent heat was given off again.

An understanding of latent heat was important in the improvement of the steam engine a few years later.

IN ADDITION

On January 5, 1762, the Russian czarina, Elizabeth, died, and was succeeded by her son, Peter III (1728–1762). A wild admirer of Frederick II, Peter III promptly deserted his allies and joined Frederick II, but within half a year he was deposed by his wife and was murdered. His wife, of German origin and infinitely more capable than her half-mad husband, reigned as Catherine II (1729–1796).

1763

POLLINATION

The existence of sexuality in plants must have seemed strange, since plants, being essentially immobile, cannot indulge in the kind of sex that animals do.

In 1763, however, a German botanist, Josef Gottlieb Kohlreuter (1733–1806), pointed out the manner in which plant pollen can be blown by the wind to reach female organs in a purely random manner. Because of this randomness, plants that depend on wind pollination must produce pollen in vast amounts.

He also pointed out that a more efficient process is to have bees or other similar animals do the job. A bee enters a flower in search of nectar (the bribe to make it enter). Pollen coats the bee's fuzzy covering, and when the bee visits the next flower, that pollen rubs off on the pistil.

IN ADDITION

The Seven Year War ended with the Treaty of Paris on Feburary 10, 1763, and the Treaty of Hubertusburg on February 15. France was the big loser. Great Britain took all of Canada from France and all of Louisiana east of the Mississippi River, together with Florida, from Spain. Spain, as compensation, received Louisiana west of the Mississippi. The result was that France, except for a few islands, was evicted from North America. As for Europe, Prussia retained Silesia, but Frederick II was so disenchanted by his narrow escape from destruction that he carefully maintained peace for the second half of his reign.

Even as France was departing from North America, the city of St. Louis, named for Louis IX, was being founded by a French fur trader, Pierre Laclede Liguest (1724?–1778).

1764

STEAM ENGINE

For half a century, the Newcomen steam engine had been used by miners despite its inefficiency. In 1764, a steam engine of this sort was given to a Scottish engineer, James Watt (1736–1819), to repair.

The repair was easy, but Watt wanted to improve it. He had learned of latent heat from his friend Black (see 1762), and he realized how wasteful it was to keep heating, cooling, and reheating the same chamber. The thought came to him, then, of having two chambers. One would always be kept hot and the other would always be kept cold. While the steam was doing its work, it would be in the hot chamber, and when it had to be condensed, a system of valves would bleed it into the cold chamber for condensation while more steam was formed in the hot chamber. That was the beginning of the first reasonably efficient steam engine.

IN ADDITION

The Seven Years War had left the British treasury with a load of debt. The taxes on the British public were already the highest in Europe and new sources of revenue were needed. The British government turned its eyes toward the North American colonies, which had, after all, benefited most from the removal of the French menace. That removal, however, had lessened colonial dependence on British protection, and the colonists were totally unenthusiastic at the prospect of taxation. The stage was set for conflict.

1765

PLUTONISM

The French geologist Nicolas Desmarest (see 1756) was interested in the changes taking place on the face of the Earth. He was the first to maintain that valleys had been formed by the streams that ran through them.

In 1765 he carried forward the ideas of Guettard (see 1752). Not only did he maintain that heat was the source of change, but he saw it as having been applied through volcanic action. He maintained that basalt was a rock of volcanic origin and that large sections of France's rocks consisted of ancient lava flows. He was a *Plutonist,* in other words (from the Greek god Pluto who was lord of the underworld). However, most geologists remained under the spell of a German geologist, Abraham Gottlob Werner (1750–1817), who was a *Neptunist* (see 1752) and insisted that water was the source of all change in the Earth's surface.

IN ADDITION

Searching for revenue, the British Parliament passed a *Stamp Act*, under which the colonists were expected to pay for revenue stamps on newspapers, legal documents, pamphlets, almanacs, playing cards, and similar items. It was the first direct tax (as opposed to tariffs) that Parliament had placed on the colonies, and they exploded with resistance. An organization called the *Sons of Liberty* sprang up to combat British interference in American affairs.

By 1765 Philadelphia had reached a population of 25,000 and was larger than any English-speaking city except London.

1766

HYDROGEN

Black's work with carbon dioxide (see 1754) had led to great interest in gases on the part of chemists. In 1766 the British chemist Henry Cavendish (1731–1810) found that some metals, when acted on by acid, liberated a gas that was highly inflammable and which he therefore called *fire air*. (We now call the gas *hydrogen*.)

Actually, early experimenters, notably Boyle (see 1661), had obtained the gas, but Cavendish was the first to study it carefully and report on its properties, so he is usually given the credit for having discovered it.

Cavendish measured the weight of particular volumes of different gases in order to determine their densities. He found this new gas to be only one-fourteenth the density of air, and no substance under ordinary conditions has ever been found to be less dense.

NERVES

From Greek times, nerves had been thought to be hollow tubes that carried some sort of subtle fluid, perhaps in analogy to veins and arteries.

A Swiss physiologist, Albrecht von Haller (1708–1777), dismissed this possibility and decided to reach no decisions on nerves that could not be demonstrated by experiment. His experimental work, which he published in 1766, showed that muscles were irritable; that is, a slight stimulus to the muscle would produce a sharp contraction. He also showed that a stimulus to a nerve would produce a sharp contraction in the muscle to which it was attached. The nerve was the more irritable and required the smaller stimulus.

Haller judged, therefore, that it was nervous stimulation that controlled muscular movement. He also showed that the tissues themselves do not experience a sensation but that the nerves channel and carry the impulses that produce the sensation.

Furthermore, Haller showed that nerves all led to the brain or to the spinal cord, which were thus clearly indicated as the centers of sense perception and responsive action. For all this, he is considered the founder of modern *neurology*.

IN ADDITION

The British Parliament rather sullenly repealed the Stamp Act but insisted that it had the right to tax the colonists.

1768

SPONTANEOUS GENERATION

A century before, Redi (see 1668) had shown that maggots were not generated spontaneously but were born of the eggs of flies. That might disprove the spontaneous generation of organisms large enough to see, but since then, scientists had learned of myriads of microorganisms. Could these arise through spontaneous generation?

The British naturalist John Turberville Needham (1713–1781) had placed mutton broth in a glass container, killed all microorganisms by heating, and then sealed the container. He reported that a few days later, the broth contained numerous microorganisms that must have developed spontaneously. This experiment, conducted in 1740, seemed impressive, but there was some suspicion that he had not heated the broth sufficiently and had not managed to kill all the microorganisms.

In 1768 the Italian biologist Lazzaro Spallanzani (1729–1799) decided to repeat the experiment and make sure. He boiled his solution for one-half to three-quarters of an hour, then sealed the flask. No new microorganisms appeared thereafter, though they did appear if he boiled the solution for shorter periods.

The conclusion was that there was no spontaneous generation even for microorganisms.

AUSTRALIA

Transits of Venus occur in pairs, eight years apart, with over a century between pairs. Since there was one in 1761, there would be another in 1769. In 1768, therefore, the English navigator James Cook (1728–1779), usually called Captain Cook, was sent out on a voyage to the Pacific Ocean. He was to observe the transit from the newly discovered island of Tahiti.

In the course of this voyage, he was the first to explore, thoroughly, the shores of Australia and to get an idea of its size. Though earlier explorers had spied its shores, it was Cook whose reports were sufficiently detailed to bring the land clearly to European attention. For that reason, he is usually considered the discoverer of Australia. This, and two more voyages in the years following, made Cook the most famous navigator since Magellan (see 1523). Cook crisscrossed the Pacific Ocean and finally demonstrated that it had no significant land areas in it. With Australia, the last major habitable land area on Earth had been discovered.

In the course of his voyages, Cook made use of Lind's dietary discovery (see 1747) and lost only one man to scurvy.

CARBONATED WATER

An English chemist, Joseph Priestley (1733–1804), grew interested in gases in 1768, partly because he lived next door to a brewery and was able to make use of quantities of the carbon dioxide generated in the brewing process.

He found that when he dissolved some of the carbon dioxide in water *(carbonated water)*, he created what he considered a pleasantly tart and refreshing drink, the one we call seltzer, or soda water, today. Since it required only flavoring and sugar to produce soda pop, Priestley may be viewed as the father of today's gigantic soft-drink industry.

IN ADDITION

Under the leadership of the British parliamentarian Charles Townshend (1725–1767) and with the enthusiastic support of George III, Parliament placed a series of tariff duties on various items. These were indirect taxes, and it was assumed the colonists wouldn't object to them. However, they did—and vigorously.

An English astronomer, Charles Mason (1728–1786), and an English surveyor, Jeremiah Dixon (d. 1777), completed the survey of the boundary between Pennsylvania and Maryland in 1767. That boundary was called the *Mason-Dixon line* and became celebrated eventually as the boundary between the free states and the slaves states.

After the French and Indian War, Parliament had decreed the land west of the Alleghenies off limits to European colonization. The colonists simply paid no attention. In 1767, for instance, the American pioneer Daniel Boone (*ca.* 1734–1820) passed through the Cumberland Gap into what is now Kentucky and settled there.

1769

QUANTITATIVE CHEMISTRY

Although Black had demonstrated the usefulness of quantitative measurements in chemistry (see 1754), what made them an integral part of the science was the work of French chemist Antoine-Laurent Lavoisier (1743–1794), who is universally considered the *father of modern chemistry*.

There were at this time some who still clung to the Greek theory of elements and their changeability. They argued that if water were boiled for a long time, a sediment appeared, and that this was clearly a conversion of water into a kind of earth, in line with Greek thinking.

Lavoisier decided, in 1769, to test this matter. He boiled water for 101 days in a device that condensed the water vapor and returned it to the flask, so that no water was lost in the process. He weighed both water and vessel before and after the boiling.

Sediment did appear, but the water did not change its weight during the boiling, so the sediment wasn't formed out of the water. The flask itself, however, had lost an amount of weight just equal to the weight of the sediment. In other words, the sediment was material from the glass, slowly etched away by the hot water and precipitating in solid fragments. Here was a clear-cut example of the way in which observation, without measurement, could be useless and misleading.

SPINNING FRAME

The textile industry was increasingly important for Great Britain, and any mechanical device that could increase production would add to the wealth of those who owned the mills and factories.

In 1769 an English inventor, Richard Arkwright (1732–1792), invented a spinning frame: a mechanical device that could produce cotton threads hard enough and firm enough to be used in textile manufacture. The frame helped produce textiles not only faster than handworkers could but so simply that Arkwright could work with relatively unskilled operators who would do the work for little money. Arkwright died a millionaire, but at a cost. Many were thrown out of work by his invention and went hungry, because the government at this time did not consider itself responsible for the welfare of its citizens.

IN ADDITION

The Spanish, in a sudden burst of energy, actively colonized the California coast. San Diego, Los Angeles, and San Francisco were all founded now. Many missions were established by Junipero Serra (1713–1784) under the sponsorship of Charles III of Spain (1716–1788) who became king in 1759.

In 1769 Captain Cook (see 1768) charted the coasts of New Zealand. Louis Antoine de Bougainville (1729–1811) completed the first circumnavigation of the globe by a Frenchman.

1770

NILE RIVER

The coasts of the continents were relatively easy to explore, but the interior represented greater toil and greater risks. This was particularly true of Africa. Its shores had been the first to be explored by the Portuguese explorers, but its interior was the last to be brought into the light of known geography, so that for a long time it was known as the *Dark Continent*.

Most paradoxical was the matter of the Nile River. On the banks of its northernmost portion, one of the two oldest civilizations of the world flourished, yet neither the Egyptians nor any of the civilized people that followed them had ever discovered the source of the Nile. This becomes less surprising when it is understood that the Nile is the longest river in the world. It flows in a fairly straight line from south to north, and the source is over 4,000 miles from the mouth. The ancient Egyptians had made their way upriver for about 1,500 miles before giving up the search.

In 1770, however, the Scottish explorer James Bruce (1730–1794) made his way upstream to Khartoum in the Sudan. There two rivers join, the White Nile coming up from the southwest and the Blue Nile from the southeast. Bruce followed the Blue Nile and finally found its source in Lake Tana in what is now northwestern Ethiopia.

This seemed at the time to solve the problem of the source of the Nile, but the White Nile, being the longer of the two, is the main river, and its source remained unknown for another century.

GULF STREAM

There are currents in the water that are as important to navigation as the winds are, though they are less noticeable.

Franklin (see 1733) had traveled from America to Europe several times and his ever-inquiring mind became aware of the difference in speed in the two directions. He was the first to study ships' reports seriously and to query the experience of whalers. As a result he found that there was a current of warm water coming up from the Gulf of Mexico (the *Gulf Stream*) and then crossing the North Atlantic toward Europe. It sped the ships sailing eastward to Europe and slowed those sailing westward to North America. Franklin mapped it and in this way showed British navigators the routes to avoid if they wished to make good time westward. This was the beginning of the scientific study of ocean currents.

Water currents did not concern only navigators. Labrador and Great Britain are at precisely the same latitude on opposite sides of the North Atlantic Ocean. However, the Gulf Stream, a warm-water current, bathes the British coasts, while a cold-water current from the Arctic bathes the Labrador coast. For that reason, Great Britain has a mild climate and is populated by tens of millions, while Labrador is frigid and is populated by tens of thousands.

SOLUBLE GASES

Priestley (see 1768) was the first to bubble gases into an upside-down mercury-filled vessel with its mouth in a tub of mercury. In this way, gases could be collected that could not be collected through water because they dissolve in the water. By 1770 Priestley had collected and studied the water-soluble gases we know as ammonia, sulfur dioxide, and hydrogen chloride.

IN ADDITION

In Boston, on March 5, 1770, British soldiers were harassed by colonials and fired in self-defense, killing five colonials. News of the *Boston Massacre* spread throughout the colonies and created a fever of anger against the British.

The Townshend duties (see 1768) were repealed, but the duty on tea remained. It was a very small one and was kept merely to maintain the principle that Parliament could legally tax the colonies.

The population of the thirteen American colonies now stood at about 2.2 million.

1771

NEBULAS

The hungry search for comets became so important to some astronomers that it seemed to consume them. One such was the French astronomer Charles Messier (1730–1817), who discovered twenty-one comets, but who was depressed whenever some other astronomer discovered one he had missed, or when he was kept away from his telescope as, for instance, when his wife was on her deathbed.

He was irritated by the false hopes aroused by the sighting of some fuzzy object that turned out not to be a comet but a nebulosity fixed to some spot in the heavens. In 1771 he published a list of 45 of these nebulas, which in later years he expanded to 103.

As it turned out, the list of *Messier objects* turned out to be far grander and more important than comets, and Messier would be completely forgotten now if his comet discoveries were his only accomplishment. It is his list of things to be disregarded that has made his name immortal, for it includes numerous star clusters and distant galaxies.

PLANTS AND CARBON DIOXIDE

Carbon dioxide was well known by this time to support neither combustion nor animal life. Priestley (see 1768) thought of checking to see whether plants, as well as animals, were unable to live in carbon dioxide.

In 1771 he placed a lit candle in an enclosed volume of air until the candle would burn no more and the air was filled with carbon dioxide. He then placed a sprig of mint in a glass of water and placed it in the enclosed air.

The plant did not die. It lived there for months and seemed to flourish. What was more, at the end of that time, a mouse could be put into that volume of air and live—and a candle would burn.

Whatever it was about air that supported combustion and animal life, and that was converted into carbon dioxide by burning candles and breathing animals, was *restored* by plant life. This was the first indication that plants and animals formed a chemical balance that kept Earth's atmosphere breathable.

IN ADDITION

Russia had been fighting Turkey for several years and making large advances in the Ukraine. In 1711, the Russians occupied the Crimean peninsula, the very last stronghold of the Tatars, who had taken Russia more than five centuries before. Austria and Prussia were alarmed at the growing strength of their eastern neighbor and sought for some way to gain territory of their own as a balance.

The first edition of the *Encyclopaedia Britannica*, in three volumes, was published in 1771.

1772

COMBUSTION

Chemists' understanding of combustion at this time was based on a theory first propounded in 1700 by a German chemist, George Ernst Stahl (1660–1734). He suggested that objects that were combustible were rich in something he called *phlogiston*, from a Greek word meaning "to set on fire."

In the process of combustion, fuel lost its phlogiston, and eventually a residue was left that lacked phlogiston and would not burn. Stahl recognized that rusting was comparable to combustion. He believed metals to be rich in phlogiston and held that they gradually lost it as they were converted to rust.

The chief flaw in the theory was that when wood burned it lost much of its weight (presumably because of the loss of phlogiston), yet when iron rusted, it *gained* weight. In Stahl's time, however, it wasn't considered important to measure quantities exactly and chemists ignored this paradox.

Lavoisier (see 1769), however, believed in weighing. In 1772 he began heating objects in enclosed volumes and weighing that volume. For instance, he burned certain elements in air and found that the material produced was heavier than the elements themselves, though nothing within the container had changed weight at all. If the elements gained weight in burning, then something else must have lost weight, and the only something else this could be was the enclosed air. If some of the air had been absorbed by the burning elements, there should be a partial vacuum in the flask. Lavoisier opened the flask, and sure enough, air rushed in. The added weight of the inrushing air was equal to the weight gained by the element that had been burned.

By such experiments, Lavoisier concluded that combustion did not come about through the loss of phlogiston, but through the combination of the substance that was burning or rusting with some portion of the air. This killed the phlogiston theory, although a few important chemists continued to accept phlogiston for some decades longer.

DIAMOND

One of the substances that Lavoisier burned (see above) was a diamond. He and some other chemists invested in one, placed it in a closed vessel, and then focused sunlight on it with a magnifying glass. The diamond, when it grew hot enough, simply disappeared, and carbon dioxide appeared within the vessel. The conclusion was that diamond consisted of carbon and, despite appearances, was chemically very closely related to coal.

NITROGEN

At this time it was known that ordinary air supported combustion and animal life and that carbon dioxide did not. If candles were burned in an enclosed volume of air until no further burning was possible, had all the air become carbon dioxide?

Black (see 1754) asked a student of his, the British chemist Daniel Rutherford (1749–1819), to investigate the matter. In 1772 Rutherford, having burned a candle to extinction in a closed container, absorbed all the carbon dioxide that had been produced by combining it with certain chemicals. He found that there was still a great deal of gas left that was *not* carbon dioxide, which also supported neither combustion nor animal life. He had, in this way, discovered a new gas that eventually came to be called *nitrogen*. This is from the Greek, meaning "niter producer," since niter, a mineral properly called potassium nitrate, contains nitrogen.

IN ADDITION

Prussia and Austria worked out a way to match Russia's increasing might by increasing their own strength at the expense of Poland, which lay between them and Russia and which was virtually helpless. Prussia took West Prussia, which formed a connection between Brandenburg and Prussia (now called East Prussia). Austria took a chunk of southwestern Poland. To keep Russia quiet, it was given a piece of eastern Poland. This has come to be called the *First Partition of Poland* and was an exercise in conscienceless might.

1773

ANTARCTIC CIRCLE

The Arctic Circle was first crossed at sea by Ottar (see 870), though unnamed primitives must surely have ventured past it on land earlier. However, the Arctic Circle passes through northern North America, Europe, and Asia, and once the glaciers had retreated, the line was open to venturesome hunters.

The Antarctic Circle, however, lies far to the south of any populated area. The closest is Tierra del Fuego at the southern tip of South America, which even at its most southerly is still about 650 miles north. It is safe to say, then, that no human being, primitive or civilized, had ever crossed the Antarctic Circle prior to 1773.

In that year, Captain Cook, on his second voyage through the Pacific in search of an important land area, ventured far to the south and on January 17 (at the height of the Antarctic summer) crossed the Antarctic Circle, a true first for humanity.

BACILLI

Bacteria had just barely been made out by Leeuwenhoek nearly a century before (see 1683). Since then, microscopists had not been able to do more than just barely see them. In 1773, however, a Danish biologist, Otto Friedrich Muller (1730–1784), managed to make them out well enough to divide them into categories. Some looked like little rods and he called them *bacilli* (from a Latin word for "little rod"); some were curved into spiral shapes and he called them *spirilla*.

He was the first to classify microorganisms generally into genera and species after the fashion of Linnaeus (see 1735), but even he couldn't do very much with the tiny specks that were bacteria.

IN ADDITION

The East India Company had a vast surplus of tea, so the British Parliament reduced the duty on tea even further and attempted to dump it on the colonies. The colonists objected not so much to paying the money as to the principle of taxation. Since they were not represented in Parliament, they felt Parliament had no right to tax them. The tea was therefore refused when it arrived in various colonial ports. In Boston, on December 16, 1773, a party of radicals organized by Samuel Adams (1722–1803) and John Hancock (1737–1793) dressed up as Indians, went aboard the tea ships, and threw 342 chests of tea overboard. This was the *Boston Tea Party*.

1774

OXYGEN

Priestley worked with mercury in his collection of gases (see 1770) and could not help using it more directly in his experiments. Mercury, when heated in air, will form a brick-red compound, which we now call *mercuric oxide*. Priestley heated some of this compound in a test-tube by using a lens to concentrate sunlight upon it. When he did this, the compound broke up, liberating mercury, which appeared as shining globules in the upper portion of the test-tube.

In addition, a gas was given off that possessed most unusual properties. Combustibles burned more brilliantly and rapidly in it than they did in ordinary air. Mice placed in an atmosphere of this gas were particularly frisky, and Priestley himself felt "light and easy" when he breathed it.

When Lavoisier heard of the experiments of Priestley and of Rutherford (see 1772), he realized in the light of his own experiments that air must consist of a mixture of two gases: one-fifth was Priestley's gas, which Lavoisier named *oxygen* (from Greek words meaning "acid producer," because it was mistakenly felt at the time that all acids contained oxygen), and four-fifths were Rutherford's gas, which Lavoisier named *azote* (from Greek words meaning "no life"), but which later came to be known as nitrogen.

Clearly, it was oxygen that supported combustion and animal life and oxygen that was involved in rusting. Animals must consume oxygen and produce carbon dioxide, and from Priestley's earlier experiment (see 1771), plants must consume carbon dioxide and produce oxygen. Between these two forms of life, the atmosphere's makeup remained stable.

CHLORINE

In a classic case of scientific misfortune, the Swedish chemist Carl Wilhelm Scheele (1742–1786) had discovered oxygen at least two years before Priestley did, and by the same method. However, Scheele's discovery was not published (through the negligence of a publisher) until after Priestley's discovery had been reported, so Priestley gets the credit.

Scheele, however, discovered many simple compounds from plants and animals, to say nothing of such poisonous gases as hydrogen fluoride, hydrogen sulfide, and hydrogen cyanide. He was also involved in the discovery of a number of elements, though he never managed to get undisputed credit for a single one of them.

Thus, by 1774 he had done most of the spadework that led to the discovery of the element manganese. His friend, the Swedish mineralogist Johan Gottlieb Gahn (1745–1818), however, took the final step and gets credit for the discovery.

Also in 1774, Scheele isolated the gas chlorine, which was unusual in not being colorless. Chlorine is greenish-yellow in color, and its name, in fact, is derived from the Greek word for "green." The trouble was that Scheele did not recognize chlorine to be an element but thought it was a combination of some substance with oxygen. Chlorine was found to be an element about thirty years later, and it was to that finder that credit for the discovery was awarded.

MIND AND DISEASE

Throughout history, diseases had been cured (so it was said) by various mystic rites, by the laying on of hands, the saying of prayers, and so on.

In 1774 the German physician and mystic Franz Anton Mesmer (1734–1815) began to apply science to this by waving magnets over his patients, and effecting cures in some cases. Later, he discovered that magnets were unnecessary and that the same happy results could be achieved by simply passing hands over the patient. He claimed to be making use of *animal magnetism*.

Naturally, in some cases he did not effect a cure, and he was driven out of Vienna, where he first practiced, as a quack. He went to Paris, where again he was first popular and then forced to leave. His methods were examined by such men as Franklin and Lavoisier.

Franklin, for one, although he condemned Mesmer's mysticism, recognized that the mind influenced the body: that it could cause disorders and that it could be used to correct them.

What Mesmer was dealing with (without understanding it as well as Franklin did) were psychosomatic ailments, where often it is only necessary for patients to believe they will be cured in order to be cured. Mesmer's methods, refined and freed of some of their mumbo jumbo, became respectable half a century later as hypnotism, and some of them entered psychoanalysis later still.

IN ADDITION

News of the Boston Tea Party (see 1773) angered George III enormously, and on March 31, 1774, Parliament passed a series of coercive measures that, among other things, closed the port of Boston, with the apparent intention of starving it out. However, food was sent to Boston from other parts of the colonies and anger against Great Britain further increased.

Louis XV of France died on May 10, 1774, and was succeeded by his grandson, Louis XVI (1754–1793), and his queen, Marie Antoinette (1755–1793).

1775

DIGITALIS

The earliest medicines were those obtained from plants of various sorts, and we have records of these herbal remedies from as far back as Dioscorides (see 50). In early modern times, physicians tended to scorn plant products, especially since those who knew most about such herbal remedies were uncivilized people or, among the civilized, old women who had picked up knowledge from old women who had preceded them. (The learned knew well how little to think of "old wives' tales.")

Nevertheless, there was something to be learned from such tales. In 1775 an English physician, William Withering (1741–1799), was persuaded to try the juices of the foxglove plant for the treatment of edema ("dropsy"), which was caused by a failing heart that lacked the power to circulate the body fluids properly. He reported on this, and added the very useful drug *digitalis* to the pharmaceutical armory of physicians.

IN ADDITION

Boston was under martial law maintained by a British general, Thomas Gage (1721–1787). When the British marched to Concord on April 19, 1775, to arrest Samuel Adams and John Hancock and confiscate the arms gathered by the colonists there, Paul Revere (1735–1818) and others warned the colonists, and the British found they would have to fight. The battles of Lexington and Concord, which the British lost, started the War of the American Revolution. The British lost heavily again at the Battle of Bunker Hill on June 17, 1775.

1776

RACES

Europeans were always aware that groups of people differed generally in appearances. The dark Mediterranean people were aware that the Germanic peoples were generally taller and blonder. Europeans noted the short height and oddly shaped eyelids of Asian invaders such as Huns and Mongols, and from ancient times they knew of the dark-hued Africans.

The first orderly classification of these varieties came in 1776, when a German anthropologist, Johann Friedrich Blumenbach (1752–1840), divided the human species into five races: Caucasian (Europeans), Mongolian (East Asians), Malayan (Southeast Asians and Pacific Islanders), Ethiopians (sub-Saharan Africans), and Americans (Native Americans). A more unsophisticated way of dividing them is to call them, respectively: the White, Yellow, Brown, Black, and Red races.

The fact remains, though, that all human beings, however superficially different in appearance they may be, are members of a single species and can freely interbreed. There is no indication that there is any *important* difference between any of these groups, either physical, mental, or psychological.

Blumenbach's divisions were far too general and simplistic, for each group can be divided into subgroups, and the Native Australians are left out altogether. Blumenbach's focus on superficial differences of skin color, hair form, and eyelid shape merely served to make it easier to express racist views in scientific-sounding language. The dark evils of slavery and of ethnic persecutions were thus made to sound lofty and biologically inevitable.

IN ADDITION

The Second Continental Congress, which met in Philadelphia on May 10, 1775, voted a Declaration of Independence, which was signed by John Hancock on July 4, 1776. This is usually taken to represent the beginning of the United States of America and from now on the colonials will be referred to as *Americans*.

The Americans were fighting with rifles against the British muskets and could inflict heavy casualties, but most of the Americans were short-term amateur volunteers who could not usually stand against the well-trained British professionals.

Philadelphia, with a population of 40,000, was now by far the largest city in the United States.

1777

TORSION BALANCE

From time immemorial, people had used two pans at equal distances from a fulcrum to weigh things (see 5000 B.C.). The object to be weighed was placed in one pan and weights were added to the other till the pans were in balance. In fact, the instrument was called a balance.

In 1777, however, the French physicist Charles-Augustin de Coulomb (1736–1806) showed that it took a certain amount of force to twist (that is, to apply torsion to) a fiber or a wire, and that the amount of torsion was proportional to the amount of force. Since weight is a force, such a *torsion balance* could be used to measure tiny weights with considerable delicacy.

IN ADDITION

The British plan was now to take the Hudson River Valley and split the United States in two. A British force headed south from Canada, reached Saratoga under great difficulties, and there, on October 4, 1977, was defeated by and surrendered to the American forces. This was the turning point of the war. The surrender of a British army in the field was an all but unprecedented event.

Meanwhile, though, the British, under William Howe (1729–1814) captured Philadelphia, the American capital, on September 26, 1777. Congress fled to York, then to Lancaster, Pennsylvania. Howe defeated Washington, just north of New York City on October 4, 1777. Washington was forced to retire to Valley Forge, west of Philadelphia, on December 14, 1777.

1778

MOLYBDENUM

Scheele, using methods that had worked in isolating manganese (see 1774), managed to isolate another metal, molybdenum. With his usual incredible run of bad luck, Scheele often loses the credit for the discovery, which is given to another friend of his, the Swedish mineralogist Peter Jacob Hjelm (1746–1813).

HAWAIIAN ISLANDS

Captain Cook, in his final voyage, explored the Pacific coast of North America north of California. This formed the basis for the later British claim to the region. In January 1778 he discovered the Hawaiian Islands, which he named the *Sandwich Islands* after John Montagu, fourth earl of Sandwich (1718–1792).

The earl of Sandwich was such an inveterate gambler that he had meat placed between two slices of bread so that he could eat with one hand and not be forced to leave the gaming table. It was in this way that the *sandwich* was invented. He was also one of the great anti-American "hawks" in the British government.

IN ADDITION

The French recognized American independence and made a treaty of alliance with the United States on February 6, 1778.

Meanwhile, the winter at Valley Forge was a dreadful one for Washington and his army, which was almost destroyed by cold and hunger. They endured, however, and a Prussian soldier, Friedrich Wilhelm von Steuben (1730–1794) drilled them in the spring and polished them into something like professional shape. When the British army, now under Henry Clinton (1738–1795), left Philadelphia and marched across New Jersey toward New York, Washington's army overtook them and fought them to a draw at the Battle of Monmouth on June 28, 1778. Washington would have won but for the treasonous behavior of his general, Charles Lee (1731–1782).

1779

FERTILIZATION

In early times, it was taken for granted that male human beings provided the "seed" and that females were merely the soil within which the seed developed. If no children were born, it was assumed that the woman involved, like desert soil, was "barren."

In 1779 Spallanzani (see 1768) studied the development of eggs. At that time, it was thought that the ovarian follicles (discovered by the Dutch anatomist Reinier de Graaf [1641–1673] in 1673 and still called *Graafian follicles* as a result) were the eggs.

Spallanzani showed that fertilization did not take place unless the sperm cells in the semen actually made physical contact with the follicles. This was a strong indication that reproduction was not a one-sided affair, but that both mother and father contributed to the birth of a child and that either side might be at fault in case children were not born.

PHOTOSYNTHESIS

Joseph Priestley had shown that plants could make air breathable after it had been filled with carbon dioxide (see 1771). In 1779 a Dutch physician, Jan Ingenhousz (1730–1799), repeated the experiments and confirmed Priestley's findings. His crucial added discovery, however, was that plants consumed carbon dioxide and produced oxygen only in the light. In the dark, plants, like animals, consumed oxygen and produced carbon dioxide.

Because of the importance of light, and because plants in the process produced not only light but also the large molecules of their tissues, the process came to be called *photosynthesis*, from Greek words meaning "to put together by light."

IN ADDITION

Spain came to the aid of the Americans (more out of enmity to Great Britain than out of friendship to the United States) and declared war on Great Britain on June 21, 1779.

Meanwhile, the American navy (such as it was) was doing well. The daring naval commander John Paul Jones (1747–1792) even raided the British coasts. In his ship *Bonhomme Richard* (*Poor Rich-*

ard in honor of Franklin—see 1733), he sank the British warship *Serapis* on September 23, 1779. When the *Serapis* had called on him to surrender in the first stage of the battle, he sent back the message, "I have not yet begun to fight."

1780

ELECTRICAL STIMULATION

An Italian anatomist, Luigi Galvani (1737–1798), noticed in 1780 that the muscles of dissected frog legs twitched wildly when a spark from a Leyden jar struck them. This was not too surprising. Electric shocks made living muscles twitch, why not dead ones, too?

Since Franklin had shown that lightning was electrical in nature (see 1752), Galvani wondered whether muscles would twitch if exposed to a thunderstorm. He therefore placed frog muscles on brass hooks outside the window so that they rested against an iron latticework.

The muscles did indeed twitch during the thunderstorm, but they also twitched in the absence of it. In fact, they twitched whenever they made simultaneous contact with two different metals.

Apparently, electricity was involved, but where did it come from, the metals or the muscle? Galvani decided it was the muscle, and he spoke of *animal electricity*. In this he was wrong, but electricity *was* involved with nerve and muscle action just the same.

IN ADDITION

Benedict Arnold (1741–1801), who had fought marvelously for the Americans and been treated poorly by them, allowed his grievances to get the better of him. He had Washington place West Point in his charge, and he then conspired with the British to give up that strong point in return for money and military advancement. The plot was uncovered, but Arnold got away. However, the name *Benedict Arnold* has become a byword for treason and treachery.

On October 7, 1780, the Americans won a spectacular, if small, victory at the Battle of Kings Mountain, and the tide began to turn in the south.

A week of anti-Catholic rioting began in London on July 2, 1780, over certain parliamentary acts giving Catholics some freedom of religion. The *No Popery* riots were put down, but the cause of religious freedom was set back.

Maria Theresa of Austria died on November 28, 1780, and her son, Joseph II (1741–1790), could now do as he wished. He was the most advanced of the "benevolent despots." He abolished serfdom in his dominions, suppressed convents, and established religious toleration. However, the people clung to their old ways and Joseph II failed completely. He had to put things back the way they were.

1781

URANUS

Since prehistoric times, human beings had been aware of the five bright starlike planets: Mercury, Venus, Mars, Jupiter, and Saturn. With the coming of Copernican principles (see 1543), Earth itself became a sixth, coming between Venus and Mars. It seemed somehow impossible that there could be another. Surely, if there were another, it would have been seen.

In the 1770s, a British astronomer (of Hanoverian birth) began studying the heavens. He was William Herschel (1738–1822). A music teacher by profession, he had grown interested in astronomy. Unable to buy a telescope he considered sufficiently good, he took to grinding his own lenses and mirrors and ended with the best telescopes of his time.

He determined to study, systematically, everything in the sky, and on March 31, 1781, he came across an object that appeared as a disc instead of as a mere point of light. He supposed it to be a comet, but the disc had sharp edges like a planet and wasn't fuzzy as a comet would be. Furthermore, when enough observations had been made to calculate an orbit, he found that orbit to be nearly circular, like a planet's, and not elongated, like a comet's. What's more, it was clear that the object's orbit lay far outside that of Saturn. It was twice as far from the Sun as Saturn was, and no comet could be seen from that distance.

The conclusion was that Herschel had discovered a seventh planet circling the Sun, which because of its great distance did not appear as bright as the others. In fact, it *was* visible to the unaided eye as a very dim star and had been observed a number of times by people who never suspected it might be a planet. The first had been Flamsteed (see 1676), who had recorded its position on his star map a century earlier and called it 34 Tauri.

After some hesitation, astronomers decided to continue naming planets for mythological characters, and the new planet was named Uranus, after the father of Saturn (Cronos) in the Greek myths.

The discovery of Uranus doubled the size of the Solar System at a stroke. It was further spectacular proof that the ancients had not known everything, and gave astronomers the exciting knowledge that there was more to be discovered in the sky than additional comets.

BINARY STARS

Astronomers were still trying to detect the parallax of the stars, the task that Bradley had failed at a half-century earlier (see 1728).

It struck Herschel (see above) that he might be able to detect parallax if he studied stars that were very close together in appearance. Both would be in the same line of sight, but if one was considerably brighter than the other, it might be considerably closer. In that case, the brighter star should show slight parallactic changes of position in the course of the year relative to the dim star.

Herschel began the study of such stars in 1781, and as time went on, he did notice that one of the stars changed position relative to the other in a number of cases. In no case, however, was this change quite what one would expect if it had resulted from the motion of the Earth. In time, astronomers were forced to the conclusion that some double stars were actually close to each other in reality and not merely in appearance and that they moved about each other. Herschel had discovered *binary stars,* from a Latin word referring to objects that exist in pairs.

Newton had presented his law of *universal* gravitation, assuming that it was a force of attraction between any two objects in the Universe, but it had actually been tested only on objects within the Solar System. Now for the first time it could be tested on distant stars, where it was eventually proved to be truly universal.

CRYSTALLOGRAPHY

Crystal is from a Greek word for "ice" or "frozen." Since ice is sometimes transparent, the word *crystal* came to be used for any transparent object. Thus, a fortune-teller uses a "crystal ball" that is a ball of ordinary glass. The planets were thought to be set in "crystalline spheres" because those spheres were transparent.

When quartz was found, it had the properties of rocky material, but it was *transparent.* Since bits of quartz often had straight edges, planar faces, and sharp angles, people began to speak of any solids with straight edges, planar faces, and sharp angles as crystals, even when they were not transparent.

In 1781, a French mineralogist, René-Just Haüy (1743–1822), was handling a piece of calcite crystal that had a rhombohedral shape (a kind of slanted cube). Accidentally, he dropped it and it shattered. Haüy noticed that the pieces all had a rhombohedral shape. He broke more pieces of calcite and found that, no matter what the original shape was, it broke into rhombohedrals.

Haüy suggested that each crystal was built of successive additions of what we now call a *unit cell,* which formed, in the absence of external interference, a simple geometrical shape with constant angles. He felt that an identity or difference in crystalline form implied an identity or difference in chemical composition.

In this way, Haüy founded the modern science of crystallography.

AXIAL INCLINATION OF MARS

The Earth's axis is tipped 23.5 degrees to the perpendicular of the plane of its orbit. It is this which gives Earth its seasons. When the Earth is in that region of its orbit in which its North Pole is tipped toward the Sun, the northern hemisphere has its summer and the southern its winter. When the Earth is in the opposite region of its orbit, with the North Pole tipped away from the Sun, the northern hemisphere has its winter and the southern its summer.

Do other planets have similar characteristics? Herschel (see Binary Stars, above) was studying the rotation of Mars—the manner in which its markings moved about it, which Cassini had used to determine the length of the Martian day (see 1665). It seemed to Herschel that the markings had to move parallel to the Martian equator and that its axis of rotation would have to be perpendicular to that. By determining the axis of rotation in this way, Herschel calculated that Mars had an axial inclination of just about 24 degrees, nearly that of Earth's. This was one more way in which Earth resembled other planets.

STEAM ENGINE

Since Watt had thought of having a hot and a cold chamber in a steam engine (see 1764), he kept introducing improvements. He arranged to have the steam push into a chamber on each side of the piston alternately so that it would be pushed rapidly in both directions rather than in only one.

In 1781 he devised mechanical attachments that ingeniously converted the back-and-forth movement of a piston into the rotary movement of a wheel, so that the steam engine could be made to power a variety of activities and machines.

The now-versatile steam engine was the first of the modern *prime movers,* the first modern device, that is, that could take energy as it occurred in nature and apply it to the driving of machinery. To be sure, wind and running water were the prime movers of ancient times, but wind is erratic, and running water only exists in certain places. The use of fuel is certain; it *always* contains energy. What's more, it can be used anywhere and in any quantity within reason.

The steam engine, bringing the use of energy to all mechanical devices in far greater quantity than anything else had offered in the past was the key to all that followed rapidly under the name of the *Industrial Revolution,* when the face of the world was changed as drastically (and far more rapidly) than at any time since the invention of agriculture, nearly ten thousand years before (see 8000 B.C.).

IN ADDITION

On October 19, 1781, the British, under Charles Cornwallis (1738–1805), were forced to surrender, and even the most stubborn of the British, even George III, saw that the war could not be won. Nevertheless, the fighting continued in desultory fashion for a considerable time.

The Articles of Confederation were accepted by the various states on March 1, 1781, and the states began to abandon their claims to western territory. This was essential to the life of the United States. Had the states perpetually quarreled over western territory, the nation would not have long endured. As the war drew to a close, the new nation had a population of about 3.5 million.

1782

ECLIPSING VARIABLES

Early on, a few stars had been noticed that were *variable;* that is, had light that dimmed and brightened instead of remaining constant.

The first variable star to be discovered was Omicron Ceti. In 1596 the German astronomer David Fabricius (1564–1617) noticed its changing light intensity. The star was eventually named *Mira* (from a Latin word meaning "wonderful") by a German astronomer, Johannes Hevel, known as Hevelius (1611–1687).

Another variable star was noted in 1672 by an Italian astronomer, Geminiano Mantanari (1633–1687). That was Beta Persei, better known as Algol (from an Arabic word meaning "the ghoul," because it represented the head of the monster Medusa in the constellation). Algol did not vary in brightness as much as Mira did, but Algol's variations, unlike Mira's, were quite regular.

In 1782 Algol was studied by a British astronomer, John Goodricke (1764–1786), a deaf-mute. To account for the regularity of Algol's variations, he suggested that it might have a dim companion that circled it in our line of sight and periodically eclipsed it, hiding most of the light.

In this, he proved to be perfectly right, but not all variable stars are *eclipsing variables.* Mira isn't. Its variations are so irregular that no eclipse can possibly be involved.

IN ADDITION

Great Britain was ready to negotiate a peace with the United States. George III's failure in North America had put an end to his hope to "be a king," and the British prime ministerial form of government was saved.

British forces had evacuated Savannah, Georgia, and Charleston, South Carolina, by the end of the year. Spanish forces took over Florida from the rather uninterested British.

Meanwhile, British forces were fighting Indian rulers in India. Led by the British statesman Warren Hastings (1732–1818), they were having rather better luck than they were having with the United States.

1783

MOTION OF THE SUN

To the ancients, the Earth was the motionless center of the Universe. To the early moderns, the Sun was.

In 1783, however, Herschel began a systematic measurement of the proper motion (see 1781) of a great many stars. Very dim and very distant stars moved so slightly they could be considered motionless and were references against which the perceptible motions of the nearer stars could be measured.

As the years passed, Herschel found that in one direction in the sky, the stars were generally moving away from each other, while in the opposite direction they were very slowly closing in toward each other. In 1805 he came to the conclusion that this could best be explained by supposing that the Sun itself was moving toward that point from which stars seemed to be separating and was moving away from the point toward which other stars seemed to be converging.

Just as Copernicus (see 1543) had maintained that the Earth was a planet like other planets, moving as they did, so Herschel now maintained that the Sun was a star like other stars and was moving as they did.

But if neither the Earth nor the Sun was the motionless center of the Universe, what was? There was no other candidate, and the question had to remain in abeyance for a time.

RESPIRATION AND COMBUSTION

Lavoisier, having worked out his theory of combustion as the combination of fuels with oxygen from the air (see 1772), thought of respiration. Animals ate food that contained carbon. They breathed air that contained oxygen and exhaled air that contained less oxygen and more carbon dioxide.

In collaboration with a French scientist, Pierre-Simon de Laplace (1749–1827), Lavoisier undertook a series of experiments designed to measure the amount of heat and carbon dioxide produced by a guinea pig. It turned out that the amount of heat was about what would be expected from the production of that much carbon dioxide, and Lavoisier concluded that respiration was a form of combustion.

The essential point was that the laws that governed combustion outside the body seemed to hold inside the body as well. This was another blow against vitalism, the system of thought that gave life special status.

BALLOON

Light objects can be borne on the air, as all of us observe in the case of small feathers, dandelion seeds, and so on. If objects are light enough, they needn't depend on wind and updrafts to remain in the air (or on muscular power as in the case of birds, bats, and insects). A light enough object can actually float in air, just as wood floats on water.

There are no known solids or liquids that are lighter than air, but some gases are. It occurred to two French brothers, Joseph-Michel Montgolfier (1740–1810) and Jacques-Étienne Montgolfier (1745–1799), that hot air expanded and was therefore lighter than an equal volume of cold air. If a quantity of hot air filled a light balloon, the hot air would float in the air and might have enough upward push to carry the balloon with it. In fact, if the balloon was large enough, there might be sufficient buoyancy to carry a human being upward.

On June 5, 1783, in the marketplace of their hometown, the brothers filled a large linen bag, 35 feet in diameter, with hot air. It lifted 1,500 feet upward and floated a distance of a mile and a half in 10 minutes. They went to Paris, and on September 19, they managed a flight of 6 miles before a crowd of three hundred thousand that included Benjamin Franklin.

The balloon carried up not only itself but a wicker basket in which were a rooster, a duck, and a sheep, which were not harmed. Finally, on November 20, a hot-air balloon carried into the air a French physicist, Jean François Pilatre de Rozier (1756–1783), and a companion. They were the first aeronauts in history.

Meanwhile, a French physicist, Jacques-Alexandre-César Charles (1746–1823), having heard of the hot-air balloons, realized that hot air had comparatively little buoyancy and lost that little as it cooled, though a fire in the basket underneath might suffice to keep it warm for a while. The gas hydrogen, studied by Cavendish (see 1766), was much lighter than hot air and had much greater buoyancy. What's more, the buoyancy was permanent. On August 27, 1783, Charles constructed the first hydrogen balloon and subsequently used one to rise nearly 2 miles in the air.

In subsequent decades, ballooning became almost a craze, and it was also used for scientific purposes.

TUNGSTEN

A Spanish mineralogist, Don Fausto D'Eluyar (1755–1833), analyzed a mineral called wolframite, which had been obtained from a tin mine, and in 1783 he obtained a new metal from it, which he called *wolfram*. The same metal is also called *tungsten* from the Swedish words meaning "heavy stone." That word derives from Scheele (see 1774), who investigated tungsten-containing minerals but, with his usual bad luck, missed spotting the new metal.

IN ADDITION

The Treaty of Paris was signed on September 3, 1783, and the War of the American Revolution came to an end. Great Britain recognized the independence of the United States, and its existence was now questioned by no one. However, the British did not evacuate New York City until November 25, 1783.

The United States consisted of thirteen colonies and all the land south of the Great Lakes and east of the Mississippi, but Florida and the coast of the Gulf of Mexico were returned to Spain, which therefore held the mouth of the Mississippi. The United States had an area of 850,000 square miles, which was nine times that of Great Britain. However, most of it was a wilderness, and its population, even including its slaves, was only half that of Great Britain.

Russia annexed the Crimea and now had a firm grip on the entire northern shore of the Black Sea.

1784

VOLCANOES

In 1783 a volcanco in Iceland erupted and killed one-fifth of the population of the island. That winter there was severe cold, and in 1784 Benjamin Franklin suggested that there might be a connection between the two events. Volcanic ash shot high into the atmosphere might reflect an abnormally high percentage of the Sun's radiation, putting the Earth in the shade, so to speak, and cooling it. This was the first speculation on a phenomenon that would be in the news beginning in the 1980s as *nuclear winter*.

BIFOCAL LENSES

Also in 1784, Franklin, who in his old age needed two sets of spectacles, one for distant vision and one for reading, grew tired of changing them frequently. He devised *bifocal spectacles,* the lenses of which were suited for distant vision above and near vision (reading) below.

HYDROGEN AND WATER

On January 15, 1784, Cavendish, still studying hydrogen (see 1766), noted that if it was burned in a container, liquid drops appeared in the cooler portions of that container. On investigation, the liquid turned out to be water. The conclusion was that hydrogen combined with oxygen to form water. When Lavoisier heard that, he gave the gas its present name of *hydrogen,* from Greek words meaning "water-former."

MARTIAN ICE CAPS

In 1784 Herschel, who had determined the axial inclination of Mars (see 1781) and therefore knew where its polar regions were, noticed that it had visible ice caps in those polar regions. Here was another similarity between Mars and Earth.

ALASKA

After Bering's discovery of the Bering Strait (see 1728), Russians from Siberia, venturing eastward, found large numbers of sea otters, whose pelts proved very profitable. In 1784 the Russians established the first European settlement in Alaska, and for the next eighty years Russian holdings expanded until all of the present-day state of Alaska was part of the Russian Empire.

TELLURIUM

An Austrian mineralogist, Franz Joseph Muller (1740–1825), was working with gold ore in 1782. In his studies, he produced a substance that he thought might be an as-yet-unknown element. Feeling unequal to the task of settling the matter, he sent the substance to a German chemist, Martin Heinrich Klaproth (1743–1817), for investigation. In

1784 Klaproth confirmed that the material was a new element, named it *tellurium* (from a Latin word for "earth"), and was careful to give full credit for the discovery to Muller.

IN ADDITION

William Pitt (1759–1806), the son of William Pitt (called the Elder Pitt) (1708–1778), who was prime minister during the French and Indian War, became prime minister in 1784. He was only twenty-four years old and stayed prime minister, with a three-year intermission, to the end of his life. He is usually considered the greatest of Britain's prime ministers.

1785

CLUSTERS AND NEBULAS

William Herschel (see 1781) studied the various fuzzy objects that Messier had listed (see 1771) and found, in 1785, that some of them were not nebulas but clusters of stars, densely packed (at least by the standard of our own uncrowded stellar neighborhood) into a roughly spherical shape. We call them *globular clusters* now and know that some consist of hundreds of thousands of stars.

There were some nebulas that Herschel couldn't resolve into constituent stars. He wondered, like Immanuel Kant (see 1755), if they might not be large collections of stars too far off to resolve.

In addition, he discovered dark areas in the Milky Way, small regions that contained no stars but that were surrounded on all sides by countless numbers of them. Herschel thought they were *holes*, which happened to be pointed in our direction so that we could see through into a cylinder of starlessness.

GALAXY

Herschel reported in 1785 on his attempt to determine the shape of the conglomeration of stars of which we were part. To count all the stars all over the sky was, of course, impractical. He therefore took samples. He chose 683 regions, well scattered, and counted the stars he could see in each one. (This was the first application of statistical methods to astronomy.)

He found that the number of stars per unit area of sky rose steadily as one approached the Milky Way, was maximal in the plane of the Milky Way, and minimal in the direction at right angles to that plane. This, he thought, would be explained if the star system were lens-shaped, with the Milky Way marking out the long diameter of the lens all around.

Such a lens-shaped star collection had been supposed by earlier astronomers, but Herschel had now made it a matter of close observation. For the first time, the *Galaxy* truly took shape, although even then no astronomer had a conception of its true size or of the vast number of stars it contained. Herschel thought it contained a hundred million stars—an enormous underestimate.

UNIFORMITARIANISM

Buffon's attempt to estimate the age of the Earth (see 1749) had been strictly speculative. Another attempt was now made by a British geologist, James Hutton (1726–1797), who wished to base it on actual observation. He spent years studying the rocks of every part of Great Britain carefully.

It seemed to him, as a result of his studies, that there had been an exceedingly slow evolution of the Earth's structure. Some rocks had been laid down as sediment and slowly compressed; some had emerged as lava and then been slowly worn down by wind and water.

The great point he made was that these changes were taking place very slowly and had been doing so in the past. Therefore, by measuring the rate of change and noting how much had been accomplished at that slow rate, one could get a notion of how long the change as a whole had taken. This belief is called *uniformitarianism* (change at a uniform rate over the eons), and it was opposed to the opinion that the Earth had been affected by sudden catastrophes such as Noah's Flood *(catastrophism)* in which huge changes could take place in comparatively little time.

Hutton published his observations and conclusions in *Theory of the Earth,* which was published in 1785. He did not attempt to estimate the age of the Earth, but he made it clear that the Earth had to be extremely old, and that there was no trace of when its beginning might have been from what he could observe.

IN ADDITION

In France, there was the *affair of the Queen's necklace,* a bizarre con game in which a very expensive diamond necklace bought for Queen Marie Antoinette ended up in the hands of crooks. In the scandal that followed, many French people believed the queen had been involved in the con. Antiroyalist fervor rose steadily, especially since France was virtually bankrupt thanks to the expense of supporting the American rebels and to the utter inefficiency and corruption of France's financial system. Every economic hardship was blamed on the idle, frivolous, and wasteful aristocracy (and with considerable justice, too).

An American physicist, John Jeffries (1745–1819), who fled to Great Britain after the Revolution, was the first person to cross the English Channel by balloon.

1786

MOUNTAIN-CLIMBING

Mountains are impressive, and because they reach toward the sky and their tops are not attainable by human beings without a great deal of trouble, they were often assumed to be the abode of gods. The connection of Mt. Olympus with the Greek gods and of Mt. Sinai with the God of the Bible are examples.

Generally, human beings admired mountains but avoided them. If it was absolutely necessary to get to the other side, they searched for comparatively accessible "passes." Some climbing had been done for scientific purposes, however. Thus, the Swiss naturalist Conrad Gesner (1516–1565) had climbed mountain peaks in the Alps to search out alpine species of plants.

By the 1700s, however, scientists had grown increasingly interested in mountains —in the plants and animals living there, in the nature of the rocks, and in the glaciers that frosted their peaks. The Alps were the mountains most available to west-European scientists, and their highest peak is Mount Blanc (French for "white," because its peak is white with snow and ice at all times). It is 15,700 feet high. No one had climbed it, of course, and it was universally regarded as the act of a lunatic to try.

However, a prize was offered for the first person who could accomplish the feat, and in 1786 a French doctor, Michel-Gabriel Paccard (1757–1827), along with a porter, Jacques Balmat, won the prize.

As often happens once one person accomplishes a feat previously considered impossible, others at once find out they can duplicate it. Paccard's climb initiated a virtual frenzy of mountain-climbing (among British aristocrats particularly). Sometimes this was for scientific purposes, but often it was merely for the excitement and adventure of it, as was the case with ballooning.

IN ADDITION

Frederick II of Prussia died on August 17, 1786, and was succeeded by his nephew, who reigned as Frederick William II (1744–1797).

In the United States there was a postwar depression and inflation and much discontent among the poor. In Massachusetts a rebellion was quickly put down, but it emphasized the weakness of the central government, which had no power to act against rebellions within a state. A meeting was called to consider improving the Articles of Confederation. It came to nothing, but Alexander Hamilton of New York (1755–1804) made the others agree to call for another meeting the following year.

1787

CHARLES'S LAW

Amontons had discovered the relationship between the volume of gas and its temperature (see 1699). For some reason, the discovery was neglected. Charles (see 1783) rediscovered the relationship in 1787, however, and the French chemist Joseph-Louis Gay-Lussac (1778–1850) again rediscovered it five years later. The relationship is sometimes called Charles's law and sometimes, Gay-Lussac's law. Amontons remains forgotten.

The amount by which a gas shrank as the temperature decreased by 1 degree Celsius was $1/273$ of its volume at 0 degrees. If the law held for all temperatures, a gas must shrink to zero volume at $-273°$ C, which would be an absolute zero. This would be true, however, only if gases remained gases as the temperature decreased, and whether this was so or not was not known.

CHEMICAL NOMENCLATURE

Language often stands in the way of scientific advance. This was especially true in chemistry, since chemists had inherited from the alchemists a bewildering array of names for various substances. The alchemists had deliberately attempted to increase their reputations by speaking metaphorically and mystically, and since no two used the same metaphors, chemists could scarcely understand each other when they reported their work.

Through the 1700s, attempts had been made to systematize the use of chemical terms and names for substances. Finally in 1787 Lavoisier (see 1769) and several coworkers published *The Method of Chemical Nomenclature,* in which a sensible and logical system was proposed. This was accepted by chemists over the next couple of decades, and finally chemistry had one language, which it retains to this day.

STEAMBOAT

Until this time, steam engines had been used only to power pumps and textile machinery. Yet if a steam engine could turn a paddle wheel, it could also serve as a mechanical oar of great power, which could drive a ship through water against both wind and current without the expenditure of human muscle (except that of feeding fuel to the engine so that steam could be generated).

The first person to achieve a workable steamboat was the American inventor John Fitch (1743–1798). On August 22, 1787, his first steamboat navigated the Delaware River for the first time. For a while Fitch kept up a regular schedule of trips from Philadelphia to Trenton and back. However, there were few passengers, the ship operated at a loss, the financial backers quit, and the ship was finally destroyed in a storm in 1792.

IN ADDITION

In the United States the Congress, sitting under the Articles of Confederation, passed the *Northwest Ordinance* on July 13, 1787. This law split the Northwest (the region between the Ohio River and the Great Lakes) into three to five states, each to have all the rights and privileges of the older states. Furthermore, the ordinance prohibited slavery in the Northwest. This reflected the increasing unpopularity of slavery in the United States, which had so recently been fighting for "life, liberty, and the pursuit of happiness."

Also in the United States, the Constitutional Convention (the one Hamilton had called for—see 1786) opened on May 25, 1787, with George Washington presiding. By September 17, it had hammered out the Constitution, under which the United States is still governed. It called for a federation in which the individual states gave up various rights to a federal government, while retaining considerable home rule. What's more, the Constitution was not to be imposed but was to be voted on and accepted voluntarily by nine of the thirteen states before it would go into effect.

1788

ALGEBRA AND MECHANICS

Geometry seemed a natural way of describing mechanics, but Descartes (see 1637) had shown that algebra could deal with geometric problems.

A French mathematician, Joseph-Louis Lagrange (1736–1813), tackled the study of mechanics in a totally nongeometric way. Using algebra and calculus, he worked out very general equations by means of which mechanical problems could be solved.

He summarized his work in *Analytical Mechanics,* published in 1788 by a very reluctant publisher. (A friend of Lagrange had to guarantee his purchase of any unsold copies.) It turned out to be a classic of science, although (as Lagrange boasted) there was not one geometric diagram in it. Geometry remains important, but Lagrange helped free the world of science from its unnecessary tyranny.

AFFINITIES

Chemists up to this time had to take chemical reactions as they found them. Substance A would react with substance B but not with substance C and that was all there was to it.

The Swedish mineralogist Torbern Olof Bergmann (1735–1784) had struggled to classify minerals and to make sense out of chemical reactions. He listed *affinities*—that is, the extent to which different chemicals interacted. He also prepared tables that made some sense out of the matter and helped predict whether certain reactions, not yet observed, would take place or not if given a chance.

His results were published posthumously in 1788, and while they represented the barest beginning in the study of chemical behavior, they were a beginning.

IN ADDITION

In the United States, New Hampshire ratified the Constitution on June 21, 1788. It was the ninth state to do so. Before the end of the year, every state but North Carolina and Rhode Island had ratified it. The old Congress held its last meeting on October 21, 1788, and the nation now waited to elect a president, a vice president, a Senate, and a House of Representatives, all for the first time under the Constitution. That would represent the true beginning of the United States of America.

Great Britain, which had been sending prisoners to North America, now sent its first load of prisoners to Botany Bay in Australia instead, near what was eventually the city of Sydney, named for Thomas Townshend, Viscount Sydney (1733–1800), who was home secretary at this time and in charge of the prison system.

1789

SATELLITES

By the end of the 1600s, ten satellites were known: Earth's moon, the four satellites of Jupiter discovered by Galileo (see 1610), and the five satellites of Saturn discovered by Huygens (see 1656) and Cassini (see 1665).

A century had passed after Cassini's discovery of Dione in 1684 without any further satellite discoveries. Then in 1787, Herschel discovered two satellites of his own planet, Uranus (see 1781). He named them Titania and Oberon, after the queen and king of the fairies in William Shakespeare's *A Midsummer Night's Dream*, thus breaking with the traditional use of the classical myths.

In 1789 Herschel discovered two more satellites of Saturn, which were nearer the planet than the others were. He named them Mimas and Enceladus, after two giants who had rebelled against Zeus (Jupiter) in the Greek myths. Fourteen satellites were now known: Earth had one, Jupiter four, Saturn seven, and Uranus two. No more were to be discovered for half a century.

ACIDS

Lavoisier (see 1774) had named the active portion of the atmosphere *oxygen* (acid-producer) because it was thought that all acids contained oxygen.

In 1789, however, the French chemist Claude-Louis Berthollet (1748–1822) showed that hydrocyanic acid and hydrosulfuric acid did not contain oxygen. To be sure, they are very weak acids, but it was eventually shown that hydrochloric acid, a strong acid, also did not contain oxygen.

CONSERVATION OF MASS

Lavoisier in 1789 wrote the best textbook of chemistry the world had seen up to that time.

The most important generalization he introduced in that book was that in any closed system (one from which no mass was allowed to leave, and into which no mass was allowed to enter), the total amount of mass remained the same no matter what physical or chemical changes went on. This is the *law of conservation of mass.* For over a century it has been central to chemistry, and when it was finally modified, it was only to make it even more fundamental.

URANIUM

In 1789 Klaproth (see 1784), who was working with a heavy black ore called *pitchblende,* obtained a yellow compound that contained a hitherto unknown element. He thought it was the element itself, and he named it after a planet, after the fashion of the medieval alchemists. Since the planet Uranus had been discovered only eight years before, Klaproth named his new element *uranium.* Neither he nor anyone else could possibly have foretold the significance it would come to have in the future.

In that same year, Klaproth also obtained a new oxide from the semiprecious jewel zircon and named the new metal contained in the oxide *zirconium.*

IN ADDITION

The situation in France had become so bad that Louis XVI was forced to call the Estates General (the French parliament) into session on May 5, 1789. The Third Estate, representing the middle class, was sure it would not be heard and so constituted itself a National Assembly, under the leadership of Honoré-Gabriel Riqueti, Comte de Mirabeau (1749–1791). The Parisians, exasperated by rumors that the king would use the army to put down those who wanted reforms, attacked the Bastille, the Parisian prison that was the very symbol of royal despotism. It fell on July 14, 1789, and that is considered the beginning of the *French Revolution.* On October 5 and 6, the Parisians marched on Versailles and brought back the royal family, which would never see Versailles again.

In the United States, elections were held for president. They were not popular votes. Instead, ten of the state legislatures chose *electors* (as the Constitution required), and on February 4, 1789, these electors unanimously chose George Washington as the nation's first president. John Adams was elected vice president. Senators and Representatives were elected in the varying states according to the state constitutions. On April 6, the first Congress convened. On April 21, John Adams was sworn in, and on April 30, George Washington reached New York and was inaugurated as the first president.

1790

INDUSTRIAL REVOLUTION

Great Britain's economic position was improving enormously with its ingenious new textile machinery and the manner of powering it by steam. It was easy for the British leaders to see that if they could retain a monopoly on this Industrial Revolution, Britain might easily become the strongest power in the world—economically, at any rate.

For this reason, what we would today call an "iron curtain" was clamped down. Blueprints of the new machinery were not allowed to leave the country, and neither were engineers who were experts in the new technology.

The new nation of the United States wanted the new technology to aid in its economic independence from Great Britain, without which its political independence wouldn't amount to much. It therefore did its best to steal the knowledge by encouraging defectors. It found its man in Samuel Slater (1768–1835).

Slater was an engineer who knew the new technology intimately, but who also knew that he could advance only so far in the class-ridden society of Great Britain. The United States was offering money for his knowledge and he decided to go for it. He couldn't take any blueprints with him, of course, so he painstakingly went about memorizing every detail of the machinery. He then disguised himself as a farm laborer and sneaked out of the country. In 1789 he arrived in the United States and made contact with rich merchants in Rhode Island.

In 1790, working from memory, Slater began building the first American factory based on the new machinery, in Pawtucket, Rhode Island.

In this way, the Industrial Revolution came to the United States, and once the process of proliferation began there was no stopping it. It continues to this day. Of course, when other nations try to make use of our technology in the same way that we made use of Great Britain's, we are, very naturally, indignant.

METRIC SYSTEM

Throughout history, every nation, sometimes every region within a nation, has developed its own system of measurements. As long as trade was slow and communication was limited, that was merely an annoyance. However, as the economic interdependence of Europe increased rapidly in modern times, the absolute jungle of different measurements was a serious handicap to progress and prosperity. Yet tradition made it almost impossible for any region to give up its time-sanctified system for others merely because the others made more sense and were easier to use.

The French, however, in the grip of their Revolution, thought it a suitable time to develop a system of sensible measurements, and they appointed a commission to work on it, a commission including such men as Laplace (see 1783), Lagrange (see 1788), and Lavoisier (see 1769).

The commission tried to base the system on natural measurements, so that the basic unit of length, for instance, the *meter* (from a Greek word meaning "to measure"), was to be equal in length to 1/10,000,000 of the distance from the North Pole to the Equator.

Other units were worked out to interconnect with the meter, when that was possible, while larger and smaller units were worked out by multiplying or dividing by ten.

This so-called *metric system* was by far the most useful and logical system of measurement ever dreamed up, and all that stood in the way of its instant adoption was, first, the dead weight of tradition, and second, the hostility of the rest of Europe to the French Revolutionaries. Even so, it gradually spread, and today the metric system is used universally, *except* in the United States. And even in the United States, the metric system is used by scientists, and increasingly by others.

The metric system represented an improvement in technique, and like other such techniques (writing, the alphabet, Arabic numerals, printing, chemical nomenclature), it may not have advanced scientific knowledge in itself, but it made such advances easier.

IN ADDITION

In the United States, Rhode Island finally ratified the Constitution on May 29, 1790, the last of the thirteen states to do so.

The Holy Roman Emperor Joseph II died on February 20, 1790, and was succeeded by his brother, Leopold II (1747–1792).

Benjamin Franklin died on April 17, 1790. He had lived long enough to see the United States well established. Its population was now almost 4 million.

1791

TITANIUM

The English minister William Gregor (1761–1817) grew interested in mineralogy. More out of curiosity than anything else, he took to analyzing as many odd minerals as he could find. In 1791 he isolated a substance from one of these minerals that he felt might be a new element. It was, and four years later it was named *titanium* by Klaproth (see 1784).

COLUMBIA RIVER

The American navigator Robert Gray (1755–1806), between 1787 and 1790, was the first American to circumnavigate the world, doing so during a voyage to obtain furs on the north Pacific coast and exchange them for tea in China. In 1791 he returned to the Northwest in the ship *Columbia.* On May 12 of that year he discovered a river, which he named Columbia for his ship, then circumnavigated the globe a second time. The United States used this trip as the basis for its eventual claim to the Oregon Territory.

IN ADDITION

Aristocrats were fleeing France and, as *émigrés* (emigrants), did their best to obtain help for their cause from other powers. What they wanted was a European invasion of France to restore the powers of the king. Louis XVI and Marie Antionette felt they would be safer if they joined the émigrés and perhaps led the foreign invasion. The escape was badly botched; Louis XVI was recognized, stopped short of the border, and returned to Paris, where he and his family were kept in virtual imprisonment.

The French Revolutionaries declared the freedom of all black slaves in the French dominions of the West Indies. The slaveholders refused to accept this and there were bloody slave rebellions.

In the United States, Vermont entered the Union as the fourteenth state on March 4, 1791. On December 15, the first ten amendments to the Constitution, the so-called *Bill of Rights,* were made part of the Constitution.

1793

COTTON GIN

The new textile industry of Great Britain, and the one that was just beginning to arise in New England, meant increasing demands for cotton, which could be grown with great profusion in the southern states. However, it was difficult to pull the cotton threads off the seeds in the cotton bolls, and that limited the quantity that could be produced.

In April 1793, however, an American inventor, Eli Whitney (1765–1825), challenged to do something about the problem, invented the *cotton gin* (*gin* is short for engine). It was a simple device in which metal wires poked through slats and entangled themselves in the cotton fibers. The wires were affixed to a wheel, and as the wheel turned, the cotton fibers were pulled off. One gin could produce 50 pounds of cleaned cotton per day.

The effect on the United States was catastrophic. The southern states began to produce cotton in great quantities, and slaves were needed, also in great quantity, to pull the bolls off the plants so that the gins would have enough to work on.

The southern states, which had been giving up on slavery, had to return to it now, defend it, and work up all sorts of excuses for its existence. They developed an agricultural economy based on slave labor, which kept them poor, while the northern states grew rich on wheat and industry. And in the end, it brought on the Civil War.

INSANE ASYLUMS

In ancient times, people who were insane were thought to be touched by some sort of divine influence and were treated, on occasion, with awe and respect. However, in Christian Europe (influenced by tales of demonic possession in the New Testament), they were thought to be infested by devils and were sometimes mistreated physically in order to force those devils to leave. The insane were also treated as figures of fun, and insane asylums were thought to be amusing places to visit, though they were dreadful nightmares of howling, demented people, imprisoned and often subjected to the most brutal treatment.

A French physician, Philippe Pinel (1745–1826), thought that psychotics were sick in mind and deserved as considerate treatment as those who were sick in body. In 1791 he published his views on *mental alienation;* that is, on minds alienated from their proper function. (For a while, physicians who specialized in mental disorders were called *alienists,* because of this.)

The French Revolutionaries, always ready to break with established custom, put Pinel in charge of an insane asylum in 1793, and there he struck off the chains of the inmates and began to make systematic studies of their conditions. He was the first to keep well-documented case histories of mental ailments.

It took half a century, though, for this civilized view of mental illness to penetrate the rest of Europe (and the United States).

ACROSS NORTH AMERICA

A British-born Canadian explorer, Alexander Mackenzie (1764–1820), worked his way into the interior of Canada, settling in what is now the Province of Alberta. From there he followed what is now called the Mackenzie River to its mouth in the Arctic Ocean, which he reached in 1789. In 1793 he crossed the Rocky Mountains to the Pacific Ocean in what is now British Columbia.

He was the first person to visit the Atlantic, Pacific, and Arctic shores—all three—of North America.

VANCOUVER ISLAND

The British navigator George Vancouver (1757–1798), who had sailed with Captain Cook (see 1768), continued to explore the lands Cook had examined—Australia, New Zealand, Tahiti, and Hawaii.

He also explored the Pacific Northwest coast of the United States and, in 1793, circumnavigated an island of moderate size off the coast of what is now British Columbia. It is called Vancouver Island in his honor.

IN ADDITION

Under émigré instigation, Prussia and Austria formed an alliance and openly threatened an invasion of France. France declared war on Austria on April 20, 1792. However, the untrained French armies didn't do well, and the more radical revolutionaries accused the moderates of showing insufficient zeal. The moderates were driven out of the country, and the radicals (the so-called *Jacobins,* because their headquarters were at a monastery on the Rue St. Jacques) seized control. Under the leadership of Georges-Jacques Danton (1759–1794), a large number of prisoners held on suspicion were lynched by a mob between September 2 and 7, 1792. This began the so-called *Reign of Terror,* which lasted for nearly two years.

Immediately after the *September massacres,* the advancing Prussians and Austrians retreated; the radicals, sensing victory, deposed Louis XVI and, on September 21, declared France a republic. The Prussians and Austrians continued to retreat, and the French took the Austrian Netherlands (now Belgium) in November.

Louis XVI was executed on January 21, 1793. As a result, Great Britain, the Dutch Republic, and Spain declared war on France. France did not flinch. Under the domination of Maximilien-François de Robespierre (1758–1794), it executed Marie Antoinette on October 16.

In the United States, Kentucky entered the Union as the fifteenth state on June 1, 1792.

In eastern Europe, meanwhile, the preoccupation of the west with France gave Russia a chance for aggrandizement. Poland was weaker than ever since its partition twenty years before. Why not a *Second Partition,* then? On January 23, 1793, Russia took a large chunk of eastern Poland and Prussia took another of western Poland. Only a small rump of Poland was left unoccupied and it was all but a corpse.

1794

METEORITES

It was the common experience of human beings that objects sometimes fell from the sky. Such falls had been reported. The Kaaba, the sacred *black stone* of the Muslims, was probably a meteorite that fell from the sky. In Ephesus, a stone was worshipped in the temple of Artemis that was probably a meteorite, and periodically reports of witnessed falls were received.

During the Age of Reason, such tales were discounted and dismissed by men of science.

In 1794, however, a German physicist, Ernst Florens Friedrich Chladni (1756–1827), published a book in which he suggested that meteorites did fall and that this happened because the space near Earth contained the debris of a planet that had once existed but had exploded.

This was the first time that a reasonable explanation had been offered for meteorites, and the tide slowly began to turn. It took a while, though.

RARE EARTHS

The term *earth* at this time was applied to any oxide that was insoluble in water and resistant to the action of heat. The Earth's crust was largely a mixture of these oxides, hence the name. Examples of common earths are calcium oxide, magnesium oxide, iron oxide, and silicon dioxide.

A Finnish chemist, Johan Gadolin (1760–1852), studied a strange mineral that had been obtained from a quarry in Ytterby (near Stockholm). Gadolin recognized it as a new earth unlike any previously known. He called it a *rare earth* to distinguish it from the common ones. From this rare earth and from others like it, a number of metallic elements (the *rare earth elements*) were eventually located, all remarkably similar in chemical properties.

IN ADDITION

In France the Terror came to an end as the Jacobins turned on each other. Danton was executed on April 6, 1794, and Robespierre on July 28. However, France remained a republic and the war continued.

In March 1794, the Poles, under the leadership of Tadeusz Kościuszko (1746–1817), who had fought in the American Revolution, rose in rebellion against the partitions.

In Haiti, the black slaves under Toussaint-L'Ouverture (*ca.* 1743–1803), rose against the slaveholders who had defied the French laws granting the slaves freedom.

1795

CANNING FOOD

The trouble with food is that much of it doesn't keep. Fairly quickly, it rots, sours, grows moldy. To preserve food for periods of time so that people don't starve over the winter, use must be made of drying, salting, smoking, and so on. A diet that depends on such foods preserves life but is monotonous.

The rising French military leader Napoléon Bonaparte (1769–1821) realized the importance of decent food to an army that had all Europe arrayed against it and offered a prize of twelve thousand francs for some way of preserving fresh food for a long period of time.

In 1795 a French inventor, Nicolas-François Appert (1750–1841), began working on the problem. He knew that Spallanzani (see 1768) had shown that meat would not rot if it was boiled for a long enough period and then sealed. Appert therefore worked out a system for applying this principle on a large scale, heating meats and vegetables and sealing them into glass or metal containers.

It took some years to perfect the process, but Appert's system represented the beginning of the canned food industry.

IN ADDITION

After the fall of Robespierre, France was ruled by a *Directory,* a group of five moderate revolutionaries under the leadership of Paul-François de Barras (1755–1829). When the Parisian mob threatened them, Barras placed Napoléon Bonaparte in charge of the armed forces in Paris, and he cleared the streets with some well-placed artillery fire. That ended the danger of the mob and accelerated Bonaparte's rise.

Meanwhile, the French army invaded the Netherlands in 1795, captured the Dutch fleet, and established the *Batavian Republic* as a French puppet. The Dutch ruler, William V (1748–1806), fled to Great Britain.

The Polish uprising failed, and what was left of Poland was divided up by Russia, Prussia, and Austria in the *Third Partition,* on October 24, 1795. Russia's territory in eastern Europe had now attained the shape it has remained, more or less, ever since.

1796

VACCINATION

Inoculation had been used to fight smallpox for some eighty years (see 1713), but it was dangerous.

The English physician Edward Jenner (1749–1823) knew that in his native Gloucestershire there were tales to the effect that anyone who caught cowpox (a very mild disease, prevalent among cows, that somewhat resembled smallpox) was thereafter immune not only to cowpox but also to smallpox.

(Since milkmaids almost invariably caught cowpox early and then never got smallpox, they retained a clear complexion. This in itself may have been enough to fuel the romantic cliché of the time concerning pretty milkmaids.)

Finally Jenner decided to test the matter. On May 14, 1796, he found a milkmaid who was undergoing an attack of cowpox. He took the fluid from a blister on her hand and injected it into an eight-year-old boy named James Phipps who, of course, got cowpox. Two months later, Jenner inoculated the boy with smallpox in the manner usual for those days. The boy did *not* get smallpox. Two years later, he found someone else with active cowpox and tried again. It worked this time also, and he felt it safe to announce his discovery.

The Latin word for cow is *vacca,* so cowpox is *vaccinia.* Jenner coined the word *vaccination* to describe his use of cowpox inoculation to create immunity to smallpox. In this way, he founded the science of *immunology.*

Such was the dread of smallpox that the new technique was instantly adopted and quickly spread all over Europe. It was the first case of a serious and frightening disease that could be reliably prevented.

NEBULAR HYPOTHESIS

Kant's suggestion (see 1755) that the Solar System had formed from the condensation of a vast nebula of dust and gas had gone mostly unnoticed.

In 1796, however, Laplace (see 1783) had published a popular nonmathematical book on astronomy and, in the appendix, had made a similar suggestion, describing it in much greater detail. As the condensing cloud spun faster and faster, he explained, first one ring of gas, then another and another were given off. The rings condensed to form the planet. The planets, in condensing, gave off smaller rings that formed the satellites.

Laplace felt that certain nebulas visible in space were themselves contracting and represented planetary systems in the process of formation. His suggestion was therefore called the *nebular hypothesis.* It proved very popular with astronomers for a time.

HEPTADECAGON

Though Greek astronomy, physics, chemistry, medicine, and geography had all been replaced since the beginning of the Scientific Revolution, Greek geometry had remained staunch and firm.

In 1796, however, a young German mathematician, Carl Friedrich Gauss (1777–1855), worked out a method for constructing a heptadecagon (a polygon built up of seventeen sides of equal lengths and sometimes called a seventeen-gon in consequence), using a compass and straightedge only. The Greeks had missed that construction, and Gauss's feat was considered the first notable addition to geometry since ancient times.

But Gauss did more than simply find a method of construction. He showed that only polygons of certain numbers of sides could be constructed with compass and straightedge alone. An equilateral heptagon (a polygon with seven equal sides) could *not* be constructed in this fashion. This was the first case of a geometric construction being proved impossible. From this point on, the proof of impossibility in mathematics grew increasingly important.

IN ADDITION

George Washington, having served two four-year terms as president of the United States, would not consider a third term, thus beginning the tradition of a two-term maximum that lasted for a century and a half. John Adams was elected the second president, with Thomas Jefferson (1743–1826) as vice president. Tennessee was admitted to the Union as the sixteenth state on June 1, 1796.

In France, Napoléon Bonaparte married Joséphine de Beauharnais (1763–1814) on March 9, 1796, and the next month took up his duties as general of a ragged French army in Italy. At once he displayed a capacity for rapid movement and instant and daring decisions that enraptured his own army and completely confused the conservative, rather dim-witted Austrian generals he faced.

Catherine II of Russia died on November 10, 1797. She was called *Catherine the Great* and was the last monarch to receive that appellation. She was succeeded by her somewhat unbalanced son, who reigned as Paul I (1754–1801).

1797

CHROMIUM

A French chemist, Louis-Nicolas Vauquelin (1763–1829), had prudently left France during the Terror but had returned after Robespierre's execution (see 1794). In 1797 he was working with a mineral from Siberia and isolated from it a new metal, which he named *chromium* from the Greek word for "color," because of the many colors of its compounds.

PARACHUTE

The principle of the parachute (a light object that presents a large surface to the air so that air resistance slows its descent) was simple enough. The French balloonist Jean-Pierre-François Blanchard (1753–1809) used one in 1785 to drop a dog in a basket safely to Earth from a balloon. The first case of a human being using a parachute successfully came in 1797 when another French balloonist, André-Jacques Garnerin (1769–1823), managed it.

IN ADDITION

Bonaparte's victories in Italy continued and on October 17, 1797, he forced the Austrians to sign the Treaty of Campo Formio, ceding Belgium to France and accepting the formation of a *Cisalpine Republic* in northwestern Italy as a French puppet. In exchange, Austria was allowed to annex the Venetian Republic, now totally moribund. Bonaparte did not bother to consult his government but treated with Austria as though he himself were the government.

Frederick William II of Prussia died on November 16, 1797, and was succeeded by his son, who reigned as Frederick William III (1770–1840).

1798

MASS OF THE EARTH

Newton's equation for the gravitational attraction of one body to another (see 1687) contained symbols for the *gravitational constant,* the masses of the two bodies, the distance between them, and the acceleration of movement toward each other.

In the case of an object falling to Earth, the mass of the object, its distance from Earth's center, and its acceleration were all known. That left two unknowns, the mass of the Earth and the gravitational constant. If one of these could be determined, the other could be calculated at once.

The gravitational constant was the same for all objects. If the gravitational attraction between two objects, each of known mass, at a known distance from each other, could be measured, that would allow the gravitational constant to be calculated and, from that, the mass of the Earth. However, two objects of known mass were sure to be so small that the gravitational attraction between them would be very tiny.

Nevertheless, Cavendish (see 1766) tried it in 1798. He suspended a light rod by a wire attached to its center. At each end of the rod was a light lead ball. The rod could twist freely about the wire, and even a small force applied to the balls in opposite directions would produce a measurable twist. Cavendish measured how large a twist was produced by various small forces.

He then brought two large balls near the two light balls, one on either side. The force of gravity between the large balls and light balls twisted the wire. Cavendish measured the twist and from that calculated the gravitational force between the two objects. That gave him the gravitational constant, and that in turn gave him the mass of the Earth. The Earth turned out to have a mass of 6,600,000,000,000,000,000,000 tons. Based on the known volume of the Earth, this made the average density of the Earth five and a half times that of water.

So precise was Cavendish's experiment that this first-time calculation was very close to the value accepted today.

COMPARATIVE ANATOMY

The greatest anatomist of his day was a Frenchman, Georges Cuvier (1769–1832). In a book published in 1798, he studied the anatomy of various animals to show how they compared with one another and did so with such excellence that he is considered the founder of *comparative anatomy.*

He introduced a new broad classification into the taxonomic scheme of Linnaeus (see 1735). Linnaeus's broadest division was that of *class,* but Cuvier grouped various classes into *phyla* (singular *phylum,* from a Greek word for "tribe").

Cuvier's eye for detail was such that he could tell whether fossil remains represented animals that belonged to one or another of the known phyla even though they were not members of any living species.

Everything he discovered seemed to imply the existence of biological evolution, but Cuvier remained firmly in the antievolution camp.

POPULATION PRESSURE

It seems obvious that population increases in times of peace, health, and prosperity and declines as a result of war, disease, and famine, but the first to try to analyze the matter dispassionately was a British economist, Thomas Robert Malthus (1766–1834). In 1798 he published a book, *Essay on Population,* in which he pointed out that population tended to increase in geometric progression (2, 4, 8, 16 . . .) while the food supply tended to increase in arithmetic progression (2, 3, 4, 5 . . .). As a result, population would always outstrip the food supply no matter what happened, and the surplus number of people would have to be stripped away by famine, war, and disease.

This imparted a certain inevitability to disaster and misery, which could only be removed by lowering the birthrate. In a second edition, Malthus suggested delayed marriage and sexual continence as a way of bringing this about. It doesn't take much of a cynic to realize that in the long run this won't work, but any suggestion that there are ways of lowering the birthrate without depriving people of the pleasures of sex is always met with strong disapproval on the part of the blue-nosed.

Malthus did not grasp the role technological advance could play in warding off disaster, even though he wrote at a time when the Industrial Revolution was in its early stages. Because of such advances, the world population is now five times what it was in Malthus's day and the Malthusian consequence has not yet been paid. But technological advance only delays, it does not stop, and the longer the delay, the more terrific the crash. Unless, of course, we lower the birthrate.

LIQUID AMMONIA

Water with substances dissolved in it melts at a lower temperature than pure water does. The French chemist Louis-Bernard Guyton de Morveau (1737–1816), by adding calcium chloride to an ice and water mixture, brought about a temperature drop to −44° C. He used this low temperature to liquefy the gas ammonia, which becomes liquid at a temperature of −33° C.

This was the first time a substance until then known only as a gas was chilled to liquid form.

INTERCHANGEABLE PARTS

In 1798 Eli Whitney, who had invented the cotton gin (see 1793), was awarded a contract to manufacture ten thousand muskets for the American government.

Up to that time, every musket (and indeed every device consisting of more than one part) had been made with each part adjusted to fit the adjoining part. If a part was broken, a new one had to be adjusted manually. A corresponding part from a similar device would not necessarily (and, in fact, would virtually never) replace the broken part without adjustment.

Whitney, however, machined his parts with such precision that a particular part could replace any other one of that part. The story is that when the muskets were done he brought some of them, disassembled, and placed them at the feet of a government official. Then, picking out parts at random, he

The development of methods for producing interchangeable parts was an important part of the developing Industrial Revolution.

BERYLLIUM

In 1798 Vauquelin, who had discovered chromium (see 1797), recognized the existence of a new element in the gems beryl and emerald. He named the element *beryllium.*

IN ADDITION

Napoléon Bonaparte, overexcited by his victories in Italy, perhaps, decided to attack Egypt and build a French Empire in the east. He managed to elude the British navy and had no trouble defeating the Egyptian forces. However, the British fleet, under Horatio Nelson (1758–1805), finally located the French ships at Abukir and destroyed them at the Battle of the Nile, on August 1, 1798. Bonaparte was thus cut off in Egypt.

1799

LAW OF DEFINITE PROPORTIONS

The French chemist Joseph-Louis Proust (1754–1826) was working in Spain at this time in order to escape the turmoil of the French Revolution and took part in a heated controversy over whether the composition of substances varied with the manner of manufacture.

Using painstakingly careful analysis, Proust showed in 1799 that copper carbonate contained definite proportions, by weight, of copper, carbon, and oxygen, no matter how it was prepared in the laboratory or how it was isolated from nature. The preparation was always five parts of copper to four of oxygen to one of carbon.

He went on to make a similar case for a number of other compounds and maintained that there was a general *law of definite proportions,* often called *Proust's law.*

In this way, Proust could distinguish between mixtures (in which different elements could exist in any proportions, as in air) and compounds (in which different elements mixed in definite proportions and no other).

STRATA

Many observers had noted that rocks existed in layers, or *strata* (the Latin word for "layers"). An English geologist, William Smith (1769–1839), who was engaged in working on canals, had frequent opportunities to see the strata at excavation sites.

He became interested in strata to the exclusion of all else and, in 1799, began writing on the subject. Smith made a new point. Each stratum had its own characteristic forms of fossils, not found in other strata. No matter how the strata were bent and crumpled—even when one sank out of view and cropped up again miles away—its fossils went with it. In fact, it seemed to Smith, one could identify a stratum from its fossil content.

Since it could be reasonably supposed that a stratum nearer the surface was younger than one deeper down, the strata offered a method for working out an orderly history of life from the fossils and even coming to some rough conclusions as to how long ago different fossils had existed as living forms.

PERTURBATIONS

In 1799 Laplace (see 1783) published the first volume of a monumental five-volume work called *Celestial Mechanics,* in which the gravitational influences on the various bodies of the Solar System were taken up in detail. Although the Sun dominates the system, and the planets move about the Sun in stately ellipses, each planet pulls at the others, and so do the satellites.

These small additional pulls introduce minor variations in the movement of the planetary bodies, called *perturbations,* and it was considered that they might gradually increase in size over time so that the Solar System would prove unstable in the long run.

Laplace showed that this was not the case. The perturbations are periodic and vary on either side of what would exist if the Sun alone had a gravitational pull. The Solar System, then, is stable.

ROSETTA STONE

While Bonaparte's army was in Egypt, a French soldier came across a black stone near a town that Europeans called Rosetta. What he found was therefore called the *Rosetta Stone.*

On the Rosetta Stone was an inscription in Greek dating back to 197 B.C. It wasn't an interesting inscription in itself, but present also were inscriptions in two different forms of Egyptian writing. If, as seemed likely, this was the same inscription in three different languages, then one had an inscription in two forms of Egyptian writing, which at that time no one could read, and a Greek inscription that many scholars could read.

In the next few decades, the Rosetta Stone was studied in order to learn the Egyptian languages, so that from inscriptions and writings left behind by the Egyptians, vast stretches of Egyptian history might come to be understood.

IN ADDITION

Bonaparte continued to win battles in Syria and Egypt, but he realized it would come to nothing as long as Great Britain controlled the Mediterranean, so on August 24, 1799, he abandoned his army and returned to France.

While Bonaparte was in Egypt, Russia joined the coalition against France, and a Russian army was sent into Italy under Aleksandr Vasilyevich Suvorov (1729–1800), Russia's greatest general. In Italy, he defeated the French in three battles. On October 22, 1799, he withdrew, however, for he failed to get the cooperation of the Austrians, who did not want the Russians to be *too* successful. Nevertheless, Italy was temporarily lost to the French.

In the United States, George Washington died on December 14, 1799. Also in 1799, the new executive mansion was completed in the new capital of the United States, which was named Washington in honor of the first president. It was located in the District of Columbia, a region on the Potomac River donated to the federal government by Maryland, and not part of any state.

1800

ELECTRIC BATTERY

Galvani, noting that dead muscle twitched when touched simultaneously by two different metals, had decided that electricity was involved and that it originated in the muscle (see 1780). The Italian physicist Alessandro Giuseppe Volta (1745–1827) thought the electricity originated in the metals.

He began to experiment with different metals in contact and was soon convinced that he was correct. In 1800 Volta constructed devices that would produce electricity continuously if it was drawn off as produced. This created an electric current, which turned out to be far more useful than the nonflowing electric charge of static electricity.

At first Volta used bowls of salt solution to produce the flow. The bowls were connected by means of arcs of metal dipping from one bowl to the next, one end of the arc being copper and the other being tin or zinc. Since any group of similar objects working as a unit may be called a *battery*, Volta's device was an *electric battery*—the first in history.

Volta then made matters more compact and less watery by using small round plates of copper and zinc, plus disks of cardboard moistened in salt solution. Starting with copper at the bottom, the disks, reading upward, were copper, zinc, cardboard, copper, zinc, cardboard, and so on. If a wire was attached to the top and bottom of this battery, an electric current would flow when the circuit was closed.

DECOMPOSITION OF WATER

The existence of Volta's electric battery (see above) was publicized on March 20, 1800. Within seven weeks, it was put to work. On May 2, an English chemist, William Nicholson (1753–1815), built a battery of his own and passed an electric current through slightly acidified water.

Bubbles of hydrogen and oxygen were produced in the water. The water had been *electrolyzed* and broken up into the hydrogen and oxygen that had been combined in its formation.

Volta had shown that a chemical reaction taking place between copper and zinc in salt water could form an electric current, while Nicholson had shown that the reverse was also true, that an electric current could produce a chemical reaction.

Later that year, the German physicist Johann Wilhelm Ritter (1776–1810) allowed the gases formed in electrolyzing water to arise from two wires and bubble into separate containers. In one container, hydrogen was collected, and in the other, oxygen. The volume of hydrogen was just twice the volume of the oxygen.

Ritter also passed a current through a solution of copper sulfate, and copper appeared on the negative electrode. (This was one of the two metal rods sticking into the solution, the other being the positive electrode.) This was the beginning of electroplating.

INFRARED RADIATION

It seemed natural to suppose that light, by its very nature, could be seen; light that could not be seen would be a contradiction in terms. Yet invisible light existed.

Herschel (see 1781) in 1800 had formed a sunlight spectrum and tested different parts of it with a thermometer to see if some colors delivered more heat than others. He found that the temperature rose as he moved toward the red end of the spectrum, and it seemed sensible to move the thermometer just past the red end in order to watch the heating effect disappear.

Except that it didn't. The temperature rose higher than ever at a spot beyond the red end of the spectrum. The region was called *infrared* (below the red).

How to interpret the region was not readily apparent. The first impression was that the Sun delivered *heat rays* as well as light rays and that heat rays refracted to a lesser extent than light rays. A half-century passed before it was established that infrared radiation had all the properties of light waves except that it didn't affect the retina of the eye in such a way as to produce a sensation of light.

GAS LIGHTING

In preparing charcoal from wood or coke from coal, little attention was paid to the materials that were heated and burned away. In 1792, however, a British inventor, William Murdock (1754–1839), began to collect the gases obtained by heating wood, peat, and coal, and found that they were inflammable. Being gases, they could be piped from place to place easily; could be easily lit and easily extinguished.

By 1800, Murdock had set up an experimental gas light, using coal gas. It was not long before gas lighting was well established, and the large cities of the industrial nations and the better-off homes were lit by flaring gas jets for most of a century. With better lighting by night, travel was safer, the crime rate fell, and the evening meal, rather than the midday meal, became the important social occasion of the day.

NITROUS OXIDE

New gases continued to be discovered. A British chemist, Humphry Davy (1778–1829), discovered *nitrous oxide* in 1800. It was his rather risky habit to smell new gases and try to study the effects on the human body of breathing them. It turned out that nitrous oxide, on being breathed, gave one a giddy, intoxicated feeling and made one suggestible, so that one would laugh or weep immoderately or make other emotional displays. (The gas is still called *laughing gas* today.)

Davy reported that under the influence of the gas he felt no pain and suggested it be used as an anesthetic. It was indeed the first true chemical anesthetic and was often so used in dentistry. It also became a fad of the day, and some people with nothing better to do would sit about a bowl of it and breathe it in order to experience the effects.

TISSUES

The French physician Marie François Xavier Bichat (1771–1802) was notable for the many postmortems he conducted in his very short professional life. His careful observations (*without* a microscope) showed him that the various organs were built up out of a mixture of different types of simpler structures, each of which might occur in many different organs. Because these structures were usually flat and delicately thin, they were called *tissues*.

In 1800 he published a book called *Treatise on Membranes* in which he listed and carefully described twenty-one different types of tissues. He is for this reason considered the founder of the science of *histology* (the study of tissues).

MALLEABLE PLATINUM

Platinum, because of its inertness and its high melting point, would be marvelously suited to laboratory equipment if only it could be easily worked and pounded into shape. A method for doing so was worked out by a British chemist, William Hyde Wollaston (1766–1828), in 1800. He kept his method secret and never allowed anyone in his laboratory. In this way he earned a fortune. He arranged to have his method published after his death.

In working with platinum, he discovered two other metals very similar in properties to it. These were *palladium* and *rhodium*.

IN ADDITION

On November 9, 1799, Bonaparte, back in France, overthrew the Directory and seized power. He established a *Consulate*, consisting of three men, to act as executives of France. He himself was First

Consul, however, and the other two were figureheads. Napoleon was dictator of France. He then went back to war and defeated Austria at the Battle of Marengo on June 14, 1800, thus regaining control of Italy. He also forced Spain to cede the portion of Louisiana west of the Mississippi, which it had gained in 1763, back to France.

In the United States, John Adams moved into the executive mansion in Washington, D.C., and Congress met there for the first time on November 17, 1800. Adams did not win reelection, however. Thomas Jefferson was elected the nation's third president.

1801

JACQUARD LOOM

To weave textiles in such a way as to introduce patterns would ordinarily require people to work in a careful manner, producing certain motions here, but not there. It is not the sort of thing that one might think could be done by machinery. A machine wouldn't have the "brains" to do what human beings do only with difficulty.

In 1801, however, a French inventor, Joseph-Marie Jacquard (1752–1834), invented what came to be called the *Jacquard loom.* In such a loom, needles ordinarily move through holes set up in a block of wood. Suppose a card containing holes in a certain pattern is interposed between the needles and the holes in the wood. Where there are holes in the card matching the holes in the wood, the needles pass through; otherwise, the needles are stopped. In this way, a card might enforce, quite automatically, just the type of needle motions that would produce a pattern. Different cards would produce different patterns.

To be sure, devising the punched card required considerable intelligence and ingenuity in the first place, but once the cards were designed and in place, the machine did not need brains; it worked automatically.

The Jacquard looms spread rapidly throughout France and eventually through Great Britain as well. The holes in the card were a primitive kind of yes-no mechanism; a century and a half later a much more subtle type of yes-no mechanism would serve as the basis for digital computers.

CERES

The German astronomer Johann Daniel Tietz, in Latin Titius (1729–1796), had suggested in 1766 that it was possible to work up a simple arithmetical series that would give the relationship of the distance of the various planets from the Sun. Six years later, the German astronomer Johann Elert Bode (1747–1826) popularized that series, which came to be known as *Bode's law.*

Bode's law didn't seem to have much scientific significance. Still, when Uranus was discovered (see 1781), it turned out to be just where Bode's law predicted a planet would be. That rather impressed astronomers. It so happened that Bode's law also predicted that a planet ought to exist in the space between Mars and Jupiter, but no such planet was known. It might be that it was a small one and had gone unnoticed.

For that reason, another German astronomer, Heinrich Wilhelm Matthäus Olbers (1758–1840), began to organize a group that would take different portions of the sky and search them for any moving object that might be that planet.

While the preparations were under way, an Italian astronomer, Giuseppe Piazzi (1746–1826), who wasn't looking for the planet, came across it while he was working at an observatory in Sicily. It was a dim object, not quite visible to the unaided eye, that changed position from night to night. The discovery was made on January 1, 1801, the first day of the nineteenth century.

Piazzi began to follow its course. Since it moved more rapidly than Jupiter and more slowly than Mars, its orbit must lie between the orbits of those two long-known planets. Since it was much dimmer than either of those planets, it must be far smaller, too, and we now know that it is about 640 miles in diameter, smaller by far than even Mercury, till then the smallest known planet. That was why it had not been noted earlier.

Nevertheless, it fit Bode's law, and Piazzi named it *Ceres,* after the Roman goddess most closely associated with Sicily.

INVERTEBRATES

Linnaeus (see 1735) and others in the past three-quarters of a century had elaborated the taxonomy of vertebrates (those animals with backbones—mammals, birds, reptiles, amphibia, and fish) in considerable detail.

The less familiar and more diverse animals without backbones *(invertebrates)* were less carefully considered. Linnaeus had lumped them all into the class *Vermes* (Latin for "worms").

The French naturalist Jean-Baptiste de Lamarck (1744–1829) tackled the problem in publications beginning in 1801. He divided the invertebrates into groups and created such divisions as the crustaceans (crabs and lobsters) and echinoderms (starfish and sea urchins). He distinguished between the eight-legged arachnids (spiders and scorpions) and the six-legged insects. In fact, Lamarck was the first to use the terms *vertebrate* and *invertebrate,* and he popularized the term *biology* for the study of life.

In founding the modern science of *invertebrate zoology,* Lamarck began the process of understanding the importance of invertebrate life. All the vertebrates are included in but a single phylum, while there are about twenty-two phyla of invertebrates. The number of species of insects alone far exceeds the number of species of vertebrates; indeed, it exceeds the number of species of all other animals combined.

ULTRAVIOLET

Herschel's discovery of infrared radiation (see 1800) naturally created a stir. Ritter (see 1800) was also studying the Sun's spectrum, but he was more interested in the chemical changes it brought about.

It had been known for nearly two centuries that light broke down the white compound silver nitrate, darkening it (by liberating tiny specks of metallic silver). This was first reported in 1614 by an Italian chemist, Angelo Sala (1576–1637).

Ritter soaked strips of paper in silver nitrate solution and placed them in different sections of the spectrum to see how rapidly they darkened. He found that the darkening was least rapid in the red end of the spectrum and took place faster and faster as one progressed toward the violet end.

Ritter, perhaps influenced by Herschel's experience, proceeded to place strips of soaked paper beyond the violet end, where nothing could be seen. There the darkening proceeded faster still. Apparently, there was radiation beyond the violet end of the spectrum as well as beyond the red. The new radiation was called *ultraviolet* (the prefix *ultra* is from a Latin word meaning "beyond").

LIGHT WAVES

The controversy over the nature of light—whether it consisted of a stream of particles or of tiny waves—had been going on for a century.

In 1801 an English physicist, Thomas Young (1773–1829), began a series of experiments that seemed to settle the matter. He showed, for one thing, that the kind of diffraction noted by Grimaldi (see 1665) did exist.

Then too, he allowed two separate beams of light emerging from two narrow slits to overlap and found *interference,* for the overlapping beams showed alternate strips of light and dark.

If light consisted of waves, then the waves of the two bands of light might move up and down in unison and reinforce each other in some places. In other places, one might move up while the other moved down, and they would cancel each other. Such interference is well known in the case of sound waves and water waves. Streams of particles, on the other hand, could scarcely produce interference effects.

It took a while for Young's demonstration to be understood and accepted, but once it was, it was seen that light was a wave phenomenon. The different colors of the spectrum marked out light with different lengths of waves (or different *wavelengths*). Short wavelengths are refracted (bent in their path) more than long wavelengths, so red light (the least refracted) has the longest waves, while orange, yellow, green, blue, and violet have successively shorter waves. From Young's work, it began to seem probable that infrared radiation had waves longer than those of red light, while ultraviolet radiation had waves shorter than those of violet light.

Because light cast sharp shadows and the diffraction effects were so small, light waves had to be very tiny. Young calculated from his interference experiments that they must be less than a millionth of a meter long.

Two kinds of wave action are known, however. There are *longitudinal waves,* in which the vibration is back and forth in the direction in which the waves are moving. Sound waves are of this type. In *transverse waves,* the vibration is up and down in the direction at right angles to that in which the waves are moving. Water waves are of this type.

Young rather suspected that light consisted of longitudinal waves, but in this he was wrong.

NIOBIUM

An English chemist, Charles Hatchett (1765–1847), analyzed an unusual mineral in the British Museum, one that had been sent there from Connecticut in pre-Revolutionary days. In 1801 he reported a new element in the mineral, an element he named *columbium* in honor of the United States, which was sometimes known by the sobriquet of Columbia. There was, however, a dispute over whether the substance was really a new element or not. By the time it was finally decided that it was, its name had become *niobium* and that is now official, so the United States lost the honor.

IN ADDITION

Bonaparte's new war with Austria ended with the Treaty of Luneville on February 9, 1801, which placed Italy once more in French hands. In addition, France annexed all the land west of the Rhine River and brought a virtual end to the Holy Roman Empire.

Paul I of Russia, more and more clearly insane, was killed in a palace coup on March 11, 1801. His son (who may have been involved in the coup) succeeded as Alexander I (1777–1825).

The United States had a population of 5.3 million now, half that of Great Britain. Russia was the most populous European power with 33 million, but India had 131 million and China 295 million. London was the largest European city with 864,000, but there were a number of cities in the Far East that had over a million in population. Canton, with 1.5 million, was the world's largest city.

1802

ASTEROIDS

Olbers and his group of German scientists were shaken in their plan to search for a planet with an orbit between those of Mars and Jupiter by Piazzi's discovery of Ceres (see 1801). Ceres was so small, however, that it scarcely seemed to be a planet. Olbers's group decided to continue the search.

In 1802 they discovered another planet between Mars and Jupiter, and they named it *Pallas,* one of the names of the goddess Athena. In 1804 they found still another, and in 1807, a fourth. These were named *Vesta* and *Juno,* after two of the sisters of Jupiter (Zeus). The three new planetary bodies, however, were even smaller than Ceres.

Herschel (see 1781) suggested they be called *asteroids* (Greek for "starlike"), because in the telescope they seemed mere dots of light, like stars (because of their small size) rather than orbs, like the larger planets.

Eventually it was realized that there were a vast number of such bodies, perhaps as many as a hundred thousand of them, between the orbits of Mars and Jupiter. This region came to be called the *asteroid belt.* Ceres was the largest of the asteroids (hence the first discovered), containing about a tenth the mass of all the other asteroids put together.

TANTALUM

In 1802 a Swedish chemist, Anders Gustav Ekeberg (1767–1813), began the analysis of minerals from Finland and located a new metal, which he named *tantalum* in honor of Tantalus (who in the Greek myths was tortured by being placed in water up to his chin, which he was never able to drink, whence the word *tantalize*). Presumably, Ekeberg found isolating the element a tantalizing task.

IN ADDITION

On March 27, 1802, the Peace of Amiens put an end to the European war after ten years. Even Great Britain, the most inveterate of the enemies of France, reluctantly suspended hostilities. On August 2, Bonaparte seized the opportunity of peace with victory to have himself declared First Consul for life with the privilege of choosing his successor. Bonaparte also established the Legion of Honor and reestablished slavery in the West Indies.

1803

ATOMIC THEORY

From the time of Robert Boyle's experiments on the compression of gas (see 1662), the evidence for the atomic nature of matter had been accumulating.

In 1803 an English chemist, John Dalton (1766–1844), advanced a summary of atomic thinking, backing it up with the evidence that had accumulated, including particularly Proust's law of definite proportions and his own investigations of the way in which gases behaved. (In 1808 he gave his arguments formally in a book entitled *New System of Chemical Philosophy*.)

Essentially, Dalton returned to the Greek notions of Democritus (see 440 B.C.) that all matter was made up of tiny, indivisible particles. Dalton even used Democritus's word *atom* for these particles. The difference was that Democritus's theory was based on speculation only, whereas Dalton's was based on a century and a half of careful chemical observation.

The Greeks, being geometers, naturally thought that atoms differed among themselves in shape. Dalton, in whose time weight and measurements had grown important, maintained that the difference was one of weight, and he pioneered the concept of *atomic weight*.

For instance, 8 grams of oxygen will combine with 1 gram of hydrogen to form 9 grams of water. Suppose that water is formed through a combination of one atom of oxygen and one atom of hydrogen (the result would be a *water molecule*). In that case, one atom of oxygen would have to be eight times as massive as one atom of hydrogen. If hydrogen was supposed to have an atomic weight of 1, oxygen would have one of 8.

Of course, water might be made up of molecules containing any number of oxygen and hydrogen atoms. Dalton supposed it to be built up of one of each merely because that was the simplest possible situation. Until such time as more was known about molecular makeup, the values obtained for atomic weights would be dubious. In Dalton's table of atomic weights (the first ever compiled), many were indeed wide of the mark.

METEORITES

Chladni had forced scientists to consider the possibility of meteorites seriously (see 1794). Continuing reports of stones falling from heaven forced investigation. A French physicist, Jean-Baptiste Biot (1774–1862), was asked to look into reports of such falls in a region 100 miles west of Paris.

After a painstaking investigation, Biot gave it as his opinion that the reports were correct and that meteorites did exist and did fall from the sky. Presumably, the discovery of the first two asteroids had forced scientists to understand that small bodies might be moving about the Sun and that some might occasionally collide with the Earth.

CERIUM, OSMIUM, IRIDIUM

New elements were being discovered rapidly. In 1803 the Swedish chemist Jöns Jakob Berzelius (1779–1848) and his friend the Swedish mineralogist Wilhelm Hisinger (1766–1852) discovered *cerium*, which they named after the newly discovered asteroid Ceres.

The British chemist Smithson Tennant (1761–1815), who had worked with Wollaston (see 1800) and was therefore interested in platinumlike elements, discovered two of them in 1803. One he named *osmium,* from the Greek word for "smell," because of the stench of one of its compounds, and the other *iridium,* from the Greek word for "rainbow," because of the different colors of its compounds.

IN ADDITION

In the United States, Ohio entered the Union as the seventeenth state on March 1, 1803. At about this time, President Jefferson sent his friend James Monroe (1758–1831) to France with instructions to offer two million dollars for New Orleans, or ten million for the entire mouth of the Mississippi. As it happened, Bonaparte had lost interest in Louisiana, and the French foreign minister, Charles-Maurice de Talleyrand-Périgord (1754–1838), blandly offered the Americans *all* of Louisiana for fifteen million dollars. Jefferson grabbed at it. On April 30, 1803, the *Louisiana Purchase* was carried through and the United States more than doubled its area, extending now to the Rocky Mountains in the north and to Texas in the south.

Meanwhile, the British were fighting steadily in India against various native rulers. There they found a capable general in Arthur Wellesley (1769–1852), who was later to become the Duke of Wellington.

In the Pacific, the Hawaiian Islands were united under the rule of Kamehameha I (1758?–1819).

1804

SCIENTIFIC BALLOONING

The first important use of balloons for scientific research came in 1804, when Biot (see 1803) and Joseph Louis Gay-Lussac (1778–1850) made a balloon ascension that reached a height of 4 miles, higher than any peak in the Alps. They took the opportunity to test the composition of the air at that height and also to study the nature of Earth's magnetic field. They found no change in either compared with measurements at sea level.

This was the beginning of high-altitude investigations that were to carry human beings beyond the atmosphere altogether a century and a half later.

STEAM LOCOMOTIVE

If steam can turn paddle wheels in the water, it can also turn wheels on land, so that one can imagine a steam *locomotive* (from Latin words meaning "moving from place to place"). To be sure, such a steam-powered land vehicle would lose too much energy crossing over rough ground. A smooth road would have to be built.

It occurred to a British inventor, Richard Trevithick (1771–1833), that what was needed was iron rails on which the wheels would fit and along which they could move, a *railroad.* He showed that, even if the rails were smooth and the wheels rolling on them were smooth as well, there was still sufficient traction for the locomotive to move itself and a train of coaches attached. Trevithick showed this by actual demonstration as early as 1801. In 1804, one of his locomotives pulled five loaded coaches for 9.5 miles at nearly 5 miles an hour.

Nevertheless, Trevithick could not manage to turn his locomotive into a commercial success.

MISSOURI RIVER

President Jefferson wanted the new territory of Louisiana to be explored. For the purpose, he chose Meriwether Lewis (1774–1809), who in turn chose William Clark (1770–1838). These two, with a party of about forty young men, made up the *Lewis and Clark Expedition*.

They went to St. Louis where they remained over the winter of 1803–1804. Then on May 14, 1804, they moved up the Missouri River and followed it back to its source. This brought them to the American border, but they went on anyway into the Oregon Territory, the only part of the American continents that had still not been effectively claimed by any one power. They followed the Columbia River to the Pacific Ocean, which they reached on November 15, 1805, and then returned to St. Louis, which they reached on September 23, 1806. This was the first trip across the United States from ocean to ocean and back. The Lewis and Clark Expedition brought back information on the Indian tribes of the region, on animal life (including huge herds of bison), on plant life, and on natural features.

IN ADDITION

In France Bonaparte had himself crowned Emperor Napoléon I on May 18, 1804. The *Code Napoléon*, the revision of French law that he sponsored, went into effect on March 21, 1804. It was the most permanent and admired product of the Napoleonic era and remained the basis of French jurisprudence, as well as influencing the law of continental Europe and Latin America.

A black republic was established in Haiti in 1804, and the island has existed under black rule ever since.

1805

MORPHINE

The use of certain plants to dull pain and discomfort and to give a feeling of well-being is old indeed. In Homer's *Odyssey*, mention is made of those who ate the lotus and forgot all else, craving only to continue eating it. There is also mention of a drug *nepenthe*, which calmed and soothed. It is very tempting to think of these as representing opium, which Dioscorides (see 50 B.C.) described. (Opium is one item that traveled from the West to the Fast East, and not vice versa.) An alcoholic extract of immature opium blooms was called *laudanum* and had first been introduced by Paracelsus (see 1556, Mineralogy).

In 1805 the German chemist Friedrich Wilhelm Adam Ferdinand Serturner (1783–1841) isolated a chemical from laudanum that he found to be the active ingredient. It was much more efficacious in dulling pain and inducing sleep than the juice itself. It was eventually called *morphine*, from the Greek word for "sleep."

Morphine has been an important adjunct to medical practice ever since, though its addictive effect was not at first understood. Its discovery initiated the study of *alkaloids,* which are important nitrogen-containing plant products with marked physiological effects even in small doses.

IN ADDITION

Austria allied itself with Russia again (with British money, as always, helping to smooth the way) and tempted fate by making war on Napoléon once more. Napoléon attacked the combined armies of Austria and Russia at Austerlitz, where on December 2, 1805, he won his greatest victory. Austria again gave up and at the Treaty of Pressburg, on December 26, 1805, was stripped of Venetia and its western provinces.

However, the redoubtable Nelson met the combined French and Spanish fleets near Gibraltar and destroyed them at the Battle of Trafalgar on October 21, 1805. After that Napoléon was penned up on the continent of Europe, while Great Britain was free to roam the world.

Egypt took advantage of the confusion and, under the leadership of Mehemet Ali (1769–1849), broke free of the Ottoman Empire.

1806

ASPARAGINE

Vauquelin (see 1797), who had earlier discovered metallic elements such as chromium and beryllium, isolated a substance from asparagus, which he called *asparagine.* In itself, it might not have seemed important, but as was eventually realized, it was the first to be isolated of a set of compounds extremely important to life, called *amino acids.*

IN ADDITION

On July 12, 1806, Napoléon organized Germany (exept for Prussia and Austria) into the *Confederation of the Rhine,* a French puppet. Francis I (1768–1835), now emperor of Austria, recognized this as the final end of the Holy Roman Empire and abdicated as Holy Roman Emperor. Prussia thereupon formed an alliance with Russia and went to war against Napoléon. Napoléon promptly struck and on October 14, 1806, crushed the Prussian army. He marched into Berlin on October 27, and Prussia lay helpless at his feet.

In Berlin, Napoléon issued the *Berlin Decree,* which was designed to put an end to all trade between Europe and Great Britain. This was an attempt to put economic pressure on the island nation whose fleet made it impossible for him to touch it in any other way. This was called the *Continental System.*

1807

SODIUM AND POTASSIUM

By this time, some thirty-eight substances were known that are today recognized as elements; most of them being metals. There were some substances that were known to be oxides—combinations of oxygen with a metal—but the combination could not be broken up and the free metal could not be isolated. None of the ordinary chemical methods used to isolate other metals would work.

It was known, however, that an electric current would break up water molecules into hydrogen and oxygen when more customary chemical methods failed (see 1800). An electric current might do the same for the recalcitrant oxides.

Davy (see 1800, Nitrous Oxide) grew interested in the problem and constructed a battery with over two hundred and fifty metallic plates, the strongest ever built to that time.

On October 6, 1807, he passed an electric current through molten potassium carbonate and liberated a metal, which he called *potassium.* The little globules of shining metal, when added to water, tore the water molecule apart as the metal eagerly recombined with oxygen, and the liberated hydrogen was heated to the point where it burst into flame. A week later, he isolated metallic *sodium* from sodium carbonate.

The next year, by similar methods, Davy isolated the elements *barium, strontium, calcium,* and *magnesium.* All were active elements that held on to oxygen tightly and would not have been isolated easily by nonelectric techniques.

These discoveries excited the scientific world enormously and greatly stimulated research into *electrochemistry*.

STEAMBOAT

Since Fitch's failure to make his steamboat succeed (see 1787), the notion had not been taken up again. But then the American inventor Robert Fulton (1765–1815) tackled the project. In 1807 he built the *Clermont,* 133 feet long. This vessel performed well, steaming up the Hudson from New York to Albany in 32 hours, maintaining an average speed of nearly 5 miles per hour. Soon he had a fleet of steamboats in operation, and unlike Fitch, he was commercially successful. For this reason, Fulton is usually considered the inventor of the steamboat.

IN ADDITION

The subjection of Prussia left its ally, Russia, still in the field. On June 14, 1807, a battle between the French and Russians at Friedland in East Prussia ended in a French victory, and Napoléon occupied the easternmost Prussian provinces. From July 7 to 9, 1807, Napoléon and Alexander I of Russia met at Tilsit on the Niemen River, which separated Prussia and Russia. By the Treaty of Tilsit, Prussia lost all its western provinces to France and gave up what it had gained in the second and third partitions of Poland. Out of that Prussian loss, the *Grand Duchy of Warsaw* was formed. Thus Poland reappeared (briefly) on the map, but only as a French puppet.

The most important neutral in the world at this time was the United States. It profited by trading with both sides in the European struggle, but each side was anxious to stop American trade with the other. Thus British ships took to stopping American ships on the high seas and searching them for deserters from the British fleet. This came to a peak on June 22, 1807, when a British warship, *Leopard,* fired on an American ship, *Chesapeake,* and removed four deserters. President Jefferson then declared an embargo on American trade with Europe, hoping that such economic pressure would force Great Britain to ease up. It did no such thing. It merely plunged New England into a deep economic depression.

1808

POLARIZED LIGHT

Bartholin's discovery that Iceland spar showed double refraction and produced two rays of light (see 1669) had never been properly explained.

In 1808 a French physicist, Étienne-Louis Malus (1775–1812), was playing idly with a crystal of Iceland spar and found that the sunlight reflected from a window produced only a *single* ray of light after passing through the crystal. What's more, as he turned the crystal, the ray of light faded and the other one appeared. When the crystal was at right angles to its original direction, only the other ray of light appeared, the first having disappeared.

Malus felt that light might have different poles, as magnets do, and that one beam of light had its poles at right angles to the other. He therefore called the rays *polarized light,* and the name stuck even though his theory was wrong.

Polarized light turned out to be enormously useful to chemists.

IN ADDITION

In March 1808, Napoléon, determined to make his Continental System work and unsure that Spain would cooperate sufficiently, deposed Charles IV of Spain (1748–1819) and made his own older brother, Joseph (1768–1844), king. This was his first serious mistake, for though Charles IV was a poor king, the Spanish people didn't want a Frenchman in his place. They rebelled on May 2. It was a *guerrilla* uprising (Spanish for "little war"), and it gave its name to all such fighting in the future. Spain continued to consume French men and money over the next four years.

In the United States, Jefferson, having served two terms, declined a third. James Madison (1751–1836) of Virginia was elected the fourth president.

1809

MECHANISM OF EVOLUTION

While the fact that biological evolution had taken place was suspected by many scientists, until now no one had suggested a mechanism that would allow it to take place. Why should some catlike creature in the course of generation after generation slowly change, some into lions, some into tigers, and some into pussycats?

The first to attempt an answer was Lamarck (see 1801) in his book *Zoological Philosophy*, published in 1809. Here he suggested that particular animals might use a part of the body steadily, or not use it, and that those parts might develop slightly or degenerate slightly as a result, and that such developed or degenerated parts could be inherited by their young, which, by use or disuse, would continue the process.

Thus some antelopes, by stretching upward to reach leaves, would gradually develop slightly longer necks and legs, which would be inherited by the young, which, also straining, would continue the process so that the giraffe would evolve. Other antelopes, by dint of constantly fleeing from predators, strengthened their leg muscles and became very fleet as the generations progressed. Water birds, by using their feet to push water backward, developed webs, while moles, who didn't need eyes underground, gradually lost them.

This sort of thing is referred to as *the inheritance of acquired characteristics*. Experiments would show that acquired characteristics were *not* inherited. Nevertheless, the advancement of a mechanism for evolution, even if a wrong one, heightened interest in the matter.

AERODYNAMICS

The dream of active flight had exercised the human imagination for millenia. Usually, however, human beings could only think of imitating the birds. The Greek myth of Daedalus had that legendary inventor building a frame to which he stuck feathers in wax and which he moved in birdlike fashion with his arms.

The first person to consider the principles that would really keep objects in the air was a British scientist, George Cayley (1773–1857). He visualized flying devices with fixed wings that presented appropriate surfaces to the air, tails with control surfaces to allow turning and braking, and propulsive mechanisms. He described all this in publications that began to appear in 1809.

In these he founded the science of *aerodynamics*. While the state of technology at that time did not allow the construction of such devices, when they did come a century later, they fulfilled Cayley's requirements.

IN ADDITION

Both Prussia and Austria, learning from defeat, began to reform their governments and economies. In Austria, Archduke Charles Louis (1771–1847) reorganized the army and once again risked war with Napoléon in April 1809. Once again, Napoléon struck rapidly. He hurried back from Spain, drove into Germany, and took Vienna on May 13. Then, on May 21, at the Battle of Aspern, just east of Vienna, Napoléon was *defeated* by Archduke Charles. It was Napoléon's first outright defeat. To be sure, he recovered, and at the Battle of Wagram, on July 5, defeated Archduke Charles (though at a high cost in casualties), and once again Austria was forced to surrender. By the Treaty of Schonbrunn on October 14, Austria ceded territory to Russia, to France, and even to the Grand Duchy of Warsaw. Napoléon's grip on Europe seemed firmer than ever.

Napoléon, feeling the need for an heir and knowing that the Empress Josephine (now forty-six years old) couldn't give him one, divorced her on December 1, 1809, and prepared for a glorious second marriage.

The Welsh philanthropist Robert Owen (1771–1858) was trying in these years to ameliorate the harsh conditions under which workers in British mills labored. He suggested that children under the age of ten not be employed and that some elementary care of children's health and education be taken. For this, of course, he was vilified.

1810

BRAIN

In 1810 a German physician, Franz Joseph Gall (1758–1828), published the first volume of a four-volume treatise on the nervous system. In it he stated that the gray matter on the surface of the brain and in the interior of the spinal cord was the active and essential part, and that the white matter, deeper in the brain and on the surface of the spinal cord, was connecting material. In this he was correct.

Gall believed that the shape of the brain had something to do with mental capacity and that different parts of the brain were involved with different parts of the human body. There was something to this, too, but Gall went much too far. He believed he could correlate the shape of the brain with all sorts of emotional and temperamental qualities and that the shape of the brain could, in turn, be deduced from the superficial unevennesses of the skull. This marks the beginning of the pseudoscience of *phrenology* (from Greek words meaning "study of the mind") in which character is supposedly analyzed by feeling the bumps on the head.

CHLORINE

Davy (see 1800) had worked with hydrochloric acid (a strong acid) and showed that it contained no oxygen. This was the final blow to the assumption that oxygen was essential to acids. However, hydrochloric acid did contain chlorine, and Scheele (see 1774) had thought chlorine to be an oxygen-containing compound. In 1810 Davy showed this was not so, and that chlorine was an element. For this reason he rather than Scheele usually receives credit for the discovery of chlorine.

IN ADDITION

On March 11, 1810, Napoléon married Marie-Louise (1791–1847), daughter of Emperor Francis I of Austria. In this way he allied himself to the proud Hapsburg family (to his satisfaction and undoubtedly to Francis I's humiliation).

Napoléon had made his younger brother, Louis (1778–1846), king of Holland in 1806. Louis opposed the Continental System because he felt the Dutch would be ruined if they adhered to it. When Napoléon insisted, Louis abdicated and fled the country on July 1, 1810. Napoléon annexed Holland to France on July 9, along with large sections of the German coast, but smuggling continued and the Continental System leaked.

Meanwhile, Napoléon, to keep Alexander I of Russia pliable, gave him a free hand in Finland, which was part of the Swedish dominions. Alexander I used it to annex Finland on September 17, 1810. Charles XIII of Sweden (1748–1818) had no heir, so the Swedes chose one of Napoléon's generals heir to the throne, as the best way of protecting themselves against further depredations from the east. Napoléon, thinking this the best way of keeping Sweden a French puppet, agreed.

The United States took advantage of Spain's troubles to annex West Florida, the portion of the gulf coast between Florida and the Mississippi River, on October 27, 1810. The population of the United States was now 7.2 million.

1811

AVOGADRO'S HYPOTHESIS

It was clear that all gases expanded by the same amount as temperature rose, provided the pressure remained constant. In 1811 an Italian physicist, Amedeo Avogadro (1776–1856), suggested that this might mean that all gases—at the same volume, pressure, and temperature—were made up of the same number of particles. This came to be called *Avogadro's hypothesis.*

If this is so, since water upon being broken up by an electric current decomposes into hydrogen and oxygen, with hydrogen having twice the volume of oxygen, then twice as many particles of hydrogen must be formed as of oxygen. This in turn makes it appear that water particles are not made up of one hydrogen atom and one oxygen atom, as Dalton had thought, but may be made up, at the simplest, of two hydrogen atoms and one oxygen atom.

In that case, since the oxygen in water has eight times the mass of the hydrogen, the oxygen atom must be eight times as massive as the two hydrogen atoms put together, or *sixteen* times as massive as a single hydrogen atom.

Again, if all gases at a given temperature, pressure, and volume are made up of the same number of particles, and if the density of one gas is twice that of another, the mass of each particle in the first gas is twice that of the other.

Thus, the density of water vapor is nine times that of hydrogen at the same temperature, but since the oxygen atom has sixteen times the mass of the hydrogen atom, then the weight of the water particle is 16 + 1 + 1, or 18. Why isn't the density of the water vapor eighteen times that of the hydrogen? It may be because the hydrogen particles are made up, not of single hydrogen atoms, but of combinations of *two* hydrogen atoms. In similar fashion, Avogadro argued that oxygen and nitrogen particles were made up of two atoms each.

Avogadro distinguished between single atoms and these combinations of atoms that made up the particles of compounds. The combinations of atoms he called *molecules* (from Latin words for "small masses"). Thus, there was an oxygen atom, and there was also an oxygen molecule made up of two oxygen atoms. There was a water molecule made up of one oxygen atom and two hydrogen atoms. And so on.

Avogadro's hypothesis, if fully applied,

would explain a great deal about atomic weights and about the atomic consitution of compounds. Unfortunately, the hypothesis was largely ignored for half a century, and chemists remained unnecessarily confused in many ways during that time.

IODINE

A French chemist, Bernard Courtois (1777–1838), was in the business of manufacturing potassium nitrate (needed in gunpowder). He got it from potassium carbonate (potash), which in turn he got from seaweed. As one of the steps to get the potassium carbonate, he heated the seaweed in acid. One day in 1811 he added too much acid and, on heating, obtained a beautiful violet vapor. On condensing the vapor, he produced dark, lustrous crystals. He suspected it might be a new element and passed it on to other chemists for confirmation. It *was* a new element, and Davy (see 1800) suggested it be named *iodine*, from the Greek word for "violet."

IN ADDITION

On March 20, 1811, Napoléon finally had his first (and only) legitimate son, François-Charles-Joseph Bonaparte, or Napoléon II (1811–1832).

In Great Britain, the cruel economic consequences of the Industrial Revolution for the lower classes led to riots, which resulted in the destruction of mills and machinery. The memory of a half-wit named Ned Ludd (fl. 1779) who had destroyed machinery some years before was evoked and the rioters were called *Luddites*. Those who are violently antitechnology are called Luddites to this day.

George III of Great Britain, who had suffered bouts of insanity (actually due to a disease known as *porphyria*, according to recent research), became permanently insane in 1811. His oldest son, George, the Prince of Wales (1762–1830), became Prince Regent.

Steamboats were continuing to make progess. In 1809 the first steamboat to appear on ocean water was the *Phoenix* under the command of Moses Rogers (1779–1821), who took it from New York around New Jersey to the Delaware River. In 1811 the *New Orleans*, the first steamboat on the Mississippi River, traveled from Pittsburgh to New Orleans.

1812

CATALYSIS

From prehistoric times, human beings have known that some substances can bring about a change without themselves being consumed. In fact the substances may increase in quantity. The best-known example is yeast, which can spread its effect through bread dough, to all intents and purposes indefinitely. But then, yeast was eventually discovered to be alive.

It would be much more surprising if something that was not alive and did not reproduce itself were to be capable of bringing about a change without being consumed.

A Russian chemist of German birth, Gottlieb Sigismund Constantin Kirchhoff (1764–1833), boiled a suspension of starch in water to which a little bit of sulfuric acid had been added. If the starch had been boiled in water *without* the sulfuric acid, nothing much would have happened. *With* the sulfuric acid, the starch was destroyed, and in its place, a substance was produced that was freely soluble in water and that tasted sweet. It was a form of sugar, and it was named *glucose,* from the Greek word for "sweet."

Several discoveries were made here. First, glucose was studied for the first time, and (as was eventually discovered) it is one of the key substances of living tissue. Second, it was the first hint that starch was built up of glucose units and could be broken down to glucose again. Third, the sulfuric acid that made the breakdown to glucose possible was not itself consumed in the reaction.

In later years, Berzelius (see 1803) gave the phenomenon of participation-without-being-consumed the name of *catalysis* (from Greek words meaning "to break down"). Sulfuric acid was an example of a *catalyst* that brought about the breakdown of starch.

CATASTROPHISM

Cuvier (see 1798) continued to find important and interesting fossils. In 1812 he reported the fossil remains of a creature that clearly must have had wings and been able to fly but whose skeleton showed it to have been a reptile. It was named *pterodactyl,* from Greek words meaning "wing-finger," because the membrane of its wing was stretched out along one elongated finger.

Cuvier could see that fossils represented creatures that were extinct and that the deeper the stratum and the older the remains, the less the fossils resembled modern organisms. Nevertheless, he couldn't bring himself to accept the idea of evolution.

Instead, in a book entitled *Inquiry into Fossil Remains,* published in 1812, he advanced the notion that there had been numerous creations, each one ended totally by some catastrophe and followed by a new creation of life closer to that of the present.

This view, called *catastrophism,* is usually considered the opposite of James Hutton's uniformitarianism, and for a long time the two were considered mutually exclusive. However, it is quite possible that Earth's history shows periods of uniformitarianism interspersed by episodes of catastrophism (though none of the catastrophes, so far, seem to have been complete).

Cuvier was the first to go into detail on the appearances of ancient forms of life, and he attempted to classify them, as far as possible, by using the same system used for living species. In this way, he is considered to have founded *paleontology,* the study of ancient extinct forms of life.

UNIVERSE AS MACHINE

In 1812 Laplace (see 1783) suggested that if the mass, position, and velocity of every particle in the Universe were known, then the entire past and future of the Universe could be calculated.

In other words, he viewed the Universe as a vast automatic machine that, once set in motion, would follow an unswervable path. On the one hand, this would explain how God, with supernatural knowledge, could know all the past and the future, but on the other hand, it would remove any need for God, except to create the whole thing and set it in motion.

This idea of a mechanistic universe dominated scientific thinking for a little over a century, but then it became clear that the Universe was more complex than a machine could be and that it might be inherently unpredictable except in a statistical sense—if that.

IN ADDITION

It was clear that Napoléon saw no way of making the Continental System work without using force on Russia. Alexander I of Russia could see that coming and hastened to prepare. Napoléon, meanwhile, built up an army that may have totaled 600,000 men, probably the largest army thrown into a single attack up to that time, and on June 22, 1812, he hurled this army into Russia. The Russians retreated before him. This was an effective policy, for even with no major battles, Napoléon was losing men and horses, which with longer and longer supply lines, he could not replace, while the Russians could replace their own losses indefinitely.

Finally, on September 7, 1812, at Borodino, 70 miles west of Moscow, the Russians were forced into battle. Moscow could not be given up without a fight. The battle was the bloodiest ever fought up to that time, with both sides losing heavily—but the Russians could replace their soldiers, and Napoléon could not.

The Russians then continued their retreat, and on September 14, Napoléon entered Moscow and took over the Kremlin. However, when Napoléon entered Moscow, the Russians simply burned it down.

Napoléon, finding that there was no move toward surrender and that half his army had melted away, had no choice but to leave, and on October 19, he began his retreat. The Russian cavalry dogged his steps, forcing him to remain on the devastated route along which he had advanced. His army disintegrated, and when they reached the Russian border there were only demoralized remnants left. Napoléon himself reached Paris on December 18, a defeated general with a vanished army.

Nor was he doing better in Spain. Wellington beat the French at Salamanca on July 22, 1812, and entered Madrid a month later.

Meanwhile, the United States, steadily harassed at sea by the British navy, declared war on Great Britain on June 18, 1812 (the *War of 1812*). To the amazement of the world, the American warships beat the British in battle after battle. In the most famous of these battles, the American warship *Constitution,* later called Old Ironsides, destroyed the British *Guerriere,* in almost no time, on August 19.

1813

CHEMICAL SYMBOLS

Once the notion of atoms was advanced, the question arose of how to represent them. Simplicity recommended the use of letters, and in 1813 Berzelius (see 1803) advanced a system that was eventually adopted. Initial letters were used: *H* for a hydrogen atom, *C* for a carbon atom, *O* for an oxygen atom, and so on. If more than one element had the same initial, a second letter from the body of the name was used, so that *Ca* was calcium, *Cl* was chlorine, and *Cr* was chromium. In the case of ancient elements that had different names in different languages, the Latin name was used in the symbol, so that gold was *Au (aurum)*, silver was Ag *(argentum)*, and so on.

Compounds were symbolized, if simple enough, by simply listing the atoms. Since water molecules are made up of two hydrogen atoms and one oxygen atom, water is H_2O. Ammonia is NH_3. Sulfuric acid is H_2SO_4. And so on.

PLANT CLASSIFICATION

The French botanist Augustin-Pyrame de Candolle (1778–1841) began a large encyclopedia of plant life in 1813, a project that outlasted him. Only seven of the twenty-one volumes had been published by the time of his death. His system of plant classification was far more scientific than that of Linnaeus and is still used today. It was Candolle who, in 1812, first used the word *taxonomy* to represent species classification.

IN ADDITION

Nothing fails like failure. Napoléon's debacle in Russia was followed at once by anti-Napoléon stirrings in Austria, Prussia, and elsewhere. Europe suddenly had the courage to combine against France.

As soon as spring came, Russian troops marched into Germany, where they were joined by Prussians and Austrians. Napoléon fought with his usual vigor and on August 26–27, 1813, won the Battle of Dresden. Twenty years of war, however, had drained the French. On October 16–19, 1813, Napoléon was smashed at the Battle of Leipzig (the *Battle of the Nations*). He had lost Germany and now had to retreat to the borders of France itself. He had had to withdraw troops from Spain to fight in Germany, and by the time of the Battle of the Nations, Spain was also lost and Wellington's army stood on the southwest border of France.

Great Britain continued to fight in North America in desultory fashion. The American officer Oliver Hazard Perry (1785–1819) had ships built to tackle the British fleet in the Battle of Lake Erie and won control of the Great Lakes.

1814

SPECTRAL LINES

Since Newton had first studied the spectrum (see 1666), nothing much had followed. Wollaston (see 1800) in 1802 had observed a few dark lines in the spectrum but had supposed them to be boundary lines between the colors and had not followed up on the matter.

Meanwhile a German physicist, Joseph von Fraunhofer (1787–1826), was meticulously manufacturing the best lenses and prisms yet made. In 1814 he was testing one by sending sunlight through a narrow slit and then through the prism. In effect, he produced innumerable lines of light, each an image of the slit and each containing a very narrow band of wavelengths. Some wavelengths were missing, however, so that the slit images at those particular wavelengths were dark. The result was that the solar spectrum was crossed by dark lines.

In theory, these should always have been visible, even to Newton, but if the prism was somewhat imperfect and the slit too wide, there would be enough fuzziness to obscure them. Where Newton had reported none and Wollaston seven, Fraunhofer, with his excellent instrument, detected nearly six hundred lines.

Fraunhofer went on to measure the position of the more prominent lines, which he marked off with letters from *A* to *K*, and showed that they always fell in the same portion of the spectrum, whether the light came directly from the Sun or was reflected from the Moon or planets. Eventually he mapped the wavelengths of several hundred of these *Fraunhofer lines*, as they were called.

Not much was done with these spectral lines for the next half-century, but eventually they proved to be major items in the research armory of chemists and astronomers.

IN ADDITION

Napoléon, still unable to conceive of defeat, refused the peace terms offered to him and fought the enemy inside France. On March 31, 1814, the Allied armies (Russian and German) marched into Paris. When his own marshals refused to fight any longer, Napoléon was forced to abdicate on April 11. The Allies exiled him to Elba, not far from his native Corsica. The younger brother of Louis XVI was then restored to the throne of France as Louis XVIII (1755–1824). The Allied leaders gathered at the *Congress of Vienna* in September to reorganize the map of Europe.

The British attempted to end the *War of 1812*, between Great Britain and the United States, once Napoléon was defeated, and on August 24, 1814, they landed near Washington, took the city easily, and burned some of its public buildings, including the executive mansion (which was later painted white to hide the burn marks and has been called the White House ever since). However, defeats at Baltimore and on Lake Champlain made Great Britain willing to sign the Treaty of Ghent on December 24, 1814, ending the War of 1812 with no gains on either side.

1815

PLANE OF POLARIZED LIGHT

Berzelius (see 1803) in 1807 had divided compounds into *organic* and *inorganic*. Organic compounds were those obtained from organisms (living or dead) or related to compounds so obtained. Those that had nothing to do with such organisms were inorganic.

Now Malus's discovery of polarized light (see 1808) gained some importance with a discovery made by Biot (see 1803). Ordinarily, the polarized light would be shining brightly in a certain plane, so that if two pieces of Iceland spar were lined up parallel, the light would pass undimmed through both.

If, however, the polarized light, en route from one piece of Iceland spar to the other, was made to pass through some organic liquid, the second piece of Iceland spar had to be turned for the polarized light to shine brightly again. This meant that the plane of the polarized light had twisted on passing through the organic liquid. For some liquids, it twisted clockwise, for others counterclockwise.

Biot suggested that this might be due to some asymmetry in the structure of the molecules of the organic substance. He was right in this, but there was as yet no way of determining what that asymmetry might be.

ORGANIC RADICALS

Gay-Lussac (see 1804, Scientific Ballooning) worked carefully with the poisonous gas hydrogen cyanide (HCN). In 1815 he discovered a related poisonous gas, *cyanogen* (C_2N_2).

He went on to show that the carbon-nitrogen combination (CN), or the *cyano group*, was very stable. In chemical changes, the two bound atoms tended to be transferred as a unit. Relatively tightly bound units that maintained their integrity through various chemical changes came to be called *organic radicals*.

This represented a major step forward in the understanding of *organic chemistry;* that is, the study of those chemicals and chemical changes characteristic of organisms.

PROUT'S HYPOTHESIS

Since Dalton had first advanced his atomic theory (see 1803), chemists had been determining the atomic weights of the chemical elements. Two facts had already emerged. One was that the hydrogen atom seemed the least massive of the atoms, and the second was that other atomic weights seemed to be even multiples of the weight of hydrogen.

An English chemist, William Prout (1785–1850), therefore suggested in 1815 that hydrogen was the fundamental atom and that all other atoms were built up of different numbers of hydrogen atoms.

This is a classic case of a scientist being far ahead of his time. *Prout's hypothesis* was not taken seriously, and as time went on, atomic weights were measured that were *not* exact multiples of hydrogen's, so that the hypothesis seemed to be disproved over and over.

Yet a century after it was advanced, it turned out that Prout's hypothesis had a large element of truth in it, although as usual in such cases, the truth was more complex and subtle than Prout could possibly have suspected.

PAVED ROADS

Through most of history and in most places, roads were simply bare earth from which vegetation and obstructions had been removed, more or less. They tended to be dusty in dry weather, muddy in damp weather, and worn into ruts by vehicle wheels, so that they were only marginally better than no roads at all. The Romans had built good roads of stone (as had other stable civilizations), and Europe had depended on them, where they existed, ever since.

A British engineer and businessman, John Loudon McAdam (1756–1836), had for years been investigating roads and experimenting with road-building methods. He recommended that roads be built higher than the fields on either side, for drainage, and that they be covered with large rocks, then with smaller rocks, the whole to be bound with fine gravel or slag.

In 1815 he finally had the opportunity to put his suggestions into practice on roads around Bristol, and their obvious superiority to what had existed before led to the rapid spread of *macadamized* highways, first in Great Britain and then in other countries. These paved roads immeasurably eased transportation and communication across land.

IN ADDITION

When the Treaty of Ghent was signed, a British force was on the way to New Orleans, and neither they nor the American defenders had any way of knowing that a peace treaty had been signed 4,000 miles away. The Battle of New Orleans was fought on January 8, 1815, and the British were mowed down. There were 2,100 British dead and wounded to 21 Americans. This reconciled the British to the peace and ended the war on a high note for the Americans.

Napoléon, unable to remain inactive on Elba, managed to make his way to the southern French coast on March 1, 1815. His overland trip to Paris quickly became a triumphal march. He entered Paris on March 20 and Louis XVIII had to flee.

The Allies, still engaged at the Congress of Vienna, quickly gathered their armies. Napoléon struck into Belgium and won some victories, but at the Battle of Waterloo, on June 18, 1815, he was defeated for the last time by Wellington. He abdicated on June 22, and this time he was sent to the distant island of St. Helena, where he spent the last six years of his life.

The Congress of Vienna ended its deliberation on June 8, 1815, just before Waterloo. Austria gained back what it had lost to Napoléon, together with the provinces of Lombardy and Venetia in northern Italy. Russia got most of the Grand Duchy of Warsaw. Prussia got most of the west bank of the Rhine so that it could serve as a stronger barrier against France. Belgium and Holland were combined as the *Kingdom of the Netherlands,* also to serve as a barrier. Sweden obtained Norway from Denmark as a reward for being anti-Napoléon toward the end. All the old royal families that Napoléon had kicked out returned, and a *Germanic Confederation,* dominated by Austria, replaced the old Holy Roman Empire.

Ferdinand VII (1784–1833) was reinstated as king of Spain, and he labored to suppress the continuing unrest in Spain's American colonies, particularly in Venezuela, where Simon Bolivar (1783–1830) led the revolt.

There was a huge volcanic eruption on an East Indian island in 1815 that sent enough dust into the upper atmosphere to affect the weather during the next year, in accordance with the suggestion of Franklin (see 1784).

The population of the United States had reached 8.3 million, or two-thirds that of Great Britain. The largest American city was still Philadelphia, with a population of 75,000. New York was second with 60,000.

1816

STETHOSCOPE

Diagnostic procedures were limited in the early days of modern medicine, but one obvious way of getting information was to place the ear to the chest wall and listen to the sound of the heartbeat. In 1816 a French physician, René-Théophile-Hyacinthe Laënnec (1781–1826), was faced with the necessity of listening to the heartbeat of a plump young woman with a heart condition. He felt that trying to hear the heart through the barrier of the breasts would be ineffective, while attempting to lift or separate the breasts for better hearing would be indelicate.

In a moment of inspiration, he bent a notebook into a cylinder and put one end between the woman's breasts and the other end to his ear. He was pleased to find that the heart sounds were actually louder than they would have been if he had placed his ear directly to the breastbone. He therefore prepared wooden cylinders for the purpose of listening to the heartbeat and thus invented the *stethoscope* (from Greek words meaning "to view the chest").

The stethoscope, steadily improved, of course, became so essential an adjunct of the medical profession that medical students with their stethoscopes became as stereotypical as engineering students with their slide rules.

IN ADDITION

James Madison, having served two terms as president of the United States, retired. James Monroe of Virginia was elected the fifth president. Indiana entered the Union as the nineteenth state.

The German philosopher Georg Wilhelm Friedrich Hegel (1770–1831) finished *The Science of Logic* in three volumes.

The year 1816 was called "the year without a summer" and "eighteen hundred and froze to death" because of the effect of the East Indian volcanic eruption of the previous year.

1817

CHLOROPHYLL

Since Priestley had first shown that plants could restore the vitality of air (see 1771), chemists had searched for the substance that gave plants this property.

Two who were particularly keen on research into plant chemistry were the French chemists Pierre-Joseph Pelletier (1788–1842) and Joseph-Bienaimé Caventou (1795–1877). Together they isolated a number of alkaloids such as brucine, cinchonine, quinine, and strychnine.

In 1817 they isolated a green compound from plants (indeed, *the* compound that made them green) and called it *chlorophyll*, from Greek words meaning "green leaf." As was eventually discovered, it was this compound that trapped the energy of sunlight and converted carbon dioxide and water into plant tissue and oxygen.

CADMIUM, LITHIUM, SELENIUM

New elements continued to be discovered.

In 1817 a German chemist, Friedrich Strohmeyer (1776–1835), analyzed a bottle in an apothecary's shop that contained zinc carbonate. He found that it turned yellow on strong heating, which it shouldn't have done. It had to contain an impurity, and when he tracked it down, he found it was a new element, which he named *cadmium* from the Latin name for a zinc ore.

In the same year, a Swedish chemist, Johan August Arfwedson (1792–1841), discovered the element *lithium* (from the Greek for "stone," because it was found in minerals, whereas sodium and potassium, which were similar, were found in plants, and Berzelius (see 1803) discovered *selenium* (from the Greek word for "moon").

IN ADDITION

James Harper (1795–1869) and his brother John (1797–1875) set up a printing office in New York in 1817 that eventually became the publishing house of Harper and Brothers and later Harper & Row (which has published this book).

Mississippi, the twentieth state, was admitted into the Union on December 10, 1817.

1818

TRANSVERSE LIGHT WAVES

Young had demonstrated that light consisted of tiny waves but thought they might be longitudinal waves, like those of sound (see 1801). In 1818, the French physicist Augustin-Jean Fresnel (1788–1827) published a treatise that considered transverse light waves in a detailed mathematical way and showed that such waves would account for reflection, refraction, and diffraction just as well as longitudinal waves would.

What's more, transverse waves could easily explain the manner in which Iceland spar divided a beam of light into two beams that were each refracted by a different amount. It also explained polarized light (see 1808), which longitudinal waves could *not* explain.

Ordinary, unpolarized light has waves that oscillate in all directions that are at right angles to its direction of propagation—up and down, side to side, and everything in between. Light that passes through certain crystals is forced to oscillate in just two directions, one at right angles to the other. This is polarized light. (This is analogous to making waves in a long rope. The wave can be made to move up and down, side to side, and everything in between. If the rope passes between the slats of a picket fence, however, the only waves that can pass through are those that move up and down.)

Fresnel's analysis settled the controversy over the nature of light, at least for the time being.

ENCKE'S COMET

Since Halley had predicted the return of Halley's comet (see 1705), no other cometary orbit had been worked out, and no other comet's return had been predicted.

In 1818, however, the German astronomer Johann Franz Encke (1791–1865) worked out the orbit of a comet that had been observed the year before by the French astronomer Jean-Louis Pons (1761–1831). The comet has been called *Encke's comet* ever since, the name going to the orbit-establisher rather than to the discoverer.

Encke's comet, the second to have its orbit established, returned to the neighborhood of the Sun every three and a third years. It was the first *short-period* comet known and, indeed, to this day no comet has been discovered with a smaller orbit. Encke's comet was the first to be visible throughout its orbit, and this went far to decrease the mystery of comets. It is, of course, a very dim comet, since repeated approaches to the Sun have drained it of the material that goes into the forming of the tail. Nowadays it has just a trace of the fuzziness that reveals it to be a comet.

ATOMIC WEIGHTS

Berzelius (see 1803) was one of those who labored to determine atomic weights, and no one in his time was as careful as he. He ran two thousand analyses on various chemicals after 1807 and used the results as a basis for working out atomic weights. In 1818 he felt justified in publishing a table of his results.

Despite the fact that he ignored Avogadro's hypothesis (see 1811) and made mistakes for that reason, many of his figures were reasonably correct. His atomic weight table was far more correct than Dalton's (see 1803) and was the first in which we can recognize many modern values.

Berzelius also presented the *molecular weights* of various compounds. These are easily obtained if you know the weight of the individual atoms and know which atoms and how many of each go to make up the molecules of a compound.

IN ADDITION

On October 20, 1818, the United States and Great Britain settled the boundary between the United States and Canada from Lake of the Woods in Minnesota westward to the Rocky Mountains at 49 degrees North Latitude. That boundary remains to this day. Illinois was admitted to the Union as the twenty-first state.

Chile declared its independence on February 12, 1818.

1819

SPECIFIC HEAT

Each substance requires a certain amount of heat to raise its temperature by 1 degree Celsius. This amount is its *specific heat*.

In 1819 the French chemists Pierre-Louis Dulong (1785–1838) and Alexis-Thérèse Petit (1791–1820) showed that the specific heat of an element was inversely proportional to its atomic weight. The higher the atomic weight, the smaller the specific heat. This means that if the specific heat of a new element is determined (which is usually easy to do), an idea of its atomic weight can at once be obtained. Since the atomic weight might be hard to determine in any other way, this discovery greatly aided Berzelius in continuing to improve his table of atomic weights (see 1818).

STEAMSHIP

The steamboats designed and built by John Fitch (see 1787) and later by Robert Fulton (see 1807) were designed for river travel. Rivers are, after all, gentler than the open sea, and banks are always close at hand in case of accident.

In 1819, however, the American steamship *Savannah* (an ocean-going vessel is a ship, not a boat), sailed from Savannah, Georgia, to Liverpool, England, in five and a half weeks. It was not much of a steamship, for it bore sails that did the major work, while the steam engine on board was at work only one-twelfth of the time. Still, it bore the promise of things to come.

IN ADDITION

Under the leadership of Klemens Wenzel von Metternich (1773–1859), Austrian minister of foreign affairs, the European powers were becoming more and more repressive in their attempt to prevent liberal notions from taking root.

Bolivar (see 1815) declared the independence of a portion of South America that included Venezuela, Colombia, and Ecuador.

In the United States, Alabama was admitted as the twenty-second state in 1819, and Florida was purchased from Spain for five million dollars.

In southeast Asia, Britain acquired an island off

the tip of the Malayan peninsula and founded Singapore.

In 1819 the American clergyman William Ellery Channing (1780–1842) denied the Trinity and founded Unitarianism.

1820

ELECTROMAGNETISM

There are many resemblances between electricity and magnetism. Both have opposites (positive and negative in electricity, north pole and south pole in magnetism). In both cases, opposites attract and similars repel, and in both cases, the strength of the attraction or repulsion declines with the square of the distance.

For that reason, many scientists thought there might be a connection between the two phenomena. The results of one experiment were published by the Danish physicist Hans Christian Ørsted (1777–1851) in 1820.

As part of a classroom demonstration, he had brought a compass needle near a wire through which a current was passing. The compass needle twitched and pointed neither with the current nor against it but in a direction at right angles to it. When Ørsted reversed the direction of the current, the compass needle pointed in the opposite direction but still at right angles to the current flow.

Ørsted himself did not follow up on his discovery, but others did. Before the year was over, the French physicist André-Marie Ampère (1775–1836) had set up two parallel wires, one of which was freely movable back and forth. When both wires carried current in the same direction, the two wires clearly attracted each other. If the current flowed in opposite directions, they repelled each other. If one wire was free to rotate, then when the currents flowed in opposite directions, the movable wire rotated through a semicircle, coming to rest in such a position that the currents flowed in the same direction in both. Clearly, wires carrying an electric current showed magnetic properties.

Ampère also demonstrated that a current flowing through a wire bent into a helix (similar in shape to a bed spring) strengthened the magnetic effect with each turn of the wire, and the helix clearly acted like a bar magnet with a north pole and a south pole.

In the same year, another French physicist, François Arago (1786–1853), showed that if a current ran through a copper wire, it could attract and pick up iron filings as easily as an ordinary steel magnet would.

Then a German physicist, Johann Salomo Christoph Schweigger (1779–1857), saw that the amount of the deflection of the needle in the Ørsted experiment could be used to measure the strength of the current. In this way, he constructed the first *galvanometer.*

All these things together meant that by the end of 1820, the phenomenon of *electromagnetism* had established itself firmly in physics.

GLYCINE

After Kirchhoff had isolated glucose from starch (see 1812), other chemists began trying to extract simple substances from complicated ones by heating in acidified water. What happened, it was eventually discovered, was that simple substances bound together by chemical bonds could be split apart, with a hydrogen atom from a water molecule added to one of the broken ends and the hydrogen-oxygen combination from the same water molecule added to the other. This process is called *hydrolysis* (from Greek, meaning "to loosen by water"). Through hydrolysis, the simple building blocks of complex compounds could be isolated.

One of those who was interested in the procedure was the French naturalist Henri Braconnot (1781–1855). He obtained glucose from sawdust, linen, bark, and other plant products, just as Kirchhoff had obtained it from starch.

Braconnot then heated gelatin, a substance derived from the connective tissue of animals, and obtained a simple substance that tasted sweet. He named it *glycine,* from the same Greek word that gave us *glucose.* (The letters *u* and *y* are interchangeable in transliterating Greek words.)

Braconnot might have thought he had a sugar, but on further investigation of glycine he obtained ammonia, so he knew its molecule must contain a nitrogen atom, which sugars do not.

As it turned out, glycine was one of the amino acids. Vauquelin had obtained *asparagine* (see 1806), and Wollaston (see 1800) had obtained *cystine* (from the Greek word for "bladder") from a bladder stone. They were amino acids, too, but glycine was the first amino acid clearly associated with substances that later came to be called *protein.*

ANTARCTIC LAND

After Captain Cook's crossing of the Antarctic Circle (see 1773), much of the exploration of Antarctic waters was conducted by sealers and whalers, as a by-product of their search for seal fur and whale blubber and oil (whale oil being a major source of illumination in the lamps of Europe and America).

On November 16, 1820, an American sealer, Nathaniel Brown Palmer (1790–1877), sighted land south of Tierra del Fuego. Also in that year, and perhaps several months earlier, a British naval commander, Edward Bransfield (*ca.* 1795–1852), sighted land in the same general area. At the time there was no way of knowing what the nature of the land was, but we now know that it is a long, curved peninsula that we call the *Antarctica peninsula.* It is the only part of the Antarctic landmass that sticks up well north of the Antarctic Circle.

Also in 1820, a Russian explorer, Fabian Gottlieb von Bellingshausen (1778–1852), discovered a small island, which he named *Peter I Island,* about 150 miles south of the Antarctic Circle.

We might list Palmer, Bransfield, and Bellingshausen as joint discoverers of Antarctic land.

DIFFRACTION GRATINGS

Since Newton's experiments with light (see 1666), scientists had used glass prisms to produce light spectra. In 1820 Fraunhofer (see 1814) was the first to find a substitute in a *diffraction grating* (closely spaced thin wires). Such gratings, the equivalent of fine, parallel scratches on glass, eventually replaced the prism in the study of spectra.

IN ADDITION

European repression simply strengthened the forces of liberalism, and in 1820 there were uprisings in Spain, Portugal, and Naples.

In Great Britain, George III finally died on January 19, 1820, and the Prince Regent became George IV.

In the United States, the question of slavery became acute for the first time as the states took opposite sides with vehemence. There were now twenty-two states in the Union, eleven where slavery was permitted *(slave states)* and eleven where slavery was illegal *(free states).* Maine had applied for entrance to the Union as a free state. The slave states didn't want to multiply the number of free states (and the number of free-state senators) without some sort of balance.

On March 3, 1820, the *Missouri Compromise* allowed Maine to enter the Union as a free state and Missouri as a slave state. That would keep the division equal at twelve to twelve. For a while it seemed the quarrel over slavery was settled—but only for a while.

The population of the United States reached 9.6 million in 1820. New York had become its most populous city at 124,000 and has remained in first place to the present day. Great Britain had a population of 14 million; France, 30 million.

1821

ELECTRICAL MOTION

The discovery of electromagnetism continued to provoke a surge of experimentation. The English physicist Michael Faraday (1791–1867) set up an electrical circuit that included two wires and two magnets. In one case, the wire was fixed and the magnet was movable. In the other, the magnet was fixed and the wire was movable. When the current passed through the wire, the movable wire revolved about the fixed magnet and the movable magnet revolved about the fixed wire.

In this way, Faraday demonstrated for the first time that electrical forces could produce motion.

This experiment led Faraday to consider magnetism a field that stretched out from its point of origin, weakening with distance. One could draw imaginary lines in the field, connecting all points of equal magnetic intensity, and call these *lines of force.* Around a wire with a current flowing through it, the lines of force were concentric circles, and it was this that caused the circular movement.

Here began the conception that today is very nearly central to physics: that the Universe consists of fields, of which particles are the origin. Lines of force, first visualized by Faraday (who had no mathematics but who had a terrific grasp of what must be) are of the utmost importance to the physics of today.

SEEBECK EFFECT

The German physicist (of Russian birth) Thomas Johann Seebeck (1770–1831) was the first to notice, in 1821, that if two different metals were joined at two places, and the two points of junction were kept at different temperatures, an electric current would flow continuously round the circuit.

This is the *Seebeck effect* and it was the first

demonstration of the phenomenon of *thermoelectricity*. Nothing much was done with it, however, for over a century.

GLACIERS

Mountain glaciers were well known to people who lived in mountainous areas such as Switzerland. It was known that they stretched down the declivities on the mountainsides like rivers of ice; that the end in the valley melted in the summer and refroze in the winter; that in advancing and retreating at the valley end, the glaciers scraped rocks, since the pebbles frozen into the bottom of the glaciers gouged striations into those rocks.

It was also noticed by some geologists that there were striations in rocks that were nowhere near glaciers. It made sense to suppose that the striations could have been caused by glaciers, but how?

A Swiss geologist, Ignatz Venetz (1788–1859), decided there was only one explanation. Glaciers must, at some time in the past, have extended far beyond their present limits. He published material to this effect in 1821, but it made no great impression at the time.

IN ADDITION

Napoléon Bonaparte died at St. Helena on May 5, 1821.

Revolutionary fervor was spreading. The Greeks, having been under Turkish rule for nearly four centuries, rebelled.

Mexico declared its independence on Feburary 24, 1821. It included California and Texas in its territory. Guatemala and Peru also proclaimed their independence. By now Spain's American colonies were irretrievably gone.

1822

HEAT FLOW

In 1807 the French mathematician Jean-Baptiste-Joseph, Baron Fourier (1768–1830) had been able to demonstrate what is called *Fourier's theorem*. This states that any periodic oscillation (that is, any variation that sooner or later repeats itself exactly, over and over) can be broken up into a series of simple regular wave motions (expressed as sines and cosines), the sum of which will be the original complex periodic variation.

Fourier used his theorem to study the flow of heat and in 1822 published a book entitled *Analytical Theory of Heat*, which amply explained that branch of physics.

In the book, he showed that it was necessary to use a consistent set of units and that, in equations involving quantities expressed as so many units, those units had to balance on both sides of the equation just as the numbers did. This began the technique of *dimensional analysis*. Fourier suggested that the basic set of units should be those of mass, length, and time, and that other units be built up of those three.

COMPUTERS

Pascal and Leibniz had constructed calculating machines (see 1642 and 1693), but these were equipped to do only the very simplest tasks.

About 1822 an English mathematician, Charles Babbage (1792–1871), began thinking of something much, much more ambitious. He wanted a machine that could be directed to work by means of punched cards as in a Jacquard loom (see 1801), that could store partial answers in order to save them for additional operations to be performed upon them later, and that could print the results.

Everything he thought of could be done, but not by purely mechanical means, using the techniques of Babbage's time. He spent virtually the rest of his life trying to build the machine, his plans growing ever more grandiose.

Babbage had conceived the modern computer, but he didn't have the necessary electronic switches. These were not to be developed for another century.

PROJECTIVE GEOMETRY

The French mathematician Jean-Victor Poncelet (1788–1867) was taken prisoner during Napoléon's invasion of Russia, and during the year and a half he spent in Russia, he meditated on geometry. The fruits appeared in 1822, when he published a book on *projective geometry* (roughly, the study of the shadows cast by geometric figures). Previously knotty problems yielded easily to the new technique. The book is usually considered to be the foundation of modern geometry.

IGUANODON

In 1822 an English geologist, Gideon Algernon Mantell (1790–1852), uncovered the bones and teeth of a large animal and sent some of them to Cuvier (see 1798). For once Cuvier made a mistake and thought the teeth were those of a rhinoceroslike mammal.

A couple of years later, Mantell came across the teeth of an iguana, a reptile that lives in desert areas of North America. The fossil teeth he had found were precisely like those of the iguana but much larger. Mantell realized then that he had uncovered an ancient reptile and named it *iguanodon* (from the Greek meaning "iguana-tooth").

As it turned out, he was the first to discover an example of what were eventually to be called *dinosaurs* (from Greek words meaning "terrifying lizards"). Dinosaurs proved to be by far the most dramatic relics of past ages and did more to acquaint the ordinary person with the fact of biological evolution than anything else.

HIEROGLYPHICS

Even with the Rosetta Stone (see 1799), nearly a quarter-century passed before important progress was made in deciphering ancient Egyptian hieroglyphic writing. The first person to make any progress at all was Young (see 1801, Light Waves).

It was not till 1822, however, that a true decipherment was made, by a French linguist, Jean-François Champollion (1790–1832), who realized that some signs were alphabetic, some syllabic, and some represented a whole word or idea. It was he who founded modern Egyptology.

IN ADDITION

After Napoléon's defeat, the victorious Allies met in a series of congresses designed to prevent revolution. Most came to nothing, but the last, which met at Verona in October 1822, took measures against the revolt in Spain, agreeing to send a French army to suppress it.

On September 7, 1822, Brazil declared itself independent of Portugal.

In 1822 the French inventor Joseph Nicéphore Niepce (1765–1833) made the first permanent photograph, but it was to be some years before photography became even faintly practical.

1823

GASTRIC ACIDITY

Vitalism—the idea that the properties of life are fundamentally different from those of nonlife, in some respects at least—seemed to have a thousand lives as it retreated from one bastion to another. Surely there had to be an important difference between the gentle chemicals that made up flesh and blood and the harsh ones that made up the inanimate world?

Among the harsh chemicals of the inanimate world are the strong acids. In 1823, however, Prout (see 1815) discovered that stomach secretions contained hydrochloric acid, one of those strong acids.

Hydrochloric acid was thus part of the animate as well as the inanimate world. But why did it not scarify and destroy the stomach lining? Sometimes it does, for it can cause ulcers, but usually it does not, and even today we are not quite sure why.

PLATINUM AS CATALYST

As early as 1816, Davy (see 1800) had noticed that certain inflammable gases seemed to ignite and burn more readily in the presence of platinum than in its absence.

In 1823 the German chemist Johann Wolfgang Döbereiner (1780–1849) found that the effect was heightened when he used powdered platinum. In fact hydrogen would ignite and burn in air without having to be heated if powdered platinum was present. Nor was the platinum consumed in the process. It was a catalyst.

Döbereiner went on to devise an automatic lighter in which a jet of hydrogen could be played upon powdered platinum, causing the hydrogen to flame at once. This could be used to light a cigar, for instance.

It wasn't practical, to be sure. The platinum was very expensive, and although it wasn't consumed, it was quickly poisoned by impurities in the hydrogen or in the air and then wouldn't work until it was cleaned up again.

In time, however, it was discovered that platinum (and also other, cheaper metals) would catalyze many reactions involving hydrogen, and such metal-catalyzed reactions became very important in industry.

ISOMERS

With the coming of the atomic view of matter, chemists routinely tried to find out the atomic composition of the molecules of the substances they were studying. In 1823 the German chemist Justus von Liebig (1803–1873) was studying a class of substances known as *fulminates.* Silver fulminate, for instance, has a molecule in which there is one atom each of silver, carbon, nitrogen, and oxygen.

At the same time, another German chemist, Friedrich Wöhler (1800–1882), was studying a class of substances known as *isocyanates.* Silver isocyanate also has a molecule containing one atom each of silver, carbon, nitrogen, and oxgyen.

Both chemists submitted their reports for publication to a journal of which Gay-Lussac (see 1804, Scientific Ballooning) was editor. He noticed that the molecular formulas were the same, yet the compounds had quite different properties. He told Berzelius (see 1803), who prepared both compounds and found that it was true: same formula, different properties. Berzelius referred to such unlike twins as *isomers* (from Greek words meaning "equal parts").

This was the first indication that counting the atoms in a molecule was not enough; it was also necessary to consider the *arrangements* of those atoms. The more complicated the molecule, the more likely it was that there would be numerous isomers. Since the molecules in living tissue tend to be far more complicated than the molecules in the inanimate world, isomerism became particularly important to organic chemists.

LIQUEFYING GASES

Generally speaking, there are two ways of making a gas condense into a liquid. You can cool it. This deprives the gas of energy, and its molecules sink toward each other and cling together. You can also put it under pressure. This forces the molecules toward each other until they cling. Naturally, if you use cold *and* pressure, you do better still.

Michael Faraday (see 1821) was the first to use cold and pressure in a systematic attempt to liquefy gases. He used a strong glass tube bent into a boomerang shape. In the closed bottom, he placed some substance that, when heated, would liberate the gas he was trying to liquefy. He then sealed the open end. He placed the bottom end in hot water. This liberated the gas in greater and greater quantities and, since the gas was in the limited space within the tube, it developed greater and greater pressure.

The other end of the tube Faraday kept in a beaker filled with crushed ice. At that end the gas would be subjected to both high pressure and low temperature and would liquefy. In 1823 Faraday liquefied the gas chlorine in this manner. (Chlorine's normal liquefaction point is $-34°$ C.) Using this method, Faraday liquefied several other gases, too.

ELECTROMAGNETS

Three years earlier, Ampère (see 1820) had shown that a wire helix (or *solenoid*, from a Greek word meaning "pipelike," because such a helix looks like a pipe with its walls made of turns of wire) acts like a bar magnet when electricity flows through the wires.

In 1823 an English physicist, William Sturgeon (1783–1850), placed an iron bar within a solenoid of eighteen turns. He found that the iron seemed to concentrate the magnetic field and make it stronger.

Sturgeon varnished the iron bar to insulate it and keep it from short-circuiting the wires. He used one that was bent into the shape of a horseshoe. His device would lift 9 pounds —twenty times its own weight—while the current was running. When the current was turned off, the magnetic properties vanished. Sturgeon had invented the *electromagnet*.

It was quickly improved upon by the American physicist Joseph Henry (1797–1878), who in 1829 wrapped *insulated* wire around an iron core. This meant that many more turns of wire could be wrapped about the core, since crisscrossing them would not produce a short circuit. The more turns of wire, the stronger the magnetic field when the current was running. By 1831, using the current from an ordinary battery, Henry could make an electromagnet lift a ton of iron.

IN ADDITION

In line with the decision of the Congress of Verona, the French sent an army into Spain. It defeated the Spanish forces on August 31, 1823, and restored Spain to the repressive rule of Ferdinand VII.

The United States was disturbed by this. If the reactionary European powers restored Spanish despotism now, they might next send armies across the ocean to restore Spanish colonies. On December 2, 1823, President Monroe therefore announced what eventually came to be known as the *Monroe Doctrine.* It declared that European powers could not interfere in matters on the American continent and that the United States in return would not interfere in European matters.

1824

PORTLAND CEMENT

The Romans used *concrete* for their structures, this being sand, gravel, or crushed rock held together by *cement,* a mixture of chemicals that hardened when water was added.

The first improvement on the Roman system came in 1824, when an English stonemason, Joseph Aspdin (1799–1855), invented a method for grinding and burning clay and limestone to produce a cement that was both cheaper and better than other cements in use at the time. He called it *Portland cement* in an attempt to emphasize its resemblance to Portland stone quarried at Portland, Dorset.

STEAM ENGINE EFFICIENCY

The steam engine, however useful, remained inefficient. With all Watt's improvements (see 1764), it still converted only 7 percent of the energy of burning fuel into work. The remaining 93 percent (or more) was simply lost as waste heat.

The first to study steam engine efficiency scientifically was a French physicist, Nicolas-Léonard-Sadi Carnot (1796–1832). In 1824 he published a book entitled *On the Motive Power of Fire* in which he was able to show that the *maximum* efficiency of a steam engine depended upon the difference in temperature between the steam at its hottest and the cold water at its coldest. It does not matter what happens at the temperatures in between; whether the steam cools or the water heats slowly, quickly, smoothly, or in stages.

Carnot was the first to consider carefully the way in which heat and work interconvert. For that reason he is considered the founder of the science of *thermodynamics* (from Greek words meaning "the flow of heat"). From his work it was possible to deduce what came to be called *the second law of thermodynamics* a quarter-century later.

DISTANCE OF THE SUN

A century and a half before, Cassini, making use of the parallax of Mars, had estimated the distance of the Sun from Earth as 87,000,000 miles (see 1672).

In 1824 Encke (see 1818), using the time that Venus entered the Sun's disk and left it during its transits, announced the Sun to be 95,300,000 miles from Earth. This was better than Cassini's figure had been. Encke's figure was only 2.6 percent too high.

QUINTIC EQUATIONS

General solutions by algebraic methods had been found for the equations of the third degree (cubic equations) and the fourth degree (quartic equations)—see 1535 and 1545. Ever since, mathematicians had been struggling to find a general solution for equations of the fifth degree (quintic equations), those that involved an x^5.

They had failed, and in 1824 the Norwegian mathematician Niels Henrik Abel (1802–1829) was able to demonstrate that a general algebraic solution of the quintic equation was impossible. Just as Gauss had shown impossibility in geometry (see 1796), so Abel was able to show impossibility in algebra.

SILICON

Silicon, as chemists now know, is, next to oxygen, the most common element in the Earth's crust. It is a component of most rocks, of sand, of glass. Nevertheless, it holds on to other atoms so tightly that it is not easy to isolate. Berzelius (see 1803) was able to manage it, however, and in 1824 was the first to obtain elementary silicon.

IN ADDITION

Louis XVIII of France died on September 16, 1824, and was succeeded by his younger brother, who reigned as Charles X (1757–1836).

In the United States, President James Monroe, having served for two terms, retired.

1825

STEAM LOCOMOTIVES

Richard Trevithick had failed in his attempt to make a commercial success of the steam locomotive (see 1804). Another English inventor, George Stephenson (1781–1848), took advantage of improved steam engines, however, and was able to build a steam locomotive that worked adequately. On September 17, 1825, one of his locomotives pulled thirty-eight cars at speeds of 12 to 16 miles an hour. For the first time in the history of the world, land transportation at a rate faster than that of a galloping horse was clearly on the way to becoming possible.

Railroads quickly began to knit nations together. It might be argued that a nation like the United States could not have been held together but for the fact that railroads made all parts reasonably accessible.

ALUMINUM

Though aluminum is more common than iron (only oxygen and silicon are commoner), it is extremely difficult to break its hold on other atoms.

Ørsted, the first person to demonstrate electromagnetism (see 1820), was also the first person to isolate aluminum. For the purpose, he made use of a still more active element, potassium, which could wrench other atoms out of aluminum's grip. In 1825 he managed to obtain the first bits of the metal.

The procedure for obtaining it was so difficult, however, that aluminum remained virtually a precious metal for some sixty years, until a cheap method for obtaining it in quantity could be worked out.

GASTRIC DIGESTION

On June 6, 1822, a nineteen-year-old Canadian, Alexis St. Martin, was accidentally shot in the side at a frontier post in northern Michigan. It was a close-range shotgun blast and he received a terrible wound.

An American army surgeon at the post, William Beaumont (1785–1853), treated him, and he recovered completely except that an opening (or *fistula*) remained in his side. It was nearly an inch across and it led into his stomach.

Through this opening, Beaumont, beginning in May 1825, was able to observe the changes in the stomach under different conditions and to extract samples of gastric juice, which he sent all over the world. It was like having the benefits of a vivisection. Beaumont's work served as a source for much early information on the process of digestion and increased interest in the field.

CANDLES

Candles had been a source of illumination for nearly five thousand years, but the most common type, tallow candles, which were all that people in ordinary circumstances could afford, had their shortcomings. For one thing, they smelled bad.

The French chemist Michel-Eugène Chevreul (1786–1889) had studied fats and worked out their chemical nature. They were combinations of glycerol with fatty acids. Each glycerol molecule could combine with three fatty acids, and each fatty acid had a long chain of (usually) sixteen or eighteen carbon atoms. Chevreul was the first to isolate the most common of these fatty acids: stearic acid, palmitic acid, and oleic acid.

In 1825 he and Gay-Lussac (see 1804, Scientific Ballooning) patented candles made out of these fatty acids. Such candles were harder than tallow candles, gave a brighter light, looked better, needed less care while burning, and didn't smell as bad. To the society of the time, this was a major advance.

ASTIGMATISM

Spectacles for the far-sighted were some five centuries old, and for the near-sighted, almost as old (see 1249 and 1451). It is possible to have difficulty seeing, however, even if one is neither, if the cornea of the eye is not perfectly curved. Such a condition is known as *astigmatism* (from Greek words meaning "no spot," because a small spot cannot be seen sharply by astimatics) and can be combined with either near- or far-sightedness.

The British astronomer George Biddell Airy (1801–1892) suffered from astigmatism, and he was the first, in 1825, to design eyeglass lenses to correct that condition.

IN ADDITION

The Erie Canal, leading from the Great Lakes to the Hudson River across central New York State, was opened on October 26, 1825. This greatly facilitated the movement of freight, increased the prosperity of the cities en route, and accelerated the growth in importance, size, and wealth of New York City.

On February 9, 1825, the House of Representatives had to select a president, since no one had gained a majority in the electoral college, and chose John Quincy Adams of Massachusetts (1767–1848) sixth president of the United States.

Alexander I of Russia died on December 13, 1825, and was succeeded by his reactionary younger brother, who reigned as Nicholas I (1796–1855).

1826

NON-EUCLIDEAN GEOMETRY

Euclid had based his massive structure of geometry, over two thousand years before, on ten axioms and propositions that could be considered so self-evident that they required no proof. (One had to start somewhere.)

One of these axioms can be stated in a number of ways, of which the simplest is "Through a given point, not on a given line, one and only one line can be drawn parallel to the given line."

This did not seem quite as self-evident as Euclid's other axioms, and the notion of parallelism implied the existence of lines of infinite length—not an easy thing to accept.

Since Euclid's time, mathematicians had desperately tried to prove that axiom from the other axioms. All failed to do so.

Nearly a century before, the Italian mathematician Girolamo Saccheri (1667–1733) had been struck with the idea of beginning by supposing that the axiom was *not* true. He could then proceed to build a geometry on that basis and, sooner or later, arrive at a contradiction. He could then conclude that the axiom *was* true. That having been proved, it would no longer be viewed as an axiom.

However, he could find no contradiction, and this made him so uncomfortable that he *imagined* he had found a contradiction and, in 1733, published a book entitled *Euclid Cleared of Every Flaw,* in which he claimed he had proved the axiom—but hadn't.

Finally, in 1826, a Russian mathematician, Nikolay Ivanovich Lobachevsky (1792–1856), got onto the same track, but he went farther. The axiom, he decided, was not an axiom because it was not needed; geometry could be made self-consistent without it. He showed that if one began with an axiom that stated "Through a given point, not on a given line, any number of lines can be drawn parallel to a given line," then that and the remaining axioms of Euclid could be used to draw up a new "non-Euclidean" geometry. It would be different from the ordinary kind, but it would be self-consistent.

Lobachevsky published his ideas in 1829 and was first in the field. A Hungarian mathematician, János Bolyai (1802–1860), had as early as 1823 worked out the same non-Euclidean geometry. However, his publication was delayed till 1832, so Lobachevsky got the credit.

Gauss (see 1797) had actually worked out the idea of non-Euclidean geometry as early as 1816 but had lacked the courage to publish it.

BROMINE

Courtois had discovered iodine in seaweed fifteen years before (see 1811). The French chemist Antoine-Jérôme Balard (1802–1876), also working with seaweed, found that at times he obtained a brown substance in solution in the liquid he was using to dissolve the ashes of the seaweed. In 1826 he tracked this color to a substance that had properties apparently just midway between those of chlorine and iodine. For a while he thought he had a compound of those two elements, but further investigation convinced him he had a new element, which he called *bromine,* from the Greek word for "smell," because of its strong odor.

IN ADDITION

John Adams and Thomas Jefferson both died on July 4, 1826, the fiftieth anniversary of the Declaration of Independence.

1827

OHM'S LAW

After Fourier had worked out the mathematical system that described the flow of heat adequately (see 1822), it seemed that the same system might be used to describe the flow of electricity. Whereas heat flow from point to point depended on the temperatures of the two points and the heat conductivity of the material between, so electric flow from point to point might depend on the electrical potential of the two points and on the electrical conductivity of the material between.

By working with wires of different thicknesses and lengths, the German physicist Georg Simon Ohm (1789–1854) found that the quantity of current transmitted was inversely proportional to the length of the wire and directly proportional to its cross-sectional area. He was in this way able to define the resistance of the wire and, in 1827, showed that, "The flow of current through a conductor is directly proportional to the potential difference and inversely proportional to the resistance." This is called *Ohm's law*.

TURBINES

Waterwheels had been in use since ancient times, with the water hitting the vanes on the outer edge of the wheel. A French engineer, Benoît Fourneyron (1802–1867), heard his teacher talk of a new kind of waterwheel that he called a *turbine* (from a Latin word meaning "whirling"). The water would come down the hub and move out to the vanes, setting the wheel turning. The faster the wheel turned, the faster the water would be thrown out against the vanes. In the end, the wheel would turn much more rapidly and deliver far more power than an ordinary waterwheel.

Such a turbine existed only in theory, but in 1827 Fourneyron built one that developed 6 horsepower. Within a few years, he built one that developed 50 horsepower. The thought of a turbine driven by steam occurred to him, but he lacked materials capable of withstanding the high temperature. The steam turbine had to wait another half-century to come into existence.

SCREW PROPELLER

Steamships were propelled by large paddlewheels on one side during the quarter-century of their existence. They worked well enough ordinarily, but in rough seas they could lift out of the water altogether if the ship listed to the other side, and that complicated steering. Furthermore, they were vulnerable to enemy fires so that it was considered completely impractical to power warships by steam.

In 1827, however, a British engineer, Robert Wilson (1803–1882), designed a screw propeller, which worked from the stern of the ship, so that it was symmetrically placed and was little affected by the ship's rolling. It was also well under water and so less vulnerable than the rest of the ship. Since it was judged to be a more efficient propulsion mechanism than the paddlewheel, it became possible to think of steam-powered warships thereafter.

MAMMALIAN OVA

De Graaf had discovered the ovarian follicles (see 1779), which were thought to be the mammalian equivalent of eggs. In 1827, however, a Russian embryologist, Karl Ernst von Baer (1792–1876), opened a follicle of a dog and examined a small yellow structure within. It was this much smaller structure, seen only through a microscope, that was the mammalian egg, or *ovum* (plural *ova*). This made it clear that mammalian development (including human development, of course) was not fundamentally different from that of other animals.

FOOD CLASSIFICATION

Prior to this time, food was seen to differ in appearance, in smell, and in taste, but anything that satisfied hunger was assumed to be as good as anything else that did the same.

However, as chemistry developed, it became clear that foods differed in chemical nature and therefore might be different in their effect on the human body. The first to make a broad classification of foodstuffs on a chemical basis was Prout (see 1815) who, in 1827, divided food into three broad classes, which today we call carbohydrates, fats, and proteins. Naturally, this classification was not exhaustive, and there are important substances that do not fall into any of these groups—some of them present in only small quantities, but vital nevertheless.

Prout's classification, however, was a good start toward understanding some of the complexities of dietetics.

BROWNIAN MOTION

In 1827 the British botanist Robert Brown (1773–1858) was studying a water suspension of pollen grains under a microscope and noted that the individual grains were moving about irregularly. This had nothing to do with water currents, for the water was still, and besides, while some grains moved in one direction, others would move in the opposite direction, and still others in all directions between.

Brown was at first not surprised. Pollen grains had sparks of life in them, of course, and the motion might be an aspect of life. However, just to check the matter, he then studied a suspension of dye particles, each about the size of a pollen grain, and unquestionably not alive. To his surprise, the dye particles moved precisely as the pollen grains did.

Brown reported the observation, and the phenomenon has been called *Brownian motion* ever since. It seemed a small thing and had no explanation at the time, but eighty years later it offered scientists the final proof of the existence of atoms.

IN ADDITION

The Turks were clearly on the point of crushing the Greek rebellion, but the British could no longer sit calmly by. They formed an alliance with the French and Russians on July 6, 1827, and demanded that Turkey accept a cease-fire. When Turkey, aided by Egypt, refused and began to land supplies on the southwestern Greek coast, the allied fleet attacked and destroyed the Turkish and Egyptian fleets at the Battle of Navarino on October 20, 1827. This insured Greek independence, although when it finally came it was within restricted boundaries.

1828

SYNTHETIC UREA

The vitalist view on the separation of chemicals into organic and inorganic was that only living tissue could create organic molecules.

In 1828, however, Wöhler (see 1825) found otherwise, even though, at the time, he was making no attempt to form an organic compound. He was simply heating ammonium cyanate, which was accepted as a definitely inorganic substance, when he found crystals forming that resembled those of urea, which was the chief mammalian waste product disposed of in urine (including human urine). He tested the crystals and found they *were* urea.

Both ammonium cyanate and urea have molecules composed of the same atoms (two nitrogen, four hydrogen, a carbon, and an oxygen) but differently arranged. They are isomers (see 1823). Yet ammonium cyanate was considered inorganic, was not found in tissue, and could be formed by the chemist in the laboratory—where urea could now be formed also.

On February 22, 1828, Wöhler announced his discovery to Berzelius (see 1803), and the latter eventually conceded the point. Now other chemists began trying to synthesize organic compounds—and succeeded. Another vitalist bastion had fallen.

NOTOCHORD

Beginning in 1828, Baer (see 1827) published a two-volume textbook on embryology. In it he pointed out that, in early stages of their development, vertebrate embryos were quite similar, even among creatures that in the end were very dissimilar.

Small structures in different embryos, scarcely distinguishable from each other at first, might develop into a wing in one case, an arm in another, a paw in a third, a fin in a fourth, and a flipper in a fifth. Baer believed that relationships among animals could be deduced more properly by comparing embryos than by comparing adult structures, so that he is considered the founder of *comparative embryology*.

In particular, Baer pointed out that early vertebrate embryos possessed a *notochord* for a short period of time. The notochord is a stiff rod running the length of the back. Some very primitive fishlike creatures possess such a structure throughout life, but in vertebrates, it is quickly replaced by a spinal cord. Nevertheless, the temporary existence of the notochord in vertebrate embryos shows their relationship to the primitive creatures.

THORIUM

In 1828 Berzelius (see 1803) discovered another element and named it *thorium* after the Norse god of thunder, Thor.

IN ADDITION

In the United States, Andrew Jackson (1767–1845) was elected the seventh president. In the same year, a fierce dispute between the northern and southern states erupted over tariffs. The north won out with a high tariff in 1828 and the south was furious.

On July 4, 1828, the United States began to build its first commercial railroad, the Baltimore and Ohio. Present at the ceremonies was Charles Carroll of Carrollton (1737–1832), the last living signer of the Declaration of Independence.

The Zulus in South Africa had come under the sway of the capable Shaka (*ca.* 1787–1828), who had reorganized their weapons and military tactics and made of them a conquering nation, though ruling with extreme cruelty. He was assassinated in 1828, and thereafter the Europeans would gradually rise to virtually complete mastery of Africa.

1829

NICOL PRISM

Biot had noted that the plane of polarized light seemed to twist when it passed through certain organic compounds that were liquid or in solution (see 1815), but the fact was of little use without some simple method of measuring this twist accurately.

In 1829 the Scottish physicist William Nicol (1768–1851) devised a useful instrument in this connection by cementing two crystals of Iceland spar together with Canadian balsam. He put them together in such a way that one of the two rays of light refracted by the first crystal was reflected out of the side of the crystal at the layer of balsam, while the other, striking the balsam at a slightly different angle, passed through to the other end of the crystal.

The one beam that appeared at the other end of the *Nicol prism* could then pass through a second Nicol prism without being weakened, if the second prism was lined up parallel to the first. If the polarized light was then allowed to pass through a container of organic liquid or solution that was placed between the prisms, and if the plane of polarization was twisted as a result, then the second prism would have to be turned to allow the beam through at maximum brightness. The degree of twist could be easily measured, and *polarimetry* became a practical technique.

IN ADDITION

Not only did the southern portion of Greece receive full independence from Turkey on November 30, 1829, but other conquered nations in the Balkans, such as Serbia and Romania, were stirring and assuming greater control of their own affairs.

On September 15, 1830, Mexico abolished slavery within its borders, but it was unable to extend this policy to the province of Texas into which Americans from the slave states were moving.

1830

ACHROMATIC MICROSCOPES

Chromatic aberration, which had been the plague of telescopists for the first century of their existence till the reflecting telescope (see 1668) and the achromatic lens (see 1733) came into use, was still plaguing microscopists. The presence of colored fuzziness effectively hid the fine details of what could be seen.

In 1830, however, an achromatic microscope was invented by a British optician, Joseph Jackson Lister (1786–1869). Using it he could see red blood corpuscles in detail for the first time and found that they were biconcave disks, rather like candy Lifesavers in which the holes had not been punched through.

It was only with achromatic microscopes that bacteria could be usefully studied.

GROUP THEORY

In mathematics it is possible to do a great deal in a short life. The French mathematician Évariste Galois (1811–1832) was killed in a duel before his twenty-first birthday. Even so, he managed to generalize the work of Abel, who had shown the impossibility of solving equations of the fifth degree by algebraic methods (see 1824).

Galois went on to show that no equation of any degree higher than the fourth could be so solved. In order to do this, he invented a mathematical technique called *group theory*, which turned out to be useful a century later in working out quantum mechanics, one of the two great physical theories developed in the twentieth century that successfully describe the Universe.

UNIFORMITARIANISM

Hutton's uniformitarianism (see 1785) was now nearly half a century old but had not made much headway. For one thing Hutton was not an inspiring writer and his chief opponent, Cuvier, the supporter of catastrophism (see 1812), was.

In 1830, however, the first volume of *The Principles of Geology* appeared. Written by a British geologist, Charles Lyell (1797–1875), it was to run to three volumes. All were clearly and attractively written and explained the uniformitarian theory so well that it finally achieved popularity. The extreme catastrophism of Cuvier and his followers thus went down to defeat. This is not to say that the pendulum doesn't swing and that certain catastrophes in the course of Earth's history did not take place. Earth's development is now viewed as mainly, but not entirely, uniformitarian.

IN ADDITION

The quarrel over tariffs heated up in the United States. South Carolina, in particular, threatened to *secede;* that is, leave the Union.

On June 26, 1830, George IV of Great Britain died and was succeeded by his brother, who reigned as William IV (1765–1837).

In France the reactionary policies of Charles X alienated the nation and on July 29, 1830, he was deposed in the course of a Parisian insurrection. Attempts to set up a republic again were defeated and Charles was succeeded by his fourth cousin, who reigned as Louis Philippe I (1773–1850).

The *July Revolution* in France inspired the Catholic population of Belgium to rise against the Protestant population of the Netherlands by which it had been dominated since the fall of Napoléon. On December 20, 1830, Belgium was recognized by the European powers as an independent nation under Leopold I (1790–1865), who began his rule in 1831.

A revolt against Russian domination also began in Poland.

Joseph Smith of New York (1805–1844) published *The Book of Mormon* in 1830, founding the Church of Jesus Christ of Latter-Day Saints on April 6, 1830. Members of the church came to be known, popularly, as Mormons.

Paintings of birds by the American artist John James Audubon (1785–1851) began to appear in 1830.

In 1830 the population of the United States was 12.9 million, about equal to that of Great Britain. The population of the world reached 1 billion.

1831

ELECTRIC GENERATORS

Since Ørsted had shown that an electric current could produce a magnetic effect (see 1820), it had seemed to Faraday (see 1821) that there ought to be some way of showing that the reverse was also true, that a magnet could induce an electric current.

To do this, Faraday made use of an iron ring. In 1831, he wound a coil of wire around one portion of the iron ring and attached it to a battery. The circuit could be opened or closed by a key. If he closed the circuit, current would flow, and a magnetic field would be set up and concentrated in the iron ring.

Suppose, then, that a second coil was wrapped around another segment of the iron ring and connected to a galvanometer. The magnetic field set up in the iron ring might produce a current in this second coil, and the galvanometer would record its presence.

The experiment worked. Faraday had devised the first *electrical transformer* and had discovered *electromagnetic induction*. However, it did not work as he had expected. There was no continuous electric current to match the continuous presence of the magnetic field. Instead, there was a momentary flash of current, marked by a jerk of the galvanometer's needle when he closed the circuit, and a second flash, in the opposite direction, when he opened the circuit.

Faraday explained this by means of the lines of force that he had visualized. When a circuit was closed and electricity was set to flowing, magnetic lines of force sprang outward and crossed the second coil, inducing an electric current. When the circuit was opened again, the magnetic lines of force collapsed inward and crossed the second coil again, inducing an electric current in the opposite direction. When the magnetic lines remained in place because the current in the first coil was flowing steadily, no lines crossed the second coil in either direction and no current was induced in it.

Faraday went on to devise a way of having metal cut across the lines of force continually. He turned a copper wheel so that its rim passed between the poles of a permanent horseshoe magnet. As long as the copper wheel turned, its rim continually cut through magnetic lines of force and an electric current flowed continually in the wheel. That current could be led off and made to do work.

Faraday had thus devised the first *electric generator.*

Until then, electric current had been produced only by batteries, which meant that the electricity was obtained by burning metals such as zinc. This meant that electricity was expensive and limited in quantity.

The turning of the copper wheel to cut across the magnetic lines of force took considerable effort, and it was this energy that was turned into electricity. If one had to turn the wheel by muscle power, little electricity could be obtained. However, the wheel could be and eventually was driven by steam power, which meant that the electricity was formed from burning fuel or from some other copious source of energy such as falling water or blowing wind.

Eventually, when the electric generator was sufficiently improved, electricity could be generated cheaply and in any quantity desired.

ELECTRIC MOTORS

Henry (see 1823) discovered electrical induction independently of Faraday, but the latter published first by a few months and gets the credit. Henry went on to study the reverse process. If the rotary motion of a copper wheel cutting across magnetic lines of force can induce an electric current, then an electric current ought to be able to produce a rotary motion.

In essence, Faraday had already shown this in a simple way (see 1821), but in 1831 Henry devised a much more practical version of such a machine, one in which a wheel would turn if electric current was supplied. This was the first practical *electric motor* (from a Latin word meaning "to move").

The importance of the motor cannot be overemphasized. A motor can be made as large or as small as desired. It can be run by electricity brought to it over a distance of many miles. Most important of all, it can (unlike a steam engine) be started in a moment and stopped in a moment.

The supply of cheap, abundant electricity made possible by Faraday's discovery of the generator (once it was sufficiently improved) would have been useless without some means of putting it conveniently to work. Henry's motor (once it was sufficiently improved) did that, so that between them, Faraday and Henry ushered in the age of electricity.

MATCHES

For thousands of years, fires had been started by means of friction—friction that entailed considerable effort.

Then, beginning with the discovery of phosphorus (see 1669), chemists began to discover chemicals so active that with very little encouragement they could burst into flame. Why not, then, simply coat the edge of a splint of wood with some appropriate chemical that could at the proper time burst into flame and set the wood on fire? You would then have a small fire that would last long enough to ignite a larger and longer-lasting one. You would then have a *match* (from an old word for the nozzle of a lamp).

Such chemical matches began to be produced in the early 1800s, but the first to be made were too hard to ignite, or too messy, or too dangerously easy to light.

In 1831, however, a French chemist, Charles Sauria, produced the first practical friction match. It contained phosphorus, which was diluted with other materials so that the matches would not start to flame until they were struck on a rough surface. The moderate amount of heat induced by the friction would then suffice to ignite it. Such matches produced flame quickly and quietly when struck, and didn't deteriorate on standing. Their use quickly spread.

There was one catch. The phosphorus used in the matches was quite poisonous, and people who worked at producing the matches would get the phosphorus into their bodies where it caused bone degeneration and killed them slowly and painfully. It took some seventy years before this was corrected.

NORTH MAGNETIC POLE

Since the time of Gilbert (see 1600), it had been understood that the Earth must have a North Magnetic Pole and a South Magnetic Pole. The general (and rather natural) feeling was that the magnetic poles would be near, or perhaps exactly at, the rotational poles. However, the Arctic and Antarctic regions of Earth, cold and desolate as they were, could only be explored with great difficulty.

It was not until June 1, 1831, that the North Magnetic Pole was actually reached. The feat was accomplished by a Scottish explorer, James Clark Ross (1800–1862). He found his compass pointing straight down, on the western shore of Boothia Peninsula, at 70.85 degrees North Latitude and 96.77 degrees West Longitude. The pole was discovered only because it was 2,100 miles from the geographic North Pole and therefore relatively accessible. In fact, it was only a few hundred miles north of the Arctic Circle.

CELL NUCLEUS

Brown, who had discovered Brownian motion (see 1827) now noted a small body within the cells that composed plant tissues. Others had observed this before Brown but had paid it little attention. Brown was the first to recognize that it was a regular feature of cells, and he gave it the name of *nucleus* (from Latin words meaning "little nut"—within the husk of the cell, so to speak).

Today it is often referred to as the *cell nucleus* in order to distinguish it from the *atomic nucleus* discovered eighty years later.

DIFFUSION

We all know that gases diffuse. If a bottle of perfume is spilled in one corner of a room, it is not long before we can smell it in the opposite corner.

A British physical chemist, Thomas Graham (1805–1869), attempted to measure the rate at which such diffusion takes place. He measured the rate at which gases diffused through a plaster of paris plug, through fine tubes, and through a tiny hole in a platinum plate.

By 1831 he was ready to announce that the rate of diffusion of a gas is inversely proportional to the square root of its molecular weight. Thus, since oxygen molecules are sixteen times as massive as hydrogen molecules, and the square root of sixteen is four, the lighter hydrogen molecules will diffuse four times as quickly as oxygen. This is still called *Graham's law,* and because of this discovery, Graham is considered one of the founders of *physical chemistry*.

CHLOROFORM

The American chemist Samuel Guthrie (1782–1848) discovered chloroform ($CHCl_3$) in 1831. In the next decade, it was to become prominent in connection with anesthesia.

CYCLONIC STORMS

An American meteorologist, William C. Redfield (1789–1857), traveled through Connecticut soon after a hurricane had ripped through New England on September 3, 1821. He noticed the manner in which the trees had fallen and deduced that, as the storm had traveled northeastward, the winds had spiraled.

He spent the next ten years studying storms and in 1831 published his evidence to the effect that storm winds whirl counterclockwise around a center that moves in the normal direction of the prevailing winds.

Eventually it was discovered that this was true only in the northern hemisphere. In the southern hemisphere, storm winds whirl clockwise.

IN ADDITION

A French expeditionary force had invaded Algeria on July 5, 1830, and by 1831 it was clear that they intended to stay there. For that purpose, the French *Foreign Legion* was established. So began the process whereby the European powers, mainly France, stripped the North African *Barbary nations* from their nominal overrule by the Ottoman Empire.

On May 26, 1831, the Russians finally succeeded in putting down the Polish rebellion. Austria easily put down revolts in various portions of northern Italy.

Emperor Pedro I of Brazil (1798–1834) abdicated on April 7, 1831, and was succeeded by his son, Pedro II (1825–1891).

In the United States a slave, Nat Turner (1800–1831), began a rebellion on August 21, 1831, in which sixty whites were killed in two days. It was quickly suppressed, but all the slave states now feared possible insurrections, and it hardened their view that abolitionists were simply trying to arrange rebellions and murders by slaves.

1832

LAWS OF ELECTROLYSIS

In his youth, Faraday (see 1821) had worked under Davy (see 1800) and carried on the elder's work in electrochemistry. The method whereby Davy had liberated a number of new metals, by passing an electric current through molten compounds of those metals, Faraday named *electrolysis* (from Greek words meaning "to loosen by electricity").

Faraday named a liquid or a solution that could conduct electricity an *electrolyte.* The metal rods inserted into the liquid or solution he called *electrodes* (from Greek words meaning "the road of electricity"). The positively charged electrode he called the *anode* (high road) and the negatively charged electrode he called the *cathode* (low road). This likened the flow of electricity to the flow of water from the height of the anode to the depth of the cathode, following Franklin's guess that electricity flowed from positive to negative (see 1752). (In fact, that turned out to be wrong, however, and electricity flowed from the negative electrode to the positive.)

All these names were suggested to Faraday by the British scholar William Whewell (1794–1866), who also coined the word *scientist* in the next decade.

In 1832 Faraday announced what are now called his *laws of electrolysis.* They are:

1. The mass of substance liberated at an electrode during electrolysis is proportional to the quantity of electricity driven through the solution.

2. The mass liberated by a given quantity of electricity is proportional to the atomic weight of the element liberated and inversely

proportional to the combining power of the element—that is, to the number of atoms that one atom of the element will combine with.

IN ADDITION

In the United States, South Carolina passed a resolution on November 19, 1832, declaring the tariff law of 1828 null and void. If this were allowed to stand, it would mean that no state need obey federal law if it didn't want to, and that would destroy the Union. President Jackson prepared to use force if necessary to suppress *nullification*.

In Italy, Giuseppe Mazzini (1805–1872) founded an organization called *Young Italy* in 1832, dedicated to the establishment of a united democratic Italy under a republican form of government.

The first *clipper ship* was built in Baltimore in 1832. It was designed for speed and for a quarter-century outraced the slowly improving steamships.

1833

DIASTASE

A French chemist, Anselme Payen (1795–1871), managed a factory engaged in the refining of sugar from sugar beets. This turned his attention to plant chemistry.

In 1833 he reported the separation of a substance from malt extract that had the property of hastening the conversion of starch to glucose. Payen called the substance *diastase*, from a Greek word for "separate," for in a sense, it separated the building blocks of starch and produced the individual glucose units.

This was an example of an organic catalyst. Yeast is an organic catalyst that was known even to prehistoric humanity, but yeast is a living organism. Diastase was the first organic catalyst to be isolated from living material that would display catalytic activity without being a living organism itself.

Diastase is an example of what later came to be called an *enzyme*, and it was the first enzyme to be prepared in concentrated form. Because of it, the suffix *-ase* eventually came to be used in the names of enzymes generally.

IN ADDITION

In a bill passed on August 23, 1833, the British Parliament provided for the abolition of slavery in all British colonies.

In the United States, the nullification crisis came to a peaceful end when Henry Clay (1777–1852) engineered a compromise tariff that gave something to the North without taking too much from the South. It was accepted even by South Carolina.

In France, Amandine-Aurore-Lucile Dudevant (1804–1876), under the name of George Sand, was both writing novels and living a life dedicated to equal rights for women.

1834

MECHANICAL REAPER

Agriculture had always been a labor-intensive occupation, particularly at harvest time, when there was often a shortage of hands to reap and gather the grain.

Attempts were therefore made to devise a mechanical reaper, and the one that finally proved successful was built by an American inventor, Cyrus Hall McCormick (1809–1884). He first built what came to be called the *McCormick reaper* in 1831, and he secured a patent for it in 1834.

It wasn't immediately successful, but McCormick pushed it with great pertinacity, and little by little, it took hold, particularly in the vast grain fields of the American Midwest.

This began a whole series of mechanical inventions that gradually reduced the number of laborers that had to work the land to produce food, until finally, in a thoroughly industrialized nation such as the United States, a work force of 4 percent of the whole suffices to grow food for itself and for the remaining 96 percent, with enough left over to export.

CELLULOSE

Having discovered diastase the year before, Payen (see 1833) went on to study the chemical composition of wood. From wood he obtained a substance that certainly wasn't starch but that could be broken down to glucose units as starch could. Because he obtained it from cell walls, he named it *cellulose*.

Sugars like glucose, and substances that can be broken down into sugars, are made up of carbon, hydrogen, and oxygen atoms, with the hydrogen and oxygen atoms in the ratio of 2 to 1, as in water. The notion arose, therefore, that these molecules consisted of carbon atoms to which water molecules had been added. For that reason they came to be called *carbohydrates* (watered carbon). The true structure of carbohydrates, it eventually turned out, was much more complex than that.

Until Payen's discovery, carbohydrates had been given ordinary names such as cane sugar, grape sugar, starch, and so on. Now use of the *-ose* suffix, as in cellulose, spread until it was used for carbohydrates generally. Cane sugar became *sucrose*, grape sugar *glucose*, starch *amylose*, and so on—at least, to chemists.

IN ADDITION

The Spanish Inquisition had been abolished under Napoléon but was restored after his fall. It was abolished again in the liberal revolution of 1820 and restored after that had been defeated. Now, at last, it was permanently abolished after six centuries of hideous existence.

A French teacher of the blind, Louis Braille (1809–1852), who had himself been blind from the age of three, devised a system of raised-point symbols that made it possible to read by touch. This is called *Braille* in his honor and is still used.

1835

DRY ICE

As Black had shown seventy years earlier, energy is consumed in converting a liquid to a vapor (see 1762). This means that if a liquid is allowed to evaporate under conditions where no outside heat can enter the system, the energy for the vaporization must be obtained from the liquid itself. As the liquid evaporates, its temperature therefore drops. (That is the purpose of perspiration. As it evaporates, the temperature of the skin drops and we remain comfortable on hot days. Increasing humidity, which reduces the rate of evaporation, makes us uncomfortable.)

The use of such evaporation to cool a liquid to the point of its solidification was demonstrated by the French chemist C.S.A. Thilorier in 1835. He began with the method introduced by Faraday to liquefy gases (see 1823) but made use of metal cylinders because they were stronger than glass and could endure higher pressures.

He prepared liquid carbon dioxide in considerable quantity and then allowed it to escape from the tube through a narrow nozzle. As it escaped, it evaporated, and the temperature of the liquid in the cylinder dropped to the point where it froze. For the first time, solid carbon dioxide was formed.

Solid carbon dioxide, exposed to ordinary pressures, will slowly *sublime*—that is, evaporate directly to gas without melting. The sublimation point is −78.5° C.

Solid carbon dioxide looks like ice, but since it doesn't form a liquid, it is popularly known as *dry ice*. Dry ice is obviously a more efficient cooling agent than ordinary ice would be. Thilorier added bits of dry ice to diethyl ether (the well-known *ether* that was soon to be used as an anesthetic and that remains liquid to a very low temperature). By allowing the mixture to evaporate, he reached temperatures as low as −110° C. For the first time, a temperature was reached in the laboratory that was lower than any temperature ever recorded on Earth's surface, even in the depths of an Antarctic winter.

CORIOLIS EFFECT

The existence of cyclonic storms, pointed out four years earlier by Redfield (see 1831) did not long remain a mystery.

In 1835 a French physicist, Gaspard-Gustave de Coriolis (1792–1843), took up the matter of motion on a spinning surface, both mathematically and experimentally. When the Earth spins, it spins all in a piece, so that a point on the Equator, forced to move a length of 25,000 miles in twenty-four hours, must move a little over 1,000 miles an hour. As one progresses farther and farther from the Equator, either north or south, a point on the surface makes smaller and smaller circles in the course of the day and moves more and more slowly. At the poles, there is no motion.

If, then, we imagine a quantity of air or water near the Equator, we can see that it must be carried west to east at just over 1,000 miles per hour. If that air or water moves north (or south) away from the Equator, it retains its speed, but the solid ground beneath it moves more slowly. The air or water gains on the ground, therefore, and curves off eastward. Similarly, if air or water moves toward the Equator, it finds the ground moving faster and gaining on it, so in effect, it curves westward.

This curving motion is the *Coriolis effect,* and it accounts for the fact that air currents and water currents take up circular paths in opposite directions north and south of the Equator.

Earth is so large that under ordinary circumstances, we don't move north or south fast enough for the Coriolis effect to come into significant play. It must be taken into account, however, in the case of artillery fire and satellite launchings.

REVOLVERS

The various handguns that had been used for nearly four centuries could only be fired once before time had to be taken to reload. Clearly this was a vulnerable time for the user, and the ability to fire more than once before reloading would provide an advantage against opponents not so equipped.

The first successful handgun of this sort held six bullets in a cylinder that automatically rotated with each shot, bringing another bullet into line for shooting. Known as a *revolver,* or a *six-shooter,* it was patented in 1835 by the American inventor Samuel Colt (1814–1862). The revolver became standard equipment in America's pioneer society. No tales of the wild West, either in print or on film, could exist without copious use of the revolver.

IN ADDITION

In South Africa, the Dutch-descended farmers *(Boers),* angered over the British freeing of slaves, moved northward beyond the Orange and Vaal rivers. Once outside British territory, they formed the *Boer Republics* of *Transvaal* and the *Orange Free State.* There they had a chance to defeat the blacks and take them as slaves.

Meanwhile, Australia was being more and more widely settled by the British. Melbourne was founded in 1835.

Emperor Francis I of Austria (who had been the last Holy Roman Emperor) died on March 2, 1835, and was succeeded by his son, who ruled as Ferdinand I (1793–1875). Ferdinand continued to allow the nation to be run by Metternich.

1836

PEPSIN

The discovery by Prout of hydrochloric acid in the stomach juices (see 1823) led to the natural assumption that it was the acid that acted to break down and digest foodstuffs in the stomach.

That this was not the case was shown by a German physiologist, Theodor Ambrose Hubert Schwann (1810–1882). He had found in 1834 that when an extract of the stomach lining was mixed with acid, it had a far greater power to dissolve meat than when acid alone was used. By 1836 he had extracted a substance from the stomach lining that proved particularly active in dissolving and digesting meat. He called it *pepsin,* from a Greek word meaning "to digest."

Pepsin, like diastase (see 1833), was an enzyme, but diastase had come from the plant world. Pepsin was the first animal enzyme to be isolated.

DANIELL CELL

The batteries that had been used since the time of Volta (see 1800), had been uncertain things at best. The amount of current produced was erratic and faded off rather rapidly. What was needed was a battery that could be relied on to deliver a constant quantity of current over a reasonably long time.

In 1836 the British chemist John Frederic Daniell (1790–1845) produced a battery, the *Daniell cell,* that made use of copper and zinc electrodes, and that filled the bill. Even though the future on a large scale belonged to generators of the type that Faraday had pioneered (see 1831), there would always be devices that were small or mobile or both, that should carry their own batteries for greater convenience.

IN ADDITION

Martin Van Buren (1782–1862) was elected eighth president of the United States. In Texas, the American settlers were in full revolt against the Mexican government. The Mexican president, Antonio López de Santa Anna (1794–1876), led an army of 4,000 against 188 Texans barricaded inside a Franciscan monastery, the Alamo. After an eleven-day siege, the Alamo was taken on March 6, 1836. All its defenders were either already dead or were killed. The Texans rallied under Samuel (Sam) Houston (1793–1863) and defeated Santa Anna at the Battle of San Jacinto on April 21. The Texans then declared their independence as the Republic of Texas and chose Houston as their president.

1837

ICE AGE

For years, Swiss geologists, notably Venetz, had been pointing to evidence that the glaciers of the Alps had been more extensive in the past (see 1821). A Swiss naturalist, Louis Agassiz (1807–1873), was among those who refused to accept these ideas—until he himself began to study the evidence.

In 1837 he began to move ahead of those who had gone before him to postulate that ice sheets had covered far more than mountainous areas, having also spread over large sections of the lowlands in the northern portions of the continents. With time he found evidence of glaciation in Great Britain, and after he went to the United States to lecture (and decided to remain there), he found evidence of glaciation in North America.

In the end he drew a convincing picture of an Ice Age, a period in the past when millions of square miles of North America, Scandinavia, and Siberia lay under a thick ice sheet. This was the first intimation that uniformitarianism (see 1830) was not one long, steady progression without surprises. The coming and going of the ice sheets had to be a kind of catastrophe, though clearly it had not put an end to life.

CHLOROPHYLL AND CELLS

When chlorophyll had been isolated (see 1817), its widespread existence in plants made it seem inevitable that it had some important function.

In 1837 the French chemist René-Joachim-Henri Dutrochet (1776–1847) was able to show definitely that photosynthesis (see 1779) took place *only* in those plant cells that contained chlorophyll. The key importance of chlorophyll to all multicellular life, animals as well as plants, was thus definitely established.

Dutrochet was a convinced antivitalist, by the way. He believed that all aspects of nature, animate and inanimate, were subject to the same physical and chemical laws.

TRISECTING AN ANGLE

The Greeks had established the principle that geometrical constructions must be carried through with only a straightedge and a compass (so that you could only draw straight lines and circular arcs) and in a finite number of steps. There was no real reason for this except that it gave geometers the very minimum of tools to work with and made their tasks more "sporting."

There were three constructions, however, the Greeks could never solve with straightedge and compass alone. The first was *squaring the circle;* that is, given a circle of any area, construct a square of the same area. The second was *doubling the cube;* that is, given a cube of any volume, construct another cube of twice the volume. The third was *trisecting the angle;* that is, given an angle of any size, divide it into three equal angles.

Since the Greeks, other mathematicians had tackled those problems with equal lack of success. The work of Gauss (see 1796) and Abel (see 1824), however, had shown the value of proving impossibility in mathematics. In 1837 a French mathematician, Pierre Wantsel (1814–1848), proved that doubling the cube and trisecting the angle were indeed impossible by the Greek rules. Eventually, squaring the circle was also shown to be impossible.

It is a queer commentary on human nature that these proofs of impossibility have not convinced any number of eager amateurs. A century and a half after these problems were shown to be impossible of solution by the Greek rules, hordes of circle-squarers and other impossibility-defiers still continue to offer the world their solutions. Needless to say, all such "solutions," without exception, have—and indeed *must* have—a fallacy somewhere.

IN ADDITION

William IV of Great Britain died on June 20, 1837. He was succeeded by his niece, who reigned as Victoria I (1819–1901). She could not succeed in Hanover, which barred female rulers, so a younger brother of William IV became king of Hanover and the connection between the two lands was broken after a century and a quarter.

In Canada, an uprising was easily suppressed, but Great Britain had learned a lesson sixty years before and set about correcting Canadian grievances.

In the United States, Michigan entered the Union as the twenty-sixth state. Since Arkansas had entered the year before, there were now thirteen free states and thirteen slave states. On March 3, 1837, the last day of Jackson's presidency, the United States recognized Texan independence.

A severe depression (they were called *panics* in those days, but nowadays, they have been downgraded to *recessions*) struck the United States, due almost entirely to Jackson's fiscal policies. The blame, of course, fell on Van Buren.

1838

DISTANCE OF STARS

Since the Earth moves around the Sun, the nearer stars should show a parallactic displacement compared with the farther ones. Bradley, when trying to detect stellar parallax, had discovered light aberration instead (see 1728). Herschel, in the same attempt, had discovered binary stars instead (see 1781).

The trouble was, the stellar parallax was so small that not till the 1830s did telescopes become sufficiently refined to detect it. At this time, the British astronomer Thomas Henderson (1798–1844) measured the parallax of Alpha Centauri from his observatory at Capetown in South Africa. (Alpha Centauri was so far south that it could not be observed from Europe.) The German astronomer Friedrich Georg Wilhelm von Struve (1793–1864), while working in Russia, measured the parallax of Vega.

Alpha Centauri is the third-brightest star in the sky, and Vega the fourth-brightest, so there was a chance they were comparatively close. The German astronomer Friedrich Wilhelm Bessel (1784–1846) chose the star 61 Cygni, which was rather dim but had the fastest proper motion (the apparent shift in position across the sky) then known, so that it was likely to be close, too.

Henderson was the first to complete the determination, but Bessel was the first to publish, in 1838, and it is he who gets the credit. His star, 61 Cygni, turned out to be about 35 quadrillion miles away. This distance is so mighty that even light takes 6 years to reach us from 61 Cygni—so that it is 6 *light-years* away. Alpha Centauri is only 4.3 light-years away, and it is actually the star closest to ourselves. Vega is 11 light-years away.

These distances made the Universe suddenly much larger than astronomers till then had dreamed. The entire Solar System shrank to a dot in space in comparison to the distance of even the nearest stars.

CELL THEORY

Hooke had first seen the empty remains of cells in cork (see 1665). Since then, biologists had often seen cells in living tissue, seen them as small bodies marked off and enclosed by *cell membranes* in animals, and by thin, cellulose-containing *cell walls* in plants. Brown had even seen the nuclei within cells (see 1831).

In 1838 the German botanist Matthias Jakob Schleiden (1804–1881) made the necessary leap of understanding and announced that all living plant tissue was made up of cells. The next year, Schwann (see 1836) extended the notion to animals as well. Both Schleiden and Schwann felt that the cell nuclei were of importance in connection with cell reproduction but could not divine the details. Those details were not to be worked out for another forty years.

The Schleiden-Schwann *cell theory* helped scientists enormously in enlarging their understanding of life.

YEAST CELLS

The discovery of substances such as diastase (see 1833) and pepsin (see 1836), which were organic catalysts but nonliving, turned the attention of chemists to yeast. Some thought that yeast might not be alive either, but it was indisputable that yeast actually grew in quantity as it was used, and that was characteristic of life.

Finally, in 1838, a French engineer, Charles Cagniard de la Tour (1777–1859), settled the matter. He studied yeast under the microscope and actually saw some of the globules budding and forming new globules. He realized he had been watching yeast cells, and that they were living things.

PROTEIN

It sometimes happens that the chief contribution of a particular scientist is the invention of an important word. Thus the Dutch chemist Gerardus Johannes Mulder (1802–1880) worked on the chemical structure of albuminous substances, which seemed to consist of molecules more complex than those of carbohydrates or fats. He came to believe that such substances were made up of a basic building block of carbon, hydrogen, oxygen, and nitrogen, to which were added varying numbers of sulfur and phosphorus atoms.

In 1838 he gave this basic building block the name *protein*, from the Greek word for "first," since it seemed the foundation of substances that were of first importance to living tissue. Eventually the word came to be used for the substances generally, and to this day, protein is what we call one of the two types of substances that are indeed of first importance to living tissue.

MORSE CODE

The idea of a telegraph had occurred to a number of people, including Henry (see 1823) and the British inventor Charles Wheatstone (1802–1875). It was just a matter of setting up a long wire and sending electricity through it in pulses by opening and closing a key. Combinations of pulses could be interpreted as letters and words.

What was needed was not so much a scientist, however, as a promoter, who could raise the necessary money to set up a sufficiently long wire, with relays to keep the signal from fading with distance. One such promoter, who had been working on the task since 1832, was the American artist Samuel Finley Breese Morse (1791–1872).

In 1838 he produced a logical list of short and long electric impulses *(dots* and *dashes)* for the various letters of the alphabet. This is called the *Morse code* to this day. It was the simple combination in Morse code of ". . . – – – . . ." that made the letter equivalents, SOS, the international distress call.

IN ADDITION

On April 23, 1838, two British steamships arrived in New York, having made the first trans-Atlantic crossing entirely by steam.

On December 16, 1838, the Boers in southern Africa defeated the Zulus and removed them as a threat.

In the United States, the *Underground Railway* began its operation. It was organized by Abolitionists who helped guide fleeing slaves northward to a haven in Canada.

Some fourteen thousand Cherokee Indians were driven westward from their homes in Georgia to the Indian territory (now called Oklahoma). About four thousand of them died on the way. The Indians called it the *Trail of Tears.*

1839

PHOTOGRAPHY

The French artist Louis-Jacques-Mandé Daguerre (1789–1851) had been working for years on a scheme for causing light to fall on a suspension of silver salts in such a way as to darken it selectively and produce a duplication of some scene. This is called *photography,* from Greek words meaning "light-writing."

The difficulty lay, first, in performing the feat after a reasonably short exposure and, second, in keeping the silver salts from continuing to darken and erasing the photograph.

By 1839 Daguerre had learned to dissolve the unchanged silver salts with a solution of sodium thiosulfate, so that what was left was permanent. It still required an exposure of at least twenty minutes, however, and the photographs that were obtained were faint.

Nevertheless, photography was born, and as others turned to it enthusiastically, the technique was very rapidly improved.

PHOTOGRAPHY OF THE MOON

With exposure times continuing to decrease rapidly, photography could be bent to scientific purposes. The British-born American chemist John William Draper (1811–1882) increased the delicacy of the photographic process to the point where, in 1839, he was able to take a photograph of the Moon. This was the first astronomical photography. He was also the first to photograph the solar spectrum. This heralded the day when astronomers could "freeze" a view and study it at leisure.

RUBBER

Rubber had come to the attention of Europeans when explorers found Native Americans using it. It was obtained from the sap of a tree originally native to tropical America but since grown in southeastern Asia and in other places as well.

Rubber was recognized as a useful waterproofing material, since it was impervious to water and wasn't rotted by it either. The trouble was, though, that when it grew cold, it stiffened and became brittle, and when it grew hot, it became soft and sticky. Attempts were made to find ways of making rubber less sensitive to temperature change, but at first there was only failure.

Then, in 1839, the American inventor Charles Goodyear (1800–1860) had a stroke of luck. He was trying to add sulfur to rubber when some of the mixture came accidentally into contact with a hot stove. To Goodyear's astonishment, those portions that weren't scorched too badly became dry and flexible, and they didn't lose flexibility in the cold or dryness in the heat. He began to heat the rubber-sulfur mixture to higher temperatures than anyone else had tried and thus obtained *vulcanized rubber* (after Vulcan, the Roman god of fire).

It was only with this discovery that rubber became truly useful, far more useful eventually than anyone might have suspected in Goodyear's day.

FUEL CELL

Ordinary electric batteries, even the Daniell cell (see 1836), obtain their energy by, in effect, burning metals. Electricity, as obtained from batteries, would be much cheaper if ordinary fuels could be burned instead, a *fuel cell*.

In 1839 the British physicist William Robert Grove (1811–1896) devised an electric cell that made use of hydrogen and oxygen, producing electricity as they combined into water. Even today hydrogen is still rather expensive, however. Using methane or coal dust along with oxygen would be close to ideal, but in all the century and a half since Grove's discovery, scientists have not been able to make the fuel cell practical. It remains a laboratory curiosity.

ANTARCTICA

The American explorer Charles Wilkes (1798–1877) commanded an exploring expedition in Antarctic waters between 1838 and 1840. He cruised along the limits of the sea to the south of the Indian Ocean. Ice everywhere prevented him from landing, but he saw enough land at a distance to realize, by 1839, that there was a continent within the Antarctic Circle. The information he brought home demonstrated this amply (and part of the Indian Ocean sector of the continent is still called *Wilkes Land* in his honor), so that Wilkes may be considered the discoverer of Antarctica.

ROSS ICE SHELF

Ross, who had located the North Magnetic Pole (see 1831), set out in 1839 to explore Antarctic waters. In the course of this exploration, he discovered a large oceanic inlet that cuts into Antarctica and is now known as *Ross Sea* in his honor. The southern portion of this sea is covered with a vast overhang of ice from the continental areas behind and is now known as the *Ross Ice Shelf*. He also located Mt. Erebus, the southernmost active volcano.

BICYCLE

The first vehicle a modern would recognize as a bicycle was designed in 1839 by a British blacksmith, Kirkpatrick Macmillan. It had two wheels, of which the rear was slightly larger, and a seat between. It had pedals, which were so arranged as to turn the rear wheel. It was heavy and clumsy and underwent a number of fundamental changes before becoming the instrument of today, but it did work. For the first time, progress had been made in allowing humans to make use of their own muscles to travel at a speed greater than they could run.

LANTHANUM

The rare earths discovered by Gadolin (see 1794) were far more complex in chemical nature than he could have expected. The Swedish chemist Carl Gustaf Mosander (1797–1858) did more to reveal that complexity than anyone else. He began, in 1839, by studying a compound of cerium, an element that had already been isolated from a rare earth mineral. In the process, he found a new element, which he named *lanthanum*, from a Greek word meaning "hidden," because it had been hidden so effectively in those minerals.

In the next few years, he isolated four other elements from the rare earth minerals: yttrium, erbium, terbium, and didymium. The first three were named from syllables of Ytterby, the quarry in Sweden where the first rare earth mineral had been obtained. The last was named from the Greek word for "twin," because it seemed so similar in properties to lanthanum. In fact, all rare earth elements are remarkably similar to each other.

IN ADDITION

As a result of the short-lived rebellion in French Canada, John George Lambton, Earl of Durham (1792–1840) came to serve as an enlightened governor-general in Canada in 1838 and, in 1839, published his recommendations that Canada be united and given a certain degree of self-government. These recommendations were eventually acted upon, so that Canada retained its ties to Great Britain.

European trade in China had taken a malignant turn. The British, in particular, encouraged trade in opium, since a drugged population was a quiet one. The Chinese objected and destroyed millions of dollars worth of illegal opium. Great Britain considered this a hostile action and so began the *Opium War.* This was the first of a series of events in which opium and other forms of degradation and despoliation were visited on China, first by Great Britain and then by other powers. It was a cruel punishment for China's failure of nerve in connection with overseas exploration four centuries before.

1840

THERMOCHEMISTRY

The study of combustion and of the heat developed in chemical reactions had languished since the time of Lavoisier (see 1769).

Then a Russian chemist, Germain Henri Hess (1802–1850), took up the matter and measured the amount of heat developed by a wide variety of chemical reactions. In 1840 he announced what is now known as *Hess's law;* that the amount of heat developed (or absorbed) in going from substance A to substance B is always fixed, regardless of the route by which the change is brought about.

The dependence on starting and ending points only, without concern for the path between, was already understood in connection with heat engines such as the steam engine. Hess's law made it seem that the laws of thermodynamics, which had been developed from the study of heat engines, might hold for chemical reactions too. In short, the laws of thermodynamics might be universal.

Hess's law founded the science of *thermochemistry* (the interrelationship of heat and chemical reactions).

OZONE

In 1840 a German chemist, Christian Friedrich Schönbein (1799–1868), working in a poorly ventilated laboratory, became conscious of a peculiar odor in the neighborhood of electrical equipment. He tracked it down and located a gas, which he named *ozone,* from the Greek word for "smell." It was later identified by an Irish physical chemist, Thomas Andrews (1813–1885), as a form of oxygen.

IN ADDITION

William Henry Harrison (1773–1841) was elected ninth president of the United States.

Frederick William III of Prussia died on June 7, 1840, and was succeeded by his son, who reigned as Frederick William IV (1795–1861).

Muhammad 'Alī Pasha of Egypt (1769–1849) was fighting the Ottoman Empire for the control of Syria and Arabia, and the rest of Europe was involved, with France on Egypt's side and the other powers on the Ottoman side. It all ended in a draw, but from then on, unrest in the Middle East would forever involve other powers, often needlessly, and often to no avail.

The population of the United States reached 17 million, still about even with that of Great Britain. New York had a population of 313,000, which was higher than that of Berlin, but Paris had nearly a million, and London had 2¼ million, making it the first city in history to pass the 2 million mark.

1841

HYPNOTISM

Mesmerism had been debunked (see 1774), but there were still those who practiced it as a form of show business. A British physician, James Braid (1795–1860), having witnessed an exhibition of mesmerism in 1841, investigated the matter and decided the phenomenon was real.

A person could indeed be put into a trancelike state resembling sleep by having the conscious mind forced into a state of weariness through repetitive stimuli. During such a state of quasi-sleep, the patient was extraordinarily open to suggestion and was relatively unresponsive to pain. Braid, avoiding the older name, called the phenomenon *hypnotism*, from the Greek word for "sleep."

It turned out to have its uses in medicine.

PHOTOGRAPHIC NEGATIVE

In the early years of photography, the photograph that was produced looked like the object photographed (a *photographic positive*) and was the only one of its type. One could not make copies of it.

In 1841, however, an English inventor, William Henry Fox Talbot (1800–1877), patented a device in which a *photographic negative* could be formed on glass, with all the naturally light areas dark and vice versa. Light could then be passed through the negative onto sensitized paper and a negative of the negative could be made, with all the light areas light again and all the dark areas dark.

The advantage of this two-stage process was that, given a negative, any number of positive prints could be made. In 1844, in fact, the first book was published that was illustrated with photographs.

NEEDLE GUN

Until now, all military small arms, whether arquebuses, muskets, or rifles, had been muzzle-loaded; that is, in reloading, the new bullet had to be forced down the length of the muzzle. Since 1836, however, a German inventor, Johann Nikolaus von Dreyse (1787–1867), had been working on a breech-loading rifle, in which bullets could be slipped in at the rear of the muzzle, making reloading much more rapid. It was perfected by 1841 and called the *needle gun* because it had a needlelike firing pin to detonate the powder and send the bullet flying. It was adopted by the Prussian army, which could then fire at a much greater rate than an opposing army equipped with only muzzleloaders. This did more than politicians and generals to insure Prussian domination of Europe in the decades to come.

SCREW THREADS

Industrial production could be increased if parts were standardized. The British inventor Joseph Whitworth (1803–1887), for instance, had worked out techniques for producing devices that matched each other not merely to the nearest sixteenth of an inch but to the nearest thousandth.

Even such improved methods wouldn't help if different manufacturers produced items of different overall dimensions, however. For instance, one factory might produce screws, all alike, and so might another, but the pitch of the screws in the two factories might be slightly different, so that one screw might fit a particular bolt and the other might not. In 1841 Whitworth suggested a standard pitch for all screws made everywhere, and the suggestion was eventually adopted.

When trade becomes first national, then international, and finally worldwide, such standardization becomes first advisable, then intensely useful, and finally indispensable.

IN ADDITION

William Henry Harrison, having become president on March 4, 1841, died of pneumonia on April 4. It was the first death of a sitting president. Vice President John Tyler (1790–1862) succeeded as tenth president of the United States.

New Zealand became a British colony in 1841, but Great Britain was having less luck elsewhere. It was fighting a war in Afghanistan that continued on and off for years, but Afghanistan stubbornly kept its independence.

In the Opium War, Great Britain seized various coastal points, including the island of Hong Kong in Canton Harbor. Hong Kong remained a British possession through the twentieth century.

1842

CHEMICAL FERTILIZER

Plants make use of minerals in the soil, and intensive cultivation deprives the soil of those minerals from year to year. If no minerals are returned to the soil, it will eventually lose fertility. The time-honored way of guarding against this is to make use of animal wastes as *fertilizer*, so that one of the important functions of domestic animals was to supply material for the manure pile, which would then have to be spread over the fields.

The manure piles, however, were not only offensively redolent, but could also be a source of disease (as was eventually discovered).

It occurred to chemists that it would be possible to determine just what elements were withdrawn from the soil and to return them in the form of odor-free, disease-free chemicals.

In 1842 an English agricultural scientist, John Bennet Lawes (1814–1900), patented a method for manufacturing what he called *su-*

perphosphate and the next year set up a factory for its production. This was the first chemical fertilizer. Such fertilizers helped sweeten the atmosphere, reduce the disease rate, and increase the food supply. (Nowadays, there is a fashion for "organically grown" food. Hiding behind that "organically" is the manure pile.)

DOPPLER EFFECT

The coming of the locomotive made a particular phenomenon much more noticeable than it had been earlier. The combination of speed and a warning whistle did the trick. People noticed that the whistle was high in pitch as the locomotive approached, and that the pitch dropped suddenly as the locomotive passed and began to recede.

An Austrian physicist, Christian Johann Doppler (1803–1853), explained the phenomenon correctly by pointing out that the sound waves partake of the motion of the source and so reach the ear at shorter intervals when the source is approaching—hence higher pitch. When the source recedes, the waves reach the ear at longer intervals—hence lower pitch.

Having done this in 1842, Doppler proceeded to check the matter experimentally a couple of years later. For two days, a locomotive pulled a flat car back and forth at different speeds. On the car were trumpeters sounding this note or that, while on the ground, musicians with a sense of absolute pitch recorded what they heard. Doppler verified his explanation in this way.

The Doppler effect turned out, in a few years, to have enormous importance in connection with astronomy.

CRANIAL INDEX

Blumenbach had divided the human species into "races," largely on the basis of skin color (see 1776). In 1842 the Swedish anatomist Anders Adolf Retzius (1796–1860) tried to make a finer division on the basis of something more subtle.

He suggested using the skull. The ratio of skull width to skull length, multiplied by 100, he called the *cranial index*. A cranial index of less than 80 was *dolichocephalic* (Greek for "long head"), while one of over 80 was *brachycephalic* (wide head).

In this way, Europeans could be divided into Nordics (tall and dolichocephalic), Mediterraneans (short and dolichocephalic), and Alpines (short and brachycephalic).

Actually, this was not a very good system for dividing the human species into smaller groups. In fact, no really satisfactory system has been devised, and every attempt has simply encouraged racism and ethnocentricity. It is safer—and better—to stop at the fact that *Homo sapiens* is a single species.

IN ADDITION

The Treaty of Nanking, on August 29, 1842, ceded Hong Kong to the British and gave them trading rights and special privileges in various coastal cities. The British, and other foreigners eventually, were not subject to Chinese law *(extraterritoriality)*. China had to pay a large indemnity and allow the opium trade to continue. It was the first of many humiliations China would have to suffer. On January 6, 1842, however, a British force of three thousand had to retreat from Kabul in Afghanistan and was slaughtered almost to a man.

In North America, the Webster-Ashburton Treaty, signed on August 9, 1842, settled the Canadian-American boundary from the Atlantic to the Rocky Mountains at the position it stands today. Only the Oregon Territory, west of the Rockies, remained in dispute.

1843

MECHANICAL EQUIVALENT OF HEAT

Some conservation laws had already been accepted by this time. Lavoisier had advanced the law of conservation of mass (see 1789) and before that there had been the law of conservation of momentum (see 1668).

There were suspicions that energy ought to be conserved also. After all, motion was a common form of energy, and according to Newton, a moving body would move on forever if not affected by an outside force. The energy would not disappear.

In real life, though, a moving body *does* stop moving after a while because of air resistance or because of friction with the ground. What happens to its energy then? Perhaps it is converted into heat. If so, however, a given amount of mechanical energy should be converted into some fixed amount of heat. Otherwise energy is not conserved.

A British physicist, James Prescott Joule (1818–1889), undertook to check this by experimentation. He expended energy in a variety of ways and measured the amount of heat produced. All his experiments showed that a fixed quantity of work ended up in a fixed quantity of heat. In 1843 he published his results: that 41,800,000 ergs of work produce 1 calorie of heat. This is called *the mechanical equivalent of heat.* Since 10,000,000 ergs are now called a *joule* in Joule's honor, we can say that 4.18 joules equal 1 calorie.

This made it appear that there might well be a law of conservation of energy if heat were included among the forms of energy. In fact a German physicist, Julius Robert von Mayer (1814–1878), had presented a figure of the mechanical equivalent of heat in 1842 (one that was far less accurate than Joule's) and deduced from it that a law of conservation of energy existed, but his work did not attract attention.

SUNSPOT CYCLE

Sunspots had been discovered by Galileo, but they had been observed only casually ever since. There seemed nothing of interest about them except for the fact of their existence.

The amateur German astronomer Samuel Heinrich Schwabe (1789–1875) had to work as a pharmacist, so he could not stay up all night to observe the sky. As something to do by day when business was slow, he began to study the neighborhood of the Sun to see if he could find some planet closer to the Sun than Mercury. Then the Sun itself caught his attention, and for seventeen years he studied its orb on every day that it was visible.

In 1843 he announced that the number of sunspots seemed to wax and wane in a ten-year cycle. (Later observations made the cycle appear eleven years long, on the average.) This discovery founded the modern study of *solar physics* and of *astrophysics* in general.

QUATERNIONS

The discovery of non-Euclidean geometries taught mathematicians that there were no *absolute* truths but that many alternative mathematics could exist depending on the nature of the axioms chosen. In other words, more than one set of axioms could give rise to self-consistent and useful consequences. This was so in geometry, and the Irish mathematician William Rowan Hamilton (1805–1865) showed that it was true in algebra as well.

Gauss had shown that complex numbers could be treated as though they were points on a plane, and that each could be represented by two numbers. Hamilton tried to deal with *hypercomplex numbers* that could be presented as points in three or more dimensions. He tried to work out a system of dealing with such points and found he could not. Then, in 1843, the thought came to him that he could do so if he were willing to abandon the commutative law of multiplication.

It had always been taken as self-evident that A × B = B × A, but if this was abandoned, Hamilton found he could devise a self-consistent algebra of hypercomplex numbers, or as he called them, *quaternions* (from the Latin word for "four," because each of Hamilton's points involved four numbers).

HIGHER ANALYTIC GEOMETRY

Descartes had worked out an analytic geometry in two dimensions, expressing curves by algebraic equations. The British mathematician Arthur Cayley (1821–1895) wanted to extend this to higher dimensions, as Hamilton had extended imaginary numbers (also planar). In 1843 Cayley succeeded in working out analytic geometry in three or more dimensions. This would be called *n-dimensional analytic geometry*.

WHEATSTONE BRIDGE

In 1843 Wheatstone (see 1838) used and made popular what came to be called the *Wheatstone bridge*. (He did not invent it—he freely admitted he did not—but there is some credit to be derived from making a device popular and introducing it to the scientific community.) The Wheatstone bridge was a device that measured the resistance of a circuit very delicately by balancing a number of currents against each other.

TRANSATLANTIC LINER

The first ship a modern might recognize as a transatlantic liner was the *S.S. Great Britain*, launched on July 19, 1843. It was 322 feet long, had a crew of 130, could seat 360 in its dining room, had an iron hull, a screw propeller, and was powered entirely by steam. It had been designed by the British builder Isambard Kingdom Brunel (1806–1859).

IN ADDITION

British dominion was spreading inexorably over India. In 1842 British forces under Charles James Napier (1782–1853) provoked a war with Sind in northwestern India, when it refused to accept the overlordship of the British East India Company. Napier defeated the Sind army at the Battle of Hyderābād on February 17, 1843, and sent a punning one-word message to London ("Peccavi," Latin for "I have sinned" [Sind]).

In New Zealand, the native Maoris began a war against the British settlers, which the settlers eventually won.

1844

TELEGRAPH

Morse (see 1838) was making his telegraph work. In order to keep the signal from fading over long lengths of wire, he made use of the *electric relay* invented by Henry (see 1823) in 1835. This is how it works. As electricity passes through a length of wire, the signal will eventually weaken but still be strong enough to produce an electromagnetic pull and attract a small iron key. This key, when attracted, can close a second circuit, which has a current from a nearby battery flowing through it. If that second circuit is long enough for its signal to weaken undesirably, a second relay can start a third circuit going. In other words, electric signals of sufficient strength to send a message can go any distance provided there are enough relays.

Morse patented his design in 1840. In 1843 he managed to persuade the United States Congress to appropriate money, and in 1844 he strung wires from Baltimore to Washington and sent the message, in Morse code, "What hath God wrought?" (a quotation from the book of Numbers in the Bible).

The telegraph spread quickly, and pretty soon nations were held together by a network of wires that meant virtually instantaneous communication from border to border.

COMPANION OF SIRIUS

Ever since Halley had discovered the proper motion of stars (see 1718), astronomers had been taking note of such motions among the nearer stars. Generally, those motions followed a straight line (allowing for parallax and other effects not due to the stellar motion itself).

Bessel (see 1838) noted, however, that the stars Sirius and Procyon had a somewhat wavy motion after all else was allowed for. In 1844 he decided that this could only be caused by the gravitational pull of another star. He suggested, therefore, that Sirius and Procyon were each binary systems. Since the companion star could in neither case be seen, Bessel proposed that they were stars near the end of their lives, with their light fading toward extinction, and were now too dim to see. They were *dark companions*.

In a way, Bessel was right, but when the truth was worked out eighty years later, it turned out that the companions were stranger than Bessel could possibly have imagined.

IN ADDITION

James Knox Polk of Tennessee (1795–1849) was elected the eleventh president of the United States.

1845

ASTRAEA

Since the four asteroids Ceres, Pallas, Juno, and Vesta had been discovered (see 1801 and 1802). No others had been found since, and it was widely assumed that four were all that existed.

The German amateur astronomer Karl Ludwig Hencke (1793–1866) began a systematic search in 1830, however, and after fifteen years of failure, he discovered a fifth asteroid in 1845. He named it *Astraea* after the Greek goddess of justice. Then, in 1847, he discovered a sixth, which he named *Hebe,* after the Greek goddess of youth.

Hencke's discoveries galvanized astronomers, who began to search for asteroids as a century earlier they had searched for comets.

SPIRAL NEBULAS

The nebulas that had been seen in the sky until now had seemed to be no more than little cloudy patches. Telescopes weren't good enough to make out much in the way of structure—or perhaps structure was simply lacking in them.

In 1827, however, an Irish astronomer, William Parsons, Earl of Rosse (1800–1867), began work on the largest telescope yet planned, and by 1845 it was finished. It had a mirror that was 72 inches across. However, the telescope, though large, was clumsy. It couldn't see much of the sky even when the sky was clear—and it was hardly ever clear.

Nevertheless, the telescope did accomplish a few things. In 1845 Rosse noted that one nebula had a distinct spiral shape, and in the following years he found fourteen others that were also spiral in appearance. These were termed *spiral nebulas,* and the time was to come, eighty years later, when they achieved considerable importance.

PERMANENT GASES

In 1845 Faraday returned to the task of liquefying gases (see 1823). He made use of a mixture of solid carbon dioxide and ether as a cooling agent, and he used higher pressures than he had in his earlier experiments.

Working in this way, Faraday managed to liquefy many gases. Indeed, there were only six gases known in 1845 that Faraday could not liquefy despite all his efforts. They were hydrogen, oxygen, nitrogen, carbon monoxide, nitric oxide, and methane.

These nonliquefiable gases were called, at least for a time, *permanent gases,* and they represented challenges that chemists tackled grimly.

IN ADDITION

On March 3, 1845, Florida joined the Union as the twenty-seventh state, and on December 29, Texas joined as the twenty-eighth. Both were slave states, the last of these to join the Union.

The United States was now involved in two areas of controversy. First, it and Great Britain both claimed all the Oregon Territory from the Spanish territory of California up to the boundary with Russian Alaska. Second, there was a dispute over the boundary between Texas and Mexico, with the United States claiming land all the way to the Rio Grande. It seemed unlikely that Great

Britain would fight over Oregon, but it looked as though Mexico would fight over Texas.

The potato crop failed in Europe in 1845 and a widespread famine resulted. This was worst in Ireland where the potatoes were particularly susceptible and where the peasantry lived on virtually nothing else. One and a half million Irish (nearly a fifth of the population) either died of starvation or emigrated (mostly to the United States).

In America, a religious zealot, William Miller (1782–1849), had predicted the end of the world in 1843 on the basis of his Biblical studies. Like all such predictions, nothing came of it. Nevertheless, his followers founded the Adventist Church in 1845, which eventually became the Seventh-Day Adventists, because they celebrated the Sabbath on Saturday. They continue to expect an imminent end to the world.

1846

ANESTHESIA

Pain, however useful as a warning signal designed to keep living organisms from damaging themselves too badly, becomes useless agony when operations must be performed.

Attempts to control pain were many. The use of alcohol or some form of what came to be called hypnotism was old. Acupuncture was used in the Orient. The new chemistry also contributed nitrous oxide, which, when inhaled, served to suppress the sensation of pain.

As time went on, substances such as diethyl ether (more commonly called simply *ether*) and chloroform were found to cause unconsciousness during which the sensation of pain disappeared. Ether came to be used by physicians during operations, the first to do so being an American physician, Crawford Williamson Long (1815–1878), who used it in 1842 to remove a tumor. He did not publish or publicize his work, however.

An American dentist, William Thomas Green Morton (1819–1868), used ether on a patient in September 1846, when extracting a tooth. The patient himself told the tale to a newspaper, and Morton was urged to demonstrate the use of ether during an operation at Massachusetts General Hospital.

It was this demonstration that effectively introduced the practice into medicine, so that Morton usually gets credit for the discovery. The American physician Oliver Wendell Holmes (1809–1894) suggested the term *anesthesia*, from Greek words meaning "no sensation."

NEPTUNE

The planet Uranus, discovered by Herschel (see 1781), was naturally observed frequently and carefully by many astronomers. In 1821 the French astronomer Alexis Bouvard (1767–1843) had shown that Uranus's position in the sky had drifted somewhat from the position to be expected if the gravitational influence of the Sun and various planets were taken into account.

One possibility was that an as-yet-unknown planet existed beyond Uranus whose gravitational influence had not been taken into account.

A British astronomer, John Couch Adams (1819–1892), attempted to calculate where such a distant planet might be in the sky based on the discrepancy in Uranus's position. He made some reasonable assumptions as to its mass and its distance from the Sun, and by October 1843, he had a possible position located. Unfortunately, he could not interest the Astronomer Royal, Airy (see 1825), in his work.

Meanwhile, a French astronomer, Urbain-Jean-Joseph Leverrier (1811–1877), was attempting the same task independently, and he ended with a position very similar to that of Adams. He wrote to a German astronomer, Johann Gottfried Galle (1812–1910), and asked him if he would check that region of the sky.

As it happened, Galle had a new map of that portion of the sky, and when he started looking, on September 23, 1846, he found the planet almost at once, for it was fairly bright (as seen through the telescope). It was not on the map.

Because of its greenish color, the new planet was named Neptune after the Roman god of the sea. Although Galle actually saw the planet first, the credit for its discovery is divided between Adams and Leverrier. The whole story is usually regarded as the greatest victory that Newton's law of gravitation was to achieve, for a tiny apparent deviation from that law was enough to lead to the discovery of a giant planet.

Later in 1846, the British astronomer William Lassell (1799–1880) discovered a satellite of Neptune, which he named *Triton* after a son of Neptune (Poseidon) in the Greek myths. It was a large satellite, larger than our Moon, and was the last large satellite to be discovered.

VULCAN

The planet Mercury has a moderately elliptical orbit. At the point in its orbit that marks its closest approach to the Sun, called its *perihelion*, Mercury moves forward very slowly because of the pull on it of other planets.

Leverrier (see Neptune, above) discovered in 1845 that, even when all planetary pulls were taken into account, Mercury's perihelion advanced just a bit more rapidly than it should.

He thought there might be a small planet closer to the Sun than Mercury was, whose gravitational pull was not being taken into account. He named it *Vulcan* after the Roman god of fire, and beginning in 1846, there were attempts to locate it.

All attempts to find Vulcan failed, however, and it was only seventy years later that a satisfactory explanation of the anomalous advance of Mercury's orbit was put forward.

CRYSTAL ASYMMETRY

Biot had shown that some substances were capable of twisting the plane of polarized light (see 1815). He had suggested at the time that some asymmetry in the system brought about the twisting. This seemed more likely to be true when it was found that some samples of a particular substance twisted polarized light clockwise, while other samples of the same substance twisted it counterclockwise.

Starting in 1846, a French chemist, Louis Pasteur (1822–1895), began his search for the possible asymmetries. He worked with small crystals of substances called *tartrates*. He studied them under the microscope and found that the crystals were rather subtly asymmetric, with one particular little facet on one side but not on the other. What's more, some crystals had the facet on the left side, while others had it on the right side, so that the two varieties were mirror images of each other.

These crystals had been obtained from a solution that had no effect on polarized light, and Pasteur suspected that one type of crystal produced one type of twist while the other produced the opposite type, the two together canceling each other. To check this, Pasteur separated the crystals painstakingly with tweezers, putting all the lefts to one side and all the rights to another. He dissolved them separately, and sure enough, one solution twisted the polarized light clockwise and the other, counterclockwise.

Crystal asymmetry could be one reason for this *optical activity*, but it could not be the sole reason. After all, the solution twisted the plane of polarized light when the crystals were dissolved and gone. There had to be a more deep-seated asymmetry present, but it took another quarter-century before that was discovered.

PROTOPLASM

A German botanist, Hugo von Mohl (1805–1872), studied plant cells assiduously. He found that the typical plant cell had a watery sap in the center, which showed no signs of life, and a granular colloidal layer rimming the cell, which did show such signs. In 1846 Mohl called this granular colloidal material *protoplasm*.

The word had been used earlier in another connection. The Czech physiologist Jan Evangelista Purkyině (1787–1869) had applied the word to the speck of living embryonic material in the egg, which was immersed in nonliving yolk that acted as its initial food supply. *Protoplasm* is from Greek words meaning "first formed," and thus referred to the living part of the egg as the first-formed part of the organism.

However, it was Mohl's more general use that caused the word to enter the scientific vocabulary.

CUNEIFORM

High on a cliff in the town of Bisitun in Persia is an inscription placed there by Darius the Great of Persia. He had achieved the throne in a dubious manner, perhaps by arranging the assassination of his predecessor. Therefore it was to his advantage to make the people of his empire think he was a legitimate ruler, and the inscription describes his version of his accession to the throne at a conspicuous height so that as many people as possible could see it. (That was the nearest an ancient could get to a television speech in prime time.) To make sure that as many of his subjects as possible understood the message, he had it inscribed in Persian, Assyrian, and Elamitic.

In 1846 a British archaeologist, Henry Creswicke Rawlinson (1810–1895), managed to work his way through the Persian bureaucracy and gain permission to investigate the inscription. He scaled the almost unscalable cliff and had himself dangled along the face of it in order that he might copy the inscription.

It turned out to be the Rosetta Stone of the ancient cuneiform writings of the Tigris-Euphrates valley. Making use of modern Persian as a guide, Rawlinson deciphered the inscription, and from this it was possible to read other inscriptions left behind by the ancient civilizations of the valley.

SEWING MACHINE

The idea of a sewing machine was natural enough, since machines that wove cloth in patterns had been around for quite a while. The trick was to make the machine small enough and convenient enough to use in the home. There were a number of near misses, but the first that really caught on—the prototype of those that quickly came to be used

—was invented by the American Elias Howe (1819–1867).

In 1846 he obtained a patent for his device, in which the eye of the needle was placed near the needle point and two threads were used, with stitches made by means of a shuttle. He proved the value of his machine by racing against five women sewing by hand and winning easily.

This was the first product of the Industrial Revolution that specifically lightened a woman's household tasks.

IN ADDITION

On January 13, 1846, President Polk ordered American forces under Zachary Taylor (1784–1850) to advance to the Rio Grande. The Mexicans crossed the Rio Grande and attacked, and the United States declared war on May 13. The *Mexican War* thus began. The Mexican War, by the way, was the first in which the telegraph, the railroad, and the revolver were used. It was also the first in which anesthesia was used for war surgery.

Though President Polk was willing to fight for every last square inch in the south, he compromised in the north. The 49 degree North Latitude boundary between the United States and Canada was extended to the Pacific Ocean on June 15, 1846, so that the Oregon Territory was divided in two. The island of Vancouver, part of which lay south of the line, was left to the British in its entirety. This boundary between the United States and Canada has remained ever since without further revision.

On December 28, 1846, Iowa became the twenty-ninth state of the Union.

Meanwhile in Ireland, the potato crop failed for the second year in a row.

1847

CONSERVATION OF ENERGY

Mayer had suggested a law of conservation of energy, and Joule had collected the data that made such a law reasonable (see 1843). Both lacked the necessary credentials as physicists to be convincing.

In 1847, however, the German physicist Hermann Ludwig Ferdinand von Helmholtz (1821–1894), whose credentials were impeccable, gathered the necessary data and published his conclusion that the law of conservation of energy existed. In other words, the total amount of energy in the Universe was constant; none could be created and none could be destroyed. Similarly, in any closed system—any portion of the Universe from which energy could not leave and into which energy could not enter—the total amount of energy was constant and could be neither created nor destroyed. (To be sure, no subsection of the Universe can be perfectly isolated from all energy loss or gain.)

Naturally, although energy can be neither gained nor lost, it can be converted from one form to another. Electricity, magnetism, chemical energy, kinetic energy, light, sound, and heat can all be interconverted.

The law of conservation of energy is also known as *the first law of thermodynamics* and it is usually viewed as the most basic of all the laws of nature.

PUERPERAL FEVER

It had long been understood that some diseases were contagious, although no one knew exactly what caused the contagion.

In the case of *puerperal fever* (from Latin words meaning "childbearing," and more familiarly known as *childbed fever*), there were suspicious signs of a certain type of contagion. A woman giving birth who was treated by a doctor who was also treating other childbearing women was much more likely to get the disease than a woman treated by a midwife who was not treating anyone else.

Some grew suspicious that doctors were somehow carrying the disease from patient to patient. The patients, being in a weakened condition and often bleeding, were much more susceptible than the doctor, who generally escaped infection. In the United States, Holmes (see 1846) took this attitude and published his opinions, but was paid little heed.

In Vienna, a Hungarian physician, Ignaz Phillipp Semmelweiss (1818–1865), thought so, too, and when he gained control of a hospital in 1847, he began to force the doctors to wash their hands in a solution of calcium chloride before touching patients. This was unpleasant for the doctors, of course, especially those older ones who were proud of the "hospital odor" of their hands.

The incidence of puerperal fever went down drastically following Semmelweiss's ruling, but this did not impress the rebellious doctors. When Hungary revolted against Austria in 1849, the Austrian doctors used Semmelweiss's Hungarian origins as a way of forcing him out. The hand-washing stopped and the incidence of puerperal fever went up again at once—which didn't disturb the doctors.

It was not until the cause of infection was finally understood, some twenty years later, that doctors began to reconcile themselves to washing their hands.

PAINLESS CHILDBIRTH

A British obstetrician, James Young Simpson (1811–1870), heard of the American experience with anesthesia (see 1846) and adopted it at once. Disliking ether, however, he made use of chloroform (much more dangerous, actually). Beginning in 1847, he was the first to administer anesthesia to women in childbirth.

There were preachers who objected to this, since God had cursed Eve when she was cast out of Eden and told her that she would bear children in "pain and sorrow." The preachers (who were all men) seemed to feel that pain and sorrow were just what women needed. In 1853, however, Simpson used chloroform to help deliver Queen Victoria's seventh child. The preachers, forced to choose between God and the queen, chose the queen, of course, and all criticism ceased.

NITROGLYCERINE

For five centuries, gunpowder had been the explosive par excellence, but now chemists were coming up with explosives that were much more powerful.

In 1845 a German chemist, Christian Friedrich Schönbein (1799–1868), accidentally spilled a mixture of nitric and sulfuric acids on his kitchen table. He snatched up his wife's apron to mop it up and then hung the wet apron over the stove to dry. It dried, and when it was dry enough, it went poof and disappeared. The astonished Schönbein experimented further and found he had discovered *nitrocellulose,* or as it came to be called, *guncotton.*

In 1847 an Italian chemist, Ascanio Sobrero (1812–1888), using the same mixture of nitric and sulfuric acids, combined it with glycerine and produced *nitroglycerine.* When he heated a single drop in a test-tube, it exploded with such force that the horrified Sobrero did no further work on it.

Nitrocellulose and nitroglycerine were entirely too explosive to be worked with safely, but each was tamed and put to work, initiating the era of modern explosives, both for good (in demolition for constructive purposes) and evil (in war and terrorism).

SYMBOLIC LOGIC

Much of Aristotle's science had been superseded during the three centuries since Copernicus, but his analysis of logic remained supreme. Even here, however, advances were finally being made.

The English mathematician George Boole (1815–1864) tried to mathematicize logical arguments. Attempts had already been made in this direction, notably by Leibniz (see 1669). It was Boole, however, who carried it through successfully. He applied a set of symbols to logical operations, choosing both symbols and operations so that they would resemble those of algebra, and showed that by algebralike manipulations, the symbols could be made to yield logical results.

In 1847 he published *The Mathematical Analysis of Logic,* thus founding what might be called *Boolean algebra* or *symbolic logic.* This was later to be of much use in studying the rigorous foundations of mathematics and, eventually, in programming computers.

SILVER FILLINGS

The American dentist Thomas Wiltberger Evans (1823–1897), who emigrated to France about 1847, introduced the use of silver amalgam for fillings in teeth whose decayed portions had been drilled away.

IN ADDITION

In Mexico, Zachary Taylor won a resounding victory at Buena Vista on February 23, 1847. On a second front, Winfield Scott (1786–1866) landed at Vera Cruz and fought his way to Mexico City, which he captured on September 14. Elsewhere, American forces swept westward from Texas, and American ships took the California port cities. Mexico had to give in and ask for peace.

On July 26, 1847, Liberia, on the western coast of Africa, became an independent republic. It has remained so ever since, even when all other African territory was occupied by European powers.

The Irish famine continued for a third year.

1848

ABSOLUTE ZERO

The work of Amagat (see 1699), on the steady decrease in the volume of gases with temperature, gave rise to the suspicion that there might be a temperature—an absolute zero—at which gas volume declined to zero. On the other hand, the volume decrease might lose validity once all gases were liquefied, and temperatures might thereafter be lowered indefinitely.

The British physicist William Thomson, later Baron Kelvin (1824–1907), pointed out in 1848 that it was not volume loss that was crucial, but energy loss; that energy loss affected all matter, liquids and solids as well as gases; and that the rate of loss was such that absolute zero should be reached at −273° C. (The modern figure is −273.15° C).

Kelvin proposed that a new *absolute scale* of temperature be created, based on absolute zero as the zero point. There would then be no negative temperatures. Each degree would be equal to a Celsius degree, so that the freezing point of water would be 273.15° A (for Absolute). Later scientists made it 273.15° K (where the *K* stands for Kelvin).

The concept of absolute temperature turned out to be essential to the science of thermodynamics.

CRAB NEBULA

Lord Rosse (see 1845), whose giant telescope had detected the spiral nebulas, studied the first nebulosity (M1) on the list Messier had prepared (see 1771). It was a curiously irregular patch of fog that marked the spot at which a bright new star had appeared in 1054, although that event had gone unnoticed in Europe.

To Rosse, the foggy patch seemed to have numerous crooked legs like those of a crab and he called it the *Crab nebula* in consequence, a name it has kept ever since. As time passed, the Crab nebula grew more and more interesting to astronomers until it was stated (with some exaggeration, perhaps) that all of astronomy could be divided into two equal parts: one, the Crab nebula, and two, everything else.

SPECTRAL LINE SHIFT

Six years earlier, Doppler had explained the *Doppler effect* with respect to sound (see 1842). Now the French physicist Armand-Hippolyte-Louis Fizeau (1819–1896) argued that the same effect would apply to any wave motion, particularly that of light.

If the light of the spectrum were unbroken, this would not be noticeable, since if the source were receding from us, ordinary visible light would move past the red boundary and become invisible, while invisible ultraviolet light would move past the violet boundary and become visible. It would work in the opposite direction if a light source were approaching us, and in neither case would there be any apparent change.

There are dark lines in the spectra, however, and these would shift their position. *That* would be noticeable. The lines would shift toward the red if the light source were receding and toward the violet if the light source were approaching. This is sometimes called the *Doppler-Fizeau effect*.

It was the *red shift* that proved of particular importance to astronomy.

IN ADDITION

Revolutions swept Europe, encouraged by the *Communist Manifesto*, which was published toward the end of 1847 by the German socialists Karl Marx (1818–1883) and Friedrich Engels (1820–1895) and came into prominence in 1848. It called for the reorganization of the world economy, with workers, rather than property owners, in charge.

In France, popular discontent forced Louis Philippe I to abdicate on February 24, 1848, and sent him into exile. He was the last of a line of French kings that had endured for nearly nine centuries. The Second Republic was proclaimed, but the leftists were totally defeated and Louis-Napoléon (1808–1873), the nephew of the Emperor Napoléon I, gained sudden popularity as a strong man on the right. On December 10, 1848, he won a landslide victory and on December 20 became the French president.

Revolutions also took place in Austria and Italy, and Metternich, the apostle of reaction, was forced to resign his post and flee Austria on March 17, 1848. Emperor Ferdinand I of Austria abdicated that same day and was succeeded by his son, who reigned as Francis Joseph I (1830–1916).

In the United States, the first woman's rights convention was held at Seneca Falls, New York, under the leadership of Elizabeth Cady Stanton (1815–1902) and Lucretia Coffin Mott (1793–1880). This represented the birth of feminism in the United States.

The Mexican War came to an end with the Treaty of Guadelupe-Hidalgo on February 2, 1848. The United States gained all of Texas to the Rio Grande, California, and the land between that now makes up the American Southwest.

President Polk was ready to retire, and Zachary Taylor, one of the heroes of the Mexican War, was elected twelfth president. Wisconsin entered the Union as the thirtieth state, and the count was now fifteen free states and fifteen slave.

1849

SPEED OF LIGHT

Roemer and Bradley had both determined the speed of light by astronomical methods (see 1675 and 1728). But until 1849 no one had been able to determine the speed of light by setting up an earthbound experiment. In that year, Fizeau (see 1848) set up a rapidly turning toothed disk on one hilltop and a mirror on another, 5 miles away. Light passed through a gap between the teeth of the disk to the mirror and was reflected. By that time, a tooth had moved into the way and the reflection could not be seen. If the disk turned rapidly enough, the reflected light passed through the next gap and could be seen again. From the speed of revolution required for the reflection to become visible, the time required for light to travel 10 miles could be calculated.

Fizeau's assistant, the French physicist Jean-Bernard-Léon Foucault (1819–1868), improved the technique the next year. Instead of using a toothed wheel, Foucault used two mirrors, one of which was rotating rapidly. Light would be reflected from the stationary mirror to the rotating one. The rotating mirror would have turned slightly in the time it took light to travel to it and would reflect that light slightly to one side as a result. From the amount of deflection, the speed of light could be calculated. Foucault's best figure for the speed of light was 185,000 miles per second. It was lower than the true value, but only by about 0.7 percent.

Foucault's method was so delicate that he only needed to let the light travel 66 feet. With only that distance to worry about, he could have it travel through water, and did. He found out that, through water, light traveled at only three-fourths its velocity through air. It turned out that the speed of light through any transparent medium is equal to the speed of light in a vacuum, divided by the index of refraction of the medium. (The index of refraction of a medium is a measure of the extent to which it will bend a light ray.)

ROCHE LIMIT

Saturn's rings had now been known for nearly two centuries, but their nature and how they came to exist remained uncertain and controversial.

In this connection, a French astronomer, Edouard Albert Roche (1820–1883), calculated the effect of one body on another when at relatively close range. His conclusions are useful in astronomy especially in studying very closely spaced binary stars.

He showed that if a smaller body circled a larger body, and if the smaller body was held together by gravitational forces only and chemical bonding was ignored, then tidal effects would break up the smaller body if it approached within 2½ times the radius of the larger body. Again, if a cloud of particles was within 2½ times the radius of a large body to begin with, that cloud could not be pulled together into a body by gravitational forces.

None of the satellites known in the Solar System at that time were within 2½ times the radius of the planet they circled. However, Saturn's rings were, in their entirety, within that distance. The conclusion was that Saturn's tidal force kept them from condensing into a satellite.

NERVE FIBERS

The cell theory, which had been advanced by Schleiden and Schwann (see 1838), grew firmer with the years. The German anatomist Rudolf Albert von Kölliker (1817–1905) had shown that eggs and sperm might be considered cells, and in 1849 he showed that nerve fibers were elongated outgrowths of cells.

IN ADDITION

The discovery of gold in California the previous year produced a mad rush to share in presumably untold wealth. Very few found wealth, but the movement helped to develop the American West.

The tide of revolution ebbed in Europe. Hungary had risen against the Austrians, but the Austrians, aided by the Russians, sent in an army, and brought Hungary under subjection again.

There was an uprising in the papal dominions against Pius IX (1792–1878), who had become Pope in 1846. It was led by Mazzini and Giuseppi Garibaldi (1807–1882) and was crushed by Austrian and other forces.

Sardinia attempted to fight Austria and gain independence for the Austrian-dominated Lombardy-Venetia region in Italy, but Austrian forces defeated the Sardinians in two battles. The Sardinian king, Charles Albert (1798–1849), abdicated and was succeeded by his son, who reigned as Victor Emmanuel II (1820–1878).

The German states outside Austria tried to form a united nation. On March 27, 1849, delegates meeting at Frankfurt established a constitution for a *German Empire* and offered to make Frederick William IV of Prussia the emperor. Lacking the nerve to stand up to Austria, he refused the crown.

The net result of the events of 1849 was that Austria remained dominant in central Europe, apparently stronger than ever.

The American inventor Walter Hunt (1796–1859) produced the safety pin in 1849, and in the same year, the French inventor Joseph Monier (1823–1906) devised reinforced concrete, whereby iron bars were embedded in the concrete, making it much stronger.

1850

SECOND LAW OF THERMODYNAMICS

The first law of thermodynamics (that is, the law of conservation of energy—see 1847) is an essentially optimistic law. Since no energy can be destroyed, it might seem that energy is always there to be used over and over.

Not all energy is equally useful, however. Carnot (see 1824) had pointed out that in a steam engine some of the energy *must* be lost as heat and so not all of it could be turned into useful work.

The German physicist Rudolf Julius Emanuel Clausius (1822–1888) found that this was true of any energy conversion: some energy was always lost as heat, and heat could never be converted completely to any other form of energy. As a result, the energy supply of the Universe was constantly being degraded to heat, and the amount of *useful* energy was constantly decreasing.

Clausius proposed that if the ratio of the heat content of a system to its absolute temperature were taken, that ratio would always increase in any process taking place in a closed system. Under perfect conditions, it might remain constant, but it would never decrease. Years later, Clausius gave this ratio the name of *entropy* for some rather obscure reason.

Clausius had thus established the *second law of thermodynamics,* that the amount of entropy in the Universe always increases and will some day reach a maximum when no useful energy is left and disorder is total. That sounds pessimistic, but since it will probably take many trillions of years to degrade all the energy in the Universe, it can't be considered a matter of immediate concern.

INFRARED WAVES

The details of light, such as the way it is refracted, the spectrum it forms, and the spectral lines it contains, can all be studied easily because light can be seen—it affects the retina. Infrared radiation cannot be seen and it is correspondingly harder to study.

An Italian physicist, Macedonio Melloni (1798–1854), had invented a *thermopile,* however, a series of strips of two different metals that produced electric currents when one end was heated. Very weak electric currents could be detected and therefore so also could very weak heating effects.

In 1850 Melloni put his thermopile to work on the infrared radiation that Herschel had discovered (see 1800). He could detect the infrared radiation by its heating effects as delicately as the eye could detect ordinary light. Melloni was able to show that infrared radiation exhibited all the properties of ordinary light, including polarization and interference, so that there was no question it consisted of waves just like those of light but longer—too long to affect the retina.

IN ADDITION

Zachary Taylor died in office on July 9, 1850. He was succeeded by his vice president, Millard Fillmore (1800–1874), who became thirteenth president of the United States.

Meanwhile, the United States faced another slavery crisis. California wished to enter the Union as a free state but there was no obvious candidate for a balancing slave state. In the end, Clay hammered out the *Compromise of 1850:* California could enter as a free state, but the southwestern territories could decide for themselves whether to be slave or free.

In China, a huge rebellion began among the southern peasants against the Manchu rulers. This *T'ai Ping Rebellion* would last fourteen years and be probably the bloodiest civil war in history, killing as many people as would be killed in World War I.

The United States in 1850 had a population of 23 million, now clearly ahead of Great Britain's 21 million, though still behind France with its 36 million. London was still the world's largest city with 2.4 million, while New York, with 700,000, was less than a third as large.

1851

ROTATION OF THE EARTH

Ever since the time of Copernicus (see 1543) it had been taken for granted that the Earth was rotating on its axis. Nevertheless, no one had actually *demonstrated* the fact. It seemed stationary, and no effect had been observed (other than the apparent rotation of the sky) that could be attributed to the rotation.

In 1851, however, Foucault (see 1849) suspended a large iron ball, about 2 feet in diameter and weighing 62 pounds, from a steel wire more than 200 feet long, hanging inside the dome of a large church. The pendulum ended in a spike that just cleared the floor but would score a mark in the sand with which the floor was sprinkled.

The iron ball was drawn far to one side and tied to the wall by a cord. To prevent vibration when the ball was released, the cord was not cut, but set on fire. The swinging pendulum would then remain in the same plane, but the Earth, as it rotated, would change its orientation. If the pendulum had been at the North Pole, for example, the pendulum would have seemed to change its plane through a complete circle in 24 hours. At the latitude of Paris, the change would have taken 31 hours and 47 minutes. Thus the spectators were actually watching the Earth rotate under the pendulum.

ARIEL AND UMBRIEL

The last large satellite, Triton, had been discovered by Lassell five years before (see 1846), but there were smaller satellites to be found. In 1848 Lassell had discovered an eighth satellite of Saturn, which he named *Hyperion* after still another of Saturn's (Cronos's) brother-Titans in the ancient myths. This satellite was discovered almost simultaneously by the American astronomer George Phillips Bond (1825–1865).

Then, in 1851, Lassell discovered a third and fourth satellite of Uranus. He followed Herschel's precedent (see 1789) by naming them after spirits in English literature. One he named *Ariel* after a spirit in Shakespeare's *The Tempest* and the other *Umbriel* after a spirit in Pope's *The Rape of the Lock.*

IN ADDITION

On May 1, 1851, the Great Exhibition opened in London. It was the first of the modern *World Fairs* and it was intended to celebrate British industry and prosperity. It was the visible evidence of the way in which the Industrial Revolution, now some three-quarters of a century old, had changed the world.

In France, in a sudden coup on December 2, 1851, Louis-Napoléon filled Paris with army units, arrested legislators, fired on unarmed protesters, and made himself absolute master of France.

A cable for telegraph wires was laid under the English Channel from Dover to Calais in 1851, connecting Great Britain with Europe.

1852

JOULE-THOMSON EFFECT

In 1852 Joule (see 1843) and Thomson (see 1848) were able to show that if a gas was allowed to expand, it consumed energy, because the molecules had to separate against a certain tendency to cling together. If energy were kept from entering the gas, the energy for expansion would have to come from the gas itself, and the temperature would drop.

This came to be called the *Joule-Thomson effect* and would later be used to liquefy some of the *permanent gases* (see 1845).

VALENCE

Chemists knew that elements had different capacities for combining with other elements. For instance, an oxygen atom would combine with two hydrogen atoms to form water, a nitrogen atom would combine with three hydrogen atoms to form ammonia, and a carbon atom would combine with four hydrogen atoms to form methane.

This was not studied closely and systematically, however, until an English chemist, Edward Frankland (1825–1899), studied organometallic compounds.

He could not help noticing the manner in which metallic atoms combined with organic compounds, and how each different kind of atom would almost invariably hook onto a fixed number of other atom groupings. In 1852 he published his theory of what came to be called *valence* (from a Latin word for "power"), pointing out that a particular atom had the power to combine with fixed numbers of other kinds of atoms according to simple rules.

The notion of valence imposed a new kind of order on the elements, because valence seemed to change regularly with atomic weight. That led to important advances in the next decade.

GYROSCOPE

Just as a pendulum has the ability to swing in an unchanging plane, so does a massive sphere in rotation have a tendency to maintain the direction of its axis of spin (as the Earth itself does). Foucault, who had made an amazing demonstration with a pendulum (see 1851), now demonstrated the matter of the rotating sphere, by setting a wheel with a heavy rim (a *gyroscope*) into rapid rotation. It not only maintained its axial direction, but when it was tipped, the effect of gravity was to set up a motion at right angles that was equivalent to the precession of the equinoxes (see 134 B.C.).

This meant that a gyroscope, maintaining its axial orientation, could be used as a steady indicator of true north, substituting for, and better than, the magnetic compass that had been in use for some six centuries.

SUNSPOTS AND EARTH

The British physicist Edward Sabine (1788–1883) showed, in 1852, that the frequency of disturbances in Earth's magnetic field paralleled the increase and decrease of sunspots on the Sun. This was the first example of a linkage between Earth and Sun involving something other than the Sun's radiation of light and heat or its gravitational pull. It was also the first hint that sunspots had magnetic properties.

ELEVATORS

As cities grow more crowded, they may spread out in area, or divide the available room into smaller and smaller living quarters, or build higher and higher structures.

High structures were originally built of stone, as the strongest available material, but the higher the building, the thicker the stone had to be at the bottom to support the structure and the less room available for living quarters. Reinforced concrete (see 1849) made higher structures possible, and steel beams would in time be even better.

Yet with the best materials and the cleverest designs, a tall building is useless if there is no way but slow climbing to lift oneself to the top floors.

In 1852 an American inventor, Elisha Graves Otis (1811–1861), built the first mechanical elevator with an adequate safety guard, one that would keep it from falling even if the cable holding it were severed completely. In 1854 Otis demonstrated the workability of the device by having himself raised to a considerable height and then having the cable cut. He descended safely.

It was the elevator more than anything else that made possible the shape of cities to come.

IN ADDITION

In the United States, Franklin Pierce (1804–1869) was elected fourteenth president.

Louis-Napoléon ordered a plebiscite in November 1852, which by rigging and force turned out to be strongly in his favor. On December 2, the anniversary of his coup, he proclaimed the *Second Empire* with himself as Napoleon III.

1853

AGE OF THE SUN

It had been assumed for thousands of years that the Sun was eternal and changeless (and would be until God wished to put it to an end). The discovery of sunspots (see 1610) had shaken this belief but not destroyed it.

Once the law of conservation of energy was well established (see 1847), however, the shining of the Sun had to be questioned. Light and heat were energy, and this could not be created out of nothing. Where, then, did the Sun obtain the energy that had kept it shining brilliantly enough to warm the Earth at a distance of nearly a hundred million miles for all the many thousands of years of history? It could not be ordinary burning, for the entire substance of the Sun would have been consumed in a mere fifteen hundred years in that case.

Helmholtz, who had been the one to establish the law of energy conservation, pondered the matter and by 1853 had decided that the only energy supply that would suffice was that of the Sun's own gravitation.

Slowly, the Sun's globe must be contracting, and the falling inward of that huge mass must be supplying the energy that was converted into light and heat. This must be the continuation of a process that had begun when a gigantic cloud of dust and gas had contracted to form the Sun in the first place (see 1796).

In order to supply light and heat at the

current rate, Helmholtz concluded, the Sun must have contracted from a size filling the orbit of the Earth to its present size in something like twenty-five million years. That meant Earth could be no older than that. Moreover, it meant that in another ten million years, the Sun would be too small and cool to support life on Earth.

This conclusion came as a great shock to geologists, who already thought the Earth must be considerably older than twenty-five million years. It was to be another half-century before the controversy was settled—in favor of the geologists.

GLIDER

Balloons had been in existence for seventy years, proving that objects that were themselves denser than air could remain aloft under proper conditions of winds and updrafts.

Nevertheless, the English engineer George Cayley (see 1809) was the first to study, scientifically, the conditions under which air might keep a heavier-than-air object aloft. He thus founded the science of *aerodynamics.* He realized that what was needed were fixed wings, like the flaps of a flying squirrel, rather than moving wings, like those of birds.

He worked out the basic shape that airplanes would eventually have—wings, tail, streamlined fuselage, and rudder. He also realized that he needed an engine and propeller to be able to proceed against the wind, but he knew that no engine then existing would be light and powerful enough to do so.

In 1853 he built the first device so constructed aerodynamically as to glide with the wind and lift with the updrafts—a *glider,* as it came to be called. Cayley, too old to take the risk, ordered his coachman to take the first glider flight. The coachman did so, vehemently objecting, flew 500 yards, and survived. Over the second half of the nineteenth century, glider-flying became a popular sport, as balloon-flying had been in the first half.

KEROSENE

In 1853 a British physician, Abraham Gesner (1797–1864), developed a process that would yield an inflammable liquid from asphalt. Because it was driven out of a waxy mixture of solid hydrocarbons, Gesner called the liquid *kerosene,* from a Greek word for "wax."

Kerosense proved ideal for lamps, but even with Gesner's process, enough kerosene could not be produced to meet the great demands represented by the lamps of Europe and America.

IN ADDITION

Japan had kept itself isolated from foreign influences for two centuries, but the Western nations desired trade. Therefore, American ships under Matthew Calbraith Perry (1794–1858) sailed into Tokyo Bay and, on July 14, 1853, delivered written communications to the rulers, making it clear that when they returned, they expected Japan to open its ports to American trade.

On December 30, 1853, the United States signed a treaty with Mexico taking possession of a strip of territory south of the Gila River (in what is now Arizona and New Mexico) in order that a railroad might be built through it. It was about 45,000 square miles in area, and the details were negotiated by James Gadsden (1788–1858). The *Gadsden purchase* fixed the boundary of the United States and Mexico to what it has been ever since.

Russia insisted that it was the natural protector of Christians inside the Ottoman Empire, particularly in the Holy Land. The Turks opposed this, and when it became clear that Great Britain and France also opposed any increase in Russian power in the Mediterranean, Turkey declared war on Russia on October 4, 1853.

1854

CHOLERA

In the early nineteenth century, Europe suffered from several cholera epidemics that arrived from India, where it was endemic. Increasingly, physicians were sure that the contagion arose from polluted water. An English physician, John Snow (1813–1858), had published his views on this matter in 1849.

When an epidemic of cholera struck London in 1854, Snow studied the geographic incidence of cholera in relation to water supply. He found five hundred cases of cholera, for instance, within a few blocks of a public water pump that drew water from a well just a few feet from a sewer pipe. He had the pump handle removed, and the incidence of cholera fell at once.

This gave a powerful impetus to improving hygiene as a means of disease prevention.

TELEGRAPH PLATEAU

With cables laid across the Hudson and Mississippi rivers in the 1840s and under the English Channel and the Irish Sea in the 1850s (see 1851), it became natural to think of a *real* job, that of laying a cable under the Atlantic Ocean to connect Europe and America.

The process of laying such a cable required some understanding of what the floor of the Atlantic Ocean was like, and the task fell to the American oceanographer Matthew Fontaine Maury (1806–1873). In the early 1850s he had prepared a chart of ocean depths, and by 1854 he had noted that the Atlantic Ocean was shallower in the center than on either side. He concluded that there was a central plateau, which he called *Telegraph Plateau*.

This was the first important physical finding concerning the ocean bottom, and little more was to be learned for another century.

NON-EUCLIDEAN GEOMETRY

The non-Euclidean geometry worked out by Lobachevsky and Bolyai (see 1826) presupposed the possibility of more than one line through a point (even an infinite number of lines), all parallel to a given line that did not include the point. As in Euclidean geometry, lines of infinite length were possible.

In 1854 a German mathematician, Georg Friedrich Bernhard Riemann (1826–1866), developed another kind of non-Euclidean geometry, one in which it was impossible for *any* two lines to be parallel, and *all* lines intersected. What's more, in this geometry all lines were finite in length. In Euclidean geometry, the three angles of a triangle add up to 360 degrees; in Lobachevskian geometry, they add up to less than 360 degrees; and in Riemannian geometry, they add up to more than 360 degrees.

Riemann's geometry is perfectly reasonable and self-consistent. Indeed, it resembles the geometry on the surface of a sphere, where all *great circles* (those that divide the surface into equal halves) are finite in length and intersect.

Riemann generalized geometry to the point where he considered it in any number of dimensions. He also considered situations in which measurements changed from point to point in space but in such a way that one could transform one set of measurements into another according to a fixed rule.

At the time, this sounded like pure abstraction, but half a century later, Riemannian geometry was shown to represent a truer picture of the Universe than did Euclid's geometry, thanks to the theory of general relativity.

IN ADDITION

Since the Russians would not stop their war against Turkey, Great Britain and France (which had not fought on the same side since the Third Crusade) declared war on Russia on March 28, 1854. This began the *Crimean War,* so-called because British and French forces landed on the Crimean peninsula jutting into the north-central portion of the Black Sea.

In the Far East, Perry returned to Japan, and on March 31, 1854, Japan signed the Treaty of Kanagawa, opening two ports to American trade and making provisions for humane treatment of shipwrecked American sailors. Almost at once, Japan saw that it dared not continue to remain weak in the face of Western strength, and it began to plan to become capable of fighting in Western ways.

In the United States, the Kansas-Nebraska Act was passed on May 30, 1854. This opened up northern territories to slavery if the population wanted it and destroyed the Compromise of 1820.

1855

LINES OF FORCE

Faraday had introduced the concept of lines of force (see 1821), but since he knew no mathematics, he could only describe them pictorially.

The British mathematician James Clerk Maxwell (1831–1879) was able to translate Faraday's concept into mathematical form in 1855 and show that the former's intuitive grasp of the subject was completely correct.

GEISSLER TUBES

Air pumps such as those devised by Guericke and Hooke (see 1645 and 1657) did not produce very good vacuums. A much better vacuum had earlier been produced by Torricelli when he allowed a mercury column to drop downward out of its closed end (see 1643).

In 1855 the German inventor Johann Heinrich Wilhelm Geissler (1815–1879) took advantage of Torricelli's discovery to devise an air pump without moving mechanical parts. By moving a column of mercury up and down, he could make the vacuum above the column suck air out of a container. In this way, Geissler produced tubes that were more thoroughly evacuated than anything seen previously. Tubes so evacuated were called *Geissler tubes,* and their use was to lead in the decades that followed to remarkable advances in the study of atomic structure.

SEISMOGRAPH

An earthquake of considerable magnitude cannot be mistaken for anything else. However, there are numerous small earthquakes that, in our busy, rumbling life, may go unnoticed.

In 1855 an Italian physicist, Luigi Palmieri (1807–1896), invented a device for the detection of such small tremors. This consisted of horizontal tubes, turned up at the ends and partly filled with mercury. Even slight quakes would cause the mercury to bob from side to side. Small iron floats were so attached that their movements could be read off on a scale and the intensity of the quake estimated.

This was the first crude *seismograph.* It was not very useful; traffic vibrations, for instance, were difficult to distinguish from minor earthquakes; but it was a beginning.

PYROXYLIN

In 1855 a British chemist, Alexander Parkes (1813–1890), found that pyroxylin (a partly nitrated cellulose), if dissolved in alcohol and ether in which camphor had also been dissolved, would produce a hard solid upon evaporation, which would soften and become malleable when heated. He found no way of doing anything commercial with it, but he had discovered the first synthetic plastic.

IN ADDITION

Nicholas I of Russia died on March 2, 1855, after a thirty-year reign and was succeeded by his son, who reigned as Alexander II (1818–1881). Matters continued to go badly for Russia in the Crimea, and it was forced to evacuate Sevastopol on September 11. Yet despite its defeat in the Crimea, Russia was beginning the conquest of central Asia, the regions north of Persia and Afghanistan.

The Italian kingdom of Sardinia joined the British and French in the Crimea in 1855, not out of any interest in the issues, but to gain British and French help in the working out of its own future plans.

Japan and Siam were both signing treaties with the Western powers and continuing with plans for modernization.

1856

GLYCOGEN

Until now, starch had been viewed as the plant's way of storing energy. Animals tend to store it as fat (as some plants do too), which is a more compact form of energy.

In 1856, however, the French physiologist Claude Bernard (1813–1878) discovered a form of starch in the mammalian liver. Because it was easily broken down to glucose (the immediate energy source in the bloodstream), he called it *glycogen* (Greek for "glucose-producer").

He showed that glycogen was built up out of blood glucose and acted as a reserve store that could be broken down to glucose again when necessary. Whether glycogen was built up or broken down depended on the exact state of the body, the energy requirements of the various tissues, the food supply in the intestines, and so on. The net result was that the glycogen balance was so maneuvered that the sugar content in the blood remained steady.

Until then, it was supposed that plants built up complex molecules *(anabolism)* and that animals tore them down *(catabolism).* Bernard's work showed that animals built up complex molecules as well, and as time went on, it appeared that both plants and animals built up and broke down substances as needed. What was different between them was that the build-up in plants was through solar energy, while in animals, the build-up was through chemical energy from food, the food being obtained eventually, if not directly, from plants.

STEEL

For over three thousand years, steel had been known to be the strongest metal, but manufacturing steel was a difficult process. First, iron came out of the smelting furnace with a great deal of carbon in it (from the charcoal or coke used to smelt it). This *cast iron* was hard but brittle. The carbon could be removed and pure iron *(wrought iron)* left behind. This was tough but too soft. Then finally, just the right amount of carbon could be added to form steel, which was both tough and hard. By then, steel was expensive.

The British metallurgist Henry Bessemer (1813–1898) sought a way of removing the carbon from cast iron more cheaply. In the old way, the carbon was burned off by oxygen from additional iron ore while the mixture was strongly heated. Bessemer wondered if the oxygen could not be supplied more simply by sending a blast of air through the molten iron. It might seem that the blast would cool the iron and spoil everything, but the combination of oxygen from the air and carbon in the iron actually heated the mix. By stopping the process at the point where the carbon level was right, Bessemer got his steel directly.

In 1856 Bessemer announced his discovery, and such *blast furnaces* began to be built. It took a while for the method to be perfected because it required phosphorus-free iron ore, something the steel manufacturers didn't understand. Eventually, however, the bugs were worked out and the era of cheap steel began. Steel, together with elevators (see 1852), helped build the cities of the next century.

SYNTHETIC DYES

Although human beings have a liking for color in their clothes, natural fibers such as cotton, linen, and wool tend to be white or off-white, and the colored dyes available to ancients tended to bleach in sunlight and wash out in water. To be sure, a few dyes were available that did neither. One was *Tyrian purple,* obtained from a snail in the harbor of Tyre, a dye so expensive it was reserved in late Roman times for royalty. There were also *cochineal,* from an insect, and *indigo* and *alizarin* from plants.

In 1856 a young British chemistry student, William Henry Perkin (1838–1907), was trying to synthesize quinine, the antimalarial drug. He was doomed to failure, for the molecular structure of quinine was far too complex for the synthetic methods of the time to handle.

In the mess that resulted, however, Perkin thought he saw a purplish glint. He added alcohol, which dissolved a substance that turned the solution a beautiful *mauve* (as the color was eventually named). Perkin wondered if it could be used as a dye. He had it checked and decided it could. He left school, invested the family savings in a factory to manufacture the dye, and the project proved most successful.

Other chemists quickly entered the field of synthetic dyes and the world of fashion became the rainbow of colors it has been ever since.

NEANDERTHAL MAN

In western Germany, in the Neander River Valley (*Neanderthal* in German), workers were clearing out a limestone cave in 1856 and came across some bones. This was not unusual, and what was commonly done in such cases was to throw away the bones.

This time, though, the word reached a professor at a nearby school, and he managed to salvage about fourteen of the bones, including a skull.

By this time, geologists were sure the Earth was very old, and biologists were sure that human beings had been alive for much longer than the Bible seemed to indicate. But however long human beings had been alive, had they always been human beings, or had they evolved from some simpler form?

The bones from the Neanderthal cave were clearly human, but the skull was rather different from those of modern humans. It had pronounced bony ridges over the eyes, a backward-sloping forehead, a receding chin, and unusually prominent teeth.

The remains were quickly dubbed *Neanderthal man* and the question arose as to whether it was a primitive form of human being or merely an individual with some bone disorder. The chief supporter of the view that it was a primitive form of humanity was a French anthropologist, Pierre-Paul Broca (1824–1880), and he won out eventually.

Nowadays, we believe Neanderthal man to be a subspecies of *Homo sapiens*. Nevertheless, the 1856 finding was the first step toward the demonstration, through fossil remains, of the evolution of the human species.

PASTEURIZATION

In 1856 France's wine industry was in a bad way. Wine often went sour as it aged, and millions of francs were lost. Pasteur (see 1846) undertook the task of investigating the problem.

He studied samples of wine under the microscope and found that, when wine aged properly, it contained yeast cells that were spherical globules. Wine that was souring contained elongated yeast cells. He decided there were two types of yeast cells, one of which created lactic acid.

He also decided that the only solution to the problem was to kill the yeast cells, good and bad alike, after the alcohol had formed but before the acid had a chance to form. The wine should be heated gently at about 50° C and then stoppered and left to age without the influence of yeast.

The vintners were horrified at the suggestion, but they were desperate enough to try it, and they found that it worked. The process of gentle heating was called *pasteurization,* and it was eventually applied to milk, too, to prevent diseases that might otherwise be spread by its contamination.

This episode turned Pasteur's attention to the world of microorganisms, with enormously important results.

IN ADDITION

Austria threatened to join the war against Russia, and that was the last straw. The Treaty of Paris, signed on February 1, 1856, ended the Crimean War. Turkey's territory was guaranteed, and Turkey promised to respect the rights of its Christian subjects. Russia, of course, gained nothing.

In the United States, James Buchanan (1791–1868), was elected fifteenth president. The slavery crisis was worse than ever. Kansas was getting ready to enter the Union, and by the Missouri Compromise, it should have been free, but the Kansas-Nebraska Bill made it either slave or free by popular vote. Both the free states and the slave states started sending in settlers, and the two groups fought each other in virtual civil war.

1857

SATURN'S RINGS

By this time it seemed very likely that Saturn's rings were composed of myriads of small particles. Since they were inside Roche's limit (see 1849), solid particles would break up under tidal influences and be unable to coalesce again. In 1857 Maxwell (see 1855) showed that this was indeed so, based on purely theoretical considerations, and no one has doubted it since.

IN ADDITION

The slave states won their greatest victory in 1857 with the Dred Scott decision, which held that slaves did not become free by entering a free state, nor could slaves bring lawsuits before the court, nor could Congress ban slavery in any territory. Naturally such a decision made many new Abolitionists and made the old more determined.

In India, the Sepoys (native soldiers who controlled India for their British masters) revolted on May 10, 1857. The *Sepoy Revolt* shook British rule, but Indian troops from the Punjab, who were willing to fight for the British, retook Delhi on September 20, 1857, and that was the turning point.

1858

EVOLUTION BY NATURAL SELECTION

The British biologist Charles Robert Darwin (1809–1882), like many biologists, believed that life forms had evolved; that some species, with time, changed into other related species, while some species became extinct. What puzzled Darwin was the mechanism that drove evolution onward.

In 1838 he read Malthus (see 1798) and realized that not only human beings multiplied past the food supply available, but that all living things did. Therefore, in each generation, there was a competition for survival among the too-many, and those tended to survive who could snatch enough food, or best evade a predator. In short, nature itself would select a few from the many for survival. The characteristics that made survival likely would be inherited by the offspring of the survivers and again there would be a natural selection. Darwin assumed that there were always variations among offspring, random and tiny variations perhaps, and that nature then used those variations as a means of selection, so to speak. The "better" among the variations would not *always* do better, since there was always the factor of chance, but on the whole, and in the long run, they would.

Darwin was theorizing, then, on *evolution by natural selection*. Knowing that this idea would create enormous controversy, and being a gentle person who could not bear to participate in such controversy, he spent some twenty years gathering evidence for his point of view, hoping that when he did publish, the evidence would be so overwhelming that it would preclude argument. (In this he was naive, for he did not count on the strength with which human beings can turn away from facts and cling to superstition.)

In 1858 another British biologist, Alfred Russel Wallace (1823–1913), was in the East Indies. He too had read Malthus and arrived at the notion of evolution by natural selection. Unhampered by Darwin's fear of controversy, Wallace wrote up his theories in three days. Then, in order that his views might be checked by an approriate expert, he sent his eleven-page paper to Darwin, who could scarcely believe his eyes when he saw himself so anticipated.

There was nothing for Darwin to do but to suggest a joint publication and this was carried through immediately. In the next year, 1859, Darwin reluctantly published a book, best known as *The Origin of Species,* in which he presented his theory in detail (but not in as much detail as he would have liked, for he had been planning a book five times as long at least).

Darwin's book was the most notable scientific work since Newton's great classic (see 1687). It was the foundation of a new biology, as Newton's had been the foundation of a new physics. It changed the very current of people's thinking, and the world has never been the same.

ORGANIC MOLECULAR STRUCTURE

Up to this point, organic molecules were still characterized by the number of atoms of each element that were present. Even when it was understood that in isomers the same number of the same sorts of atoms must be arranged differently, it was not clear what the difference in arrangement might be.

However, Frankland's notion of valence (see 1852) offered an important handle, which the German chemist Friedrich August Kekule von Stradonitz (1829–1896) made use of. He pointed out that since hydrogen had a valence of 1, oxygen of 2, nitrogen of 3, and carbon of 4, this meant that each hydrogen atom could be attached to other atoms by means of a single "hook," each oxygen by means of two "hooks," each nitrogen by means of three, and each carbon by means of four.

A British chemist, Archibald Scott Couper (1831–1892), working independently, suggested that each atom have its valence indicated by dashes. Thus, the hydrogen molecule, the oxygen molecule, and the nitrogen molecule would be written H—H, O═O, and N≡N. Water would be H—O—H, carbon dioxide would be O═C═O, and so on.

Kekule's second great point was that carbon atoms could hook onto each other freely, so that molecules might be made up of chains of carbon atoms, with other atoms latching onto the valence bonds not taken up by the formation of those chains.

This system finally made sense of many organic molecules. The different arrangements of isomers could often be worked out, and a great deal of the mystery of organic chemistry vanished.

CELLULAR PATHOLOGY

The cell theory (see 1838) continued to grow firmer. The German pathologist Rudolph Virchow (1821–1902) published the results of his studies of diseased tissues in a book called *Cellular Pathology* in 1858. There he showed that abnormal cells that marked one disease or another arose from normal cells. There was no sudden change, but a smooth development of abnormality. This served to found the science of *cellular pathology*.

Virchow went on to oppose the idea of spontaneous generation. He pointed out that "all cells come from cells," and since the cell is already a complex organization of simpler parts, it is unlikely that it can originate easily from dead matter.

REFRIGERATORS

Reducing food to a low temperature kept it from spoiling, and in early times, natural ice was used for the purpose. Even in summer, snow or ice could be brought from nearby mountainous areas. And ice gathered from ponds in the winter could be insulated with straw and kept in underground icehouses through the summer.

In the early nineteenth century, a number of people tried to design mechanical refrigerators. The efforts of scientists to liquefy gases had made it clear that, if a gas were liquefied and then allowed to evaporate, it would lower its own temperature and therefore its surroundings. If the vapor were then condensed by pressure and allowed to evaporate again, over and over, heat would, in effect, be pumped out of the refrigerator into the surrounding air.

The first such device that we would recognize as similar in essence to those we have today, and that was a reasonable commercial success, was designed by a French inventor, Ferdinand Carré (1824–1900). His first device, built in 1858, used water, but in 1859 he switched to ammonia, which was much more efficient.

Of course, the early refrigerators were bulky and inconvenient, and ammonia was a corrosive and poisonous substance. They were used only industrially, for ice-making or in meat-packing plants, for instance, to preserve the meat. It took some three-quarters of a century of improvement before refrigerators became convenient and almost universal household appliances.

ELECTRICITY IN A VACUUM

Occasionally, scientists would try to force an electric current through a vacuum, hoping, perhaps, that they would be able to study the electric fluid itself, which would be forced out into the open if there were no matter obscuring it. Faraday (see 1821) had tried it, for instance, and noticed that the glass vessel enclosing the vacuum showed a greenish glow. This glow was given the name of *fluorescence* in 1852 by the British physicist George Gabriel Stokes (1819–1903) (a term now applied to any visible light that arises from the energetic collision of radiation with matter).

However, the vessels were never sufficiently evacuated to make possible the study of the electric fluid itself, at least not until the Geissler tube (see 1855) was devised.

In 1858 the German physicist Julius Plücker (1801–1868) forced an electric current through a Geissler tube. He detected the fluorescence, described it, and noted that the position of the fluorescence shifted in response to an electromagnetic field. Whatever was happening in the tube, an electric charge was involved. This was the first tentative advance toward understanding that atoms were more than tiny little balls.

IN ADDITION

In 1858 the Sepoy Rebellion was finally quelled. On August 2, 1858, the East India Company was relieved of its rule over India, and India was made a direct possession of Great Britain. The last of the Mogul emperors, Bahādur Shāh II (1775–1862), was deposed and the Mogul dynasty came to an end after two and a quarter centuries. India was to be ruled instead by a British viceroy.

Austria was now at the height of its post-Napoléonic power, having apparently recovered entirely from the disorders of 1848–1849.

Frederick William IV of Prussia was declared insane in 1858, and his younger brother, William (1797–1888), was made regent.

Minnesota entered the Union in 1858 as the thirty-second state, and the count became seventeen free states to fifteen slave states. What's more, it looked as though the free settlers in Kansas were winning over the slave settlers. Increasingly, the slave states could see no peaceful resolution of the crisis that would allow them to keep their slaves.

1859

OIL WELLS

Petroleum (from Latin words meaning "rock oil"), sometimes called simply oil, is a complex mixture of hydrocarbons, formed, it is believed, from the fat content of myriads of microorganisms in the distant past.

In the Middle East, where petroleum is particularly common, some was found even on the surface. There the smaller molecules evaporated, leaving behind a tarry substance variously called *pitch, bitumen,* and *asphalt.* This was used in waterproofing. From pitch one could also sometimes extract an inflammable liquid called *naphtha* (from a Persian word meaning "liquid"), which could be used in lamps.

There was a limit to what could be obtained on the surface, however. It was also possible to dig for it. Two thousand years before, the Chinese had dug for brine and occasionally obtained oil.

An American railway conductor, Edwin Laurentine Drake (1819–1880), had invested in a firm that gathered oil for medicinal purposes, from seepages near Titusville, Pennsylvania. It occurred to Drake, who knew about the brine-drilling, that oil could be drilled for as brine was.

He studied drilling methods and, in 1859, set about using those methods at Titusville. He drilled 69 feet into the ground and, on August 28, struck oil. He had drilled the first oil well, which was soon producing 400 gallons per day.

Other drillings followed, and other oil wells. The first consequence of this was that kerosene could be obtained in great quantities from petroleum, and the kerosene lamp became almost universal in the United States and elsewhere. Kerosene served to replace whale oil and to cut down somewhat on the carnage that humanity was visiting on those inoffensive animals.

Much more than that, however, lay in the future.

STORAGE BATTERY

All the electric batteries used in the six decades since Volta had invented the first one (see 1800) were one-shot affairs. The chemical reaction that gave rise to the electric current eventually proceeded to a point where it could no longer support current. It then had to be discarded, for the chemical reaction could not be reversed.

Yet some chemical reactions are easily reversible. In 1859 a French physicist, Gaston Planté (1834–1889), took two sheets of lead, with an insulating sheet of rubber between them, rolled the lead sheets into a spiral, and upended it into dilute sulfuric acid. He found that a chemical reaction resulted that would produce an electric current. Furthermore, when the battery was discharged, an electric current could be forced through it in the opposite direction, and the chemical reaction would be reversed, so that the battery could eventually produce an electric current again.

Naturally, we are not getting something for nothing. The second law of thermodynamics (see 1850) would not allow that. The electrical energy required to charge a storage battery is always greater than the amount it can then deliver. It would be a losing proposition, therefore, to charge a discharged storage battery from one that was fully charged. To charge a storage battery, you must get your electricity from a generator that makes use of fuel energy or some other nonelectrical source of energy.

SPECTRAL LINES AND ELEMENTS

Since their discovery by Fraunhofer nearly half a century before (see 1814), spectral lines had not been connected with chemistry.

The German physicist Gustav Robert Kirchhoff (1824–1887), however, had carefully studied the spectra of light produced when various elements were heated to incandescence. He found that each element produced light of only certain wavelengths, so that the spectrum consisted sometimes of only a few lines of light, which were sometimes well separated.

In 1859 Kirchhoff announced that every element produced characteristic spectral lines, which it also absorbed when its vapors were cooler than the light source. The pattern of lines was different for each element; no two elements shared spectral lines in exactly the same positions. In essence, each element had a spectral "fingerprint."

This meant that if any sample of ore, on being heated to incandescence, produced even a single spectral line in a position not recorded for any known element, then some new element must be present in that ore.

Making use of such spectroscopic data, Kirchhoff discovered *cesium* in 1860. The name is derived from the Latin word for "sky-blue," since that was the color of the spectral line that had given away cesium's existence. The next year Kirchhoff discovered *rubidium,* a related element, its name coming from the Latin for "red."

Kirchhoff also pointed out that the dark lines in the solar spectrum were the result of light being absorbed by the gases in the relatively cool atmosphere of the Sun. From those lines, one could be sure that sodium existed in the Sun's atmosphere, and half a dozen other elements as well.

This was the first actual observation indicating that the same chemical elements that existed on Earth existed also on other astronomical bodies, and therefore presumably throughout the Universe.

SOLAR FLARES

The British astronomer Richard Christopher Carrington (1826–1875) measured the Sun's rotation by the motion of the sunspots. Galileo had done that two and a half centuries before, but Carrington could work with more delicate instruments. He found that the Sun did not rotate all in a piece, which meant that it was not a solid body, but that the outer regions, at least, might well be gaseous. (In view of the temperature of the Sun's surface, this was not surprising.) A point on the solar equator took 25 days to make one turn around the Sun, while at solar latitude 45 degrees, a point that had to travel a considerably smaller distance to make the turn, it took 27½ days.

On September 1, 1859, Carrington, watching the sunspots, observed a starlike point of light burst out of the Sun's surface. It lasted 5 minutes and subsided. Carrington speculated that a large meteor had fallen into the Sun, but it was eventually realized that he had been the first to see a *solar flare*, a brief but violent explosion that was usually associated with sunspots. This was the beginning of an understanding that when the number of sunspots was high, we had an "active Sun" and when it was low, a "quiet Sun."

KINETIC THEORY OF GASES

Maxwell, having worked on the small particles that made up Saturn's rings (see 1857), turned in 1859 to other, even smaller particles—the molecules that made up gases. Maxwell treated the situation statistically, assuming that the molecules moved in random directions at random speeds and bounced off the walls of a container, and off each other, with perfect elasticity.

Eventually he worked out mathematical relationships that showed the distribution of velocities among the molecules of a gas at a particular temperature. A few molecules moved very slowly and a few very quickly, but larger percentages moved at intermediate velocities, with the most common velocity somewhere in the middle. A rise in temperature caused the average velocity to rise, while a drop in temperature caused it to fall. In fact temperature—and heat itself—could be pictured as molecular movement and nothing else. This is called the *kinetic theory of gases*, where *kinetic* is from the Greek word for "motion."

From the random motion of the molecules, the various laws involving gases could be deduced (Boyle's law—see 1662; Charles's law —see 1787), but they were obeyed statistically. That is, gases might act quite differently from the way they do by chance deviation from the normal. However, there are so many molecules that statistical deviations involving more than a vanishingly small percentage of them are enormously improbable.

In the same way, as Maxwell pointed out, gases can defy the second law of thermodynamics if all the molecules happen to move in the same direction, or all the faster molecules gain still more speed from the slower ones—but the chance of that happening is so small that in the entire lifetime of the Universe, it is not likely to have happened in any cubic centimeter of it.

IN ADDITION

Austria had no doubt that Sardinia planned the unification of Italy by force and demanded on April 23, 1859, that Sardinia demobilize its troops. Sardinia refused, and Austria invaded the little country on April 29. This was a mistake because it made Austria the aggressor and France could now join Sardinia on the pretext of defending a small neighbor. Thus on May 12, France declared war on Austria. However, Napoléon III, unlike Napoléon I, did not enjoy battles. He was depressed at the casualties and it suddenly occurred to him that a united Italy might be an uncomfortable neighbor. He therefore met with Francis Joseph of Austria on July 11 and came to a quick agreement. Sardinia would get Lombardy, but Austria could keep Venetia. Naturally, Sardinia denounced it as a betrayal. What Napoléon had done was make both Austria and Italy his enemies. But as for Austria, this was the beginning of a fifty-year decline.

Oregon entered the United States as the thirty-third state. Now there were eighteen free states and fifteen slave states.

1860

SPONTANEOUS GENERATION

Pasteur (see 1846) now took up the matter of spontaneous generation. It was already clear that nutrient broth that had been strongly heated till all microorganisms within it were dead would not develop microorganisms if it were sealed away from air. Vitalists, however, maintained that the heating destroyed some *vital principle* in the air.

Pasteur had shown that dust in the air carried microorganisms. In 1860, he therefore boiled meat extract and then left it in a flask exposed to air, but only by way of a long narrow neck that was bent down, then up. Although unheated air could thus freely penetrate into the flask, dust particles settled to the bottom of the curved neck and did not enter the flask. The meat extract did not spoil. No decay took place. No organisms developed. There was no question now of a destroyed vital principle. When Pasteur knocked the neck off the flask, microorganisms developed in the meat extract at once.

That was the final destruction of the notion of spontaneous generation—at least under present conditions.

Darwin's theory of evolution, however, had opened the question as regards the past. If species had evolved, might there not have been a time when life itself evolved from nonliving precursors? This was a most uneasy thought to those who had always assumed life to be a creation of God, but Darwin himself had raised the possibility—and that possibility has concerned science ever since.

ORGANIC SYNTHESIS

When Wöhler had synthesized urea (see 1828), it might very well have been considered an anomaly, but the techniques of organic chemistry were advancing, and other organic molecules were being synthesized from the elements.

In particular, Pierre-Eugène-Marcelin Berthelot (1827–1907), a French chemist, went about such syntheses and, by 1860, had synthesized such well-known and important organic molecules as methyl alcohol, ethyl alcohol, methane, benzene, and acetylene. He even synthesized molecules that fulfilled all the requirements of organic molecules in terms of structure and properties but that did not occur in living tissue, which put an end once and for all to the idea that only living tissue could manufacture organic molecules.

Indeed, in 1861, when Kekule (see 1858) published a textbook of organic chemistry, he defined it as the chemistry of carbon compounds, with no mention of life. After that, when a term was needed for the chemistry of compounds that occur in living tissue specifically, the term chosen was *biochemistry* (with *bio-* from the Greek word for "life").

INTERNAL-COMBUSTION ENGINE

For a century and a half, steam engines had produced heat outside the system they were powering. Steam formed by the heat entered a cylinder and moved a piston.

What if a mixture of inflammable vapor with air could be exploded within a cylinder?

The force of this internal explosion would then move the piston directly. (The fuel would have to be a gas to begin with, or an easily evaporated liquid.)

If such an *internal-combustion engine* were developed, it would be much smaller than a steam engine and so much more readily set into motion. After all, a vapor-air mixture will explode at the touch of a spark, while the initial boiling of water over a fire, in the case of the external-combustion steam engine, is a slow process.

A working internal-combustion engine was first constructed by a Belgian-born, French inventor, Jean-Joseph-Étienne Lenoir (1822–1900). In 1860 he hitched such an engine to a small conveyance and had a "horseless carriage." Horseless carriages run by steam had existed previously, but Lenoir's was more compact and more responsive to controls.

Nevertheless, Lenoir's internal-combustion engine was very inefficient. It was not till the next decade that an engine of this sort was built that was efficient enough for widespread use.

SOLAR PROMINENCES

Photography of astronomical phenomena was becoming more common. In 1858 the English astronomer Warren De la Rue (1815–1889) had devised a telescope that was particularly adapted to taking photographs of the Sun. After that, solar photography became an almost daily routine for him.

In 1860 he traveled to Spain to observe a total eclipse of the Sun. The photograph he took of the Sun during the eclipse showed gouts of flame about its edge. These were called *solar prominences* and, like solar flares (see 1859), were evidence of violent activity on the Sun. This was the first astronomical discovery to be made as the result of photography.

AVOGADRO'S HYPOTHESIS

Despite everything, despite even Kekule's system for presenting organic structures (see 1858), there was still ample confusion over the structure of many organic molecules.

For that reason, Kekule called for an international conference of chemists to consider the matter. In 1860 the conference took place in Karlsruhe, just across the Rhine from France. (It was the first international scientific conference.)

Attending was an Italian chemist, Stanislao Cannizzaro (1826–1910), who, two years before, had come across Avogadro's hypothesis (see 1811), now half a century old. Cannizzaro saw that from Avogadro's hypothesis the molecular weight of various gases could be determined with certainty and that this would reduce much of the confusion that attended molecular structure.

He made a strong speech, introducing the hypothesis and explaining how to use it. He carried conviction to some immediately, and to others afterward. Avogadro's hypothesis was accepted at last, and general agreement over molecular structure followed.

BLACK BODIES

Kirchhoff, having pointed out that materials emitting certain wavelengths when hot will absorb the same wavelengths when cool (see 1859), drew a natural conclusion.

He suggested in 1860 that a body that absorbed all light and reflected none (and was a *black body* as a result) would, when heated, emit all wavelengths of light.

This may not have seemed very important at the time, but the question of how the wavelengths were given off, how they were distributed across the total spectrum, and how this changed with temperature turned out to be a major puzzle. Within four decades, it had led to a revolution in physics.

IN ADDITION

The Republican antislavery nominee, Abraham Lincoln, was elected sixteenth president of the United States. South Carolina, always the most extreme of the slave states, feared an antislavery backlash and on December 20, 1860, seceded from the Union.

As the United States seemed to be falling into disunion, Italy was becoming unified. Small states in northern Italy asked to join Sardinia-Lombardy and Naples joined the northern states.

On October 12, 1860, British and French forces occupied Peking in response to the imprisonment of a British diplomat. This was a pattern. Any attempt by China to assert its rights within its own borders was met by an overwhelming response from the Western military. China was then forced to pay indemnities and make further economic concessions. This was to continue for decades.

The population of the United States was now over 31 million, well beyond that of Great Britain and approaching the population of France. New York City had 800,000 people, but it then consisted only of the island of Manhattan. With Brooklyn (which is now part of the city), the population was 1.25 million.

1861

ARCHEOPTERYX

Although the giant dinosaurs, whose fossil relics had been dug up over the past forty years, were the most dramatic remnants of ancient life, the most important single fossil was not a giant, but a relatively small lizardlike animal that was discovered in 1861. It is estimated now to be about 140 million years old.

It left a clear impression in a rock, which showed that it had a head possessing teeth and no beak, a long neck, a long tail, and a flat breastbone—all very lizardlike. There was, however, one all-important added feature. The "lizard" had *feathers*. The imprints of those feathers are unmistakable. They are in a double row down the length of the tail and are also present along the forelimbs.

In the world today, every known bird has feathers, and no living thing that is not a bird has them. Therefore, it is necessary to think of this fossil as representing a very primitive bird. It is called *archeopteryx* (from Greek words meaning "ancient wing").

The archeopteryx is the best-known example of an ancient life form that seems to lie exactly between two major groups of today's life forms. It is half reptile and half bird and is therefore a perfect example of a reptile in the process of becoming a bird. No other fossil discovered before or since is such a clear example of evolution at work.

BROCA'S CONVOLUTION

Gall had been of the opinion that different parts of the brain controlled different parts and functions of the body (see 1810). Gall's conclusions were strictly speculative, however. The first person to present conclusive evidence of a specific region of the brain in charge of a specific function was Broca (see 1856).

Broca had a fifty-one-year-old patient who had lost the ability to speak. When the patient died in 1861, a postmortem revealed damage to the third convolution on the left frontal lobe of the cerebrum (now known as *Broca's convolution*). Gall's insight, which had been misdirected into phrenology, was thus put right.

THALLIUM

Kirchhoff (see 1859) was not the only one to discover new elements by means of spectroscopy. The British physicist William Crookes (1832–1919) was working with selenium ores, and in 1861 he came upon a sample of such ore that when heated showed a bright green line characteristic of no known element. He analyzed the ore in search of the new element, found it, and named it *thallium,* from the Greek word for "green twig" after the color of its line.

IN ADDITION

Although elected president in November 1860, Lincoln was not inaugurated until March 4, 1861. In that interval, ten more slave states seceded. On February 4, 1861, the states that had already seceded met in Montgomery, Alabama, and set up what they called the Confederate States of America. Jefferson Davis of Mississippi (1808–1889) was chosen president.

Four slave states (the Border States) did not secede. They were Delaware, Maryland, Kentucky, and Missouri. What's more, Kansas became the thirty-fourth state on January 29, 1861, as the ninteenth free state.

The Confederate States took over American military bases in their territory. Fort Sumter, in the harbor of Charleston, South Carolina, refused to surrender, however. On April 12, 1861, the Confederates bombarded the fort and the small garrison was forced to surrender. That bombardment began the American Civil War.

On March 17, 1861, the Kingdom of Italy was proclaimed. It included all of Italy except Venetia, which was still Austrian, and the westernmost portion of the old Papal States, which remained independent under the protection of French forces.

Frederick William IV of Prussia died on January 2, 1861, and William I became king.

In Russia, the liberation of the serfs by order of Alexander II was completed.

1862

GERM THEORY OF DISEASE

More and more biologists were coming to suspect that contagious diseases were caused by microorganisms.

In 1862 Pasteur (see 1846) took up the matter and issued a publication that gathered all the evidence. His prestige was such that the *germ theory of disease* then had to be taken seriously. There is no question but that this was the most important single advance in the history of medicine.

Using the theory, Pasteur, and others as well, located the specific microorganisms that caused certain diseases and could then work out logical ways of preventing the disease in the first place or of curing it once it struck.

This was the beginning of modern medicine. It led to a fall in the death rate, a doubling in life expectancy, and an accelerated population explosion that has more than tripled world population since Pasteur's time and presented humanity with enormous problems in consequence.

DIM COMPANION OF SIRIUS

Bessel had suggested the existence of a dark companion of Sirius, one that could not be seen but that betrayed its existence by its gravitational influence (see 1844).

On January 31, 1862, an American astronomer, Alvan Graham Clark (1832–1897), had ground a new 18-inch lens and was testing it in a telescope by examining Sirius. He saw a tiny spot of light near Sirius and thought at first that it might represent an imperfection in the lens. Since other bright stars did not show such an effect, however, he concluded that he was actually seeing the companion of Sirius. It was not truly dark, merely dim.

Clark did not know that he was looking at anything more than a dim star. It was to be six decades before the true and most unusual nature of the companion of Sirius was to be demonstrated.

HYDROGEN IN THE SUN

Once Kirchhoff had shown the use of spectral lines in determining the constitution of the solar atmosphere (see 1859), astronomers began to compare the position of the dark lines in the solar spectrum with the lines produced by the elements.

In 1862 the Swedish physicist Anders Jonas Ångström (1814–1874) announced his discovery of hydrogen in the Sun. He went on, in later years, to publish a map of the spectrum in which he located about a thousand lines, measuring the wavelengths represented by each in units equal to a ten-billionth of a meter. That unit is still referred to as the *angstrom*.

CHLOROPLASTS

Plants seem to be uniformly green, and it is easy to suppose that chlorophyll is spread uniformly throughout the cells. That this is not so was discovered in 1862 by the German botanist Julius von Sachs (1832–1897). He noted that plant cells had discrete bodies in which the chlorophyll was concentrated. These bodies were eventually named *chloroplasts*. Furthermore, starch grains appeared within the chloroplasts, and Sachs was thus able to show that starch was the product of photosynthesis.

SOURCE OF THE WHITE NILE

Bruce had discovered the source of the Blue Nile in northwestern Ethiopia (see 1770), but the White Nile was the main stream, and its source was still unknown.

In 1857 two British explorers, Richard Francis Burton (1821–1890) and John Hanning Speke (1827–1864), started from Zanzibar in east central Africa and probed westward in search of the great lakes that Arab traders had reported. In February 1858 they reached Lake Tanganyika, a long, narrow body of water 620 miles from the African coast.

There, Burton had had enough and left. Speke, however, moved northward on his own and, on July 30, reached Lake Victoria, in area the second-largest body of fresh water in the world. (Only Lake Superior is larger.)

In 1862 Speke confirmed that a stream issuing from the northern rim of Lake Victoria was the Nile River. It was necessary, then, to trace back the longest river that flowed *into* Lake Victoria. That turned out to be the Luvironza, which is 715 miles long and flows into the lake from the west. The source of the Luvironza is therefore the true source of the Nile, and it is located in the modern nation of Burundi about 35 miles east of Lake Tanganyika.

IRONCLAD WARSHIPS

In 1862 the American Confederacy raised a scuttled Union ship, the *Merrimack,* from its resting place near Norfolk, Virginia, renamed the ship *Virginia,* covered her with iron plates, placed a cast-iron ram under the waterline, and outfitted her with ten guns.

On March 8, the ironclad *Merrimack* attacked the wooden Union ships that were blockading the port and sank them. She was absolutely untouchable by the Union guns. For a while it seemed as though the *Merrimack* could single-handedly take on and destroy the entire Union navy and break the blockade. The Confederacy could then, with British help, win its independence.

However, an American engineer of Swedish birth, John Ericsson (1803–1889), had already built an ironclad vessel, the *Monitor,* for the Union. It was small, floated very low in the water, and had two guns. It sailed south just as the *Merrimack* was launching its attack. On March 9, only one day after the *Merrimack*'s, triumph, the *Monitor* arrived, and there was a 5-hour fight in which neither ship could damage the other. However, the *Merrimack* sprang a leak, moved into dry dock, and never left again, so the Union was saved.

It was clear to the whole world, however, that wooden warships were now obsolete, and every navy (including especially the British) began the process of converting itself into an ironclad force.

MACHINE GUNS

The Colt revolver, which had come into use two decades before, was not the last word, of course. Repeating rifles, called *carbines,* had come into use in 1860. More destructiveness was wanted, though—an indefinite stream of bullets.

The American inventor Richard Jordan Gatling (1818–1903) sought to devise a gun that could fire bullets out of a chain of cartridges and that wouldn't stop until the chain ran out. By November 1862 he had developed a rapid-fire gun that could shoot nearly six bullets per second, though it had to be hand-cranked.

This *Gatling gun* was the first *machine gun* and it was used by the Union forces toward the end of the Civil War. Gatling's name lives today in the slang term *gat,* used for any handgun.

HEMOGLOBIN

One of the most familiar proteins is the one in red blood corpuscles that combines with oxygen in the lungs and carries it to the cells of the tissues, where it gives it up again. One of those who analyzed the protein carefully was the German biochemist Felix Hoppe-Seyler (1825–1895). He crystallized it in 1862 and gave it its name of *hemoglobin* (*hemo-* from the Greek word for "blood" and *globin* as a brief version of *globulin,* which is the class of proteins to which it belongs). When it combines with oxygen, it becomes *oxyhemoglobin.*

IN ADDITION

The Union navy nibbled away at the Confederate coastline in the course of 1862. On March 4, 1862, Confederate forces took Santa Fe, but Union forces drove them back and the West remained solidly Union thereafter. In the East, however, the Union forces met only disaster. The Confederates had three great generals, Joseph Eggleston Johnston (1807–1891), Robert Edward Lee (1807–1870), and Thomas Jonathan (Stonewall) Jackson (1824–1863). The Union had only George Brinton McClellan (1826–1885). Lincoln issued the *Emancipation Proclamation* on September 22, 1862, declaring all slaves in rebellious states to be free as of January 1, 1863, and making it clear that total emancipation was a war aim.

Mexico was in trouble with Great Britain, France, and Spain, because it had defaulted on its debts. All three had sent troops in order to collect, but Great Britain and Spain thought better of it. France, however, remained, planning to set up a French-dominated empire, since the United States, wrapped up in its Civil War, had no way of implementing the Monroe Doctrine except by verbal protests.

In Prussia, Otto Eduard Leopold von Bismarck (1815–1898) became prime minister, and for the first time since Frederick II Prussia had a strong leader.

1863

GREENHOUSE EFFECT

In 1863 the Irish physicist John Tyndall (1820–1893) pointed out that such gases as carbon dioxide and water vapor are transparent to the visible light that reaches Earth's surface from the Sun, but rather opaque to the infrared radiations that the Earth gives off to space when it cools down at night.

This means that the presence of even a small amount of carbon dioxide and water vapor in the air will keep the Earth's surface temperature higher than it would otherwise be. This resembles the situation in a greenhouse, where light enters through the glass and warms the atmosphere within, but heat escapes only with difficulty, so that the net effect is to warm the greenhouse. For this reason, the action of carbon dioxide and water vapor is referred to as the *greenhouse effect*.

Since human activity has somewhat increased the carbon dioxide of the atmosphere and is continuing to do so, the greenhouse effect has become a serious threat.

LAW OF OCTAVES

By this time, over sixty elements were known, and they seemed to represent an unruly jungle of characteristics. Chemists were uneasy with that, so they tried to group elements into families. Several attempts had already been made, but an English chemist, John Alexander Reina Newlands (1837–1898), tried something a bit different.

He listed the elements in order of atomic weight and found that the second seven repeated the properties of the first seven quite closely. He tried to extend this further, referring to it as the *law of octaves* (as in music, where the same seven notes are repeated over and over, with every eighth note resembling the first octave higher, *octave* coming from the Latin word for "eight").

Newlands's work was not taken seriously, because the list of elements fit the law of octaves very imperfectly. Nevertheless, it helped put the listing of elements in vogue, and in the course of the following decade a much superior table was established.

CONSTITUTION OF THE STARS

Granted that the elements in the Sun seemed to be the same as those on Earth (see 1859), might it be argued that the Solar System would naturally be made of one set of materials but that other stars might be made of other elements?

The English astronomer William Huggins (1824–1910) studied the spectra of some of the brighter stars and announced in 1863 that their spectral lines were those of the old familiar elements. Presumably, then, the entire Universe was made up of the same elements.

BARBITURATES

In 1863 the German chemist Adolf von Baeyer (1835–1917) discovered *barbituric acid.* (There is a story that he named it for a girlfriend of the moment whose name was Barbara.) Barbituric acid is the parent substance of a family of compounds known as *barbiturates,* which are well known today for their use in *sleeping pills.*

INDIUM

A German mineralogist, Ferdinand Reich (1799–1882), suspected that a yellow precipitate he had obtained from a zinc ore might contain a new metal. Because he himself was color-blind, Reich had his assistant, Theodor Richter (1824–1898), examine it spectroscopically. Richter did so and spotted an indigo-colored line that was not characteristic of any known element. The new element was therefore named *indium.*

IN ADDITION

In the United States Lee took the offensive and marched his army north. In Gettysburg, Pennsylvania, he collided with the Union army. The Battle of Gettysburg raged for three days, July 1–3, 1863, and now at last the Union forces stood firm and their superior artillery inflicted heavy damage on the Confederates. Lee was forced to retreat, and that was the turning point of the war.

In the west, the Confederacy held on to Vicksburg, Mississippi, blocking the passage of Union troops along the Mississippi River. The Union general Ulysses Simpson Grant (1822–1885) kept attacking, however, until on July 4, 1863, just as news of the victory at Gettysburg was reaching Washington, he took Vicksburg and all of the Mississippi was in Union hands.

The western section of Virginia, sympathetic to the Union, had declared itself a separate state and, as West Virginia, entered the Union on June 20, 1863, as the thirty-fifth state.

In Mexico, French troops took Mexico City on June 7, 1863, and Napoléon III shopped around for someone who would be willing to be a puppet emperor on his behalf.

On January 10, 1863, the first subway system in the world opened for use of the public, in London.

1864

NATURE OF ORION NEBULA

Some patches of nebulosity, including the Milky Way itself (see 1609), had turned out to be clusters of faint stars. A number of Messier objects (see 1771) had turned out to be globular clusters (see 1785). Was any nebulosity what it appeared to be—simply a cloud of gas?

In 1864 Huggins (see 1863), studying the Orion nebula, found that its spectrum was typical of what would be expected of a luminous gas. It is, indeed, a large cloud of gas (though we know today that stars are embedded in it and that they are what heat the gas to luminosity.)

IN ADDITION

Grant was given command of the Union armies on March 10, 1864. On May 3, Grant's army marched into Virginia and engaged Lee in a bloody series of battles. Grant lost more than he won, but he could replace his losses and Lee could not. Unlike his Union predecessors in the east, Grant did not retreat when defeated, but continued to advance.

Further south, the Union general William Tecumsah Sherman (1820–1891) invaded Georgia. He entered Atlanta on September 2, then marched through Georgia to the sea, deliberately wrecking whatever he encountered. He reached Savannah on December 20.

The Confederacy's only hope was the election. There were many in the Union who were tired of the war. Lincoln was running for reelection against McClellan. Fortunately, the taking of Atlanta came at the right time, and Lincoln was reelected, the first president to be reelected since Andrew Jackson in 1832.

Nevada entered the Union as the thirty-sixth state in 1864.

It was a bad year for rebellions generally. The Russians crushed a Polish rebellion, and the T'ai Ping Rebellion came to an end as the Manchu forces took Nanking with the help of the British.

In Mexico, Ferdinand Maximilian Joseph (1832–1867), the younger brother of Francis Joseph I of Austria, was persuaded by Napoléon III, on April 10, 1864, to rule over Mexico as the Emperor Maximilian.

1865

GENETICS

There was a flaw in Darwin's theory of evolution by natural selection (see 1858). Granted that in every generation of a particular species there are random variations, there are also (to some extent, at least) random matings. Therefore, the variations should tend to vanish into an average, since extremes are not likely to mate with similar extremes. (It might even be argued that the second law of thermodynamics requires this tending toward an average.)

The flaw was corrected by an Austrian botanist and Augustinian monk, Gregor Johann Mendel (1822–1884), who experimented with peas that he grew in the monastery garden.

Carefully, Mendel arranged for various plants to self-pollinate, wrapping them to guard against accidental pollination by insects. Carefully, he saved the seeds produced by each self-pollinated plant, planted them separately, and studied the new generation.

Mendel found that if he planted seeds from dwarf pea plants, only dwarf pea plants sprouted. The seeds produced by this second generation also produced only dwarf pea plants. The dwarf pea plants "bred true."

Seeds from tall pea plants behaved in a more complicated fashion. Some bred true, but some did not. Those that did not breed true gave rise to tall plants three-quarters of the time and to dwarf plants one-quarter of the time.

Mendel then crossed dwarf pea plants with the tall pea plants that bred true. All the peas produced grew into tall pea plants. The characteristics of dwarfness seemed to have disappeared.

Next Mendel had each of this new generation of tall plants self-pollinate and found that they produced peas that grew into tall plants and dwarf plants in a 3-to-1 ratio.

Dwarfness had been submerged in one generation but had then shown up in the next.

In other words, tallness was *dominant* and dwarfness *recessive,* so that tallness overwhelmed dwarfness—but only temporarily.

Mendel found that other types of traits in pea plants worked the same way. There was *no* mixing of extremes.

Mendel also found that male and female contributed equally. It was as though each organism had two factors for a particular trait and each contributed one of them to the offspring. If the factors were different in the offspring, and one was dominant, the recessive characteristic didn't show but was still there. It might be the one handed on to the next generation, and if the other parent also handed on a recessive, so that the offspring had two recessives and no dominant, the recessive trait would appear again.

Mendel thus worked out the *Mendelian laws of inheritance,* as they were later to be called, and founded the science of *genetics* (from a Greek word meaning "to give birth to"). He published his first paper on this work in 1865 (a second followed in 1869), but it was totally ignored for thirty-three years.

Since the Mendelian laws showed that extremes did *not* mix into a blind average, but continued to show up, they provided the mechanism by which natural selection could bring about a gradual change in species. Thus, Mendel corrected the flaw in Darwin, but both Mendel and Darwin were dead before the world of science came to realize what had happened.

BENZENE RING

Kekule's way of writing chemical formulas (see 1858) did not solve all problems. There was one compound, benzene, which was extremely important, being involved in the molecular structure of the new artificial dyes, for instance (see 1856), that did not yield to his system.

The molecule of benzene is made up of six carbon atoms and six hydrogen atoms. Given a chain of six carbon atoms, there seemed to be no way of adding six hydrogen atoms without producing a very unstable compound—and benzene was quite stable.

It was Kekule himself who found the answer in 1865. According to his own story, he was in a semidoze in a horse-drawn bus and was visualizing carbon chains when suddenly the tail end of one chain attached itself to the head end and formed a spinning ring. At once he saw that if he imagined the six carbon atoms forming a hexagonal ring, with one hydrogen atom attached to each carbon, a stable atom might be the result. Once the notion of carbon rings was added to that of carbon chains, many structural problems were solved.

AVOGADRO'S NUMBER

The hydrogen molecule is composed of two hydrogen atoms. The hydrogen atom has an atomic weight of just about 1, so that a hydrogen molecule has a molecular weight of 2. Hydrogen gas is made up of hydrogen molecules, and at 0° C and normal atmospheric pressure at sea level, 22.4 liters (5.9 gallons) of hydrogen gas weigh 2 grams. This is the molecular weight in grams, or 1 *mole.*

Since equal volumes of gases are made up of equal numbers of molecules, according to Avogadro's hypothesis (see 1811), and since each oxygen molecule has a weight of 32, then 22.4 liters of oxygen gas would weigh 32 grams—or 1 mole.

In fact, 22.4 liters of any gas are likely to weigh 1 mole.

The question is, how many molecules are there in 22.4 liters of a gas? In 1865 the Austrian chemist Johann Joseph Loschmidt (1821–1895) used Maxwell's kinetic theory of gases (see 1859) to calculate what that number ought to be. It turned out to be about 600,000,000,000,000,000,000,000, or six hundred billion trillion. Since all this was based on Avogadro's hypothesis, the number was called *Avogadro's number.*

From Avogadro's number, you could calculate the actual mass of a hydrogen molecule, since it would be 2 grams divided by six hundred billion trillion. A hydrogen atom would have a mass half that, and the mass of other atoms and molecules could be calculated as well.

For the first time, scientists had a notion, and a rather good one, as it turned out, of the mass of the tiny atoms and molecules that made up matter.

ANTISEPTIC SURGERY

Anesthetics had come into use nearly twenty years before (see 1846), but if the process had become more nearly painless, it still remained deadly. The operation might be successful, yet the patient might develop inflammation and die.

In 1865 the British surgeon Joseph Lister (1827–1912) learned of Pasteur's germ theory of disease (see 1862) and it occurred to him that death after operations might result from a germ infection to which the traumatized tissues were particularly susceptible. The germs producing the infection might come from the doctors themselves, or from their instruments.

Lister therefore instituted the practice of using phenol solutions to clean hands and instruments, and the death rate after operations dropped at once. Semmelweiss had attempted the same thing seventeen years earlier (see 1847), but without the justification of Pasteur's theory, and physicians refused to listen. This shows the importance of theory in the most practical of affairs.

Eventually, chemicals less irritating and more effective than phenol were used, and *antiseptic* (from Greek words meaning "against putrefaction") *surgery* became the rule.

MAXWELL'S EQUATIONS

The culminating work of Maxwell (see 1855) came in 1865 when he began to devise a set of equations *(Maxwell's equations),* simple in form, that expressed all the varied phenomena of electricity and magnetism and bound them together indissolubly, as Newton had done for gravitation.

Maxwell showed that electricity and magnetism did not exist separately, but that each was an inevitable aspect of the other. There was a single electromagnetic force.

Maxwell showed that the oscillation of an electric charge produced an electromagnetic field that radiated outward from its source at a constant speed. This speed could be calculated from the equations and turned out to be just the speed of light. Maxwell therefore maintained that light was a form of electromagnetic radiation and that the wavelengths of such radiation depended on the oscillation rate of the charge and could be anything—far shorter than the ultraviolet and far longer than the infrared, for example. (Two decades later, this view was to be confirmed.)

In this way, Maxwell brought about the first *unification* in physics, or the first bringing under the umbrella of a single set of mathematical relationships such apparently disparate phenomena as electricity, magnetism, and light. Further unifications of this sort were to involve major efforts by later physicists.

MÖBIUS STRIP

In 1865 the German mathematician August Ferdinand Möbius (1790–1868) presented what came to be called a *Möbius strip.* This is a long, flat strip of paper (or other flexible material) that is given a half twist and the two ends then pasted together to give a circular figure. The resulting construction has but one edge and one side.

This made Möbius one of the founders of *topology,* the branch of mathematics that deals with those properties of figures that are not altered by deformations (as long as there is no tearing).

CYLINDER LOCKS

Human behavior being what it is, there has always been a demand for locks. A lock becomes more useful if it is small and not easily picked; if the key is small and not easily duplicated. No lock can be perfect, but some are obviously better than others.

In 1865 an American locksmith, Linus Yale (1821–1868), patented a cylinder lock with tumblers that had to be brought into a certain alignment if the lock was to open. The key had a serrated edge that sufficed to bring the tumblers into line. A vast number of combinations were possible, so that every key could be unique, even if there were millions of locks. The familiar locks and keys that have controlled house and apartment doors ever since are based on this principle.

IN ADDITION

Between Grant in Virginia and Sherman in Georgia, what was left of the Confederacy crumbled. Lee surrendered to Grant at Appomatox Court House on April 9, 1865, four years almost to the day after the bombardment of Fort Sumter. The last Confederate army surrendered at Shreveport, Louisiana, on May 26, 1865. The war had cost the embattled states over a million casualties.

Meanwhile, though, a more personal tragedy struck. On the night of April 14, 1865, President Lincoln was shot in a theater box by an actor, John Wilkes Booth (1838–1865), and died the next day. He was the first American president to be assassinated.

The Thirteenth Amendment to the Constitution was ratified in 1865, outlawing slavery.

As the United States reunited, the Prussian chancellor, Bismarck, was carefully engineering the unification of Germany, using a system of fighting one enemy at a time while carefully refraining from frightening the next.

1866

DYNAMITE

During the twenty years since nitroglycerine had been discovered by Sobrero (see 1847), it had been used in blasting roads through mountains, digging canals, and laying foundations. It fulfilled such tasks admirably, but it was such a touchy explosive that it sometimes blew up at the wrong time, destroying factories and killing people as well.

The Swedish inventor Alfred Bernhard Nobel (1833–1896) was a member of a family that produced nitroglycerine and had lost a brother in an explosion. He therefore sought a way of taming nitroglycerine. In 1866 he came across a cask of nitroglycerine that had leaked. The liquid had been absorbed by the packing, however, which consisted of diatomaceous earth, or *kieselguhr* (made up of the siliceous skeletons of myriads of microscopic diatoms).

The soaked kieselguhr remained dry, and Nobel experimented with the mixture. He found that the nitroglycerine could not be set off without a detonating cap once it had been mixed with kieselguhr. Short of that, the mixture could be handled virtually with impunity. Once set off, though, the nitroglycerine in the kieselguhr retained its shattering quality.

Nobel called the combination *dynamite* (from the Greek word for "power"), and it replaced free nitroglycerine, initiating the era of safe use of explosives in construction. (At his death, Nobel left his estate of nearly ten million dollars for the establishment of the annual Nobel prizes.)

CLINICAL THERMOMETER

Doctors had long recognized the importance of knowing a patient's temperature, but the thermometers they were forced to use were long, clumsy, and required up to 20 minutes to register results. A British physician, Thomas Clifford Allbutt (1836–1925), devised a small thermometer, no more than 6 inches long, which took only 5 minutes to produce results. With this *clinical thermometer*, taking a patient's temperature became truly routine.

KIRKWOOD GAPS

By now, nearly ninety asteroids had been discovered and their orbits calculated. It was clear that they were not evenly distributed, and the American astronomer Daniel Kirkwood (1814–1895) showed in 1866 that there were definite gaps, since called "Kirkwood gaps."

He explained these gaps by postulating that any asteroids that would have been in those gaps would revolve around the Sun in a period that bore a simple ratio to the period of Jupiter. That meant that every two or three turns of the asteroid would bring it into the same position with respect to Jupiter. Jupiter's gravitational pull would then be cumulative, and the asteroid would be forced either farther from the Sun or nearer to it, leaving a gap.

He pointed out that the gaps in Saturn's rings existed for similar reasons. If there were ring particles in Cassini's division, for instance, those particles would revolve about Saturn in just half the period of Saturn's satellite, Mimas.

EARTH'S IRON CORE

Ever since Cavendish had determined the Earth's mass (see 1798), it had been known that the Earth's average density was about twice that of its rocky crust. The conclusion was that Earth's center must be made of material that was denser than rock. Metal seemed to be the only reasonable alternative.

The French geologist Gabriel-Auguste Daubrée (1814–1896), considering the fact that a number of meteorites were composed of a nickel-iron alloy (though most were rocky), assumed that if meteorites were remnants of an exploded planet, the rocky ones came from the outermost portions of the planet, and the nickel-iron ones came from its core. He reasoned that this would be true of Earth as well, so that Earth had a nickel-iron core. This has been accepted by geologists ever since.

COMPOSITION OF A NOVA

Since the exceedingly bright novas spied by Tycho Brahe and by Kepler nearly three centuries before, further novas of the sort had been seen. Nevertheless, dim novas had been sighted, and they still attracted notice because they had brightened suddenly and without warning.

In 1866 Huggins (see 1863) managed to study the spectrum of a nova and showed from its spectral lines that it was surrounded by a cloud of hydrogen. This was the first indication that hydrogen might be a dominating constituent of the Universe.

IN ADDITION

Bismarck's plan of German unity under Prussian domination was working out beautifully. He made an alliance with Italy, then picked a quarrel with Austria, and in June 1866, the *Seven Weeks War,* in which Prussia and Italy fought against Austria, began.

Italy was defeated almost at once, but the Prussians were another matter. They used the telegraph and the radio, having learned that lesson from the American Civil War, and they also had the needle gun (see 1841), which the Austrians did not. The result was that on July 3, 1866, the Austrians were crushed at the Battle of Sadowa and gave up. The Treaty of Prague was signed on August 23, 1866. By its terms, Austria gave up Venetia to Italy. Prussia annexed Schleswig-Holstein and other territories in northern Germany, including Hanover.

The big loser was France, which had expected a long war that would wear out both parties. Prussia was now the dominant force in Germany, and the next item on Bismarck's agenda was taking care of France. He therefore treated defeated Austria gently to prevent it from seeking revenge by helping France when that nation's turn came.

The Atlantic cable, connecting Europe and America, was finally laid. (An earlier cable laid in 1858 only worked for a brief period.) This new cable was laid by the *Great Eastern,* the first of the really large liners (692 feet long), which had been designed by Brunel (see 1843), and launched in 1858. Too large to be commercially successful, it was ideal for laying the cable. The financial end was taken care of by the American businessman Cyrus West Field (1819–1892).

1867

DRY CELL

Batteries in use up to this time, including storage batteries, were made up of liquids inside a container. They had to be, since the chemical reactions required to produce an electric current proceeded in solution. Batteries containing fluid had to be handled with care. They could be easily tipped and spilled, and the fluid was usually corrosive.

In 1867, however, a French engineer, Georges Leclanché (1839–1882), devised a cell and, experimenting over the course of twenty years, gradually converted its fluid into a stiff paste by adding flour and plaster of Paris.

In the end, while such a cell was not really dry, it was not so wet that it would spill. It could be thrown about, placed on its side, and turned upside down without being adversely affected. It came to be called a *dry cell.* It was these dry cells that came to be used in myriads of small devices from flashlights to children's toys.

TYPEWRITERS

Although printing had come into use four centuries before, letters or stories still had to be written by hand. (The text of a piece of writing before it is published is still called a *manuscript,* from Latin words meaning "handwritten.") Attempts had been made to devise machines that could print letters when certain levers were struck, but they usually proved extremely cumbersome and worked much more slowly than a person could write.

The first writing machine that was reasonably compact and, with practice, could be made to print at least as quickly as people could write was built by an American inventor, Christopher Latham Sholes (1819–1890). He constructed his first machine in 1867 and patented it the next year. Within a few years it had gained the name of *typewriter*.

LAW OF MASS ACTION

Chemists found that chemical reactions could often proceed in either of two directions. By increasing the mass of one set of reagents, the reaction could be forced to move in the direction that consumed that set.

The Norwegian chemists Cato Maximilian Guldberg (1836–1902) and his brother-in-law, Peter Waage (1833–1900), showed that it was not mass alone that did this but concentration; that is, the quantity of mass in a given volume. The rule, known as the *law of mass action*, was promulgated in 1867 and helped lead the way toward the application of thermodynamics to chemical reactions.

IN ADDITION

After the end of the Civil War, the American demands that Napoléon III pull his troops out of Mexico became more strident and insistent, and on March 12, 1867, Napoléon III withdrew. Maximilian remained behind, but Mexican troops captured him on May 14, 1867, and he was executed on June 19.

Alaska was proving a drain on the Russian Empire. It was too far away to be administered successfully, and Alexander II was glad to sell it to the United States on March 30, 1867, for 7.2 million dollars.

On August 28, 1867, the United States acquired its first overseas territory when it annexed the Midway Islands west of Hawaii.

Nebraska entered the Union as the thirty-seventh state. The southern states were undergoing a harsh *Reconstruction* period at the hands of vengeful Republicans who controlled the Congress. Upon Lincoln's assassination, his vice president, Andrew Johnson (1808–1875), had become the nation's seventeenth president. He tried to soften Reconstruction, but he was not a skilled politician and the congressional leaders easily thwarted him.

The Seven Weeks War had weakened Austria to the point where it could no longer dominate Hungary by force. A compromise was worked out in October 1867, by which Hungary and Austria would have separate governments but common ministers for war, foreign affairs, and finance, while Francis Joseph I would be both emperor of Austria and king of Hungary. The nation would henceforth be called *Austria-Hungary*.

Mutsuhito (1852–1912) became emperor of Japan in February 1867. Under him, the shogunate came to an end and the modernization of Japan was accelerated.

1868

AIR BRAKE

Advances in technology invariably pose problems in safety. For instance, a locomotive with a number of carriages following it has a huge mass and consequently a huge inertia. Traveling, as it might, at a high speed, it would take an enormous force to decelerate and bring it to a halt. Failure to do so could well make collisions inevitable and extremely deadly.

In 1868 the American engineer George Westinghouse (1846–1914) invented the air brake, which used compressed air rather than human muscle. Such an air brake could be applied simultaneously to every wheel on the train. Cornelius Vanderbilt (1794–1877), the American train magnate, dismissed the idea as a "fool notion" because he did not believe that one could stop a train by the use of air. However, the air brake, after some improvements, caught on quickly and greatly increased rail safety.

HELIUM

Improvements in transportation had made it possible for astronomers to travel all over the world if they had to. The most spectacular astronomical phenomenon that might be confined to a small area almost anywhere in the world was a total eclipse of the Sun. When an eclipse was due to be visible at certain areas in India in 1868, European astronomers didn't hesitate to make the journey.

A French astronomer at the site, Pierre-Jules-César Janssen (1824–1907), studied the spectra of the solar prominences (see 1859). Meanwhile, a British astronomer, Joseph Norman Lockyer (1836–1920), showed that by allowing light from the edge of the Sun to pass through a prism, he could obtain spectra of prominences even without an eclipse.

Janssen observed a spectral line that did not match any of the known lines in position. He sent a report on this to Lockyer, who decided that the line represented a hitherto unknown element. The element was named *helium,* from the Greek word for "sun." (Since then, a number of strange lines have been found in astronomical light sources, and other new elements have been postulated, but helium—the first of these—remains the only one to have proven truly a new element, though this was not proven until a quarter-century later.)

CRO-MAGNON MAN

In 1868 a French paleontologist, Édouard-Armand-Isidore-Hippolyte Lartet (1801–1871), discovered four skeletons of human beings in a cave called Cro-Magnon. These came to be viewed as relics of *Cro-Magnon man.* They were some thirty-five thousand years old, but human in every respect. It was clear then that the human race was far older than Adam and Eve were thought to be by the Biblical literalists.

DEEP-SEA LIFE

It was easy to assume that life in the sea was confined to the uppermost layers. After all, light could only penetrate about 250 feet below the surface of the ocean, and since plants depend on light, they couldn't exist lower than that. Since animals can't live without plants to feed on, it seemed they couldn't exist any lower than that either.

When cables were being laid under the Mediterranean Sea and under the Atlantic Ocean, however, life forms were occasionally brought up from great depths. Scientists were reluctant to believe this. Then in 1868 a British zoologist, Charles Wyville Thomson (1830–1882), began a series of deep-sea dredging operations that, after eight years, had led him some 70,000 zigzag miles over the oceans. He made 372 deep-sea soundings and showed once and for all that life inhabits the ocean from top to bottom.

What happens is that when plant life near the surface dies, it drizzles downward, and some escapes being eaten all the way to the bottom. The animals that live on this drizzle also die and themselves add to the drizzle. Deep-sea life does not live in the profusion that surface life does, but it exists.

IN ADDITION

President Johnson's continuing quarrel with Congress resulted in his impeachment on February 21, 1868, by the House of Representatives. This remains the only occasion on which a president has actually been impeached. Johnson was then tried before the Senate, beginning on March 30, but the vote on May 16 was one short of the two-thirds majority needed for conviction. Johnson continued as president, therefore, but a new election followed and there was no chance that he could run for reelection. Grant, the victorious general, was elected the eighteenth president of the United States.

Russian forces occupied Samarkand in central Asia. It had been the capital of Tamerlane nearly five centuries before.

The capital of Japan was transferred from Kyoto to Tokyo, and Mutsohito was now in full charge.

1869

PERIODIC TABLE OF THE ELEMENTS

Newlands had tried to set up a table of elements that would allow them to be grouped into natural families (see 1863). A Russian chemist, Dmitry Ivanovich Mendeleyev (1834–1907), now tried his hand at the task.

Like Newlands, Mendeleyev arranged the elements in order of increasing atomic weight. However, he did not try to be too simple and arrange them in a fixed gridwork of seven elements per row. Instead, letting himself be guided by the valence of each element, he allowed the length of the period (row) to increase. He had hydrogen all by itself, then two periods of seven elements each, then two periods of seventeen elements each. He had prepared what is now called the *periodic table of the elements*.

He published his table on March 6, 1869, beating out others who were attempting the same task, notably the German chemist Julius Lothar Meyer (1830–1895).

In 1871 Mendeleyev went a step further, one which put him in a class by himself. In order to keep the valences and other properties of elements similar as one looked down the columns of his table, he had to leave empty spaces. He did not view this as an imperfection in his table. He merely announced that the empty spaces represented elements that had not yet been discovered.

He picked out three gaps in particular, one

under boron, one under aluminum, and one under silicon, and called them *eka-boron, eka-aluminum,* and *eka-silicon.* (*Eka* is "one" in Sanskrit, so that each element is one below the indicated element in the periodic table.)

Mendeleyev went on to predict the properties of the missing elements according to their places in the table. Naturally, no one paid this serious attention at the time, but Mendeleyev turned out to be right.

NUCLEIC ACID

The classification of foodstuffs into carbohydrates, fats, and proteins (see 1827) still seemed quite satisfactory.

In 1869, however, a Swiss biochemist, Johann Friedrich Miescher (1844–1895), isolated a substance from the remnants of cells in pus that did not belong to any of these three classes. It contained both nitrogen and phosphorus.

Miescher took his new discovery to Hoppe-Seyler (see 1862), who investigated the matter himself. When he discovered a similar substance in yeast, the discovery was announced. Since the material seemed to originate in cell nuclei, it was called *nuclein* at first, and later, because it had acidic properties, *nucleic acid.*

It was not till three-quarters of a century later that the true and overwhelming importance of nucleic acid was appreciated.

CRITICAL TEMPERATURE

The Irish physical chemist Thomas Andrews (1813–1885) worked on the liquefaction of gases, particularly carbon dioxide, a gas that can be turned into a liquid at ordinary temperatures by pressure alone.

Andrews carefully raised the temperature of carbon dioxide and found that, as the temperature went up, more and more pressure was required to liquefy it. Finally, at 31° C, no pressure he could apply would turn the trick. The gas might become as dense as a liquid, but it would still retain gaseous properties.

Andrews therefore suggested, in 1869, that for every gas there was a *critical temperature* above which pressure alone could not liquefy it. This was useful for chemists to know, because in their efforts to liquefy the so-called permanent gases, they would clearly have to lower the temperature below the critical temperature first, and only then apply pressure.

BIOGEOGRAPHY

Wallace (see 1858) noticed that the animal species of Australia were quite different than those of Asia. In 1869 he drew a line separating the two types of animals (a line still called *Wallace's line*), which followed the deep-water channel between the large islands of Borneo and Bali to the west and Celebes and Lombok to the east.

Clearly, the animals in Australia and Asia had been separated for a long time and had gone their separate evolutionary ways. Out of this grew the notion of dividing animal species into large continental and supercontinental blocks. This founded the science of *biogeography.*

ISLETS OF LANGERHANS

The German physician Paul Langerhans (1847–1888), while working for his medical degree, studied the microscopic structure of the pancreas, a digestive gland second in size only to the liver. In 1869, in his dissertation, he noted that it contained numerous small groups of cells, different from the cells in the body of the gland. These groups came to be called the *islets of Langerhans.*

At the time, they may have seemed a detail of no great importance beyond making it possible to gain a degree, but the time was to come when they proved of enormous value.

CELLULOID

Something is *plastic* if it is capable of being molded or shaped with relative ease. In that sense, clay, plaster, wood, rubber, and glass are all plastic. However, the noun *plastic* these days has come to be applied to a group of synthetic organic substances that possess plastic properties. The first of these was pyroxylin, prepared by Parkes seventeen years earlier (see 1855).

Pyroxylin had no commercial use at the time. Then a prize of ten thousand dollars (a fortune in those days) was offered for a substance cheaper than ivory but just as good, for the manufacture of billiard balls.

An American inventor, John Wesley Hyatt (1837–1920), wanted the prize money and had heard of pyroxylin. Hyatt improved on Parkes's method of preparing it and in 1869 patented his method of manufacturing billiard balls out of this material, which he named *celluloid*.

He did not win the prize, but celluloid enjoyed a minor boom as a material for baby rattles, shirt collars, photographic film, and other products. It was light, flexible, waterproof, and easily cleaned. Its one great flaw was that it was flammable. Celluloid was the first commercially successful plastic.

STOCK TICKER

In 1869 an American telegrapher, Thomas Alva Edison (1847–1931), came to New York seeking employment. While he was in a broker's office waiting to be interviewed, a telegraph machine transmitting gold prices broke down. Edison was the only one present who saw how to fix it.

He went on to devise something better, and invented a stock ticker of a type that was used for many decades. This he proceeded to sell to a large Wall Street firm. He wanted to ask five thousand dollars but lost his courage and asked the president to make him an offer. The president offered forty thousand.

That launched Edison on his career as probably the greatest inventor the world has ever seen.

IN ADDITION

The Union Pacific Railroad and the Central Pacific Railroad were joined at Promontory Point near Ogden, Utah, with the symbolic hammering of a golden spike on May 10, 1869. The United States now had a transcontinental railroad that could carry a person from the Atlantic to the Pacific in eight days. Later in the year, a horde of Wall Street speculators brought on a stock market crash on September 24.

In Africa, the Suez Canal, connecting the Mediterranean and Red seas, was opened to shipping on November 17, 1869. The long trip around Africa to India that the Portuguese explorers had pioneered four centuries before was shortened by 5,000 miles.

The term *ecology*, signifying the interrelationships of life forms among themselves and with the inanimate environment, was first used by the German biologist Ernst Heinrich Philipp August Haeckel (1834–1919) in 1869.

1870

TROY

The ancient non-Biblical site of most interest to Europeans was surely Troy, in what is now the northwestern part of Asian Turkey. It was there that the semilegendary Trojan War was fought about 1200 B.C., and many who had read Homer's *Iliad* in the last twenty-five centuries had thought there was a historical basis to it.

One of the believers was a German businessman, Heinrich Schliemann (1822–1890). Born poor, he worked hard to make a fortune, for the sole purpose of using it to locate the ruined city of Troy.

In 1870 he went to Turkey, managed to choose the right spot (following the description in the *Iliad*), and uncovered a series of ancient cities, built one on top of the other. He obtained various fascinating artifacts, many of them made of gold.

Although Schliemann was not the first archaeologist, and was a very unscientific one, ruining a great deal more than he found, his findings were sensational and rang through the world. His work was a great stimulant to archaeological research in general.

IN ADDITION

It was now France's turn. With consummate skill, Bismarck escalated a small problem over the succession to the Spanish throne into a deadly confrontation. He even maneuvered the French into declaring war on July 19, 1870, and thus made France look like the aggressor. Bismarck had made sure that Austria-Hungary and Russia were friendly, and the *Franco-Prussian War* that followed was totally one-sided.The Prussians outmatched the French in every way.

It was all over in no longer than it had taken to break Austria, and Prussia was avenged for its defeat in 1806. The credit for a spectacular six-year march to triumph belonged not only to Bismarck's diplomatic skill but to the reorganization and direction of the army by one of history's great strategists, Helmuth Karl Bernhard von Moltke (1800–1891).

To fight the war, France had to withdraw its troops from the Pope's remaining dominions. Italian troops immediately marched in and annexed the region, making Rome capital of the Italian kingdom. Pius IX (1792–1878) remained in the Vatican palace, which was left to him, and considered himself a "prisoner of the Vatican," as did the Popes who succeeded him for the next half-century. Italian unification was now virtually complete.

Chicago was now the largest city in the American West, with a population of 300,000. The population of the United States was 39 million, now well ahead of France and Great Britain, and close to the population of the German nation that was now about to be unified. Canada, which had been unified into its present form by 1870, had a population of about 3.3 million.

1871

HUMAN EVOLUTION

Darwin, when writing his book on evolution (see 1858), had deliberately refrained from discussing human evolution in order to avoid controversy. It was impossible, however, to suppose that biological evolution had shaped life in general but had somehow refrained from touching human beings. It quickly turned out, then, that there was no point in ignoring the most important (to human beings) of all aspects of evolution.

Now, in 1871, Darwin published *The Descent of Man,* in which he applied the principles of evolution to humanity specifically. He pointed out the vestigial organs in man as evidence of descent from nonhuman animals: the trace of a point to human ears, the existence of useless muscles that once moved those ears, the four small bones at the bottom of the spine that were once tailbones, and so on. There were still no fossil remnants, however, that could give information as to the details of human ancestry. (Even the Neanderthals were too like ourselves to be useful in this respect.)

DRY PLATES

Until now, photography had had to make use of solutions smeared on plates or films. These solutions were very touchy and one had to be an expert to get good photographs.

In 1871 an English chemist, Joseph Wilson Swan (1828–1914), found that heat increased the sensitivity of the solutions, and that even if they were dried, the result would be beneficial. This made possible the use of dry plates, which greatly increased the ease with which photographs could be taken.

It also turned out that if the sensitive silver compound was mixed with gelatin rather than with collodion, exposure times could be reduced to seconds or even to fractions of a second.

PLANT STRAINS

In 1871 an American naturalist, Luther Burbank (1849–1926), began his life's work of cultivating plants and watching for new strains. He specialized in producing new strains of fruit and, for example, developed sixty varieties of plums in his lifetime. He also developed new varieties of flowers. His work, which caught the public eye, beautifully illustrated the capacity for variation in organisms and helped support Darwin's theory of evolution by natural selection.

IN ADDITION

On January 18, 1871, the German Empire came into being in the Hall of Mirrors at Versailles, where two centuries before Louis XIV had ruled. The significance was unmistakable. Germany, with William I of Prussia as emperor and Bismarck as chancellor (prime minister), was now the lord of Europe. It was the birth of a stronger version of the Holy Roman Empire, which had come to an end sixty-five years earlier. The new empire was called the *Second Reich* in consequence.

Paris surrendered to the surrounding German troops on January 28, 1871, and France agreed to a peace whereby the provinces of Alsace and part of Lorraine were ceded to Germany.

The French deposed Napoléon III on March 1, 1871, and he went into exile in Great Britain. He was the last monarch of France. A *Third Republic* was established, with Louis-Adolphe Thiers (1797–1877) as the first president.

1872

GILGAMESH

The last great king of the Assyrian Empire had been Ashurbanipal (see 2500 B.C.). He had patronized art and literature and had established a large library in his capital city of Nineveh (probably the largest before the time of the Greeks).

In the 1860s, British archaeologists excavated the site of ruined Nineveh (which had been destroyed fourteen years after Asshurbanipal's death) and brought cuneiform tablets from the ruins of the library to the British Museum. There, the English archaeologist George Smith (1840–1876), who was an expert at reading cuneiform languages thanks to the work of Rawlinson (see 1846), was startled to read on one tablet the tale of a flood, much as it appeared in the Bible.

He announced his results in 1872. He had uncovered the epic of Gilgamesh, the oldest surviving literary product of humanity, and it contained a flood story that the Biblical writers had obviously used as a source. This made a huge sensation, and following, as it did, hard upon the discovery of Troy, it further increased popular interest in archaeology.

BACTERIOLOGY

Bacteria had been known for two centuries but were so small that till now they had not been studied in detail. Pasteur's germ theory of disease (see 1862), however, had brought them into lurid prominence. It turned out that a number of different bacteria were pathogenic. They were the cause of certain diseases and of their contagious character.

A German botanist, Ferdinand Julius Cohn (1828–1898), inspired by Pasteur's work, was the first to treat bacteriology as a special branch of knowledge.

In 1872 he published a three-volume treatise on bacteria that may be considered to have founded the science. He made the first systematic attempt to classify the bacteria into genera and species. He was also the first to describe bacterial spores (the form in which bacteria survive periods of desiccation and other unfavorable conditions by retreating behind a thick cell wall). Bacterial spores are so tenacious of life that they may even survive boiling.

STELLAR PHOTOGRAPHY

The advantage of taking a photograph of something you can see perfectly well by eye is that the photograph represents a permanent record. This is valuable in the case of spectra of various light sources, which can then be studied with care and at leisure.

The American astronomer Henry Draper (1837–1882) was the first to photograph a star's spectrum (that of Vega), in 1872. As he adopted the use of dry plates (see 1871), stellar photography became easier, and eventually Draper photographed the spectra of over a hundred stars.

EXPERIMENTAL PSYCHOLOGY

It might seem that psychology is one science that we would know as a matter of course. We know how we think, what our motivations and emotions are, and, perhaps, feel that others are like ourselves in these respects. However, being sure we know something is no substitute for observation and measurement.

It seemed to a German psychologist, Wilhelm Wundt (1832–1920), that there were facets of human behavior that could be profitably measured. There was, for instance, the manner in which the human brain handled sense impressions. He thus initiated *experimental psychology*, writing a textbook on the subject in 1872. He went on in later years to establish the first laboratory devoted to experimental psychology and the first journal devoted to publishing findings on the subject.

IN ADDITION

Railroads began to be built in Japan in 1872.

The Ebers papyrus (see 1550 B.C.) was located in the ruins of ancient Thebes in Egypt.

1873

GAS LAWS

Ever since Boyle (see 1662), it had been known that gases obeyed certain rules, which interrelated volume, pressure, and temperature.

In 1873, however, the Dutch physicist Johannes Diderik van der Waals (1837–1923) showed that gases would obey those simple laws exactly only if the volume of the molecules themselves were zero and if there were no attraction between them. Actually, molecules *do* have a small volume and there *is* a small attraction between them. If the pressure on the gas is low and the temperature is high, these facts scarcely matter, but as pressure goes up and temperature goes down, the facts grow increasingly important.

In 1873 van der Waals introduced a modified equation that gases followed much more closely than they do the simpler, unmodified one. Only *perfect gases* (which have zero-volume molecules and no intermolecular attraction—and which don't exist in reality) follow the unmodified equations.

As a result of van der Waals's work, it could be seen that the Joule-Thomson effect—that gases cool as they expand (see 1852)—only holds below a certain temperature. For most gases, that doesn't matter, because the temperature below which the effect holds is quite high. In the case of hydrogen, however, that temperature is quite low, so that attempts to liquefy hydrogen by means of the Joule-Thomson effect would fail if the gas was not sufficiently cooled to begin with.

In 1910 van der Waals received a Nobel Prize in physics for this work.

LEPROSY

Leprosy, a deadly and disfiguring but slow disease, had been known and dreaded from ancient times. A Norwegian physician, Gerhard Henrik Armauer Hansen (1841–1912), was in charge of a leper hospital in Norway in 1873 when he found that a specific bacterium was the cause of the disease. As a result, leprosy came to be called *Hansen's disease*.

This was the first occasion on which a particular microorganism was found to be the cause of a particular disease. Many other cases followed.

TRANSCENDENTAL NUMBERS

An algebraic number is one that can serve as a solution to a polynomial equation made up of x and powers of x. All integers and all fractions can serve as solutions to such equations and so can certain irrational numbers.

Those irrational numbers that cannot serve as solutions are called *transcendental numbers* (from Latin words meaning "to climb beyond," since they climb beyond the algebraic numbers to further heights). The trick is, however, to *prove* that a particular number is transcendental and that no conceivable polynomial equation can yield it as a result.

In 1873 a French mathematician, Charles Hermite (1822–1901), succeeded in showing that e (a very important quantity in mathematics, equal to 2.71828 . . .) is transcendental. This was the first transcendental number to be identified.

PLATELETS

Red blood corpuscles had been known since Swammerdam had discovered them over two centuries before (see 1658). White blood cells, or *leucocytes* (from the Greek for "white cell"), are larger, but there are far fewer of them in the blood, and they had not been studied carefully till the British physician Thomas Addison (1798–1866) had done so three decades before.

A third type of formed object in the blood had first been reported in 1842. Even smaller than red blood corpuscles and not as numerous (though more numerous than white blood corpuscles), they were first studied carefully by a Canadian physician, William Osler (1849–1919), who reported on them in 1873. They were called *platelets* from their shape. Eventually they were found to be intimately involved with blood clotting and also came to be called *thrombocytes* (from the Greek for "clot cells").

IN ADDITION

Napoléon III died in exile in England on January 9, 1873, and since France had paid off its indemnity, German occupation troops evacuated the county, the last soldier leaving on September 16, 1873. The French elected Marie-Edme-Patrice-Maurice de MacMahon (1808–1893) as their new president. MacMahon was a monarchist and France prepared once again to abandon a republic in favor of a monarchy.

Charles X had a grandson, Henri Dieudonné d'Artois, comte de Chambord (1820–1883). Louis Philippe I also had a grandson, Louis-Phillippe-Albert, comte de Paris (1838–1894). Since Chambord had no heirs, it was agreed he would rule first and Paris would succeed. However, Chambord insisted on restoring the white fleur-de-lis flag of the monarchy, while the country would not abandon the tricolor.

Cable cars (the world's first streetcars) went into action in San Francisco in 1873.

1874

GALLIUM

The discovery of new elements continued after Mendeleyev announced his periodic table of elements (see 1869). In 1874 the French chemist Paul-Émile Lecoq de Boisbaudran (1838–1912) found a zinc ore that displayed hitherto unknown spectral lines. He extracted the new element and named it *gallium,* from the old Latin word for the area that became France after the fall of the Roman Empire.

Once the discovery was announced, Mendeleyev pointed out at once that the element was his eka-aluminum. He was right. The characteristics and properties possessed by gallium were exactly what he had predicted for eka-aluminum. The validity of the periodic table could not thereafter be denied.

TETRAHEDRAL CARBON ATOM

Kekule's way of writing the formulas of organic molecules (see 1858), now fifteen years old, was essentially two-dimensional. The four valence bonds of the carbon atom were directed toward the four angles of a square. That was inadequate for some purposes. For instance, some organic molecules, in solution, rotated the plane of polarized light. That meant the molecules must be asymmetric in one way or another, but the asymmetry wasn't visible in the Kekule formulas.

In 1874 a Dutch physical chemist, Jacobus Hendricus van't Hoff (1852–1911), advanced a three-dimensional representation of organic molecules. In his system, the four bonds of the carbon atom pointed toward the vertices of a tetrahedron, so that the atom could rest on three of the bonds as though it were a splayed-out three-legged stool, with the fourth bond pointing straight upward, each bond equidistant from the other three.

This *tetrahedral carbon atom* produced the necessary asymmetry. If four different groups were attached to the four valence bonds of a particular carbon atom, two distinct molecules could be formed, one being the mirror image of the other.

If one of these compounds twisted the plane of polarization clockwise, the other would twist it counterclockwise. Indeed, any compound shown by the tetrahedral carbon atom to be asymmetric was in fact optically active when tested, twisting the plane of polarized light in one direction or the other. Any compound not shown to be asymmetric was not.

The way the tetrahedral carbon atom explained optical activity was so useful that the new outlook was adopted quickly. Van't Hoff's way of looking at formulas was considered *stereochemistry,* from the Greek meaning "solid chemistry," since the molecules were pictured in three dimensions.

A French chemist, Joseph-Achille Le Bel (1847–1930), advanced the tetrahedral carbon atom at about the same time, independently.

TRANSFINITE NUMBERS

The notion of *infinity,* of endlessness, is always troublesome. The number series 1, 2, 3, 4 . . . is endless, but so is the series 2, 4, 6, 8. . . . You can match up every even number to an ordinary number half its size so that the total number of even numbers is just as great as the total number of all numbers. (This had been pointed out by Galileo two and a half centuries before.)

The German mathematician Georg Ferdinand Ludwig Philipp Cantor (1845–1918) used this sort of one-to-one correspondence to show that all fractions are *denumerable* and can be counted by the integers.

However, all real numbers (rational plus irrational) cannot be counted by the integers. Whatever system one uses to count them will always leave out an infinite number. The group of real numbers represents a higher infinity, a *transfinite number*. Furthermore, all real numbers correspond to the points in a line, so that points are also *nondenumerable*.

Cantor showed that transfinite numbers themselves exist in an endless number, which is larger than the ordinary infinity of the integers.

CURRENT AND CRYSTALS

Every once in a while a scientist will note something that might come under the heading of "curious-but-so-what?"

Thus, in 1874, a German physicist, Karl Ferdinand Braun (1850–1918), noted that in the case of some crystals, an electric current would pass through in one direction but not in the other. He didn't know the reason for it, or what could be done with it, but as it turned out, it was an enormously important finding leading to amazing consequences in later years.

IN ADDITION

Germany was now quiet. Bismarck had the rare ability in a conqueror of knowing when enough was enough. He had attained his goal. Germany was unified and dominant in Europe, and now he seemed anxious to leave it that way.

Japan, on the other hand, was learning the worse part of European ways as quickly as the better. In April 1874, it invaded the Chinese-ruled island of Taiwan on a flimsy excuse and then left, but not before extracting an indemnity from China in European fashion.

1875

FERTILIZATION

Egg cells and sperm cells were both known, and it was clear that the birth of young required a union of the two. Nevertheless, such a union was not observed until the German embryologist Oskar Wilhelm August Hertwig (1849–1922) saw a sperm cell actually enter the egg cell of a sea urchin. What's more, although sperm cells are produced in swarming plenty, Hertwig could see that but a single sperm cell entered the egg cell and that it sufficed for fertilization.

RADIOMETER

In 1875 Crookes (see 1861) devised the *radiometer*, consisting of a set of pivoted vanes in a vacuum. One side of each vane was blackened so that it would absorb heat, and the other side was shiny so that it would reflect it. In the presence of sunlight, the vanes turned steadily. This was not due to solar radiation, for if the container was evacuated particularly well, the motion ceased. Actually, the air molecules in the imperfect vacuum rebounded from the heated side more strongly than from the cooler, shiny side, thus "kicking" it around. This required a partial vacuum to work, as otherwise air resistance would muffle the motion. The radiometer was only a toy, but it offered vivid support for the kinetic theory of gases.

IN ADDITION

The American religious leader Mary Baker Eddy (1821–1910) published *Science and Health*, founding *Christian Science*.

Ismā'īl Pasha (1830–1895), viceroy of Egypt, found himself mired in debt and sold a controlling share in the Suez Canal to Great Britain. The shrewd prime minister of Great Britain, Benjamin Disraeli (1804–1881), engineered the coup.

1876

TELEPHONE

The telegraph, now over thirty years old, transmitted only signals. The British-born American inventor Alexander Graham Bell (1847–1922) wanted something better than that. He wanted to send actual speech over the wires, by turning sound waves into a fluctuating electric current that waxed and waned as the sound waves compressed and decompressed air. The electric current could then be reconverted into sound at the other end.

He finally invented a device capable of doing that and first made use of it accidentally. He had spilled battery acid on his pants and automatically cried out to his assistant, "Watson, please come here. I want you." Thomas Augustus Watson (1854–1934), at the other end of the circuit on another floor, heard the instrument speak and ran downstairs.

On March 7, 1876, Bell patented the telephone. Edison improved it almost at once by devising a mouthpiece that contained carbon powder. When the carbon powder was compressed, it carried more current than when not compressed. As the sound waves compressed and decompressed the carbon powder, the electric current waxed and waned.

The telephone utterly revolutionized human communication.

FOUR-STROKE ENGINE

The Lenoir internal-combustion engine had been devised sixteen years before, but its inefficiency was recognized (see 1860).

A German engineer, Nikolaus August Otto (1832–1891), built a version in which the piston made four strokes. As the piston moved outward (first stroke), a mixture of air and inflammable vapor would be drawn into the cylinder. As the piston moved inward (second stroke), the mixture would be compressed, and at the height of the compression, a spark would set off an explosion. The explosion would drive the piston outward (third stroke, which would supply the power that did the work). As the piston moved inward (fourth stroke), the waste gases would be forced out. The cycle would then be repeated.

Otto built such a four-cycle engine in 1876. The *Otto engine*, as it was called, was such a vast improvement that it caught on at once and was the basis for the internal-combustion engines of today.

CHEMICAL THERMODYNAMICS

Thermodynamics, although worked out originally through the study of heat (see 1850), applies to all forms of energy. An American physicist, Josiah Willard Gibbs (1839–1903), applied it to chemical change.

He wrote a series of papers on the subject that totaled four hundred pages and appeared, over a two-year period beginning in 1876, in *The Transactions of the Connecticut Academy of Sciences.* In the course of his papers, he evolved the modern concepts of *free energy* and *chemical potential* as the driving force behind chemical reactions.

In the papers, he also considered equilibria (singular, equilibrium: the point at which a system comes to rest and there is no further change) between different phases (liquid, solid, or gas), where one or more components of a system are involved. The number of ways (*degrees of freedom*) in which temperature, pressure, or concentration can be varied in such cases can be expressed by a simple equation, which Gibbs called the *phase rule.* By the time he was through, Gibbs had left little more to do in what is now called *chemical thermodynamics.*

BACTERIAL CULTIVATION

When an anthrax epidemic struck the cattle in Silesia (eastern Germany), a local physician, Robert Koch (1843–1910), grew interested. In 1876 he located the particular bacterium that caused anthrax in the spleen of infected cattle and transferred it to mice, carrying the infection from mouse to mouse and recovering the same bacilli in the end. More important still, he learned to cultivate the bacteria outside the living body, using blood serum at body temperature.

Then he learned to make use of solid media for growing bacteria—gelatin, or a complex carbohydrate called agar, which could be obtained from seaweed. When grown in such solid media, the bacteria could not move about easily, and if the bacteria happened to be isolated in one spot, they would, by division and redivision, give rise to a patch of descendant bacteria with no admixture of outside varieties. Bacteria could then be transmitted to animals or allowed to start new cultures with the certainty that only a particular strain of bacteria was being worked with.

In short, Pasteur's germ theory was given practical application by Koch. Koch showed how the causative bacteria could be isolated, then used to produce the disease, then regained from the diseased animal, and finally worked with to find a prevention or cure.

CATHODE RAYS

There was increasing interest in the passage of electric currents through a vacuum. In 1876 a German physicist, Eugen Goldstein (1850–1930), repeated the work of Plücker, nearly two decades before (see 1858). Goldstein thought the fluorescence in the evacuated tube seemed to form as though a stream of radiation had traveled from the cathode (the negatively charged electrode) to the spot on the glass where the fluorescence appeared. He therefore described the phenomenon as *cathode rays.*

This was the first indication that Franklin might have been wrong in his decision that electricity flowed from what he had arbitrarily considered positive to what he had arbitrarily considered negative (see 1752). In the evacuated tube, certainly, the electric current seemed to be flowing from the negative.

IN ADDITION

In the United States, Grant, having served two full terms as president (the first to do so since Andrew Jackson, forty years earlier), retired, and Rutherford Birchard Hayes (1822–1893) was elected the nineteenth president.

Ever since the end of the Civil War, the American army had been busy subduing the Indian tribes of the West and with no light hand, either. Occasionally, the Indians would manage to get in a blow of their own. The most remarkable of these moments came in 1876, when the American general George Armstrong Custer (1839–1876) led a force of 264 men against a much superior Indian contingent under the Sioux leader Sitting Bull (*ca.* 1831–1890). Custer and all his men were wiped out at the Battle of the Little Big Horn on June 25 of that year.

In April 1876, Queen Victoria was declared Empress of India, a title the British monarchs were to hold for seven decades.

In Mexico, Porfirio Díaz (1830–1915) gained power and kept it for thirty-five years, bringing to the land order and foreign investments but ruling dictatorially and doing little for the mass of the population.

The American librarian Melvil Dewey (1851–1931) devised the *Dewey decimal system* for classifying and arranging the books of a library.

1877

PROTEIN SIZE

The fact of osmosis—that some substances could go through certain membranes while others could not—had been known for two decades. In 1877 a German botanist, Wilhelm Friedrich Philipp Pfeffer (1845–1920), explained the phenomenon by showing that large molecules would not go through, while small ones would. If all the large molecules were on one side of the membrane, they would block the passage of some of the small ones. That meant that more small molecules would pass through from the side that lacked the large molecules. The side with the large molecules would then swell, as more small molecules entered than left, and it would experience an *osmotic pressure.*

Pfeffer showed how one might measure this osmotic pressure and related that pressure to the size of the molecules that could not pass through the membrane. If a particular protein was on one side of the membrane, then from the osmotic pressure developed on that side Pfeffer could calculate the molecular weight of the protein. Pfeffer was also the first to determine the molecular weights of proteins and to show that they were *macromolecules* made up of hundreds, even thousands, of atoms.

LIQUID OXYGEN

The work of Andrews (see 1869) and of van der Waals (see 1873) had emphasized the fact that gases that resisted liquefaction had to be cooled as far as possible before pressure could be usefully applied, or the Joule-Thomson cooling-by-expansion effect attempted.

In 1877 a French physicist, Louis-Paul Cailletet (1832–1913), cooled oxygen intensively and then let it expand. He got a fog of liquid droplets of oxygen. By using similar techniques, he also managed to get small quantities of liquid nitrogen and liquid carbon monoxide as well.

A Swiss chemist, Raoul-Pierre Pictet (1846–1929), managed to do this independently in the same year. Using a more elaborate method, he produced greater quantities of the liquids.

As it turned out, oxygen liquefied at −183° C (90° K); carbon monoxide at −191° C (82° K); and nitrogen at −196° C (77° K). The only known gas that remained a gas at the temperature that sufficed to liquefy nitrogen was hydrogen. (Two other gases, as yet unknown, would also have remained unliquefiable at this time.)

PHONOGRAPH

In 1876 Edison had opened the first industrial research laboratory, at Menlo Park, New Jersey. In 1877, the same year in which he improved the telephone mouthpiece in a crucial manner, he also made what he always said was his favorite invention—the *phonograph* (from Greek words meaning "sound-writing").

He put tinfoil on a cylinder, set a free-floating needle skimming over it as the cylinder turned, and connected a source that would carry sound waves to the needle. The needle, vibrating in time to the sound waves, impressed a wavering track on the tin. Afterward, following that track, the needle reproduced the sound waves in a form that was a bit distorted but recognizable.

We all know what the phonograph, and its much improved descendants, have done to bring music (good and bad) into the home.

MARTIAN CANALS

Every thirty years or so, Mars and Earth happen to pass each other where their orbits are closest. Mars is then only 35,000,000 miles from Earth, and astronomers get ready to study it carefully. One of these close approaches (or *conjunctions*) took place in 1877, and one of those interested was the Italian astronomer Giovanni Virginio Schiaparelli (1835–1910).

For one thing, he tried to map Mars. Even 35,000,000 miles is a large distance, and the Martian atmosphere tends to obscure things (to say nothing of our own atmosphere), so that earlier attempts to map the shadowy markings on Mars had not been very successful; different astronomers saw different markings. Schiaparelli, however, recorded what he saw with so good a telescope and so clear an eye that, for once, other astronomers saw the same markings he did. Schiaparelli gave the markings classical names, and others were content to use those names.

Schiaparelli did more. At this time, and at later, less favorable conjunctions of Mars, he detected rather narrow, dark markings. He thought they were bodies of water, so he called the narrow ones *channels*. The Italian word for *channels* was *canali*, and this was mistranslated back into English as *canals*.

The difference is that *channels* refers to natural bodies of narrow water, while *canals* are made by intelligent beings. The notion came into being, then, that Mars was a dying world, with its water slowly leaking into space; and that it was home for a superintelligent race, for whom the canals were a huge engineering development designed to bring water from the polar ice caps to the agricultural tropics. It was nearly a century before that notion was to be quashed once and for all.

MARTIAN SATELLITES

By now, Jupiter was known to have four satellites, Saturn seven, Uranus four, and Neptune one. Of the inner planets, Earth had one satellite, but no satellites were known for Mercury, Venus, or Mars.

At the 1877 conjunction of Earth and Mars, the American astronomer Asaph Hall (1829–

1907) seized the opportunity to make certain that Mars had no satellites. If it did have, they would have to be small and dim, and might well be very close to Mars, whose light would tend to obscure them. Without all three of those qualifications, they would certainly have been seen by now.

Hall therefore began to search the neighborhood of Mars for any signs of little sparks moving about the planet. He worked inward toward Mars's surface and, by August 11, his telescope was trained so close to Mars that its glare was beginning to make it impossible to see anything else. He decided to give it up and assume that satellites were not present, but his wife, Angelina Stickney Hall, urged him to try for one more night.

Hall agreed, and the next night he spotted a small satellite. Five nights of clouds followed, and then on August 17 he noted a second satellite. The satellites were small indeed, the smallest that had yet been discovered, but they were there. Hall named them *Phobos* (Greek for "fear") and *Deimos* (Greek for "terror"), after the two sons of Mars (Ares), the god of war.

IN ADDITION

Russia once again declared war on Turkey, and again Great Britain determined to keep Russia from gaining too much.

In the United States, the Reconstruction era came to an end as the last federal troops left the ex-Confederacy.

In Japan, the feudal samurai rebelled against the modernization efforts of the emperor. They were defeated by an army of ordinary Japanese—with modern weapons. Modernization from this point on proceeded unchecked.

1878

ENZYME

At this time biological catalysts were called *ferments,* whether they existed in intact cells or could be isolated as nonliving substances. The German physiologist Wilhelm Friedrich (Willy) Kuhne (1837–1900) clung to a vitalist position and felt that "ferment" belonged to living systems only.

In 1878 he suggested calling substances that could exert a catalytic effect when separated from living tissue *enzymes* (from Greek words meaning "in yeast," because enzymes acted like materials in the living cells of yeast).

In less than two decades, however, the distinction between biological catalysts inside and outside cells was to be lost, and the word *enzymes,* originally implying an inferior status, came to be applied to all such catalysts indiscriminately.

VARVES

Forty years before, Agassiz had shown that an Ice Age had existed in the past (see 1837). Careful studies since had indicated the presence of several ice ages separated by warm interglacial periods. However, the times when the ice had come and gone, or the lengths of these ice ages and interglacial periods, were shrouded in uncertainty.

In 1878 a Swedish geologist, Gerard Jakob de Geer (1858–1943), studying the sediments in glacier-fed lakes, found that layers of fine silt and coarser sediment alternated, depending on whether the layers were laid down in the winter or in the summer.

By counting the *varves* (Swedish for "layers"), and counting each fine/coarse accumulation as one year, de Geer estimated that the process had been going on for twelve thousand years. That meant it had been twelve thousand years ago that the glaciers of the most recent ice age had retreated. (This was about the time agriculture was coming into use, in the Middle East—see 8000 B.C.)

By studying the thickness of the varves, some idea of changing climate could also be obtained.

This was the first technique for counting thousands of years with considerable accuracy. A number of other such techniques were to be developed later.

YTTERBIUM

Six elements had been isolated thus far from the rare earths, but the list was not complete. In 1878 the Swiss chemist Jean-Charles-Gallissard de Marignac (1817–1894) discovered another. He named it *ytterbium* from the quarry of Ytterby where the first rare earth had been discovered by Gadolin (see 1794). This was the fourth element to be named for that quarry, the other three being yttrium, erbium, and terbium.

IN ADDITION

The Russians defeated the Turks and, on March 3, 1878, forced upon them the Treaty of San Stefano, in which Turkish dominions in the Balkans were virtually dismantled and Russia dominated the peninsula instead. This, however, the British opposed, and some sections of the British public wanted to use force.

The British did not intervene militarily but forced a European congress in Berlin, with Bismarck as host and Disraeli as the chief deal-maker. It met on June 13, 1878, and in a month hammered out a new settlement, in which Serbia, Romania, and Montenegro became independent nations for the first time in six centuries. Turkey was left with a small strip of Balkan territory north of Greece, running from the Black Sea to the Adriatic.

Victor Emmanuel II of Italy, in whose reign the nation had been united, died on January 9, 1878, and was succeeded by his son, Umberto I (1844–1900).

1879

ELECTRIC LIGHT

With a source of energy like electricity, there seemed a chance of having light without flame. A current forced across an air gap can produce a bright spark, and arc lights had been introduced by Davy (see 1800) nearly three-quarters of a century before, but the light was harsh and the danger of fire was great. Edison now tackled the problem of an alternate form of electric lighting.

Electricity passing through a wire warmed it because of the wire's resistance. If the resistance was made great enough by making the wire thin enough, the wire would heat to incandescence. If it did so, however, it would either melt or burn.

To prevent burning, the wire filament

would have to be in an evacuated glass bulb. That would also soften the light and lessen the danger of fire to nearly nothing.

That still left the problem of finding a conductor that would not melt. Edison spent a considerable time experimenting with platinum wire, but platinum was too expensive—and wouldn't work.

Finally, after thousands of experiments, Edison found what he wanted, and it wasn't a metal at all. He used a scorched cotton thread, which proved to be equivalent of a carbon wire. On October 21, 1879, Edison sent a current through such a filament in an evacuated glass bulb. It burned for 40 continuous hours. Edison promptly obtained a patent, and on the following New Year's Eve, he illuminated the main street of Menlo Park using his electric light bulbs before a crowd of three thousand.

To make the light bulb useful, Edison had to develop a generating system that would supply electricity as needed and in varying amounts as lights were switched on and off. He accomplished that, too.

The age of electric lighting had come, and the dark of night was on the road to vanishing.

ORIGIN OF THE MOON

Newton had thought that tides were the result of gravitational attraction by the Moon. The British astronomer George Howard Darwin (1845–1912), the second son of Charles Darwin, went into the matter with particular care.

He realized that the flow of water around the Earth as it turned swept that water across shallow arms of the seas and up sloping shorelines. This produced a certain friction and converted some of the energy of Earth's rotatory motion into heat.

This meant that Earth's rotation was slowing and its period of rotation (the length of the day) was lengthening at a very slow rate. That meant, in turn, that some of the Earth's angular momentum was disappearing. However, the angular momentum could not disappear into nothingness, for there was a law of conservation of angular momentum. If the Earth lost some, then it seemed most likely that the Moon would have to gain just the amount Earth lost. The simplest way for the Moon to gain the angular momentum was for it to move very slowly away from the Earth.

If this scenario were played backward in time, we would see the Moon coming closer and closer to Earth and Earth's day growing shorter and shorter.

George Darwin suggested that the Moon was once part of the Earth, in fact, and because Earth was spinning very rapidly in primordial times, part of its outermost region whirled off and became the Moon.

Although this suggestion is not accepted today, it was very dramatic and interesting in its time.

SACCHARIN

Discoveries can be made entirely by accident. In 1879 the American chemist Ira Remsen (1846–1927) and a student of his, Constantine Fahlberg, synthesized a compound named orthobenzoyl sulfimide. Ordinarily it would have ended there, for new organic compounds are manufactured endlessly every year. This time, however, Fahlberg happened to put his fingers to his mouth without knowing that a few grains of the new compound had adhered to them. He was astonished by an intensely sweet taste.

The compound was eventually named *saccharin* (from a Latin word for "sweet"). It was the first of the commercial sugar substitutes and is still a major one today.

SCANDIUM

Since Gadolin had discovered the rare earths (see 1794), about a dozen elements had been extracted from them that all closely resembled each other. An element that was not exactly one of the rare earth series, but that rather resembled them in its properties, was discovered in 1879 by the Swedish chemist Lars Fredrik Nilson (1840–1899). He named it *scandium,* for Scandinavia.

The Swedish chemist Per Teodor Cleve (1840–1905) pointed out soon afterward that its properties were exactly those of Mendeleyev's "eka-boron" (see 1869) and that it fit into the periodic table just where eka-boron had been placed. A second of Mendeleyev's predicted elements had thus been discovered.

THULIUM, HOLMIUM, AND SAMARIUM

If scandium was not quite a rare earth element, other elements were discovered that were. Cleve, who identified scandium as eka-boron (see above), isolated two new rare earth elements in 1879. He gave both of them names of Scandinavian aura. One was *thulium,* after Thule, which in classic legend was the land farthest to the north, and which was interpreted by later Europeans as representing Scandinavia. The other he named *holmium,* after Stockholm.

In the same year, Lecoq de Boisbaudran, who had discovered gallium (see 1874), identified a new rare earth element, which he named *samarium,* because he had found it in a mineral called *samarskite,* which, in turn, bore the name of an otherwise obscure Russian mining engineer named Samarski.

HEAT AND RADIATION

An Austrian physicist, Josef Stefan (1835–1893), was particularly interested in how hot bodies cooled and, therefore, in how much radiation they emitted. He studied hot bodies over a considerable range of temperature and, in 1879, was able to show that the total radiation of a body was proportional to the fourth power of its absolute temperature *(Stefan's law).*

For example, if a body's temperature tripled, from 1,000° K to 3,000° K, its radiation output would increase $3 \times 3 \times 3 \times 3$, or 81 times. This very rapid increase of radiation with temperature was more of an increase than would have been expected and turned out to be of great use in deducing the evolution of stars. More immediately, from the total radiation of the Sun, its surface temperature was deduced from the rule, and turned out to be about 5,700° K.

IN ADDITION

The Zulus, who had fought valiantly in South Africa against the British and the Boers, were finally defeated, and their land absorbed by the British. Among the casualties on the British side was the "Prince Imperial," Eugène Louis Jean Joseph Napoléon (1856–1879), the son of Napoléon III. He died on June 1, 1879, and with him went all hope of a future Napoléonic restoration.

Bismarck meanwhile was busily engaged in building up a network of alliances intended to keep the peace in Europe and stymie any hope of revenge the French might have. On October 7, 1879, Germany and Austria-Hungary signed a pact of alliance, which was to hold for nearly forty years.

1880

MALARIA

Malaria is perhaps the most widespread and debilitating disease in the world. It was only the discovery of quinine nearly two and a half centuries before that had made it possible for Europeans to maintain themselves in the tropics.

In 1880 a French physician, Charles-Louis-Alphonse Laveran (1845–1922), isolated the microorganism that caused malaria and, to everyone's astonishment, found it to be not a bacterium but a protozoon, a one-celled animal. It was the first pathogenic organism to be found that was anything but a bacterium.

For this discovery, Laveran received the Nobel Prize in physiology and medicine in 1907.

Other bacteriologists were identifying pathogenic microorganisms, too. In 1880, for instance, the German bacteriologist Karl Joseph Eberth (1835–1926) identifed the bacillus that caused typhoid fever.

NEBULAR PHOTOGRAPHY

As the art of photography advanced, it grew steadily more important in astronomy. In 1880 Draper (see 1872), who had learned the use of dry plates from Huggins (see 1863), used the new technique to photograph the Orion nebula and to study its spectrum. It was the first time a nebula had been photographed.

From its spectrum, it was possible to deduce that the Orion nebula was a cloud of dust and gas lit by stars that must be embedded in it. It would be natural to suppose, as a result, that all other nebulas might be of such composition, but within forty years it turned out that there were all-important exceptions.

SEISMOGRAPH

The first attempt to develop a device that would measure the strength of earthquakes had been made by Palmieri (see 1855). In 1880 the first modern *seismograph* (from Greek words meaning "to record earthquakes") was devised by a British geologist, John Milne (1850–1913). This was, essentially, a horizontal pendulum, one end of which was fixed in bedrock. When the ground moved as a result of a quake, the motion was recorded on a turning drum by a pen (eventually by a ray of light). Milne established a chain of seismographs in Japan and elsewhere, which marked the beginning of modern *seismology*.

GADOLINIUM

Marignac, who had discovered ytterbium (see 1878), discovered yet another rare earth element in 1880. This time Gadolin, who had discovered the rare earths (see 1794), was honored, and the element was named *gadolinium*.

UNCONSCIOUS MIND

In 1880 an Austrian physician, Josef Breuer (1842–1925), acquired a new patient whom he referred to as "Anna O." She suffered from various psychological disturbances, including occasional paralysis and a dissociated personality.

Breuer found that if she could be induced to relate her fantasies, sometimes with the help of hypnosis, her symptoms were alleviated. He decided that important causes of such ailments were embedded in the unconscious mind, and that verbalizing them consciously offered a chance of a cure.

ELECTROMECHANICAL CALCULATOR

The American census was growing more and more elaborate. There were more and more people, and more and more questions were being asked of each person. The information gathered was so voluminous that it took literally years to sort it.

An American inventor, Herman Hollerith (1860–1929), who worked for the census, thought there might be a better way of handling the data, and beginning in 1880, he set about the task.

He made use of punch cards after the fashion of Jacquard (see 1801) and Babbage (see 1822). Each card could be punched to represent data gathered in the census: holes in appropriate places could represent sex, age, occupation, and so on.

In order to add up and analyze all this information, the cards were placed on a stand and a metal device was pressed down against them. The device had many pins, which would be stopped by the cardboard. Wherever there was a hole, however, a pin would go through and reach a pool of mercury underneath. Electricity would pass through that pin and control the pointer on a dial. As the punch cards were sent rapidly through the machine, it was only necessary for people to record the numbers indicated on the dial.

What made all the difference between Hollerith and Babbage was that Hollerith had the use of electricity. He had developed an *electromechanical* calculator, and not merely a mechanical one.

Eventually, Hollerith founded a company devoted to making all kinds of machines that could handle and analyze information. That company developed into the *International Business Machines Corporation*, usually known simply as *IBM*.

CHARGED CATHODE RAYS

There was some question about the nature of the cathode rays to which Goldstein had given a name (see 1876). Most of the reported observations could be explained equally well by supposing cathode rays to be either a form of electromagnetic radiation, like light, or a stream of particles.

The matter was investigated by Crookes (see 1861), who in 1875 had devised a vacuum tube that was even better than Geissler's (see 1855).

In 1880, using his own *Crookes tube*, he showed that cathode rays traveled in straight lines and cast sharp shadows. The radiation could even turn a small wheel, which it struck on one side. This still left the nature of the rays uncertain.

But then Crookes showed that a magnet would cause cathode rays to curve in their path, and from the manner of curving, it seemed quite certain that they carried a negative electric charge. Since there seemed no way electromagnetic waves could carry such a charge, Crookes concluded that cathode rays were actually a stream of electrically charged particles.

That decision held up in all subsequent studies.

HIGH PRESSURE

Scientists starting with Boyle (see 1662) had used pressure to study the compression of gases or, as Faraday had, to bring about the liquefaction of gases (see 1823).

A French physicist, Emile Hilaire Amagat (1841–1915), was the first, though, to attempt to reach really high pressures and study the behavior of materials under those conditions. He began his experiments in 1880 and soon managed to attain a pressure equal to 3,000 atmospheres. It set a record and laid the groundwork for further work in decades to come at still more extreme pressures.

PIEZOELECTRICITY

In 1880 a French chemist, Pierre Curie (1859–1906), noted that if a crystal of quartz was placed under pressure, an electric potential appeared across it. If, in reverse, the crystal was subjected to an electric potential, it would compress as though under pressure. If the electric potential changed rapidly, the crystal would compress and expand in time with the change, and the vibration of the crystal would set up sound waves that oscillated too rapidly to be heard. Curie had thus discovered a way of producing *ultrasonic vibrations.*

The manner in which this is done—the interaction of pressure and electric potentials—is known as *piezoelectricity,* where *piezo* comes from the Greek word for "pressure." Crystals with piezoelectric properties came to form an essential portion of sound-electronic devices such as microphones and record players.

IN ADDITION

In the United States, James Abram Garfield (1831–1881) became the twentieth president.

Africa continued to be carved up by the European powers. The French took over west central Africa in 1880 and called it French Equatorial Africa. The western Sahara and northwestern Africa were also becoming French. On July 3, 1880, a meeting of European powers, with the United States included, decided that Morocco was to be independent, and so it was—for a time.

The Boers established the independence of their republics north of the British dominions in South Africa in 1880, and Great Britian allowed this—for a time.

By now, Manhattan had three elevated railway lines, London had a telephone directory, and newspapers were beginning to reproduce photographs.

In 1880 China was the most populous nation with 400 million people, while India was in second place with 240 million, and Russia in third place with 100 million. The United States had climbed to fourth place now with a population of 53 million. The four nations have remained in this rank order ever since.

1881

INTERFEROMETER

In 1881 a German-born American physicist, Albert Abraham Michelson (1852–1931), devised an *interferometer,* with the financial help of Bell, who had invented the telephone five years before (see 1876).

This device acted to split a beam of light in two, send the parts along different paths, then bring them back together—a way of treating light that Maxwell (see 1855) had suggested six years before. Then, no one had had an instrument capable of the job; now Michelson had one.

If the two portions of the light traveled precisely the same distance at the same speed, they would then come back together in phase, and the light would be unchanged. If their distance or speed were slightly different, however, the two beams would be slightly out of phase when they rejoined, and interference fringes should be set up such as those Young had detected and used to prove the wave nature of light (see 1801).

At this time, it was thought that light, being a wave, had to be a wave of something. Consequently, it was supposed that all space was filled with a *luminiferous ether*. (The word *luminiferous* is from Latin, meaning "light-carrying," and *ether* harks back to Aristotle's *aether*—see 350 B.C., Five Elements.) This luminiferous ether was thought to be in a state of absolute rest, and the Earth was thought to be moving at some particular speed relative to it, called Earth's *absolute motion*.

Michelson used his interferometer to measure Earth's absolute motion by sending two halves of a light beam traveling at right angles to each other. Light sent in the direction of Earth's motion through the ether and back again ought to complete its trip a little sooner than light traveling at right angles to the motion and back. Therefore the two halves of the light should rejoin out of phase, and by measuring the width of the interference fringes, one should be able to measure the speed of Earth relative to ether. Once that was known, the absolute motion of all other bodies should follow.

However, the experiment failed. No interference fringes were found. Michelson assumed there was something wrong with the experiment and set about refining it. It took years before he was satisfied that there truly were no interference fringes. That observation helped revolutionize science.

ANTHRAX INOCULATION

It had been three-quarters of a century since Jenner had successfully prevented smallpox by inoculating people with cowpox, a much milder disease that conferred immunity to the more serious one (see 1796). This feat could not be repeated in other cases because other serious diseases did not seem to have milder cousins. Pasteur decided, however, that one might be able to manufacture those milder cousins in the laboratory.

Pasteur studied the deadly disease anthrax, which ravaged herds of domestic animals. Koch (see 1876) had reported observing the bacteria responsible for the disease. Pasteur confirmed Koch's finding and showed that the germs sometimes survived in the ground as heat-resistant spores for long periods of time. It was therefore necessary to kill all infected animals and bury them deep.

If an animal did survive anthrax, however, it was immune thereafter. Pasteur therefore prepared cultures of anthrax germs and heated them. In this way, he destroyed the activity and virulence of the germs but left an "attenuated" preparation that could still produce immunity against anthrax.

In 1881 he carried through a dramatic experiment. Some sheep were inoculated with his attenuated germs; others were not. After a time, all the sheep seemed healthy, but when they were injected with deadly anthrax germs, those that had been inoculated lived; those that had not been inoculated died.

Pasteur, in tribute to Jenner, called the process *vaccination,* even though the disease vaccinia was not involved.

PNEUMOCOCCUS

The physician George Miller Sternberg (1838–1915) was the American pioneer in bacteriology. In 1881 he isolated the bacterium that caused pneumonia. It was a small spherical bacterium of the kind called *coccus* (from a Greek word meaning "seed"). The pneumonia-causing bacterium was therefore called *pneumococcus.*

VENN DIAGRAM

In 1881 the British mathematician John Venn (1834–1923) extended the work of Boole on symbolic logic (see 1847), by representing logical statements as intersecting circles *(Venn diagrams).*

Such circles made it easier to handle statements like "All A are B" or "Some A are B" or "A can be B or C but not both." If Boole's work was a kind of algebraic logic, Venn's was geometric logic.

IN ADDITION

On July 2, 1881, President Garfield was shot by a frustrated office-seeker, Charles Guiteau (1840?–1882). Garfield died on September 19 and was succeeded by his vice president, so Chester Alan Arthur (1829–1886) was twenty-first president of the United States. Guiteau was eventually hanged for his crime.

On March 13, 1881, Alexander II of Russia was assassinated, and his son succeeded as Alexander III (1845–1894). The reforms carried out in Alexander II's reign did not continue. Alexander III returned to the repressive attitudes of Nicholas I and, in particular, encouraged massacres (*pogroms,* a Russian word meaning "devastation") of Jews.

In 1881 France annexed Tunisia in northern Africa and made it part of the growing French African empire.

The population of London stood at 3.3 million in 1881; Paris was over 2 million; while Berlin and Vienna were 1 million each.

1882

CHROMATIN

Studies of the intimate details of cellular structure through the microscope were hampered by the fact that the cell is transparent, so that the inner structure can ordinarily only be seen as gray on gray.

Perkin, however, had produced the first synthetic dye a quarter-century before (see 1856), and since then many others had been produced. It was possible that a dye might combine with some intracellular structures but not with others. In that case, the affected structures would stand out in bright colors.

It was in this way that the German botanist Eduard Adolf Strasburger (1844–1912) was able to observe some of the changes that took place during cell division among plants. In 1882 Strasburger divided protoplasm into two regions, one inside the nucleus, which he called *nucleoplasm,* and the other outside the nucleus, which he called *cytoplasm* (the prefix *cyto-* is from the Greek word for "cell"). These words, especially *cytoplasm,* are now common among biologists.

Much more thorough studies were made by the German anatomist Walther Flemming (1843–1905). He used a dye that combined with a material inside the nucleus that he called *chromatin,* from the Greek word for "color."

When he dyed a section of growing tissue, cells were caught at different stages of division. The cells were killed by the dye, of course, so that all cell divisions were stopped. What he saw, therefore, was a series of "stills" all jumbled together. Given enough observations, however, he could arrange the various stages in order and deduce what was happening.

As the process of cell division began, the chromatin collected into short, threadlike objects, which eventually came to be called *chromosomes* (from Greek words meaning "colored bodies"). Because these threadlike chromosomes were so characteristic a feature of cell division, Flemming named the process *mitosis,* from a Greek word for "thread."

As cell division proceeded, the chromosomes doubled in number. Then came what seemed the crucial step: the chromosomes, entangled in the fine threads of a structure that Flemming named the *aster* (Greek for "star"), were pulled apart, half going to one end of the cell and half to the other. The cell then divided, and the two offspring were each left with an equal supply of chromatin. Because the chromosomes had doubled before the division, each offspring had as much chromatin as the original undivided cell.

Flemming summarized his observations in *Cell Substance, Nucleus, and Cell Division,* which was published in 1882.

All this was of the highest significance, but scientists failed to draw the proper conclusions, because they didn't know of Mendel's genetic work (see 1865).

SPEED OF LIGHT

Since Foucault's determination of the speed of light (see 1849), no one had succeeded in bettering the figure.

However, Michelson (the inventor of the interferometer—see 1881), measured the speed of light in 1882 and reported it to be 186,320 miles per second. This was over a thousand miles per second faster than Foucault had reported, and Michelson's value was the more accurate. In fact, it was less than 40 miles per second above the value that is currently accepted.

DIFFRACTION GRATINGS

Fraunhofer had been the first to use diffraction gratings for the formation of light spectra (see 1820). Diffraction gratings would form far better spectra, which much sharper lines, than prisms would, if the parallel scratches on glass or metal were fine enough and numerous enough.

The American physicist Henry Augustus Rowland (1848–1901) had devised a method for preparing gratings of unexampled fineness, and in 1882, he prepared a grating in which there were almost 15,000 lines per inch. Eventually, using gratings such as these, Rowland could prepare a map of the solar spectrum in which the precise wavelengths of some 14,000 spectral lines were given.

TUBERCULOSIS

Tuberculosis was a very common disease in the nineteenth century. It did not disfigure as smallpox did, and it did not kill quickly, but it progressed inexorably and took a heavy toll, of young people particularly.

In 1882 Koch (see 1876) isolated the *tubercle bacillus,* the bacterium that caused tuberculosis. This led him to search for a cure, and for a while he thought he had one, but was disappointed. Still, such was the dread of the disease that merely identifying the germ seemed a feat important enough to warrant granting Koch a Nobel Prize in medicine and physiology in 1905.

PI AS TRANSCENDENTAL

In 1882 the German mathematician Ferdinand von Lindemann (1852–1939) studied *pi,* the ratio of the circumference of a circle to its diameter. Its value is 3.14159 . . . and so on, forever, in a nonrepeating decimal, for pi is an irrational number.

Lindemann was able to show that it was more than irrational. It was transcendental, as Hermite had shown *e* to be (see 1873). Therefore, pi could not serve as a solution for any conceivable polynomial equation, and that meant that the problem of *squaring the circle* by straightedge and compass in a finite number of steps (see 1837) was impossible.

IN ADDITION

The world continued to be parceled out by the European powers. Great Britain felt the need to protect the Suez Canal from Egyptian nationalists and so bombarded Alexandria on July 11, 1882, and occupied Cairo on September 15. Egypt became part of the still-growing British Empire.

Meanwhile, little Belgium, under Leopold II (1835–1909), who had been king since 1865, was expanding its holdings over a vast tract of central African land that eventually came to be called first the Congo Free State and then the Belgian Congo. France was extending its hold over Madagascar and what was later called French Indo-China (which included the nation now known as Vietnam). Italy seized a port on the Red Sea coast of Africa, which eventually became the center of an Italian colony named Eritrea.

Meanwhile Bismarck ignored overseas acquisition in favor of strengthening his position in Europe. He continued to build his system of alliances, and on March 20, 1882, Italy joined Germany and Austria-Hungary to form the *Triple Alliance.*

1883

ALLOY STEEL

For three thousand years, steel (carbonized iron) had been the strongest material known for tools, structures, and weapons, but that didn't mean it couldn't be improved.

Attempts were made to improve the qualities of steel by adding other metals. Manganese was one of the metals tried, but it seemed to make the steel more brittle. A British metallurgist, Robert Abbott Hadfield (1858–1940), added more manganese than others had thought advisable, however, and when the steel was 12 percent manganese, it was no longer brittle. If it was then heated to 1,000° C and quenched, it became superhard. Where ordinary steel used for railroad rails had to be replaced every nine months, rails made of manganese-steel lasted twenty-two years.

Hadfield patented his manganese-steel in 1883, and that marked the beginning of the triumph of *alloy steel.* Other metals were then added to steel in varying quantities and mixtures—chromium, tungsten, molybdenum, vanadium, niobium, and so on—in search of alloys with new and useful properties.

ALTERNATING CURRENTS

Electric currents, as used in the first half of the nineteenth century, flowed from one point to another in one direction. That is the kind of current one gets from batteries, and it is *direct current* (dc). It is easier, in working with electric generators, however, to get a kind of current that goes first in one direction, then in the other, alternating very rapidly, with the current intensity rising and falling like a sine wave. This is *alternating current* (ac).

Alternating current didn't seem as useful as direct current, but then in 1883 a Croation electrical engineer, Nikola Tesla (1856–1943), constructed an induction motor that could make use of alternating current to do useful work.

Edison had committed himself to direct current, and he fought the use of alternating current, but he lost out eventually.

EDISON EFFECT

Once Edison had invented the electric light (see 1879), he naturally endeavored to improve it. In particular, he wanted to make the filament last longer. In 1883 he sealed a metal wire into a light bulb near the hot filament. Perhaps he thought this might absorb some of the remaining air in the tube and lessen its destructive effect on the filament.

To Edison's surprise, electricity flowed from the hot filament to the cold wire across the gap that separated them. This is called the *Edison effect*.

Edison wrote up the effect meticulously and patented it, as a matter of course, but he could think of no use for it. This was his only purely scientific discovery, and he didn't follow it up. The Edison effect proved the basis for the science of electronics that was eventually to come.

RAYON

The French chemist Louis-Marie-Hilaire Bernigaud de Chardonnet (1839–1924) had begun some five years earlier to produce fibers by forcing solutions of nitrocellulose through tiny holes and allowing the solvent to evaporate. He perfected the process in 1883 and found he had a fiber that strongly resembled silk in its fineness and shininess. He called it *rayon* (from the French word for "a ray of light") because of its shininess. He used only partly nitrated cellulose, so rayon wasn't explosive, but it was still uncomfortably inflammable.

Rayon was the first synthetic fiber, and it was to be followed by many others.

MEIOSIS

The discovery of the mechanism of cell division by Flemming (see 1882) stimulated many biologists into investigating the matter further.

The Belgian cytologist Edouard Joseph Louis-Marie van Beneden (1846–1910) found that the number of chromosomes in the cells of a particular species was always constant, though the number varied from species to species. (It is now known that there are forty-six chromosomes in human cells.)

Furthermore, he discovered that in the formation of the sex cells (that is, the ova and spermatozoa), the division of chromosomes during one of the cell divisions was *not* preceded by a doubling. Each egg and sperm cell, therefore, ended with only half the usual number of chromosomes. (This halving of number was called *meiosis*, from a Greek word meaning "to make smaller"). Consequently, when a sperm cell entered an egg cell in the process of fertilization, the chromosomes reached their normal number for the species, half of them coming from the mother and half from the father.

Meiosis fit in perfectly with Mendel's genetic discoveries (see 1865) but those discoveries were still being ignored.

PHAGOCYTES

A Russian-born French bacteriologist, Élie Metchnikoff (1845–1916), found that in simple animals there were semi-independent cells that were capable of ingesting small particles. Any damage to the animals brought these cells to the spot at once.

In 1883 Metchnikoff followed up this lead and studied more complicated animals. He was able to show that the white cells in blood were also semi-independent and were capable of ingesting bacteria. The white cells flocked to the site of any infection, and what followed was a battle between bacteria and what Metchnikoff called *phagocytes* (Greek for "eating cells"). When the phagocytes lost heavily, their disintegrated structures made up pus.

The white cells, Metchnikoff held, were an important factor in resistance to infection and disease. For this work, he received a share of the Nobel Prize in physiology and medicine in 1908.

DIPHTHERIA

The German pathologist Edwin Klebs (1834–1913) discovered the bacterium responsible for a serious children's disease, diphtheria, in 1883.

MAXIM GUN

The Gatling gun (see 1862) had to be cranked. In 1883 an American-born, British inventor, Hiram Stevens Maxim (1840–1916), went one step further, with a fully automatic machine gun. It made use of the energy of a fired bullet's recoil to eject the spent cartridge and load the next.

The Maxim gun gave European armies a still greater advantage over native levies in Africa and Asia. One popular jingle of the time went:

"Whatever happens, we have got
The Maxim gun, and they have not!"

However, the time was to come when non-Europeans found out how to use advanced weapons, and the jingle then lost its luster.

EUGENICS

From earliest times, human beings have bred their animals in such a way as to enhance desired characteristics, so that larger and speedier horses were obtained, sheep with more wool, cows with more milk, hens with more eggs. It must have occurred to many that similar tactics might improve the human species.

One who thought this was the British anthropologist Francis Galton (1822–1911), a first cousin of Charles Darwin. In 1883 he coined the term *eugenics* (from Greek words meaning "good breeding") for the study of the improvement of human qualities by careful breeding.

However, eugenics is not easy to put into practice. In the first place, one can't guide human breeding with quite the ease that one can guide the breeding of animals. Second, we are not as sure of what we want in human beings as we are in the case of animals. Third, when Mendel's discovery of recessive traits (see 1865) was understood, it came to be seen that it was hard to get rid of what one might consider undesirable characteristics. Fourth, the loudest supporters of eugenics were unsavory characters who wished to place it at the service of prejudice and racism.

IN ADDITION

The Brooklyn Bridge, connecting New York City and Brooklyn, then two separate cities, was opened to the public on May 24, 1883. It was the first of the great modern suspension bridges, the longest bridge built up to that time (1,595 feet, or 0.3 mile), and the first to use steel cables. It was designed by the German-born American engineer John Augustus Roebling (1806–1869), who died in an accident during the early stages of work on the bridge. It was carried to completion by his son, Washington Augustus Roebling (1837–1926), who was badly crippled by caisson disease, or the "bends", from which he suffered during the construction.

Railroads had to run according to schedule, and the use of local time in every town, based on longitude, reduced matters to chaos. In 1883 the railroads adopted standard time zones, which the United States divided into Eastern, Central, Mountain, and Pacific zones. The nation generally was so dependant on railroads that it had to follow, and the idea eventually spread to all but the most primitive portions of the world.

A non-European conqueror arose in Africa, a Sudanese Muslim, Muhammad Ahmad (1844–1885), who called himself the *Mahdi* (Arabic for "the guided one"—guided by Heaven, that is). By the end of 1883, he had defeated three Egyptian armies and taken over full control of the Sudan (theoretically an Egyptian dependency at the time).

In a fearsome natural cataclysm, the island of Krakatoa, an apparently dormant volcano lying between Sumatra and Java, exploded on August 27, 1883. The sound was heard 3,000 miles away; thirty-six thousand people were killed by tsunamis *(tidal waves);* and the upper atmosphere was filled with dust that created brilliant sunsets for a long time. It was the worst such explosion in over three thousand years.

1884

HEAT AND TEMPERATURE

Stefan had related heat emission to the fourth power of the temperature (see 1879), and in 1884, the matter was taken up further by the Austrian physicist Ludwig Eduard Boltzmann (1844–1906), who had once served Stefan as a laboratory assistant. Boltzmann had earlier increased the rigor of the mathematical treatment of thermodynamics and had emphasized the statistical interpretation of the second law of thermodynamics, so that he is considered the founder of *statistical mechanics*.

Boltzmann showed that Stefan's finding could be derived from thermodynamic considerations, so that the rule is sometimes called the *Stefan-Boltzmann law.*

IONIC DISSOCIATION

A substance dissolved in water lowers the freezing point of water somewhat. This lowering is in proportion to the quantity of the substance dissolved.

For different substances, the amount of lowering depends on the number of molecules present. If substance A has molecules only half the mass of substance B's molecules, and if equal masses of both are dissolved in separate quantities of water, there will be twice as many molecules of the half-sized A in solution as of the full-sized B. Substance A will therefore be more effective in lowering the freezing point of water than substance B.

That goes for substances that don't conduct electric current in solution *(nonelectrolytes)*. For substances that do conduct electric current *(electrolytes)*, the situation is different. For instance, when sodium chloride is dissolved in water, one can calculate how many of its molecules are in solution and therefore how far the freezing point will be depressed. However, the freezing point is depressed twice as much as the calculated figure. The same is true for potassium bromide and sodium nitrate. With substances like barium chloride and sodium sulfate, the depression of the freezing point is three times what is calculated.

A Swedish chemistry student, Svante August Arrhenius (1859–1927), decided that the logical way to explain this was to suppose that sodium chloride molecules broke up into two fragments of opposite electrical charge (such fragments are called *ions*, a word first used by Faraday—see 1821). The ions carried the current, and since there were twice as many particles as there would have been if the molecule had remained intact, this accounted for the doubled depression of the freezing point. In the same way, barium chloride must break up into three ions, and so on.

In 1884 Arrhenius presented his theory of *ionic dissociation* as his Ph.D. thesis. It was not greeted with enthusiasm, for chemists had assumed for a century that atoms were featureless, indestructible, and unchanging. The thought of atoms carrying electric charges lent them features and a structure, and that was hard to accept. However, Arrhenius's theory made sense in every other way, so he was given a passing grade—but the lowest possible one. In 1903, when key discoveries had been made concerning the atom's structure, this same thesis obtained for Arrhenius a Nobel Prize.

SUGAR STRUCTURE

Though sugars had been studied throughout the nineteenth century, and though their atomic content was more or less known, the actual three-dimensional arrangement of the atoms was not known. The work of van't Hoff, ten years before, had shown the importance of the three-dimensional arrangement of atoms in a molecule and the way this could be responsible for optical activity (see 1874).

In a long series of experiments beginning in 1884, the German chemist Emil Hermann Fischer (1852–1919) isolated pure sugars and studied their structures. He showed that the best-known sugars contained six carbon atoms and could exist in sixteen varieties, depending on how the carbon bonds were arranged. Each different arrangement was reflected in the way the plane of light polarization was twisted.

Fischer showed that there were two series of sugars, mirror images of each other, which he called the *D-series* and the *L-series*. He had to pick which mirror image belonged to which method of writing the formula, and he did so arbitarily. He had a fifty-fifty chance of guessing correctly, and (unlike Franklin on the electric charge—see 1752), he guessed right. The natural sugars all happen to belong to the D-series.

While engaged in this work, Fischer also studied a group of substances called *purines*, with molecules that consisted of a double ring of atoms made up of five carbons and four nitrogens. They were eventually found to make up an important part of certain key biochemical substances.

For his work on sugar structure and on purines, Fischer received the Nobel Prize for chemistry in 1902.

COCAINE

Cocaine is an alkaloid obtained from the leaves of the coca bush, which originally grew wild in Peru and Bolivia. The Incas chewed the leaves to suppress feelings of pain and weariness and so be able to endure the strains of life better.

Europeans eventually discovered cocaine without understanding at first that it was strongly addictive and therefore dangerous. One of the first to study it was an Austrian physician, Sigmund Freud (1856–1939). He did not follow it up but passed on his information to an Austrian colleague, Carl Koller (1857–1944). Freud suggested that cocaine might be a pain-relieving agent, but Koller went further and studied its uses as a local anesthetic (one that would deaden pain in a particular area of the body without producing unconsciousness).

In 1884, after having experimented on animals, Koller performed an eye operation on a patient—the first using local anesthesia. It was successful. Eventually, local anesthetics were developed that were more effective, and safer, than cocaine.

BACTERIAL STAINING

Bacteria can be stained with synthetic dyes, just as Flemming had shown that ordinary cells can be (see 1882).

In 1884 a Danish bacteriologist, Hans Christian Joachim Gram (1853–1938), stained bacteria with a dye. He found that, when treated with iodine and an alcohol wash, the stain could be removed from some kinds of bacteria but not from others.

The bacteria that retained the dye were called *Gram-positive*, while those that lost it were *Gram-negative*. The importance of this became apparent once antibacterial agents were developed and it was found that some attacked the Gram-positive varieties and some the Gram-negative ones.

STEAM TURBINE

Turbines had been devised (see 1827), but none were capable of withstanding the high temperatures to which they would be subjected if they were to be turned by jets of steam and of withstanding the simultaneous stress of rapid rotation while preventing steam from escaping prematurely.

In 1884, however, the British engineer Charles Algernon Parsons (1854–1931) constructed the first successful steam turbine. Such devices could be used to power ships, which could then attain unusually high speeds. They could also be used in electric generators.

LINOTYPE

Ever since the invention of printing (see 1454), advances had been made in the speed with which it could be carried through. The greater the speed, the more printed matter would be produced, the more would be read, and the more would be desired, especially as literacy increased the world over.

One bottleneck, however, was the setting of type. Plucking out an individual letter and setting it in line was time-consuming.

In 1884 a German-born American inventor, Ottmar Mergenthaler (1854–1899), patented a machine that, when directed by an operator at a keyboard, could set a whole line of type at one time. It was therefore called a *Linotype* machine. For the next three-quarters of a century it was a mainstay of publishing—especially newspapers, which increasingly had to be turned out steadily and in quantity.

FOUNTAIN PEN

Through most of history, those in the West have been writing with pen and ink. The pens advanced from feather quills to steel, but always they had to be dipped into the inkwell every few moments. This was tedious and led to a great number of ink blots and ink spills and tended to turn hands, face, and clothes inky as well.

In 1884 an American inventor, Lewis Edson Waterman (1837–1901), patented a pen with a reservoir that could be filled with ink. The pen could then write for a long time before another filling was called for. The reservoir held a virtual fountain of ink, and before long the *fountain pen*, with further improvements, had replaced the dip pen almost everywhere. Even the typewriter (see 1867) did not completely remove the need for writing, after all.

IN ADDITION

Great Britain and France both established footholds on the Somalian coast east of Ethiopia, while Russia took the city of Merv in central Asia and advanced to the northern border of Afghanistan. Even Bismarck was forced by German public opinion to join in the scramble, picking up scraps of territory the other powers had overlooked or had not had time for. Germany established footholds in Togo, the Cameroons, and southwest Africa.

In the United States, Grover Cleveland (1837–1908), the governor of New York, was elected twenty-second president.

The first roller coaster was built in Coney Island. In Chicago, a ten-story building was erected with an internal metal frame of iron supporting the floors so that there was no need of thick walls and massive foundations. This might be viewed as the first *skyscraper*.

1885

RABIES

Rabies, or hydrophobia, is a severe disease of the central nervous system. It can occur in any warm-blooded animal. The microorganism that causes it can be present in the salivary glands and may be communicated by a bite. For human beings, dogs can be particularly dangerous. They become excitable and vicious ("mad dogs") when they have the disease, and bite at the slightest cause, or for no cause at all.

Once infected, it may take quite a while for a human being to show symptoms, because the microorganism has to penetrate the nervous system, but once it is established there, death is fairly quick, almost certain, and agonizing.

Pasteur tried to treat the disease as he had treated anthrax (see 1881), by developing an attenuated and feeble preparation of the causative agent that would give immunity but not the disease. He showed that an attenuated agent could be obtained by passing a rabies infection through a variety of different species of animals until its virulence had abated.

In 1885 Pasteur made the first use of his attenuated preparation to prevent a case of rabies in a boy, Joseph Meister (1878–1940), who had been bitten by a mad dog. The treatment worked, and the boy was saved.

PURINES AND PYRIMIDINES

Since the nucleic acids had been discovered by Miescher (see 1869), little had been done to determine their molecular structure.

Then the German biochemist Albrecht Kossel (1853–1927) took up the matter. He got rid of the proteins associated with the nucleic acids and worked on the material itself. By 1885 he had obtained substances from it that included the double-ring purines Emil Fischer had studied a few years earlier (see 1884). He also obtained *pyrimidines,* whose molecules were made up of a single ring of atoms, four carbons and two nitrogens.

He isolated two different purines, *adenine* and *guanine,* and three different pyrimidines, *uracil, cytosine,* and *thymine.* He also decided that a sugar molecule was present but couldn't tell which one.

Much remained to be done, but it was a good start and far more important than anyone could tell at the time. This was some of the work for which Kossel received a Nobel Prize in physiology and medicine in 1910.

PRASEODYMIUM AND NEODYMIUM

Didymium was a rare earth element that had been discovered by Mosander some forty years before. Its name is derived from the Greek word for "twin" because it was so like other rare earth elements. However, the name turned out to be more apt than was thought, for the element was actually twins, a mixture of two very similar elements.

In 1885, after much careful work, an Austrian chemist, Carl Auer, Freiherr von Welsbach (1858–1929), managed to separate the two elements. One he named *praseodymium* ("green twin," because of the color of a prominent line in its spectrum) and the other he named *neodymium* (new twin).

WELSBACH MANTLE

Von Welsbach (see above) was the first to find an important use for the rare earth elements in which he was so interested.

It occurred to him that gas flames might be made to give more light if they were allowed to heat up some compound that would itself then glow brightly.

He tried many substances that would glow at high heat without melting and finally found that if he impregnated a cylindrical fabric with thorium nitrate, to which a small amount of cerium nitrate (a compound of one of the rare earth elements) had been added, he would obtain a brilliant white glow in a gas flame.

This *Welsbach mantle,* when used in a kerosene lamp, gave such a bright light that it allowed kerosene lamps to compete with electric light for another thirty years or so.

TRANSFORMER

In the battle between alternating current and direct current (see 1883), victory went to alternating current, because it could be transformed into another current of much higher voltage and of correspondingly lower amperage (to get something, you have to give up something in the real world). While it is in this high-voltage state, the current can be transported long distances with comparatively little loss. Once it arrives at its destination, it can be reconverted into a low-voltage, high-amperage current again, in which form it can do its work best.

For these transformations from low to high voltage and back again, a device was needed that was invented in 1885 by an American electrical engineer, William Stanley (1858–1916), who was working for Westinghouse (see 1868). His *transformer* could shift voltage and amperage, but only with alternating current and not with direct current.

AUTOMOBILE

Until the invention of the steam engine (see 1712), the dream of a carriage that would move without a horse pulling it (a "horseless carriage") had belonged to the world of myths and legends. The steam-engine is supposed to have been put into action as early as 1769, but such steam-powered vehicles were bulky, clumsy, and slow. Even fairly advanced ones, built much later, took time to start, because water had to be heated and boiled first.

The coming of the internal-combustion engine, especially the Otto four-stroke engine (see 1876), offered a much better hope. What was needed now was an appropriate fuel, and eventually that turned out to be *gasoline*, a petroleum fraction with smaller molecules than those of kerosene, so that it vaporized more easily and burned more readily.

The first working automobile with a gasoline-burning internal-combustion engine was built in early 1885 by a German mechanical engineer, Carl Friedrich Benz (1844–1929). Its wheels looked like bicycle wheels, and there were three of them, a smaller one in front and two larger ones in back. It ran at a speed of 9 miles per hour, and it was the forerunner of all that was to follow.

FINGERPRINTS

In 1885 Galton (see 1883) pointed out the individuality of fingerprints. No two people (barring identical twins) had identical fingerprints, and Galton worked out a thoroughgoing system of classifying and identifying them.

In handling smooth surfaces, people always left sweaty, greasy fingerprints, even when these were unnoticeable unless the surface was appropriately powdered. Eventually, fingerprints proved a useful way of showing that a given person had been present at a given place and had handled a given object. This added a new dimension to *forensic medicine* (from a Latin word referring to a public place such as a courtroom).

IN ADDITION

Germany annexed what is now Tanzania and established it as German East Africa.

Alfonso XII of Spain (1857–1885) died on November 24, 1885. His wife was pregnant and later gave birth to a son who reigned as Alfonso XIII (1886–1941).

1886

ALUMINUM

Aluminum is the most common element in the Earth's crust. It was first isolated by Ørsted (see 1825), but the isolation was so difficult that it was virtually a precious metal. Napoléon III had his cutlery and a baby rattle for the Prince Imperial made out of aluminum, and the capstone on the Washington Monument is a slab of aluminum.

In 1886 an American student of chemistry, Charles Martin Hall (1863–1914), heard his teacher say that anyone who discovered a cheap way of making aluminum would grow rich and famous. In his home laboratory, using homemade batteries, Hall devised a method of preparing aluminum by the use of an electric current, as Davy had prepared sodium and potassium nearly eighty years before (see 1807). He used aluminum oxide dissolved in a molten mineral named cryolite, and into it he stuck carbon electrodes.

Oddly enough, that same year a French metallurgist, Paul-Louis-Toussaint Héroult (1863–1914), with the same last initial and the same birth and death years, independently devised precisely the same system, which is therefore called the *Hall-Héroult process*.

Aluminum became cheaper almost at once, and it is now second only to steel as a structural material. A combination of lightness and strength makes it ideal for aircraft, for instance.

GERMANIUM

A German chemist, Clemens Alexander Winkler (1838–1904), analyzed a silver ore and, when he had completed his work, found that the elements he had located added up to only 93 percent of the whole. Puzzled, he searched out the remaining 7 percent and, in 1886, found a hitherto unrecognized element, which he named *germanium,* after Germany.

It turned out to be eka-silicon and filled the one remaining gap for which Mendeleyev had predicted a new element (see 1869). What's more, the properties that Mendeleyev had predicted for it once again turned out to be absolutely correct. Three predictions made, three predictions fulfilled. It was as amazing a feat as the discovery of Neptune (see 1846).

FLUORINE

For three-quarters of a century, chemists had known that a certain element must exist. They had even given it a name—*flourine.* However, it was a particularly active element, the most active one known, more active than oxygen and chlorine, and nothing seemed capable of forcing it out of combination with other elements.

A number of chemists tried to isolate it and found the matter not only difficult but dangerous, for the materials they had to work with were poisonous.

Finally, a French chemist, Henri Moissan (1852–1907), tried. He used platinum for all his equipment, because it was one of the very few substances that fluorine would not attack and combine with instantly. If he isolated some fluorine in platinum, that fluorine would stay isolated.

He placed a solution of potassium fluoride in hydrogen fluoride in his platinum equipment, chilled it to −50° C to tame the fluorine a bit, and passed an electric current through it on June 26, 1886. He obtained a pale yellow gas that was the long-sought fluorine.

For this, he obtained the Nobel Prize in chemistry in 1906 (receiving, according to report, one vote more than Mendeleyev did, which if true was an injustice, for Mendeleyev was more deserving).

DYSPROSIUM

Lecoq de Boisbaudran, who had already isolated gallium (see 1874) and samarium (see 1879), was working on a rare earth ore containing holmium when he discovered that it contained a small amount of still another rare earth element, which he named *dysprosium,* from a Greek word meaning "hard to get at."

CANAL RAYS

Goldstein, who had given cathode rays their name (see 1876), continued to work with them. In 1886 he used a perforated cathode, one with small holes, or channels, penetrating it. When he did this, he found that not only did cathode rays emerge as usual, but radiation passed through the channels and streamed in the opposite direction.

He called the radiation *Kanalstrahlen,* which is German for "channel rays," but they were usually called *canal rays,* by the same mistranslation that had given rise to the notion of Martian canals (see 1877).

RAOULT'S LAW

The French physical chemist François-Marie Raoult (1830–1901) showed in 1886 that the partial pressure of solvent vapors in equilibrium with a solution is directly proportional to the ratio of the number of solvent molecules to solute molecules. This is called *Raoult's law.*

To the nonchemist, the law may seem unclear, but it provided chemists with a new way of calculating the molecular weight of dissolved substances. It also showed that the freezing point was depressed, and the boiling point elevated, in proportion to the number of particles present in a given volume of solution. Arrhenius had assumed this when he had worked out his theory of ionic dissociation two years earlier (see 1884).

NITROGEN FIXATION

Plants must have nitrogen atoms to build their tissues, and they get them from nitrates in the soil. There are enormous amounts of nitrogen in the atmosphere, but nitrogen itself is not active, and it is not easy to force it into combination with other substances.

The German chemist Hermann Hellriegel (1831–1895) discovered in 1886 that some leguminous plants (peas, beans, and related species) had root nodules that contained *nitrogen-fixing bacteria* (that is, bacteria that could affix nitrogen to other kinds of elements, eventually forming nitrates). Since there is always a danger that plants will consume the nitrates in soil, causing it to decline in fertility, it was useful to know that the planting of leguminous plants would restore the fertility of the soil—at least with respect to nitrogen.

IN ADDITION

The Indian wars in the United States came to an end after two and a half centuries, with the capture in August 1886 of the Apache leader Geronimo (1829–1909), the last formidable Indian adversary.

Great Britain had now absorbed all of Burma. At home, however, Prime Minister William Ewart Gladstone (1809–1898) tried to give Ireland a certain autonomy (*home rule*). He was defeated, but the Home Rule movement would agitate Great Britain for the next few decades.

1887

MICHELSON-MORLEY EXPERIMENT

Michelson, who had hoped to determine the Earth's speed of motion through the stationary ether (see 1881), continued to improve the delicacy of his procedure and of his instruments, and to repeat the experiment. Finally, in 1887, with the help of a colleague, the American chemist Edward Williams Morley (1838–1923), Michelson ran a definitive experiment and, once again, found no interference fringes.

There had to be an explanation. Either the

Earth was motionless with respect to the ether, or the Earth dragged the ether with it, or *something*. All possible explanations seemed highly unlikely, and for nearly a quarter of a century, the world of science was completely puzzled. It took a scientific revolution to explain the matter, so that the *Michelson-Morley experiment* is perhaps the most important "failure" in the history of science.

PHOTOELECTRIC EFFECT

The German physicist Heinrich Rudolph Hertz (1857–1894) was interested in Maxwell's equations (see 1865). As part of his investigation of electromagnetic effects in the light of those equations, he set up an electrical circuit that oscillated. It surged alternately, first into one, then into the other of two metal balls separated by an air gap. Each time the potential reached a peak in one direction or the other, it sent a spark across the gap.

In the course of these experiments, Hertz noted that when ultraviolet light shone on the negative terminal of the gap, the one from which the spark issued, the spark was more easily elicited. This did not seem to have anything to do with what he was trying to observe, so he noted the matter but did not follow it up.

This was the first observation of the effect of light on electrical phenomena, however. This *photoelectric effect* turned out to be extremely important.

MACH NUMBER

With advancing technology, human beings were traveling faster than they had in the past, and they were likely to travel faster still in the future. The faster human beings traveled, the more important air resistance would be. The Austrian physicist Ernst Mach (1838–1916) studied the conditions that occurred when a solid object and air were in rapid motion relative to each other.

The natural rate at which air molecules can move is the speed of sound through air. When an object moves through air at higher speeds, the air molecules can no longer move aside naturally but must be shoved aside (so to speak) faster than they want to go. This produces new conditions, which Mach studied.

For instance, faster-than-sound motion compresses air and sets up a bunching of sound waves that then expands to produce a sudden clap. Thunder is the best example of such a *sonic boom*, with the heat of lightning expanding the air at greater than the speed of sound. Another example of a sonic boom is the crack of a bullwhip.

A speed equal to that of sound is now called *Mach 1* in Mach's honor. Twice that speed is *Mach 2*, and so on.

RUBBER TIRE

Since wheeled transportation had been invented five thousand years before, wheels had had wooden or metal rims. These were noisy and offered no spring, so that travel on carts and wagons was a jolting affair.

In 1887 a British inventor, John Boyd Dunlop (1840–1921), decided to rim the wheels of his son's tricycle with a rubber tire (and patented the notion the following year). Rubber, though soft, actually wears better than wood or iron does. What's more, Dunlop made it a pneumatic tire: what went round the rim was an air-filled tube, which was covered by a tire with a rubber tread.

This gave a vehicle spring and cut down the noise tremendously, so that tires were soon used for automobiles and other vehicles as well.

IN ADDITION

Bismarck signed a secret treaty with Russia on June 18, 1887, and thus succeeded in isolating France entirely.

A transcontinental rail link across Canada was completed on May 23, 1877.

A Polish philologist, Ludwik Lejzer Zamenhof (1859–1917), designed an artificial language, which he called *Esperanto* (hope), because he dreamed that a common language for the world might foster international understanding and peace. Unfortunately, it never caught on, and neither has any other artificial language.

The German-born American inventor Emile Berliner (1851–1929) improved the phonograph immeasurably in 1887. Where Edison had used cylinders with a needle moving up and down, Berliner introduced flat disks with the needle moving from side to side as it followed a spiral groove. The disk rapidly replaced the cylinder.

In 1887 the Huang (Yellow) River in China flooded its banks and, directly or indirectly, killed nine hundred thousand people. It was the worst flooding disaster in history.

1888

RADIO WAVES

When Hertz was performing the experiments that gave him the first glimpse of a photoelectric effect (see 1887), he expected that the oscillating current he was working with would produce an electromagnetic wave. Each oscillation should produce a wave, and the wave should be very long. After all, since light travels at a bit over 186,000 miles per second, a wavelength formed in an oscillation of a mere hundred-thousandth of a second will still be nearly 2 miles long. In 1888, he actually observed such waves.

Hertz used, as a device for detecting the possible presence of such long-wave radiation, a simple loop of wire, with a small air gap at one point. Just as the current in his first coil gave rise to radiation, so the radiation produced ought to give rise to a current in the second coil. Sure enough, Hertz was able to detect small sparks jumping across the gap in his detector coil.

By moving his detector coil to various parts of the room, Hertz could tell the shape of the waves by the intensity of the spark formation and could calculate a wavelength of 2.2 feet. This is a million times the size of a wavelength of ordinary light. He also managed to show that the waves were electromagnetic in nature. At first these long-wave radiations were called *Hertzian waves,* but later the name *radio waves* came into use instead.

In this way, Hertz verified the usefulness of Maxwell's equations and showed that light was only a tiny section of the electromagnetic spectrum.

LE CHÂTELIER'S PRINCIPLE

The French chemist Henri-Louis Le Châtelier (1850–1936) stated a rule in 1888 that became known as *Le Châtelier's principle.* The rule is this: Every change in one of the factors of an equilibrium brings about a rearrangement of the system in such a direction as to minimize the original change.

If a system in equilibrium is placed under increased pressure, for instance, the system rearranges itself to take up less room, so as to minimize the increased pressure. Again, if the temperature is raised, the system undergoes a change that absorbs some of the additional heat, so as to minimize the rise in temperature.

This very general statement included the law of mass action of Guldberg and Waage (see 1867) and fit in well with Gibbs's chemical thermodynamics (see 1876). It also worked as a guidepost to direct scientists in how best to bring about desired changes in a system.

CHROMOSOME

Flemming's book describing chromatin and the changes it underwent during cell division had appeared six years before (see 1882).

In 1888 the German anatomist Heinrich Wilhelm Gottfried von Waldeyer-Hartz (1836–1921) suggested that the stubby threads of chromatin that appeared during cell division be called *chromosomes.* This has become one of those scientific terms that is well known to the public at large.

GREENLAND ICE CAP

Though the Greenland coast had first been seen by Europeans nine centuries before, no one up to this point had explored the interior. It was the last significant Arctic landmass that remained to be probed.

In 1888 the Norwegian explorer Fridtjof Nansen (1861–1930) landed on the eastern shore of Greenland with five others. All six managed to cross to the inhabited western shore in a six-week trek. It was the first time Greenland had ever been crossed.

The crossing seemed to indicate that the entire interior of Greenland was buried under a huge ice cap, a last remnant of the Ice Age. That one ice cap is now known to contain about 8 percent of all the ice in the world.

HOME PHOTOGRAPHY

Photography was now about half a century old, yet it remained almost entirely the province of experts and specialists, because it took considerable knowledge and skill to take and develop a photograph.

The American inventor George Eastman (1854–1932) simplified the process. He substituted a flexible film for the large and inflexible piece of glass on which photographic emulsion was spread and dried. With a film that could be rolled compactly, cameras could be made much smaller.

In 1888 Eastman developed a camera that weighed only about 2 pounds. He called it the *Kodak,* a meaningless name but one Eastman thought would be catchy and easily remembered. It had the film rolled up in it, and all the owner had to do was to point the camera, press a button, then send the camera to Rochester. Eventually, the photograph and the newly loaded camera were sent back. The Kodak slogan was "You press the button—we do the rest."

There were many improvements to come, but the Kodak started the process that allowed anyone and everyone to take photographs. The camera entered the home and began to make itself universal.

IN ADDITION

Benjamin Harrison (1833–1901) was elected twenty-third president of the United States.

William I of Germany died on March 9, 1888, just before his ninety-first birthday. He was succeeded by his son, who reigned as Frederick III (1831–1888) but who died of throat cancer on June 15, to be succeeded in turn by his son, who reigned as William II (1859–1941).

The electric chair became part of the American scene in 1888 when New York State adopted electrocution as the method of executing criminals.

The "Blizzard of '88" struck New York City and other parts of the Atlantic coast between March 11 and 14, 1888. New York was isolated from the world for days, with snowdrifts up to 20 feet high.

The Irish inventor John Robert Gregg (1867–1948) devised a method of shorthand in 1888, a system of abbreviated writing that made it possible to transcribe words as quickly as they could be spoken. There were other systems in use, but *Gregg shorthand* replaced them all and became the indispensable tool of stenographers for three-quarters of a century.

1889

NEURON THEORY

Of all the cells, the nerve cells seem most complex, and of all the organs and systems of organs, the brain and nervous system seem most complex. Moreover, of all the parts of a human body, the brain and nervous system are, or should be, the most interesting, since it is they that make us human.

Waldeyer-Hartz (see 1888) was the first to maintain that the nervous system was built out of separate cells and their delicate extensions. The delicate extensions, he pointed out, approached each other closely but did not actually meet, much less join, so that the nerve cells remained separate. He called the nerve cells *neurons,* and his thesis that the nervous system is composed of separate neurons is the *neuron theory.*

An Italian histologist, Camillo Golgi (1843 or 1844–1926), fifteen years before, had devised a system of staining with silver compounds that brought out the structure of neurons in fine detail. Using this stain he could demonstrate that Waldeyer-Hartz's views were correct. He could show that fine processes did indeed issue from the neurons and that those from one neuron approached but did not touch those of neighboring neurons. The tiny gaps between one neuron and the next are called *synapses* (oddly enough, from a Greek word meaning "union," which they appeared to be at a casual glance but were not in actuality).

A Spanish histologist, Santiago Ramón y Cajal (1852–1934), improved on Golgi's stain, and by 1889 had worked out the cellular structure of the brain and spinal cord in detail, firmly establishing the neuron theory.

For their work on the neuron theory, Golgi and Ramón y Cajal shared the Nobel Prize in medicine and physiology in 1906.

TETANUS

The modernization of Japan included an orientation of Japanese scholars toward western science, for which they showed a ready aptitude. (This is true of scholars in all nations. There is no such thing as "western" science in the sense that it can only come out of western minds.)

Thus, a Japanese bacteriologist, Shibasaburo Kitasato (1856–1931), came to Germany to study under Koch (see 1876) and in 1889 isolated the bacillus that caused tetanus. (After he returned to Japan, he isolated the microorganism that caused bubonic plague and another that caused dysentery.)

AXIOMATICS

The work of Lobachevsky and others in non-Euclidean geometry (see 1826) had highlighted the importance of choosing different systems of axioms to develop different types of geometry. The work of Boole on symbolic logic (see 1847) seemed to offer a tool for testing the necessary axioms of mathematics generally.

In 1889 an Italian mathematician, Giuseppe Peano (1858–1932), published *A Logical Exposition of the Principles of Geometry,* in which he applied symbolic logic to the fundamentals of mathematics. He built up a system beginning with undefined concepts for *zero, number,* and *successor,* and developed them symbolically into the arithmetic on which all mathematics is based.

ENERGY OF ACTIVATION

It was the experience of early humans that fires were hard to start but that once started, they kept right on going with no further trouble as long as they were fed fuel.

This is also true of many chemical reactions. Nevertheless, a reaction that ordinarily yields energy when it proceeds will not take place spontaneously until energy is added to it. The reaction has to be *activated,* by breaking molecules apart, perhaps, so that individual atoms or molecular fragments can react with each other more freely. The energy required is called the *energy of activation.*

Once the reaction starts, it yields energy that can be used to activate neighboring parts of the initial substance. Thus, when you heat a hydrogen and oxygen mixture and a small amount begins to react, the energy it produces will cause a wave of reaction to spread to other parts of the mixture so rapidly that it explodes. This is a *chain reaction,* since each step leads to the next, as each link in a chain leads to the next.

Arrhenius (see 1884) analyzed the energy activation concept systematically in 1889 and brought about new understanding of chemical reactions, of chain reactions, and of explosions.

SPECTROSCOPIC BINARIES

Binary stars had been known since Herschel had discovered them nearly a century before (see 1781). However, they could be detected optically only if they were far enough apart to be seen as separate objects in a telescope. What if the pair were so far away or so close together (or both) that they could not be distinguished?

In 1889 the American astronomer Edward Charles Pickering (1846–1919) noted that Mizar, the middle star in the handle of the Big Dipper, had dark lines that divided in two. They moved apart, then came together, joined, and moved apart again in the opposite direction. He suspected that what he was viewing was a binary star with components so close together they could not be seen separately.

However, as they circled each other in the plane of his line of sight, one was advancing and the other receding, so that the former showed a violet shift and the latter a red shift. After they had swung about each other, the one that was receding began to advance and vice versa.

His assistant, Antonia Caetana Maury (1866–1952), worked out the period of this *spectroscopic binary*—104 days—and then went on to discover a second such star, Beta Aurigae, with a period of 4 days.

The following year, a German astronomer,

Hermann Karl Vogel (1842–1907), independently discovered spectroscopic binaries.

MERCURY'S ROTATION

Schiaparelli, who had been the first to see the markings on Mars that were interpreted as canals (see 1877), went on to study the markings on Mercury. Mercury was a smaller world than Mars and farther away. It was often lost in the Sun's glare, and when it could be seen best, only a crescent phase was visible. Nevertheless, Schiaparelli did make out some markings, and since he always saw the same markings when the planet was in a particular position, he concluded, in 1889, that Mercury always presented the same side to the Sun.

Since Mercury was close enough to the Sun for tidal influences to produce this effect, Schiaparelli's suggestion was accepted without serious question for the next three-quarters of a century.

CORDITE

For four centuries, gunpowder had ruled the battlefield. It produced a smoke and a stench, however, and battlefields hidden under a thickening pall of gunpowder smoke made adequate generalship difficult.

An English chemist, Frederick Augustus Abel (1827–1902), worked on the problem of developing explosives that would not produce smoke. In 1889, with the help of another British chemist, James Dewar (1842–1923), he developed *cordite.*

Cordite was a mixture of nitroglycerine (see 1847) and nitrocellulose (see 1834) to which some petroleum jelly was added. The resulting gelatinous mass could be squirted out into cords (hence the name of the material), which after careful drying could be measured out in precise quantities.

The use of cordite (and other such smokeless explosives) freed the battlefield from its cloud. While it might seem small comfort that the scene of carnage and atrocity could be seen clearly, it was important militarily, for it allowed generals to survey a battle's progress and better avoid the casualties that come about when soldiers blunder about uselessly.

MOTION PICTURES

Once photography had been invented (see 1839), it seemed natural to use it to simulate motion. If photographs are taken of a moving object in quick succession and flashed before the eyes rapidly, the eye retains an image of each for a brief while, so that all the flashings melt together and are not seen as separate images. Instead, the object photographed seems to move.

Such devices were constructed, but generally they showed very brief movements that might be repeated over and over again. Primitive as such devices were, they amused the audience.

Edison thought of a way to extend such illusions and perpetuate them for a considerable length of time. He used a strip of film, of the type brought into production by Eastman (see 1888), and took a series of closely spaced photographs of a moving object, which he placed along its length. The film could then be moved by means of perforations along the sides over sprocket wheels, which could push the film in front of a flashing light at a carefully regulated speed.

In effect, Edison had invented motion pictures. With a steady stream of improvements, they developed into a giant industry as indispensable to modern life as the automobile.

IN ADDITION

In the United States, North Dakota, South Dakota, Montana, and Washington entered the Union, which now consisted of forty-two states. Oklahoma, which had been reserved for Indian settlement (and was thus called Indian Territory), was opened to whites on April 22, 1889. They entered in a mad race for land.

In Austria, the Archduke Rudolf (1858–1889), only son of Emperor Francis Joseph I and heir to the throne, was found dead in his hunting lodge, along with his mistress, on January 30, 1889. Francis Joseph's nephew, Francis Ferdinand (1863–1914), became heir.

In Brazil, Emperor Pedro II (1825–1891), who had been reigning for forty-nine years, was deposed. Brazil, which for sixty-seven years had been the only monarchy in Latin America, became a republic.

The Eiffel Tower, designed by the French engineer Alexandre-Gustave Eiffel (1832–1923), was completed in Paris on May 6, 1889. It was the tallest structure in the world for nearly half a century and became the symbol of Paris. On September 27, 1889, New York completed its first skyscraper, which was thirteen stories tall.

1890

ANTITOXIN

A microorganism itself may not be the immediate cause of a disease, but some substance it produces may be (a *toxin,* from a Greek word for "poison"). An organism stricken with the disease then produces a substance (an *antitoxin*) capable of neutralizing the toxin. If the organism recovers, the antitoxin present in its bloodstream confers immunity to the disease thereafter.

A German bacteriologist, Emil Adolf von Behring (1854–1917), decided in 1890 that it might be possible to produce an immunity against tetanus, in an animal, by injecting into it graded doses of blood serum from another animal suffering from tetanus. Enough of the serum would be given to induce antitoxin formation, but not enough to cause it to die.

It was found that the animal thus immunized could then be used as a source of antitoxin that could confer at least temporary immunity on still another animal, or in a human being.

Behring tried this also in the case of diphtheria, a disease that was in those days common among children and often fatal. The diphtheria antitoxin, once marketed, not only conferred a certain immunity, but helped defeat the disease even after it had established itself.

For this work, Behring received the Nobel Prize for physiology and medicine in 1901, the first year in which Nobel Prizes were awarded.

JAVA MAN

The skeletal remains of Neanderthal man had been uncovered thirty-four years earlier (see 1856), but although the Neanderthals possessed some primitive features, they had brains as large as our own. A truly primitive human ancestor had not yet been discovered.

By this time, though, the anthropoid apes were well known, and a Dutch paleontologist, Marie Eugène François Thomas Dubois (1858–1941), believed that primitive human beings would be found where the anthropoid apes now were: in sections of Africa and southeast Asia. He could not easily reach Africa, but the Indonesian islands were Dutch territory, and Dubois, who was serving in the army, wangled an assignment to Java.

He had incredible luck, for in 1890, not long after reaching Java, he discovered a skullcap, a thighbone, and two teeth of what was undoubtedly the most primitive humanlike organism (hominid) that had yet been found. It was clear, for instance, that the brain was only three-fifths the size of a modern human being's.

Dubois called the organism *Pithecanthropus erectus* (erect ape-human), since the thighbone was sufficiently human to signify that the hominid had walked erect as we do. The existence of this hominid was a powerful argument in favor of evolution.

SPECTROHELIOGRAPH

For three-quarters of a century, the solar spectrum had been studied in increasing detail, but always the Sun had been photographed by light from the entire spectrum.

In 1890 an American astronomer, George Ellery Hale (1868–1938), perfected and put into use the *spectroheliograph,* a device that made it possible to photograph the light of a small band of wavelengths of the Sun. Thus he was able to photograph the wavelengths around a line emitted by calcium and the result was a clear indication of the distribution of calcium in the solar atmosphere. It now became possible to study the chemistry of the Sun's outermost layer in considerable detail.

SURGICAL GLOVES

As the danger of infection became better and better understood, surgeons went to increasing lengths to prevent it, and in some cases very simple precautions turned out to be enormously useful.

An American surgeon, William Stewart Halsted (1852–1922), suggested that nurses wear rubber gloves to protect their hands against contact dermatitis. It then occurred to him that rubber gloves were more easily sterilized than human hands were, and that if the gloves were thin enough, they would not seriously cut down on the hand's delicacy of touch.

In 1890 Halsted became the first surgeon of importance to wear rubber gloves during an operation, and this innovation was rapidly adopted. It marked the transition from *antiseptic surgery* (killing the germs that were present) to *aseptic surgery* (in which the germs weren't allowed to get there in the first place).

Also in 1890 Halsted perfected the technique of radical mastectomy (cutting off the breast and the muscles underneath) to prevent the spread of breast cancer to other parts of the body, where it would surely be fatal.

IN ADDITION

On March 18, 1890, William II of Germany forced Bismarck to retire after serving as the virtual ruler of Germany for a quarter of a century. Bismarck was seventy-five at the time and could not rule forever, and William II wished to direct the nation himself. Since William II's abilities and intellect were quite limited, he proceeded to lead his nation into disaster.

William III of the Netherlands (1817–1890) died on November 23, 1890, after a forty-one-year reign, and was succeeded by his daughter, who reigned as Wilhelmina (1880–1962).

In the United States, Idaho and Wyoming were admitted as the forty-third and forty-fourth states of the Union. Sioux lands in South Dakota were taken over by the American government, and when the Sioux tried to resist, they were massacred at the Battle of Wounded Knee on December 29, 1890. This was the last whisper of the Indian wars, and it also marked the end of the *frontier* in the main territory of the United States.

The population of the United States reached 63 million in 1890.

1891

ASTEROID PHOTOGRAPHY

It had been almost a century now since the first asteroid was discovered by Piazzi (see 1802), and others had been discovered in astonishing numbers. By 1891, 322 asteroids had been discovered and their orbits calculated.

Each one, however, had been discovered by eye; an object would be seen that looked like a rather dim star but would be shifting position against the starry background. If it moved at a certain rate, it was almost certainly an asteroid.

In 1891 it occurred to a German astronomer, Maximilian Franz Joseph Cornelius Wolf (1863–1932), to make the discoveries by photography. If a telescope is set to turning in time with the movement of the vault of the sky (a movement that arises because of the rotational motion of the Earth), then all the stars in the telescopic view will show up on the film as sharp points. An asteroid, however, will move with respect to the stars and so will show up as a small dash. The object responsible for the dash can then be put under observation and its orbit eventually calculated.

In this fashion, Wolf discovered the 323rd asteroid, which he named Brucia, and went on to discover others. In the course of his life, he discovered 500 asteroids. Nowadays, the orbits of nearly 2000 asteroids are known, and it is estimated that there may be as many as 100,000 asteroids that are at least a mile across.

GRAVITATIONAL AND INERTIAL MASS

Newton had defined mass in terms of the amount of acceleration produced in a body through the application of a force of a given magnitude. This is called the *inertial mass*, because the greater the mass of a body, the less acceleration a given force produces and the greater the inertia (the resistance to a change in velocity).

Newton also found that the intensity of an object's gravitational field at a given distance depended upon its mass. That is the *gravitational mass*.

These two kinds of mass were determined by two entirely different kinds of observations and would seem to have no necessary connection, yet the mass determined by inertial effects always seemed to be the same as the mass determined by gravitational effects.

A Hungarian physicist, Roland Eötvös (1848–1919), saw that if the gravitational mass and the inertial mass were truly identical, then objects in a given gravitational field would always drop (in a vacuum) at the same rate, regardless of mass. He made particularly delicate measurements in 1891 and found that objects of different mass fell at the same rate within five parts per billion. If there was a difference between the two kinds of mass, it was extremely tiny.

This measurement turned out to be very important later in the development of a new and better way of looking at gravity.

FUNDAMENTAL UNIT OF ELECTRICITY

Arrhenius's theory of ionic dissociation (see 1884) made it look as though atoms or groups of atoms could carry electric charges. What's more, it looked as though atoms or atom groups would carry charges of different sizes that were related to each other in ratios of exact whole numbers.

An Irish physicist, George Johnstone Stoney (1826–1911), suggested that electricity existed in fundamental particles, as matter did, and that all these particles carried the same electric charge. A particular atom or atom group would then carry one of these particles, or two or three, and that was why the ratios would be in whole numbers.

He suggested in 1891 that the fundamental particle be called an *electron*. This didn't seem to make much of an impression at the time, but four years later it was to come into its own.

CARBORUNDUM

People from the earliest times have needed *abrasives;* that is, hard substances that can grind down softer substances, removing unevennesses and imparting a smooth polish to the surface. The diamond is made up of carbon atoms symmetrically placed in tetrahedral patterns, and they are very close to each other as well since the carbon atoms are so small. Therefore, the atoms in diamonds cling together more tightly than in any other substance, so that diamond is by far the hardest material known. Perfect diamonds are too beautiful to use for anything other than show and ornamentation, but imperfect diamonds, or diamond dust, can be used as abrasives. Even imperfect diamonds, however, are in short supply.

Graphite is also made up of carbon atoms, but in a different, looser arrangement. It is so far from being an abrasive that it can be used as a lubricant. By Le Châtelier's principle (see 1888), however, if graphite is placed under great pressure, its atoms should take up a more compact arrangement and it should become diamond (as it does under great pressure deep underground).

Various scientists had tried to supply the necessary pressure, but up to this time none had succeeded. An American inventor, Edward Goodrich Acheson (1856–1931), made the attempt, too.

He did not succeed in forming diamond either, but in 1891, he found that when he heated carbon strongly with clay, he obtained what he eventually found to be a compound of silicon and carbon, or *silicon carbide.* Silicon is much like carbon in its properties, so that the atoms of silicon carbide have a diamond arrangement with a silicon atom taking the place of every other carbon. Silicon is a larger atom and doesn't hold together quite as tightly as carbon atoms do, so silicon carbide is not quite as hard as diamond. Still, it was harder than anything else known at the time, and much cheaper than diamond.

Acheson called the new compound *carborundum,* and it became extremely useful in industry as an abrasive.

GLIDERS

Four decades had passed since Cayley had built the first glider capable of carrying a human being (see 1853). Now a German aeronautical engineer, Otto Lilienthal (1848–1896), made them into things of grace and ability. As early as 1877 he had shown that curved wings were superior to flat wings as far as gliding was concerned. In 1891 he launched himself on his first glide. He died a few years later in a crash landing, but he made gliding popular, and it turned out to be not too long a step from a glider to an airplane.

1892

AMALTHEA

Galileo had discovered the four large satellites of Jupiter (see 1610). In the interval since, no other Jovian satellite had been discovered, even though Saturn, a smaller and more distant planet, was known to have eight satellites by this time.

The American astronomer Edward Emerson Barnard (1857–1923) reasoned, as Hall had in connection with Mars (see 1877), that a fifth satellite, if it existed, would have to be small and close to Jupiter. He searched the neighborhood of Jupiter and in 1892 spotted a satellite only 112,500 miles from Jupiter's center and only 68,000 miles above Jupiter's cloud surface. We now know it to be only 125 miles in diameter.

It was variously called *Barnard's satellite* and *Jupiter V* (because it was the fifth satellite to be discovered), but the French astronomer Camille Flammarion (1842–1925) suggested it be named *Amalthea*, after the goat (or nymph) that served as wet-nurse for Jupiter (Zeus) during the god's infancy.

Amalthea was the twenty-first satellite to be discovered and the last to be discovered without photography.

LIGHT PRESSURE

From Maxwell's equations (see 1865) one could deduce that light ought to exert a pressure, albeit a very feeble one. The Russian physicist Pyotr Nikolayevich Lebedev (1866–1912) checked the point by making use of very light mirrors in a vacuum. In 1892 he was able to observe and measure the pressure exerted by light.

FITZGERALD CONTRACTION

For five years the negative results of the Michelson-Morley experiment had puzzled physicists. The Irish physicist George Francis FitzGerald (1851–1901) offered an explanation in 1892.

He suggested that distances contracted with speed. If a light source sped along toward point A at a certain velocity, then the light it emitted in the direction of that velocity had to cover a slightly smaller distance to reach point A than light emitted in other directions. The faster the source moved, the shorter the distance light would have to travel to reach a point that would seem to be a fixed distance away to a motionless observer. That change in distance would keep the light moving in different directions in phase, so that there would be no interference fringes when it rejoined. Furthermore, if the source traveled at the speed of light, then the distance from the source to any other point in the direction of travel would be zero, because a speed faster than light was impossible.

FitzGerald derived a simple equation, involving the ratio of the speed of the light source to the speed of light, that would reduce distances in the direction of travel just enough to produce a negative result in the Michelson-Morley experiment. This was called the *FitzGerald contraction*, but it seemed *ad hoc* (that is, an explanation made up just to account for a particular observation and nothing else). It would take thirteen more years for a more fundamental solution to be found.

DEWAR FLASK

Heat can be transported from one place to another in three ways: by conduction (travel through matter), convection (carriage by the movement of matter itself, as by air and water currents), and radiation.

Of these, only radiation can transport heat in a vacuum. Realizing this, Dewar (see 1889) in 1892 constructed a double-walled flask with a vacuum between the walls. He was interested in low-temperature work, and very cold liquids such as liquid nitrogen, if kept in such a double-walled flask, would gain heat from the outside only slowly. Dewar cut down on transport by radiation even more by covering the interior of the double walls with a mirror surface so that radiation would be reflected rather than absorbed.

These were called *Dewar flasks* and proved indispensable to low-temperature work. They also entered the commercial home market, where they sold as *thermos bottles* for keeping cold drinks cold and hot drinks hot.

IN ADDITION

Ex-president Grover Cleveland was elected twenty-fourth president of the United States. This was the only time in American history that a man was elected president for two nonconsecutive terms.

The first automobile constructed in the United States appeared in Springfield, Massachusetts. In France, the first automobile equipped with pneumatic rubber tires was built.

1893

PSYCHOANALYSIS

Breuer had started to use hypnotism in the treatment of mental diseases such as hysteria (see 1880). The method had later been taken up by Freud (see 1884), but he eventually abandoned hypnotism for free association, allowing the patient to talk randomly and at will with a minimum of guidance. In this fashion, the patient was gradually put off guard, and matters came to be revealed that in ordinary circumstances would have been kept secret from the patient's conscious mind.

The advantage of this method over hypnotism was that the patient was aware of what was going on at all times and did not have to be informed afterward of what had been said.

In 1893 Freud and Breuer published *The Psychic Mechanism of Hysterical Phenomena.* This is considered to have laid the foundations of the medical technique of *psychoanalysis.*

CATTLE FEVER

In 1893 the American pathologist Theobald Smith (1859–1934) reported that Texas cattle fever was caused by a protozoan parasite, as had been found to be true of malaria thirteen years before.

Smith went on to show that the parasite was spread from infected animals to healthy ones by blood-sucking ticks. This was the first definite indication that disease could be spread by an arthropod. Ticks were among the arachnids (the spider family), but it would soon be shown that blood-sucking insects could also be responsible for the spread of disease.

WAVELENGTHS AND TEMPERATURE

At any temperature above absolute zero, objects give off electromagnetic radiation. In general, very long wavelengths are rare, as are very short wavelengths. Somewhere in between there is a peak wavelength, which is radiated to a greater extent than any other.

The German physicist Wilhelm Wien (1864–1928) demonstrated this by actually measuring the wavelengths. In 1893 he was able to show that the peak wavelength varies inversely with absolute temperature. As the temperature rises, not only does the total amount of radiation increase, in line with Stefan's finding (see 1879), but the peak wavelength decreases.

Thus, warm bodies radiate chiefly in the infrared, of which we are unaware except as a feeling of warmth. As temperature goes up, the peak slides toward the visible red, and eventually enough red is produced to give the body a deep red glow. It is then "red-hot." With further temperature rise, the red brightens, grows orange, yellowish, and then white, by which time the peak is in the yellow and all the wavelengths of light are well represented, as in the case of the Sun.

Some stars are so hot that the peak wavelength is in the ultraviolet. Such stars gleam with an intense blue-white light.

For his work on radiation, Wien received the Nobel Prize in physics in 1911.

ALTERNATING CURRENT

Tesla had made alternating current useful (see 1883). By 1893 a German-born American electrical engineer, Charles Proteus Steinmetz (1865–1923), had worked out the intricacies of alternating current circuitry in complete mathematical detail, making use of complex numbers.

This made it possible to design alternating current equipment with increased efficiency. Steinmetz's mathematics spread among the electrical engineering profession and completed the victory of alternating current over direct current. (The house current we draw on whenever we plug an electrical device into a wall socket is invariably alternating current.)

ARCTIC OCEAN

Nansen, who had crossed Greenland (see 1888), now made ready to explore the Arctic Ocean.

He designed a strong ship that would be lifted, rather than crushed, when the ocean about it froze. He named the ship *Fram (Forward)* and set sail in 1893 with thirteen men aboard. His idea was to let himself be frozen in by the sea ice and carried along with it in its slow swirl about the Arctic Ocean, perhaps reaching the North Pole itself.

He remained on the ship for a year and a half, and although he never reached the North Pole, he got closer to it (86.23 degrees North) than anyone before him ever had.

IN ADDITION

Without Bismarck at hand to guide him, William II of Germany allowed the alliance with Russia to lapse. France seized the chance at once, and in the last days of 1893, negotiations between France and Russia were clearly producing a military alliance aimed primarily against Germany.

The Hawaiian Islands had been a monarchy since 1795, but on January 14, 1893, American settlers overthrew the queen of Hawaii and established a provisional government. Attempts to secure annexation by the United States failed, however.

1894

ARGON

Ever since Prout had suggested that all atoms were built up of hydrogen atoms (see 1815), chemists had been checking the atomic weights of various elements with greater and greater accuracy. The fact that so many atomic weights were *not* multiples of hydrogen's seemed to disprove Prout's hypothesis.

Twelve years before, for instance, the British physicist John William Strutt, Lord Rayleigh (1842–1919), had shown that the atomic weight of oxygen, although usually considered to be 16 times the atomic weight of hydrogen, is actually 15.882 times the atomic weight of hydrogen.

Rayleigh went on to measure the atomic weight of nitrogen and then encountered a puzzle. Whereas oxygen always had the same atomic weight no matter how it was prepared, nitrogen did not. Nitrogen prepared from the atmosphere consistently showed a slightly higher atomic weight than nitrogen prepared from a variety of nitrogen-containing compounds.

Rayleigh could not find a suitable explanation for this and wrote a letter to the journal *Nature,* asking for suggestions. The British chemist William Ramsay (1852–1916) rose to the challenge. He remembered having read that Cavendish (see 1766) had tried to combine the nitrogen of air with oxygen and had found that a small bubble of gas remained behind, which would simply not combine with oxygen. Cavendish had thought there might be some small quantity of gas in the atmosphere that was more dense than nitrogen, and more inert, too, but had not pursued the matter.

Ramsay repeated the experiment, and he too found himself with a small bubble of gas left over. He had spectroscopic techniques, however, which Cavendish had not had. Ramsay heated the bubble of gas, studied the spectral lines it emitted, and found them to be in positions that fitted no known element.

Clearly here was a hitherto-unknown gaseous element, which made up about 1 percent of the atmosphere. It was completely inert and would not react with any substance. It was also denser than nitrogen. The presence of this new gas as an impurity in the nitrogen obtained from air gave the nitrogen an abnormally high atomic weight, whereas nitrogen obtained from chemicals without any admixture of this impurity gave the true atomic weight.

The discovery was announced on August 13, 1894, and the new gas was named *argon*, from the Greek word for "inert." As a result, Rayleigh received the Nobel Prize in physics and Ramsay the Nobel Prize in chemistry in 1904.

IN ADDITION

Japan and China were at odds over Korea, which lay between the two. Korea, although nominally independent, had been more or less under Chinese cultural and political domination through most of its history. Aggressive moves by Japan forced a war with Korea on July 27, 1894, and with China on August 1. Before the end of the year, the modernized Japanese army had defeated China in two battles and was clearly going to win the war.

In France, an army officer, Alfred Dreyfus (1859–1935), was accused of selling military secrets to the Germans. Dreyfus, who was Jewish, was convicted on December 22, 1894, and sentenced to exile for life on Devil's Island off French Guiana. This was accompanied by a wave of anti-Semitism in France.

A Hungarian-Jewish journalist, Theodor Herzl (1860–1904), began to advocate a new Jewish state in Palestine, the ancient home of the Jewish people—thus founding the Zionist movement.

In Russia, Alexander III died on November 1, 1894, and was succeeded by his son, who reigned as Nicholas II (1868–1918).

On August 7, 1894, the United States recognized the Republic of Hawaii.

1895

X RAYS

Work on cathode rays by Goldstein (see 1876) and by Crookes (see 1861) had come to interest a number of physicists in the subject. One of them was a German physicist, Wilhelm Conrad Röntgen (1845–1923), who was particularly interested in the ability of cathode rays to make materials fluoresce.

He placed certain chemicals, known to fluoresce easily, inside a cathode ray tube, surrounded it by dark paper, and darkened the room to observe the pale fluorescence that would result.

On November 5, 1895, he set his cathode-ray tube to working, and in the dimness, a flash of light that did not come from the tube caught his eye. A sheet of paper coated with barium platinocyanide (one of the chemicals he was planning to use) was glowing. The glow ceased when the cathode-ray tube was turned off. The coated paper glowed even in the next room once the cathode-ray tube was turned on.

Radiation was clearly emerging from the tube when the cathode rays were streaming, and was penetrating matter to some extent. Röntgen didn't know what the radiation might be, so he referred to it as *X rays, x* being the usual symbol for an unknown quantity in mathematics. He published his findings on December 18, 1895.

The news of these X rays roused the world of physics to a furor not seen since Ørsted had discovered electromagnetism (see 1820). So much work was done, and so many revolutionary findings were made (most of them as a direct result of Röntgen's finding) that Röntgen is often viewed as having set off a second Scientific Revolution, as Copernicus had set off the first (see 1543).

For this work, Röntgen received the Nobel Prize in physics (the first one) in 1901.

CATHODE RAY PARTICLES

There were still some who thought that cathode rays were a wave form and doubted Crooke's observation that they carried electric charges.

Hertz, the discoverer of radio waves (see 1888), had found that cathode rays could pass through thin sheets of metal, and this seemed to favor their being a wave form. In 1892 one of Hertz's assistants, the German physicist Philipp Eduard Anton Lenard (1862–1947), devised a cathode ray tube with a thin aluminum "window" through which cathode rays could emerge into open air. Lenard studied these open-air cathode rays and received a Nobel Prize in physics in 1905 as a result. He, too, thought cathode rays to be waves.

In 1895, however, a French physicist, Jean-Baptiste Perrin (1870–1942), showed that when cathode rays were made to impinge upon a cylinder, the cylinder gradually gained a large negative charge. That finally settled the matter. Cathode rays were streams of negatively charged particles, and no one has doubted it since.

VELOCITY AND MASS

The negative results of the Michelson-Morley experiment (see 1887) still attracted the attention of physicists. The Dutch physicist Hendrik Antoon Lorentz (1853–1928) came to the same conclusions that FitzGerald had come to concerning the shortening of distance with speed (see 1892) but went farther. It seemed to Lorentz that mass would have to increase with velocity and do so at a steadily increasing rate. At a speed of 160,000 miles per second, the mass of any object should have doubled, and at the speed of light (a little over 186,000 miles per second), it should become infinite.

This idea made the speed of light appear to be an absolute maximum. As a result of Lorentz's work, physicists often speak of the *Lorentz-FitzGerald contraction*.

HELIUM ON EARTH

The time had passed when a new element could be discovered and then treated as though it existed all by itself. Mendeleyev with his periodic table (see 1869) had shown that elements existed in families.

Argon, discovered the year before by Rayleigh and Ramsay (see 1894), did not fit in with any existing family, but it seemed to be in the neighborhood of chlorine and potassium as far as its atomic weight was concerned. Since the periodic table was based largely on valence (see 1852), it made sense to put argon between chlorine and potassium. Chlorine and potassium each had a valence of 1, and since argon formed no combinations with other atoms at all, it had to have a valence of 0. The valence progression of 1, 0, 1 made sense, and argon would be the first member of a new family of elements of 0 valence.

In that case, where were the others? Ramsay made it his business to look for them. In 1895 he learned that in America, samples of a gas taken for nitrogen had been obtained from a uranium mineral. That looked hopeful, for the 0-valence gases could be mistaken for the rather inert nitrogen but for no other gas.

Ramsay repeated the work with a uranium mineral and got a gas that resembled nitrogen in its inertness. He tested the gas spectroscopically, however, and it did not produce the spectral lines of nitrogen. Instead it produced the spectral lines that Ramsay recognized as those detected by Janssen in sunlight (see 1868).

Janssen's solar element had been named *helium,* and now Ramsay had identified it on Earth, and found that it would lie between hydrogen and lithium in the periodic table. As to what helium might be doing in the uranium mineral, that was a puzzle, but the answer was not long in coming.

MAGNETISM AND HEAT

People must have noticed that if a magnet was made red hot it would lose its magnetism. Curie (see 1880) showed that this was not a gradual effect. There was a specific temperature (now called the *Curie temperature*) at which the magnetism of iron disappeared. This happens to be 770° C.

Other metals that have the same strong response to magnetism that iron does (and are therefore said to be *ferromagnetic*, from the Latin word for "iron") also have Curie temperatures. For nickel it is 358° C, and for cobalt, 1131° C.

RADIO ANTENNAS

Once Hertz had demonstrated the existence of radio waves (see 1888), the possibility of using such waves in signaling across long distances occurred to a number of people. If one could use such radiation for the purpose, one could do away with total dependence on telegraph wires and cables. Indeed, the British call communications by radio waves *wireless telegraphy* or for short *wireless.* Americans call it *radiotelegraphy* or for short *radio.*

Of course, to make such communication possible, one needs a far better detector than the simple loop Hertz used. In 1890 a French physicist, Édouard-Eugène Branly (1844–1940), invented a detector that consisted of a container of loosely packed metal filings. It ordinarily conducted little current, but it conducted quite a bit when radio waves fell upon it. Using this detector, Branly could detect radio waves 150 yards from the source.

The British physicist Oliver Joseph Lodge (1851–1940) improved the device in 1894, called it a *coherer,* and used it to detect radio waves half a mile from the source. He also sent out the radio waves in dots and dashes so that he could transmit and pick up a message in Morse code.

In 1895 two men made a particularly crucial discovery. They found that a long vertical wire attached to the source, and another attached to the receiver, made the signals much stronger and easier to detect. The long wires were called *antennas,* because they seemed to resemble the long feelers that insects wear on their heads.

One of the discoverers was a Russian physicist, Aleksandr Stepanovich Popov (1859–1905), the other an Italian electrical engineer, Guglielmo Marconi (1874–1937). It was these antennas that really made radio communication possible.

IN ADDITION

The Sino-Japanese war ended on April 17, 1895, with the Treaty of Shimonoseki. The Chinese ceded the island of Taiwan to Japan, which in this way began a career of expansion beyond its home islands that was to continue successfully for nearly half a century. Korea was to be "independent," which meant in actuality that it would be under Japanese domination.

Cuba, which was one of the very few remnants of Spain's once-vast dominions in Latin America, rose in rebellion against Spain in 1895. The rebellion was put down, but it continued to simmer—and to attract American attention.

The independent Boer Republics north of the British dominions at the southern tip of Africa were a continuing irritation to some of the British. Cecil John Rhodes (1853–1902), who had been prime minister of the Cape Colony since 1890, plotted to overthrow the Boers, and sent his friend Leander Starr Jameson (1853–1917) on a raid northward on December 29, 1895. Jameson was captured and imprisoned for a short while, and Rhodes was forced to resign. Nevertheless, the bad feeling provoked by the incident was to have serious consequences.

In St. Petersburg, Russia, Vladimir Ilyich Ulyanov (1870–1924) began to work toward the overthrow of the czars and the establishment of a socialist form of government. Later he would take the pseudonym of *Nikolai Lenin*.

1896

URANIUM RADIATIONS

The discovery of X rays by Röntgen (see 1895) fascinated a French physicist, Antoine-Henri Becquerel (1852–1908), who had been working on fluorescent substances as his father had before him. He wondered if perhaps the radiation given off by fluorescent substances might include X rays.

A fluorescent substance both he and his father had been interested in was potassium uranyl sulfate. In February 1896 Becquerel wrapped photographic film in black paper and put it in sunlight with a crystal of potassium uranyl sulfate upon it. He reasoned that sunlight would make the crystal fluoresce, and any X rays it produced would penetrate the black paper (as ordinary light would not) and fog the photographic plate.

Sure enough the plate was fogged and Becquerel decided that fluorescence did produce X rays. But then came a series of cloudy days and Becquerel could not continue his experiments. He had a fresh plate neatly wrapped in the drawer with a crystal resting upon it, but there was no sunlight to expose it to. Finally, unable to bear doing nothing, he developed the film anyway, just to make sure that nothing happened in the absence of sunlight.

To his amazement, the film was strongly fogged. Whatever radiation was passing through the paper did not depend on either sunlight or fluorescence.

For this discovery, Becquerel received a share of the Nobel Prize in physics in 1903 and rightly so, for it had enormous consequences.

DIETARY DEFICIENCY DISEASES

People in the Dutch East Indies commonly suffered from *beriberi*, which produced weakness and death. Naturally it was assumed to be a germ disease, since Pasteur (see 1862) and others had been so successful at combating disease on that assumption. However, no one could find the germ that caused the disease.

A Dutch physician, Christiaan Eijkman (1858–1930), who had gone to the East Indies to study the disease, was nonplussed. But then an ailment broke out in 1896 among the chickens being used at the laboratory for bacteriological researches. The *chicken polyneuri-*

tis showed symptoms similar to beriberi, and Eijkman was busily studying the disease and checking on its contagiousness—when it suddenly disappeared and all the chickens got well.

Eijkman investigated and found that during the period when the chickens had had the disease, they had been feeding on rice ordinarily meant for the human patients. The disease disappeared when a new cook put them back on commercial chicken feed. Eijkman found he could produce the chicken disease at will when he fed the chickens polished rice. By feeding them unpolished rice, he cured them.

Eijkman was the first to correct a specific disease by diet since Lind had connected citrus fruits and scurvy (see 1747). Although Eijkman missed the point at first, it became clear that beriberi was a dietary deficiency disease. It was caused by the *absence* of some substance (present in the hulls of unpolished rice and not present in polished rice from which the hulls had been removed) that seemed necessary to health in small traces.

For this discovery, Eijkman received a share of the Nobel Prize in physiology and medicine in 1929.

LIGHT AND MAGNETISM

Maxwell had maintained that an oscillating electric charge could produce radiation (see 1865), and Hertz had shown that to be correct with his discovery of radio waves (see 1888). Maxwell had also maintained that light was an electromagnetic radiation, but if so, what was the electric charge that was oscillating and therefore producing it?

Arrhenius had maintained that atoms or groups of atoms could carry electric charges (see 1884), and Lorentz (see 1895) had wondered if it might be electric charges within the atom that did the oscillating. If that were the case, then placing a light source under the influence of a strong magnetic field ought to affect the oscillating charges and introduce changes in the spectral lines.

Lorentz had a student, the Dutch physicist Pieter Zeeman (1865–1943), who undertook the experiment, and indeed, the magnetic field split the spectral lines into three components. This *Zeeman effect* could be used to study the fine details of atomic structure and to yield information on the structure of stars as well.

As a result, Lorentz and Zeeman shared the Nobel Prize in physics in 1902.

FERMENTS AND ENZYMES

Kuhne had suggested that the term *ferments* be restricted to catalysts within living cells while those that could be isolated as nonliving molecules be considered *enzymes* (see 1878).

In 1896 a German chemist, Eduard Buchner (1860–1917), wondered if the enzymes in yeast would work if they were extracted from the living cell.

To do this, he ground up yeast cells with sand until not one of them was left alive. He then filtered the material, obtaining a clear, cell-free liquid. He added sugar to preserve it against bacterial contamination and found that before long carbon dioxide was forming vigorously. The clear extract from the yeast was doing the job of fermenting sugar just as the intact cells would have, and Buchner had demonstrated the opposite of what he had expected.

From this point on, the term *enzymes* was applied to all biochemical catalysts, in or out of a cell, and another bastion of vitalism had fallen.

For this work, Buchner was awarded the Nobel Prize in chemistry in 1907.

ACOUSTICS

In 1896 an American physicist, Wallace Clement Ware Sabine (1868–1919), completed the study of a new lecture room that had been built at Harvard University the previous year. The lecture room had a flaw; the lecturer could not be heard because of excessive reverberation. Sabine had investigated every angle of the problem, even photographing sound waves that were made visible by the changes in light refraction they produced.

As a result of his studies, Sabine founded the science of architectural *acoustics*. In the process he showed how to design a hall that would make the sounds of voice and music clear. For the purpose, he made use of mathematical equations that related absorptivity of sound by various materials and the volume of the room to the amount of reverberation.

IN ADDITION

The Mahdists were still in control of the Sudan, and in 1896 the British sent a punitive expedition south from Egypt under Horatio Herbert Kitchener (1850–1916).

In southern Africa, the British were still feeling the bad results of the Jameson raid. William II of Germany, with characteristic bad judgment, gratuitously sent a telegram of congratulations to the Boer leader, Paul Kruger (1825–1904), on the capture of Jameson. This made Great Britain aware for the first time of possible German hostility.

Italy had worse troubles. It had been advancing southwestward from the Red Sea, penetrating Ethiopia. Some twenty-five thousand Italian troops encountered a much larger force of Ethiopians at Adowa in northeastern Ethiopia on March 1, 1896, and were wiped out. The Ethiopian victory put an end to Italian advance in the region for forty years.

In the United States, William McKinley was elected twenty-fifth president and Utah entered the Union as the forty-fifth state. Gold was discovered in the Klondike region of northwestern Canada near the Alaskan border. Another gold rush began like the one in California a half-century before.

1897

ELECTRON AS SUBATOMIC PARTICLE

By now Perrin's work had convinced everyone that cathode rays consisted of negatively charged particles (see 1895). The last bit of evidence was supplied by the British physicist Joseph John Thomson (1856–1940), who showed that cathode rays could be deflected not only by a magnet but by an electric field.

From the amount of the deflection, Thomson could work out the ratio of the electric charge of the cathode ray particle to its mass. This ratio turned out to be quite high, so that either the charge was high or the mass was low or both. Thomson supposed that the charge would be the unit charge worked out from Faraday's laws of electrolysis (see 1832). If so, the mass of the cathode ray particle would be only a small fraction of the hydrogen atom's, the smallest atom known. (We now know that the particle has a mass only $1/1837$ of hydrogen's.)

Thomson called the particle the *electron*, making use of Stoney's suggested name for the fundamental unit of electric charge (see

1891), since he suspected that the particle carried that fundamental unit. And to be sure, no smaller electric charge has actually been observed since then.

The electron was the first *subatomic particle* (particle smaller than an atom) to be discovered.

URANIUM RADIATION

One of those who followed up instantly on Becquerel's discovery of the radiation from potassium uranyl sulfate (see 1896) was a Polish-born French chemist, Marie Sklodowska Curie (1867–1934). She was the wife of Pierre Curie (see 1880).

In 1897 she made use of her husband's discovery of piezoelectricity to measure the intensity of radiation given off by a variety of uranium compounds. She showed that the intensity was always in proportion to the quantity of uranium present.

This demonstrated that the radiation did not come from the compound as a whole, but from the uranium atom specifically. It was an atomic phenomenon and not a molecular one.

ALPHA RAYS AND BETA RAYS

The radiations given off by uranium were of more than one kind. Some were deflected only slightly in a direction that indicated them to be positively charged. Some were deflected much more sharply in the opposite direction and were therefore negatively charged. Both had to consist of streams of particles, and the particles of the former were clearly the more massive.

The British physicist Ernest Rutherford (1871–1937) noted this in 1897. He called the massive, positively charged radiation *alpha rays,* and the lighter, negatively charged radiation *beta rays,* after the first two letters of the Greek alphabet. Those names have remained in use ever since.

NICKEL AS CATALYST

The metal nickel combines with carbon monoxide to form a compound that is liquid at ordinary temperatures and boils at 43° C. In 1897 the French chemist Paul Sabatier (1854–1941) wondered if other volatile nickel compounds could be formed by combining the metal with other gases. He chose ethylene for his attempt, since like carbon monoxide, it has a double bond.

He failed, but found that some of the ethylene had, in the presence of nickel, changed to ethane. The double bond in ethylene had thus added two hydrogen atoms. That happens easily if a metal such as powdered platinum is used as a catalyst. Sabatier had now shown that powdered nickel, which was much cheaper, would do the same.

Sabatier went on to perfect the technique, and nickel catalysis went on eventually to make possible the formation of edible fats, such as margarine and shortenings, from inedible plant oils, such as cottonseed oil.

For this work Sabatier was awarded a share of the Nobel Prize in chemistry in 1912.

MOSQUITOES AND MALARIA

Smith's discovery of the manner in which ticks served to communicate disease (see 1893) turned the minds of pathologists in the direction of stinging insects as possible disease-spreaders.

The British physician Ronald Ross (1857–1932) devoted himself to collecting, feeding, and dissecting mosquitoes until in 1897 he discovered the malarial parasite that had been identified by Laveran (see 1880) in the anopheles mosquito. This meant that malaria might well be fought by wiping out the breeding places of these mosquitoes, by making use of mosquito netting, and so on.

For this, Ross received the Nobel Prize in physiology and medicine in 1902.

OSCILLOSCOPE

In 1897 a German physicist, Karl Ferdinand Braun (1850–1918), modified the cathode-ray tube in such a way that the spot of green fluorescence formed by the stream of speeding particles shifted in accordance with the electromagnetic field set up by a varying current.

The device was called an *oscilloscope*, because the spot could follow and reveal the oscillations of the field. Braun's device was the ancestor of the present-day television screen.

LARGE REFRACTING TELESCOPE

The first telescope that Galileo used was a refracting telescope, making use of lenses only. In the nearly three centuries since, such telescopes had grown larger and more elaborate.

In 1897 Clark, the discoverer of Sirius's dim companion (see 1844), supervised the construction of a refracting telescope with a lens 40 inches across. It was the largest and best refractor built up to that time, but it had reached the limits of the art. No larger refractor has been built since then, or is likely to be built. All larger telescopes are of the reflecting variety that Newton invented (see 1668).

DIESEL ENGINE

The four-stroke engine invented by Otto (see 1876) used low-boiling gasoline for fuel and ignited the vapor-air mixture with an electric spark.

A German inventor, Rudolf Diesel (1858–1913), tried to eliminate the complexities that resulted from running an electrical system in conjunction with an engine. By 1897 he had perfected a *Diesel engine* that ignited the vapor-air mixture by the heat developed through compression. This allowed him to use higher-boiling fuel such as kerosene, which was cheaper and less inflammable (hence safer) than gasoline.

However, the compression had to be great, so the Diesel engine had to be considerably larger and heavier than the Otto engine if the higher pressures were to be brought about and maintained. Diesel engines therefore found their use in heavy transport vehicles such as trucks, buses, locomotives, and ships.

IN ADDITION

The island of Crete, which was still under Turkish rule, had been in a state of revolt, and conflicting ambitions in the Balkans threatened to embroil Great Britain and Russia in the matter. In the western hemisphere, Cuba continued in a state of revolt against Spain, and the United States was increasingly on the side of Cuba.

When two German missionaries were killed in China's Shantung province, Germany occupied the port of Tsingtao in that province, initiating a new scramble for economic concessions by other European powers.

Queen Victoria of Great Britain celebrated her Diamond Jubilee, the sixtieth anniversary of her accession. Great Britain was now at the very peak of its strength, with a worldwide empire, a navy that controlled the seas, and unrivaled prosperity.

The first Boston marathon race was run on April 19, 1897.

1898

POLONIUM AND RADIUM

Marie and Pierre Curie continued to work on the radiations produced by uranium (see 1897).

In 1898 Marie Curie showed that thorium, another heavy metal, also produced radiations, and she coined the term *radioactivity* for the phenomenon, so that it could be said that both uranium and thorium were radioactive.

She also discovered that although pure uranium compounds were always radioactive only to the extent that uranium was present, some uranium ores produced far more radioactivity than could be accounted for by the uranium present. It seemed to her that the ores must contain other elements (in small quantities or they would have been discovered earlier) that were much more intensely radioactive than uranium.

In July 1898 the Curies detected such an element, which they called *polonium* after Marie Curie's native land. In December 1898 they detected another, which they called *radium* because of its intense radioactivity.

For their work on radioactivity, the Curies shared the Nobel Prize for physics with Becquerel (see 1896) in 1903. For the detection of polonium and radium, Marie Curie (her husband by then being dead) received the Nobel Prize for chemistry in 1911.

NEON, KRYPTON, AND XENON

Polonium and radium were not the only elements discovered in 1898. In the previous four years, Ramsay had discovered argon (see 1894) and helium (see 1895). There had to be several more elements in the zero-valence family, and he searched for them with the assistance of a British chemist, Morris William Travers (1872–1961).

A British inventor, William Hampson (1854–1926), had developed a method for producing liquid air in sizable quantities. He gave some to Ramsay and Travers, who carefully distilled it and, in the argon fraction, discovered *neon* (from the Greek word for "new"), *krypton* (hidden), and *xenon* (stranger). The new elements were all gases and all zero-valence. The five elements (including helium and argon) were called, as a group, the *inert gases,* or more recently, the *noble gases.*

LIQUID HYDROGEN

Although nitrogen and other gases had been liquefied twenty years before, hydrogen still remained recalcitrant. In 1895, however, the German chemist Carl Paul Gottfried von Linde (1842–1934) had developed a technique of cooling gases in stages, by expansion, each cooled fraction being returned to cool another sample, which could be cooled still further by expansion, and so on. In this way, Linde could produce commercial quantities of liquid air.

Dewar (see 1889) adopted Linde's system, improved it, and used it on hydrogen. The result was that in 1898 he liquefied hydrogen at a temperature of only 20° K.

Even so, that did not represent the final victory over gases. The new group of inert gases remained. Those with the higher atomic weights were easily liquefied, to be sure. Even neon, with the next-to-lowest atomic weight, could be liquefied at 27° K.

Helium, however, with the lowest atomic

weight of the inert gases, remained a gas even at liquid-hydrogen temperatures.

PHOEBE

In 1898 the American astronomer William Henry Pickering (1858–1938) discovered a ninth satellite of Saturn, which he named *Phoebe,* after still another of the siblings of Saturn (Cronos) in the Greek myths. It was far more distant from Saturn than the satellites discovered earlier, and it revolved around the planet in a retrograde direction (clockwise rather than counterclockwise, if viewed from high above Saturn's north pole), so that it is considered to be a captured asteroid.

EROS

Since the time of Kepler, it had been thought that, except for the Moon, no sizable object approached Earth more closely than the planet Venus, which at its closest can be only 25,000,000 miles away from us. As for the swarming asteroids, they seemed all to be circling between the orbits of Mars and Jupiter, and none were expected to approach more closely than 35,000,000 miles.

But then, on August 13, 1898, the German astronomer Gustav Witt discovered asteroid number 433. When its orbit was plotted, it proved to be a real stunner. At aphelion (when it is most distant from the Sun), it is well within the asteroid belt. At perihelion, however, it is only 105,000,000 miles from the Sun—well within the orbit of Mars. At its closest to the Earth, the two could be only a little over 14,000,000 miles apart.

Since it approached Earth more closely than Mars or Venus did, Witt gave it the name of the child of Mars and Venus in the Greek myths—*Eros.* That began the habit of giving masculine names to asteroids whose orbits extend outside the asteroid belt.

Since then, other asteroids have been discovered with orbits that intersect that of Mars. All those that can approach Earth more closely than Venus does are called *Earth-grazers.* Eros was the first of these to be discovered and is the largest, being at least 15 miles in its longest diameter.

FILTRABLE VIRUS

Pasteur had not been able to find a microorganism that caused rabies (see 1885). Rather than suppose there was something wrong with the germ theory of disease, however, he suggested that the causative agent existed but was too small to see even by microscope.

Another disease with no known causative agent was tobacco mosaic disease, so called because a mosaic pattern forms on the leaves of the tobacco plant. It was easier to work with a plant disease than with something as virulent as rabies, and in 1892, a Russian botanist, Dmitri Iosifovich Ivanovsky (1864–1920), mashed up infected leaves and passed the fluid through a fine filter that would keep back all bacteria.

The fluid that passed through the filter could still infect healthy tobacco plants with the disease, but Ivanovsky could not bring himself to believe there were causative agents smaller than bacteria and supposed simply that the filters were defective.

In 1898 the Dutch botanist Martinus Willem Beijerinck (1851–1931) did a similar experiment, but he had no hesitation in maintaining that there was a causative agent smaller than bacteria. He did not know what it might be, so he simply called it a *filtrable virus* (*virus* is the Latin word for "poison").

It turned out that a considerable number of human diseases were caused by viruses, including the common cold, influenza, chickenpox, mumps, and poliomyelitis. Such diseases were not as easily handled as those caused by larger organisms such as bacteria and protozoa.

MITOCHONDRIA

As microscopes improved, it was clear that the cell did not consist of a homogeneous fluid but had a complex structure, even outside the nucleus. In 1898 the German histologist Carl Benda (1857–1933) made out tiny bodies in the cytoplasm. He called them *mitochondria,* from Greek words meaning "threads of cartilage," which is what he thought they were. They turned out to be far more than that.

EPINEPHRINE

What we now call the adrenal glands (from Latin words meaning "at the kidney") are small lumps of tissue above each kidney. They first came into prominence in 1855, when a British physician, Thomas Addison (1793–1860), showed that the deterioration of the adrenal glands gave rise to a serious condition (known as *Addison's disease* to this day).

In 1894 a British physiologist, Edward Albert Sharpey-Schafer (1850–1935), showed that a substance extracted from the adrenals would raise the blood pressure if injected into an animal's bloodstream.

An American pharmacologist, John Jacob Abel (1857–1938), was able to study this substance in 1898, and he named it *epinephrine* (from Greek words meaning "above the kidney"). Three years later, a Japanese chemist, Jokichi Takamine (1854–1922), working in the United States, isolated the substance in pure, crystalline form. He called it *adrenaline.* Both names are still used.

Actually epinephrine/adrenaline was the first hormone to be isolated, but the hormone concept had not yet been elaborated.

SUBMARINE

Ships made to go under the water have been a human fancy for centuries, but the first to attempt to build one was a Dutch inventor, Cornelis Jacobszoon Drebbel (1572–1633), who piloted one in the Thames River between 1620 and 1624. An American inventor, David Bushnell (1742–1824), devised simple submarines that were used unsuccessfully against British ships in harbor during the Revolutionary War and the War of 1812.

It was not till 1898, however, that an American mechanical engineer, Simon Lake (1866–1945), succeeded in devising a submarine that could actually go out to sea. In that year, his submarine *Argonaut I* sailed from Norfolk, Virginia, to New York. This marked the true beginning of modern submarines.

IN ADDITION

The American battleship *Maine* sailed to Havana, and there, on February 15, 1898, she blew up, killing 260 men. It seemed impossible to suppose that the Spaniards would have done anything this foolish, but the United States declared war on April 11, despite Spain's frantic efforts to avoid it.

The *Spanish-American War* didn't last long. The American navy, modern and efficient, destroyed the Spanish fleet, and what land battles took place were won by the Americans. The Treaty of Paris ended the war on December 10, 1898. The United States annexed the former Spanish colonies of Puerto Rico, the Philippine Islands, and Guam, though twenty million dollars were paid to Spain for the Philippines. Cuba became independent.

In a matter unconnected with the war, the United States annexed the Republic of Hawaii on July 7, 1898.

In Africa, Kitchener defeated the Mahdists at the Battle of Omdurman on September 2, 1898, and took Khartoum. Meanwhile, though, a French expedition had reached the Nile and occupied

Fashoda, 400 miles south of Khartoum. Kitchener reached it on September 19, and for a while it looked as though those old enemies, Great Britain and France, might go to war again for the first time since Waterloo.

However, Great Britain was directly threatened by Germany's growing military strength and did not wish to get embroiled with France. Nor did France want to do Germany the favor of fighting Great Britain. France therefore ordered the evacuation of Fashoda on November 3.

In France, it became quite obvious that the conviction of Dreyfus had been a miscarriage of justice carried through by corrupt army officers. A pamphlet called *J'Accuse (I Accuse)* by Émile Zola (1840–1902) forced a new trial.

1899

ACTINIUM

There was more in uranium ores than the polonium and radium that the Curies had detected (see 1898). In 1899 the French chemist André-Louis Debierne (1874–1949), a close friend of the Curies, isolated still another element from the ores. He called it *actinium*, from a Greek word for "ray," so that the name was the Greek equivalent of the Latin *radium*.

LOGIC AND GEOMETRY

In 1899 the German mathematician David Hilbert (1862–1943) published *Foundations of Geometry*, in which he proposed a set of axioms for geometry that was the most satisfactory yet. He began with points, lines, and planes as undefined concepts. It was not necessary to define them, merely to describe certain properties they possessed. He also used certain relationships, such as *between*, *parallel*, and *continuous*, without defining them. Again, provided the consequences of using those words were clearly set forth, it didn't matter what they actually meant. Hilbert proved his system of axioms to be self-consistent, and that was crucial.

SOLID HYDROGEN

Dewar, who had liquefied hydrogen the year before, took another step toward absolute zero by producing solid hydrogen in 1899. This brought the lowest temperature ever obtained down to 14° K. At this temperature, all known substances but one were solid. The only exception was helium, which even at this temperature remained gaseous.

IN ADDITION

In southern Africa, the Boer Republics were convinced that the British were planning a takeover, and so they were. The Boer War broke out on October 12, 1899. The British troops on the spot were outnumbered by the Boers, who were well equipped with German arms. Almost at once the British suffered humiliating defeats.

The Philippine Islands, which had fought with the United States against Spanish rule, expected the independence Cuba had received. When they found they were only changing masters, a rebellion arose, led by Emilio Aguinaldo (1869–1964).

The United States, under the guidance of Secretary of State John Milton Hay (1838–1905), proposed an *Open Door Policy* for China. Fearful that the race of the European powers for trade advantages in China might shut out the United States, Hay suggested that trade be free for all and that all powers have a fair and equal chance at the Chinese market.

1900

QUANTA

Kirchhoff had pointed out that a black body (one that absorbed all electromagnetic radiation that fell on it) would radiate at all wavelengths if heated (see 1860). Thus, a hollow body with a small hole in it would absorb all the radiation that entered through the hole, for virtually none of it would be reflected and find its way out again. If such a body was heated, radiation would therefore emerge from the hole at all wavelengths, with very little at the extremes and with a peak at some intermediate value. The higher the temperature, the shorter the wavelength of the peak value.

A number of physicists tried to work out mathematical equations for the distribution of wavelengths in such *black body radiation.* Rayleigh and Wien (see 1894) both advanced equations in 1900, but Rayleigh's worked well only for the long-wavelength half, and Wien's only for the short-wavelength half. Neither one could work out an equation that gave the distribution across the board.

Then a German physicist, Max Karl Ernst Ludwig Planck (1858–1947), produced an equation that did just that. In order to derive that equation, he had to assume that energy was given off not continuously but in discrete pieces, and that the size of the piece was inversely proportional to the wavelength. Thus, since violet light had half the wavelength of red light, violet light would be delivered in pieces that were twice the size, and therefore twice the energy content, of red light.

Planck called the bits of energy *quanta* (singular, *quantum,* a Latin word meaning "how much?"). He worked out the relationship between energy and wavelength (or energy and frequency, since frequency is 1 divided by the wavelength), making use of a very small value called *Planck's constant,* which represents the "graininess" of energy. Since Planck's constant is exceedingly small, energy has a very fine grain indeed, and it is not noticeable in most circumstances, so that the laws of thermodynamics could be deduced as though energy were a continuous fluid without grain. The problem of black body radiation was the first in which the graininess had to be taken into account.

There was no evidence, at first, for the existence of quanta, except for the fact that it made the equation for black body radiation possible. Even Planck himself suspected that quanta might be only a mathematical device that had no physical meaning.

Nevertheless, the *quantum theory,* as it is now called, proved so fundamental that all physics prior to 1900 is called *classical physics* and all physics after 1900 *modern physics.*

For his work, Planck received the Nobel Prize for physics in 1918.

MASS INCREASE

Lorentz had maintained that mass increased with velocity (see 1895). It did not seem at the time that this could ever be checked, for the velocities needed to make the change in mass measurable were too huge to deal with in the laboratory.

As physicists began to study the speeding electrons involved in cathode rays, and in other phenomena, however, it turned out that electrons might be moving at respectable fractions of the speed of light—up to 90 percent in some cases. This meant the mass increase ought to be measurable, since fast electrons, becoming more massive than slow ones, would bend less sharply in response to electromagnetic fields. This turned out to be so, and the increase in mass, measured in 1900 for the first time, matched quite closely the theoretical predictions of Lorentz.

The Lorentz-FitzGerald contraction was borne out, but physicists had to wait five more years before a comprehensive physical theory that explained it was advanced.

BETA PARTICLES

Becquerel, who had discovered uranium radiations (see 1896) continued to study them. The beta rays were clearly composed of negatively charged *beta particles,* and from the way these curved in a magnetic field of a particular strength, they seemed to resemble electrons. In 1900, after a close study of the properties of beta particles, Becquerel decided that they *were* electrons.

Until then, electrons had seemed to be associated only with electric currents, as in the case of cathode rays. Now it seemed they were associated with atoms as well—at least with radioactive atoms.

GAMMA RAYS

In 1900 the French physicist Paul Ulrich Villard (1860–1934), studying the radiations from uranium that Becquerel had discovered, noted that in addition to alpha rays and beta rays, there was some radiation that was totally unaffected by magnets.

This, it was decided, consisted of electromagnetic radiation. Its properties were very much like those of X rays, but it was even more penetrating and therefore even shorter in wavelength. It came to be called *gamma rays,* after the third letter of the Greek alphabet.

RADON

A German physicist, Friedrich Ernst Dorn (1848–1916), studying the radium that Curie had discovered (see 1898), found in 1900 that it gave off not only radiations but a gas that was itself radioactive. The gas was called *radium emanation* at first, but on closer study it turned out to be a noble gas (see 1898), the sixth one, and was named *radon.*

ATOMIC CHANGE

In 1900 Crookes (see 1861) found that it was possible to treat a solution of uranium compound in such a way that part of it would be removed as an insoluble material. The removed portion was an impurity, the uranium compound itself remaining in solution.

But almost all the radioactivity was in the insoluble impurity and very little in the uranium compound. Crookes suspected that uranium itself might not be radioactive at all.

However, as Becquerel speedily pointed out, the purified uranium compound, which was only slightly radioactive if at all, steadily gained in radioactive intensity if allowed to remain standing. It was then suggested that uranium was at least slightly radioactive and that as it gave off its radiations, its atoms were converted into other kinds of atoms, which were much more intensely radioactive than uranium.

This was the first suggestion that radioactivity might represent the spontaneous change of one kind of atom into another. If this was true, it would imply that atoms had structure, and that rearrangement of the still-smaller particles that made up atoms took place in the course of radioactive breakdown.

ELECTRON EMISSION

Edison had discovered the Edison effect—that an electric current seemed to jump the gap between the heated filament of an electric light bulb and a cold metal rod (see 1883).

Beginning in 1900, the British physicist Owen Willans Richardson (1879–1959) studied the phenomenon and realized that heated metals tended to emit speeding electrons, which carried the electricity. His observation made it possible to apply the Edison effect to the developing electronic technology.

For his work, he received the Nobel Prize in physics in 1928.

MUTATIONS

The American evening primrose had been introduced into the Netherlands in the 1880s, and in 1886 a Dutch botanist, Hugo Marie De Vries (1848–1935), came across a colony of them growing in a clump in a meadow. He could see that some were very different from others, even though it seemed likely that all had grown from a common batch of seeds.

He dug them up, took them to his garden, and bred them separately and together. By 1900 he had evolved the laws of genetics that Mendel had worked out (see 1865).

Unknown to De Vries and to each other, two other botanists were making the same rediscovery. One was a German, Karl Erich Correns (1864–1933), and the other an Austrian, Erich Tschermak von Seysenegg (1871–1962).

Each of the three independently decided to publish his discovery, and each of the three found, on searching the literature, that he had been anticipated by Mendel. Each of the three (in a remarkable display of scientific ethics) reported Mendel as the discoverer and presented his own work merely as confirmation.

De Vries, however, went beyond Mendel. He had observed that the primroses didn't always breed true, that some characteristics appeared that didn't seem to exist in earlier generations. He suggested, therefore, that evolution did not always proceed in small, nearly microscopic steps, but sometimes in changes large enough to be clearly visible. He called these visible changes *mutations* (from a Latin word for "change"), and indeed the concept of mutations now forms an integral and indispensable part of the theory of evolution.

BLOOD TYPES

Although the usual eighteenth-century practice was to remove blood from a sick patient, there were occasional physicians who tried to add blood to patients, taking it from the veins of healthy humans or even from animals. Sometimes it was helpful, but sometimes it hastened death, so that most European nations by the end of the nineteenth century had prohibited such *blood transfusion*.

In 1900, however, an Austrian physician, Karl Landsteiner (1868–1943), was able to demonstrate some specific properties of human blood. Plasma (the liquid portion of blood) from one donor might clump red cells from person A but not from person B. Serum from another donor might clump red cells from person B but not from person A. Still other samples of plasma might clump both—or neither.

Clumped red cells could block blood vessels and lead to death. Therefore, it was necessary in performing a blood transfusion to know that the blood of the donor would not clump the red cells of the receiver. Landsteiner showed that human blood fell into four classes: O, A, B, and AB. It was always safest if both donor and receiver were in the same class. In an emergency, O blood could be given to any receiver, but A blood could be given only to A and AB receivers, B blood only to B and AB receivers, and AB blood only to AB receivers.

Landsteiner made blood transfusion rational and safe and added an important weapon to the medical armory. For this discovery, he received the Nobel Prize in medicine and physiology in 1930.

YELLOW FEVER

Yellow fever was a terrifying scourge of coastal cities. New York and Philadelphia, for instance, were periodically hit, and many deaths followed.

The Spanish-American War made the United States particularly disease-conscious, as far more American soldiers died from disease and from tainted meat than died from enemy bullets.

The American military surgeon Walter Reed (1851–1902) was sent to Cuba in 1899 to see if he could do something about yellow fever. He had already showed, in 1897, that it was not caused by a bacterium, as had been earlier proposed.

His studies in Cuba showed that the disease was not transmitted by bodily contact, by clothing, or by bedding. He suspected mosquito carriers, as Ross had shown was the case with malaria (see 1897). The case was proved for yellow fever, too, in 1900, when doctors allowed mosquitoes to sting patients and then themselves. One doctor, Jesse William Lazear (1866–1900), died as a result.

Yellow fever was brought under control by fighting mosquitoes and destroying their breeding grounds. The last yellow fever epidemic in the United States was in New Orleans in 1905.

DREAMS

Dreams had always mystified human beings. Dreaming that dead people were alive helped lead to a belief in the spirit world. Erotic dreams led to belief in incubi and succubi. Dreams seemed to be doorways into a different world, to be messages from the gods, to be revelations of events at a distance or in the future.

All this was dismissed as superstition by rationalists, but Freud (see 1884) gave new meaning to dreams.

In 1900 he published *The Interpretation of Dreams*. Dreams, he maintained, might represent truths about human beings that they were not willing to accept in their waking hours, so that psychoanalysis might be hastened and made more effective if the analyst carefully considered the literal and symbolic meaning of dreams.

TRYPTOPHAN

By now, some thirteen different amino acids had been isolated, each of which were among the building blocks of protein molecules. Still another was isolated in 1900 by the British biochemist Frederick Gowland Hopkins (1861–1947). He called it *tryptophan*, from Greek words meaning "appearing through trypsin," because he had obtained it from protein molecules that had been broken down by the digestive enzyme trypsin.

A French physiologist, François Magendie (1783–1855), had shown, in 1815, that animals could not be kept alive if gelatin was the only protein fed them. Hopkins found that tryptophan did not occur in gelatin, and he suggested that tryptophan could not be formed in the body from other substances but had to be present as such in the diet.

All the amino acids that are found as part of protein molecules are essential to life, but only some are essential as such in the diet. It is to these latter that the term *essential amino acids* refers. Hopkins originated this concept, and it proved an important finding in the field of dietetics and food chemistry.

FREE RADICALS

Organic chemists find it interesting to try to synthesize compounds of unusual structure. The Russian-born American chemist Moses Gomberg (1866–1947) tried to attach a carbon atom to four different benzene rings. There is scarcely room around a single carbon atom for four such bulky groupings, and a number of chemists had tried and failed. Gomberg, however, succeeded in forming small quantities of this compound, called *tetraphenylmethane*.

He next tried to attach six benzene rings to two carbon atoms hooked together, thus forming *hexaphenylethane*. This he could not do, but he did get a strongly colored compound that did not have the properties to be expected of hexaphenylethane.

Gomberg studied the colored compound carefully and, in 1900, came to the conclusion that he had a half-molecule. The hexaphenylethane tended to break in two and leave him with two single carbon atoms, each with three benzene rings attached—*triphenylmethyl*.

Ordinarily the carbon atom is attached to four different groups, but in this case it was attached to only three, while the fourth bond was unoccupied. When compounds form, there may be moments when a carbon atom loses a fourth attachment and has not yet gained a substitute attachment, but this usually lasts such a brief time that it isn't noticeable.

Such groupings with a carbon bond unattached are called *radicals*. Something like triphenylmethyl, which lasts for a perceptible time, is called a *free radical*. The explanation for the existence of such free radicals was not advanced for over three decades more.

DIRIGIBLE

Balloons had been in operation for over a century, and it was easy to see that if one could place a steam engine in a balloon's gondola and attach a propeller to it, the balloon might be guided in any desired direction, even dead against the wind. However, steam engines are very heavy, and supporting one by balloon would not be easy.

The coming of Otto's internal-combustion engine (see 1876) offered a much lighter energy source, but even so it was difficult to force a balloon through the air against air resistance.

It occurred to a German inventor, Ferdinand Adolf August Heinrich von Zeppelin (1838–1917), to streamline a balloon so that it would meet with less air resistance. This could be done by confining it within a cigar-shaped metal envelope. The Hall-Heroult method had made aluminum cheap (see 1886), and aluminum had the combination of lightness and strength needed to make such a streamlined balloon feasible.

On July 2, 1900, one of Zeppelin's cigar-shaped vessels rose into the air. Beneath it was a gondola bearing an internal-combustion engine and a propeller. For the first time, an aircraft was not at the mercy of the wind but could move in any direction at will. It was called a *dirigible balloon;* that is, "a balloon that could be directed." This was soon abbreviated to *dirigible,* and the device was also sometimes called a *zeppelin* after its inventor.

KNOSSOS

During classical times, Crete had lain outside the current of history and was little regarded. Yet Homer, describing the Trojan War, gave Crete an important role, and in the Greek myths, Crete, under its king Minos, was described as dominating Greece in early times.

A British archaeologist, Arthur John Evans (1851–1941), felt there might be some historical fact in that myth and, beginning in 1894, he engaged in archaeological excavations in Greece. By 1900 he had dug up the site of Knossos, the capital of Crete in Minos's legendary time. (Evans called it the *Minoan* civilization after the king.) He showed that a complex and advanced civilization had indeed existed for nearly two thousand years before the Trojan War, and that it had indeed dominated the islands of the Aegean and parts of the Greek mainland for a long period.

IN ADDITION

The Chinese, driven frantic by the way the European powers were plucking the nation clean, formed a *righteous harmony band.* This was mistranslated by the Europeans as *righteous harmonious fists* and the group was therefore called the *Boxers* in derision. When they rose, it was called the *Boxer Rebellion.* The Boxers seized legations and killed the German minister on June 20, 1900. An international force, which included American troops, was organized under German leadership. It defeated China, and the Chinese court was forced to flee Peking on August 15, 1900.

In South Africa, the British now defeated the Boers, relieved besieged cities, annexed the Boer Republics, and drove Kruger to exile in Europe. The Boers, however, continued a guerrilla war that kept the British busy for quite a while. The Boer War might be considered the beginning of the decline of Great Britain.

Umberto I of Italy was assassinated on July 29, 1900, and was succeeded by his son, who reigned as Victor Emmanuel III (1869–1947).

In the United States, McKinley was reelected president by a bigger margin than he had won with in 1896.

Galveston, Texas, was struck by a hurricane on September 8, 1900. Up to eight thousand people may have died, the greatest number of casualties ever in a natural disaster in the United States.

In 1900 the population of the United States reached 76 million, twice that of Great Britain. New York City was now the second-largest city in the world with 3.6 million, compared with London's 6.6 million. Chicago was the second-largest city in the United States with 1.7 million.

1901

RADIOACTIVE ENERGY

In 1901 Pierre Curie measured the heat given off by radium as it emitted radiation. He determined that each gram of radium gave off 140 calories per hour.

This becomes much more surprising when one realizes that radium continues to emit energy at this rate hour after hour after hour for years and centuries. It falls off with time, but only slowly, and even after 1600 years, the rate of energy emission is still half of what it was at the start.

The total energy emitted by radium (or by radioactive substances generally) was far beyond anything humanity had experience with in ordinary chemical changes, such as the burning of fuel or the shattering of explosives.

Curie's finding was the first indication that a new, hitherto unknown, but enormous energy source existed somewhere inside the atom. But until the details of the internal structure of the atom were discovered, and the nature of the changes that gave rise to radioactivity were worked out, scientists could only refer to this new energy source as *atomic energy.*

RADIO

The sending of signals by radio waves reached a climax on December 12, 1901. Marconi (see 1895) broadcast radio waves from the southeastern tip of England, using balloons to lift his antenna as high as possible. The signals were received in Newfoundland. This day is usually considered the one on which radio was invented, and Marconi is given credit as the inventor.

EUROPIUM

Eleven rare earth elements were now known, but that did not exhaust the list. The French chemist Eugène-Anatole Demarcay (1852–1903) detected a twelfth, which he named *europium* in honor of Europe.

GRIGNARD REAGENTS

The French chemist Victor Grignard (1871–1935) was searching for methods of attaching carbon-containing groups to organic molecules. He needed a catalyst. Zinc shavings and magnesium shavings did some good but not much. Frankland (see 1852) had reported on zinc combinations with organic compounds using diethyl ether as the solvent. Grignard tried it with magnesium instead, in 1901, and found he had the catalyst he needed.

Magnesium-organic compounds dissolved in diethyl ether are called *Grignard reagents* and proved extraordinarily useful to organic chemists who were trying to build up relatively complicated compounds. For this work, Grignard shared with Sabatier (see 1897) the Nobel Prize in chemistry in 1912.

IN ADDITION

On January 22, 1901, Victoria of England died after a reign of nearly sixty-four years. She was succeeded by her son, who reigned as Edward VII (1841–1910).

On January 1, 1901, Australia was created a commonwealth within the British Empire, with home rule akin to that of Canada. In southern Africa, Boer guerrillas were still holding out but without much success.

On September 7, 1901, the Boxer Rebellion ended. China had to pay an enormous indemnity and grant further political and economic concessions to the European powers.

On September 6, 1901, United States President McKinley was shot by an anarchist named Leon F. Czolgosz (1873–1901) and by September 14, the president was dead. His vice president, Theodore Roosevelt (1858–1919), succeeded as twenty-sixth president.

Nobel Prizes were awarded for the first time. They have continued ever since to be the scientific awards of greatest prestige.

1902

CHROMOSOMES AND INHERITANCE

Mendel, in working out the laws of genetics (see 1865), had suggested that for every characteristic there was a pair of factors. Mother and father each contributed one of that pair to the offspring, who in this way inherited characteristics from each parent.

By the time Mendel's work had been rediscovered by De Vries and others (see 1900), the role of chromosomes in cell division had been worked out by Flemming (see 1882) and their role in the formation of sex cells had been researched by Beneden (see 1883).

In 1902, then, an American geneticist, Walter Stanborough Sutton (1877–1916), made a suggestion that in hindsight seems obvious. He pointed out that chromosomes were (or contained) the genetic factors Mendel spoke of. He proved to be correct.

SECRETIN

The pancreas begins to secrete its digestive juice as soon as the acid food contents of the stomach enters the small intestine. How does the pancreas "know" that food, requiring digestion, is making its appearance? The natural assumption is that the entering food stimulates a nerve that in turn stimulates the pancreas. The Russian physiologist Ivan Petrovich Pavlov (1849–1936) suggested that this was so.

Two British physiologists, Ernest Henry Starling (1866–1927) and his brother-in-law, William Maddock Bayliss (1860–1924), tested the matter. They cut all the nerves leading to the pancreas—yet it still performed on cue.

They then discovered that the lining of the small intestine secreted a substance (which they named *secretin*) under the influence of stomach acid. It was this secretin that stimulated the pancreatic flow. In short, then, Starling and Bayliss had discovered that it was possible for chemical messages as well as nerve messages to exist in the body.

Eventually other chemical messengers were discovered and Starling suggested they be called *hormones*, from Greek words meaning "to rouse to activity." Secretin was the first hormone to be recognized as such, but epinephrine, isolated by Abel (see 1898), turned out to be a hormone, also.

ANIMAL INHERITANCE

Mendel's work, once rediscovered, created an enormous stir. The British biologist William Bateson (1861–1926) was a strong supporter of Mendelian views and translated his papers into English.

Mendel, together with the rediscoverers of the laws of genetics, had worked with plants, in which breeding was easier to control and inheritance easier to study than in animals. Bateson showed that the same laws of genetics that applied to plants applied also to animals.

ANAPHYLACTIC SHOCK

The French physiologist Charles Robert Richet (1850–1935) worked on immune sera, the sort of thing that Behring had used successfully on diphtheria (see 1883). To his surprise, Richet discovered that if he caused an animal to produce an immune serum to a particular foreign protein (an *antigen*) and then injected the antigen, the animal died. In 1902 Richet named this phenomenon *anaphylaxis*, from Greek words meaning "overprotection."

From then on, physicians were warned. Serum therapy had to be conducted in such a way as to prevent the possibility of sensitization, which would produce serum sickness. It came to be understood that people might be sensitized to foreign proteins in the environment—in plant pollen, in dust, in food—and exhibit unpleasant reactions. These reactions came to be called *allergies* (from Greek words meaning "other work," because the mechanisms of the body do something other than the work they are supposed to do).

Richet's work on anaphylaxis, and the understanding of allergies that it led to, brought him the Nobel Prize in medicine and physiology in 1913.

SUTURES

The French surgeon Alexis Carrel (1873–1944) was particularly deft in the field of blood-vessel repair. He developed a technique by which blood vessels could be delicately sutured (sewn together) end to end. He did this successfully in 1902, requiring as few as three stitches for the job.

The usefulness of such techniques in surgery led to Carrel's being awarded the Nobel Prize in medicine and physiology in 1912.

RADIOACTIVE SERIES

Crooke's discovery that much of uranium's radioactivity could be precipitated and that the radioactivity would spontaneously return thereafter (see 1900) led Rutherford and an assistant, the English chemist Frederick Soddy (1877–1956), to investigate the matter more thoroughly. By subjecting uranium and thorium to chemical manipulations and following the fate of the radioactivity, they demonstrated that the two elements broke down in the course of radioactivity into a series of intermediate elements.

This meant that one could speak of a *radioactive series*.

PHOTOELECTRIC EFFECT AND ELECTRONS

Hertz had first noted the photoelectric effect—the greater ease which which electric sparks cross small gaps when ultraviolet light falls upon those gaps—fourteen years before (see 1887). Now that electrons were known, the photoelectric effect could be studied more effectively.

Lenard (see 1895) showed in 1902 that the electrical effects produced by light falling upon certain metals were the result of electrons being emitted from the metal surface. He showed that only light of a certain wavelength or shorter could bring about electron emission from a particular metal, and that the crucial wavelength was different for different metals.

Increasing intensity of light of a given wavelength resulted in the emission of a greater number of electrons, but the energy of the individual electrons remained the same. If the wavelength decreased, the electrons emitted had higher energy; if the wavelength increased, the electrons emitted had lower energies. There was no way of explaining this by nineteenth-century physics.

Even unexplained, however, the photoelectric effect made it appear that electrons existed in matter even when electric currents were not involved. The fact that identical electrons were emitted by various metals made it seem that electrons were a constituent of all atoms without exception.

KENNELLY-HEAVISIDE LAYER

The fact that Marconi's message, generated in southwestern England, was picked up in Newfoundland (see 1901) presented a puzzle. Electromagnetic radiation, including radio waves, travels in straight lines. Even if the radio waves were traveling parallel to the ground to begin with, they should have headed off into outer space as the Earth's surface curved downward. (Earth is a sphere, after all.) Instead, the signal apparently clung to the neighborhood of Earth's surface, working its way around the Earth's spherical curve across the Atlantic Ocean.

A British-born American electrical engineer, Arthur Edwin Kennelly (1861–1939), suggested that there might be a layer of electric charges in the upper atmosphere, which would serve to reflect radio waves. In that case, the radio waves would bounce between earth and clouds, thus explaining why they clung to the bulge of the Earth.

A British electrical engineer, Oliver Heaviside (1850–1925), independently advanced the same idea, so that the supposed layer of charged particles came to be called the *Kennelly-Heaviside layer.* It was to be twenty years before these speculations were proved correct.

STRATOSPHERE

Ever since balloons had been invented (see 1783), scientists had used them to explore heights that they could otherwise reach only by climbing the highest mountains. However, by the time a balloon was 6 miles in the air, the cold and lack of oxygen could be fatal.

The French meteorologist Léon-Philippe Teisserenc de Bort (1855–1913) therefore took to sending up balloons bearing only instruments that could be read when the balloon returned to Earth.

He found in this way that the temperature of the atmosphere dropped steadily up to a height of about 7 miles. At higher altitudes, the temperature remained constant as high as he could reach.

In 1902 Teisserenc de Bort suggested that the atmosphere be divided into two parts. The lower part, below the 7-mile mark, had temperature variations and therefore produced winds, clouds, and in short, the weather changes so familiar to us. This would be called the *troposphere* (from Greek words meaning "sphere of change"). Above it would be the constant-temperature *stratosphere* ("sphere of layers," because Teisserenc de Bort felt that constant temperature meant the gases would lie in undisturbed layers). The names have remained with us ever since.

SYMBOLIC LOGIC AND MATHEMATICS

The German mathematician Gottlob Frege (1848–1925) improved and extended Boole's system of symbolic logic (see 1847). In fact he spent almost twenty years putting together a system of symbolic logic that would serve as the basis of all mathematics, reducing it to complete rigor by minimizing assumptions and proving each step.

In 1902, when the second volume was in galleys, Frege received a letter from the British mathematician Bertrand Arthur William Russell (1872–1970). The letter contained a question about an apparent self-contradiction in Frege's system and asked Frege to remove it. Frege, after deep thought, realized that his system of logic could not handle the contradiction, so that he found his project worthless at the moment of completion.

This failure was to have important ramifications in mathematics.

ULTRAMICROSCOPE

Substances such as salt and sugar, which dissolve in water, separate into single ions or molecules of the same order of size as the water molecules among which they are dispersed. Sometimes substances are made up of giant molecules, such as those of proteins, or disperse as clusters of ordinary small molecules, the clusters being of comparatively large size.

In 1861 the Scottish physical chemist Thomas Graham (1805–1869) noted that small molecules in solution could pass through membranes with fine holes, such as a sheet of parchment. Large molecules or molecular clusters in solution could not. Graham called the former *crystalloids,* because crystalline solids usually dissolved as small molecules. The latter he called *colloids,* from the Greek for "glue," because a solution of glue contained large molecules (proteins, usually) that did not pass through the parchment.

The Irish physicist John Tyndall (1820–1893) pointed out that light passing through a crystalloidal solution was not affected by the molecules of water or dissolved material. Such solutions were *optically clear.* The larger molecules or molecular clusters in colloids scattered light, on the other hand, and did so more effectively for the shorter wavelengths. This is the *Tyndall effect.*

Tyndall pointed out that dust particles in the air scattered shortwave light particularly, so that the sky was blue and the Sun at setting (when its light passed through a great thickness of dusty air) was red—the long waves being less scattered.

The Austrian-born German chemist Richard Adolf Zsigmondy (1865–1929) took advantage of this scattering of light by colloidal particles in 1902. He sent a beam of light through a colloidal solution and viewed the light scattered at right angles to the beam through a microscope. In this way, colloidal particles, invisible to ordinary microscopes, could be seen, in what Zsigmondy called an *ultramicroscope.*

For his work on colloids with his ultramicroscope, Zsigmondy was awarded the Nobel Prize for chemistry in 1925.

IN ADDITION

Great Britain, recognizing the dangers of isolation in the face of an ever-more-powerful Germany, sought allies. On January 20, 1902, Britain formed an alliance with Japan, thinking that Japan's special interest in Korea would keep the Far East quiet if Britain got involved in Europe.

The Boer War finally came to an end on May 31, 1902, with the Treaty of Vereenigung. The Boers accepted British sovereignty, but in return the British didn't press too hard on the Boers and their way of life.

The United States pulled its troops out of Cuba, which became theoretically independent, though in fact it remained an American protectorate under the Platt Amendment. By the terms of this amendment, Cuba could not take any action the United States disapproved of, and the United States was free to intervene militarily at will.

1903

AIRPLANE

Once Lilienthal had devised his gliders (see 1853), it seemed natural to think of putting an internal-combustion engine on a glider as Zeppelin had done for balloons (see 1900). The American astronomer Samuel Pierpont Langley (1834–1906) made the attempt on three separate occasions between 1897 and 1903 and almost succeeded but not quite.

Then two brothers, Orville (1871–1948) and Wilbur (1867–1912) Wright took up the task. They made shrewd corrections in design and invented "ailerons," movable wing tips that enabled the pilot to control the plane. In addition, they built a crude wind tunnel to test their models, and they designed new engines of unprecedented lightness for the power they could deliver.

On December 13, 1903, at Kitty Hawk, North Carolina, Orville Wright made the first powered flight in a heavier-than-air machine. It remained in the air for almost a minute and covered 850 feet. Such devices soon came to be called *airplanes*.

SPACE FLIGHT

Oddly enough, in the same year that air flight became a reality, space flight began to receive true scientific attention.

In 1903 a Russian physicist, Konstantin Eduardovich Tsiolkovsky (1857–1935), began a series of articles for an aviation magazine in which he went into the theory of rocketry quite thoroughly. He wrote of space suits, satellites, and the colonization of the solar system. He was the first to suggest the possibility of a space station.

ELECTROCARDIOGRAM

That muscles gave rise to tiny electric potentials had been known since Galvani's time (see 1780). It seemed natural to suppose, then, that the heart, beating rhythmically, might give rise to rhythmic electric potentials. Perhaps a departure from the natural rhythm might be used to diagnose pathological conditions before they could be discovered in any other way. The problem was to detect the small currents with sufficient accuracy.

In 1903 a Dutch physiologist, Willem Einthoven (1860–1927), developed the first *string galvanometer*. This consisted of a delicate conducting thread stretched across a magnetic field. A current flowing through the thread would cause it to deviate at right angles to the direction of the magnetic lines of force. The delicacy of the device was sufficient to make it possible to record the varying electrical potentials of the heart.

The result was an *electrocardiogram*. The abbreviation is *EKG* because in German *-cardio-* is spelled with a *k*. For the development of electrocardiography, Einthoven was awarded the Nobel Prize for medicine and physiology in 1924.

IN ADDITION

The United States decided to dig a canal through the Isthmus of Panama, which was part of the South American nation of Colombia, and arranged a treaty with Colombia on January 22, 1903. The Colombian legislature rejected in on August 12. The United States therefore sent a warship to Panama and arranged to have the Panamanians declare their independence, which they did on November 3. The United States recognized Panamanian independence on November 6, and the treaty that Colombia had rejected was signed with Panama on November 18.

After the Boxer Rebellion, Russia occupied Manchuria. The Japanese, who felt that Manchuria was *theirs* to steal, sent letters of protest to the Russians, who ignored them. Relations between Russia and Japan began to deteriorate.

An automobile crossed the United States for the first time, taking 52 days. President Roosevelt sent the first message to go around the world by wire and cable, which took 12 minutes.

1904

ELECTRONIC RECTIFIER

A British electrical engineer, John Ambrose Fleming (1849–1945), studied the Edison effect on a hot filament and a cold plate enclosed in an evacuated glass vessel and separated by a gap. He noted that the electric current only flowed across the gap if the hot filament was the negative electrode (the cathode) and the cold plate was the positive electrode (the anode). In that case, electrons poured into the hot filament and were driven off by the heat. If the current was reversed and electrons poured into the cold plate, there was insufficient energy to send them flying outward.

If an alternating current was sent through the vessel, the anode and cathode changed place many times a second, and there was a spurt of electrons each time the filament was a cathode but nothing when it was an anode. The current went in as an alternating current but came flowing out in only one direction, as a direct current, even if the flow was in little gushes.

In this way, the vessel containing the filament and plate acted as an electronic *rectifier,* because it let electricity through in only one direction. Fleming called it a *valve,* which is descriptive. In the United States, however, it was for some reason called a *tube,* which is not descriptive. Because it contains two electrodes, it can also be called a *diode.*

Fleming's rectifier was the first of a long line of *radio tubes* (so called because they became most familiar to the general public in radios) that made electronic devices work.

ATOMIC STRUCTURE

With the discovery of the electron and its emission from many different metals by way of the photoelectric effect, there was no use in continuing to pretend that the atom was a featureless, indivisible particle. It had to have a structure involving electrons, and the first to speculate on that structure was Thomson, who had discovered the electron (see 1897).

In 1904 he suggested that the atom was positively charged electricity in which negatively charged electrons were embedded (like raisins in a pound cake) in just sufficient quantity to neutralize the positive charge. Naturally, under stress, such as that produced by electric currents or light, the electrons could be jarred loose from the atom.

The suggestion was interesting but quickly proved inadequate. However, it did set people to thinking seriously about atomic structure.

COENZYME

Once Buchner had shown that yeast enzyme could be extracted from yeast and still be active (see 1896), it became a popular substance for experimentation.

In 1904 the British biochemist Arthur Harden (1865–1940) placed an extract of yeast in a bag made of a semipermeable membrane (see 1748) in order to dialyze the extract and remove any small molecules that might be present.

Harden found to his surprise that when he did this, the material left behind in the bag lost its activity—it no longer fermented sugar. Did the enzyme have so small a molecular size that it passed through the semipermeable membrane? No, for the material outside the bag showed no activity either. However, if the material outside the bag was poured back into the bag, the combined mixture showed enzyme activity.

Harden concluded that an enzyme might consist of two parts, a large part that could not go through the semipermeable membrane and a small part that could and was so loosely held that it easily broke away and *did* go through the membrane.

The material that remained inside the bag, if it was boiled, did not regain activity when the material outside the bag was added. If the material outside the bag was boiled, it could still restore activity if added to unboiled material inside the bag. The conclusion was that the large molecule was a protein, since protein molecules are routinely destroyed by boiling. The small molecule was not.

Harden called the small molecule a *coenzyme,* and the name came to be applied to any nonprotein structure of relatively small size that proves essential to the activity of an enzyme. Not all enzymes require coenzymes, but coenzymes, when they do exist, eventually explained a great deal about the workings of enzymes and the requirements of diet.

ORGANIC TRACERS

The body is a "black box." We know what goes in, in the way of food and air; we know what comes out, in the way of waste products. But what happens in between? All the changes from beginning to end are referred to as *metabolism.* Those changes that are hidden inside the body are called *intermediary metabolism.*

In 1904 a German biochemist, Franz Knoop (1875–1946), tried an ingenious trick that he thought might give some hint of what went on inside the body. He attached benzene rings to the long carbon chains of the fatty acids that make up fats. The benzene ring is not easily destroyed in the body, and Knoop expected it to show up in the urine.

Knoop found that if the fatty acids had an even number of carbon atoms, the waste product was a benzene ring attached to a two-carbon fragment. Fatty acids with an odd number of carbon atoms produced a benzene ring attached to a one-carbon fragment.

From this, Knoop decided that fatty acids were broken down in the body by having two-carbon fragments snipped off, one after the other, to the point where the next snip would involve a benzene ring. They were probably also built up two carbons at a time, so that the fatty acids that occur naturally in living things always have even numbers of carbon atoms. (Those with odd numbers that Knoop worked with were synthesized in the laboratory.)

The benzene ring acts like a label attached to the molecule, one that allows the final fragment of the molecule to be identified.

It is considered a *tracer*, because it enables scientists to trace the route of a particular compound as it works its way through intermediary metabolism.

Of course the benzene ring is not a natural component of the fatty acids, and there was always the chance that the presence of this unnatural grouping might distort the workings of intermediary metabolism. Ideally, a tracer should be perfectly natural and yet perfectly identifiable at the same time. That seemed too much to ask for, but such tracers were eventually found.

NOVOCAINE

Alkaloids such as cocaine and morphine are extremely useful as painkillers, but they have serious physiological effects and are addictive. Plants, after all, do not make them for the sake of animals, but for self-protection.

Organic chemists strove to modify the alkaloid molecule in an effort to find a substitute that kept some of the good aspects while losing some of the bad ones. In 1904, for instance, a molecule named *novocaine*, or *procaine*, was found, with a structure something like part of a cocaine molecule. It acted as a local anesthetic but was much safer to use than cocaine, so that it became highly important to dentists.

Chemists weren't always so lucky. In 1898 a modified molecule of morphine was found to be even more effective as a pain-reliever than morphine itself. After a few years, it was recognized to be even more dangerous and addictive. It was *heroin*.

STAR STREAMS

When Halley first noted that stars moved (see 1718), astronomers strove to determine the motions of as many stars as possible, and the usual conclusion was that they moved randomly.

The Dutch astronomer Jaçobus Cornelis Kapteyn (1851–1922) found out differently. Observing numerous stars, he was able to conclude by 1904 that there were two huge streams of stars moving in opposite directions; three-fifths of the stars were in one, and two-fifths in the other. The reason for this wasn't clear, but at least it showed that there was order to the movement of the many millions of stars that made up the Milky Way Galaxy. The nature of that order became clear a quarter-century later.

JUPITER'S OUTER SATELLITES

The large satellites had by now all been discovered, but there still remained small ones to find. In 1904 the American astronomer Charles Dillon Perrine (1867–1951) discovered a small satellite circling Jupiter. It was the sixth to be discovered and was no more than 110 miles across. The next year he discovered a seventh, even smaller, satellite, only 50 miles across. Both circled Jupiter at average distances of 7,000,000 miles or so, far outside Jupiter's large satellites. They may be captured asteroids.

For a long time they were not given names but called *Jupiter VI* and *Jupiter VII*. The sixth is now known as *Himalia* and the seventh as *Elara*, after obscure nymphs in the Greek myths.

IN ADDITION

Russia was taking over Manchuria, and Japan did not intend to allow that. On February 8, 1904, the recently westernized Japanese navy bombed Port Arthur, the Russian port in Manchuria, without warning. They damaged the Russian ships there and blockaded the port. Then on February 10, Japan declared war, and the *Russo-Japanese War* was on.

The Russians were badly outnumbered in the Far East; they were incompetently led; supplies and reinforcements could only come by way of a single 6000-mile railroad line just being completed; and there was seething discontent and rebellion at home. The Japanese won a series of victories, and by the end of the year, the Russian armies had been driven out of Korea and southern Manchuria. The world was astonished.

In the United States, President Theodore Roosevelt ran for election in his own right and won easily.

Great Britain, still searching for friends, signed an *entente cordiale* (friendly agreement) with France, in which all the points of dispute between them were ironed out. France agreed to let Great Britain have Egypt, and Great Britain agreed to let France have Morocco.

1905

SPECIAL RELATIVITY

The Michelson-Morley experiment (see 1887) was still troublesome. The work of FitzGerald (see 1892) and Lorentz (see 1895) got rid of the difficulty in a way, but the notion of decrease of distance and increase of mass seemed to hang in the air without an overall physical theory for support.

The German-born physicist Albert Einstein (1879–1955) supplied that in 1905. He began with the assumption that the speed of light in a vacuum would always be the same, regardless of the motion of the light source relative to the observer. This was what Michelson and Morley had observed, but Einstein maintained that he was unaware of the Michelson-Morley results when he worked out his theory.

From this assumption it was possible to deduce length contraction and mass increase with velocity. It was also possible to deduce that the speed of light in a vacuum was an absolute speed limit and that the rate of time-flow would decrease with velocity.

This is Einstein's *theory of special relativity*. It is *relativity*, because velocity has meaning only as relative to an observer, there being no such thing as "absolute rest" against which an "absolute motion" can be measured. There is also no such thing as "absolute space" or "absolute time," since both depend on velocity and therefore have meaning only relative to the viewer. Nevertheless, despite this absence of absolutes, the laws of physics still held for all "frames of reference." In particular, Maxwell's equations (see 1865) still held, though the much older and more revered laws of motion as worked out by Newton had to be modified.

The theory is *special*, because it confines itself to the special case of objects that are moving at constant velocity. Under those conditions, the theory does not take into account the effect of gravitational interactions, which are everywhere present and which force accelerations on motion.

Einstein's view of the Universe seemed to go against common sense, but that was only because the average person deals with a Universe of small distances and small velocities. Under those conditions, Newton's theories hold almost perfectly. In fact, Einstein's equations reduce to Newton's under such conditions. However, where large distances and large velocities are involved, Einstein's equations hold and Newton's do not.

In the eight decades since special relativity

was advanced, endless tests and observations have upheld it completely. No divergences between reality and the Einsteinian view have been found.

MASS-ENERGY

Another consequence of Einstein's theory of special relativity (see above) is that mass must be viewed as a highly concentrated form of energy. Einstein's equation representing this is the famous $e = mc^2$, where e is energy, m is mass, and c is the speed of light. The speed of light is so huge that to square it and multiply it by even a small amount of mass is to represent a large amount of energy (1 gram of mass equals 900 billion billion ergs of energy).

Whenever any process gives off energy, it loses a little mass; when it absorbs energy, it gains a little mass. The amount of mass lost or gained under ordinary conditions is so minute it had never been detected. That is why Lavoisier could consider mass conserved independently of energy (see 1769) and Helmholtz could consider energy conserved independently of mass (see 1847).

With the study of radioactivity, much larger energy changes per unit mass were involved, as Pierre Curie had found (see 1901). Mass-energy equivalence could then be measured and was found to be precisely as Einstein's theory required it to be. The law of conservation of energy was thus extended and made more precise by the inclusion of mass as one more form of energy. The law of conservation of mass became obsolete, or rather, was included in what is sometimes known as *the law of conservation of mass-energy*.

PHOTOELECTRIC EFFECT AND QUANTA

In 1905 the photoelectric effect, as observed by Lenard (see 1902), was combined with quantum theory (see 1900) by Einstein.

He showed that if light consisted of quanta, with energies proportional to frequency (inversely proportional to wavelength), then the atoms in a metal surface could only absorb intact quanta. Furthermore, long-wavelength quanta would not supply enough energy to eject an electron from the metal, no matter how intense the light might be. But as wavelength grew shorter, energy quanta grew larger, and a point would be reached where the energy was just sufficient to eject an electron. The shorter the wavelength beyond that, the more energetic the ejected electron would be and the more speedily it would travel. Since some metals may hold electrons more firmly than others, the critical wavelength must be shorter in some cases than in others.

Einstein's analysis had several consequences.

1. It explained the photoelectric effect completely. There have not had to be any extensions or additions to the explanation since.

2. It made use of quantum theory to explain a phenomenon unexplainable otherwise, and a phenomenon moreover that had not been in Planck's mind when he worked out the theory. If quantum theory were simply a mathematical device needed to make black-body radiation come out right, it wouldn't be likely to be applicable without modification to an altogether different phenomenon. Einstein's work therefore established quantum theory as legitimate and not merely a mathematical trick.

3. It showed that light could be treated as particles in some respects. Newton's particles and Huygens's waves (see 1678) were

thus combined into a whole that was far more complex and useful than could have been imagined on the basis of seventeenth-century knowledge. The particle aspects of light, and of electromagnetic radiation generally, are referred to nowadays as *photons.*

It was for this work on the photoelectric effect (not on his still greater discoveries in connection with relativity) that Einstein gained a Nobel Prize in physics in 1921.

BROWNIAN MOTION AND ATOMIC SIZE

Brownian motion had remained somewhat of a puzzle since Brown had discovered it (see 1827). In 1902 the Swedish chemist Theodor Svedberg (1884–1971) had suggested that the unequal bombardment of small particles by molecules from all sides impelled them to move randomly, this way and that.

Einstein, in 1905, analyzed the possibility of molecular bombardment thoroughly. He reasoned that any sizable object immersed in water (or any fluid) is bombarded from all sides, and surely more from one side at one moment and more from another side at another moment. However, the countless trillions of molecules involved means that any small differences will be so small as to be indetectable.

If we consider a smaller and smaller object, the total number of molecules striking it at a given moment will be smaller, and little deviations will loom larger. By the time objects approach the microscopic in size, an additional molecule from this side or that should be sufficient to give it a noticeable push in this direction or that.

Einstein worked out an equation to describe Brownian motion, one from which it was possible to work out the size of molecules and of the atoms making them up, provided one found a way to measure certain other variables that occur in the equation.

It was not long before Einstein's equation was put to good use.

COLOR AND STELLAR LUMINOSITY

We all know that some stars are brighter than others; astronomers measure that brightness as *magnitude.* A star can appear bright for either of two reasons. It may radiate a large amount of light (be of high *luminosity*), or it may be unusually close to us so that it appears bright even though it has low luminosity.

The Danish astronomer Ejnar Hertzsprung (1873–1967) suggested that if a star's distance were known, one could calculate what magnitude it would have if it were some standard distance away. The distance chosen was 10 parsecs, or 32.6 light-years. The brightness of the star at that distance would then be its *absolute magnitude.* Thus, if our Sun were 10 parsecs away from us, it would seem to have a magnitude of 4.86 (a rather dim star), and that would be its absolute magnitude.

In studying the absolute magnitude of various stars, Hertzsprung could calculate their relative luminosities. In 1905 he noticed that there were two kinds of red stars: red stars with very high luminosity (which we now call *red giants*) and red stars with very low luminosity (which we now call *red dwarfs*).

The most interesting aspect of the findings was that there were no red stars of intermediate luminosity. Hertzsprung's report did not attract much attention at first (he published it in a journal of photography), but it represented the first step toward understanding the evolution of stars.

PLANETESIMAL HYPOTHESIS

Laplace's nebular hypothesis of the origin of the Solar System (see 1796) had held sway for a century, even though astronomers had grown steadily more dubious about it. It turned out that most of the angular momentum in the Solar System was concentrated in the planets. (Jupiter alone has 60 percent of all the angular momentum in the Solar System because of its rapid rotation and the whirling of its major satellites.) There seemed no way in which the Solar System could have been formed by the slow condensation of a vast nebula without virtually all the angular momentum ending up at the core —in the body that became the Sun.

Despite this concern, no alternate hypothesis was advanced till the American geologist Thomas Chrowder Chamberlin (1843–1928) and the American astronomer Forest Ray Moulton (1872–1952) came along with one.

Chamberlin had been working on it since 1900 and, in 1905, he and Moulton advanced the suggestion that the Solar System began when the Sun, already in existence, passed near another star. The gravitational influence of each star tore gouts of matter out of the other, and the gravitational pulls of the stars as they separated gave those gouts of matter a sidewise yank so that they became planets with high angular momentum.

In the process of planet formation, the gouts of solar matter condensed into small solid pieces called *planetesimals,* and these came together little by little to form planets. This *planetesimal hypothesis* retained a certain popularity for nearly half a century.

If it were true, it would mean that there were very few planetary systems in the Universe, since the close approach of two stars is an excessively rare phenomenon.

METABOLIC INTERMEDIATES

Harden, who had demonstrated the existence of coenzymes the year before (see 1904), continued to study the behavior of yeast enzyme on the fermentation of glucose. The enzyme breaks down glucose rapidly at first and produces carbon dioxide, but as time goes on, the level of activity drops off. The natural assumption was that the enzyme broke down with time.

In 1905, however, Harden showed that this could not be so. If he added inorganic phosphate to the solution, the enzyme went back to work as hard as ever. This was a strange finding, for neither the sugar being fermented, nor the alcohol and carbon dioxide produced, nor the enzyme itself contained phosphorus.

Since the inorganic phosphate disappeared, Harden searched for some organic phosphate formed from it, which he located in the form of a sugar molecule to which two phosphate groups had become attached. This sugar phosphate was formed in the course of fermentation; then, after other reactions had taken place, the phosphate groups were removed again. The sugar phosphate was a *metabolic intermediate.* Harden was the first to isolate such a metabolic intermediate, but other biochemists were soon hot on the trail of such things.

Harden was also the first to indicate the important role that phosphate groups play in metabolism. For this work, he was awarded a share of the Nobel Prize in chemistry in 1929.

HORMONES

Starling, who had shared in the discovery of secretin (see 1902), suggested the name *hormone* in 1905. He also suggested that other hormones would be found to exist and that they were produced by various small glands in the body. In this he proved to be quite correct.

LINKAGE OF CHARACTERISTICS

When Mendel had studied his pea plants (see 1865), he had followed the course of seven characteristics and found them all to be inherited independently. It was natural to suppose, therefore, that every characteristic had its own factors, which all reached the fertilized egg cell independently.

When Sutton pointed out that the chromosomes were Mendel's factors, however (see 1902), there was a problem: there weren't enough different chromosomes to account for all inherited characteristics.

Bateson, who first applied the laws of heredity to animals (see 1902), pointed out that indeed not all characteristics were independently inherited—some were transmitted together—so it might be supposed that a single chromosome contained more than one factor, perhaps many more than one factor. It might be that Mendel, by pure chance, had chosen seven factors that were each on a different chromosome.

The thought that a chromosome was not *a* factor but a collection of factors was a crucial one in the development of genetics. (It was Bateson, incidentally, who introduced the term *genetics*.)

HIGH PRESSURE

Quite early, human beings learned how to produce temperatures considerably higher than normal. Producing high pressures was a more difficult task.

The American physicist Percy Williams Bridgman (1882–1961), while a doctoral student, wanted to work with high pressures but found his equipment inadequate. In 1905, therefore, he turned to the problem of devising better high-pressure apparatus. He designed seals that squeezed more tightly together as the pressure increased so that there would be no leakage, and he quickly attained pressures equal to 20,000 atmospheres (128 tons to the square inch).

This opened up the realm of truly high pressures to investigation.

INTELLIGENCE QUOTIENT

While abnormal psychological phenomena have long interested physicians and scientists, the normal manifestations of the mind were what interested the French psychologist Alfred Binet (1857–1911).

He strove to devise tests that would measure the ability of the human mind to think and reason, independently of learning and education in one field or another.

To do this, he asked children to name objects, follow commands, rearrange disordered things, copy designs, and so on. In 1905 he and his associates published the first batteries of tests designed to measure intelligence.

Standards were set empirically. If a particular test was passed by some 70 percent or so of the nine-year-olds in the Paris school system, then it represented the nine-year-old level of intelligence.

The phrase *intelligence quotient* (often abbreviated *IQ*) became popular. It represents the ratio of the mental age to the chronological age, with 100 considered average. Thus, a six-year-old who can pass a ten-year-old's test has an IQ of $10/6 \times 100$, or 167.

Binet's initial efforts gave rise to batteries of tests designed to measure personalities, attitudes, aptitudes, and potentialities as well as intelligence—and their value is almost certainly overestimated.

IN ADDITION

The Russo-Japanese War continued to be disastrous for the Russians. By September 5, 1905, the Russians had had enough, and agreed to a peace treaty, giving up Korea and Manchuria, and the southern half of the island of Sakhalin (just north of Japan), together with Port Arthur, to the Japanese. Russia refused to pay an indemnity, however, and the Japanese felt cheated.

A strong factor in the Russian defeat were the disorders at home. On January 22, 1905, a peaceful demonstration in St. Petersburg was countered by troops who fired into the crowd, killing 70 and wounding 240. This produced strikes and demonstrations all over Russia, and there was an intensifying demand for an end to autocracy—for a constitution and for representative government. The Russian court was forced to yield and to promise liberalization.

Norway, ruled by Sweden, wanted its independence. In this case, a plebiscite was held, and Sweden let Norway go in peace. The separation was made final on October 26, 1905, and a Danish prince was chosen to rule over Norway as Haakon VII (1872–1957).

Germany responded to the *entente cordiale* of the previous year by expressing its displeasure at the assignment of Morocco to France. On March 31, 1905, William II declared for Moroccan independence. From this point on, Europe was divided into two armed camps: Germany and her allies, and France and her allies.

1906

RADIO WAVES AND SOUND

Radio communication first came into use only as a wireless telegraph, forming the dots and dashes of the Morse code in appropriate bursts of radio waves.

It occurred to the Canadian-born American physicist Reginald Aubrey Fessenden (1866–1932) to send out a continuous signal with an amplitude (the height of the waves) varying in such a way as to follow the irregularities of sound waves. The radio wave was said to have *amplitude modulation* (AM) imposed on it.

At the receiving end, the modulation could be reconverted into sound waves, with the result that you could use a radio for speaking and hearing thanks to modulated radio waves the same way you could use a telephone for speaking and hearing thanks to modulated electric currents.

On December 24, 1906, the first such message was sent out from the Massachusetts coast, and wireless receivers picked up music.

TRIODE

The rectifying diode worked out by Fleming (see 1904) was a useful tool but limited in its range. That range was extended in 1906 by the American inventor Lee De Forest (1873–1961). He inserted a third element, called a *grid,* into the tube to make it a *triode* (three electrodes).

The grid is an electrode with holes in it, so that electrons can move from the hot filament through the holes of the grid to the plate. Even a weak charge placed on the grid can have a relatively enormous effect on the electron current. It can increase the intensity of the current if the grid is slightly positively charged, since it then attracts electrons from the heated filament; and it can decrease the intensity if slightly negatively charged, since it then repels electrons. By placing a small varying charge on the grid, you get a much larger variation in the electron flow.

A triode therefore acts as an *amplifier,* and it can be modified to perform a great variety of tasks. Radio became more than ever adapted to the transmission of sound through Fessenden's modulation (see above).

TROJAN ASTEROIDS

By 1906, no fewer than 587 asteroids had been spotted (thanks to Wolf's photographic technique—see 1891) and had their orbits calculated.

In 1906 Wolf discovered an asteroid that moved unusually slowly and must therefore be unusually distant. In fact it moved in Jupiter's orbit at the outer extreme of the asteroid belt, keeping step with Jupiter at a distance of 60 degrees.

This meant that the Sun, Jupiter, and the new asteroid were located at the vertices of a vast equilateral triangle. Lagrange (see 1788) had shown back in 1772 that such an arrangement was gravitationally stable, but this was the first case of such a gravitational triangle that had been actually found in space.

Wolf named the asteroid *Achilles,* after the hero of Homer's *Iliad,* the tale of the Trojan War. In giving an asteroid with an unusual orbit a masculine name, he followed the precedent of Witt (see 1898).

Other asteroids were eventually discovered that accompanied Achilles in its triangular position, and still others at a point 60 degrees from Jupiter in the other direction, so that there were two equilateral triangles adjacent to each other. All were named after various heroes of the Trojan War, so that they could be lumped together as the *Trojan asteroids.* The position at the third apex of a triangle where the other two are occupied by larger bodies is now called the *Trojan position.*

ALPHA PARTICLES

By now it was understood that beta rays were streams of speeding electrons (beta particles), while gamma rays were electromagnetic radiation of still shorter wavelength and higher frequency than X rays. The nature of the alpha particles that made up the alpha rays remained to be determined.

In 1906 Rutherford (see 1897), working with a German assistant, Johannes Hans Wilhelm Geiger (1882–1945), managed to determine the ratio of the electric charge to the mass in the case of the alpha particles. This ratio turned out to be equal to that in a helium atom from which two electrons had been removed.

Later Rutherford fired alpha particles at a double wall of glass with a vacuum between. The alpha particles had energy enough to penetrate the first partition but, in the process of penetration, lost so much energy that they were unable to penetrate the second. They therefore remained in the vacuum between, and after enough had accumulated, Rutherford found that the thin gas that had appeared in the vacuum was indeed helium, judging from the spectrum.

Alpha particles and helium were thus related, but not identical. After all, streams of helium atoms would not penetrate glass.

CHARACTERISTIC X RAYS

X rays had been discovered eleven years earlier and were still of profound interest to physicists. The British physicist Charles Glover Barkla (1877–1944) studied the manner in which X rays were scattered by gases and found that the higher the molecular weight of the gas, the greater the scattering of the X rays. From this he deduced, in 1904, that the more massive the atoms and molecules, the more charged particles they contained, since it was the charged particles that did the scattering. This was the first indication of a connection between the number of charged particles in an atom and its position in the periodic table.

He further showed, from the manner of scattering, that X rays were transverse waves like light, not longitudinal waves like sound. This was the final proof that they were examples of electromagnetic radiation.

In 1906 Barkla went on to something still more important. He showed that when X rays were scattered by particular elements, they produced a beam with a particular degree of penetration. The higher the atomic weight of an element, the more penetrating the *characteristic X rays* they produced. He went on to describe two types of such X rays: the more penetrating he called *K radiation* and the less penetrating he called *L radiation*.

At the time it was difficult to know what to make of this, but before long, characteristic X rays were to be important in rationalizing the periodic table.

For his work on X rays, Barkla was awarded the Nobel Prize in physics in 1917.

THIRD LAW OF THERMODYNAMICS

By now, temperatures as low as 14 degrees above absolute zero had been reached with Dewar's solid hydrogen (see 1898). It seemed that the race to reach absolute zero might be soon completed.

In 1906, however, the German physical chemist Walther Hermann Nernst (1864–1941) advanced thermodynamic reasons for supposing that absolute zero could not be reached by any technique. He proposed that, like the speed of light in a vacuum, absolute zero was a limit that could be approached more and more closely but never actually reached. This is sometimes called *the third law of thermodynamics*, and for this Nernst was awarded the Nobel Prize in chemistry in 1920.

VITAMIN CONCEPT

Since Eijkman had found that beriberi could be cured by dietary means (see 1896), biochemists had become aware of other examples of *dietary-deficiency diseases*.

Hopkins (see 1900) was convinced that various organic compounds were present in the diet that were essential to life but needed only in tiny quantities. In 1906 he made this point in a lecture and suggested that scurvy and rickets were two diseases that arose through the lack of trace substances.

When these trace substances became known as *vitamins* a few years later, Hopkins was given credit for having helped give rise to the *vitamin concept*, so that he and Eijkman shared the Nobel Prize in medicine and physiology in 1929.

MAGNESIUM AND CHLOROPHYLL

Ever since Pelletier discovered chlorophyll (see 1817), it had been understood that the substance was of the utmost importance. After all, it brought about the production of food and oxygen on which animal life, including human beings, subsisted. The chemical structure of chlorophyll was therefore of devouring interest to biochemists, but it was still poorly understood for all that.

The German chemist Richard Willstätter (1872–1942), however, managed to supply a key item of information about chlorophyll structure in 1906. He showed that each chlorophyll molecule contained a magnesium atom, held in much the same way that an iron atom was held by hemoglobin.

For this and for other work on planet pigments, Willstätter was awarded the Nobel Prize for chemistry in 1915.

CHROMATOGRAPHY

The Russian botanist Mikhail Semenovich Tsvett (1872–1919) worked with plant pigments, which are made up of a large number of rather similar organic compounds, difficult to separate into individual substances that can be studied singly. (This is a difficulty that frequently arises in biochemistry.)

In 1906 Tsvett found a convenient means of separation. He let a solution of a mixture of the pigments trickle down a tube of powdered aluminum oxide. The different substances in the pigment mixture held onto the surface of the powder particles but with different degrees of strength. As the mixture was washed downward, the substances began to separate, those holding with less strength being washed down farther.

If the tube of aluminum oxide was long enough, the substances in the mixture would be completely separated by the bottom of the column, and they would be washed out individually. The separation could be judged by the appearance of different shades of color on the column, so the technique was called *chromatography,* from Greek words meaning "writing in color." The name was retained even in the case of mixtures of colorless substances.

Chromatography, modified in many ways, became one of the most important techniques for the study of complex mixtures.

RADIOACTIVITY AND EARTH

In 1906 the American geologist Clarence Edward Dutton (1841–1912) suggested that pockets of radioactivity in the Earth's crust delivered enough heat over time to activate volcanic action.

This was the beginning of the understanding that radioactivity added substantial heat to the Earth's crust, enough to balance that lost by radiation, so that any attempt to judge the Earth's age by calculating the time it took Earth to "cool down" from an initial high temperature was far off base. Earth could be billions of years old and still retain a heated interior.

Dutton also developed methods for determining the depth of earthquake origins and the velocity with which earthquake waves traveled through the Earth. This opened the way for a technique that finally offered strong evidence concerning the physical and chemical nature of the Earth's deep interior.

IN ADDITION

In Europe, Great Britain escalated the naval rivalry with Germany by launching the *Dreadnaught,* the most powerful battleship in the world.

The Dreyfus case finally came to an end in France. Twelve years after his conviction, Dreyfus was fully exonerated and restored to his rank.

Germany's population was now the largest in western Europe at 62 million. The United States, however, had 85 million and Russia 120 million.

1907

RADIOACTIVE DATING

Once it was understood that radioactivity involved the change of uranium or thorium to some other radioactive atom, which then broke down into still another, and so on (see 1900, Atomic Change, and 1902, Radioactive Series), the question that naturally arose was, Where did it all end?

The American chemist Bertram Borden Boltwood (1870–1927) was convinced that lead was the final product of the radioactive series that began with uranium and thorium. As early as 1905 he had noted that lead was always present in uranium and thorium ores. In 1907 he pointed out that from the quantity of lead in uranium ores and from the known rate of uranium disintegration, it might be possible to determine with reasonable accuracy how long a portion of the Earth's crust had been solid and had existed essentially undisturbed.

It had been about a century and a quarter since Hutton had said that he saw no sign of a beginning to Earth's history (see 1785), but now at last radioactive dating looked like a technique that might yield such a sign. Boltwood's suggestion has born fruit ever since.

LUTETIUM

The rate of discovery of elements had been falling off—at least of elements that were not involved in radioactive series. Up to this point, thirteen elements had been isolated from the rare earth minerals, which might have exhausted the supply, but there turned out to be room for another. In 1907 the French chemist Georges Urbain (1872–1938) discovered a fourteenth and named it *lutetium,* from an old name for the Roman town that had stood on the site of what is now Paris.

SYNTHETIC PEPTIDES

It was well known that protein molecules were built up out of amino acids, but it was not known in just what manner those amino acids connected with each other. It seemed most likely that the amino group of one amino acid combined with the acid group of another, but that was not certain.

In 1907, however, Fischer, who had determined the structure of sugar molecules (see 1884), combined amino acids by chemical techniques that insured a combination between the amino group of one amino acid and the acid group of the next. He continued this until he had built up a chain consisting of eighteen amino acids.

He compared such a chain with those resulting from the breakdown of protein molecules by digestive enzymes. The fragments of a protein molecule so obtained are called *peptides,* from the Greek word for "to digest." Fischer found that his *synthetic peptides* matched the properties of the natural peptides obtained from protein in all important ways. Indeed, the synthetic peptides could also be broken down by digestive enzymes.

The basic structure of the protein molecule was thus known. What remained was to determine the precise order of amino acids in given protein molecules, but that was not to happen for another half-century.

CHEMOTHERAPY

The medieval alchemists had attempted to cure disease by the use of various chemicals. They were not truly successful (except occasionally by accident), because they did not know the cause of disease and had no way of testing particular chemicals in a rational manner before using them. Their techniques were largely abandoned, therefore.

The German bacteriologist Paul Ehrlich (1854–1915) returned to the notion of chemical cures (he coined the word *chemotherapy* in this connection) on a much more knowledgeable basis. It was understood, thanks to the work of Flemming (see 1882), that synthetic dyes could combine with some parts of cells and not others and could more markedly affect some cells than others. It occurred to Ehrlich that if a dye could be found that combined with some pathogenic organism but not with human cells, it would be a "magic bullet" that could kill the pathogen while leaving the human patient largely unaffected.

By 1907 he had located a dye called *Trypan red* that combined with and killed trypanosomes, a type of protozoa that caused sleeping sickness. Trypan red was therefore a possible cure for the disease.

This development of chemotherapy earned Ehrlich a share of the Nobel Prize in physiology and medicine in 1908.

FRUIT FLIES

Mendel had worked out the laws of inheritance by studying pea plants (see 1865), and these laws had been verified for animals by Bateson (see 1902). However, animals are much harder to work with on the whole than plants are.

In 1907, however, the American geneticist Thomas Hunt Morgan (1866–1945) began to use a tiny insect called *Drosophila,* or *fruit fly.* They have only four chromosome pairs in each cell, are simple to feed, and breed readily and copiously at brief intervals.

In studying them, Morgan found that there were characteristics that were linked and inherited together, but that the linkage was not necessarily permanent. Every once in a while, two characteristics that had previously been inherited together were suddenly inherited independently. He was able to correlate this with the fact that chromosomes sometimes interchanged parts so that two characteristics ordinarily on the same chromosome came to be on different chromosomes.

Fruit fly research greatly hastened the pace at which knowledge of genetic mechanisms increased. For his work with them, Morgan was awarded the Nobel Prize for medicine and physiology in 1933.

TISSUE CULTURE

If intact animals are hard to study, might it be possible to take small sections of tissues and keep them alive and growing? In this way, it might be possible to study the growth, development, and functioning of cells, as well as changes in their organization. Eventually, this technique offered a method for the cultivation and study of microorganisms that would not grow on grosser, simpler media.

The first to culture tissues successfully was an American zoologist, Ross Granville Harrison (1870–1959). He cultivated tadpole tissue and found that nerve fibers grew from it. He was able to study the protoplasmic movements within the fibers, and this served as the foundation for later studies on nerve physiology.

CONDITIONED RESPONSE

Salivation at the sight of food is an *unconditioned response.* It is brought about by the construction of the nerve network with which the organism is born. In 1907 Pavlov (see 1902) began an attempt to see if he could impose a new pattern upon such inborn ones.

Thus, a hungry dog that is shown food will salivate. If a bell is made to ring every time the dog is shown food, the dog will eventually salivate when the bell rings even when food is not shown. The dog has associated the sound of the bell with the sight of food and reacts to the first as though it were the second. This is a *conditioned response.*

Studies of the conditioned response led to the thought that a good part of learning and of the development of behavor is the result of conditioned responses of all sorts picked up in the course of life.

SPACE-TIME

Einstein's theory of special relativity (see 1905) forced many physicists to reconsider their view of the Universe. It was clear from Einstein's work that an ordinary three-dimensional view of the Universe was insufficient.

The Russian-born German mathematician Hermann Minkowski (1864–1909) published *Time and Space* in 1907. In it he demonstrated that relativity made it necessary to take time into account as a kind of fourth dimension (treated mathematically somewhat differently from the three spatial dimensions). Neither space nor time had existed separately, in Minkowski's view, so that the Universe consisted of a fused *space-time.*

Einstein adopted this notion as he continued to work on his theories. He was trying to extend them to accelerated motion in order to take gravitational interactions into account.

IN ADDITION

Great Britain, increasingly aware of the dangers of a Germany that was trying to be strong both on land and sea, continued to search for allies. On August 31, 1907, Great Britain concluded a *rapprochement* (French for "reconciliation") with Russia.

Europe was now divided into two armed camps of roughly equal strength. There was the *Triple Entente* (Great Britain, France, and Russia) on one side and the *Triple Alliance* (Germany, Austria-Hungary, and Italy) on the other. The powder train was laid; all that was needed was the spark.

In the United States, Oklahoma joined the Union as the forty-sixth state.

Oscar II of Sweden (1829–1907), who had begun his reign in 1872, died and was succeeded by his son, who reigned as Gustav V (1858–1950).

A record number of immigants—more than one and a quarter million—entered the United States in 1907. That number has never been surpassed.

1908

ATOMIC SIZE

Einstein had devised an equation showing how one might calculate the size of atoms and molecules from Brownian motion (see 1905). In 1908 Perrin, who had shown that cathode rays consisted of particles carrying a negative electric charge (see 1895), set about making the calculation.

Through a microscope, he counted the number of small particles of gum resin suspended at different heights in water. That they were suspended at all was the result of recoil from collisions with water molecules; in other words, from Brownian motion. From his observations and from Einstein's equations, Perrin could make his determination.

For the first time, the approximate size of atoms could be deduced from an actual observation. Atoms, it turned out, are about a hundred-millionth of a centimeter in diameter. In other words, 250,000,000 of them, lined up side by side, would stretch about an inch.

This was the final demonstration of the real existence of atoms. They were *not* merely a convenient hypothesis designed to make chemical calculations easier.

LIQUID HELIUM

It had been ten years since Dewar liquefied hydrogen (see 1898) and left helium as the only unliquefied gas.

In 1908 the Dutch physicist Heike Kamerlingh Onnes (1853–1926) set about the task of liquefying helium. He built an elaborate device that would cool helium intensively by means of evaporating liquid hydrogen. When the helium had reached a very low temperature under compressed conditions, it would be allowed to expand so that it would cool still further.

In this way, liquid helium was finally collected in a flask set within a larger flask of liquid hydrogen, which was in turn contained in a still larger flask of liquid air—so that the liquid helium would gain heat and vaporize only very slowly.

It turned out that helium liquefied at a temperature of only 4 degrees above absolute zero. By allowing some of it to evaporate, Kamerlingh Onnes reached a temperature of only 0.8 degrees above absolute zero, but even then he could not solidify it.

For his liquefaction of helium, Kamerlingh Onnes was awarded the Nobel Prize in physics in 1913.

GEIGER COUNTER

Rutherford, who had recently shown the relationship of alpha particles to helium (see 1906), was deeply engaged in investigations involving the energetic particles emitted by radioactive substances. His labors were made easier by the invention of a device to detect and, eventually, record such particles.

The invention was that of his assistant, Geiger (see 1906), who in 1908 produced the first rather primitive version of his device. Essentially, it was a cylinder containing a gas under a high electric potential, but one not quite high enough to overcome the resistance of the gas and set up a spark of discharge.

If a high-energy subatomic particle entered, it would rip some of the electrons off a gas molecule, and what was left of the molecule would be a positively charged ion, which would be pulled toward the negatively charged cathode with great energy. In the process, as a result of collisions, the ion would produce further ions, each of which would do the same. In short, a single subatomic particle, on entering, would set off an avalanche of ionization, producing a momentary electric discharge, which could be arranged to make a clicking sound. The clicking of such a *Geiger counter* gives us information that our senses cannot give us.

SUNSPOTS AND MAGNETISM

For some three centuries, astronomers had observed sunspots, counted them, noticed the manner in which they increased and decreased cyclicly in number, but knew nothing more about them than they could see. Hale, who had invented the spectroheliograph (see 1890) and supervised the construction of the 40-inch refractor (see 1897), changed that.

In 1908 he was able to show from the spectrum of sunspots that they exhibited the Zeeman effect (see 1896). That indicated that they were subjected to a strong electromagnetic field. It was the first time such a field had been detected for any astronomical object but Earth itself.

RICKETTSIA

The American pathologist Howard Taylor Ricketts (1871–1910) was investigating the serious disease of Rocky Mountain spotted fever. In 1906 he had shown that it was spread by cattle ticks.

In 1908 he finally located the pathogenic agent and found it to resemble a rather small bacterial cell. It was not, however, a complete cell, capable of independent life and growth. Apparently it lacked some essential components of life, and like a virus, could only grow inside a cell, where it could make use of the cellular machinery to eke out its own deficiencies.

These agents, which came to be called *rickettsia,* in honor of the discoverer, were therefore intermediate between viruses and bacteria.

ASSEMBLY LINE

During the first twenty years of its existence, the automobile had been improved greatly and manufactured in greater numbers. It remained very largely a toy of the rich, however, rather as yachts are today.

The man who changed that was the American industrialist Henry Ford (1863–1947). He had built his first automobile in 1893 and started a car-manufacturing company in 1899. His aim was to produce cars in quantity *(mass production)* and to make them cheaply enough to put them within reach of middle-class Americans.

His key innovation came in 1908, when he thought of dividing the manufacture of cars into steps, each of which could be performed simply by a single worker. He then placed the future car on a moving belt that brought it to different workers in succession, each of whom performed the assigned task over and over, with all necessary tools and parts within reach. What started at one end of the *assembly line* was a mere skeleton of a car. What rolled off the other end was a complete, functioning automobile, including a gasoline supply so that it could be driven away.

Ford produced a series of models identified by letters of the alphabet and finally considered the *Model T* suitable for mass-production. It cost only $950 to begin with, but the price dropped in succeeding years, eventually reaching a low of $290. For the first time, the average man could afford to buy a car, and the automobile age—still in full swing today—began.

HABER PROCESS

Nitrogen is absolutely essential to life, and also to explosives. Hellriegel had discovered that leguminous plants could fix atmospheric nitrogen and keep soil fertile (see 1886), but that was a slow process and would not suffice for the manufacture of explosives in the quantities that a war economy required.

Nitrogen in its most usable form exists in nature as *nitrates* in the soil. Nitrates are therefore sought after, both for fertilizers and for explosives. Nitrates, however, are uniformly soluble and tend to be leached out of the soil by rainfall. One can therefore expect a reliable source of nitrates to be found chiefly in desert areas such as those in northern Chile.

Germany, anticipating possible war with Great Britain, knew that British control of the seas would prevent Chilean nitrates from reaching Germany, reducing Germany's ability to fight a long war. German chemists were therefore encouraged to find an alternate source for nitrates.

The German chemist Fritz Haber (1868–1934) sought a way of fixing atmospheric nitrogen in the laboratory. He found that if he placed a mixture of nitrogen and hydrogen under high pressure, in the presence of iron as a catalyst, he could manufacture ammonia. From ammonia, it was easy to obtain nitrates. By 1908 he had perfected the method and provided a source of home-grown (so to speak) affordable nitrates.

This *Haber process* made it possible for Germany to fight long wars—unfortunately.

IN ADDITION

The Ottoman Empire continued to fall apart. Bulgaria declared itself independent in 1908. Crete joined itself to Greece. Austria-Hungary annexed Bosnia and Herzegovina in the northwestern corner of the Balkans. All that was left of Turkish dominions in Europe was a strip of territory running westward from Constantinople to the Adriatic. Stung by the humiliation of Turkey, a group of revolutionaries, the *Young Turks,* forced Abdülhamid II (1842–1918), the Turkish sultan, to grant a constitution and establish a parliament.

In the United States, William Howard Taft (1857–1930) became the twenty-seventh president.

In central Siberia, on June 30, 1908, the most devastating meteor strike of historic times knocked down every tree for miles about and destroyed a herd of reindeer. Through the sheer chance of having struck uninhabited land, it created no destructive tidal waves and cost not a single human life. Since no crater was ever found, it was probably a small comet that struck, its icy substances heated to explosion before it ever quite reached the ground.

The population of New York City reached 4.4 million.

1909

SYPHILIS

Ehrlich had won a Nobel Prize for finding a chemical that would kill trypanosomes and might therefore serve as a cure for sleeping sickness (see 1907). The chemical involved contained nitrogen, and it occurred to Ehrlich that if the similar but much more poisonous element arsenic were substituted for the nitrogen, it might prove a more effective chemotherapeutic agent.

He tried every organic arsenic compound he could obtain or synthesize, and the 606th was a compound now called *arsphenamine.* It did not seem unusually promising as an agent against trypanosomes, but in 1909 an assistant of Ehrlich found that it was effective against spirochetes, the agents that caused syphilis.

Syphilis had been a dreaded disease for four centuries, all the more so since it was spread by sexual activity so that it was considered shameful and could not be mentioned in polite society. Naturally, that just multiplied its incidence.

Arsphenamine cut down the incidence of syphilis by half in England and France over the next five years, but bluenoses were not wanting who thought it a wicked invention because it encouraged "immorality" by lessening the fear of medical consequences.

TYPHUS

Typhus, a contagious disease with a high mortality rate, periodically struck in epidemic form.

In Tunis, a French physician, Charles-Jean-Henri Nicolle (1866–1936), noted that typhus was very contagious outside the hospital but not inside. Something must happen upon entrance to the hospital, and Nicolle wondered if it might be the fact that patients' clothes were removed and that they were washed and put into clean hospital garments.

Nicolle decided that the contagious agent must be connected with the old clothes, and he suspected the body louse, which exists everywhere among human beings where the chance for frequent washing of clothes and body does not exist.

By 1909 he had proved the case. The body louse, by biting someone with typhus and then biting someone without it, spread the disease. This meant the disease could be controlled if some practical way could be found of removing the louse, in societies where for one reason or another washing was infrequent. That was not to come for another third of a century.

RIBOSE

Kossel had isolated the nitrogenous bases of nucleic acid (see 1885) but had not been able to go further. It was clear, though, that nitrogenous bases were not all there was in the molecule.

In 1909 the Russian-born American chemist Phoebus Aaron Theodore Levene (1869–1940) extracted a sugar from nucleic acid and identified it as *ribose.* It had a five-carbon molecule. Not all nucleic acid molecules possessed it, but those that did came to be known as *ribose nucleic acid,* almost invariably abbreviated as *RNA.*

The sugar, if any, in the samples of nucleic acid that did not contain ribose was not identified for another twenty years.

GENES

Thanks to the work of Morgan and his fruit flies (see 1907), it was established that chromosomes contained chains of many units of inheritance. It would obviously be convenient to have some way of referring to those units in a concise way. In 1909 the Danish botanist Wilhelm Ludvig Johannsen (1857–1927) suggested that they be called *genes.* The suggestion was adopted.

TUNGSTEN WIRE

When Edison introduced the electric light bulb (see 1879), he used carbon fibers as filaments. These were brittle, hard to handle, and didn't last long. Obviously some sort of metal wire would be better. However, it would have to be of a metal with a high melting point so that it could withstand white-hot temperatures, and such metals were for the most part expensive, hard to draw into wires, or both.

The metal with the highest melting point is tungsten, which melts at about 3410° C. It is not inordinately expensive, but it is brittle. In 1909, however, the American physicist William David Coolidge (1873–1975) managed to perfect a method of drawing tungsten into fine wires.

As a result, tungsten filaments became universal in light bulbs, radio tubes, and other devices. Light bulbs lasted for considerably longer before having to be replaced, therefore, though further improvements remained to be made.

BAKELITE

Hyatt's celluloid had been the first important synthetic plastic (see 1869), but in the forty years since it had been introduced, there had been no flood of additional plastics on the market.

The real beginning came with the work of a Belgian-born American chemist, Leo Hendrik Baekeland (1863–1944).

It often happened in organic chemistry that hard tarry residues fouled chemical equipment and then could not be removed. Baekeland tried to find some solvent that would remove them. For the purpose, he deliberately reacted phenol with formaldehyde to form such a residue, and then searched for a solvent. He couldn't find one.

It occurred to him that if the residue was so resistant to solvents, it might have a useful application in its own right as an inert, strong, and cheap material, so he began to concentrate on forming the resinous mass more efficiently and making it still harder and tougher. By using the proper heat and pressure, he obtained a liquid that solidified and took the shape of the container it was in. Once solid, it was hard, water-resistant, solvent-resistant, and an electrical insulator. Yet it could be cut without trouble and was easily machined.

In 1909 he brought this substance (which he named *Bakelite*, after himself) on the market. Bakelite was the first of the *thermosetting plastics* (one that once set will not soften under heat) and has been useful ever since. It sparked the modern development of plastics.

MOHOROVIČIĆ DISCONTINUITY

There was an earthquake in the Balkans in 1909, and a Croatian geologist, Andrija Mohorovičić (1857–1936), studied the waves that traveled through the Earth's crust as a result. He found that waves that penetrated deeper into the Earth arrived sooner than waves traveling along the surface.

Mohorovičić maintained that the Earth's outermost crust rested on a layer that was more rigid and in which earthquake waves therefore traveled more quickly. Furthermore, the separation between the two layers did not seem to be gradual but sharp. It was a *discontinuity* and came to be known as the *Mohorovičić discontinuity*.

This was the first indication that the Earth was composed of more than one layer, with widely different properties.

NORTH POLE

Ever since the search for the Northwest Passage had begun, some three and a half centuries before, the North Pole had been a distant goal for explorers, but the ice defeated everyone until the early years of the twentieth century.

The American explorer Robert Edwin Peary (1856–1920) made it his goal. Beginning in 1886, he began a detailed exploration of Greenland and in 1891 explored its northern coast (relatively ice-free and still called *Peary Land* in his honor). He proved that Greenland was an island that did not extend to the North Pole, although the northernmost bit of Greenland is closer to the pole than any other land area.

In 1909 Peary organized an elaborate travel party, of which successive members were to turn back at periodic intervals until at the end only Peary and his black associate, Matthew Alexander Henson (1866–1955), reached the North Pole in a final dash on April 6, 1909.

There was some controversy over this. An erstwhile associate of Peary's, Frederick Albert Cook (1865–1940), claimed to have reached the Pole in 1908. The controversy has never been settled and perhaps never will be, but the general opinion is that it was Peary who first attained the goal.

IN ADDITION

On April 26, 1909, Abdülhamid II of the Ottoman Empire was forced to abdicate, and his brother succeeded to the sultanate, reigning as Mohammed V (1844–1918).

Leopold II of Belgium died on December 17, 1909, and was succeeded by his nephew, reigning as Albert I (1875–1934).

1910

NEON LIGHTS

Beginning in 1910, the French chemist Georges Claude (1870–1960) showed that electric discharges through the noble gases could be made to produce light. Most spectacular was the red light produced in this manner by neon, so that light produced in this manner by any gas came to be called *neon lights*.

The fact that tubes filled with neon or other gases could be twisted into any shape —so that they spelled out words, for instance —made it inevitable that they would replace ordinary incandescent bulbs in advertising signs.

SEX-LINKED CHARACTERISTICS

In 1910 Morgan, still working with fruit flies (see 1907), noted a white-eyed male fly among a mass of ordinary red-eyed ones. It was a mutation such as De Vries had observed among plants (see 1900).

Morgan crossed the white-eyed male with a red-eyed female, and all the offspring were red-eyed (red was dominant). In the next generation, however, there were both red-eyed and white-eyed flies, and all the white-eyed ones were males.

This was the first observation of *sex-linked characteristics*. It meant that males and females had to be differentiated in chromosome makeup. The way they are is that not all the chromosomes make up perfect pairs. In the case of one of them, the female fruit fly did indeed have a pair (an X chromosome and an X chromosome), but the male had one normal chromosome and a stub (an X chromosome and a Y chromosome). A white-eye gene on the female X chromosome could be overbalanced by a red-eye gene on the pair, but a white-eye gene on the male X chromosome had nothing to balance it on the Y-chromosome stub.

The chromosome pairs of male and female human beings show a similar differentiation.

MATHEMATICS AND LOGIC

Russell (see 1902) and the British mathematician Alfred North Whitehead (1861–1947) collaborated on a monumental three-volume work, *Principia Mathematica,* the first volume of which appeared in 1910. It was another effort to establish mathematics as a branch of logic, building all of it out of basic definitions and processes. It was the most nearly definitive accomplishment of this sort.

IN ADDITION

On May 6, 1910, Edward VII of Great Britain (1841–1910) died and was succeeded by his son, who reigned as George V (1865–1936). Elsewhere in the British Empire, the Union of South Africa was established on May 31, 1910, uniting the British and the Boers in a virtually independent dominion. Its first prime minister was Louis Botha (1862–1919), who had fought for the Boers during the Boer War.

The Balkans continued to see a diminution of Turkish power as the little nation of Montenegro, just north of Albania, declared its independence on August 28, 1910, under Nicholas I (1841–1921). A revolt in Albania, however, was suppressed by the Turks.

A revolution in Portugal put an end to the nearly eight-century-old Portuguese monarchy. Portugal's last king, Manuel II (1889–1932), who had begun his reign in 1908, fled the country on October 4, 1910.

In Asia, Japan continued its steady program of expansion by annexing Korea on August 22, 1910.

The population of the United States reached 92 million.

Halley's comet approached Earth on its return from around the Sun (its third return since Halley's prediction—see 1705). Its tail enveloped Earth, the prospect of which induced panic, but the tail is so utterly thin that it had no effect whatever.

1911

NUCLEAR ATOM

For some years, Rutherford had been firing alpha particles at sheets of matter (see 1906), thinking that even if they could penetrate the matter, they might be deflected and scattered by the atoms making it up. From the manner of the deflection, Rutherford hoped to gain some knowledge of atomic structure.

In 1908, for instance, he fired alpha particles at a sheet of gold only 1/50,000 of an inch thick. Most of the alpha particles passed through, unaffected and undeflected, recording themselves on the photographic plate behind.

Since the gold represented a barrier that was two thousand atoms thick, the fact that alpha particles could pass through them all as though there were nothing there seemed to mean that atoms were mostly empty space. Yet some alpha particles *were* deflected, and struck the photographic plate some distance away from the central spot formed by the main stream of alpha particles. Some were deflected through quite an angle.

This meant that part of the atom contained considerable mass. From the fact that so few alpha particles were deflected, it could be concluded that the part with the mass must make up a very small fraction of the atom.

By 1911 Rutherford had gathered enough evidence to put forward his theory of the nuclear atom. The atom, it would seem, has virtually all its mass squeezed into a tiny, positively charged *atomic nucleus* (which we now know is only 1/100,000 the diameter of the atom). The outer regions of the atom contain enough electrons, each with a unit negative charge, to neutralize the nuclear charge and leave the atom as a whole electrically neutral.

This theory was adopted quickly, for it answered major questions. For example, the relationship between alpha particles and helium was now understood. The alpha particles were helium *nuclei,* not helium atoms. This accounted for the electric charge of the alpha particle and, thanks to its subatomic size, for its penetrating power.

CLOUD CHAMBER

Since the discovery of radioactivity by Becquerel (see 1896), the use of speeding subatomic particles had been increasing. This made it important to have devices that could yield information concerning them. The Geiger counter (see 1908) could detect their presence, but more was needed.

A Scottish physicist, Charles Thomson Rees Wilson (1869–1959), had been studying clouds and labored to produce small artificial clouds in the laboratory that he could also study.

In 1896 he had allowed moist air to expand within a container so that the expansion lowered the temperature and not all the moisture could be retained, the excess coming out as water droplets to form a tiny cloud. In this way he determined that the presence of dust or of electrically charged ions encouraged the formation of water droplets and therefore of clouds.

It occurred to Wilson eventually that energetic radiation would produce ions as they spread through the atmosphere. If he could prepare air that was dust-free, he could make it so moist that water drops would only be kept from condensing by lack of the dust that would serve as condensation-seeds.

If an energetic particle then passed through the chamber, and if the chamber was expanded, droplets of water would form around the ions produced by the passage of the particle, and not only would the presence of the particle be detected but its route of travel as well. If the *cloud chamber* was then placed in a magnetic field, the curvature of the path of the particle would indicate the nature of its electric charge and give information concerning its mass. It would also indicate collisions of particles with molecules and with other particles and offer a guide to events that took place before and after the collision.

Wilson perfected his cloud chamber in 1911 and it quickly became an important adjunct of nuclear research. For this work, Wilson was awarded a Nobel Prize in physics in 1927.

ELECTRON CHARGE

The ratio of the electric charge of the electron to its mass had been worked out and compared with that of ordinary ions by Thomson (see 1897). The size of the electric charge in an absolute sense, however, was not known.

The American physicist Robert Andrews Millikan (1868–1953) tackled the job. Beginning in 1906, he had followed the course of tiny electrically charged water droplets falling through air, under the influence of gravity, against the pull of a charged plate above. The evaporation of the water confused the results, and in 1911 he began to use tiny oil droplets instead.

Every once in a while, such an oil droplet attached itself to an ion, which Millikan produced by passing X rays through the chamber. With the ion added, the effect of the charged plate above was suddenly strengthened and the droplet would fall more slowly or perhaps even rise. The minimum change in motion was due, Millikan felt, to the addition of a single electronic charge. By balancing the effects of the electromagnetic attraction upward and the gravitational attraction downward, both before and after such an addition, Millikan was able to calculate the charge on a single electron. The figure we now have is sixteen-quintillionths of a coulomb.

For this work, Millikan was awarded the Nobel Prize in physics in 1923.

COSMIC RAYS

One way to detect the presence of energetic radiation is by the use of a gold-leaf electroscope. This device consists of two pieces of gold leaf, joined at the upper end, and contained in a sealed jar. They can be electrically charged from the outside, and since both leaves have the same charge, they repel each other, forming an inverted V. Any energetic radiation entering the jar produces ions, which will carry off the electric charge and allow the gold leaves to come together slowly.

There seemed no way of keeping the leaves permanently apart, however, even in the absence of any known source of energetic radiation. Some radiation was apparently coming from an unknown source.

An Austrian physicist, Victor Franz Hess (1883–1964), felt that the source must be somewhere in the ground, so in 1911 he took electroscopes up on balloon flights to get them out of range of the ground radiation.

He made ten flights and observed quite to his astonishment that the gold leaves came together up to eight times as rapidly at considerable heights as they did at ground level. The radiation seemed to be coming from above, after all—to be coming from outer space, from the cosmos generally. Millikan (see above) suggested, therefore, that the radiation be called *cosmic rays,* a name that stuck.

For his discovery, Hess was awarded a share of the Nobel Prize in physics in 1936.

SUPERCONDUCTIVITY

Kamerlingh Onnes, having liquefied helium and obtained temperatures of 4 degrees above absolute zero (4° K) and even lower (see 1908), was eager to study the properties of matter at such exceedingly low temperatures.

For instance, it was known that metals tended to lessen their resistance to an electric current as the temperature went down. It seemed to Kamerlingh Onnes that this lessening must continue all the way down to absolute zero, where it must disappear altogether.

He tested the matter on mercury, and resistance dropped more or less as expected till he reached a temperature of 4.2° K. There, to his surprise, the resistance dropped suddenly to zero.

This phenomenon—the perfect conductivity of an electric current at temperatures close to absolute zero—was called *superconductivity.* It was soon found that other metals (though not all) also showed the phenomenon, the resistance dropping to zero at some very low temperature that was characteristic for each metal.

CHROMOSOME MAPS

Morgan had shown that chromosomes could cross over from one gene to another, which allowed them to be inherited separately where previously they had been linked, or inherited together. Obviously, the farther two genes were from each other on a particular chromosome, the greater the chance that a crossover somewhere along the chromosome would separate them.

Morgan and his assistant, the American geneticist Alfred Henry Sturtevant (1891–1970), investigated the frequency of separation by crossover in an attempt to locate the genes governing particular characteristics on a chromosome. The first such *chromosome map* was presented in 1911.

TUMOR VIRUS

Cancer, one of the most dreaded diseases, is not obviously contagious. On the other hand, it is not a single disease but a whole family of diseases involving unrestrained growth, with characteristics that may differ from one to another.

An American physician, Francis Peyton Rous (1879–1970), had occasion to examine a chicken with a tumor. When it died, Rous decided to test among other things whether it might contain a virus. He was sure it didn't but felt it wise to make sure by experiment.

He mashed up the tumor and passed it through a filter fine enough to keep out anything larger than a virus. To his surprise, he found that this filtrate was infectious and would produce tumors in healthy chickens. He published the report in 1911, and the disease came to be called the *Rous chicken sarcoma virus*. It was the first of a family of tumor viruses.

For this work, Rous was awarded a share of the Nobel Prize for medicine and physiology in 1966, fifty-five years after his report.

EARTHQUAKES AND FAULTS

It was known that there were *faults* in the Earth's crust, regions where two dissimilar sets of rocks came together. It was as though there had been a smooth stretch of rock originally, which had cracked, with one side sliding along the other to produce a mismatch. The general feeling was that such faults were produced by earthquakes.

The American geologist Harry Fielding Reid (1859–1944) studied the San Francisco earthquake and came to the conclusion, in 1911, that the reverse was true. The fault existed first, and the earthquake resulted when pressure upon the fault caused it to slip further.

This idea has been accepted ever since.

SEAPLANES

The American inventor Glenn Hammond Curtiss (1878–1930) was deeply involved in the early airplane flights. He was the first to fly a mile in the United States, in 1908, and he flew from Albany to New York in 1910.

In 1911 he finally built a practical *seaplane*, or *hydroplane*, with pontoons rather than wheels, that could take off and land on water.

SOUTH POLE

Peary's attainment of the North Pole (see 1909) sharpened the race to reach the South Pole. This was a more difficult objective, for the South Pole was more distant from important population centers, and it was at the center of a landmass, so that it was considerably colder than the North Pole.

The Norwegian explorer Roald Amundsen (1872–1928) had already made his way by sea across the northern border of North America in 1903 (achieving the Northwest Passage at last). Now he prepared for a race to the South Pole.

In October 1911 he set off, with dogs (who could eat meat—and each other, in case of need), reaching the South Pole on December 14 and returning safely.

The British explorer Robert Falcon Scott (1868–1912) was also trying to reach the pole, but he didn't plan his expedition as well. He ran into enormous misfortune, reached the South Pole a month after Amundsen did, and perished with his party on the way back.

SELF-STARTER

The automobile still had to be started manually by inserting a crank into the front of the car, where it could grip the rotor of the engine, forcing it to turn and "catch." That took a great effort, and sometimes when the engine caught, the crank began to turn at great speed, pulled out of the cranker's hand, and broke his arms.

The American inventor Charles Franklin Kettering (1876–1958) invented an electric starter in 1911 that would do the job at the turn of a key. It was used in the 1912 Cadillac and quickly grew popular. With the crank gone, automobiles could be started and driven by far more people, which greatly spread the automobile way of life.

IN ADDITION

In 1911 a revolution in China led by Sun Yat-sen (1866–1925) overthrew the child emperor, Hsuantung (1906–1967), and brought the Manchu dynasty to an end after two and a half centuries. For the first time in thousands of years, there was no emperor in China. The Chinese Republic was established.

There was a revolution in Mexico, and continuing unrest in Russia, too.

Europe, meanwhile, continued its imperialist ways. The only portion of North Africa that was still under the control of the Ottoman Empire was Libya. On September 29, 1911, Italy declared war on the Turks and took Tripoli, the Libyan capital, on October 5. The Turks were in no position to resist and the Ottoman Empire was finally driven out of Africa after four centuries.

A much more serious crisis took place farther west. Despite the fact that Morocco's independence had been guaranteed, France entered Fez, in northern Morocco, and was clearly planning to annex the country. An indignant Germany sent a gunboat, the *Panther*, to the Moroccan port of Agadir, and for a while it looked as if there might be war. By November 4, 1911, however, Germany agreed to let France have Morocco in return for a portion of French territory in west central Africa.

1912

CEPHEID VARIABLES

There is a group of variable stars in which the rise and fall of brightness follows a distinctive pattern, although the period—the time taken by one rise and fall—varies from star to star. These stars are called *Cepheid variables* because the first one discovered was in the constellation Cepheus.

The American astronomer Henrietta Swan

Leavitt (1868–1921) was interested in Cepheid variables and studied a number of them in the Magellanic Clouds, two large groups of stars that lie beyond the Milky Way (see 1678).

Her observations, beginning in 1904, showed that the brighter the Cepheid, the longer the period. This fact was obscured among the stars nearer to us, because some Cepheids seem dim only because they are quite far away, while others seem bright only because they are near. In the Magellanic Clouds, however, all the Cepheids are about the same distance from us, so that their apparent brightness reflects their *real* brightness, or *luminosity*.

By 1912 Leavitt had worked out a method for determining the luminosity of a Cepheid variable from its period. Once its luminosity was known, its distance could be calculated from its apparent brightness. What was needed to make this method work was a reliable estimate, by some other method, of the absolute distance of at least one Cepheid variable. This was a knotty problem, for even the nearest Cepheid is too far away for its absolute distance to be determined easily.

Once this was accomplished, however, the Cepheid variables could be used as a yardstick to determine distances far greater than those that could be determined by the method of parallax (see 150 B.C.).

NEBULAR VELOCITIES

The Andromeda nebula, which had first been observed telescopically exactly three centuries before, was still rather a puzzle for astronomers. It seemed to be a cloud of dust and gas to the eye, but its light was similar to starlight in characteristics, even though no stars were visible within it.

One thing that could be done was to study its spectrum and, from the position of its dark lines, see whether it was approaching us or receding from us. The American astronomer Vesto Melvin Slipher (1875–1969) made the measurement in 1912 and reported that it was approaching Earth at a speed of 125 miles per second.

This did not seem very important at the time, but Slipher went on to determine the radial velocities of other nebulae, and such results led, in the next decade, to a revolution in our notions of the structure of the Universe.

CONTINENTAL DRIFT

As soon as the shoreline of South America had been mapped, three and a half centuries before, people had noticed that South America and Africa gave the impression that they would fit together neatly if they were moved together.

In 1912 the German geologist Alfred Lothar Wegener (1880–1930) proposed that Africa and South America had indeed once formed a single landmass, which had broken in two, and that the two parts had then moved apart in a sort of *continental drift*.

In fact, he suggested that all the continents had originally formed a single mass (*Pangaea*, Greek for "all-earth") surrounded by a continuous ocean (*Panthalassa*, Greek for "all-sea"). This large granite mass of Pangaea had broken into chunks that slowly separated, floating on a basalt ocean-floor, and over hundreds of millions of years, produced the pattern of fragmented continents that we now have.

Unfortunately, the notion of granite floating on basalt did not seem a tenable hypothesis, and on the whole, few people took Wegener's notions seriously at the time.

X-RAY DIFFRACTION

Once Barkla had shown that X rays were electromagnetic radiation (see 1906), the question arose of what their wavelengths might be. The ordinary way of measuring the wavelength of electromagnetic radiation was to pass that radiation through a fine grating. However, to measure very short wavelengths, very fine gratings were required, and no grating fine enough to measure wavelengths as short as those of X rays could be manufactured.

It occurred to a German physicist, Max Theodor Felix von Laue (1879–1960), that it was unnecessary to manufacture such gratings; nature had already done the job. A crystal consisted of layers of atoms that were spaced just as regularly as, but far more closely than, the ruled scratches of any grating. A beam of X rays aimed at a crystal ought then to be diffracted as ordinary light would be by an ordinary grating. The results would be complicated, because ordinary gratings consist simply of parallel lines, while crystals contain layers of atoms extending in various directions. Still, it could be done.

In 1912 Laue tried the experiment of passing X rays through a crystal of zinc sulfide and recording the diffraction pattern on a photographic plate. It worked perfectly. It was clear that such a diffraction pattern could be used to calculate X-ray wavelengths, and that once the wavelengths were known, the technique could be used to study the fine structure of crystals.

For his work on X-ray diffraction, Laue was awarded the Nobel Prize for physics in 1914.

NEON VARIETIES

Thomson (see 1897) found himself interested in the canal rays that Goldberg had discovered a quarter-century earlier (see 1886). By 1912 it was clear that they were composed of positively charged subatomic particles, so Thomson called them *positive rays.* In the light of Rutherford's nuclear atom (see 1911), the positive rays appeared to be streams of atomic nuclei.

In 1912 Thomson studied the manner in which positive rays were deflected by magnetic and electric fields. He balanced those fields in such a way that particles of different charge-to-mass ratio curved differently and fell in different spots on a photographic plate.

When he deflected streams of neon nuclei, he found to his surprise that they landed on two different spots, as though there were two varieties of neon atom that differed in charge, mass, or both. Observations such as this were soon to introduce new concepts of atomic structure that proved to be of the greatest importance.

DIPOLE MOMENTS

Now that it was understood that electrons formed part of atoms, it followed that when atoms joined to form molecules, electrons ought to be distributed over the various atoms of the molecules.

Such distribution might be symmetrical, so that the molecule was electrically uncharged, or unsymmetrical, so that one part of the molecule might have a surplus of electrons and thus a slight negative charge while another part might have a deficit of electrons and a slight positive charge. These would represent positive and negative *poles* of the molecules, which would thus be *polar molecules,* or molecular *dipoles.*

Naturally, polar molecules and nonpolar molecules would act differently with respect to electric fields. In addition, polar molecules would attract each other, positive toward negative, and therefore have higher melting and boiling points than nonpolar molecules of similar size.

In 1912 a Dutch physical chemist, Peter Joseph William Debye (1884–1966), worked out a set of equations that would represent the behavior of such polar molecules. This gave rise to the concept of *dipole moments* and greatly helped chemists understand the behavior of molecules.

For this work, Debye was awarded the Nobel Prize for chemistry in 1936.

VITAMINS

The Polish-born biochemist Casimir Funk (1884–1967) strongly supported the notion advanced six years earlier by Hopkins (see 1906) that diseases such as beriberi, scurvy, pellagra, and rickets were caused by the absence of vital trace substances in the diet.

It was Funk's notion (an erroneous one) that such substances contained *amine groups* (combinations of a nitrogen atom and two hydrogen atoms), and he therefore called the trace substances *vitamines* (Latin for "life-amines").

When it turned out a few years later that not all such substances contained amine groups, the *e* was dropped to dilute the reference and the substances became known as *vitamins*.

COAL HYDROGENATION

The Haber process for hydrogenating nitrogen to form ammonia (see 1908) set off a flurry of activity. The German chemist Carl Bosch (1874–1940) improved the process and supervised the construction of large plants to make use of it. In 1912 another German chemist, Friedrich Bergius (1884–1949), applied the principles in treating coal and heavy oil with hydrogen to form gasoline.

For their work on high-pressure processes, Bosch and Bergius shared the Nobel Prize for chemistry in 1931.

IN ADDITION

In the United States, Woodrow Wilson (1856–1924) became the twenty-eighth president. New Mexico and Arizona became the forty-seventh and forty-eighth states and the entire area between Canada and Mexico was now divided into states.

Emperor Mutsohito of Japan, who had overseen Japan's rise to become a modern power, died on July 30, 1912, and his son Yoshihito (1879–1926) succeeded him.

Frederick VIII of Denmark died on May 14, 1912, and was succeeded by his son, who reigned as Christian X (1870–1947).

A war broke out in the Balkans on October 18, 1912. Serbia, Bulgaria, and Greece in alliance fought against Turkey. The Balkan powers won victory after victory, but Austria-Hungary had no intention of letting Serbia grow too strong (there were too many Serbians and allied peoples in its own southeastern territories).

Piltdown man was discovered in southern Great Britain, consisting of a thoroughly human skull and a thoroughly apelike jaw. It turned out eventually to be a fraud, the most notorious perhaps in scientific history. British anthropologists were fooled because so little was known at that time concerning hominid fossils, and because they were inspired with nationalistic fervor (till then, France and Germany had had all the good prehuman fossils).

1913

ISOTOPES

During the intense study of radioactive phenomena that had been going on for the previous seventeen years, some forty to fifty different elements had been reported (as judged by differences in radioactive properties—the kinds, intensities, and energies of particles emitted). However, there were only ten to twelve places available for them in the periodic table. Either the periodic table did not apply to radioactive elements or there was something subtle about those elements that had been missed.

The British chemist Frederick Soddy (1877–1956) worked on the problem. He made clear what is now called the *radioactive displacement law*. (The Polish chemist Kasimir Fajans [1887–1975] made these same suggestions independently at about the same time.) When an atom gives off an alpha particle, the alpha particle has a positive charge of 2 and a mass of 4. The atom that gave it off therefore has to turn into another atom with a smaller charge (by 2) on its nucleus and a smaller mass (by 4).

When a beta particle is given off, with a negative charge of 1, the loss of the negative charge is equivalent to the gain of a positive charge. An atom that gives off a beta particle therefore turns into another atom with a larger nuclear charge (by 1). Since an electron has only a tiny mass, the mass of an atom is virtually unchanged in beta-particle emission.

A gamma ray, having neither electric charge nor mass, does not affect the nature of an atom when given off; it merely decreases the atom's energy content.

By following these changes, Soddy suggested that a given place in the periodic table might be occupied by two or more different substances, distinguishable from each other by their differing radioactive properties. These occupants of the same place would be equal in nuclear charge but different in mass. Soddy called them *isotopes*, from Greek words meaning "same place."

For his advancement of the isotope concept, Soddy was awarded the Nobel Prize for chemistry in 1921.

LEAD ISOTOPES

Soddy had worked out the isotope concept in connection with radioactive elements (see above), but these were available in such tiny quantities that atomic weights could not be determined, and the existence of isotopes could not be checked in that way. However, the radioactive displacement law made it clear that uranium and thorium each broke down to a different isotope of lead, and that could be checked.

The American chemist Theodore William Richards (1868–1928) had worked out unprecedentedly accurate methods for determining atomic weights, and he used these methods to determine the atomic weight of lead obtained from ores containing uranium or thorium and those containing neither.

In 1913 Richards found definite variations in the atomic weight of lead, and this strongly supported the isotope concept. For his work on atomic weights, Richards was awarded the Nobel Prize for chemistry in 1914.

QUANTIZED ATOM

Now that Rutherford had formulated the nuclear atom (see 1911), it was possible to view the hydrogen atom as consisting of a nucleus (bearing a charge of +1) and a single electron (with a charge of −1) circling it.

One could argue that the electron, as it circled the nucleus, in effect oscillated from side to side. This oscillation, according to Maxwell's equations, should result in electromagnetic radiation. But if this were so, the electron would lose energy as it circled and would spiral into the nucleus.

The Danish physicist Niels Henrik David Bohr (1885–1962) tried to solve this problem by applying quantum theory to the atom. The electron, he decided, could not radiate energy except in intact quanta, each of which represented a large amount of energy on the atomic scale. Consequently the electron, when it radiated, would lose a large packet of energy and would not spiral into the nucleus gradually but drop very suddenly to a lower orbit nearer the nucleus. It would do this each time it radiated a quantum of energy. Eventually it would reach the lowest orbital state, below which it couldn't fall, and it would then emit no more energy.

In reverse, if the atom absorbed energy, the electron would suddenly rise to a higher orbit, and this would continue with further absorption of energy until it left the atom altogether, at which time the atom would be *ionized*, becoming a fragment with a positive charge equal in size to the number of electrons that had been boiled off, so to speak.

As electrons rose to higher orbits and fell to lower orbits, they would radiate only certain wavelengths, and under other conditions, absorb those same wavelengths, as Kirchhoff had shown over half a century before (see 1859).

The presence of many electrons rising and falling in orbits might confuse the issue, but hydrogen, with its single electron, should be easier to handle. Indeed, hydrogen has a simple spectrum, giving off radiation at a series of wavelengths that can be related to each other by a rather simple equation. This equation had been worked out by a Swiss physicist, Johann Jakob Balmer (1825–1898), in 1885. It had not seemed to have much significance at the time, but now Bohr could choose orbits for a hydrogen electron that would yield just those wavelengths that the hydrogen spectrum displayed.

Bohr's suggestion wasn't perfect. There were fine details of the hydrogen spectrum that it couldn't account for. There was also no explanation of why the electron, when it was in a particular orbit and oscillating back and forth, did not lose energy. If it couldn't give off an entire quantum, why didn't it stop oscillating?

Bohr's suggestion, however, was the first application of quantum theory to the atom and was enormously important for that reason. The imperfections were gradually removed in succeeding years, and for his work, Bohr was awarded the Nobel Prize for physics in 1922.

COOLIDGE TUBE

Coolidge, who had pioneered the use of tungsten filaments in electric light bulbs (see 1909), made a further advance involving that metal in 1913: he used a block of tungsten as an anode in a cathode-ray tube to produce X rays. This increased the ease and efficiency with which X rays could be produced. Whereas hitherto X rays had been almost entirely a laboratory phenomenon, the new *Coolidge tube* made it possible to use them in industry, medicine, and dentistry.

NITROGEN-FILLED LIGHT BULBS

Even with the use of Coolidge's tungsten filaments (see above), light bulbs were not immortal. The high-melting tungsten very slowly vaporized at the temperatures required to produce adequate light. The filament thinned and eventually broke.

The American chemist Irving Langmuir (1881–1957) argued that by keeping hot metal in a vacuum, vaporization was encouraged, but the presence of gas pressure would cut down the rate. Naturally, one couldn't use air, for then the tungsten would burn, but why not use pure nitrogen? Nitrogen-filled bulbs did indeed decrease the rate of evaporation and provide longer-lasting light bulbs.

Eventually, argon was used in place of nitrogen. Hot tungsten reacted slowly even with nitrogen, but not at all with the just about totally inert argon. The use of gas in light bulbs was also a safety measure. If light bulbs were accidentally dropped, the breakage was not as shattering when there was gas pressure inside as when there was a vacuum.

STARK EFFECT

In 1913 the German physicist Johannes Stark (1874–1957) demonstrated that strong electric fields caused a multiplication in spectral lines. The *Stark effect* was the analog of the Zeeman effect involving magnetic fields (see 1896). For this finding, Stark was awarded the Nobel Prize for physics in 1919.

MAGELLANIC CLOUD DISTANCES

In 1913 Hertzsprung, who had earlier noted the difference between red giant stars and red dwarf stars (see 1905), was able to work out the actual distances of some Cepheid variables. Once he had done this, he could make use of Leavitt's *period-luminosity law*, worked out the previous year (see 1912), to determine the distances of Cepheids in the Magellanic Clouds and therefore the distance of the Magellanic Clouds themselves. This distance proved to be something over 150,000 light-years.

This was the first distance to be determined for any object lying outside our own Milky Way Galaxy.

OZONOSPHERE

Although oxygen is a major component of Earth's atmosphere, the three-atom molecule ozone (see 1840) is not found in more than trace amounts in the air around us. That is just as well, of course, for it is poisonous.

In 1913, however, the French physicist Charles Fabry (1867–1945) was able to demonstrate the presence of significant quantities of ozone in the upper atmosphere, between heights of 6 and 30 miles. This region is sometimes called the *ozonosphere* in consequence. There, the ozone is extremely useful, for it absorbs and blocks the more energetic ultraviolet light from the Sun, which is dangerous to life, and keeps it from reaching the Earth's surface.

VITAMINS A AND B

As research on vitamins continued, it became clear that there were several of them. The American biochemist Elmer Verner McCollum (1879–1967) found, in 1913, that there was some factor essential to life in some fats. It had to be fat-soluble, and this alone had to make its molecular structure considerably different from the trace substance, essential to life, that cured beriberi, since that was water-soluble.

In the absence of any real information about the molecular structure of these compounds, McCollum called them *fat-soluble A* and *water-soluble B.* Eventually, this was changed to *vitamin A* and *vitamin B.*

The use of letters has persisted, at least to some extent, ever since. For instance, the factor that Lind had unwittingly worked with in curing scurvy (see 1747) came to be called *vitamin C* and the one that prevented rickets was called *vitamin D.*

MICHAELIS-MENTEN EQUATION

Although catalysts had been known and used for a century (even if we don't count the prehistoric use of fermentation), the manner of their action remained unknown. In fact there seemed something almost magical about them. How could they hasten a chemical reaction and yet not take part in it themselves? How could enzymes, present in so small a quantity that no chemist knew what their molecular structure might be, so greatly hasten a reaction?

The German chemist Leonor Michaelis (1875–1949) and his assistant Maud Lenora Menten derived an equation that would describe the rate at which enzyme-catalyzed reactions took place. In order to do that, they supposed that the enzyme combined with the substance whose reaction it catalyzed and that after the reaction took place the two separated again.

This supposition allowed the *Michaelis-Menten equation* to be derived, which showed how the rate of reaction varied with the concentration of the substance undergoing the reaction. It seemed that enzymes and catalysts generally *did* take part in the reaction. The enzyme (or the catalyst, more usually) offered a surface to which the reactant could attach under circumstances that allowed it to undergo the reaction much more readily.

To be metaphorical, the catalyst played the part of a hard flat surface in the writing of a note. It is difficult to write on a piece of paper in the air, but place the paper on a desk and writing is easy. The desk takes no part in the writing, but it does offer an appropriate surface. Catalysts began to lose their mystery.

GLYCOLYSIS

The British physiologist Archibald Vivian Hill (1886–1977) was particularly interested in the relationship between muscle contraction and heat development. To measure the small and transient heat effects, he made use of thermocouples, which swiftly and delicately recorded heat changes in the form of tiny electric currents. He refined his methods till he could measure a rise of three-thousandths of a degree for a few hundredths of a second.

In 1913 he found that heat was not developed nor was oxygen consumed during muscular contraction, but that both took place *after* contraction when the muscle was at rest.

The German biochemist Otto Meyerhof (1884–1951) independently demonstrated the same fact in chemical terms. While the muscle is contracting, glycogen disappears and lactic acid appears. Six-carbon units, in other words, are split into three-carbon units without the consumption of oxygen or the development of heat. Eventually, accumulating lactic acid prevents further muscular contraction. (We feel worn out.) After the contraction, lactic acid is oxidized (consuming oxygen and developing heat), thus paying off the *oxygen debt* that piled up during the preceding reaction (*anaerobic glycolysis*, Greek for "sugar-splitting without air").

As a result of this work, Hill and Meyerhof were awarded shares of the Nobel Prize for medicine and physiology in 1922.

IN ADDITION

In Mexico, the democratic President Francisco Indelecio Madero (1873–1913) was overthrown and killed in February 1913 by Victoriano Huerta (1854–1916), who took over the presidency.

George I of Greece (1845–1913) was assassinated on March 18, 1913, after a fifty-year reign and was succeeded by his son, who reigned as Constantine I (1868–1923). This took place less than two weeks before the end of the Balkan War. By the Treaty of London on May 30, Turkey gave up all European territory except for that in the immediate neighborhood of Constantinople.

The Balkan nations quarreled over the spoils and there was a month-long *Second Balkan War* in which Bulgaria was quickly defeated by the others. A new treaty brought peace on August 10. Serbia, Montenegro, Greece, and Bulgaria all gained territory at Turkish expense. Albania was made an independent nation, largely because Austria-Hungary and Italy did not want Serbia to reach the Adriatic.

1914

ATOMIC NUMBER

Once Laue had shown that X rays could be diffracted by crystals (see 1912) and Barkla had shown that elements could be made to emit characteristic X rays (see 1906), it became possible to measure the wavelengths of those characteristic X rays precisely by crystal diffraction.

This work was done by the British physicist Henry Gwyn Jeffreys Moseley (1887–1915), who completed it in 1914. He showed that the characteristic X rays decreased in wavelength and increased in frequency with the increasing weight of the elements emitting them. This Moseley attributed to the increasing positive charge on the atomic nucleus as the atomic weight of the element increased.

This discovery led to a major improvement of Mendeleyev's periodic table (see 1869). Mendeleyev had arranged the elements in order of increasing atomic weight but had to place some slightly out of order to keep them in their proper families. Moseley showed that if the elements were arranged in order of increasing nuclear charge, no element would have to be placed out of order.

The amount of positive charge on the nucleus was called the *atomic number*. It was 1 for hydrogen, the smallest atom, and 92 for uranium, the most complex atom then known. For the first time, chemists could now be sure how many new elements remained to be discovered, and where in the periodic table they would fall.

At the time Moseley advanced his atomic number suggestion, only seven numbers between 1 and 92 remained unrepresented by elements. They were numbers 43, 61, 72, 75, 85, 87, and 91.

Moseley would undoubtedly have earned a Nobel Prize for physics for this work, but he died in action a year later in World War I.

X-RAY WAVELENGTH

Laue's discovery of X-ray diffraction by crystals (see 1912) led almost at once to efforts to determine the wavelengths of X rays by this method.

This was accomplished by a father-and-son team of British physicists. William Henry Bragg (1862–1942) and William Lawrence Bragg (1890–1971). They worked out the mathematical details involved in the diffraction and showed how to calculate wavelengths from it.

As a result, the two Braggs were awarded the Nobel Prize for physics in 1915.

IONS AND CRYSTALS

Thirty years earlier Arrhenius had advanced the notion that electrolytes in solution dissociated into ions (see 1884). The idea was that a substance such as sodium chloride existed in solid form as a molecule, symbolized as NaCl, but on solution split up into the positively charged *sodium ion* (Na^+) and the negatively charged *chloride ion* (Cl^-).

When the Braggs were studying X-ray diffraction, however (see above), they found that that phenomenon could best be understood if it was supposed that in a solid crystal of sodium chloride there were no intact molecules, merely sodium ions and chloride ions positioned with geometric regularity.

Sodium chloride and many other compounds did not exist as molecules in the older sense, then, but as arrays of ions held together by electromagnetic interaction.

BETA-PARTICLE ENERGIES

If a particular atom broke down to emit an alpha or beta particle, it would seem that a definite energy source had been broached, and particles of definite energies ought to be given off.

As early as 1904, W. H. Bragg (see above), studying radium, had shown that alpha particles were emitted with several different sharply delineated ranges. Presumably there were several different processes going on within the radium nucleus, and each gave off alpha particles of a particular energy.

In 1914, however, the English physicist James Chadwick (1891–1974) showed that this was not true of beta particles. They came off in a continuous range of energies, from a sharply defined maximum down to zero. This was a puzzle that was not to be explained for a number of years.

PROTON

Thomson felt that positive rays consisted of streams of high-velocity atomic nuclei (see 1912). Rutherford studied them and came to the conclusion, in 1914, that the positive rays involving hydrogen nuclei were the smallest of all and that no smaller positively charged particles existed. He therefore called the hydrogen nucleus a *proton* (from the Greek word for "first").

The proton is not the positive analog of the electron, however, even though the size of their electric charge is precisely the same. The proton is a much more massive particle (now known in fact to be 1836.11 times as massive as the electron).

After Rutherford had made his suggestion, it seemed that the massive nucleus at the center of the atom must be made up of protons. (This meant that Prout's hypothesis that all elements were made up of hydrogen —see 1815—was in a manner of speaking correct.)

Of course it didn't make too much sense that the atomic nucleus should consist of protons. After all, the protons, being positively charged, would repel each other. The natural suggestion, then, was that electrons were also present in the nucleus and that their negative charges acted as a kind of cement. This seemed all the more logical since radioactive changes caused atoms to eject electrons in the form of beta particles, and these seemed to come from the nucleus.

Furthermore, only in hydrogen were the mass and charge of the nucleus equal. Thus an alpha particle (a helium nucleus) had four times the mass of a proton but a positive charge only twice that of a proton. It seemed logical to explain this by supposing that an alpha particle consisted of four protons plus two electrons. The electrons neutralized the charge of two of the protons without affecting the mass much. The remaining two positive charges on the nucleus were neutralized by the two *planetary electrons* circling outside the nucleus.

It seemed, then, that all atoms were made up of equal numbers of protons and electrons, with some of the electrons inside the nucleus and some circling around it.

This picture of the atom *seemed* simple and satisfactory but proved quite wrong, although the matter wasn't straightened out for sixteen years.

MAIN SEQUENCE

Hertzsprung had noted that some red stars were giants and some were dwarfs (see 1905), with nothing in between. Independently, the American astronomer Henry Norris Russell (1877–1957) noted this in 1914.

Russell also went into greater detail. He plotted the temperatures of stars against their luminosity and obtained a diagonal line that showed that the stars grew steadily dimmer as they grew cooler. This line, which extended from the hot luminous stars to the cool dim ones, was called the *Main Sequence,* because about 95 percent of the stars, including the red dwarfs, fell upon it. The red giants did not fall on it, because although they were quite cool (that was why they were red), they were very luminous, because of their enormous size.

The first thought was that the Main Sequence represented the line of evolution of stars; that they started as large, cool conglomerations of gas (red giants) that condensed and grew hotter till they reached the top of the Main Sequence as the hottest stars there were. They then gradually cooled and grew dimmer, ending as red dwarfs, finally becoming too cool to glow altogether.

This view turned out to be incorrect, but Russell's diagram, based on observation, was correct. It merely had to be interpreted in better fashion. Giving Hertzsprung credit for priority, it is usually called the *Hertzsprung-Russell diagram,* or the *H-R diagram* for short.

WHITE DWARFS

A star that didn't fit into the Main Sequence (see above) was the dim companion of Sirius, which had first been postulated by Bessel (see 1844) and first observed by Clark just over half a century before (see 1862).

From the strength of the companion's attraction to Sirius, the companion had to be about the mass of the Sun. At the distance of Sirius, a star with the mass of the Sun that was as dim as the companion was should be very cool indeed and therefore red in color. Yet the companion didn't seem to be red, but white.

The American astronomer Walter Sydney Adams (1876–1956) managed to get its spectrum in 1914, despite the blaze of nearby Sirius. The companion turned out to be a hot star, as hot as Sirius and hotter than our Sun. To be our Sun's mass and to be hotter than the Sun, too, the companion should have been blazing out as a first-magnitude star, but it wasn't. In fact it is too dim to see without a telescope. The only way of explaining that was to suppose that Sirius's companion was a very *small* star, more nearly the size of the Earth than of the Sun—even though it had the mass of the Sun.

Had the discovery been made a few years

earlier, it would have been dismissed as preposterous. However, now that science had accepted Rutherford's nuclear atom (see 1911), it seemed possible that under some conditions atoms might break down and the nuclei approach each other more closely than in ordinary matter. In the case of broken-atom *degenerate matter*, densities might easily be a million times that of ordinary matter.

Sirius's companion (now more commonly known as *Sirius B*, while Sirius itself is sometimes called *Sirius A*), being both tiny and white-hot, was called a *white dwarf*. It was the first of this class to be discovered, but white dwarfs were eventually found to be quite common, though only those relatively close to us can be seen because of their low total luminosity.

JUPITER IX

The tale of Jupiter's satellites did not end with the discovery of Jupiter VI and Jupiter VII (see 1904).

In 1908 the French astronomer Philibert Jacques Melotte discovered Jupiter VIII, which was still farther from Jupiter than VI and VII were. Jupiter VIII circled Jupiter at an average distance of 14,600,000 miles.

Then in 1914 the American astronomer Seth Barnes Nicholson (1891–1963) discovered Jupiter IX. Its average distance from Jupiter is 14,700,000 miles and it circles Jupiter in two years and one month. No other known satellite anywhere in the Solar System is so far from its planet or has so long a period of revolution.

Jupiter VIII and Jupiter IX are both about 25 miles in diameter and are known as *Pasiphae* and *Sinope*, respectively.

ACETYLCHOLINE

A fungus called ergot produces a variety of alkaloids with powerful effects on animal tissues. The eating of fungus-ridden grain can produce a disease called *ergotism*, and in the centuries before the cause was understood, there were terrible epidemics of ergotism.

Among those who worked on ergot was a British biologist, Henry Hallett Dale (1875–1968). In 1914 he isolated from it a compound called *acetylcholine*, which seemed to produce effects on organs similar to those produced by certain nerves. The significance of this was not fully understood for several more years.

EARTH'S MANTLE AND CORE

The waves set up by earthquakes apparently did not reach all parts of the Earth's surface even when they were strong enough to be detected everywhere. There was a *shadow zone* within which no waves were detectable.

The German-born American geologist Beno Gutenberg (1889–1960), studying this phenomenon, suggested in 1914 that there was a core at the center of the Earth, about 2100 miles in radius, that was markedly different in density and chemical composition from the material outside that radius. Earthquake waves entering the core would be refracted in such a way as to be sent beyond the shadow zone. Based on the fact that transverse waves did not enter the core at all, Gutenberg suggested that the core was liquid.

Thus the Earth is divided into two parts, an inner *core*, which (from its high density, and the fact that iron meteorites are common) seems likely to consist of liquid nickel-iron in a 1-to-9 ratio, and an outer *mantle* made up of rocky material. These two are in about the proportion of yolk to white in an egg, and the outermost *crust* is about the proportion of the eggshell.

The sharp dividing line between the mantle and the core is called the *Gutenberg discontinuity*.

BEHAVIORISM

Freudian psychology (see 1893 and 1900, Dreams) had grown extremely popular by this time, but those who favored other views were not wanting. In 1914 the American psychologist John Broadus Watson (1878–1958) developed the thesis that human behavior was explainable in terms of conditioned responses, such as those Pavlov had demonstrated (see 1907). Watson relegated even heredity to a minor role.

Animals, including the human being, he viewed as intensely complicated machines, which reacted according to their nerve-path *wiring*, those nerve paths being altered, or conditioned, by experience. This view of Watson's was called *behaviorism*.

IN ADDITION

On June 28, 1914, Archduke Francis Ferdinand of Austria-Hungary (1863–1914) was assassinated by a Serbian terrorist. Austria-Hungary, determined to break Serbia, demanded satisfaction. Russia backed Serbia, and Germany backed Austria-Hungary.

Austria-Hungary declared war on Serbia on July 28. Russia began to mobilize, whereupon Germany did too. On August 1, Germany declared war on Russia and on August 3, on Russia's ally, France. Planning a quick victory, Germany drove westward, violating Belgium's neutrality, whereupon Great Britain declared war on Germany on August 4. World War I was on full blast.

In the east, Russia launched an immediate offensive, which was met by capable German generals, who wiped out a Russian army at the Battle of Tannenberg and the Battle of the Masurian Lakes. The Germans then marched into Poland and, for the rest of the war, were not in serious danger on the east.

Turkey, anxious to profit by a Russian defeat, entered the war on Germany's side on October 29, while Japan, with eyes on German possessions in the Pacific, joined the war on the British side on August 23.

On the western front—the really crucial area—the Germans plunged westward and reached the Marne River, less than 20 miles from Paris but were stopped at the Battle of the Marne. There, for the rest of the year, the two sides built a line of trenches and fell into a bloody stalemate.

The United States maintained neutrality.

Mexico had dissolved into a civil war between Venustiano Carranza (1859–1920) and Francisco (Pancho) Villa (1877–1923), who were both anti-American.

Mohandas Karamchand (Mahatma) Gandhi (1869–1948), who had spent twenty-one years in South Africa fighting for equal rights for non-whites, returned to India in 1914, determined to begin fighting British domination at home through a program of nonviolent civil disobedience.

On a peaceful note, the Panama Canal was opened for use on August 3, 1914, even as World War I started.

The American social reformer Margaret Louise Sanger (1879–1966) coined the term *birth control*.

The American inventor Clarence Birdseye (1886–1956) pioneered the production of quick-frozen foods.

The last of the passenger pigeons, which had once flown across American skies by the tens of millions, died in the Cincinnati Zoo on September 1, 1914.

1915

PELLAGRA

Pellagra was a disease that was endemic in the American South after the Civil War. It didn't seem contagious, and Funk had speculated that it might be a vitamin-deficiency disease (see 1896).

The Austrian-born American physician Joseph Goldberger (1874–1929) noted that it struck wherever the diet was monotonous and limited and did not include much in the way of milk, meat, or eggs.

In 1915 he conducted a dramatic experiment on prisoners in a Mississippi jail. Volunteers (who were promised pardons in return) were placed on a limited diet lacking meat or milk. After six months, they developed pellagra, which was relieved by adding milk and meat to their diet.

During this time Goldberger's study group went to great lengths to try to contract pellagra by contact with the patients, their clothing, and their excretions. They failed. Pellagra was definitely not contagious. Goldberger spoke of a P-P *(pellagra-preventive)* factor in the diet, but its chemical nature was as yet unknown.

THYROXINE

About a quarter of a century earlier, the thyroid gland in the throat had been shown to be responsible for the overall rate of metabolism of the body, so that the human engine raced, so to speak, when the thyroid was overactive and slowed to a crawl when it was underactive.

Once Starling had advanced the hormone concept (see 1902 and 1905), it was clear that the thyroid must do its work by way of a hormone. It was known that the thyroid contained a characteristic protein called *thyroglobulin,* which was unique in possessing iodine, an element not till then known to be essential to life.

The American biochemist Edward Calvin Kendall (1886–1972) studied thyroglobulin looking for some simple component that would do the work of the thyroid gland in trace quantities and that would therefore qualify as the hormone involved. In 1915 he isolated what he called *thyroxine.* In the course of subsequent years, it turned out to be an iodine-containing amino acid related to the tyrosine found in proteins. It was the hormone.

BACTERIOPHAGES

No type of cell is immune to the ravages of the subcellular parasites called viruses. Even the smallest cells, those of bacteria, can be victimized. The British bacteriologist Frederick William Twort (1877–1950) discovered, in 1915, a type of virus that infested and killed bacteria. The discovery was made again, independently, some time afterward by a Canadian-born bacteriologist, Felix d'Hérelle (1873–1949). D'Hérelle named the viruses *bacteriophages* (Greek for "bacteria-eaters").

ELLIPTICAL ELECTRON ORBITS

Bohr's quantized atom did not explain all the fine details of spectra. Some dark lines that seemed simple, on closer investigation proved to consist of groups of narrow, closely spaced lines.

The German physicist Arnold Johannes Wilhelm Sommerfeld (1868–1951) felt the answer lay in the fact that electron orbits were more complicated than Bohr had suggested. Bohr had made use of perfectly circular orbits, but the orbits (like planetary orbits) might also be elliptical. Sommerfeld used Einstein's theory of relativity to calculate the elliptical orbits and quantum theory to show that only certain kinds of ellipses were possible. The combination of circular and elliptical orbits explained some of the details of the spectra that Bohr's original treatment had left unexplained. For this reason, people sometimes speak of the *Bohr-Sommerfeld atom*. It made use of the two great physical theories of the twentieth century, relativity and quanta.

HYDROGEN-HELIUM CONVERSION

Pierre Curie had shown that radioactivity implied the existence of enormous amounts of energy somewhere deep in the atom (see 1901). Physicists had not yet straightened out the matter in detail, but in 1915 the American chemist William Draper Harkins (1875–1951) noted that the helium nucleus was not quite four times as massive as the hydrogen nucleus. Therefore, if four hydrogen nuclei could somehow be converted to a helium nucleus, the excess mass not needed for the helium nucleus would represent a great quantity of energy that could be released. He was perfectly correct, but it was to be forty years before scientists learned how to bring about that conversion.

IN ADDITION

World War I continued. In northeastern France (the *western front*), huge, bloody, and useless battles were fought that did not shift the line significantly. On April 22, 1915, the Germans introduced the use of poison gas, liberating quantities of chlorine. The Allied troops broke and fled, but the Germans, caught unprepared by their own success, didn't exploit the victory.

On the *eastern front*, the Russians lost heavily but fought tenaciously and did not break. The Germans took Warsaw on August 7, 1915, and by year's end most of Poland was in their hands.

At sea, the Germans made use of submarines in an attempt to destroy British shipping and starve out the British Isles. On May 7, 1915, the British liner *Lusitania* was sunk by a German submarine with the loss of 1198 lives, of which 139 were American. The sinking of the *Lusitania* swung American opinion heavily in favor of the Allies.

Italy joined the war on the Allied side, and Bulgaria joined the war on the side of Germany.

On April 25, 1915, the British landed in Gallipoli, a peninsula south of Constantinople, in an attempt to force Turkey out of the war and to open a route whereby arms and supplies could be sent to the beleaguered Russians. The strike was the brainchild of the British statesman Winston Leonard Spencer Churchill (1874–1965). The Gallipoli campaign proved a fiasco, and Churchill lost his job.

Allied forces landed in Salonika, Greece, and advanced in Mesopotamia, which was then under Turkish control. German dirigibles bombarded London, not very effectively, and airplanes came to be used, first for reconnaissance and then, outfitted with machine guns, for battle.

In Africa the Germans quickly lost all their colonies except for German East Africa. In the Pacific, German colonies were lost to Japan.

1916

GENERAL RELATIVITY

Eleven years before, Einstein had advanced his special theory of relativity, in which he showed that the laws of physics remained unchanged in systems moving at constant velocity relative to each other (see 1905). In 1916 he extended this to systems moving relative to each other at any velocity, however changing. This was his *general theory of relativity,* often shortened to *general relativity.*

To make this possible, he assumed that inertial mass (mass derived from measurements of acceleration) and gravitational mass (mass derived from measurements of gravitational intensity) were identical. He also assumed that space was curved in the presence of mass, and that gravitation was not a force but merely the result of moving objects following the shortest possible path in curved space.

He advanced a set of equations to cover all this, equations that allowed grand conclusions to be drawn about the universe as a whole, so that he founded the science of *cosmology.* Einstein pointed out that Newton's laws of gravitation produced results quite close to those of general relativity but that there were three kinds of differences that might be measured in order to decide which was a closer approach to reality.

First, Einstein's theory allowed for a shift in the position of the perihelion of a planetary orbit slightly beyond that which would be expected in a Newtonian Universe. This *advance of the perihelion* had been discovered by Leverrier in the case of Mercury (see 1846, Vulcan) seventy years earlier. General relativity explained it without having to call upon an intra-Mercurian planet, which had never been found. The fact that this effect was already known rather diminished its importance, however.

Second, Einstein showed that light moving upward against a strong gravitational field should show a red shift. This involved something that had not been expected earlier, but even the Sun's gravitational field wasn't intense enough to make the measurement practical. Nothing much could be done with the prediction, therefore.

Third, and most dramatic, Einstein showed that light would be deflected by a gravitational field substantially more than would be expected in a Newtonian Universe, and the difference could be measured. To do so, however, it would be necessary to locate stars in the neighborhood of the Sun during a total eclipse. Their light would have reached Earth after passing close to the Sun and bending in its path very slightly as it did so, so that each star would be seen a bit farther from the Sun than it really was. This would appear if the same region of the sky was photographed when the Sun was in another part of the sky.

However, World War I was raging and there was no way any eclipse expedition could be safely organized. In fact, it was difficult to get word of Einstein's views out of Germany. So the world had to wait.

BLACK HOLES

After Einstein's equations for general relativity were published, the first to work out solutions for them was a German astronomer, Karl Schwarzschild (1873–1916). He also calculated the gravitational phenomena in the neighborhood of a star with all its mass concentrated to a point.

In order for an object to move infinitely far from a body as the result of a single initial impulse, that initial impulse must produce a speed that will carry the object away so rapidly that its speed will not decline as quickly as does the gravitational pull (which declines with the square of the increasing distance). In that case, the gravitational pull will never be intense enough to bring the object to a complete halt. This *escape velocity* equals 7 miles per second for Earth but only 1.5 miles per second for the Moon.

In general, the escape velocity from an object's surface increases with the mass of the attracting object and also with its density. Over a century earlier, Laplace (see 1783) had pointed out that if an object was sufficiently massive and dense, even light wouldn't have sufficient velocity to escape.

Schwarzschild studied the case of a star with a mass compressed more and more strongly till the star's volume sank to zero, so that the gravitational pull at its surface got higher and higher without limit. Schwarzschild calculated the distance from such a point-mass at which light would barely have the speed to escape. This is the *Schwarzschild radius.* Once anything approached closer to the star than the Schwarzschild radius, it could never escape again. Not even light could.

Since nothing could escape, not even light, such a star would behave like a bottomless hole in space, so to speak. It would be a *black hole,* a name given to it half a century later.

ELECTRONS AND CHEMICAL BONDS

Once Moseley had advanced his notion of atomic numbers (see 1914), it was understood that the neutral atom contained as many electrons in its outskirts as the value of the atomic number. A hydrogen atom had one planetary electron, a uranium atom had ninety-two, and the other elements had numbers in between.

What's more, study of the characteristic X rays of elements, which had been initiated by Barkla (see 1906), gave rise to the notion that the electrons existed in successive shells. Clearly, it was the electrons in the outermost shell that were most exposed and were therefore most capable of being removed, or of being shifted from one atom to another. The German chemist Richard Wilhelm Heinrich Abegg (1869–1910) had already suggested in 1904, even before the details of electron shells had been worked out, that chemical reactions were the result of electrons being transferred from one atom to another.

An American chemist, Gilbert Newton Lewis (1875–1946), went into the matter in greater detail. He pointed out in 1916 that the electron arrangements seemed to be particularly stable when the outermost shell contained eight electrons (or two, in the special case of helium). An atom with eight electrons in an outer shell and then one outside that (as in sodium or potassium) would so readily lose that one superfluous electron that it would be a very active element. An atom with seven electrons in its outermost shell (as in chlorine or bromine) would so readily accept an additional one that it would be reactive also. And of course, atoms that had eight electrons in their outermost shell (as in the noble gases) or two (as in the special case of helium) would have little or no tendency to lose or gain an electron and so would be chemically inert.

Moreover, two chlorine atoms might each contribute an electron to a shared pool, so that each would have eight electrons in its outermost layer (its own seven plus one of its neighbor's which it was sharing). Naturally, in order to be able to share electrons, the two atoms would have to remain in close contact, so that chlorine exists as two-atom molecules, and it takes energy to pull those atoms apart.

This sort of reasoning also accounted for other bonds routinely found in organic molecules among carbon, oxygen, hydrogen, and nitrogen atoms. And Lewis showed from a consideration of electron arrangements why various elements had the valences (see 1852) they did and why valence varied regularly over the periodic table.

Langmuir (see 1913) worked out a similar system of chemical bonding independently, and the English chemist Nevil Vincent Sidgwick (1873–1952) soon pointed out that the Lewis-Langmuir notions applied to complicated inorganic molecules as well.

SUPERHETERODYNE RECEIVER

Until now, radio operation had been a rather complicated thing that had to be left to radio engineers. In 1916, however, the American radio engineer Edwin Howard Armstrong (1890–1954) worked out a system for lowering the frequency of electromagnetic waves and then amplifying them. He called this a *superheterodyne receiver*.

The addition of such devices to radios made them far easier to use. The turn of a dial was all that was needed to get good reception, or to transfer reception from one wavelength to another. It was only after such devices were widely adopted that radios could be operated by anyone, so that they entered the home and became a vehicle for mass entertainment and information.

IN ADDITION

On the western front, there was another year of appalling and indecisive bloodletting. The Germans attacked at Verdun, and the British attacked on the Somme. Neither battle achieved anything but mass murder.

The British introduced the use of tanks on September 15, 1916, during the Battle of the Somme. These had the potential for breaking the stalemate of trench warfare, but they were used timidly and against the will of the uninspired generalship that marked the war.

On the eastern front, the Russians launched a successful offensive against Austria-Hungary, but Russian unrest at home was rising and crippled the Russian war effort.

The battlefields in Italy and the Balkans were marked by indecisive fighting, too. Romania entered the war on the Allied side on August 27, 1916, but was promptly defeated by the Germans, who took Bucharest on December 6.

The British and German fleets met in the North Sea in the Battle of Jutland on May 31. The Germans did surprisingly well, outfighting the British, ship for ship. The British, however, could afford to lose more ships than the Germans could. The German fleet eventually retreated to port and did not venture to emerge again.

The war offered opportunities to oppressed peoples. The Arabs rebelled against their Turkish overlords, and the British came to their support.

On the other hand, on April 24, the *Easter Rebellion* took place in Ireland, and the British managed to suppress it by May 1.

The United States continued to remain neutral, though public opinion was becoming ever more strongly anti-German. Nevertheless, President Wilson was reelected for a second term largely because he had maintained neutrality.

The Mexican Civil War was spilling over into the United States. On March 9, 1916, Pancho Villa raided Columbus, New Mexico, killing seventeen American citizens. An American force was sent into Mexico in pursuit of him, but the expedition was a total failure.

Francis Joseph I of Austria-Hungary died on November 21, 1916, after having ruled for sixty-eight years. He was succeeded by a grandnephew, who reigned as Charles I (1887–1922).

1917

EXPANDING UNIVERSE

It had always been assumed by the Greeks that the Universe was changeless, and even the astronomers of the modern period, while aware that stars varied and moved and came into being and dimmed to death, felt that such changes canceled each other out and left the Universe as a whole unchanged.

Einstein, in formulating his general relativity equations, understood that if he was to have a static Universe, something would have to be added to the equations. He therefore added an arbitrary constant designed to make the equations come out right, so to speak. He later said this was the greatest scientific mistake of his life.

A Dutch astronomer, Willem de Sitter (1872–1934), however, was content to travel wherever Einstein's equations took him. He pointed out, in 1917, that if the equations were solved as they stood, the implication was that the Universe was expanding. Such a picture of an *expanding universe* seemed grotesque, for at the time, nothing about the Universe seemed to give a hint of this expansion, but de Sitter's suggestion took on a great deal of importance in the next decade.

MICROCRYSTALLINE DIFFRACTION

The Braggs had shown how crystal structure could be worked out by X-ray diffraction (see 1914). Intact and reasonably large crystals are hard to come by, however, and it represented a great advance when the Dutch-born American physicist Peter Joseph William Debye (1884–1966) showed, in 1917, that useful results could be obtained when X rays were diffracted by powdered solids that consisted of tiny crystals oriented in every direction.

The technique was also discovered independently, and at about the same time, by the American physicist Albert Wallace Hull (1880–1966).

100-INCH TELESCOPE

In no science is the advance of instrumentation as important and as dramatic as in astronomy. In 1917 a new reflecting telescope was put into action on Mount Wilson, near Pasadena, California. Its mirror was 100 inches across, which made it the largest telescope in the world at that time, and it was to remain the largest for three decades.

PROTACTINIUM

Few of the new substances being detected as radioactive breakdown products of uranium and thorium were truly new elements. Once Soddy's isotope concept had been advanced (see 1913), most of the new substances were recognized as isotopes of elements that were already known.

In 1917, however, the German physical chemist Otto Hahn (1879–1968) and a coworker, Austrian physicist Lise Meitner (1878–1968), discovered a truly new element that disintegrated into actinium (see 1899). They therefore named the new element *protactinium* (meaning "before actinium").

It turned out to be element number 91, one of the seven missing elements at the time Moseley had advanced his atomic number concept (see 1914). That left only six elements to go.

SONAR

Pierre Curie, in discovering piezoelectricity (see 1880), had shown how to produce ultrasonic sound vibrations. These were put to use by the French physicist Paul Langevin (1872–1946).

Waves are more easily reflected the shorter their length. Ordinary sound waves are long enough to bend around ordinary obstacles and are not efficiently reflected. The much shorter sound waves of ultrasonic vibration can easily be reflected by relatively small objects. The direction from which the reflection is received indicates the direction of the object, and the time it takes between emission and reflection (knowing the speed of sound) gives the distance, or *range*, of the object.

Since light cannot penetrate large thicknesses of water and ultrasonic vibrations can, the latter can be used to detect underwater objects such as submarines when light is impotent. Since World War I was raging and German submarines were a deadly danger to the Allies, it is not surprising that Langevin worked with that purpose in mind.

Langevin's system of *echolocation* was called *sonar*, an acronym (combination of initial letters) of *sound navigation and ranging*. While Langevin had the technique worked out in 1917, it could not be put into effective use before the end of the war.

Sonar is now used not only for the detection of underwater objects such as submarines and schools of fish, but also to study the conformation of the ocean floor. Sonar utterly revolutionized oceanography.

IN ADDITION

On the western front the slaughter continued.

On the eastern front, Germany advanced rapidly and Russia collapsed into the *Russian Revolution*. Russian troops mutinied on March 10, 1917, and Nicholas II abdicated on March 15, bringing the Russian monarchy to an end. A new democratic government under Aleksandr Fyodorovich Kerensky (1881–1970) attempted to continue the war, but the Russian soldiers and people had been tried beyond even their phenomenal endurance.

On November 6, 1917 (October 24, by the Russian calendar, which was thirteen days behind the Gregorian calendar), what was called the *October Revolution* took place. The provisional government was overthrown, and a more radical Bolshevik (better known nowadays as Communist) government took its place. Under the leadership of Lenin, the new government asked for peace.

In northeastern Italy, German troops inflicted a catastrophic defeat on the Italians at Caporetto on October 24, 1917, and all of Venetia was lost.

In the Middle East, the British took Jerusalem on December 9, 1917, and Palestine came under Christian control for the first time in six and a half centuries.

In order to break the British blockade, Germany resorted to unrestricted submarine warfare, and the United States declared war on Germany on April 6, 1917. American troops arrived in France under the command of John Joseph (Black Jack) Pershing (1860–1948) and went into action for the first time on October 27.

The population of the United States reached 100 million.

1918

CENTER OF THE GALAXY

The concept of the Galaxy had entered the astronomical mainstream with Herschel (see 1781). Since then, astronomers had assumed the Sun to be near the center of the Galaxy since the Milky Way encircled us more or less evenly, in a great arc around the sky.

There was one important asymmetry, however. Globular clusters were not evenly spread over the skies. (Globular clusters are closely packed spherically shaped groups of up to a hundred thousand stars. They were first noted by Herschel.)

These globular clusters were to be found almost entirely in one hemisphere of the sky, something first pointed out by Herschel's son, John Frederick William Herschel (1792–1871). Indeed, about one-third of the globular clusters were to be found in the single constellation of Sagittarius, in which the Milky Way is rather brighter and richer in stars than it is anywhere else.

Once the Cepheid yardstick (see 1912) had been worked out by Leavitt and Hertzsprung (see 1905), it became possible to find Cepheids in the globular clusters and determine their distances and the distances of the clusters themselves.

This task was undertaken by the American astronomer Harlow Shapley (1885–1972), making use of the new 100-inch telescope (see 1917). By 1918 he was able to make a three-dimensional model of the globular clusters and could see that they themselves formed a loose sphere around a point in Sagittarius far distant from the Solar System.

Shapley assumed (correctly, as it turned out) that the globular clusters were distributed around the center of the Galaxy. His estimate of the distance to that center was a trifle overlarge, something that was later corrected. We now know that the center of the Galaxy is 30,000 light-years away and that the Galaxy is about 100,000 light-years across. Our Solar System, far from being at the center of the Galaxy, is about 20,000 light-years from one end and 80,000 from the other. Shapley had dethroned the Sun from its position at the center of the Universe, as Copernicus had dethroned the Earth (see 1543).

We cannot see the center of the Galaxy (let alone the far half of the structure), because dark clouds of dust and gas in the Milky Way obscure the view. We are, indeed, at the center of what we can see—which is not surprising.

With Shapley, astronomers finally got a more or less correct picture of the size of the Galaxy and of our own position in its outskirts. It was far larger than anyone had thought—it contained at least 100 billion stars and perhaps twice that number. It is not surprising, then, that astronomers thought our Galaxy and its two attendant and much smaller *satellite galaxies,* the Magellanic Clouds, made up the Universe.

This seemed large enough, but as it happened, astronomers had not yet even begun to grasp the true size of the Universe.

SPECTRAL CLASSES

Not all stellar spectra are alike. Some resemble the Sun's, but most stars have spectra that are altogether different. The first to point this out was an Italian astronomer, Pietro Angelo Secchi (1818–1878), who in 1867 had divided stellar spectra into four classes.

The notion of *spectral classes* came of age, however, with the work of the American astronomer Annie Jump Cannon (1863–1941) who in 1918 began the meticulous and time-consuming task of studying and classifying the spectra of many thousands of stars.

In so doing, she worked up a classification that is still used today. She used a system of letters, planning that there would be a smooth transition from A to B to C and so on. However, as the stars were arranged in order of decreasing surface temperature, the letters fell out of order, and some had to be omitted.

The spectral classes of the main sequence, in order of decreasing temperature, turned out to be O, B, A, F, G, K, and M.

Each spectral class can be divided into ten subdivisions from 0 to 9, so that our Sun is a G2 star, Sirius is A1, Rigel is B5, Arcturus is K2, Betelgeuse is M2, and Proxima Centauri M5. All this turned out to be useful in dealing with stellar evolution.

RADIOACTIVE TRACERS

Fourteen years before, Knoop had used benzene rings as a tracer in fat metabolism (see 1904). It occurred to the Hungarian-born chemist Georg Karl von Hevesy (1885–1966) that it might be possible to use radioactive atoms as tracers. The advantage would be that their concentration could be accurately determined even when they were present in trace quantities, thanks to the radiations they emitted.

In 1918 Hevesy decided to make use of a radioactive isotope of lead, formed by uranium breakdown. This *radiolead* was identical in chemical properties to ordinary stable lead. Suppose, then, he were to add a small quantity of radiolead to lead and use the mixture to produce certain lead compounds.

These lead compounds are only slightly soluble in water, so that one could not measure the concentration of the dissolved stable lead with anything approaching accuracy. If, however, a lead compound with radiolead added is stirred in water so that a tiny fraction of the molecules dissolve, that same tiny fraction of the molecules containing radiolead also dissolves. The amount of radio-lead in solution can then be easily and accurately determined, and the percentage of radiolead compound dissolved will be the same as that of the ordinary lead compound.

In later years, Hevesy followed the manner in which plants absorbed and distributed water by spiking it with a tiny quantity of radiolead compound that he could follow accurately.

As long as radioactive tracing was confined to lead, it could only play a minor role, of course. Nevertheless, Hevesy had pointed out the potential of the technique, and the time was to come when it was to be a superb and essential tool for biochemists and others.

For this, and for other work, Hevesy was awarded the Nobel Prize for chemistry in 1943.

ORGANIZER

The German zoologist Hans Spemann (1869–1941) was interested in the development of embryos. Thirty years later, it had been shown that if the fertilized ovum of a test animal was divided in two, and if one of the resulting cells was killed with a hot needle, then the remaining cell would evolve into a longitudinal half of an embryo. The act of dividing in two had established the plane of bilateral symmetry.

If, however, a fertilized ovum divided in two and the two resulting cells were separated and each allowed to develop, each would form a complete (though smaller than normal) embryo. (It is this which results in the birth of identical twins among human beings, for instance.)

To Spemann, the difference between a cell that developed from a divided fertilized ovum and one that developed while attached to a dead partner indicated that the individual cells of an embryo affected each other.

In a series of experiments, Spemann showed that even after an embryo had begun to show definite signs of differentiation, it could still be divided in half, with each half producing a whole embryo. This showed that cells remained plastic until quite late in the game.

Spemann found that an embryo develops according to the nature of neighboring areas. An eyeball develops originally out of brain material and is joined by a lens that develops out of nearby skin. If the eyeball is placed near a distant section of the skin, one that would never in the course of nature develop a lens, it nevertheless begins to develop one.

There were apparently *organizers* in the embryo that brought about certain developments nearby. For his work, Spemann was awarded the Nobel Prize for medicine and physiology in 1935.

IN ADDITION

Germany seemed within reach of victory now. The eastern front had been cleared up. On March 3, 1918, the Russians were forced to sign the Treaty of Brest-Litovsk and give up any claim to their border areas: Poland, Finland, the Baltic States, the Ukraine, Transcaucasia. Romania also made peace on May 7.

The Germans undertook a vast spring offensive, beginning on March 21, to crush the Anglo-French allies before American reinforcements could become significant. But the German gamble failed, for by July American forces were flooding into the front in increasing numbers. The Germans had to fall back, and by the end of August they were back at the line from which their spring offensive had started.

Bulgaria gave up on September 30, Turkey gave up on October 30, and Austria-Hungary gave up on November 3.

Germany, unable to stand in the west and with all its allies gone, could not avoid defeat. On November 9, William II abdicated, and the German monarchy came to an end. On November 11, Germany signed an armstice and World War I was over. It had killed ten million, wounded twenty million, and cost perhaps three hundred billion dollars.

As if this were not enough, the *Spanish flu* killed twenty million people all over the world before the end of the year, twice as many as were killed in four years of war.

Russia was in chaos. There was a civil war in which the Communists *(Reds)* fought against the counterrevolutionaries *(Whites)* who wanted to restore Russia more or less to what it had been.

New nations appeared on the map of Europe. Finland and Poland established their independence. Serbia and Montenegro combined and absorbed the Slavic southeastern provinces of Austria-Hungary to form Yugoslavia. The northern provinces of Austria-Hungary formed the independent nation of Czechoslovakia.

Charles I of Austria-Hungary, his realm having disintegrated, abdicated on November 11, 1918, and Austria and Hungary became independent and separate republics.

1919

MASS SPECTROMETER

Thomson had shown that neon atoms seemed to exist in two different varieties (see 1912). Soon afterward, Soddy had advanced the isotope concept (see 1913), but it had at first seemed to apply only to radioactive elements and their breakdown products. Could perfectly stable elements also exist in the form of different isotopes? Was that the significance of Thomson's findings with respect to neon?

The British chemist Francis William Aston (1877–1945) improved on Thomson's apparatus and in 1919 developed a *mass spectrometer,* which could separate ions so delicately that all ions of a particular mass would focus in a fine line on a photographic film. Working with neon ions, Aston showed that there were two lines, one indicating a mass of 20 and the other a mass of 22. From the comparative darkness of the two lines, Aston calculated that the ions of mass 20 were 10 times as numerous as those of 22. If all the ions were lumped together, they would have an average mass of 20.2, and that was indeed the atomic weight of neon. (Later, a third group of neon ions of mass 21, occurring in only a tiny concentration, was discovered.)

Working with chlorine atoms, Aston found again two types of atoms, with masses of 35 and 37, in the ratio of 2 to 1. A weighted average came out to 35.5, which was the atomic weight of chlorine.

The use of the mass spectrograph eventually showed that most stable elements (but not all) consisted of two or more stable isotopes. The atomic nuclei of a particular element all had the same positive electric charge, but the isotopes differed in mass. Each had its own *mass number*.

For this work, Aston was awarded the Nobel Prize for chemistry in 1922.

NUCLEAR REACTION

Rutherford, in his work with alpha-particle bombardments of matter (see 1906), had made gases his target. For instance, when he bombarded hydrogen, he got bright scintillations on a zinc sulfide screen (which gave off a flash of light whenever struck by an energetic subatomic particle). The alpha particles themselves produced scintillations, but the new scintillations, which appeared when hydrogen was present, were particularly bright.

Rutherford decided (correctly) that the alpha particles occasionally struck a hydrogen nucleus (consisting of a single proton) and hurled it forward. It was these speeding protons, he felt, that produced the bright scintillations.

In 1919, when nitrogen was introduced into the cylinder, occasional bright scintillations like those of protons were produced. Rutherford knew that the nitrogen nucleus had a charge of $+7$, so that it had to contain at least seven protons. He felt that the alpha particle must knock one of these protons out of the nitrogen nucleus every once in a while.

The number of alpha-particle scintillations produced under these circumstances gradually decreased. Rutherford reasoned that some of the particles must be absorbed by the nitrogen nuclei. If a nitrogen nucleus absorbed an alpha particle with its charge of $+2$ and lost a proton with its charge of $+1$, the net change would leave the nitrogen nucleus with a total charge of +8. This is the characteristic charge of an oxygen nucleus.

In short, what Rutherford had done was to combine a helium nucleus (an alpha particle) and a nitrogen nucleus to form a hydrogen nucleus (a proton) and an oxygen nucleus. He had converted one type of atom into another by subatomic bombardment.

Ordinary chemical reactions involve the transfer or sharing of electrons, as Lewis had pointed out (see 1916). Now Rutherford had produced reactions that involved the transfer of particles inside the nucleus. In other words, he had brought about the first humanly engineered *nuclear reaction*.

GRAVITATIONAL DEFLECTION OF LIGHT

Einstein had advanced the theory of general relativity, which among other things predicted that light rays would bend slightly and follow a gently curved path when in a gravitational field (see 1916). The one way of testing that would be to study the stars in the immediate neighborhood of the Sun during a total eclipse, but astronomers had to wait for the end of World War I to make an eclipse expedition feasible.

On May 29, 1919, a solar eclipse was scheduled to take place at just the time when more bright stars would be in the vicinity of the eclipsed Sun than would be there at any other time of the year.

The Royal Astronomical Society of London, under the leadership of Arthur Stanley Eddington (1882–1944), who was a great enthusiast of Einstein's theories, made ready two expeditions, one to northern Brazil and one to Principe Island in the Gulf of Guinea off the coast of West Africa. The positions of the bright stars near the Sun were measured relative to each other at the time of the eclipse. If light was bent in its passage near the Sun, those stars would all seem to shift away from the Sun and would appear to be slightly farther apart from each other than they would six months earlier or six months later when they rode high in the midnight sky.

The positions of the stars proved to be in line with what Einstein had predicted, and this was considered a grand verification of general relativity.

However, the measurements were borderline and a trifle uncertain, and none of the tests of general relativity were entirely conclusive for an additional forty years, so that a number of other types of cosmological theories were advanced to compete with that of Einstein. (On the other hand, special relativity was confirmed over and over, and there has been no reasonable doubt concerning it for three-fourths of a century.)

BEE COMMUNICATION

Pavlov's conditioned responses (see 1907) could be used to elicit answers from animals about what they sensed.

The Austrian-born German zoologist Karl von Frisch (1886–1982), for instance, conditioned bees to go to certain locations to pick up nectar, making certain that the location was of a certain color. They would then fly to other places of the same color, since they had been conditioned to react to that color as they would to a food source.

Frisch then changed the color to see what would happen to the conditioning. Thus, if he conditioned the bees to black and then substituted red, they flew to it anyway, indicating that they could not see red—red was black to them. However, if they were conditioned to black and that was changed to ultraviolet (which would still look black to human eyes), the bees no longer flew to it. They could see ultraviolet.

By 1919 Frisch had also interpreted the manner in which the bee communicated its findings to its colleagues of the hive. Having obtained honey from a new source, the returning bee would "dance," moving round and round or side to side. The number of the evolutions and their speed gave the necessary information about the location of the new source. Frisch also showed that bees could orient themselves in flight by the direction of light polarization in the sky.

For this work, Frisch was awarded a share of the Nobel Prize for physiology and medicine in 1973.

IN ADDITION

A peace conference opened at Versailles in France on January 18, 1919. The victorious nations planned to work out a peace treaty that would prevent future wars.

Almost the first thing done at the conference was to institute a *League of Nations,* at which disputes and disagreements could be argued out and settled in a way short of war.

The Treaty of Versailles was signed on June 28. Germany was forced to give up Alsace-Lorraine to France, West Prussia to Poland, and all its colonies to Great Britain, France, and Japan. It was expected to pay a heavy indemnity besides.

Germany's ex-allies were forced to sign treaties later. Austria-Hungary was broken up into Austria, Hungary, and Czechoslovakia, with outlying provinces going to Italy, Romania, Yugoslavia, and Poland. Turkey lost all its territory outside Asia Minor, Syria being transferred to France, Palestine and Iraq to Great Britain.

However, the United States would neither ratify the Treaty of Versailles nor join the League of Nations, and this condemned the league to certain impotence.

German leaders met at Weimar and officially set up a republic (thereafter called the *Weimar Republic*) on July 31, 1919.

In Russia, the civil war continued, and short-lived Communist revolts erupted in central Europe but were quickly quelled.

In the United States there was a "Red scare" accompanied by a feeling that harsh repression ought to be the order of the day against any sign of serious dissent. The first dial telephones were introduced in Norfolk, Virginia.

1920

STELLAR DIAMETER

Throughout history, stars had been studied only as points of light, and all information about them had to be obtained in ways that required only points.

In 1920, however, Michelson, who had first used his interferometer to compare the speed of light in different directions (see 1881), used it for another purpose. He built a 20-foot interferometer and attached it to the new 100-inch telescope (see 1917). With it, he could measure light emerging from either side of the star Betelgeuse. (Betelgeuse is a relatively near red giant, so that its diameter was likely to prove more measurable than that of smaller or more distant stars.)

The two rays from the sides of Betelgeuse made a very tiny angle, but from the interference fringes they produced, Michelson could measure the angle, and from that angle, knowing the distance of Betelgeuse, he could calculate its diameter. He worked it out to be about 260 million miles, or 300 times the width of our Sun.

It was news that made the first page of the *New York Times.*

NOVAS IN ANDROMEDA

The Andromeda nebula was a matter of dispute among astronomers in the early decades of the twentieth century. There were some who thought it was a cloud of dust and gas that was part of our own Galaxy (and that the same was true of other nebulas that resembled it). Others pointed out that the spectrum of the Andromeda and other such nebulas was that of starlight and not at all like that of the Orion nebula, which was clearly a cloud of dust and gas. The Andromeda nebula and others like it might therefore be vast assemblages of stars that were independent galaxies but so far away from us that there was little hope of seeing their stars individually.

The leading exponent of a nearby Andromeda nebula was Shapley, who had worked out the shape and size of the Galaxy and determined the position of the Solar System within it (see 1918). The leading exponent of a far-off Andromeda nebula was the American astronomer Heber Doust Curtis (1872–1942).

Curtis reasoned that, although the Andromeda nebula might be too far away for ordinary stars in it to be made out, unusually bright stars such as novas ought to be visible. He therefore observed the nebula carefully and did see numerous very faint stars that appeared and disappeared as though they were exceedingly faint novas. There were far more novas in the patch of light represented by the Andromeda nebula than there were in any other similarly sized patch of the sky. To Curtis, this meant that the Andromeda nebula contained a vast number of extremely faint stars and that it was an independent Galaxy (as Kant had suggested—see 1755).

In 1920 Curtis and Shapley met in a great debate before the National Academy of Sciences and discussed the matter of Andromeda. Majority opinion among astronomers was on the side of Shapley, but Curtis made an unexpectedly strong showing and won a moral victory. However, the matter could not be decided by debate. It had to await the necessary observations.

HIDALGO

Asteroids with atypical orbits continued to be discovered. In 1920 the German astronomer Walter Baade (1893–1960) detected an asteroid that he named *Hidalgo.* At its perihelion, it is within the asteroid belt, but its orbit balloons outward, so that at aphelion it is as far from the Sun as Saturn is.

Of all the asteroids that spend at least part of their time in the asteroid belt, none is known to this day that recedes as far as Hidalgo does.

DENDROCHRONOLOGY

Ancient wood is well preserved in Arizona's dry climate, and the American astronomer Andrew Ellicott Douglass (1867–1962) grew interested in it.

The wooden trunk of a tree contains rings that mark annual growth, wide ones in good years, narrow ones in bad years. Since all trees in a region share the same good years and bad years, all of them have similar tree-ring patterns. These patterns are characteristic and do not ever quite repeat themselves over time. Therefore, if a piece of old wood is placed overlapping the tree-ring pattern of a freshly produced tree stump, you can determine the year in which the old wood was cut by counting rings of the new wood back to where the overlap begins. You can then follow its pattern back farther than you could that of the fresh wood, still older pieces of wood can be fitted on, and so on.

By 1920, when Douglass convinced himself that such *dendrochronology* (Greek for "time-telling by trees") was a practical way of determining age where wood was concerned, he had traced his tree-ring calendar back for about five thousand years. The technique was particularly useful in dating objects of Native American prehistory.

CLIMATIC CYCLES

Weather is so erratic that even the most modern devices have trouble predicting it long in advance. Nevertheless, very general cycles of weather may exist, and these may explain the periodic ice ages that have occurred in the last million years of Earth's history.

In 1920 the Yugoslavian physicist Milutin Milankovich (1879–1958) suggested that astronomic factors played a part. Slow oscillations in the eccentricity of Earth's orbit and in the tilting of the Earth's axis, together with the precession of that axis, seemed to suggest a 40,000-year cycle, which could be divided into a *Great Spring, Great Summer, Great Autumn,* and *Great Winter,* each about 10,000 years long.

Milankovich's suggestion was ignored at the time, but after more than half a century, it came to be considered seriously.

ANEMIA

Anemia, from Greek words meaning "no blood," is the name of a family of diseases in which the blood performs its functions with less than normal efficiency. One common reason for anemia is a lack of iron in the diet. This reduces the level of hemoglobin in the blood and makes the red corpuscles less efficient at picking up oxygen from the lungs. Such anemics are pale and easily tired.

The American pathologist George Hoyt Whipple (1878–1976) induced an artificial anemia in dogs by bleeding them, then followed the manner in which new red blood corpuscles were formed. He kept the dogs on various kinds of diets to see what effect would be produced on corpuscle formation and found that, of the various foodstuffs he tried, liver was the most potent in correcting such anemia. This eventually led the way to the cure for a kind of anemia much more dangerous than simple iron-deficiency anemia.

For this reason, Whipple was awarded a share of the Nobel Prize for medicine and physiology in 1934.

AIR MASSES

A father-and-son team of meteorologists, Vilhelm Friman Koren Bjerknes (1862–1951) and Jacob Aall Bonnevie Bjerknes (1897–1975), had set up weather-observing stations all over Norway during World War I.

By 1920 they had shown that the atmosphere is made up of large *air masses* and that there is a sharp differentiation in temperature between warm tropical air masses and cold polar air masses. The sharp boundaries between them they called *fronts,* from an analogy to the battle lines that had so preoccupied Europe in recent years.

This simplified the technique of weather prediction.

IN ADDITION

The Russian civil war turned in favor of the Reds. The Red forces were even able to take the offensive and drive into Poland on July 17, 1920. The Polish forces defeated the Russians on August 16, however. Poland thereupon absorbed large sections of territory inhabited by Byelorussians and Ukrainians. The Baltic nations, too—Estonia, Latvia, and Lithuania—established their independence.

Turkey was still at war, for Greece claimed

Smyrna, a district on the Turkish Aegean shore, and Greek forces invaded Turkey.

Warren Gamaliel Harding (1865–1923) was elected twenty-ninth president of the United States. Two Italian anarchists, Nicola Sacco (1891–1927) and Bartolomeo Vanzetti (1888–1927), were arrested on May 5, 1920, for the murder of two people in the course of a robbery. They were very likely innocent, but they were anarchists, and under the leadership of Attorney General Alexander Mitchell Palmer (1872–1936), who was the Red scare leader, that meant automatic guilt.

The population of the United States reached 105.7 million. Russia's population, despite its catastrophic losses in war, stood at 137 million. The world population reached 1.8 billion.

1921

INSULIN

By this time it seemed clear that the serious disease of diabetes mellitus had something to do with the pancreas; removal of the pancreas from experimental animals invariably produced a diabeteslike condition.

Once the hormone concept had been propounded by Starling (see 1902 and 1905), it was easily assumed that some hormone produced by the pancreas exerted control over carbohydrate metabolism. In the absence of the hormone, carbohydrate metabolism went out of control, with the result that glucose built up in the blood and spilled out in the urine. A variety of unpleasant symptoms made themselves manifest—with death as the end.

To be sure, it was well established that the chief function of the pancreas was to produce protein-digesting enzymes, but there were portions of the pancreas that seemed different from the rest. These portions, scattered through the pancreatic background, were the Islets of Langerhans (see 1869). Some suspected that the islets produced the necessary hormone, and there were suggestions that the hormone be called *insulin*, from the Latin word for "island."

No one had managed to isolate the hormone from pancreatic tissue, however. This was not surprising, since if insulin were protein in nature (as indeed it turned out to be), the digestive enzymes in the pancreas would digest and destroy it before it could be isolated.

The Canadian physician Frederick Grant Banting (1891–1941) had read that if the pancreatic duct was tied off in a living animal, the pancreatic tissue degenerated. The Islets of Langerhans, however, whose hormone product would not then be discharged through the duct but secreted directly into the bloodstream, were not affected. Then why not attempt to isolate insulin from a degenerated pancreas, in which no digestive enzymes would exist to spoil the extraction?

In 1921 Banting managed to obtain space at the University of Toronto and to get the services of an assistant, the American-born Canadian physiologist Charles Herbert Best (1899–1978). Together Banting and Best tied off the pancreatic ducts in a number of dogs and waited seven weeks for the pancreas to degenerate. They then extracted a solution that quickly stopped the symptoms of diabetes. They had insulin.

As a result, Banting was awarded a share of the Nobel Prize for medicine and physiology in 1923.

VAGUSSTOFFE

It was well known by this time that the nerve impulse was electrical in nature. The German-born American pharmacologist Otto Loewi (1873–1961) felt that chemicals were involved, too, especially where the nerve impulse had to jump the tiny gap (or synapse) from one nerve cell to another.

In 1921, working with the nerves attached to a frog's heart, particularly the *vagus nerve,* he showed that chemical substances were indeed set free when the nerve was stimulated.

The idea for testing to see whether the chemical so obtained could be extracted and made to stimulate another heart directly, without the intervention of nerve activity, came to Loewi at 3 A.M. one night. He jotted down notes and went back to sleep but in the morning had forgotten the details and couldn't read his own writing. When the following night the same idea occurred to him, he took no chances. He went to the laboratory and got to work. By 5 A.M. he had proved his point.

Loewi called the substance *Vagusstoffe* (German for "vagus material"). Dale, who had discovered acetylcholine a few years before (see 1914), recognized that its effects were similar to those described by Loewi and suggested that Vagusstoffe was acetylcholine. So it was, and Loewi shared the Nobel Prize for medicine and physiology with Dale in 1936.

RICKETS

By now McCollum had distinguished between fat-soluble vitamin A and water-soluble vitamin B (see 1913). Vitamin C, the antiscurvy factor, was also water-soluble but had no effect on beriberi and was therefore a substance different from vitamin B.

The disease of rickets, thought to be a vitamin-deficiency disease at this time, was not affected by any of these three vitamins. The British biochemist Edward Mellanby (1884–1955) undertook to track down some additional vitamin that might be involved.

By 1921 he had located a rickets-inhibiting substance in such animal fats as cod-liver oil, butter, and suet. The vitamin was fat-soluble, but its distribution was not that of vitamin A, and its absence did not produce the symptoms that absence of vitamin A did. Clearly, then, there were two fat-soluble vitamins, and the rickets-inhibiting one was labeled vitamin D.

Also in 1921, other researchers discovered that sunshine had a rickets-inhibiting effect. Sunshine itself couldn't contain a vitamin, but it might convert some substance in skin into a vitamin. This, in fact, turned out to be the case.

GLUTATHIONE

In 1921 Hopkins (see 1900, Tryptophan) isolated *glutathione* from tissues. It is a combination of three amino acids (a *tripeptide*) and switches easily between a reduced and oxidized state. That is, it can easily be made either to give up a pair of hydrogen atoms or to take them up again.

Hopkins demonstrated this ability on the part of glutathione and argued for its importance in tissue chemistry. First, such compounds can act as protectors for delicate compounds. By giving up their hydrogen atoms readily, they make it unnecessary for other compounds to give up theirs when that loss might do some damage harder to repair. Also, by bouncing back and forth between the two states, glutathione (and other such compounds) can bring about the easy transfer of atomic groupings from one substance to another.

MAGNETRONS

By this time, many different varieties of radio tubes were being developed. In 1921 the American physicist Albert Wallace Hull (1880–1966) developed a diode (see 1904) that could produce bursts of short radio waves *(microwaves)* of high intensity. He called it a *magnetron*, because an external magnet was used to apply a magnetic field to the electrodes inside the tube.

In the next decade, tubes of this sort played a key role in the development of radar.

TETRAETHYL LEAD

One of the difficulties in automotive engineering was that of getting the gasoline vapors to burn smoothly within the cylinder. If they burned too rapidly, there was too great an explosion and a resulting "knock" in the engine. This was hard on the engine, disturbing to the ear, and wasted energy.

In 1921, however, the American chemist Thomas Midgley, Jr. (1889–1944) discovered that if the compond *tetraethyl lead* was added to the gasoline, it inhibited the burning just enough to prevent knock. It was an *antiknock compound*. From that time on, one could speak of "ethyl gas" or of "leaded gasoline."

A bromine compound was also added to the gasoline, to prevent accumulation of lead in the cylinder. It meant that the relatively volatile compound lead bromide was formed and discharged through the exhaust. This added another factor to the air pollution produced by automobiles.

INTROVERT AND EXTROVERT

Freud, who devised psychoanalysis (see 1893 and 1900, Dreams), had a number of colleagues, with whom he always managed to quarrel. Those colleagues went on to found psychoanalytic schools of their own and to add refinements to psychoanalytic theory. Thus the Austrian psychiatrist Alfred Adler (1870–1937) had popularized the notion of the *inferiority complex* in 1911.

In 1921 the Swiss psychiatrist Carl Gustav Jung (1875–1961) popularized the concepts of *introvert*, for a person whose thoughts and interests are directed inward, and *extrovert*, for a person whose thoughts and interests are concentrated on other people and the outside world.

RORSCHACH TEST

Psychoanalysis generally involves conversation between the patient and the psychiatrist, with the patient usually doing most of the talking. In 1921, however, a Swiss psychiatrist, Hermann Rorschach (1884–1922), invented a nonconversational device for diagnosing psychopathological conditions.

This involved the use of ten symmetrical inkblots, which patients were asked to interpret—to give their notion of what the abstract images represented. This *Rorschach test* became well known to the general public, although as with psychoanalytic techniques generally, it is difficult to tell objectively how useful it is.

IN ADDITION

The Middle East was now regaining a kind of equilibrium after the dislocations of World War I. Persia (a Greek name that was soon to be replaced by the native Iran) expelled all Russian officers and regained its full sovereignty.

Iraq was placed under Faisal I (1885–1933) as king, after a plebiscite.

Turkey made peace with Russia and settled its boundaries.

In Russia, the civil war came to an end and the

nation remained intact except for its western strip, which now made up the nations of Finland, Estonia, Latvia, Lithuania, and Poland. It also lost the southwestern province of Bessarabia to Romania.

Peter I of Yugoslavia died on August 16, 1921, and was succeeded by his son, who reigned as Alexander I (1888–1934).

Great Britain granted southern Ireland dominion status and a large measure of self-government, and it became the Irish Free State on December 6, 1921. The six northeasternmost counties, largely Protestant, came to be known as Northern Ireland and remained under British rule.

In 1921 airmail had its beginnings in the United States, with airplanes crossing the nation from the Pacific to the Atlantic in a little over 33 hours.

1922

SUMERIA

From the Greek histories and from Biblical accounts, modern historians knew something of Babylonian and Assyrian history, but archaeological investigations had to be depended on to delve back earlier still.

In 1922 the English archaeologist Leonard Woolley (1880–1960) began his investigations along the course of the lower Euphrates River, particularly at what seemed to be the site of Ur, a very early city mentioned in chapter 11 of Genesis. His excavations introduced moderns to the ancient civilization of Sumeria in what is now southeastern Iraq. Sumeria was probably the first civilization on Earth; it is the Sumerians who invented writing.

Woolley's most startling finding was geological evidence of a great flood that must have spread devastation through Sumeria about 2800 B.C. It was this flood that gave rise to the flood story in the epic of Gilgamesh (see 2500 B.C.) and to the biblical account of Noah's flood.

Woolley's findings were a great stimulus to the further study of early civilizations.

TUTANKHAMEN'S TOMB

The ancient Egyptian pharaohs had magnificent burials, and much in the way of gold and other precious materials was interred with them. Every effort was made to keep the tombs from being rifled and the contents stolen, even to the extent of placing the tomb at the center of a solid pyramid.

All efforts failed. All tombs were burglarized—and a good thing, too. If all the gold had remained buried, it would have ruined the economy of the ancient world. The tomb-robbers did civilization a remarkable favor by restoring the tomb contents to circulation.

By 1000 B.C., the great days of the pharaohs were over, and every last tomb was empty—except one. From 1361 to 1352 B.C., the pharaoh ruling over Egypt had been Tutankhamen. He was only about 21 when he died, but he was given the usual sumptuous burial. His grave was at once robbed, but for a wonder, the robbers were caught in the act and forced to return the loot. Perhaps the fact that the grave had been looted had gotten out but the return had been kept quiet. At any rate, no further effort at looting was made for two centuries. Then, while a grave was being excavated for a later pharaoh, the resulting showers of stone chips covered the entrance to Tutankhamen's tomb, hiding it so effectively that it came down to the twentieth century intact.

A British archaeological expedition under George Edward Stanhope Molyneux Herbert, Earl of Carnarvon (1866–1923), and Howard Carter (1873–1939) found the first sign of the entrance to Tutankhamen's tomb on November 4, 1922. Three days later they reached the sealed burial chamber, and a rich treasure trove of ancient Egyptian artifacts was uncovered. It gave tremendous impetus to Egyptian studies.

It also gave rise to the silly tale of the "Pharaoh's curse," when Lord Carnarvon died five months later of an infected mosquito bite complicated by pneumonia, but surely no sane man could believe Tutankhamen had anything to do with it. Carter lived on for seventeen years after opening the tomb.

VITAMIN E

Nutritionists were carefully placing rats, guinea pigs, and other animals on limited diets in order to find disorders that could be eliminated by the addition of particular foods to that diet. When diet could correct one disorder without having much effect on other disorders known to be caused by lack of a vitamin, the presence of a new vitamin could be suspected.

In 1922 the American anatomist Herbert McLean Evans (1882–1971), who four years earlier had determined that human cells contained twenty-four pairs of chromosomes (actually twenty-three, it was later shown), found a limited diet that rendered rats sterile and that could be corrected by fresh lettuce, wheat germ, or dried alfalfa but not by already known vitamins. As a result, it looked as if another vitamin (later known as vitamin E) existed, one that was fat-soluble, as were vitamins A and D.

GROWTH HORMONE

Evans, the discoverer of vitamin E (see above), also showed in 1922 that an extract of the pituitary gland (which is attached to the lower surface of the brain) served to promote *gigantism* in rats, making them grow to a significantly larger size than normal. That indicated the presence of a *growth hormone* in the pituitary.

LYSOZYME

In 1922 the Scottish bacteriologist Alexander Fleming (1881–1955) isolated an enzyme called *lysozyme* from tears and mucus. He found it to have bactericidal (bacteria-destroying) properties. It was the first example of a human enzyme that had such properties, and it represented the as-yet pale dawn of a time when more powerful natural products of this sort would be found.

ORIGIN OF LIFE

Darwin, who had advanced his theory of evolution six decades earlier (see 1858), had not taken up the matter of the origin of life. Not enough was known, and the subject was too touchy.

Spontaneous generation (life derived from nonlife) had been disproven by Pasteur (see 1860), but only under present conditions. In earlier times, when life first made its appearance, the Earth's environment was radically different from what it is today (there was no oxygen in the atmosphere, for instance), and no life existed that might promptly consume the material that represented the first tentative approaches to life.

Even so, there was some reluctance to consider the matter of a naturalistic origin to life, one that did not require the intervention of a Creator. The first person to make a significant study of the matter was a Russian biochemist, Alexander Ivanovich Oparin (1894–1980), who, living under a government that was officially atheist, had no inhibitions in doing so. He began his investigation of the subject in 1922, suggesting that life developed through the slow buildup of organic substances from the simple compounds present in the primordial ocean and atmosphere.

NERVE FIBERS

The electric currents in nerves are tiny, so tiny that it was difficult to study them in detail. The American physiologists Joseph Erlanger (1874–1965) and Herbert Spencer Gasser (1888–1963) developed delicate methods of detection using Braun's oscilloscope (see 1897) to aid in the studies.

Beginning in 1922, they determined the rates at which nerve fibers conducted their impulses and showed that the velocity of the impulse varied directly with the thickness of the fiber. For this, the two men shared the Nobel Prize for medicine and physiology in 1944.

EXPANSION OF THE UNIVERSE

Five years earlier, Sitter had suggested that Einstein's equations of general relativity implied a spontaneously expanding Universe (see 1916). Sitter's work, however, applied to a Universe empty of matter.

In 1922 the Russian mathematician Alexander Alexandrovich Friedmann (1888–1925) went further. Solving the equations for a Universe containing mass, he showed that it too would naturally expand.

IN ADDITION

On December 30, 1922, Russia organized itself as a union of republics. It became the *Union of Soviet Socialist Republics,* commonly abbreviated as *USSR,* and often referred to as the *Soviet Union.*

Under the Washington Conference, which met at the invitation of the United States, Chinese independence was guaranteed, the Open Door policy (all powers having an equal chance to loot China) was adopted, and naval armaments were limited.

Egypt became a British puppet kingdom. In Turkey, the sultanate came to an end after six centuries, and a republic was established under Kemal Atatürk (1881–1938).

In Italy, a right-wing organization known as the *Fascists* arose under Benito Amilcare Andrea Mussolini (1883–1945) and took control of the government on October 28.

In Germany, there was economic hardship, which included a high inflation rate.

1923

WAVES AS PARTICLES

Einstein had pointed out that electromagnetic radiation could be viewed as having particle aspects, but it was still necessary to observe phenomena that could best be explained in that way. The shorter the wavelength of electromagnetic radiation, the greater the energy content of the quantum and the more pronounced the particle aspect is likely to be. It would seem a good idea, then, to work with X rays in this respect.

In 1923 the American physicist Arthur Holly Compton (1892–1962) showed that X rays scattered by matter tended to lengthen their waves. This was called the *Compton effect*.

Compton was able to account for this by presuming that when a quantum of X rays struck an electron, the electron recoiled, subtracting some energy from the quantum and therefore increasing its wavelength. This was a clear demonstration of the particle aspects of an energetic wave, and indeed, it was Compton who now began to refer to such waves in their particle aspects as *photons*.

For this work Compton was awarded a share of the Nobel Prize for physics in 1927.

PARTICLES AS WAVES

Even as Compton was demonstrating that waves showed particle properties (see above), the French physicist Louis-Victor-Pierre-Raymond de Broglie (1892–1987) was maintaining that, from theoretical considerations, every particle ought also to have an associated *matter-wave* and therefore show wave properties.

The wavelength of such matter-waves would be inversely related to the momentum of the particle (that is, its mass times its velocity). A massive particle such as a baseball, or even a proton, would have matter-waves of such ultrashort wavelengths that they would be difficult or impossible to detect. Electrons, however, should have matter-waves with wavelengths similar to those of X rays.

Increasingly, after the work of Compton and de Broglie, physicists began to take the view that all objects had both wave and particle aspects. Where energy was low (and mass is a form of energy), the wave aspect would predominate, and where energy was high, the particle aspect would predominate.

To be sure, de Broglie's work was strictly theoretical. An actual demonstration of matter-waves would not come for several years, after which de Broglie would be awarded a share of the Nobel Prize for physics in 1929.

DEBYE-HUCKEL EQUATIONS

When Arrhenius had worked out the theory of electrolytic dissociation (see 1884), it seemed clear that some compounds dissociated only partially on solution. As crystal structure was elucidated by X-ray scattering (see 1912), however, it appeared that many compounds existed in totally dissociated form even in crystals. Why, then, should they seem to be only partially dissociated in solution?

Debye, who had developed the notion of dipole moments (see 1912), tackled the problem. He suggested that dissolved electrolytes were indeed completely dissociated in solution but that each positive ion was attended by a cloud of negative ions, while each negative ion was attended by a cloud of positive ions. The two types of ions therefore tended to insulate each other to some extent, and this gave the appearance of incomplete dissociation. Debye, with an assistant, the German chemist Erich Huckel, worked out equations to represent the situation. The *Debye-Huckel equations* are the key to modern interpretation of the properties of solutions.

ACID-BASE PAIRS

After the notion of ionic dissociation was established by Arrhenius, an acid came to be defined as a substance that split up to yield hydrogen ions, while a base was one that split up to yield hydroxyl ions (OH^-). These neutralized each other, because hydrogen ions and hydroxyl ions combined to form neutral water molecules.

The Danish chemist Johannes Nicolaus Bronsted (1879–1947) suggested a more general definition in 1923. An acid didn't simply give up a hydrogen ion (which could be viewed as a proton), because as a proton, it could not exist loose in solution. The proton, once it broke free of the acid molecule, must promptly attach itself to another molecule. Therefore, chemists ought to speak of *acid-base pairs.* Whenever a proton transferred from one molecule to another, the one that gave up the proton was an acid, the one that accepted it was a base. This broadened the concept and made it more useful.

COENZYME STRUCTURE

Harden had shown that the yeast enzyme that fermented sugar had a nonprotein portion, which he called a coenzyme (see 1904). The chemical structure of that coenzyme was unknown, however.

It was not until 1923 that a German chemist, Hans Karl August Simon von Euler-Chelpin (1873–1964), solved that problem and worked out the structure of Harden's coenzyme. The structure was related to that of the nucleotides that formed the building blocks of nucleic acid molecules, and it received the name of *diphosphopyridine nucleotide.*

The most interesting thing about the molecule was that a portion of it, when broken off, proved to be the well-known chemical compound *nicotinamide,* which could easily be changed to *nicotinic acid.* Both possessed a six-atom ring made up of five carbon atoms and a nitrogen atom. This type of ring did not occur in living tissue except in this coenzyme and in other closely related compounds.

For this work, Euler-Chelpin was awarded a share (with Harden) of the Nobel Prize for chemistry in 1929.

CEPHEIDS IN ANDROMEDA

Three years earlier, Curtis had debated Shapley over whether the Andromeda nebula was a distant galaxy (see 1920).

The still-new 100-inch telescope offered a way of settling the matter. The American astronomer Edwin Powell Hubble (1889–1953) used it in 1923 to study the Andromeda nebula and managed to make out some ordinary stars (not novas) in it. Some of these stars were Cepheids, which meant that Hubble could determine the distance of the nebula using the technique worked out by Leavitt (see 1912).

Hubble's calculations showed that the Andromeda nebula was 750,000 light-years away, although eventually this turned out to be a substantial underestimate. Even so, the distance was great enough to indicate that the Andromeda nebula was *not* part of our galaxy but existed far beyond it and must be an independent galaxy. From that time on, it was termed the Andromeda Galaxy, and it came to be understood that the Universe consisted of numerous galaxies, as many as a hundred billion perhaps.

For the first time, the general structure of the Universe came to be understood, and it was realized that its size was far, far greater than had been thought.

HAFNIUM

Hevesy, who had introduced the notion of radioactive tracers (see 1918), managed to reduce further the small number of still-missing elements. In collaboration with a Dutch physicist, Dirk Coster (1889–1950), Hevesy used an X-ray analysis method worked out by Coster and discovered *hafnium,* from the Latin name for Copenhagen. It is not a particularly rare element, but it is very like zirconium, which is just above it in the periodic table, and it is never found except in association with zirconium, which is fifty times as common, so that hafnium is hard to separate out.

Hafnium has an atomic number of 72, and its discovery reduced the number of still-missing elements between 1 and 92 to five.

ULTRACENTRIFUGE

Materials that ordinarily remain in suspension in water, due to the incessant battering of surrounding molecules, can be made to settle out by means of a centrifugal effect. Instruments called *centrifuges* whirl suspensions and force the suspended material to the side of the vessel away from the center of rotation. In this way, red blood corpuscles can be separated from blood, and cream from milk. (Since cream is less dense than the watery portion of milk, it collects on the inner side of the vessel, closer to the center of rotation.)

Ordinary centrifuges do not produce effects large enough to force colloidal particles smaller than red blood corpuscles or droplets of cream to settle out. But the Swedish chemist The Svedberg (1884–1971) developed an *ultracentrifuge* in 1923, which spun so quickly it developed effects equivalent to a gravity hundreds of thousands of times normal.

At such high rates of spin, the centrifugal effect can force ordinary protein molecules to settle out, and since mixtures of different types of proteins settle at different rates (the greater the molecular weight, the faster the rate of settling), different proteins can to some extent be separated in this way.

The rate of settling allowed the molecular weights of proteins to be determined with reasonable accuracy, and for this work Svedberg was awarded the Nobel Prize for chemistry in 1926.

IN ADDITION

Germany was in deep trouble. Inflation had gone completely out of control because of war reparations. The savings of many middle-class Germans vanished and they were reduced to penury. Humiliation over defeat and resentment over their treatment since made many Germans look for violent solutions to their problems and for scapegoats at home.

Taking advantage of this dangerous mood, an Austrian-born German demagog, Adolf Hitler (1889–1945), was bringing the National Socialist *(Nazi)* party to the fore by capitalizing on the discontent of the middle classes and using anti-Semitism to give them an outlet for their anger. (The Jews were few enough and weak enough to attack safely.)

In the United States, President Harding died suddenly on August 2, 1923, and Vice President Calvin Coolidge (1872–1933) succeeded as the thirtieth president.

1924

AUSTRALOPITHECUS

Until this point, the most primitive hominid known was *Pithecanthropus erectus,* discovered by Dubois (see 1890). Although this fossil had a brain only half the size of a modern human being's, it was still comparatively advanced. Therefore, it was considered very likely that there were earlier and more primitive hominids.

In 1924 a small skull that, except for its size, looked human was discovered in a limestone quarry in South Africa. An Australian-born South African anthropologist, Raymond Arthur Dart (1893–1988), examined it and recognized it as a primitive hominid. He called it *Australopithecus,* which is Greek for "southern ape."

It was not an ape, however, for subsequent discoveries of several species of such *australopithecines* showed that they walked upright and were closer to modern humans than to any apes, modern or past. They are the earliest hominids, as far as we can now tell.

BOSE-EINSTEIN STATISTICS

In 1924 the Indian physicist Satyendra Nath Bose (1894–1974) worked out a statistical method of handling certain subatomic particles. Einstein was enthusiastic about this and generalized Bose's work the next year.

The resulting *Bose-Einstein statistics* may be used with any of a group of subatomic particles called *bosons* in Bose's honor. The best-known example of a boson is the photon.

IONOSPHERE

Heaviside and Kennelly had predicted the existence of regions in the upper atmosphere that contained charged ions and that reflected radio waves (see 1902).

Details concerning this layer were worked out by an English physicist, Edward Victor Appleton (1892–1965). There was a problem at this time with the fading of radio signals, and Appleton found that this tended to take place at night. He felt it might be due to reflection from the charged layers in the upper atmosphere, which might be particularly prominent in the night sky.

If this were so, such reflection might set up interference, since the same radio beam would reach a given spot by two different routes, one direct and the other by bouncing off the charged layers.

Appleton experimented by using a transmitter and receiver that were about 70 miles apart. He altered the wavelength of the signal and noted when the two beams, direct and reflected, were in phase, so that the signal was strengthened, and out of phase, so that it was weakened. From this he could calculate the minimum height of reflection. In 1924 he found that the Kennelly-Heaviside layer was some 50 miles high.

At dawn, the Kennelly-Heaviside layer broke up, and the phenomenon of fading was no longer particularly noticeable. However, there was still reflection from charged layers higher up. These are called *Appleton layers* and are 150 miles high.

Because of this content of reflecting ions, the layer of air above the stratosphere came to be called the *ionosphere.*

CYTOCHROME

It had been known for a long time that oxygen from the lungs was absorbed by hemoglobin in the red blood corpuscles and carried to the cells of the body. What happened to the oxygen within the cells remained unknown, however.

In 1924 the Russian-born British biochemist David Keilin (1887–1963) was studying the absorption spectrum of the muscles of the horse botfly, and he noticed four absorption bands that disappeared when the cell suspension was shaken in air but reappeared afterward.

Keilin suggested, therefore, that there was a substance in the cells that absorbed oxygen. This substance showed the absorption bands when it had not yet absorbed oxygen but did not show them after it had absorbed oxygen.

He called this substance *cytochrome* (Greek for "cell color") and showed, eventually, that the substance was actually a chain of enzymes that passed hydrogen atoms from one to another until the final enzyme combined them with oxygen.

IRRADIATION

Vitamin D is not often to be found, as such, in food. However, it was known that sunshine converted an inactive precursor in the skin to vitamin D (see 1921). It followed that similar inactive precursors of vitamin D might exist in food, which exposure to sunlight *(irradiation)* might produce.

The American biochemist Harry Steenbock (1886–1967) showed in 1924 that this was, in fact, the case, and the use of irradiated food grew common.

IN ADDITION

In the Soviet Union, Lenin died on January 21, 1924, and a power struggle began among possible successors. The two leading candidates for control were Leon Trotsky (1879–1940) and Joseph Stalin (1879–1953).

Mussolini slowly tightened his control over Italy.

In the United States, President Coolidge won reelection in his own right in 1924.

1925

PACKING FRACTION

Six years earlier, Aston had used his mass spectrograph to determine the mass and relative occurrence of the isotopes of various stable elements (see 1919). Out of the 257 stable isotopes we now know to exist, he obtained the masses of 212.

By 1925 Aston had brought his mass spectrograph to such a pitch of accuracy that he was able to show that the mass numbers of the individual isotopes were actually very slightly different from integers, sometimes a little above, sometimes a little below. These slight discrepancies occurred because forming the nuclei from the individual subatomic particles that made them up absorbed or produced energy, and this energy was the equivalent of tiny bits of mass lost or gained in accordance with Einstein's equation relating mass and energy (see 1905, Mass-Energy). The energy change produced by packing together subatomic particles in a nucleus was called the *packing fraction.* Since it represented the energy binding the particles together, it could also be called the *binding energy.*

This meant that if one type of nucleus was converted into another kind of nucleus with a tighter packing, mass would be lost and energy produced on a far larger scale per particle than was produced in ordinary chemical reactions involving the outer electrons of an atom.

This made sense out of Harkins's contention that hydrogen-to-helium conversion would produce a great deal of energy (see 1915) and generalized it to nuclear reactions as a whole.

It also explained the energy of alpha particles. When an atom broke down radioactively to another atom, comparative packing fractions reduced the total mass somewhat, and this made a certain amount of kinetic energy available for the emerging alpha particles. The alpha particles invariably came out with the energy to be expected of the mass loss.

Beta particles remained puzzling. The maximum energies they possessed were always those to be expected of the mass loss, but there were beta particles that invariably emerged at lesser energies—all the way to zero—and the explanation for that still eluded physicists.

EXCLUSION PRINCIPLE

Bohr and Sommerfeld had worked out the energy levels of the electrons in an atom (see 1913). These could be expressed as *quantum numbers,* which followed certain simple rules. Three quantum numbers were known at this time.

The Austrian-born American physicist Wolfgang Pauli (1900–1958) considered the matter and felt that there was need for a fourth quantum number. If that were allowed according to certain rules, then it would be possible to show that no two electrons in a particular system of electrons could have all four quantum numbers alike. In other words, if a particular electron in an atom had one of the four quantum numbers, all the other electrons would be excluded from having that number. This was called the *exclusion principle,* and it allowed the arrangement of electrons in any atom to be worked out. It also explained why Mendeleyev's periodic table (see 1869) took the form it did.

For the exclusion principle, Pauli was awarded the Nobel Prize for physics in 1945.

PARTICLE SPIN

Once Pauli had enunciated the exclusion principle (see above), two Dutch physicists, George Eugene Uhlenbeck (1900–1988) and Samuel Abraham Goudsmit (1902–1978), at once pointed out that the fourth quantum number that Pauli had decided was required could be interpreted neatly as *particle spin.* Each particle, such as an electron, could spin either clockwise or counterclockwise at a rate that could be expressed as $+1/2$ or $-1/2$.

Eventually, similar spins (equal to $1/2$ or some multiple thereof) were found to exist for almost all other particles.

MATRIX MECHANICS

Beginning with Bohr (see 1913), physicists had tried to interpret spectral lines (which represented energy given off or taken up as electrons passed from one state to another) using images similar to those used for planets circling a star. They spoke of circular orbits, elliptical orbits, tilted orbits, rotation about an axis (particle spin), and so on.

The German physicist Werner Karl Heisenberg (1901–1976) considered all this useless and misleading. He preferred to take the numbers representing the energy level and manipulate them without regard to their pictured significance. In 1925 he developed a form of manipulation called *matrix mechanics* for the purpose.

MAGNETISM AND ABSOLUTE ZERO

The Dutch physicist Willem Hendrik Keesom (1876–1956) had managed to reach a temperature of 0.5 degrees above absolute zero (0.5° K), but hope of getting lower temperatures by methods such as gas expansion, which had worked till then, seemed over.

In 1925, however, Debye (see 1912) suggested something new. He pointed out that a paramagnetic substance (one that concentrates magnetic lines of force) can be placed almost in contact with liquid helium, separated from it by helium gas, and the temperature of the whole system reduced to about 1° K. If the system is then placed within a magnetic field, the molecules of the paramagnetic substance will line up parallel to the field's lines of force and, in doing so, give off heat. If this heat is removed by further slight evaporation of the surrounding helium, and the magnetic field is then removed, the paramagnetic molecules will immediately fall into a random orientation, absorbing heat as they do so. Since the only source of heat will be the liquid helium, its temperature will consequently drop below 0.5° K.

The same suggestion was made independently, soon afterward, by the American chemist William Francis Giauque (1895–1982), but it was not until another decade had passed that the idea was put into practice.

GRAVITATIONAL RED SHIFT

Einstein had predicted that light rising against a gravitational field would lose energy so that it would redden slightly (see 1916). The Sun's gravitational field, though intense, was not intense enough to show a measurable effect.

Ten years before, however, W. S. Adams had shown that Sirius's companion star, Sirius B, was tiny and extraordinarily dense. Its combination of large mass and small size meant it had a gravitational field more than ten thousand times as intense as the Sun's. It should show a gravitational red shift if, in fact, one existed.

In 1925 Adams managed to study the spectrum of the tiny star and detected a red shift close enough to that predicted by Einstein to be considered evidence of the validity of the theory of general relativity. As in the case of the bending starlight, however (see 1919), it was a borderline observation.

RHENIUM

In 1925 two German chemists, Walter Karl Friedrich Noddack (1893–1960) and Ida Eva Tacke (b. 1896), detected a new element with the atomic number of 75. They named it *rhenium,* after the Latin name of the Rhine river.

Although they didn't know it at the time, Noddack and Tacke (they were married the next year) had discovered the eighty-first and last element that possessed stable isotopes. Four elements between 1 and 92 now remained to be discovered, atomic numbers 43, 61, 85, and 87. Of these, elements 85 and 87 could be assumed to be radioactive, but there seemed no reason to suppose that elements 43 and 61 were. In fact, Noddack and Tacke announced the discovery of element 43 at the same time as that of rhenium. They called element 43 *masurium,* after a district in eastern Germany. In the case of this element, however, their observations were mistaken.

MORPHINE SYNTHESIS

Organic chemists were working out ever-subtler methods for placing atoms exactly where they wanted to and building up molecules of intricate form. The alkaloids present in plant tissues (see 1805) were particularly complex for compounds that were not simply chains of simple and repeated units.

An English chemist, Robert Robinson (1886–1975), was particularly skillful at synthesizing complex molecules, and in 1925 he synthesized morphine. From the methods of the synthesis, he could deduce its exact structure, atom for atom.

For his work in this field, Robinson was

awarded the Nobel Prize for chemistry in 1947.

PARATHORMONE

The thyroid gland was known to produce a hormone that regulated metabolic activity (see 1915). Embedded within it were four small glands called the *parathyroids,* which regulated calcium metabolism. In 1925 the Canadian biochemist James Bertram Collip (1892–1965) isolated from the parathyroids an extract that contained the hormone, which was named *parathormone*.

IRON AND CYTOCHROME

Keilin had earlier demonstrated the existence of cytochromes within the cell—a series of enzymes that combined oxygen atoms with pairs of hydrogen atoms (see 1924).

A German biochemist, Otto Heinrich Warburg (1883–1970), was studying the cytochromes and noticed that carbon monoxide attached itself to them in the same way that they attached themselves to hemoglobin. In 1925 Warburg showed that the cytochromes possessed the same iron-containing heme group that hemoglobin did.

IN ADDITION

On December 1, 1925, a series of treaties were signed, in Locarno, Switzerland, by Belgium, France, Great Britain, Italy, and Czechoslovakia on one side and Germany on the other. These *Locarno Treaties* guaranteed the postwar boundaries in the west and provided for arbitration of disputes. Seeming to heal the wounds of war, finally, and to provide against a repetition, they gave western Europe a feeling of security. Nevertheless, France began construction of a heavily fortified *Maginot Line* at its border with Germany, so-called after André Maginot (1877–1932), the French minister of war.

Despite the "Spirit of Locarno," Germany continued to rumble. Hitler published the first part of *Mein Kampf (My Battle),* a hysterical farrago of his hatreds.

In the United States, fundamentalist religious groups, particularly powerful in the rural South, forebade the teaching of evolution in the public schools. A test case arose over the fact that a biology teacher in Tennessee, John Thomas Scopes (1900–1970), had taught evolution. In 1925 the *Scopes trial* that resulted attracted international attention. It made the United States a laughingstock before the civilized world, but although the fundamentalists were defeated in the court of intelligent opinion, they did not give in and in fact haven't yet.

1926

WAVE MECHANICS

Three years earlier, de Broglie had suggested that particles such as electrons might have wave aspects (see 1923).

In 1926 an Austrian physicist, Erwin Schrödinger (1887–1961), decided that if the electron was viewed as a wave rather than as a particle, Bohr's electron orbits (see 1913 and 1915) might make more sense.

Schrödinger imagined an atom in which the electron could be in any orbit, provided that its matter-waves would extend around that orbit in an integral number of wavelengths. This would produce a *standing wave* and therefore would not represent an electric charge in oscillation. The electron, as long as it remained in such an orbit, need not radiate light and would not violate the conditions of Maxwell's equations (see 1865).

Furthermore, the permissible orbits worked out by Bohr and others all involved whole wavelengths, and the lowest orbit was one that involved a single wavelength. In this way, Bohr's suggestions came to make solid sense.

This view of Schrödinger's was called *wave mechanics,* and it was soon shown to be mathematically equivalent to Heisenberg's matrix mechanics of the year before (see 1925). However, Schrödinger's waves seemed the more attractive idea of the two, because they did offer the mind a picture of the atom.

Schrödinger worked out the mathematical underpinnings of wave mechanics; a key relationship that he presented is known as the *Schrödinger wave equation.*

For his work on wave mechanics, Schrödinger was awarded a share (along with Dirac—see 1930) of the Nobel Prize for physics in 1933.

WAVE PACKETS

Like Schrödinger (see above), the German physicist Max Born (1882–1970) tried to work out the implications of the electron viewed as a wave. He gave electron waves a probabilistic interpretation: the rise and fall of waves could be taken to indicate the rise and fall in probability that the electron behaved as though it was a particle existing at those points in the *wave packet.*

He also worked on the mathematical basis of such a view. He, Schrödinger, and Heisenberg (see 1925) may be viewed as the inventors of *quantum mechanics,* which ever since has been used to interpret chemistry and subatomic physics with remarkable success. Quantum mechanics and Einstein's relativity (see 1905 and 1916) are the two great theoretical foundations of twentieth-century physics.

For his work on quantum mechanics, Born was awarded a share of the Nobel Prize for physics in 1954.

FERMI-DIRAC STATISTICS

Bose and Einstein had worked out the Bose-Einstein statistics a year earlier, but the statistics now turned out to hold only for those particles (like the photon) that had spins of integral values: 0, 1, 2, and so on. Particles like the proton and electron had spins of *half-values:* ½, 1½, and so on.

Once Pauli had worked out the exclusion principle (see 1925) for particles with half-value spins, it was clear that Bose-Einstein wouldn't work for particles with such spins. A new set of statistics would therefore have to be worked out.

Taking the lead in working out that problem, in 1926, was an Italian physicist, Enrico Fermi (1901–1954). Dirac (see 1930) also contributed, so that the result is known as the *Fermi-Dirac statistics.* All particles subject to these statistics, like the proton and electron, are called *fermions* in honor of Fermi.

GALACTIC ROTATION

Twenty-two years before, Kapteyn had observed that there were two star streams moving in opposite directions (see 1904). In 1926 the Swedish astronomer Bertil Lindblad (1895–1965), analyzing these motions carefully, showed that they were just what would be expected if we viewed the Galaxy as rotating about its center.

Shortly thereafter, the Dutch astronomer Jan Hendrik Oort (b. 1900) came to the same conclusion.

LIQUID-FUEL ROCKETS

The Chinese had made use of rockets in the Middle Ages, and Isaac Newton, with his law of action and reaction (see 1687), had shown that rockets were one way of traveling through outer space. Until the twentieth century, however, rockets had made use of gunpowder as the fuel and had depended on the atmosphere for oxygen.

A crucial change was introduced by the American physicist Robert Hutchings Goddard (1882–1945). He had been interested in rocketry since his teens, and it occurred to him that a rocket would be more powerful and under better control if it made use of a liquid fuel such as gasoline and carried its own oxidizer in the form of liquid oxygen.

On March 16, 1926, Goddard fired his first liquid-fuel rocket. It was about 4 feet high, 6 inches in diameter, and was held in a frame like a child's jungle gym. It rose 200 feet into the air. It didn't seem like much, but that little rocket marked the first step in the expansion of the human range beyond the atmosphere.

ENZYME CRYSTALLIZATION

Although Payen had isolated an enzyme for the first time nearly a century before (see 1833), scientists still didn't understand the chemical nature of enzymes.

From the fact that enzyme action could be destroyed by gentle heating, it might be supposed that enzymes were proteins. However, Willstatter (see 1906) had purified enzyme solutions as far as he could while still retaining the enzyme action and had then tested the solution for protein content and found none. From this he deduced that enzymes were not proteins, but actually the test was not conclusive. Enzymes might be so active that they need be present in only trace amounts to do their work—in amounts too small for Willstatter's tests to be positive.

The solution might be to prepare more concentrated preparations of enzymes and then test them. This was done by the American biochemist James Batcheller Sumner (1887–1955). In 1926 he was extracting an enzyme from jackbeans, one that catalyzed the breakdown of the waste product urea to ammonia and carbon dioxide. The enzyme was therefore called *urease.*

In performing his extraction, Sumner found that a number of tiny crystals precipitated out of one of his fractions. He isolated the crystals, dissolved them, and found that not only did they show strong urease activity, but he could not separate the activity from the crystals. He concluded that the crystals *were* the enzyme. If he was correct, it would be the first time anyone had prepared an enzyme in crystalline form and therefore in relatively pure condition.

When he tested the crystals, he found them to be incontrovertibly protein. For this, he was awarded a share of the Nobel Prize for chemistry in 1946.

PERNICIOUS ANEMIA

One form of anemia is particularly deadly and is therefore called *pernicious anemia.* This disease was investigated by an American physician, George Richards Minot (1885–1950). Recalling that Whipple had found liver to be helpful in treating ordinary anemia (see 1920) and believing that pernicious anemia might be a dietary deficiency disease caused by the lack of some essential vitamin, Minot wondered if this vitamin might be found in liver. He and his assistant, William Parry Murphy (b. 1892), put pernicious anemia patients on a liver diet and by 1926 had demonstrated success. It wasn't pleasant, eating quantities of liver for indefinite periods, but it delayed death.

As a result, Minot and Murphy (along with Whipple) were awarded the Nobel Prize for medicine and physiology in 1934.

IN ADDITION

In the Soviet Union, Stalin had established himself as the heir of Lenin and from this point on he ruled the land with an increasingly firm grip.

In China, Chiang Kai-shek (1887–1975) established himself as the nearest approach to a Chinese ruler in that chaotic nation.

In Japan, the emperor Yoshihito died and was succeeded by his son, Hirohito (1901–1989).

1927

UNCERTAINTY PRINCIPLE

It had always been assumed by scientists that given enough patience, enough delicacy, and the proper instruments, any and all properties of anything observable could be measured to any degree of accuracy. It seemed axiomatic.

In 1927, however, Heisenberg (see 1925) showed that a careful consideration of quantum mechanics proved this to be wrong. One could determine, for instance, the momentum of a subatomic particle to any degree of precision, and the position of a subatomic particle to any degree of precision, but one could not determine *both at the same time* to any degree of precision. The more closely you determined a particle's momentum, the less certain you could be of its position, and vice versa. (The same was true if you tried to determine both the energy content of a particle and the exact time at which you were trying to observe it.) The uncertainty in the knowledge of momentum multiplied by the uncertainty in the knowledge of position was equal to Planck's constant (see 1900).

Planck's constant represented, so to speak, the graininess of the Universe. It seemed that if you tried to inspect the Universe very closely indeed, you would come up against the grain and could detect nothing finer than that.

It was rather like trying to magnify a photograph made up of tiny patches of dark and light. Under ordinary conditions, the picture looks smooth and well defined. Magnify it sufficiently, though, and the tiny patches expand till the picture seems to be a chaos of meaningless light and dark. You have reached the limits of magnification and of further information.

Heisenberg had discovered the *uncertainty principle.* At first glance, it would seem to negate any hope scientists might have of discovering "truth" to the finest detail, but looked at in another way, the uncertainty principle is how the Universe works, and its existence as a limitation explains many facets of the Universe that would be meaningless if that limitation did not exist. For example, to explain why helium doesn't freeze at ordinary pressures, even at absolute zero, requires a line of reasoning in which the uncertainty principle plays an essential part.

It was for this principle that Heisenberg was awarded the Nobel Prize for physics in 1932.

ELECTRON DIFFRACTION

De Broglie had suggested that electrons and indeed all particles had wave aspects (see 1923). No one since, however, had actually caught a particle behaving like a wave.

An American physicist, Clinton Joseph Davisson (1881–1958), however, was studying the reflection of electrons from a metallic nickel target enclosed in a vacuum tube. The tube shattered by accident, and the heated nickel promptly developed a film of oxide that made it useless as a target. To remove the film, Davisson heated the nickel for an extended period. Once this was done, he found that the reflecting properties of the nickel had changed.

Whereas the target had contained many tiny crystal surfaces before heating, it contained just a few large crystal surfaces afterward. Davisson therefore carried matters to a logical extreme and, in 1927, prepared a single nickel crystal for use as a target.

Once that was done, he found that an electron beam was not only refracted, but also diffracted just as a beam of X rays would be. Diffraction was a characteristic property of waves, so in this way the wave aspect of electrons was detected.

In 1927 the British physicist George Paget Thomson (1892–1975), the son of J. J. Thomson, who had discovered the electron (see 1897), sent a beam of fast electrons through thin gold foil and also demonstrated electron diffraction.

Thus de Broglie's theory was amply confirmed, and for doing so Davisson and Thomson were awarded shares in the Nobel Prize for physics in 1937.

SPEED OF LIGHT

In his last years, Michelson, who had performed the fateful Michelson-Morley experiment (see 1887), grew interested in measuring the speed of light with new precision. In the California mountains, he surveyed a 22-mile distance between two mountain peaks to an accuracy of less than an inch. He made use of a special eight-sided revolving mirror to reflect a light beam in the fashion used by Foucault (see 1849).

In 1927 he obtained a value of 199,798 kilometers per second for the speed of light. This was only about 6 kilometers per second faster than the figure determined nowadays by much more sophisticated equipment.

COSMIC EGG

Friedmann had developed the theoretical concept of an expanding universe (see 1917), and in 1927 the Belgian astrophysicist Georges-Henri Lemaître (1894–1966) drew what seemed a natural conclusion.

If the Universe was expanding as time went forward, then if we imagined the situation reversed and looked back in time, we should see the Universe contracting. (It would be as though we had taken a film of the expanding Universe and were running it backward.)

If we looked forward, the Universe might well expand forever, but if we look backward, the contraction had to be limited. Eventually, at a point far enough back in time, all the matter of the Universe would be compressed into one relatively small body, which Lemaître called the *cosmic egg*.

This cosmic egg apparently exploded in what came to be called *the big bang* and started the expanding Universe that now exists. Of course, Lemaître could offer no scientific explanation of where the cosmic egg came from and just how its explosion led to the present Universe. Physicists have been trying to work that out ever since.

ELECTRON BONDS

Lewis had viewed chemical bonds as resulting from the transfer or sharing of electrons (see 1916). Now, immediately after Schrödinger and Born had worked out the mathematical underpinnings of quantum mechanics (see 1926), two German physicists, Fritz Wolfgang London (1900–1954) and Walter Heitler (b. 1904), tried to apply quantum mechanics to chemical bonds. In 1927 they took the simplest case, that of the hydrogen molecule. It consisted of two hydrogen atoms, each contributing its single electron to a shared pair.

It turned out that quantum mechanics explained the properties and behavior of the hydrogen molecule neatly. This was the beginning. Eventually quantum mechanics was applied to every facet of chemistry, greatly advancing our understanding of the subject and making it, to some degree, a branch of physics.

PEKING MAN

A Canadian anthropologist, Davidson Black (1884–1934), was convinced that humanity had its origins in Asia. In 1920 he took up a post at the Peking Union Medical College so that he might investigate the region for hominid remains.

In 1927, in a cave at Cho-K'ou-tien, 25 miles west of Peking, he located a single human molar. From this tooth alone, he deduced the existence of a small-brained hominid, which he called *Sinanthropus pekinensis* (Greek for "Chinese human of Peking") and popularly known as *Peking man*.

Later discoveries showed that Peking man was very similar to Java man, which Dubois had discovered (see 1890). Both are now considered examples of *Homo erectus,* a species that preceded *Homo sapiens* (ourselves and Neanderthal man), and which in turn was preceded by the australopithecines, which Dart had discovered (see 1924).

Naturally the line of descent was not complete. Other discoveries remained to be made.

X RAYS AND MUTATIONS

Morgan (see 1907) had been working with the fruit fly for two decades, and one of his students was the American biologist Hermann Joseph Muller (1890–1967).

To make the studies on fruit flies worthwhile, mutations had to occur, and so they did, but randomly and not often enough for Muller. He found, in 1919, that keeping the fruit flies at a higher temperature increased the mutation rate, and he decided that this might be so because at a higher temperature, the molecules in the genes vibrated more strongly and were therefore more likely to undergo spontaneous random change.

If this were so, and if mutations were the result of molecular change, was there something that would bring them about faster than raising the temperature would?

It occurred to him to try X rays. They were more energetic than gentle heat, and if they struck a gene they would certainly affect some atomic group. He made use of X rays, and by 1927 it was quite clear that they worked. The mutation rate went up sharply.

This was not only useful from the standpoint of genetics. It became plain that most X-ray-induced mutations (and indeed, mutations induced in any fashion) were deleterious. This meant that X rays and radioactive radiations had the capacity to harm human beings who worked with them carelessly. Muller publicized this and subsequently fought for the safe handling of radiation.

M AND N BLOOD GROUPS

Landsteiner had discovered the A, B, O series of blood groups and demonstrated their importance in blood transfusion (see 1900). He thought it was possible that other blood groups might exist that were not of importance to transfusion but that might still be of interest in studying heredity, distinguishing between geographically separated human groups, determining paternity questions, and so on.

Thus in 1927 Landsteiner and his group discovered blood groups they designed as M, N, and MN.

TALKING PICTURES

For a quarter-century, motion pictures had been an increasingly important amusement for the world, but in all that time they had been silent, accompanied only by the conventional use of a tinkling piano. By and large, the silent movies consisted of a dumb show, broken occasionally by title plates.

Attempts had been made now and then to add sound to film, but the first true success came on October 6, 1927, with the opening of *The Jazz Singer*, starring Al Jolson. This new development spread with amazing rapidity, and in two or three years the silent movie was dead.

IN ADDITION

On May 20–21, 1927, the American aeronaut Charles Augustus Lindbergh (1902–1974) flew from New York to Paris. Others had flown across the Atlantic ocean before, but Lindbergh did it nonstop and alone in a small single-engine plane, *Spirit of St. Louis*, in a flight that kept him awake for 33½ hours. With this feat, aeronautics came of age.

In the Soviet Union, Stalin further consolidated his power. He expelled Trotsky from the Communist Party in November 1927.

In China, Chiang Kai-shek consolidated his power as well. In addition, he took a sharp turn to the right and abandoned the leftist allies he had inherited from Sun Yat-sen.

Sacco and Vanzetti were finally executed on August 23, 1927, for a crime that many people were sure they did not commit and in the face of worldwide protest.

Oil was discovered in northern Iraq on October 14, 1927, which began the gradual development of the Middle East as the world's richest oil reserve.

1928

PENICILLIN

Some discoveries are made by accident. In 1928 Fleming, who had discovered lysozyme (see 1922), left a culture of staphylococcus germs uncovered for some days. He was through with it, actually, and was about to discard the dish containing the culture when he noticed that some specks of mold had fallen into it. Around every speck, the bacterial colony had dissolved away for a short distance. Bacteria had died and no new growth had invaded the area.

Fleming isolated the mold and eventually identified it as one called *Penicillium notatum*, closely related to ordinary bread mold. He decided that it liberated some compound that, at the very least, inhibited growth, and he called the substance *penicillin*.

Fleming tested the mold on various types of bacteria and found that some were affected and some were not. Human cells were not affected. He did not go further, and it was to be over a decade before scientists returned to the problem. Nevertheless, for this discovery, Fleming received a share of the Nobel Prize for medicine and physiology in 1945.

DIELS-ALDER REACTION

Chemists who specialize in organic synthesis are delighted when they can find a chemical reaction that will put together atoms in some fashion and that will work under a variety of conditions.

In 1928 two German chemists, Otto Paul Hermann Diels (1876–1954) and Kurt Alder (1902–1958), found a reaction that would join two compounds in such a way as to form a ring of atoms. This was properly a *diene synthesis* but came to be called the *Diels-Alder reaction*. The reaction was useful in synthesizing many compounds of biological interest, and as a result, the two chemists were awarded the Nobel Prize for chemistry in 1950.

RAMAN SPECTRA

After Compton had discovered the Compton effect—that X rays tended to lengthen their waves when diffracted (see 1923)—Heisenberg (see 1925) pointed out, in 1925, that this ought to be true of any electromagnetic radiation, including visible light. That it was so in practice was shown in 1928 by the Indian physicist Chandrasekhara Venkata Raman (1888–1970).

By demonstrating that scattered light had weak components of changed wavelengths, he showed that photons of visible light had particle aspects. Furthermore, the exact wavelengths produced in the scattering depended upon the nature of the molecules doing the scattering. For this reason, *Raman spectra* proved to be useful in determining some of the fine details of molecular structure.

For his discovery, Raman was awarded the Nobel Prize for physics in 1930. He was the first Asian to win a Nobel Prize.

GAME THEORY

A new branch of mathematics was opened by a Hungarian-born American mathematician, John von Neumann (1903–1957), in 1928. He began devising the principles of what came to be called *game theory* because it dealt with the best strategies to follow when playing simple games with fixed rules, such as coin-matching.

The principles so developed could then be applied to far more complicated games, such as business or war, and an attempt made to work out the best strategy to beat a competitor or an enemy. Even scientific research can be considered a game, one in which scientists pit their wits against the impersonal Universe.

HEXURONIC ACID

In 1928 the Hungarian-born American biochemist Albert von Nagyrapolt Szent-Györgyi (1893–1986), while working at Cambridge University under Hopkins (see 1900,

Tryptophan), isolated a substance from adrenal glands that easily lost and regained a pair of hydrogen atoms and was therefore, like glutathione (see 1921), a *hydrogen carrier.*

Since its molecules seemed to have six carbon atoms and since it had the properties of a sugar, Szent-Györgyi named it *hexuronic acid.* (The *hex* is from the Greek word for "six" and the *uronic* is a common suffix for sugar-related compounds.)

Szent-Györgyi also isolated the substance from cabbages and oranges, both rich in vitamin C, but it took him a while to decide that hexuronic acid was itself the vitamin.

IN ADDITION

The peace movement that had arisen as a reaction to the horrendous bloodletting of World War I reached its peak with the Pact of Paris, popularly called the *Kellogg-Briand Pact.* On August 17, 1928, sixty-three nations signed the pact, outlawing war as a national policy. Various nations made exceptions, however, and there was no arrangement for the nations to unite and impose economic penalties *(sanctions)* of one sort or another on any nation breaking the pact. The result was that the pact proved an empty flood of words and never accomplished anything to prevent or even inhibit the war-making tendencies of nations.

In the United States, Herbert Clark Hoover (1874–1964) was elected the thirty-first president.

1929

RECEDING GALAXIES

Slipher had measured the radial velocity of the Andromeda nebula even before it had been found to be an independent galaxy (see 1923). After that he measured the radial velocity of other galaxies and found that all but two were receding from us.

The American astronomer Milton La Salle Humason (1891–1972) continued these labors, working with Hubble, who had first proved the Andromeda to be an independent galaxy. Humason measured the radial velocity of many more galaxies and also found that all were receding from us, some at unusually large rates.

Hubble went over all the work carefully and used various methods for estimating the distances of the various galaxies. By 1929 he felt sufficiently confident of his conclusions to announce them. Galaxies, he maintained, receded from us at a rate proportional to their distance from us *(Hubble's law).*

What is so important about our Galaxy that all the others (except for two of the very closest) should be receding from us, and why should those farther off be receding more quickly than those nearer? The explanation that seemed most logical was that the Universe was expanding in accordance with the suggestion of Friedmann (see 1917), so that the galaxies (or clusters of galaxies) were all receding from each other, and not merely from us. An observer in *any* galaxy would see the other galaxies receding at a rate proportional to their distance.

With Hubble, then, the expanding Universe ceased being a matter of theory and became an observed fact.

SOLAR COMPOSITION

Ångström had demonstrated the existence of hydrogen in the Sun two-thirds of a century before (see 1862). It was not till 1929, however, that the Solar spectrum was used to give a fairly detailed picture of the overall composition of the Sun.

Russell, who had earlier helped work out the matter of the Main Sequence (see 1914), showed that the Sun consisted almost entirely of hydrogen and helium, in a 3-to-1 mass ratio. The most important of the minor components were oxygen, nitrogen, neon, and carbon. (This turned out to be similar to the composition of the Universe as a whole, as nearly as astronomers can determine.)

SOLAR ENERGY

Three-quarters of a century before, Helmholtz had suggested that the Sun obtained its energy from gravitationally induced shrinkage (see 1853). This had given an impossibly small age for the Earth, but no suitable substitute source of energy had been found until the existence of nuclear energy was demonstrated by Pierre Curie (see 1901). Then it seemed very likely that some sort of nuclear reaction was the solar power source.

In 1929 the Russian-born American physicist George Gamow (1904–1968) suggested that the nuclear source was the conversion of hydrogen nuclei into helium nuclei, since these two elements made up nearly all the Sun, as Russell had shown (see above). As four hydrogen nuclei had to fuse together to form one helium nucleus, this process was called *nuclear fusion*, or more specifically, *hydrogen fusion*.

However, not enough was known about nuclear fusion as yet for Gamow to flesh out his suggestion in detail.

COINCIDENCE COUNTER

In 1929 the German physicist Walther Wilhelm Georg Franz Bothe (1891–1957) devised a new method of studying cosmic rays. He placed two Geiger counters (see 1908) one above the other and set up a circuit that would record an event only if both counters recorded it virtually simultaneously.

This would happen only if a cosmic ray particle, streaking down from above, shot vertically through both counters. Particles coming from some other direction would pass through one counter and not the other. And particles that were not cosmic rays, though coming from the right direction, would be insufficiently energetic to pass through both.

Such a *coincidence counter* turned out to be very useful in measuring the short intervals of time between passage through one counter and the next. These ultrashort intervals were still long enough for events to take place on the subatomic scale. Bothe used this technique to show that the laws of conservation of energy and of momentum were as valid for atoms as for billiard balls.

For his coincidence counters, Bothe was awarded the Nobel Prize for physics in 1954.

PARTICLE ACCELERATOR

In the quarter-century since the discovery of radioactivity, the most energetic particles easily available to nuclear physicists had been alpha particles. The shorter the half-life of a particular radioactive isotope, the more energetic the alpha particles they produced. Rutherford used alpha particles to bombard atoms and induce nuclear reactions (see 1906), but even the most energetic alpha particles could only do so much. Of course, cosmic ray particles were more powerful still, much more powerful, but they were not under scientists' control.

What was needed was some way of beginning with ordinary nuclear particles, say protons obtained by ionizing hydrogen atoms, and then accelerating them, perhaps by means of an electromagnetic field.

This was first accomplished by the British physicist John Douglas Cockcroft (1897–1967) and his coworker, the Irish physicist Ernest Thomas Sinton Walton (b. 1903). In 1929 they devised a *voltage multiplier* that would build up high electrical voltages capable of accelerating protons to the point where they contained more energy than the alpha particles occurring in nature.

This was the first *particle accelerator* (better known to the public for a time as an *atom-smasher)*. For this work, Cockcroft and Walton were awarded the Nobel Prize for physics in 1951.

OXYGEN ISOTOPES AND ATOMIC WEIGHTS

Aston's work with the isotopes of stable elements (see 1925) had not uncovered them all. In 1929 Giauque (see 1925) found that of every 10,000 oxygen atoms, 9,976 indeed had a mass of 16, which was the atomic weight of oxygen. This very common isotope was therefore called oxygen-16. However, of the remaining 24 atoms, 4 had a mass of 17 and 20 a mass of 18. These were called oxygen-17 and oxygen-18.

This was upsetting news, since oxygen atoms had been used for a century as the standard against which atomic weights could be determined. If oxygen was a mixture of isotopes, with the mixture varying a bit (perhaps) from place to place and from time to time, it wouldn't serve as a standard. There were suggestions that oxygen-16 be used as a standard, even though that meant that all atomic weights would have to be slightly altered. Eventually, the commonest isotope of carbon, carbon-12, was accepted as standard, since this would entail less change in the value of atomic weights generally.

DEOXYRIBOSE

Levene had identified the sugar in some molecules of nucleic acid as ribose (see 1909). There were other nucleic acids from which a sugar that was *not* ribose could be obtained, but it was not till 1929 that Levene could identify that other sugar. It turned out to be *deoxyribose,* which had a molecule just like that of ribose except for a missing oxygen atom.

That meant that there were two types of nucleic acid: ribosenucleic acid and deoxyribonucleic acid, usually abbreviated as RNA and DNA, respectively. It was DNA that was found in chromosomes.

HEME

For ten years the German chemist Hans Fischer (1881–1945) had been puzzling out the structure of *heme,* the complex molecular group that joined with protein to form hemoglobin. It was heme that supplied blood's color, that contained the iron atom, and that did the work of picking up oxygen at the lungs and giving it up at the cells.

Heme was not composed of amino acids, as proteins were. It was a nonprotein adjunct of the molecule, which consisted of a *porphyrin ring,* four small rings of atoms connected into a larger ring. Attached to this *ring of rings* were eight *side chains,* four of one kind, two of another, and two of still another.

Once Fischer had the general structure worked out, he realized that the side chains could be arranged in any of fifteen different ways. He organized his students into groups and had them each set about synthesizing a different arrangement. Fischer then checked them all to see which arrangement had the properties of the natural substance.

By 1929 he had the structure of heme worked out to the last of its seventy-five atoms, and for this he was awarded the Nobel Prize for chemistry in 1930.

ESTRONE

Males and females of the same species develop differently. The sex organs arise out of similar structures, but though the penis and the clitoris are homologous, they are very different in appearance and function. The male grows a larger larynx, the female larger breasts, the subcutaneous fat distribution and the hair pattern are different in the two sexes, and so on.

It wasn't very difficult to suppose that hormones were involved and that, like the organs whose development they controlled, they were similar but different in the two sexes.

In 1929 the American biochemist Edward Adelbert Doisy (1893–1986) and, working independently, the German chemist Adolf Friedrich Butenandt (b. 1903) isolated a female sex hormone, which came to be called *estrone,* from the Greek word for "sexual heat."

INTRINSIC FACTOR

Minot and Murphy had found that some vitamin in liver relieved pernicious anemia (see 1926). The American physiologist William Bosworth Castle (b. 1897) noted that although diet could prevent pernicious anemia from developing in most patients, it did not cure the few who actually developed the disease. He reasoned that, if a vitamin was involved, those who had trouble absorbing it even when it was present in the diet were the ones who came down with pernicious anemia.

Since patients with pernicious anemia had a characteristic lack of hydrochloric acid in their gastric juice, Castle suggested, in 1929, that it was some component of normal gastric juice that brought about the absorption of the antipernicious-anemia vitamin. He called that missing component *intrinsic factor*.

ELECTROENCEPHALOGRAPHY

Einthoven had developed methods for detecting the rise and fall of electric potentials involved in the heartbeat and devised the electrocardiogram (see 1903).

The German psychiatrist Hans Berger (1873–1941) thought the same might be done for the brain. During the 1920s he devised a system of electrodes that, when applied to the skull and connected to an oscillograph, would give a recording of the rhythmic shifting of electric potentials commonly called *brain waves*.

In 1929 he published his results, describing *alpha waves* and *beta waves*. In this way, *electroencephalography* (Greek for "the writing of brain electricity") was developed. It offered a technique for the diagnosis of such serious brain disorders as tumors and epilepsy.

CARDIAC CATHETER

A German surgeon, Werner Theodor Otto Forssmann (1904–1979), was the first to work out a practical system of cardiac catheterization. He inserted a catheter (a long, thin, flexible rod that was opaque to X rays) into a vein in his own elbow and pushed it along the vein until it reached the heart. This made it possible to study the structure and function of an ailing heart and to make accurate diagnoses without the necessity of exploratory surgery.

The technique was not used clinically for over a decade, but Forssmann was awarded a share of the Nobel Prize for physiology and medicine in 1956.

CHILD DEVELOPMENT

By 1929 the Swiss psychologist Jean Piaget (1896–1980) had worked out his observations on child development. By observing and asking questions, he learned much. He also developed what he called *conservation tasks:* he made changes in simple objects and studied whether children could tell what had *not* changed. For instance, if one pours water from a broad low vessel into a tall thin vessel, the height of the liquid increases. This may be taken to imply that there is more water, but after a certain stage of development the child recognizes that the quantity of water has not changed.

Piaget described four phases of mental growth, paralleling the physical growth taking place, and maintained that all children went through the same phases in the same order.

IN ADDITION

In the United States, the stock market crashed on October 24, 1929, and billions of dollars in paper profits vanished. This eventually sparked the worldwide Great Depression.

In the Soviet Union, Trotsky was forced into exile. Stalin's domination was unquestioned thereafter.

The papacy had never agreed to the end of the Papal States in Italy during the course of Italian unification in the 1860s, and ever since 1870, the pope had considered himself a prisoner in the Vatican, which was the papal palace in Rome. On June 7, 1929, however, Pope Pius XI (1857–1939) came to an agreement with the Italian government of Mussolini. The pope's secular power was restored by recognizing 109 acres around the Vatican as an independent territory under the pope's rule. The Italian government paid the Vatican a large indemnity, and the pope no longer considered himself a prisoner.

In 1929 the American sociologists Robert Staughton Lynd (1892–1970) and Helen Merrell Lynd (1896–1982), his wife, published *Middletown,* in which they applied techniques usually employed in studying primitive cultures to the analysis of a midwestern American city.

1930

PLUTO

Neptune had been discovered by Adams and Leverrier (see 1846) because Uranus was not moving in its orbit exactly as it should, so that the gravitational field of a more distant planet was suspected.

Neptune's presence reduced but did not entirely wipe out the discrepancies in Uranus's orbit, and some astronomers thought that a still more distant but fairly large planet might exist beyond Neptune.

Lowell, who was so enthusiastic about Martian canals (see 1877), was also enthusiastic about this "Planet X" and spent much of his time calculating its possible orbit and then searching for it. He died without having found it, but at Lowell Observatory, which he had founded, the search went on.

The American astronomer Clyde William

Tombaugh (b. 1906) continued the search methodically. His technique was to take two pictures of the same small part of the sky on two different days. Each of these would show from 50,000 to 400,000 stars. Despite all those stars, the two plates would be identical if the spots of light were stars, and only stars. If the two plates were projected on a screen in rapid alternation, none of the stars would seem to move. If one of the "stars" was really a planet, however, one that had moved against the starry background during the interval between photographs, as the plates were alternately thrown upon the screen, that one object would seem to dart back and forth.

On February 18, 1930, Tombaugh found such a flicker in the constellation Gemini. From the smallness of the shift, the object had to be moving very slowly and must therefore lie beyond Neptune. On March 13, 1930, the seventy-fifth anniversary of Lowell's birth, the discovery was announced. The new planet was named Pluto, first because that god of the nether darkness was appropriate to a planet swinging farthest out from the light of the Sun, and second because the first two letters were the initials of Percival Lowell.

In time, though, Pluto turned out to be a small planet incapable of influencing Uranus's orbit measurably. The possibility of another large planet beyond Neptune therefore remains to this day.

SURFACE TEMPERATURE OF THE MOON

Nicholson, who discovered Jupiter IX (see 1914), also developed sensitive heat-measuring devices that could determine the surface temperature of an astronomical body such as the Moon from the amount of heat it delivered to the instrument. By using it he could show that the surface temperature of the Moon dropped nearly 200° C during the period of its eclipse by the Earth's shadow. The Moon had no ocean or atmosphere to conserve and circulate heat, and its solid body was so poor a heat-conductor that the surface lost a great deal of heat before any new supply could work its way upward from the deeper layers.

In 1930 Nicholson and Edison Pettit obtained the first reasonably accurate measurement of the temperature of the lunar surface. As corrected by work since, it seems that the sunlit side of the Moon reaches a maximum temperature of 117° C, which is well above the boiling point of water. The night side of the Moon, after two weeks without sun, sinks to a temperature, just before dawn, of −169° C, far colder than Antarctica at its coldest.

Since the Moon is at Earth's average distance from the Sun, we see how fortunate it is that Earth has a faster spin and shorter day, an ocean to conserve heat, and an atmosphere to distribute it.

CORONAGRAPH

For two centuries, astronomers had been traveling the world in order to witness rare astronomical events that were not easily visible in other places and at other times. Examples are the far southern stars, total eclipses of the Sun, and transits of Venus and of Mercury.

One of the rare sights, from the point of view of both beauty and scientific interest, was the Sun's pearly upper atmosphere, or *corona,* in which helium was first discovered (see 1868).

In 1930 the French astronomer Bernard-Ferdinand Lyot (1897–1952) devised the *coronagraph,* a telescope that focused the light of the Sun on an opaque disk, cutting out all scattered light from the atmosphere and from the lens itself. Mounting a telescope in the clear air of the Pyrenees, Lyot managed to observe the inner corona, at least, of the uneclipsed Sun. This meant that astronomers no longer had to wait for total eclipses to study the corona and its spectrum.

SCHMIDT CAMERA

One of the difficulties with the large telescopes of the twentieth century was that they could focus on such a tiny part of the sky. It was like looking at the Universe through a keyhole. Any attempt to enlarge the view resulted in inadmissable distortion.

In 1930 the Estonian-born German optician Bernhard Voldemar Schmidt (1879–1935) devised a special *corrector plate,* a small glass object with a complicated shape that could be placed near the focus of a spherical mirror. The corrector plate bent the light waves in such a way as to eliminate distortion, so that even wide fields could be enlarged steadily.

An instrument outfitted with such a mirror and corrector plate is called a *Schmidt telescope,* or, since it is invariably used with a photographic plate rather than an eye receiving the light, a *Schmidt camera.* Used in conjunction with a conventional telescope, it can locate the interesting spots, on which the keen peephole of the telescope can then be focused.

INTERSTELLAR MATTER

Three centuries earlier, it had come to be understood that there was a vacuum between the astronomical bodies. It was all too easy then to assume that the vacuum was perfect; that there was really nothing at all once one got outside any atmosphere that might be clinging to the immediate surface of a body.

In 1930, however, the Swiss-born American astronomer Robert Julius Trumpler (1886–1956) noted that the light of the more distant globular clusters was dimmer than would be expected from their sizes. The more distant the cluster, the more marked this departure from the expected brightness. What's more, the more distant the cluster, the redder the light.

The easiest way of explaining this was to suppose that space, even far from sizable bodies, was *not* a perfect vacuum. (Indeed, a perfect vacuum does not exist and probably cannot exist in the Universe.) There are thin wisps of gas and dust throughout interstellar space, and over vast distances, there is enough of this—of dust, particularly—to dim and redden light. By taking the dimming effect of this *interstellar matter* into account, the size of the Galaxy was shown to be somewhat smaller than Shapley's too-large estimates (see 1918).

ANTIMATTER

The demonstration by Davisson and Thomson that electron waves really existed (see 1927) encouraged the British physicist Paul Adrien Maurice Dirac (1902–1984) to develop the mathematics of electron waves.

From the equations he derived, it seemed to him that electrons and protons (the only two subatomic particles then known) had to exist in two energy states, one positive and one negative. At first Dirac assumed that the electron and proton themselves represented the two energy states and were fundamentally the same particle.

That would have been a great simplification of subatomic physics, but it didn't stand up. The proton and electron were too different in too many ways, notably in mass.

In 1930 Dirac suggested that both the electron and proton, as known, were in the positive state, but that each could exist in a negative state as well. That is, there should be a particle precisely like the negatively charged electron in every way except that it had a positive electric charge, and there should be a particle precisely like the positively charged proton in every way except that it had a negative electric charge.

These negative-state particles came to be thought of eventually as *antiparticles*—an *antielectron* and an *antiproton*—and it was not difficult to imagine that, just as protons and electrons made up matter, so antiprotons and antielectrons, if they actually existed, would make up *antimatter*.

Dirac was correct, they did exist, and he was awarded a share (along with Schrödinger—see 1926) of the Nobel Prize for physics in 1933.

CYCLOTRON

The particle accelerator developed by Cockcroft and Walton (see 1929) forced particles to travel in a straight line, faster and faster, thus piling up more and more energy, while being pushed on ahead by a magnetic field. The difficulty with this was that in order to get really high energies, the device had to be inconveniently long.

It seemed to the American physicist Ernest Orlando Lawrence (1901–1958) that, instead of pushing the particles ever onward, it might be handier to make them travel in circles, giving them a magnetic-field push each time they came around.

In 1930, therefore, he built a small device in which protons were made to travel between the poles of a large magnet that deflected their paths into circles. At each turn, they received another push of electric potential. This made them travel faster and therefore in a path that, under the constant force of the magnet, curved less sharply. The path was a sort of spiral that brought them closer and closer to the rim of the instrument. By the time the charged particles finally shot out of the instrument altogether, they had accumulated high energies indeed.

Lawrence called the device a *cyclotron*, because the particles moved in cycles. His first device was quite small, but it produced more energetic particles than a very long voltage multiplier would have.

For this, Lawrence was awarded the Nobel prize for physics in 1939.

COMPUTER

Babbage had tried to build a machine that would solve complicated mathematical problems by purely mechanical means (see 1822). He was defeated by the fact that mechanical methods of the day were, by and large, too coarse to do the work.

By the 1920s, engineers had at their disposal electric currents, together with radio tubes to control those currents. These cut down on the number of moving parts needed and supplied a much more delicate way of controlling the parts of what came to be called a *computer*.

In 1930 the American electrical engineer Vannevar Bush (1890–1974) produced the first machine capable of solving differential equations and the first one that Babbage would have recognized as fulfilling his design. However, Bush's computer was only partly electronic.

CRYSTALLINE ENZYMES

Sumner had crystallized urease and shown that one enzyme at least was a protein (see 1926). He had little reputation in the field, however, and biochemists generally hesitated to accept his results in the face of contrary views by much better-known biochemists such as Willstatter (see 1906).

In 1930, however, the American biochemist John Howard Northrop (1891–1987) managed to crystallize, not a rather off-beat enzyme such as urease, but the very well-known digestive enzyme pepsin and showed it to be a protein. That really settled the matter (especially when Northrop went on to crystallize other well-known enzymes as well).

The result was that Northrop was awarded a share (along with Sumner) of the Nobel Prize for chemistry in 1946.

VITAMIN A STRUCTURE

Vitamins had been known and worked with and used in nutrition and medicine for a third of a century, but they still remained mysterious, because their molecular structure was not known.

In 1930, however, the Swiss chemist Paul Karrer (1889–1971) showed that vitamin A was related to the carotenoids, of which the most familiar is *carotene,* the coloring matter of carrots. Other carotenoids are found in sweet potatoes, egg yolk, tomatoes, lobster shells, and human skin.

Vitamin A rather resembled half a molecule of carotene. This was finally proved when Karrer eventually synthesized it, which led to a rash of molecular determinations and syntheses of other vitamins, and the mystery (of their structure, at least) was removed.

FREON

By now, refrigerators and air-conditioners existed, and general use was made of liquids that could be evaporated and so draw heat from the surrounding environment, thus lowering the temperature, as in the method used for liquefying gases.

The main trouble was that the gases used for the purpose were ammonia and sulfur dioxide, for the most part, and that these had choking odors and were actually poisonous, so that leaks in the system were bound to be uncomfortable and might be dangerous.

It would be ideal if one could find a liquid that was odorless, easily evaporated, nonpoisonous, and stable.

Just such a liquid was discovered by the American chemist Thomas Midgley, Jr. (see 1921). In 1930 he prepared difluorodichloromethane, with a molecule consisting of a carbon atom to which two chlorine atoms and two fluorine atoms were attached. It had all the properties needed for refrigeration.

It belonged to the class of *chlorofluorocarbons,* and other molecules of the class had uses too. For example, such liquids could be used in spray cans to force other liquids out in a fine mist.

Trademarked preparations of the liquid, like Freon, rapidly went into use. They made air-conditioners common and, indeed, in developed societies, nearly universal equipment. Eventually, however, the chlorofluorocarbons were found to have dangers that at the time of their discovery could not have been foreseen, and their use is now being increasingly discouraged.

IN ADDITION

In the United States, stock market prices recovered a substantial part of their losses from the October 1929 crash, but in May they broke again and began a long decline. Congress passed the Hawley-Smoot Tariff Act, which raised duties to new highs, thus forcing other nations to retaliate and choking off trade. This made the Depression worse and insured that it would be worldwide. Banks started closing, and the accumulated savings of millions disappeared. Unemployment began to rise.

An Ethiopian prince, Ras Tafari (1892–1975), became the emperor of Ethiopia, ruling as Haile Selassie.

In Germany, Hitler and his National Socialists gained power rapidly as the Great Depression intensified.

The population of the United States reached 123 million, but for the first time, more people were leaving Depression-sunk America than were entering. The world population reached 2 billion.

1931

GÖDEL'S PROOF

Thirty years earlier, Frege had attempted to place all of mathematics on a formal logical basis, making it fully rigorous, and had failed (see 1902). Other, more elaborate attempts followed.

An Austrian mathematician, Kurt Gödel (1906–1978), put a final end to such schemes in 1931 by advancing what is now called *Gödel's proof.* He translated the symbols of symbolic logic into numbers in a systematic way and showed that it was always possible to construct a number that could not be arrived at from the other numbers of his system.

What it amounted to was this: Gödel showed that if you began with any set of axioms, there would always be statements within the system governed by those axioms that could be neither proved nor disproved on the basis of those axioms. If the axioms were modified in such a way that the statement *could* be either proved or disproved, then another statement could be constructed that could be neither proved nor disproved in the new system, and so on forever.

Gödel had thus ended the search for certainty in mathematics by showing that it did not and could not exist, just as Heisenberg had in physics (see 1927).

However, Gödel's proof does not affect the nitty-gritty of ordinary mathematics. Two plus two is still four.

NEUTRINO

For over a decade, physicists had been plagued with the problem of beta-particle emission. Beta particles might be fired out of a nucleus with all the energy to be expected from the loss in mass as one nucleus broke down into another. Generally, though, they came out with less energy, and to an unpredictable degree. They sometimes emerged with very little energy, in fact, and some physicists, in despair, felt that the law of conservation of energy simply didn't hold in connection with beta-particle emission.

In 1931, however, Pauli, who had worked out the exclusion principle (see 1925), suggested an explanation that did not violate energy conservation. He suggested that, along with the electron, another particle was given off, and that the energy was divided between the electron and the other particle in a random manner.

Since the electron had all the electric charge available, the other particle had to be electrically uncharged, or *neutral.* Since all the kinetic energy of an electron could be converted into only a tiny quantity of mass, the other particle would have little or no mass.

The next year Fermi, who had devised a mathematical treatment for electron distribution (see 1926), named the other particle the *neutrino* (Italian for "little neutral one").

Since it lacked both mass and electric

charge, the neutrino was sure to be very difficult to detect, assuming it existed at all, and for a quarter of a century it remained a kind of "ghost particle," with theoretical reasons for existing but backed by no observational evidence.

DEUTERIUM

As more and more stable elements proved to be made up of mixtures of isotopes, there was considerable feeling that even the lightest and simplest of the elements, hydrogen, ought to consist of isotopes. Hydrogen has an atomic weight of very close to 1, so that if it did consist of isotopes, hydrogen-1 must be overwhelmingly the most common. Still, there might be very small quantities of hydrogen-2 present.

The American chemist Harold Clayton Urey (1893–1981) tackled the problem in 1931. He reasoned that hydrogen-2, being the more massive atom, would be less easily evaporated than hydrogen-1. If, then, he slowly evaporated a large quantity of liquid hydrogen, the final bit of liquid ought to have a percentage of hydrogen-2 larger than that in the supply he began with.

Now if hydrogen-2 were present, its spectral lines ought to have slightly different wavelengths than those of hydrogen-1, and the ordinary hydrogen spectrum ought to have, accompanying each spectral line, a very faint one nearby—too faint to be sure of.

However, when most of the liquid hydrogen had evaporated, leaving an unusually high concentration of hydrogen-2, the hydrogen-2 lines ought to become more prominent and should be unmistakable.

This turned out to be so, and Urey announced the discovery at once. Hydrogen-2 came to be called *heavy hydrogen* or *deuterium* (from the Greek word for "two"). For this discovery, Urey was awarded the Nobel Prize for chemistry in 1934.

RESONANCE

Four years earlier, London had applied quantum mechanics to the electron-sharing between hydrogen atoms in the hydrogen molecule. In 1931 the American chemist Linus Carl Pauling (b. 1901) extended this to the electron-sharing in organic compounds generally.

Consider the benzene molecule, for instance. It consists of six carbon atoms in a hexagonal ring, with one hydrogen atom attached to each carbon atom. Such a ring must have single bonds and double bonds in alternation. Double bonds in ordinary organic compounds represent a region of activity where two hydrogen atoms can easily be added. In the case of benzene, however, the double bonds are stable and are difficult to add to. The great stability of the benzene ring was a puzzle. Some suggested that the double bonds switched endlessly, so that any given adjacent pair of atoms were connected by a single bond and a double bond in rapid alternation.

Pauling showed in 1931, however, that if all the atoms of a molecule were in a single plane (as was true of benzene) and symmetrically placed (as was again true of benzene), then the electron waves would spread out over the carbon atoms generally, so that what connected the carbon atoms was neither an ordinary single bond nor an ordinary double bond but something intermediate between the two. This spreading out of the electrons *(resonance)* represented a very stable conformation, and any molecule increased in stability where there was an opportunity for resonance.

The concept of resonance helped greatly in explaining the manner of chemical reactions and made them far more predictable.

For this, and for his work on chemical structure made possible by the concept of resonance, Pauling was awarded the Nobel Prize for chemistry in 1954.

ANDROSTERONE

Butenandt had isolated estrone, a female sex hormone (see 1929). It seemed obvious that if female sex hormones existed, male sex hormones must also exist.

In 1931 Butenandt obtained a small quantity of a hormone that was named *androsterone* (from the Greek word for "man"). It is produced by cells of the testicle and stimulates the type of development required to produce the characteristics of the adult male. Butenandt isolated only 15 milligrams of the hormone (a two-thousandth of an ounce), but by delicate microanalytical methods, he was able to make two analyses of the compound.

For his work on the sex hormones, Butenandt was awarded a share of the Nobel Prize for chemistry in 1939.

NEOPRENE

Rubber had become a vital resource in an age of automobiles. Its use in tires was essential, and it had a myriad of other uses as well. The major producer of rubber was originally Brazil, but production had been transferred to Malaya. The possibility of such a distant source being cut off in time of war or political unrest made it advisable, even imperative, that the industrial nations of Europe and America devise some sort of synthetic rubber that could be made at home.

One of those involved in the search was a Belgian-born American chemist, Julius Arthur Nieuwland (1878–1936). His researches uncovered the fact that acetylene, a compound with a molecule made up of two carbon atoms and two hydrogen atoms, could *polymerize;* that is, add other atoms onto itself to form a long chain of atoms. At some stage, this chain gained some of the properties of rubber.

Nieuwland found that if, at the four-carbon stage, chlorine atoms were attached, the resulting chain was far more like rubber. By 1931 he had what was called *neoprene.* This kept essential facets of the American economy rolling later when the rubber supply was indeed cut off.

NYLON

One of the most important natural polymers is silk, which is a protein fiber composed of a long string of relatively simple amino acids. Making silk from the cocoon of the silkworm (a caterpillar) is a tedious job.

It occurred to the American chemist Wallace Hume Carothers (1896–1937) that an artificial polymer might be prepared that had the properties of silk. He had worked on polymers along with Nieuwland and had helped synthesize neoprene (see above).

Carothers began working with compounds called *diamines* and *dicarboxylic acids.* These combine alternately to form the same types of connections that amino acids do. In 1931 Carothers found a polymeric fiber that, after stretching, proved even stronger than silk. It was eventually called *nylon,* and the time was to come when it would replace silk to a very large extent.

VIRUS PARTICLES

Since the existence of viruses had first been demonstrated by Beijerinck (see 1898), some forty diseases (including measles, mumps, chickenpox, influenza, smallpox, poliomyelitis, rabies, and the common cold) had been found to be caused by viruses, but their nature remained a mystery.

In 1931, an English bacteriologist, William

Joseph Elford (1900–1952), managed to trap some in filters. He used fine collodion membranes, graded to keep out smaller and smaller particles. From the fineness of the membrane that could filter out the agents capable of producing a given disease, he could judge the size of that particular virus.

He found that viruses were smaller than even the smallest bacteria, but that even the smallest viruses were larger than most molecules. Viruses, in other words, were objects that ranged in size from that of large protein molecules to nearly that of tiny bacteria.

VIRUS CULTURE

A great deal of the success in dealing with bacterial disease came about because it was possible to grow pure colonies of bacteria in glassware and study them in detail. Virus diseases presented a more intricate puzzle, because viruses would only multiply inside living cells.

In 1931, however, the American pathologist Ernest William Goodpasture (1886–1960) devised a technique for culturing viruses in eggs, which are, after all, living cells. This technique made it possible eventually to develop vaccines for a number of viral diseases, notably poliomyelitis.

STRATOSPHERIC BALLOONS

Human beings had reached heights of up to 6 miles in balloons but had done so at great peril, for the air is not dense enough at such heights to be thoroughly capable of supporting life (see 1902). Uncrewed balloons were used to reach greater heights.

This was not satisfactory to the Swiss physicist Auguste Piccard (1884–1962), who wanted to study the ionosphere and cosmic rays more intimately than the instruments of the day made possible. What had to be done, he felt, was to construct a sealed aluminum gondola within which a normal atmosphere could be maintained however high the balloon climbed.

In 1931, making use of such a sealed gondola, Piccard ascended to a height of nearly 10 miles, half again as high as human beings had ever risen before. This paved the way for other and even higher penetrations of the stratosphere by himself and others. Eventually, heights of nearly 20 miles were achieved.

IN ADDITION

The Great Depression gathered steam. On May 11, 1931, the Kreditanstalt, a banking institution in Vienna, went bankrupt, and financial chaos spread through Europe.

In the Far East, the situation grew ominous. The Japanese charged into Manchuria and took it over, renaming it Manchukuo.

Chiang Kai-shek of China couldn't do anything about it. He had begun a deadly rivalry with Mao Tse-tung (1893–1976), who had remained leftist after Chiang had turned rightist. For years afterward, Chiang was far more interested in fighting Mao than in fighting Japan.

Alfonso XIII of Spain (1886–1941) was ousted from his throne on April 14, 1931, after the republican parties won a great election victory. A Spanish Republic was proclaimed.

The Jehovah's Witnesses came into being, led by Joseph Franklin ("Judge") Rutherford (1869–1942).

1932

NEUTRON

For twenty years, it had been assumed that the atomic nucleus was made up of protons and electrons, since these were the only two subatomic particles known. Thus the nitrogen nucleus had a mass of 14, so it must contain fourteen protons. On the other hand, the nitrogen nucleus had an electric charge of +7, whereas fourteen protons would have an electric charge of +14. Therefore, the nucleus must have seven electrons with a charge of −1 each to neutralize the charge of seven of those protons. The net result was that the nitrogen nucleus was thought to be made up of 14 protons and 7 electrons, or 21 particles altogether.

But seven years earlier, when Uhlenbeck and Goudsmit worked out the concept of particle spin (see 1925), it had become clear that something was wrong with the proton-electron notion of nuclear structure. Protons and electrons both had a spin of either +½ or −½. If twenty-one such spins, or in fact any odd number of such spins, were added up (regardless of the distribution of pluses and minuses), the total spin should be one of the half-integers: ½, or 1½, or 2½, and so on. The actual measured spin of the nitrogen nucleus, however, was an integer. The nucleus, therefore, could not contain an odd number of particles but had to contain an even number. There were other nuclei that shared this anomalous characteristic.

If the proton-electron combination could be counted as a single particle, however, then the nitrogen nucleus would contain seven protons and seven proton-electron combinations, for fourteen particles—an even number.

A proton-electron combination would have the mass of a proton but would carry no electric charge. If such a neutral particle existed, however, it would be difficult to detect, for the various devices for detecting subatomic particles all depended on the electric charge these particles carried.

In 1930 Bothe, who had devised the coincidence counter (see 1929), became aware of strange radiations emerging from beryllium that had been exposed to bombardment with alpha particles, but he couldn't tell what the radiation consisted of.

In 1932, however, the English physicist James Chadwick (1891–1974) repeated the experiments and maintained that the best way of explaining the observations was to suppose that the alpha particles knocked neutral particles out of the beryllium nucleus. These particles could not be detected directly, since they lacked an electric charge, but they succeeded in knocking protons out of the atomic nuclei in paraffin, and those protons could be detected. A radiation capable of ejecting protons must consist of particles with a mass in the proton range, and a particle possessing the mass of a proton and no charge was just the thing to represent a proton-electron pair and yet be one particle and not two. The new particle was called a *neutron,* and it proved to be by far the most useful particle for initiating nuclear reactions.

For his discovery of the neutron, Chadwick was awarded the Nobel Prize for physics in 1935.

PROTON-NEUTRON NUCLEUS

As soon as the neutron was discovered by Chadwick (see above), Heisenberg (see 1925) pointed out that the atomic nucleus must be made up of protons and neutrons rather than protons and electrons.

Thus the nitrogen nucleus with a charge of +7 must contain seven protons. Since it had a mass of 14, it must also contain seven neutrons. That would mean fourteen particles altogether, and fourteen particles, each with with a spin of +½ or −½, would add up to a total spin that was an integer no matter how you distributed the pluses and minuses. Given a proton-neutron nucleus, all nuclear spins turned out to match what was expected, and all spin anomalies disappeared.

What's more, the proton-neutron picture clearly explained the existence of isotopes. The nuclei of all atoms of oxygen, for instance, must contain eight protons, but the common oxygen-16 nucleus must contain eight protons and eight neutrons, the oxygen-17 nucleus eight protons and nine neutrons, and the oxygen-18 nucleus eight protons and ten neutrons. In the case of hydrogen, the hydrogen-1 nucleus must contain a proton only, with the hydrogen-2 nucleus containing one proton and one neutron.

This new view of the nucleus solved the spin problem but introduced a new problem of its own. If the tiny nucleus were filled with protons, all of which were positively charged, there must be an intense repulsion among them. The neutrons, being uncharged, would do nothing to reduce the repulsion. How then account for the fact that the nucleus clings together strongly? (As long as it was thought that electrons existed in the nucleus, they might be viewed as a kind of cement, but that possibility was now gone.)

Heisenberg suggested that *exchange forces* might exist. In other words, the proton and neutron might exchange particles in such a way that a strong attraction would be produced, since the particles could not be exchanged unless the proton and neutron remained very close together. It took a few years for this notion to be properly developed.

POSITRON

Dirac had predicted, from theoretical considerations, that there must be a particle just like the electron but with a positive charge rather than a negative one (see 1930). However, such an *antielectron* had not been observed in nature.

The American physicist Carl David Anderson (b. 1905) was working with Millikan (see 1911) on cosmic rays. To carry out his studies, he had devised a cloud chamber (see 1911) with a lead partition running across it. A cosmic ray particle entering an ordinary cloud chamber was so energetic that a magnetic field scarcely managed to deflect it at all, and from the virtually straight-line trace that resulted, not much information about it could be derived. If a lead partition was placed in the chamber, the cosmic ray particle had energy enough to smash through it but lost enough energy in the process to allow it to be easily deflected on the far side of the partition.

In 1932, while conducting research with his lead-partitioned cloud chambers, Anderson noted tracks emerging from the lead that were clearly electron tracks (experienced researchers could read tracks at a glance). The only peculiarity was that the electron tracks curved in the wrong direction. Anderson had found the positively charged particle that Dirac had predicted—the antielectron. Anderson, however, thought of it as a *positive electron* and gave it the shortened name of *positron,* which it has carried ever since.

For his discovery, Anderson was awarded the Nobel Prize for physics in 1936, sharing it with Hess, who had discovered cosmic rays (see 1911).

PARTICLE ACCELERATOR AND NUCLEAR REACTION

The first human-made nuclear reaction had been carried through by Rutherford (see 1919). For the purpose, he had used alpha particles produced naturally by radioactive sources. Other nuclear reactions brought about since then had also made use of such alpha particles.

In 1932, however, Cockcroft and Walton, who had devised the first particle accelerator (see 1929), used protons from such an accelerator to bombard lithium nuclei. This made alpha particles come streaking out of the lithium nuclei.

The lithium nucleus is made up of three protons and four neutrons. When accelerated protons are fired at such a nuclei, a proton will occasionally enter a nucleus, giving it four protons and four neutrons. This instantly breaks up into two alpha particles containing two protons plus two neutrons. In other words: hydrogen + lithium yields helium + helium.

This was the first nuclear reaction produced by human beings as a result of bombardment with artificially accelerated particles. It was to be the first of many.

RADIO WAVES FROM SPACE

As radio came more prominently into use for communication and home entertainment, the problem of *static* (a crackling interference that made communication uncertain and music unpleasant) cried out for correction. Static had a number of causes, including thunderstorms, nearby electric equipment, and aircraft passing overhead.

The Bell Telephone Company, which needed radio for ship-to-shore calls, among other things, set one of its employees, Karl Guthe Jansky (1905–1950), to exploring the problem.

Jansky detected a new kind of weak static from a source that at first he could not identify. It came from overhead and moved steadily. At first it seemed to Jansky that it must be the Sun. However, the source gained slightly on the Sun to the extent of 4 minutes a day. This is just the amount by which the vault of the stars gains on the Sun, so that the source must lie beyond the Solar System.

By the spring of 1932, Jansky had decided that the source was in the constellation of Sagittarius, the direction in which Shapley had placed the center of our galaxy (see 1918).

This represented the birth of radio astronomy, in which astronomers learned to receive and interpret radio waves rather than light waves.

Light waves, of course, are easily perceived by the retina of the eye and by photographic film. In 1932, however, perceiving radio waves with any precision was extremely difficult, for instruments were lacking. Therefore, the development of radio astronomy was delayed for some twenty years.

ELECTRON MICROSCOPE

Davisson had demonstrated that electrons did indeed have their wave aspects (see 1927). Therefore they might be manipulated in much the same way that light waves were. They might be focused as light waves were in microscopes, for instance.

The sharpness with which tiny objects can be seen varies inversely with the wavelength of the viewing medium. The shorter the wavelength, the smaller the objects that can be made out. Since electron waves have a wavelength equal to that of X rays (but are much easier to focus than X rays), an *electron microscope* should be much more powerful than an ordinary optical microscope.

Electron waves cannot be focused by passing them through lenses, of course, but they can be focused by appropriate magnetic fields. The German electrical engineer Ernst August Friedrich Ruska (1906–1988) built the first such instrument in 1932. It was a crude affair, to be sure, but it was capable of magnifying four hundred times. It was rapidly improved and became an essential item in the microscopists' armory.

For his work, Ruska was belatedly awarded a share in the Nobel Prize for physics in 1986.

PRONTOSIL

It had been a quarter-century since Ehrlich had found chemicals that attacked pathogenic microorganisms without damaging higher animals (see 1907), and since then the matter had languished.

It was clear that possible chemicals for the purpose lay among the dyes. Some dyes combined with cells and not with others, so surely there should be some dyes that combined with germs, killing them in the process, without affecting human cells.

The German biochemist Gerhard Domagk (1895–1964) conducted systematic tests of new dyes synthesized since the time of Ehrlich to see if he could find something appropriate. One of them was a newly synthesized orange-red compound with the trade name *Prontosil*. In 1932 Domagk found that injections of the dye had a powerful curative effect on streptococcus infection in mice.

As it happened, Domagk's young daughter, Hildegard, had been infected by streptococci following the prick of a needle. When other treatment failed, Domagk injected large quantities of Prontosil. She recovered, and the world had another item in its chemotherapeutic armory. Prontosil was the forerunner of a large number of drugs that would cause many infectious diseases to lose their terrors.

For this discovery, Domagk was awarded the Nobel Prize for medicine and physiology in 1939. (Hitler refused to allow Germans to accept Nobel Prizes, however, because a Peace Prize had been awarded to Carl von Ossietzky [1889–1938], an imprisoned German pacifist. Domagk was therefore not able to accept his prize formally till 1947.)

ASCORBIC ACID

In 1932 the American biochemist Charles Glen King (1896–1988) concluded his investigations of vitamin C, in the course of which he had isolated the vitamin and determined its structure. The substance had a six-carbon molecule closely resembling those of the sugars and was named *ascorbic acid*, from Greek words meaning "no scurvy." The work of Lind (see 1747) was thus successfully concluded.

Two weeks later, Szent-Györgyi reported that his hexuronic acid (see 1928) was vitamin C, and he was right. This set off a violent and bitter controversy over precedence, which continued throughout the long lives of both men (each lived into his 90s).

UREA CYCLE

As biochemists learned more and more about metabolism, it became clear that certain reactions fit into the pattern of a chain. Often the chain moved from its starting point back to that same point after a number of steps (forming a *cycle*), achieving some result with each turn of that cycle.

Thus the German-born British biochemist Hans Adolf Krebs (1900–1981) showed that when the amino acid arginine broke down and was reconstituted, it produced in the process a molecule of urea (so that it was called the *urea cycle*). Urea is the chief nitrogen-containing waste in mammals generally.

POLAROID

Nicol had made use of Iceland spar to produce an instrument that could be used in the study of polarized light (see 1829). Some substitute for Iceland spar that would be cheaper and easier to work with would have been welcome, but none was known.

To be sure, there were certain organic crystals that behaved as Iceland spar did, but they did not readily produce crystals of the proper size, and the crystals would have been far too fragile to work with in any case.

An American inventor, Edwin Herbert Land (b. 1909), had the idea that a single crystal was not necessary. A myriad of tiny crystals would do if all were oriented in the same direction.

In 1932 he devised methods of aligning the crystals and of then embedding them in clear plastic. When set, this served nicely to keep them from drifting out of alignment and made their fragility irrelevant. The result was given the trade name *Polaroid,* and it quickly replaced Iceland spar in scientific instruments. It also came to be used in automobile windshields and in spectacles.

QUINACRINE

Malaria, a widespread disease in the tropics and perhaps the most debilitating disease, on the whole, that was suffered by the human species, had been treated with quinine for three centuries (see 1642). Quinine comes from the bark of a tropical tree, however, and the supply to the industrial nations might easily be cut off in wartime. The search was on for a substitute.

The first satisfactory substitute to be found was *quinacrine* (also called *Atabrin*), first developed in Germany and established as a successful antimalarial in 1932. When not too many years afterward, war did indeed disrupt the quinine supply, the existence of quinacrine made it possible for troops to operate in tropical regions.

IN ADDITION

In the United States, Franklin Delano Roosevelt (1882–1945) was elected the thirty-second president. Wall Street hit an all-time low on July 28, 1932. Unemployment stood at something like sixteen million. The unemployed and their dependents made up nearly a quarter of the nation.

In Europe, Fascism grew steadily more powerful. In Italy, Mussolini's power was absolute, and in Germany, Hitler was becoming ever stronger. In Portugal, the government came under the power of a Fascist dictator, Antonio de Oliveira Salazar (1889–1970). Even in the countries that remained under democratic institutions, Fascist parties were founded and grew more prominent.

In China, Japanese troops landed in Shanghai on January 28, 1932, where they met with little resistance.

In India, Mohandas Karamchand (Mahatma) Gandhi was carrying on a relentless campaign of civil disobedience against British rule.

1933

SYNTHETIC VITAMIN C

King and Szent-Györgyi had succeeded in determining the molecular structure of vitamin C (see 1932).

In 1933 the Polish-born Swiss chemist Tadeus Reichstein (b. 1897) succeeded in synthesizing the vitamin. Eventually other vitamins were analyzed and then synthesized. This meant that it was no longer necessary to obtain small quantities of vitamins from huge quantities of foodstuff. Instead they could be synthesized in the laboratory by the pound and even by the ton. Nor were the synthetic vitamins in any way different from the natural ones.

The ability to synthesize vitamins led to the development and popularity of vitamin pills, and when these became available at affordable prices, the nutritional problem of vitamins in the diet was diminished or even abolished.

The English chemist Walter Norman Haworth (1883–1950) independently synthesized vitamin C not long after Reichstein did, and it was Haworth who gave the vitamin the name of ascorbic acid. For Haworth's work on sugars generally, he was awarded a share in the Nobel Prize for chemistry in 1937.

MOLECULAR BEAMS

If gases are allowed to escape from a container through a tiny hole into a high vacuum, the escaping molecules meet virtually no molecules with which they can collide. They therefore form a tight beam of moving particles. These *molecular beams* are made up of particles that are neutral but that are themselves made up of charged particles: nuclei and electrons.

As a result, in accordance with Maxwell's equations (see 1865), they ought to behave like tiny magnets.

The German-born American physicist Otto Stern (1888–1969) studied these molecular beams for years and by 1933 had conclusively demonstrated that they *did* behave like magnets and in ways, moreover, that also supported quantum mechanics. Furthermore, Stern showed that the molecular beams possessed wave aspects.

For his work on molecular beams, Stern was awarded the Nobel Prize for physics in 1943.

APPROACH TO ABSOLUTE ZERO

Debye and Giauque had suggested a way of approaching absolute zero more closely by using magnetic techniques (see 1925).

In 1933 Giauque managed to turn theory into practice. He imposed a magnetic field on gadolinium sulfate, then let it retreat into disorder on the removal of the field. The retreat into disorder required heat and absorbed it from the surrounding helium. In this way he lowered its temperature to 0.25 degrees above absolute zero. By the end of a year, others using the same technique had reached 0.0185° K.

For this and for later work on ultracold temperatures, Giauque was awarded the Nobel Prize for chemistry in 1949.

IN ADDITION

Roosevelt introduced his New Deal, taking the attitude that the federal government was responsible for the welfare of the people and strongly controlling the direction and manner of business affairs. This did not end the Depression, but it restored public confidence and eased human suffering.

On January 30, 1933, Hitler became chancellor of Germany. At once, he proceeded to wipe out opposition parties, to institute a persecution of the Jews, and to bend all of Germany's strength toward the buildup of a new military machine. On October 14, Germany withdrew from the League of Nations.

Germany was preceded in this respect by Japan, which announced on May 27 that it would leave the league.

In Austria, the chancellor Engelbert Dollfuss (1892–1934), established himself as Fascist dictator of that land.

In the Soviet Union, Stalin instituted the first of a series of purges of the Communist Party, designed to remove anyone he suspected of less than complete loyalty.

1934

NEUTRON BOMBARDMENT

As soon as Chadwick had discovered the neutron (see 1932), it was clear that it would be an extremely useful tool for the bombardment of nuclei. Protons and alpha particles, being positively charged, are repelled by the positively charged nuclei, and much of their energy must be expanded in overcoming this repulsion. Neutrons, being uncharged, can drift into a nucleus even when low in energy.

When a neutron is absorbed by the nucleus of a particular atom, the new nucleus may be unstable and may emit a beta particle, which by subtracting a negative charge converts a neutron in the nucleus into a proton. The net result is that the nucleus ends up with one proton more than it started with, which makes it an element that is one higher in atomic number.

In 1934 it occurred to Fermi (see 1926 and 1931) that it would be particularly interesting to bombard uranium atoms with neutrons and perhaps form atoms with an atomic number of 93, which are not known in nature.

The results of the bombardment turned out to be confusing, and the confusion was not straightened out for five years—but then with portentous results. For his work on neutron bombardment generally, Fermi was awarded the Nobel Prize for physics in 1938.

WEAK INTERACTION

Pauli had suggested that every time a beta particle was emitted by a nucleus, a neutrino (without charge or mass) was also emitted (see 1931).

In 1934 Fermi worked out the theoretical basis on which the two particles were formed. He showed that there was an interaction involved in connection with neutrinos that was like an electromagnetic interaction but much weaker. It therefore came to be called the *weak interaction.* (It was stronger than the gravitational interaction, however.)

The electromagnetic interaction and the gravitational interaction decreased in intensity with the square of the distance. This was a rather slow decrease, so that both interactions could make themselves felt over long distances. (This was particularly true of the gravitational interaction, which was purely attractive, whereas the electromagnetic interaction had attractive and repulsive components that tended to cancel each other.)

The weak interaction, however, fell off so rapidly with distance that it was confined entirely to distances the size of an atomic nucleus or less. It might therefore have been called the *nuclear interaction* were it not that another interaction of small range was soon discovered.

ARTIFICIAL RADIOACTIVITY

Ever since Rutherford had brought about nuclear reactions by bombardment with subatomic particles (see 1919), physicists had been inducing more and more such reactions.

In 1934 the French physicists Frédéric Joliot-Curie (1900–1958) and Irène Joliot-Curie (1897–1956), his wife, who was also the daughter of Pierre and Marie Curie (see 1897), were bombarding aluminum atoms with alpha particles.

In the process of the bombardment, the nucleus of the aluminum atom would absorb an alpha particle and give off a proton. The aluminum nucleus contains thirteen protons and fourteen neutrons, so that it is aluminum-27. By taking up an alpha particle (two protons and two neutrons) and then giving up a proton, it ended up with fourteen protons and sixteen neutrons, so that it was *silicon-30,* which occurs in nature.

However, after the bombardment ceased, so that alpha particles were no longer absorbed and protons were no longer given off, another form of radiation continued. The Joliot-Curies investigated and decided that in some cases the aluminum nucleus, after absorbing the alpha particle, gave off a neutron. The result was a net gain of two protons and one neutron, for a total of fifteen protons and fifteen neutrons, which made it *phosphorus-30.*

Phosphorus-30, however, does not occur in nature, and it is radioactive, breaking down with a half-life of just under 3 minutes (which is why it doesn't occur in nature). In breaking down, it gives off a stream of positrons (particles that had been discovered by Anderson—see 1932). Each escaping positron converts a proton to a neutron, so that phosphorus-30 becomes stable silicon-30.

The Joliot-Curies were thus the first to observe radioactive breakdowns that emitted positrons and were the first to produce a radioactive isotope of an ordinarily stable element. This is called *artificial radioactivity,* since it comes about as the result of bombardment of nuclei in the laboratory. Eventually it turned out that every element that possessed one or more stable types of nuclei could also possess radioactive nuclei *(radioisotopes).*

For the discovery of artificial radioactivity, the Joliot-Curies were awarded the Nobel Prize for chemistry in 1935.

CHERENKOV RADIATION

Light travels at a speed of 299,792.5 kilometers per second (186,282 miles per second), and nothing can go faster than that according to special relativity (see 1905). However, light travels more slowly when passing through matter, and the decrease in speed is more marked as the index of refraction of the transparent medium increases. In water, light travels at 224,900 kilometers per second, and only 124,000 kilometers per second in diamond. The speed of light is even a bit below maximum when it is passing through air.

A rapidly moving particle can never travel faster than the speed of light in a vacuum, but it can travel faster than light does in water, let us say. And if the particle is moving extremely close to the speed of light in a vacuum, it may move faster in air than light does. When particles move faster than light in some medium other than a vacuum, they leave a wake of light trailing behind.

A Soviet physicist, Pavel Alekseyevich Cherenkov (b. 1904), was the first to observe this wake of radiation, which came to be called *Cherenkov radiation* as a result. The reason for the wake was explained by the Russian physicists Igor Yevgenyevich Tamm (1895–1971) and Ilya Mikhaylovich Frank (b. 1908).

From the angle at which Cherenkov radiation is emitted, the speed of ultrafast particles can be calculated, and for this finding, Cherenkov, Tamm, and Frank were awarded the Nobel Prize for physics in 1958.

SUPERNOVAS

Tycho Brahe had observed a very bright nova (see 1572) and so had Kepler in 1604. Since then, three and a quarter centuries had passed without such a bright nova being seen. To be sure, from time to time new stars had been seen that were fairly bright and that qualified as novas, but none had attained a brightness surpassing that of Jupiter, or even Venus, as the novas of Tycho and Kepler had.

In 1885 a nova had appeared in the Andromeda nebula that had attained a brightness of magnitude 7, almost bright enough to see with the unaided eye, but this had not been made much of at the time. When Hubble, however, showed that the Andromeda nebula was really a far distant galaxy (see 1923), the brightness of the 1885 nova had to be reassessed. To be nearly visible to the unaided eye from the vast distance of the Andromeda galaxy, it had to be enormously more luminous than the run-of-the-mill novas that had been seen since Kepler's time.

The Swiss astronomer Fritz Zwicky (1898–1974) pointed this out in 1934 and suggested that novas such as those Tycho and Kepler had seen and the one in the Andromeda galaxy were *supernovas.* Since Kepler's time, none had been seen in our own galaxy, but Zwicky, watching other galaxies, spotted a number. Since a supernova shines at its peak with a brilliance about equal to the total light of an average galaxy, a supernova can be seen as far as a galaxy can.

NEUTRON STARS

It seemed reasonable that when a star used up its nuclear fuel, its center would cool down. Without fierce central heat to keep its huge body bloated, it might collapse to a white dwarf (see 1914) under the pull of its own huge gravity.

Zwicky (see above) suggested that in the case of supernovas, collapse might be extreme. After expending its energy, a supernova might collapse under its own gravitational pull till its subatomic particles (protons and electrons) were crushed together to form neutrons and the neutrons were forced into contact.

Such a *neutron star* might possess all the mass of a full-sized star and yet be only 8 miles across. Naturally, a neutron star would be difficult to detect, and it was not until thirty-five years after Zwicky's suggestion that the detection was accomplished.

SEX HORMONES

Butenandt had isolated the male sex hormone androsterone and worked out its structure (see 1931). In 1934 the Croatian-born Swiss chemist Leopold Stephan Ružička (1887–1976) synthesized androsterone and showed that Butenandt's analysis was correct. For this he shared the Nobel Prize for chemistry in 1939 with Butenandt.

Meanwhile, also in 1934, Butenandt isolated *progesterone,* a female sex hormone of vital importance to the chemical mechanisms involved in pregnancy. Butenandt, like Domagk (see 1932, Prontosil), could not accept the Nobel Prize because of Hitler's ruling. It was not until 1949 that he could accept it formally.

BATHYSPHERES

Since earliest time, people had dived beneath the ocean surface, either for pleasure or to gather such objects as sponges and pearl oysters. Naturally they could not go very deep or stay very long.

Later they could stay underwater for extended periods in caissons or diving suits, but only at the cost of having the air they breathed compressed to the pressure of the surrounding water. This limited the depth to which they could penetrate and the time they could remain.

At high pressures, nitrogen dissolves in the blood and bubbles out when the pressure is relieved, causing the *bends,* or *caisson disease,* which results in agonizing pain and sometimes paralysis and death. To prevent this, pressure has to be released very slowly, a procedure first suggested in 1878 by the French physiologist Paul Bert (1833–1886).

To achieve real depth for indefinite periods, it was clearly necessary to provide a sealed gondola capable of maintaining an interior at ordinary pressures, as Piccard had done for high-altitude balloons (see 1931). The task was more difficult at sea, for in the heights the difference between inside and outside pressure was less than one atmosphere, while in the depths the sea pressed in with many atmospheres of pressure.

The American naturalist Charles William Beebe (1877–1962) devised the necessary steel vessel. It had thick quartz windows and was a sphere (at the suggestion of President Roosevelt) rather than a cylinder for greater strength.

In 1934 Beebe used this *bathysphere* (Greek for "sphere of the deep") to descend to a record depth of 3,028 feet, nearly three-fifths of a mile. It was terribly risky, for he had to be suspended from a ship, and if the lifeline had snapped, no rescue was possible. However, no untoward incidents took place during his explorations.

IN ADDITION

In Germany, President Hindenburg died on August 2, 1934, and Hitler assumed the post, coming to be styled *Der Fuehrer* (The Leader) by his National Socialist (Nazi) followers.

Nations that might have opposed Germany, including France, Belgium, and Yugoslavia, grew steadily weaker as Germany grew steadily stronger.

The Soviet Union, aware that Germany to its west and Japan to its east were determined on war, no longer wished to remain isolated and therefore joined the League of Nations on September 18, 1934, but the League was too moribund to matter.

In China, the Communists under Mao Tse-tung embarked on a *Long March* deep into the recesses of China, where they reestablished themselves and prepared for what they felt would be their ultimate victory.

1935

URANIUM-235

Since the concept of isotopes had been advanced by Soddy (see 1913), almost all the stable isotopes had been discovered, thanks to the work of Aston and his mass spectograph (see 1919 and 1925).

Nevertheless, it was not till 1935 that uranium was found to consist of two isotopes in its natural state. The common form had a nucleus made up of 92 protons and 146 neutrons, so that it was uranium-238. In 1935 the Canadian-born American physicist Arthur Jeffrey Dempster (1886–1950) showed that there was also another isotope, making up only 1 out of 140 uranium atoms, that had a nucleus containing 92 protons and 143 neutrons, so that it was uranium-235.

There was no way of telling at the time of its discovery how enormously important uranium-235 was, but the demonstration came within a few years.

ISOTOPIC TRACERS

Hevesy had first used radioactive atoms as tracers in biochemical work (see 1918). However, he had used radioactive lead, and lead is not a natural component of tissue, so that its presence might well have interrupted the natural flow of chemical reactions there.

Since then, however, it had been found that elements that were not radioactive were also made up of isotopes. In particular, the four most common and important elements of tissue each existed also as an isotope that was relatively rare, quite stable, and capable of being differentiated from the common isotope. Thus there was carbon-13, which was 8.5 percent more massive than the common carbon-12; nitrogen-15, 7.1 percent more massive than the common nitrogen-14; oxygen-18, 12.5 percent more massive than the common oxygen-16; and hydrogen-2, 100 percent more massive than the common hydrogen-1.

The greater the percentage difference in mass, the easier it was to analyze for the unusual isotope, so that hydrogen-2, discovered by Urey (see 1931, Deuterium), should be particularly useful.

The German biochemist Rudolf Schoenheimer (1898–1941) obtained a quantity of hydrogen-2 from Urey and used it to form fat molecules that contained a considerably higher percentage of hydrogen-2 than would be found in natural fat.

Samples of this *isotope-enriched* fat were fed to rats. It was assumed that the tissues would not differentiate between the two hydrogen isotopes but would treat hydrogen-2 exactly like hydrogen-1. After a time, a rat could be killed and its body fat analyzed for hydrogen-2, which would have acted as an *isotopic tracer*, indicating the types of reactions that fat molecules underwent. By 1935 he had achieved surprising results.

For instance, until then it had been believed that the fat stores of an organism were usually immobile, remaining where they were until such time as famine demanded their use. Under ordinary conditions, it was thought, the body would make use of fat obtained from the immediate diet.

However, Schoenheimer found that when isotope-enriched fat was fed to rats, half of what had been fed was to be found in the fat stores. In other words, ingested fat was stored, and stored fat was used. There was a rapid turnover, and body constituents were not static but changed constantly and dynamically.

Schoenheimer later made use of nitrogen-15 to label amino acid molecules and show that there was constant action there, too—molecules rapidly changing and shifting even though the overall movement might be small.

CRYSTALLINE VIRUSES

The first enzyme had been crystallized by Sumner (see 1926), and other enzymes had been crystallized since by Northrop and others (see 1930), so that the mystery of the chemical nature of enzymes was removed.

The American biochemist Wendell Meredith Stanley (1904–1971) thought similar techniques might clarify some of the mysteries of virus structure. He prepared a quantity of tobacco mosaic virus (the first virus to be recognized as such by Beijerinck—see 1898) by growing tobacco and infecting it. He then mashed up the infected leaves and put the mash through the usual procedures used by chemists to extract proteins and crystallize them, since he felt that viruses were probably protein in nature.

In 1935 he obtained fine needlelike crystals, which he isolated and found to possess all the infective properties of the virus in high concentration.

This raised problems, since it had always been thought that crystals were a form of matter characteristic of nonliving atoms and molecules, and since viruses could multiply, they had been thought to be living. The answer, quickly arrived at, was that crystallization was not a dividing line between life and nonlife, and that viruses were so simple a life form that they could crystallize.

As a result of this work, Stanley was awarded a share of the Nobel Prize for chemistry in 1946, along with Sumner and Northrop.

STRONG INTERACTION

Heisenberg had tried to explain why the nucleus remained intact despite the repulsion between protons by supposing the existence of exchange forces (see 1932, Proton-Neutron Nucleus). Fermi had made use of that concept in working out his theory of the weak interaction (see 1934).

The Japanese physicist Hideki Yukawa (1907–1981) used the ideas of Heisenberg and Fermi to build up a theory that would explain the nucleus. There had to be a short-range force like that of the weak interaction, one that could be felt only within the nucleus. This short-range force would have to be much stronger than the weak interaction and indeed much stronger than the electromagnetic interaction in order to overcome proton repulsion. For that reason, it came to be called the *strong interaction*.

Yukawa worked out his mathematical treatment in 1935. He showed that in order for the nucleus to exist, protons and neutrons had to exchange a particle possessing mass. The shorter the range of the force, the more massive the particle. He calculated that the particle exchanged ought to be about two hundred times as massive as an electron, or one-ninth as massive as a proton. Such intermediate-sized particles were unknown at the time, but eventually they were discovered. Yukawa was then awarded the Nobel Prize for physics, in 1949. He was the first Japanese to win a Nobel Prize.

SULFANILAMIDE

Domagk had found that the dye Prontosil had certain antibacterial properties (see 1932). Prontosil had a fairly complicated molecule, however, and there was a reasonable hope that some smaller fraction of that molecule might have the same desirable properties but be easier to synthesize and so more available in quantity.

In 1935 Domagk managed to break up the molecule of Prontosil into several fragments, one of which was *sulfanilamide,* a compound already well known to organic chemists. Sulfanilamide turned out to have very pronounced antibacterial properties.

This heralded the synthesis of a whole family of similar compounds, collectively called the *sulfa drugs,* which were used to combat infections of various kinds.

RIBOFLAVIN

Since the work of Eijkman on beriberi (see 1896), biochemists had discovered a number of vitamins that, like the vitamin B that cured beriberi, were water-soluble and contained rings made up of carbon and nitrogen molecules. They came to be labeled vitamin B-1 (which cured beriberi), vitamin B-2, and so on. The whole was the *B-vitamin complex*.

These letter-number combinations turned out to be unsatisfactory, however, because some reported vitamins turned out to be false alarms and others were simply not given such combination names. The B vitamins therefore came to be known by chemical names. Thus vitamin B-1 came to be called *thiamin,* vitamin B-2 *riboflavin,* and so on.

Karrer, who had determined the structure of vitamin A (see 1930), synthesized riboflavin in 1935 and presented the final proof of its structure. For this work, he was awarded a share of the Nobel Prize for chemistry in 1937.

CORTISONE

The first hormone to be isolated was adrenaline. It had been obtained by Takamine (see 1898, Epinephrine) from the adrenal glands. The adrenals consist of two parts, however, which turned out to be two different glands. The inner part, or *medulla* (the Latin word for "core"), is what manufactures adrenaline. The outside part, which wraps about the medulla, is the *cortex* (from a Latin word for "bark"). It too manufactures hormones.

The American biochemist Edward Calvin Kendall (1886–1972) isolated no fewer than twenty-eight different cortical hormones, or *corticoids,* of which four showed effects on laboratory animals. He named the corticoids by letter, and the four effective ones were Compound A, Compound B, Compound E, and Compound F.

Of these, Compound E, which Kendall isolated in 1935, proved the most useful. It came to be called *cortisone* and was widely used as an anti-inflammatory drug.

For his work on corticoids, Kendall was awarded a share of the Nobel Prize for medicine and physiology in 1950.

PROSTAGLANDINS

In 1935 the Swedish physiologist Ulf Svante Von Euler (1905–1983) isolated a hormone-like substance from semen, which he took to represent a product of the prostate acting as a gland. He therefore called it *prostaglandin.*

Prostaglandin and related compounds referred to collectively as the *prostaglandins* have since been found in other tissues and have proved to have a wide variety of physiologic effects on the body.

RADAR

To send out a beam of light and record the time it took to be reflected back across a given distance had first been used to determine the speed of light by Fizeau (see 1849). Once the speed of light was determined with good accuracy, the time it took for a beam of light to strike an object and return could be made to yield both the direction and the distance of the reflecting object. Such a scheme, making use of ultrasonic sound, had been used by Langevin in devising sonar (see 1917).

The use of light for such a purpose is impractical, because light is too easily stopped by obstacles and too easily absorbed and scattered by mist, dust, and fog. Radio waves are much more penetrating, but those most commonly used in communication are so long that they bend around obstacles rather than being reflected. However, if the shortest radio waves *(microwaves)* are used, they are reflected by moderately sizable objects and are sufficiently penetrating to pass through cloud and fog.

The Scottish physicist Robert Alexander Watson-Watt (1892–1973) worked on devices for the emission of microwaves and for the detection of the reflected beam. By 1935 he could use his device to follow the path of an airplane by the microwave reflections it sent back.

The system was called *radio detection and ranging,* abbreviated to *ra. d. a. r.,* or *radar.* This development was to prove of life-and-death importance in just a few years.

IMPRINTING

In 1935 the Austrian-born German zoologist Konrad Lorenz (1903–1989), a student of bird behavior, described *imprinting.* He showed that at a certain critical point in early life, soon after hatching, young birds learned to follow a moving object, usually their mother, who was sure to be in the neighborhood after all. If this for some reason was not the case, the hatchlings would follow some other adult bird or a human being or even a pulled inanimate object.

Once imprinting had taken place, it affected their behavior to some extent all their lives.

Lorenz thus established the science of *ethology,* the study of animal behavior in natural environments, and initiated the study of how learned behavior in very early life affected later events. For his work on animal behavior, he was awarded a share in the Nobel Prize for physiology and medicine in 1973.

RICHTER SCALE

Earthquakes, as everyone who has studied, experienced, or heard of them knows, come in various intensities. Some are perceptible only to instruments. Some are capable of leveling a city.

In 1935 the American geophysicist Charles Francis Richter (1900–1985) established the *Richter scale* for the measurement of earthquake intensity.

This scale measured the extent of Earth movement by a series of numbers, each one of which represented a movement ten times as great as the number before. The damage done rises even faster than the extent of movement. The strongest earthquakes so far recorded have reached 8.9 on the Richter scale.

IN ADDITION

On October 3, 1935, Italian forces invaded Ethiopia.

As for Germany, Hitler's plans proceeded smoothly. On January 13, 1935, the Saar Basin, a coal-producing region on Germany's western border, which had been under jurisdiction of the League of Nations since the end of World War I, voted to return to Germany. It was Hitler's first territorial acquisition. Hitler denounced the Versailles Treaty on March 16, 1935, and began rearming Germany, including an air force. He pushed through the infamous *Nuremberg laws,* depriving Jews of all rights. By agreeing to keep his navy at only 35 percent of Great Britain's, he insured that Great Britain would not interfere with him otherwise.

On March 21, 1935, Persia declared itself *Iran.*

1936

NEUTRON ABSORPTION

After Chadwick's discovery of the neutron (see 1932), the use of neutron bombardment to induce nuclear reactions had achieved considerable importance, especially in the work of Fermi (see 1934).

In 1936 the Hungarian-born American physicist Eugene Paul Wigner (b. 1902) worked out the mathematics that governed the manner in which neutrons were absorbed by atomic nuclei. It showed how the likelihood of neutron absorption varied with the energy of the neutron. It also introduced the concept of *nuclear cross section:* the greater the cross section of a particular type of nucleus, the greater the likelihood of its absorbing a neutron.

For this and other work, Wigner was awarded a share of the Nobel Prize for physics in 1963.

THIAMINE

Thiamine (also known as vitamin B-1) is the name given to the vitamin that prevents beriberi, and whose existence had come to be suspected as a result of Eijkman's work (see 1896).

The American chemist Robert Runnels Williams (1886–1965) brought this work to fruition. He had isolated about a third of an ounce of the vitamin from a ton of rice polishings, and this was enough to enable him to determine its molecular structure. In 1936 he synthesized thiamine—the final proof that his structure was correct.

It was then no longer necessary to work with huge quantities of rice polishings. Thiamine by the ton could be prepared in the laboratory.

PERFUSION PUMP

Carrel (see 1902) was much in the news in the 1930s through his demonstrations that it was possible to keep organs or tissues alive by means of *perfusion;* that is, by passing blood or blood substitutes continuously through the tissues by way of its own blood vessels. He kept a piece of embryonic chicken heart alive and growing (it had to be periodically trimmed) for over thirty-four years—much longer than the normal life expectancy of a chicken—before the experiment was deliberately terminated.

In 1936, to make the process more efficient, Carrel put into use a perfusion pump that Lindbergh (see 1927) had helped him design. It drove the blood and was germ-proof. It was called an *artificial heart,* though it was not the kind of artificial heart that could substitute for the natural one in the human chest.

IN ADDITION

Italy's Ethiopian War came to a successful conclusion with the capture of Addis Ababa, the Ethiopian capital, on May 5, 1936. Italy annexed Ethiopia to its empire on May 9, and Victor Emmanuel III of Italy was declared Emperor of Ethiopia.

George V of Great Britain died on January 20, 1936, and was succeeded by his son, who ruled as Edward VIII (1894–1972). He fell in love with an American divorcée and so mishandled the matter that he was forced to abdicate on December 10, 1936. He was succeeded by his younger brother, who reigned as George VI (1895–1952).

Hitler took advantage of the Ethiopian War and the British abdication crisis to send his army into the Rhineland (the German area west of the Rhine river).

On October 25, 1936, Germany signed an alliance with Italy. Mussolini grandiosely described the two nations as "an axis" about which other nations might assemble and collaborate, so that Germany and Italy were thereafter known as the *Axis powers.* On November 25, 1936, Germany signed a pact with Japan as well.

In Spain, there was a revolt of army generals against the liberal anti-Fascist government. Under the leadership of Francisco Franco (1892–1975), they quickly established their power over much of Spain.

In the United States, Roosevelt was triumphantly reelected.

Fuād I of Egypt (1868–1936) died on April 28, 1936, and was succeeded by his son, who reigned as Farouk I (1920–1965).

The British economist John Maynard Keynes (1883–1946), in a book published in 1936, argued strongly for the kind of government intervention in economics typified by Roosevelt's New Deal, as a way of preventing depressions. His thinking has had a strong effect on governments ever since.

The population of the United States was 127 million, growth having slowed because of the Depression.

1937

TECHNETIUM

By now, only four gaps remained in the periodic table of elements between number 1 (hydrogen) and number 92 (uranium). They were numbers 43, 61, 85, and 87. Noddack and his colleagues, who had discovered rhenium (see 1925), thought they had also discovered number 43, but it turned out they were mistaken.

The Italian physicist Emilio Gino Segrè (1905–1989) tackled the problem in a different way. Thinking the chances small that he could find the element in the Earth's crust, he wondered if he might not manufacture it. After all, Fermi had been bombarding elements with neutrons to produce elements higher by one atomic number (see 1934). If Segrè bombarded molybdenum (element 42), might he not then manufacture detectable quantities of element 43?

In 1937 Segrè bombarded molybdenum with *deuterons,* the atomic nuclei of hydrogen-2, which had been discovered by Urey (see 1931, Deuterium). A deuteron consists of one proton and one neutron, rather loosely bound. If such a particle approached a nucleus, the positive charge of the nucleus would repel the positively charged proton, breaking it from the neutron and deflecting it away. The uncharged neutron, unaffected, would strike the nucleus.

It was an American physicist, Robert Oppenheimer (1904–1967), who had demonstrated that deuteron bombardment was equivalent to neutron bombardment. Why not use neutrons in the first place? Because the deuteron, carrying an electric charge, could be accelerated in a cyclotron and made very energetic. A neutron by itself could not.

Segrè analyzed the bombarded molybdenum and did indeed isolate very small quantities of an element that, from its chemical properties, could be seen to be number 43. It was the first element to be indisputably formed in the laboratory, rather than discovered in nature, so it was named *technetium,* from the Greek word for "artificial."

As it turned out, no technetium isotope was stable; all were radioactive. It is, indeed, the simplest element to have no stable isotopes. The most nearly stable isotope, technetium-97, has a half-life of 2,600,000 years. That is not a long enough half-life for it to have endured in the soil since the formation of the Earth. It is for that reason that it was never detected in the soil, and why Noddack had to be mistaken.

MUON

Anderson, who had discovered the positron as a result of studying cosmic ray bombardment of atoms within Earth's atmosphere (see 1932), was still investigating such bombardments. In the process, he discovered the track of a particle that, from its curvature in the magnetic field, appeared to be more massive than an electron but less massive than a proton.

By 1937 there was no doubt about it, for the same phenomenon had been observed by others as well. It seemed to be the type of particle that Yukawa had postulated (see 1935, Strong Interaction), when trying to work out what held protons and neutrons together in the atomic nucleus.

The new particle was called a *mesotron,* from a Greek word for "intermediate," and this was eventually shortened to *meson.* Once it was found that there were different varieties of these intermediate particles, different mesons had to be identified and differentiated in some way. Anderson's meson was therefore called a *mu-meson, mu* being a Greek letter equivalent to our *m.*

It turned out, however, that the mu-meson was not the particle Yukawa had postulated. It showed no tendency to interact with nuclei, and Yukawa's particles would have had to. Its name was therefore contracted further to *muon,* since it was not a meson in the sense that other intermediate-sized particles were.

ELECTROPHORESIS

Proteins carry electric charges here and there on their large molecules, some positive, some negative. If an electric field is imposed on a solution of proteins, those with a positive charge outweighing the negative charge will travel toward the cathode, and those with a negative charge outweighing the positive charge will travel toward the anode. The rate at which a protein molecule will travel in either direction depends on how great the overall charge is and, to some extent, on the pattern of the charges over the molecular surface. Enough variation exists so that every protein molecule travels at a different speed. A technique that takes advantage of this to separate and isolate protein components of a mixture is called *electrophoresis.*

A Swedish biochemist, Arne Wilhelm Kaurin Tiselius (1902–1971), developed the technique in 1937. He devised a special tube in the form of a rectangular U within which the proteins could move and separate. The Tiselius tube consisted of portions fitted together at specially ground joints, which could be separated to isolate one of a mixture of proteins in a particular chamber.

In addition, by the use of proper cylindrical lenses, it was possible to follow the process of separation by observing changes in the refraction of light passing through the suspension. These changes could be photographed, and the resulting wavelike pattern could be used to calculate the quantity of each protein present in the mixture. Failure to separate a mixture into components by electrophoresis was good evidence of the purity of a protein preparation, particularly if there continued to be no separation when the acidity of the solution was changed.

For his work on electrophoresis, Tiselius was awarded the Nobel Prize for chemistry in 1948.

ELECTRON MICROSCOPES

The first electron microscope had been constructed by Ruska (see 1932). It was not till 1937, however, that an electron microscope was constructed that could clearly outperform the best optical microscopes. The Canadian physicist James Hillier (b. 1915) accomplished this feat. His microscope could magnify seven thousand times, whereas the best optical microscope could only manage a magnification of two thousand times.

Eventually, Hillier and others devised electron microscopes with still further capabilities, until magnifications of two million times became possible.

FIELD-EMISSION MICROSCOPE

A magnifying device even more impressive than the electron microscope, although of only limited use, is the *field-emission microscope,* first devised in 1937 by the German-born American physicist Erwin Wilhelm Mueller (1911–1977).

This involves a very fine needle tip in a high vacuum. The tip can be made to emit electrons that shoot outward in diverging straight lines and strike a fluorescent screen. What appears on the screen, then, is a vastly magnified picture of the needle tip.

Eventually, the image of single atoms could be made out.

RADIO TELESCOPE

Radio waves from outer space had been detected by Jansky (see 1932), but nothing had been done about it because the devices necessary for detecting and analyzing the radiation were not yet in existence.

In 1937, however, an American radio engineer, Grote Reber (b. 1911), built the first radio telescope in his back yard. It was a parabolic reflector 31 feet in diameter, and the radio waves he received could be reflected to a focus so that their intensity could be measured.

In this way, Reber discovered points in the sky that emitted radio waves more strongly than did the general background, and he was the first to prepare a *radio map* of the sky. Naturally, Reber's work was primitive compared to what was to come in succeeding decades, but for several years he was the only radio astronomer in the world.

VIRUS NUCLEIC ACID

Stanley had succeeded in crystallizing tobacco mosaic virus and proving that it was protein in nature (see 1935). The question was whether the virus was *entirely* protein. Viruses are indisputably living, and if they consisted only of protein, it would follow that protein molecules (if sufficiently complex) were the essence of life. Everything else, even simpler proteins, would be merely auxiliary material.

In 1937, however, a British plant pathologist, Frederick Charles Bawden (b. 1908), was able to show that tobacco mosaic virus contained, in addition to protein, a small quantity of ribonucleic acid (RNA). Eventually, viruses generally were found to contain either RNA or DNA, which means they are nucleoprotein rather than merely protein.

The fact that chromosomes are also nucleoproteins (containing DNA) makes it seem that viruses are chromosomes "on the loose" and that it is nucleoprotein that is the essence of life.

CITRIC ACID CYCLE

Many biochemists had contributed insights concerning the metabolism of carbohydrate, including Harden (see 1904), Meyerhof (see 1913), and Warburg (see 1926).

Szent-Györgyi (see 1928, Hexuronic Acid) had found that any of four different four-carbon acids, if added to tissue slices, stimulated oxygen uptake. He suspected that they must play some role in carbohydrate metabolism.

Beginning in 1937, the German-born British biochemist Hans Adolf Krebs (1900–1981) found two six-carbon acids, including the familiar *citric acid,* that also played a role. He worked out the details of a cycle that began and ended with citric acid. Sugar molecules entered the cycle, and carbon dioxide molecules emerged at the other end, along with several pairs of hydrogen atoms that, through another chain of reactions including cytochromes (see 1924), were united with oxygen to yield the energy used by the body.

This *citric acid cycle* is often called the *Krebs cycle* in honor of its discoverer, and for this work, Krebs was awarded a share of the Nobel Prize for physiology and medicine in 1953.

NIACIN

Goldberger had shown that pellagra was a vitamin-deficiency disease (see 1915). A disease of dogs called *blacktongue* was the canine equivalent of pellagra.

In 1937 the American biochemist Conrad Arnold Elvehjem (1901–1962) considered the problem. Euler-Chelpin had shown that Harden's coenzyme included nicotinamide (see 1923). The enzyme wouldn't work without the coenzyme, and the coenzyme wouldn't work without the nicotinamide portion. Human beings, dogs, and other animals can manufacture all parts of the coenzyme from simpler substances—except the nicotinamide. That has to be present in the diet. Since enzymes are needed in trace quantities only, so are coenzymes, and so, in this case, is nicotinamide. Animals can risk depending on the diet for something needed in so small a quantity. Plants can make nicotinamide out of simpler substances.

But suppose the diet lacks even the trace quantities of nicotinamide that are needed? In that case, the enzyme won't work, carbohydrate metabolism will proceed limpingly or not at all, and serious symptoms will arise.

Nicotinic acid is simpler than nicotinamide and is easily converted to nicotinamide in mammalian tissue. Elvehjem therefore added a very small quantity of nicotinic acid to the diet of a dog with blacktongue, and it improved markedly and very quickly. It was at once apparent that nicotinic acid and nicotinamide were the vitamins that prevented pellagra, or cured it if it occurred.

Because physicians didn't want the lay public to confuse nicotinic acid with nicotine (a poisonous alkaloid) and assume there were vitamins in cigarettes, they used the names *niacin* and *niacinamide* for nicotinic acid and nicotinamide.

It turned out that other vitamins also worked because they were essential parts of coenzymes, parts that could not be manufactured in the body, were needed only in trace amounts, and could be picked up ready-made from the diet.

YELLOW FEVER VACCINE

A vaccine that conferred immunity to one or another serious disease was first used by Jenner in connection with smallpox (see 1796). Pasteur had added other vaccines against such diseases as cholera, anthrax, and rabies (see 1881).

The South African-born American microbiologist Max Theiler (1899–1972) developed a vaccine against yellow fever. By 1937, the vaccine was quite safe and effective, which did much to remove the dread of this terrible disease.

For this, Theiler was awarded the Nobel Prize for medicine and physiology in 1951.

EVOLUTION AND MUTATION

Darwin had advanced the theory of evolution by natural selection a century before (see 1858). He assumed that selection took place because in every generation there were small variations among the offspring of a particular species. How those variations arose, he didn't know.

Soon after Darwin, Mendel had developed the laws of genetics (see 1865), and some decades later, De Vries had shown the existence of mutations (see 1900). It seemed possible that mutations provided the variations that allowed natural selection to function as a mechanism for producing evolutionary changes. The exact rationale of this was not quite understood, however.

In 1937 the Russian-born American geneticist Theodosius Dobzhansky (1900–1975), who worked with fruit flies after the fashion of Morgan (see 1927), published a book entitled *Genetics and the Origin of Species* in which mutation and evolution were neatly joined together. As a result, evolution was understood on a molecular level as well as on an organismic level.

IN ADDITION

Japanese aggression against China took a strong leap forward. Before the end of July, the Japanese had taken Peiping and Tientsin, and by the end of the year they held all of northeastern China.

On August 8, 1937, the Japanese attacked Shanghai again. On November 8, they took the city and drove up the Yangtse River. They took the Chinese capital of Nanking on December 13, 1937, sacking it with appalling cruelty and atrocity. The Chinese government had to retreat far upriver to Chungking.

The Western powers continued to confine themselves to condemnation by words alone, which had no effect at all on the Japanese.

In Spain, the civil war continued bloodily with the Rebels making slow gains, thanks to major help from the Axis powers. On March 18, 1937, the Spanish Loyalists inflicted a crushing defeat on Italian troops who were helping the Rebels.

Stalin, meanwhile, virtually destroyed the Soviet army in his continuing purge.

The dirigible, which had been invented by Zeppelin thirty-seven years earlier, came to an end as a major aeronautic device on May 6, 1937, when the German dirigible *Hindenburg*, the largest ever built, exploded and burned at Lakehurst, New Jersey.

1938

SOLAR ENERGY SOURCE

Gamow had suggested that hydrogen fusion was the source of the Sun's energy (see 1929) but could not supply the details.

By 1938, however, much had been learned about nuclear reactions—the energy they delivered and the speed with which they took place—from experiments in the laboratory. This, together with reasonable estimates of the pressures and temperatures deep in the Sun's core, enabled the German-born American physicist Hans Albrecht Bethe (b. 1906) to outline in detail the mechanism by which hydrogen fusion could take place in the Sun's core. (The German astronomer Carl Friedrich Weizsacker [b. 1912] independently advanced similar ideas at about the same time.) For the first time, a suitable answer was given to Helmholtz's question about the source of the Sun's energy (see 1853).

For this, and for other work in nuclear physics, Bethe was awarded the Nobel Prize for physics in 1967.

MAGNETIC RESONANCE

The Austrian-born American physicist Isidor Isaac Rabi (1898–1988) continued the work of Stern on molecular beams (see 1933). Rabi, in 1938, developed the technique of *magnetic resonance* whereby it was possible to measure the energies absorbed and given off by particles in the beam in an extremely accurate manner. For this, he was awarded the Nobel Prize for physics in 1944.

VITAMIN E SYNTHESIS

Vitamin structure was rapidly being worked out in full detail. Karrer, who had synthesized vitamin A (see 1930) and riboflavin (see 1935), synthesized vitamin E in 1938 and thus established its structure.

PHASE-CONTRAST MICROSCOPE

In 1938 the Dutch physicist Frits Zernike (1888–1966) constructed a microscope based on a phenomenon he had been investigating for years. When it is diffracted, light slightly alters its phase so that different objects in a cell appear to take on color even though they are colorless ordinarily. Making use of such diffraction allows objects within the cell to be made out clearly and easily without killing the cell by adding a dye. For this *phase-contrast microscope,* Zernike was awarded the Nobel Prize for physics in 1953.

ICONOSCOPE

The existence of the cathode ray tube (see 1876) had raised a new possibility. If an electron beam could be made to pass over every part of a screen as the result of an appropriately varying magnetic field, and if the screen could be made to fluoresce appropriately, the electron beam could paint a picture, so to speak. The result would be what was eventually called a television screen.

The first practical television camera was constructed in 1938 by the Russian-born American engineer Vladimir Kosma Zworykin (1889–1982). He called it an *iconoscope.* The rear of the iconoscope was coated with a large number of tiny cesium-silver droplets. Each emitted electrons as a light beam scanned it, in proportion to the brightness of the light. The electrons in the television tube were controlled by the electrons in the iconoscope, so that the screen showed the same scene that entered the iconoscope.

Eventually, with improvements, the iconoscope made television a practical reality.

XEROGRAPHY

In 1938 the American physicist and lawyer Chester Floyd Carlson (1906–1968), aware of the constant need for copies of patent specifications, labored to devise some new way of providing them. The means he found involved attracting carbon black to paper by localized electrostatic forces.

This technique he called *xerography* (from Greek words meaning "dry writing," since no ink was used), and his first successful attempt came on October 22, 1938.

It took years to secure a patent and interest a corporation in the process. Eventually, though, the process led to the modern technique of photocopying, which has virtually eliminated the mimeograph machine and greatly reduced the need for carbon paper.

BALLPOINT PEN

In 1938 two Hungarian brothers, Ladislao Biro and Georg Biro, designed a ballpoint pen: ink from an internal reservoir coated a tiny ball at the end of the pen, and the ball rolled, depositing ink on the paper.

When the design was sufficiently improved and a high-viscosity ink was developed that would not blot, streak, or stain the fingers and that dried almost at once, the ballpoint pen took over. Fountain pens and inkwells became almost obsolete, and even pencils and erasers became less prominent.

COELACANTH

Finding a new kind of fossil is exciting enough, but finding a living specimen of a species thought to exist only as a fossil is far more exciting.

On December 25, 1938, a trawler fishing off South Africa brought up an odd fish about 5 feet long. The oddest thing about it was that its fins were attached to fleshy lobes rather than directly to the body.

A South African zoologist, J. L. B. Smith, who had the chance of examining it, recognized it as a *coelacanth,* a primitive fish that zoologists thought had been extinct for seventy million years—to have vanished from the Earth while dinosaurs still existed.

Eventually, other specimens of the fish were obtained as well. What is most interesting about this (from our own viewpoint) is that the coelacanth is a close relative of the kind of fish that, hundreds of millions of years ago, clambered onto land, became amphibious, and thus became our ancestors.

IN ADDITION

On March 12, Hitler sent his army into Austria and declared it part of Germany. This was the so-called *Anschluss* (German for "annexation"). He then began a campaign of propaganda and terror against Czechoslovakia. Prime Minister Neville Chamberlain (1869–1940) of Great Britain and Premier Édouard Daladier (1884–1970) of France quickly chose the path of *appeasement*. This consisted of giving Hitler what he wanted, hoping that he would then settle down and be a good fellow. On September 19, Great Britain and France, in an agreement reached at Munich, handed over the border regions of Czechoslovakia to Germany and left it defenseless. Chamberlain declared the Munich agreement to be "peace with honor."

Inside Germany, anti-Semitism went into high gear in response to the assassination of a German embassy official in France by a Jew. On the night of November 9, 1938, Jewish synagogues, shops, and homes were smashed and the Holocaust began as tens of thousands of Jews were carted off to death in concentration camps.

Meanwhile, Japan continued its advance in China, taking coastal cities in May and June. In July and August, however, Japanese and Soviet forces clashed at a point where the borders of the Soviet Union, Manchuria, and Korea met. The Japanese forces were defeated by the Soviets, and Japan decided that in future its sphere of expansion would be southward, not northward.

1939

NUCLEAR FISSION

Fermi had bombarded uranium with slow neutrons in the hope of obtaining element number 93, but the results had been confusing (see 1934).

Hahn and Meitner, who had discovered protactinium (see 1917), were investigating the matter. Among other things, they dissolved the bombarded uranium and added a barium compound. When they precipitated the barium, they found that a fraction of the radioactivity came down with it.

This was what they had hoped for. Barium is chemically very like radium, so that anything that precipitated barium would also precipitate radium. In fact, the bombardment of uranium (element number 92) with neutrons might somehow result in the emission by uranium of *two* alpha particles, which would reduce the atomic number by four and so produce an isotope of radium (element number 88). Then, since barium is similar to radium but not identical, the two could be separated, and the idea of double alpha-particle emission would be proved. (That was the idea behind adding the barium in the first place.)

However, although the barium did precipitate radioactivity, the radioactivity simply could not be separated from it. The problem was therefore more puzzling than ever.

In 1938 Meitner, an Austrian national, had to flee the country because, with its annexation by Germany, it became subject to Germany's anti-Semitic laws. She escaped to Sweden.

Hahn, now working with the German chemist Fritz Strassman (1902–1980), finally began to suspect that the radioactivity could not be separated from the barium because it consisted of a radioactive isotope of that very element. But how could a radioactive isotope of barium have been formed in the neutron bombardment of uranium? The atomic number of barium is 56. For uranium to form barium, the uranium nucleus would have to break in two (something that later came to be called *nuclear fission*).

Nuclear fission was unheard of, and Hahn hesitated to make such a suggestion. He published his findings in January 1939 but said nothing about fission.

Meitner, in Sweden, thinking about the matter, also came to the conclusion that uranium had undergone nuclear fission and had produced a radioactive isotope of barium, but she decided to advance the possibility publicly. With the help of her nephew, the physicist Otto Robert Frisch (1904–1979), she prepared a paper dated January 26, 1939, and sent it to the British scientific journal *Nature*.

Frisch was working for Bohr (see 1913) and told him of the contents of the paper before it was published. Bohr was going to the United States to attend a physics conference in Washington, D.C., on January 26, 1939, and he spread the word there, also before the paper was published.

In the United States, nuclear fission as a result of the neutron bombardment of uranium was quickly confirmed. Bohr had suggested that it was the rare isotope, uranium-235, that underwent fission, and that was confirmed, too.

For his discovery of nuclear fission, Hahn was awarded the Nobel Prize for chemistry in 1944, but it was not until 1946 that he could accept it.

NUCLEAR CHAIN REACTION

Chain reactions are familiar in chemistry. A reaction may produce heat, which encourages further reaction, which produces more heat, and so on. The manner in which a lightning stroke, or a single match, can burn down an entire forest is an example of a chemical chain reaction. A chemical reaction can also produce a molecular product, which brings about more of the chemical reaction, which produces more of the molecular product, and so on. The formation of polymers is sometimes the result of such a chain reaction.

As early as 1932, it had occurred to the Hungarian-born physicist Leo Szilard (1898–1964) that a nuclear chain reaction was also possible. By that time, atomic nuclei were being bombarded by neutrons, and in some cases, the entry of one neutron into a nucleus meant the ejection of two neutrons. Szilard envisaged the case of a neutron disrupting a nucleus and producing two neutrons, which would disrupt two nuclei, producing four neutrons, which would disrupt four nuclei, and so on. Each disruption would deliver a tiny quantity of energy, but each link in such a nuclear chain reaction would proceed so quickly that the energy produced would mount in a fraction of a second to titanic proportions. The result would be a *nuclear bomb*, tremendously more powerful than ordinary bombs based on chemical chain reactions.

Szilard was one of many scientists fleeing Central Europe because of German anti-Semitism, and the brains that Germany thus rejected and the allies absorbed were eventually to prove of inestimable help to the allies. Szilard patented the notion of a nuclear chain reaction and offered it to Great Britain.

The nuclear reactions known in 1932 and for several years afterward were not suitable for nuclear chain reactions, however. It took a fast, energetic neutron to eject two neutrons, but those two ejected neutrons were too slow, too lacking in energy, to continue the chain.

In 1939, however, when Szilard heard of nuclear fission, which was brought about by *slow* neutrons, he realized that a nuclear chain reaction and a nuclear bomb were finally possible, and he promptly began a campaign to persuade American scientists to keep their researches into such things secret. To a large extent, he succeeded.

FRANCIUM

Only three elements remained undiscovered: numbers 61, 85, and 87. In 1939 the French physicist Marguerite Perey (1909–1975), working with the radioactive element actinium, discovered a type of beta activity that was not quite like that of any known isotope. She tracked it down and found that it resulted from the breakdown of an isotope of element number 87. She named it *francium*, for her native country.

As it turned out, the most nearly stable isotope of the element is francium-223, which has a half-life of only 22 minutes. Of all the elements between 1 and 92, francium is the only one that possesses no isotope with a half-life of as much as half an hour.

NEUTRON STARS

Zwicky had suggested the possible existence of a neutron star (see 1934). In 1939 Oppenheimer (see 1937) analyzed the possibilities mathematically, in the light of what was known of nuclear reactions. Nevertheless, the matter remained strictly theoretical, for no such object had actually been observed, or would be for another thirty years.

MAGNETIC MOMENTS

The magnetic properties of atoms and molecules had been determined by the studies of molecular beams undertaken by Stern (see 1933) and by Rabi (see 1938).

The Swiss-born American physicist Felix Bloch (1905–1983) worked out methods for determining the magnetic properties of molecules within liquids and solids. From this work, he calculated the magnetic moment of the neutron. This was important, because the neutron is not an electrically charged particle and should therefore have no magnetic field. Yet Bloch demonstrated that it had.

This was the first indication that the neutron might be a composite particle, made up of more fundamental particles that *did* have electrical charges.

The fact that the neutron had a magnetic moment, incidentally, showed that such a thing as an *antineutron* could exist, which would have a magnetic field pointed in the direction opposite to that of the neutron.

The magnetic moment of the neutron was independently worked out at about the same time in another way by the American physicist Edward Mills Purcell (b. 1912). As a result, Bloch and Purcell shared the Nobel Prize for physics in 1952.

VITAMIN K

A Danish biochemist, Carl Peter Henrik Dam (1895–1976), fed hens a synthetic diet and noticed that on some diets they developed small hemorrhages under the skin and within the muscles. Vitamin C, which cures some conditions like that, didn't help.

He finally decided that the problem stemmed from lack of some fat-soluble vitamin not yet known, the presence of which was necessary for proper coagulation of the blood. He named it vitamin K (for *koagulation,* the German spelling of the word).

This was done in 1934, but it was not till 1939 that the American biochemist Edward Adelbert Doisy (1893–1986) worked out the structure of the vitamin and synthesized it. For this work, Dam and Doisy shared the Nobel Prize for medicine and physiology in 1943.

RH FACTOR

After the A, B, and O blood factors were discovered by Landsteiner (see 1900), other blood factors were discovered that did not interfere with transfusion and did not involve health problems (see 1927).

A Russian-born American immunologist, Philip Levine (b. 1900), was studying erythroblastosis fetalis, a disease of fetuses and newborn infants that was marked by severe destruction of the red blood cells. (They were therefore called *blue babies*.)

Levine noted, in 1939, that the mothers in these cases lacked a blood component called the *Rh factor* (because it had first been recognized in the blood of rhesus monkeys used as experimental animals). The mothers were *Rh-negative,* but the fathers were *Rh-positive,* which was genetically dominant, so that their offspring were Rh-positive also. However, the fetus's blood apparently brought about the production of antibodies to the Rh-positive factor in the mother's blood. These antibodies filtered into the fetus's blood and destroyed the red corpuscles.

Routine testing for Rh factor prepared physicians for this possibility, and the replacement of the infant's blood supply by new blood reduced the death rate.

PENICILLIN

Fleming had discovered penicillin (see 1928), but nothing much came of it until 1939. Then an Australian-born British pathologist, Howard Walter Florey (1898–1968), in collaboration with a German-born British pathologist, Ernst Boris Chain (1906–1979), set about isolating the actual antibacterial agent from the mold.

They reached their goal quickly and the groundwork was laid for penicillin's use during the dark days of war that were to follow. As a result, Florey, Chain, and Fleming were awarded the Nobel Prize for medicine and physiology in 1945.

TYROTHRICIN

Most pathogenic bacteria do not survive well in the soil, which is why burying people who have died of disease does not necessarily keep epidemics going. It is possible to reason that this may be because antibacterial substances are produced by bacteria that do commonly live in the soil and are not pathogenic. (One might suppose that there is an evolutionary benefit for these bacteria in preventing the invasion of other species of bacteria.)

The French-born American microbiologist René Jules Dubos (1901–1982) investigated the matter and, in 1939, isolated an antibacterial substance from a bacterium called *Bacillus brevis*. He named it *tyrothricin*. This was found to be a mixture of several polypeptides (amino acid chains rather smaller than those found in proteins generally).

Tyrothricin was not a very effective antibacterial agent, but it and penicillin ushered in an era characterized by the most powerful weapons against infection the medical profession had yet possessed.

ESSENTIAL MINERALS

In 1939 Keilin, who had demonstrated the existence of the cytochromes (see 1924), showed that the enzyme carbonic anhydrase contained a small quantity of zinc that was essential to its working. This meant that since the enzyme was essential to life, so was zinc.

A number of elements, usually associated with minerals rather than with life, have since been shown to be essential to life in trace quantities because of their association with enzymes. Included are manganese, molybdenum, and copper. These may be grouped together as *essential minerals*, or *essential trace elements*.

DDT

The most serious enemies of humanity, next to pathogenic microorganisms, are the insects. Not only do they carry and spread diseases such as yellow fever, malaria, typhus, and encephalitis, but they also eat crops and make serious inroads into the human food supply. Through the ages, they have been feared and fought, and as knowledge of chemistry grew, poisons had been used to kill them. Unfortunately, the inorganic poisons used, such as Paris green, were deadly to mammals, including human beings.

The Swiss chemist Paul Hermann Müller (1899–1965) began a search for organic substances that might be poisonous to insects but not to other forms of life, and that would also be cheap, stable, and without unpleasant odor.

In September 1939 he tried *dichlorodiphenyltrichloroethane* (of which the common abbreviation is *DDT*), a compound known to chemists since 1873, and it seemed to fulfill all the requirements. DDT proved exceedingly valuable in the years to come, particularly in fighting lice-spread typhus, so that Müller was awarded the Nobel Prize for medicine and physiology in 1948.

In time, to be sure, DDT turned out to have its harmful aspects after all, and its use dwindled. Nevertheless, it was the forerunner of a great variety of *pesticides* that have served humanity.

HELICOPTER

One problem with airplanes is that they must move quickly in order to produce aerodynamic lift under their wings. If they slow up, the lift dwindles and they crash. A device that exerted a force straight upward instead of merely forward as propellers do would eliminate this necessity for speed-born lift.

The obvious solution was a large propeller directly overhead. Since the ends of the propeller would mark out a helix as the vehicle lifted upward, such a device was called a *helicopter,* from Greek words meaning "helical wing."

The Russian-born American aeronautical engineer Igor Ivan Sikorsky (1889–1972) had been working with helicopters for thirty years and finally in 1939 produced a satisfactory model. On September 14, a helicopter with Sikorsky himself at the controls flew successfully. The time was to come when it would be a primary weapon in wars fought in Vietnam and Afghanistan, to say nothing of its uses in tracking traffic, rescue work, and intra-urban transportation.

FREQUENCY MODULATION

Static in radio broadcasts was proving a rather intractable problem. Radio transmission for the first forty years after Marconi's discovery (see 1901) was carrried out by systematically altering the amplitude (the height of the wave) of the carrier signal to match the variation in the amplitude of the sound waves being transmitted. This was called *amplitude modulation,* or AM.

Unfortunately, thunderstorms and electrical appliances also modulate the amplitude, doing it randomly and producing the irritating noise of static.

In 1939, however, Armstrong, who had invented the superheterodyne receiver (see 1916), devised a method of transmitting a signal by systematically altering the frequency (the length of the wave) of the carrier signal. This was called *frequency modulation,* or FM. Thunderstorms and electrical appliances have no effect on frequency, so FM transmission is largely static-free. Unfortunately, FM will work only for carrier waves of high frequency, and these cannot be transmitted much beyond the horizon.

IN ADDITION

Hitler broke the Munich agreement and invaded what was left of Czechoslovakia, incorporating most of the land into Germany. On March 21, he forced Lithuania to cede him the German-speaking Memel area on its western border. He then began demanding annexation of the German-speaking Free City of Danzig and threatened Poland.

As a way of showing its own military might, Italy invaded Albania on April 7, 1939. It fell without resistance.

In Spain, the Rebels finally won, taking Madrid on March 28. This meant that France was surrounded by three Fascist powers: Germany, Italy, and Spain.

Stalin decided to protect himself by coming to an accommodation with Hitler, and on August 23, Germany and the Soviet Union signed a nonaggression pact.

On September 1, Hitler sent German forces into Poland. Great Britain and France declared war on Germany on September 3, and World War II had begun.

While all this was happening, the Japanese advance in China had virtually stopped. The Japanese had had as much of China as they could conveniently digest, and the war in Europe turned their eyes toward European possessions in southeastern Asia.

In the United States, public opinion favored

Great Britain and France, but the winds of isolationism still blew hard.

Szilard and his fellow refugees from Hungary, Eugene Paul Wigner and Edward Teller (b. 1908), persuaded Albert Einstein to send a letter to President Roosevelt, urging him to proceed with the development of a nuclear fission bomb so that Hitler would not get one first.

1940

NEPTUNIUM AND PLUTONIUM

Fermi had tried to form element number 93 by bombarding uranium with neutrons (see 1934). Hahn and Meitner had shown that the result was nuclear fission (see 1939). The two were not mutually exclusive, however. It was possible that some uranium nuclei might undergo fission and others might undergo the kind of changes that would produce element number 93.

In 1940 the American physicists Edwin Mattison McMillan (b. 1907) and Philip Hauge Abelson (b. 1913), studying uranium that had been bombarded with neutrons, detected a beta particle with a half-life of 2.3 days. When they had tracked this down, they announced on June 8, 1940, that they had located traces of element number 93.

Since uranium had been named for the planet Uranus (see 1789), the new element, lying beyond uranium, was named *neptunium,* for Neptune, the planet lying beyond Uranus.

Since the neptunium isotope that had been located emitted beta particles, it had to gain another unit in atomic number and produce element number 94, which was named *plutonium,* for the planet Pluto. In this part of the work, the American physicist Glenn Theodore Seaborg (b. 1912) was prominent.

Neptunium and plutonium were the first of the *transuranium elements* to be discovered. There would be others.

Seaborg recognized that the transuranium elements were part of a series analogous to the rare earth elements. There were fifteen elements running from lanthanum (number 57) to lutetium (number 71) inclusive (with number 61 *still* undiscovered), and they were now called the *lanthanides* after the first member. The second group would be the fifteen elements running from actinium (number 89) to element number 103 inclusive, and they were named the *actinides.* Six of the actinides were now known, and nine remained to be discovered.

For their work on transuranium elements, McMillan and Seaborg were awarded the Nobel Prize for chemistry in 1951.

URANIUM HEXAFLUORIDE

One of the problems in preparing a nuclear bomb was that only uranium-235 would participate in a slow-neutron nuclear chain reaction, but uranium-238 was present to the extent of 140 atoms for every atom of uranium-235. The uranium-238 atoms would absorb neutrons and undergo reactions other than fission, which dampened and even aborted the chain reaction.

It would be necessary to prepare *enriched uranium,* with a larger percentage of the uranium-235 atoms. Separating two isotopes is difficult, however. If uranium were a gas, which could be made to pass through long narrow passages, or very fine holes, the uranium-235, being lighter by 1.26 percent, would move a little faster, so that the first sample to arrive at the other end would be slightly richer in uranium-235 than the natural element would be. If this was done over and over *(gaseous diffusion),* satisfactorily enriched uranium would result. But uranium is not a gas.

In 1940 Abelson, who was helping to isolate neptunium (see above), suggested that uranium hexafluoride, with each molecule made up of one uranium atom and six fluorine atoms, was a liquid that was easily evaporated, and the vapors might then be made to pass through tubes or holes.

Uranium hexafluoride containing a uranium-235 atom had a molecular weight of 349 compared to 352 for uranium hexafluoride with a uranium-238 atom. The difference in weight is nearly 1 percent, and that was enough. Gaseous diffusion sufficed to produce enriched uranium.

ASTATINE

In 1940 Segrè, who had isolated technetium (see 1937), bombarded bismuth (element number 83) with alpha particles. If an alpha particle struck the bismuth and remained, or even if a neutron were emitted thereafter, the bismuth would have gained two protons and the result would be the undiscovered element number 85. This was accomplished in 1940, but World War II interrupted, and it wasn't until the war was over that they could confirm their finding.

The new element was quite unstable. Its most long-lived isotope had a half-life of only 8.3 hours. It was therefore named *astatine,* from the Greek word for "unstable." It belonged to the same group as fluorine, chlorine, bromine, and iodine, which is why it received the *-ine* ending.

With the discovery of astatine, only one gap remained in the entire periodic table from element number 1 (hydrogen) to element number 94 (plutonium), and that was element number 61.

BETATRON

Cyclotrons were used to accelerate protons, which are massive particles and energetic enough to get results even when they are not moving at very high speeds (see 1930).

It might be useful to accelerate electrons, but they are so light that to obtain useful energies they would have to be brought to speeds quite close to that of light. That would introduce a mass increase according to special relativity (see 1905) and cause the accelerating pulse to go out of synchronization, setting a too-low limit on the speeds achieved.

In 1940, however, the American physicist Donald William Kerst (b. 1911) devised an accelerator that whirled the electrons in circles rather than spirals. This new device, called the *betatron* (because it accelerated beta particles), made useful electron bombardment possible.

STREPTOMYCIN

Dubos's discovery of tyrothricin (see 1939) had galvanized one of his past teachers, the Russian-born American microbiologist Selman Abraham Waksman (1888–1973).

Using Dubos's methods, Waksman began to search for bacteriocidal compounds in microscopic fungi. In 1940 he located one he called *actinomycin* because it was found in fungi of the *Actinomycetes* family. Not long after, he found one in fungi of the *Streptomycetes* family and called it *streptomycin.*

Streptomycin was quite effective against bacteria that were not affected by penicillin but was considerably more toxic to human beings than penicillin and had to be used very cautiously.

It was Waksman who coined the term *antibiotic* (from Greek words meaning "against [microscopic] life"), and for his work on them, he was awarded the Nobel Prize for medicine and physiology in 1952.

COLOR TELEVISION

Although television existed as a laboratory exercise only and wouldn't enter the American home till after World War II, methods were already being developed to transmit television in full color.

The first to work out such a system was the Hungarian-born American engineer Peter Carl Goldmark (1906–1977), who used a rotating three-color disk for the purpose in 1940. The method was not, in the end, used. More effective methods were finally made commercial some fourteen years later.

IN ADDITION

The year opened with active fighting only in Finland. The Finns defended themselves tenaciously, but by March 12, 1940, the Soviet Union had worn them down, and they accepted a peace that gave the Soviets territory and advantages.

Then, on April 9, Germany struck northward. Denmark fell in a day, and German troops landed in Norway. By the end of April, Norway was effectively in German hands.

The situation forced Chamberlain to resign, and Winston Churchill took his place on May 7, 1940.

Things continued to grow worse for the Allies, however. By May 14 the Netherlands had surrendered, and on May 26 Belgium capitulated also. Northeastern France was overrun, and by the end of May the entire British expeditionary force, along with some French and Belgian troops who were still fighting, were surrounded and pinned against the English Channel at Dunkirk.

For some reason, Hitler called his army to a halt and left the task to the German airforce, which proved unequal to the task, and the British army was rescued. It was Hitler's first great mistake.

Mussolini, feeling that an Axis victory was now assured, declared war on France and Great Britain on June 10, 1940.

Paris was declared an open city and abandoned to the Germans on June 14. The French premier, Paul Reynaud (1878–1966), resigned on June 16 and was replaced by Philippe Pétain (1856–1951), who gave in at once, and an armistice was signed by the French on June 24. The Germans established themselves in northern and western France, while Pétain's puppet regime made its capital at Vichy in central France. Pierre Laval (1883–1945), who also assumed Germany's victory to be complete, was the real leader of what passed for a government in Vichy. However, General Charles-André-Marie-Joseph de Gaulle (1890–1970) managed to make his way to London and there he held high the banner of what came to be called the *Free French*.

Now Great Britain stood alone against Germany. When Great Britain refused to surrender, the Germans began a systematic course of bombing—of London in particular—thus fighting what came to be called the *Battle of Britain*.

Again, however, the German airforce, under its bombastic leader Hermann Göring (1893–1946), was unequal to the task. By the end of the year, Germany had clearly lost the Battle of Britain.

Meanwhile, the Soviet Union had converted Estonia, Latvia, and Lithuania into Soviet Socialist Republics and had also annexed the Romanian province of Bessarabia (in each case taking back territory it had lost in 1918).

In Asia, Japan moved into French Indo-China and formed a military alliance with Germany and Italy. Before the end of the year, Hungary and Romania had joined the alliance, too.

In the United States, Roosevelt decided to run for an unprecedented third term because of the unprecedented situation abroad. He was successful, the only president ever to win a third term.

The population of the United States was 132 million and that of the Soviet Union 180 million. Germany and its controlled territory totaled 110 million, and the population of the world had reached 2.3 billion.

1941

HIGH-ENERGY PHOSPHATE

Harden had discovered the existence of phosphate esters in tissues (see 1905, Metabolic Intermediates). Since then, Meyerhof (see 1913) and others had learned how the phosphate esters formed and how the phosphate groups were transferred from compound to compound in the course of metabolism.

In 1941 the German-born American biochemist Fritz Albert Lipmann (1899–1986) showed that there were two kinds of phosphate bonds. In one kind, the loss of the phosphate bond liberated a relatively small quantity of energy; in the other, the loss liberated a relatively large quantity. The latter esters possessed a *high-energy phosphate.*

In the course of carbohydrate metabolism, phosphate groups are added to sugar molecules forming a low-energy phosphate. The sugar molecule then undergoes the kind of change that concentrates the energy in the phosphate group, so that a high-energy phosphate is formed. These high-energy phosphate groups then serve as the "small change" of energy. In the simplest terms, food and oxygen combine to form high-energy phosphate bonds, which then deal out energy for all the energy-consuming functions of the body.

The most versatile of the high-energy configurations is a compound called *adenosine triphosphate (ATP),* each molecule of which contains two high-energy phosphates and which has been found to be concerned with body chemistry at almost every point where energy is required.

POLARIMETRY

The Czechoslovakian physical chemist Jaroslav Heyrovsky (1890–1967) had worked for years on a device that contained a mercury electrode so arranged that a small drop of mercury repeatedly fell through a solution to a mercury pool beneath. An electric current flowed through the solution and, as the potential was increased, the current reached a plateau, the height of which depended on the concentration of certain ions in the solution. In this way one could analyze a solution of unknown composition.

By 1941 Heyrovsky had perfected this technique, which he called *polarimetry,* and for this work, he was awarded the Nobel Prize for chemistry in 1959.

CARDIAC CATHETERIZATION

Forssmann had introduced the principle of cardiac catheterization (inserting a catheter into a vein and maneuvering it to the heart—see 1929). In 1941 it was introduced into clinical practice by the French-born American physiologist André Frédéric Cournand (1895–1988) and the American physician Dickinson Woodruff Richards (1895–1973). As a result, Cournand and Richards were awarded the Nobel Prize for physiology and medicine in 1956, sharing it with Forssmann.

DISTANCE OF THE SUN

The earliest reasonable estimate of the distance to the Sun had been based on Cassini's measurement of the parallax of Mars (see 1672). Measurement of the parallax had improved with time, but it was always difficult to deal with Mars in this respect, since it showed a small orb in the telescope, which forced a certain ambiguity on the measurement of its exact position.

Nearly a century before, the German astronomer Johann Gottfried Galle (1812–1910) had suggested that the parallax of an asteroid be used for determining the scale of the Solar System and the distance of the Sun, since its size and starlike appearance would make its positioning more accurate. However, asteroids were farther off than Mars and their parallax was correspondingly smaller and harder to measure.

But then Eros had been discovered by Witt (see 1898), and Eros could on occasion approach more closely than any planet.

In 1931 Eros approached within 16,000,000 miles of Earth, and a long, detailed program was set up in advance. Fourteen observatories in nine countries took part under the leadership of the English astronomer Harold Spencer Jones (1890–1960). Seven months were spent on the project, nearly three thousand photographs were taken, and the position of Eros was determined on each.

Ten years of calculation followed, and by 1941 Jones was able to announce that the distance of the Sun was 93,005,000 miles. That was the most accurate determination yet and was probably the best that could be done until methods transcending the accuracy of parallax determination were devised.

JET PLANES

During the forty-year history of airflight, planes had been propelled through the air by the aptly named *propeller*. There was no question, though, that a plane could also be made to move through the air, perhaps even more quickly and efficiently, by means of the rocket principle—by burning fuel and ejecting a jet of exhaust gas at high speed (such planes are therefore called *jet planes*).

The advantage jet planes had over rockets such as those developed by Goddard (see 1926) was that they traveled through the atmosphere, so they needed to carry only fuel and could make use of the oxygen in the surrounding air as the oxidizer.

Plans for engines that made some use of the jet principle can be traced back to 1921, but the first patent for a jet engine of the type used today was obtained by a British aeronautical engineer, Frank Whittle (b. 1907), in 1930.

The first jet plane making use of Whittle's engine was flown in May 1941. Jet planes were developed too late to play much of a role in World War II, but they came into their own afterward.

NEUROSPORA

Often genetics takes a step forward with the introduction of a simpler organism for study, as was the case, for instance, when Morgan began to study fruit flies (see 1907).

The American geneticist George Wells Beadle (1903–1989), in collaboration with the American biochemist Edward Lawrie Tatum (1909–1975), began to work in 1941 with a mold called *Neurospora crassa*.

In the wild state, this mold will grow on a nutrient medium in which sugar is the only significant organic compound. For its supply of elements not present in sugar, such as nitrogen, phosphorus, and sulfur, Neurospora can make do with inorganic compounds.

If Neurospora are bombarded with X rays, using Muller's technique (see 1927), however, mutations take place. Some of the mutated Neurospora lose the ability to form a particular organic compound necessary for growth, so that it has to be added to the nutrient medium.

Beadle found that it was not always necessary to add the missing compound itself to the medium—a different but similar compound might do. This meant that the similar compound could be converted into the necessary one. By trying a variety of similar compounds and noting which would promote growth and which would not, Beadle could deduce the sequence of chemical reactions and locate the step that the mutated mold could not manage.

From his work, Beadle concluded that the characteristic function of the gene was to supervise the formation of a particular enzyme (one gene to each enzyme). A mutation took place when a gene was so altered that it could no longer form a normal enzyme.

For this work, Beadle and Tatum were awarded a share of the Nobel Prize for medicine and physiology in 1958.

IN ADDITION

The Germans turned east, where they forced Bulgaria and Yugoslavia to join the Axis. They raced into Greece as well and by April 27 controlled the entire Balkan peninsula.

In North Africa, the British faced the Italians and swept through all of eastern Libya. Hitler sent a brilliant general, Erwin Johannes Eugen Rommel (1891–1944), to Libya along with some German tank troops who were trained in desert fighting. They had cost the British their Libyan conquests by the end of the year.

On the other side of the world, Japan consolidated its hold on Indo-China and on April 13, 1941, signed a nonaggression pact with the Soviet Union.

At this point, Hitler made his second mistake. He decided to invade the Soviet Union. On June 22, Germany struck without warning. The Soviet Union was caught completely by surprise and suffered catastrophic losses but did not collapse. On November 22, the Germans took Rostov, at the mouth of the Don River, but on November 29 the Soviets drove them out again. It was the first time in over two years that the Germans had been forced to give up a position they had taken. In addition, the German army was stopped 20 miles west of Moscow and, unlike Napoleon's French army a century and a third earlier, never reached the city.

In the United States, President Roosevelt signed a secret order on December 6, 1941, that ordered the start of the so-called Manhattan Project to develop a nuclear fission bomb. On December 7, Japan attacked Pearl Harbor. The United States at once declared war on Japan, and with that, Hitler declared war on the United States.

1942

NUCLEAR REACTOR

Szilard's notion of a nuclear chain reaction (see 1939) could not be made practical. Once the Manhattan Project had been put in motion (see 1941), Fermi was placed in charge of producing such a chain reaction. Uranium and uranium oxide were piled up in combination with graphite blocks in a structure called an *atomic pile*. Neutrons colliding with the carbon atoms in graphite did not affect the carbon nuclei but bounced off, giving up their energy and moving more slowly as a result, thus increasing their chance of reacting with uranium-235.

Making the pile as large as possible made it more likely that neutrons would strike uranium-235 before blundering out of the pile altogether and into the open air. The necessary size was the *critical mass*. Naturally, the more enriched in uranium-235 the pile was, the smaller the critical mass needed to be.

Cadmium rods were inserted into the pile, because cadmium would soak up neutrons and keep the pile from becoming active prematurely. When the pile was large enough, the cadmium rods were slowly withdrawn and the number of neutrons produced slowly increased. At some point, the increase would reach the level where more were being produced than were being harmlessly consumed by the cadmium. At that time the nuclear chain reaction would begin and the whole thing would go out of control in a moment.

What stopped that from happening was that some neutrons didn't come out of the bombarded nuclei at once. In other words, after the pile went critical, there was a slight pause before the *delayed neutrons* were emitted and everything went awry. During that pause, the cadmium rods could be shoved in again.

At 3:45 P.M. on December 2, 1942, in the squash court of the University of Chicago, the chain reaction became self-sustaining and was choked off. The *atomic age* had begun. The atomic pile was the first *nuclear reactor*.

BIOTIN

At this time a new vitamin could only be detected by noting that certain foods or food extracts could correct symptoms that known vitamins would not touch. One such new vitamin was called *vitamin H*. The American biochemist Vincent du Vigneaud (1901–1978) isolated tiny quantities of vitamin H in reasonably pure form, as judged by its powerful effect on the symptoms it could alleviate, and by 1942 had worked out its rather complicated two-ring structure. The compound was now called *biotin*. It was synthesized and its structure proven.

BACTERIOPHAGE STRUCTURE

The electron microscope (see 1932) had reached a pitch of excellence by now and could show viral structure as ordinary light microscopes could show cellular structure.

In 1942 the Italian-born American microbiologist Salvador Edward Luria (b. 1912) managed to get good photographs of bacteriophages. These were unusually large viruses, to be sure, but they were considerably smaller than bacterial cells. Luria showed that the bacteriophage consisted of a rounded head and a thin tail, rather like an extremely small sperm cell. This was the first time a virus had been seen as anything more than a vague dot.

IN ADDITION

Most of the year remained grim for the Allies. By June, all of the western Pacific was under Japanese control. In Europe, the Germans began their second massive offensive, confining it this time to the south, and in August they approached Stalingrad. In North Africa, Rommel drove eastward. A turning point then occurred in each of these three areas of conflict.

In the ferocious three-day Battle of Midway, in the Hawaiian Islands, the United States broke the Japanese fleet, and the days of Japanese conquest were brought to an end. On August 7, 1942, American forces landed on Guadalcanal in the Solomon Islands, and the long American counteroffensive began. (Meanwhile, in the United States, 110,000 people of Japanese descent were placed in concentration camps.)

In North Africa, the British forces had come under the leadership of Bernard Law Montgomery (1887–1976). On October 23, 1942, Montgomery struck and fought the Battle of El Alamein, which sent the Germans on a long retreat.

For three months in the Soviet Union, the tremendous battle over Stalingrad continued. On November 19, the Soviets launched a terrific counterattack that sent the Germans fleeing in disorder.

On December 24, 1942, the German rocket engineer Wernher von Braun (1912–1977) fired the first rocket we would now call a guided missile.

In the Soviet Union, the biologist Trofim Denisovich Lysenko (1898–1976) was at the peak of his power. Backed by Stalin, his fallacious genetic theories set back Soviet genetics for years and demonstrated the danger of allowing scientific thought to be influenced by political power and ideology.

1943

ADRENOCORTICOTROPHIC HORMONE

It was becoming more obvious that the pituitary gland was a particularly important source of protein hormones, some of which activated and controlled other glands, such as the thyroid and the gonads.

In 1943, the Chinese-born American biochemist Choh Hao Li (b. 1913) isolated, from the pituitary, a hormone that stimulated the adrenal cortex, using it to produce and release the cortical hormones (see 1935, Cortisone). This pituitary hormone was named *adrenocorticotrophic hormone,* a name usually abbreviated *ACTH.* It had the same effect on the body as cortisone, though more indirectly.

LYSERGIC ACID DIETHYLAMIDE

In 1913 a Swiss chemist, Albert Hoffman (b. 1906), was working with lysergic acid, which is obtained from ergot, a mold that produces serious and sometimes deadly disorders in the human body. Hoffman modified it to form the diethyl amide of the compound and apparently absorbed some of the substance. He was overcome by strange sensations, vivid fantasies, and brilliant colors. He deliberately swallowed a tiny bit more of the material—and the results were even more weird.

He had clearly suffered from hallucinations, so that lysergic acid diethylamide (usually abbreviated *LSD*) came to be called a *hallucinogen,* or a *psychedelic drug.* Other hallucinogens occur in nature, in certain mushrooms, in peyote cactus, and elsewhere. Even alcohol in sufficient quantities becomes hallucinogenic. Hallucinogens have been widely used in religious ceremonies, presumably because they seem to offer a vision of another world.

LSD was a particularly effective hallucinogen, and in time its use became a fad among young people, which helped fasten the *drug culture* on America.

SEYFERT GALAXIES

Over twenty years had passed since it had become clear that there were innumerable galaxies lying deep in space, but there didn't seem much hope of learning any significant details about the inner structures of objects that were millions of light-years distant.

In 1943, however, the American astronomer Carl K. Seyfert (1911–1960) detected an odd galaxy with a very bright spot at the center. Other galaxies of the sort have since been observed, and the entire group is known as *Seyfert galaxies.* Altogether perhaps 1 percent of all galaxies are Seyfert galaxies.

This was the first case of what came to be called *active galaxies,* those with centers that seem to be the site of activity beyond the normal. Much more remained to be discovered about such galaxies when it became possible to observe them outside the range of visible light.

AQUALUNGS

Until now, if you wanted to spend a reasonable period of time under water with sufficient mobility to allow exploration, you needed a diving-suit, which was heavy and required a lifeline.

In 1943, even while he was with the French underground fighting the occupying Germans, the French oceanographer Jacques-Yves Cousteau (b. 1910) invented the *Aqualung.*

This was a device that supplied the diver with air under pressure. It was self-contained and light, so that with an Aqualung and finned devices on their feet, divers could probe under the water's surface with freedom and mobility. This made it possible to observe coral reefs and ocean life in the uppermost layers of the ocean and also introduced a new sport, *scuba diving.* (*Scuba* is an acronym for "self-contained underwater breathing apparatus.")

IN ADDITION

The Soviets ended the long siege of Leningrad on January 3, 1943, and forced the surrender of the Germans at Stalingrad on February 2. On July 5 the Germans began their third summer offensive in the Soviet Union, limited this time to the salient near the city of Kursk. In the greatest tank battle in history, the Soviets stopped the Germans cold.

In North Africa, Roosevelt and Churchill met at Casablanca on January 17, 1943, and settled on a policy of requiring *unconditional surrender* from Germany and Japan. By May 12, Tunisia was taken and the Axis forces were driven out of Africa.

Allied bombing of Germany steadily intensified and the Germans could respond only against the Jews, which they did savagely. The Jews in the Warsaw ghetto rose in hopeless rebellion on April 18, 1943, and few survived.

On July 10, 1943, British-American forces invaded Sicily. They entered southern Italy on September 2 and advanced northward against stubborn German resistance.

In the Pacific, Allied forces began the slow task of advancing from island to island against fanatical but in the end always futile Japanese resistance.

On November 28, 1943, Roosevelt, Churchill, and Stalin met in Teheran, Iran, for the purpose of planning an invasion of France.

1944

DNA AS GENETIC MATERIAL

By now it had been understood for some forty years that chromosomes carried the genetic material. It was also known that chromosomes were nucleoprotein in character, containing both protein molecules and deoxyribonucleic acid (DNA) molecules.

It was assumed that the key portion of the genetic material in chromosomes was the protein, since at every point proteins seemed to be the key to living tissue. They were giant molecules of enormous variety and versatility; the enzymes, which controlled the chemistry of the body, were proteins. The nucleic acid was thought to consist of relatively small molecules that behaved as adjuncts, like the heme in hemoglobin or the coenzymes in enzymes—though what the function of the nucleic acid adjunct might be was not known.

Then biochemists began discovering that nucleic acids were not small molecules after all. Methods of isolating DNA had been so harsh that the molecules were recovered in small fragments. Gentler isolation retrieved larger molecules. It was also noticed that in sperm cells, where chromosomes were condensed to minimum size, the protein portion seemed unusually simple while the DNA seemed present in the usual quantity and complexity. Nevertheless, faith in proteins remained unshaken.

At this time, a Canadian-born American bacteriologist, Oswald Theodore Avery (1877–1955), was working with pneumococci—bacteria that caused pneumonia. Two different strains were grown in the laboratory—one with a smooth coat made up of a complicated carbohydrate molecule and one with a rough surface. The strains were called S (for smooth) and R (for rough).

Clearly the R strain lacked a gene for the formation of the carbohydrate surface. It was possible to prepare an extract from the S strain that contained no cells and was clearly nonliving, which on addition to the R strain would convert it into an S strain. The extract must contain the gene (or *transforming principle*) that catalyzed the production of the carbohydrate. But what was the chemical nature of the transforming principle?

In 1944 Avery and his associates purified the transforming principle as far as they could while still leaving it functional and showed that it was DNA in nature and nothing else. There was *no* protein present. This was the first indication that the genetic material in cells was DNA and *not* protein. The discovery revolutionized genetics, which now began a quick advance.

Avery's discovery was clearly worthy of a Nobel Prize, but he died too soon to receive one.

PAPER CHROMATOGRAPHY

Chromatography, which had been devised by Tsvett (see 1906), worked well in separating complex mixtures, but it was slow and required a considerable quantity of solution.

In 1944 two British biochemists, Archer John Porter Martin (b. 1910) and Richard Laurence Millington Synge (b. 1914), invented *paper chromatography*. Instead of dripping the mixture through a column of absorbent powder, this technique allowed it to creep up a column of absorbent *filter paper*. The paper could then be turned at right angles and dipped into another solvent, so that if the first did not separate the mixture entirely, the second, working on an already partially separated mixture, would complete the job.

Paper chromatography was faster than ordinary chromatography and worked on mere drops of solution. It was an excellent technique for separating the complex mixtures that were formed when a protein molecule was broken into fragments. For this work, Martin and Synge were awarded the Nobel Prize for chemistry in 1952.

TEFLON

The need to make use of uranium hexafluoride in the development of the nuclear bomb (see 1940) led to an intensive study of fluorine compounds and in particular of *fluorocarbons.* These were molecules in which all the free valences of a carbon chain or ring were occupied by fluorine atoms.

It was possible to develop long chains of carbon atoms with fluorine atoms attached, which would form polymers something like polyethylene (in which all the carbon valences were attached to hydrogen atoms). Fluorine atoms held much more tightly to the carbon chain than hydrogen atoms did, however. For that reason, long-chain *polytetrafluoroethylene* (or *Teflon* for short) resisted change. It didn't burn, dissolve, or stick to anything.

In 1944 Teflon was first produced for commercial use as a lining for frying pans. Since material would not stick to it, oil did not need to be used, and the frying pans were easily cleaned after use. Nor was the material in any way toxic.

SYNTHESIS OF QUININE

By now the techniques and methods of organic synthesis were so advanced that any organic molecule, no matter how complex, could be synthesized.

Perkin had attempted to synthesize quinine (see 1856) but was bound to fail, since its molecule was too complex for the methods then available.

In 1944, however, the American chemists Robert Burns Woodward (1917–1979) and William von Eggers Doering (b. 1917) began with simple compounds that could be easily synthesized from their constituent elements, and from these they synthesized quinine.

Woodward went on to synthesize other complicated molecules and was eventually awarded the Nobel Prize for chemistry in 1965.

2,4-D

The chemical *2,4-dichlorophenoxyacetic acid* (*2,4-D* for short) was introduced in 1944 and was the first effective chemical herbicide (Latin for "plant killer"). Naturally, killing plant life indiscriminately is not something one would want to do, but 2,4-D is selective in its effects. It does not seriously affect grasses (including the various grains so important to humanity) but prevents the growth of broad-leaved plants, usually regarded as *weeds.*

NEW NEBULAR HYPOTHESIS

For nearly two centuries, astronomers had been trying to work out some reasonable mechanism that would account for the formation of the Solar System. Laplace's *nebular hypothesis* (see 1796) had broken down over the fact that 98 percent of the angular momentum of the Solar System was concentrated in the planets, which made up only 0.1 percent of the total mass of the system.

Chamberlin had advanced the *planetesimal theory* (see 1905), which required a near collision that would draw out solar matter by gravitational pull and form the planets. Eddington (see 1919), however, showed the interior of stars to be so incredibly hot that matter pulled out of the Sun would simply disperse and not collect into planets.

In 1944 Weizsacker (see 1938) worked out a new version of the nebular hypothesis. He introduced the notion of turbulence in the outer layers of the condensing nebula and showed how, as a result of such turbulence, planets would form in their actual observed orbits, more or less.

Furthermore, the development of *magnetohydrodynamics* at just about this time, by the Swedish astronomer Hannes Olof Gösta Alfvén (b. 1908), showed how thin gases moved when immersed in magnetic fields and how they could carry energy and angular momentum outward. This solved the problem of the concentration of angular momentum in the planets.

The Weizsacker mechanism, with minor modifications, is now viewed as a likely explanation of the formation of the Solar System.

RADIO WAVES FROM HYDROGEN

In German-occupied Europe, scientific work could proceed only with extreme difficulty. The Dutch astronomer Hendrik Christoffel van de Hulst (b. 1918) was forced to do what work he could with theoretical notions that required no more in the way of instrumentation than pen and paper.

He considered the behavior of cold hydrogen atoms and worked out how the magnetic fields associated with the proton and the electron in the hydrogen atom were oriented to each other. They could line up in the same direction or in opposite directions. Every once in a while a particular atom could flip from one configuration to another, and in so doing, it would emit a radio wave 21 centimeters in length.

Any single hydrogen atom ought to do this only once in eleven million years or so on the average, but there were so many such atoms in space that a continuing drizzle of 21-centimeter radiation should be emitted, a drizzle that might be intense enough to be detectable.

Jansky's discovery had shown that radio waves were emitted by objects in the sky (see 1932), but beyond Reber's simple radio telescope (see 1937) there was nothing with which to observe such radio waves in any detail. Van de Hulst's calculations had to wait for corroboration.

AMERICIUM AND CURIUM

After Seaborg had helped McMillan isolate plutonium (see 1940), it was clear that other elements might exist beyond that element. Seaborg devoted himself to the task of preparing such elements by bombarding the massive atoms already known with subatomic particles.

In 1944, by bombarding plutonium with neutrons and alpha particles, Seaborg and his associates prepared *americium,* with an atomic number of 95, and *curium,* with an atomic number of 96. The former was named for America and the latter for the Curies (see 1897).

V-2

Since Goddard flew the first liquid-fueled rocket (see 1926), his work had been taken most seriously in Germany, where a group of 870 enthusiasts, including Wernher von Braun (see 1942), had begun experimenting with rockets in 1930.

Hitler favored such work for military use, and in 1936 von Braun headed a research project for the purpose of developing military rockets. In 1942 the first true missile, carrying its own fuel and oxygen, was shot off and reached a height of 60 miles.

On September 7, 1944, such missiles, called V-2s, with the *V* standing for *vergeltung* (vengeance), were first fired at London. (The V-1 had been a pilotless airplane wired with explosives, not nearly as damaging as the V-2.)

In all, 4,300 V-2s were fired; 1,230 hit London. Casualties included 2,511 dead and 5,869 seriously wounded. These missiles were armed with only chemical explosives, however, and came too late in the war to save Germany.

IN ADDITION

The fortunes of the Axis powers were now declining precipitously. The Soviet forces advanced steadily westward and by mid-1944 had virtually driven the Germans from Russian territory. Romania surrendered to the Soviets on August 24, 1944; Bulgaria on September 16. Belgrade fell to the Soviets on October 20.

In Italy British and American forces took Rome on June 4 and Florence on August 12.

Farther north, a full-scale invasion of France began successfully when British and American forces landed in Normandy on June 6, 1944 (D-day). By the end of August, almost all of France was cleared. Paris was taken on August 25 and Brussels on September 2.

One last German offensive opened in Belgium on December 16, 1944, but it was crushed before the end of the year.

In the Pacific the United States wiped out what remained of the Japanese navy on October 21 in the Philippine Sea.

In the United States Franklin Roosevelt was reelected to a fourth term.

1945

NUCLEAR FISSION BOMB

For a nuclear chain reaction to take place (see 1939), it is necessary to accumulate enough fissile material (uranium-235 or an appropriate plutonium isotope manufactured from uranium-238) to bring about an explosion. The amount of fissile material has to be large enough (the critical mass) for enough of the neutrons produced to strike other nuclei before wandering out into open air.

If two pieces of fissile material, each below the critical mass and therefore nonexploding, are fired into each other to form a single piece above the critical mass, any stray neutron (and there are always some around) will start the chain reaction and the material will explode in a fraction of a second.

By mid-1945 enough fissile material had been collected to carry through a test. On July 16, 1945, at a site 60 miles northwest of the town of Alamogordo, New Mexico, a nuclear fission bomb (popularly called an *atomic bomb*, or an *A-bomb*) made of plutonium was detonated before dawn. The scientists in charge had expected an explosive force equivalent to 5,000 tons of TNT. What they got was the equivalent of 20,000 tons of TNT.

At a stroke, the face of war had changed totally. It was even possible that the fate of humanity had been sealed.

SYNCHROCYCLOTRON

Since the invention of the cyclotron by Lawrence (see 1930), such devices had grown larger and larger and had produced particles with more and more energy. By the time the particles had reached energies of about 20,000,000 electron-volts (20 MeV), however, they had gained so much mass, in line with special relativity (see 1905), that they were curving less sharply and their circlings within the instrument fell out of phase with the periodic alternations of the magnetic field. They could gain no further energy.

In 1945 McMillan (see 1940) worked out a method for synchronizing the magnetic field in such a way that the alternations of the magnetic field slowed and kept time with the mass increase of the particles. With such a *synchrocyclotron,* particle energies far in excess of 20 MeV could be reached, and it became possible to look to the day when particle accelerators might produce particles with energies in the range of cosmic ray particles.

The synchrocyclotron was independently devised at about the same time by the Soviet physicist Vladimir Iosifovich Veksler (1907–1966).

PROMETHIUM

By now, four elements beyond uranium had been discovered, but a single empty spot still remained in the periodic table below uranium. This belonged to element number 61, one of the lanthanides (see 1940, Neptunium and Plutonium).

In 1945 a group under the direction of the American chemist Charles DuBois Coryell (b. 1912) discovered element number 61 among the fission products of uranium. Its most nearly stable isotope had a half-life of 17.7 years.

It was eventually named *promethium,* because just as the Greek god Prometheus had snatched fire from the nuclear furnace of the Sun, so promethium had been snatched from the nuclear furnace of the fissioning uranium atoms.

With the discovery of promethium, the periodic table had no further gaps, and all that remained was to discover further elements beyond the now-known curium (element number 96).

VIRAL MUTATIONS

It was well known that mutations occurred in both plants and animals, and they had been studied with care for over half a century.

In 1945 Luria (see 1942) and, independently, the American microbiologist Alfred Day Hershey (b. 1908) showed that bacteriophages underwent mutations as well. It is for this reason that it is so hard to develop immunity to viral diseases like flu and the common cold. Antibodies might be developed against a particular strain of the virus, but then a mutation will produce a new strain against which the old antibodies are not effective.

For their work, Luria and Hershey received shares of the Nobel Prize for physiology and medicine in 1969.

JET STREAMS

During World War II, both Japanese and American pilots in high-flying planes became aware of a high-altitude wind blowing from west to east. The Japanese, making the discovery as early as 1942, used the wind in an effort to blow balloon-bombs across the Pacific Ocean to the United States. The Americans became aware of the wind when flying to bomb Japan in 1944.

By 1945 these high winds had been recognized as a permanent feature of the atmosphere. There were several bands of such winds blowing at the troposphere-stratosphere boundary. These bands were hundreds of miles wide and several miles deep. Speeds of up to 300 miles per hour were measured.

These were called *jet streams* and were studied by the Swedish-born American meteorologist Carl-Gustaf Arvid Rossby (1898–1957). The course they follow is a strong determinant of weather on Earth.

ARTIFICIAL KIDNEYS

The era of modern artificial organs was initiated in 1945, when the Dutch-born American inventor Willem J. Kolff designed an artificial kidney. People suffering from kidney failure could then be kept alive by periodically filtering urea out of their bloodstream.

IN ADDITION

Germany was collapsing rapidly now. By February 20, 1945, the Soviets had advanced to within 30 miles of Berlin. At the end of February, American armies were entering Germany from the west, and on March 7 they crossed the Rhine River.

Roosevelt, Churchill, and Stalin met in Yalta, in the Crimea, from February 7 to 12, in order to plan the postwar world.

On April 11, American and Soviet forces met at the Elbe River, and on April 20 the Soviets battered their way into Berlin. Hitler committed suicide on April 30. In Italy, Mussolini was captured by anti-Fascist Italians and executed on April 18.

On May 8 *(Victory in Europe,* or *V-E, Day)*, Germany surrendered unconditionally and the war in Europe was over. Roosevelt did not live to see the day, however. He had died of a cerebral hemorrhage on April 12, 1945, and Vice President Harry S Truman (1884–1972) became the thirty-third president of the United States.

Meanwhile, in the Pacific, American forces took the island of Iwo Jima on March 17, 1945. On April 1, the Americans invaded Okinawa, one of Japan's home islands, and after nearly three months of bitter fighting, took that too.

On August 6, 1945, the United States dropped a nuclear fission bomb on the Japanese city of Hiroshima and on August 9, another bomb on Nagasaki. The Japanese surrendered formally on September 2, 1945 *(Victory in Japan,* or *V-J, Day)*.

World War II was over, six years and one day after it had begun. It had killed 55 million people and left ten million displaced and homeless. Hitler's Holocaust had killed one-third of the world's Jews.

Between July 17 and August 2, Truman, Churchill, and Stalin met at Potsdam, Germany, to confer on the fate of that defeated nation. In the midst of the conference, Churchill was replaced by a new prime minister, Clement Richard Attlee (1883–1967). Of the six leaders at the start of World War II (Roosevelt, Churchill, Stalin, Hitler, Mussolini, and Tojo), only one, Stalin, was still in power on V-J Day.

At a conference in San Francisco from April 25 to June 26, 1945, the *United Nations* came into being as a successor to the League of Nations.

1946

ENIAC

Bush had devised a computer that made use of radio tubes as electronic switches in addition to the usual mechanical parts (see 1930).

The obvious next step was to make an entirely electronic computer, with no moving mechanical parts at all. This was done by the American engineers John William Mauchly (1907–1980) and John Presper Eckart, Jr. (b. 1919), who devised the first practical electronic digital computer in 1946. It was called *ENIAC (electronic numerical integrator and computer)*.

It was an enormous, energy-guzzling device, weighing 30 tons and taking up 1,500 square feet of space. Though it was a wonder of its time, ENIAC was retired some nine years after it had been set up. It was by then hopelessly obsolete, for its descendants were growing steadily smaller, cheaper, more efficient, and much more capable.

MICROWAVE REFLECTION FROM THE MOON

Thanks to radar, it was possible to detect microwave reflections from airplanes, and from the reflections to calculate the distance and direction of the airplanes. In principle, there was no reason this could not be done for an astronomical body.

In 1946 a Hungarian scientist, Zoltan Lajos Bay, sent a beam of microwaves to the Moon and detected the reflection. The time it took for the reflection to return could be used to determine the distance of the Moon at the time of reflection with far greater precision than had ever been possible before.

NUCLEAR MAGNETIC RESONANCE

It is possible for chemical substances to absorb certain frequencies of microwave radiation when those substances are in a strong, steady, and homogeneous magnetic field. The frequencies that are absorbed depend on the magnetic properties of the atoms involved.

In particular, atomic nuclei act as tiny spinning magnets, and these can be lined up in a magnetic field, to strengthen the microwave absorption. The Swiss-born American physicist Felix Bloch (1905–1983) and the American physicist Edward Mills Purcell (b. 1912) independently developed this technique in 1946, and for it they shared the Nobel Prize for physics in 1952.

The technique, named *nuclear magnetic resonance* (NMR), provided a noninvasive way of studying the interior of living organisms. Since microwaves are far less energetic than X rays, they are correspondingly less likely to do damage to tissues. Then too, X rays are best at detecting massive atoms, of which there are few in the body, while NMR detects light atoms, particularly hydrogen, and these are plentiful in the body.

The use of NMR is growing more common, but because of public fears of the word *nuclear* (thinking that it implies the existence of destructive radiation—definitely not so in this case), physicians commonly speak of it simply as *magnetic resonance.*

NORADRENALINE

It was already known that acetylcholine was a chemical that served to transmit a nerve impulse from one neuron to a neighboring neuron (see 1921, Vagusstoffe).

In 1946 the Swedish physiologist Ulf Svante von Euler (1905–1983) showed that in that part of the nervous system called the *sympathetic nerves,* the transmitting chemical was *noradrenaline* (more properly, *norepinephrine*), which was very much like the hormone adrenaline in structure but was missing a carbon atom.

Von Euler received a share of the Nobel Prize for medicine and physiology in 1979 for this discovery.

BACTERIAL GENETICS

It had long been customary to think of reproduction in the simplest organisms as essentially simpler than that in multicellular ones. Organisms such as ourselves reproduce by means of sex and the intermingling of genetic material from male and female. A one-celled organism, however, seems able simply to divide, making use of only its own genetic material. It would seem that evolution should proceed much faster in multicellular than in unicellular organisms over a given number of generations because of this intermingling of genetic material. The shuffling of genes might represent the great biological advantage of sexual reproduction.

In 1946, however, the American geneticist

Joshua Lederberg (b. 1925), working with Tatum (see 1941), showed that bacterial reproduction was not entirely asexual. Different strains of bacteria could be crossed in such a way as to make the genetic material intermingle. It was therefore possible for organisms even as simple as bacteria to engage, on occasion, in what is essentially sexual reproduction.

VIRUS GENETICS

While Lederberg was showing the unexpected complexity of bacterial reproduction (see above), the German-born American microbiologist Max Delbrück (1906–1981) and Alfred Day Hershey (see 1945) were independently doing the same for those even simpler organisms, the viruses.

Both showed that the genetic material of different strains of viruses could be combined to form a new strain different from either. This too is a form of sexual reproduction, and for this Delbrück and Hershey received shares of the Nobel Prize for physiology and medicine in 1969.

CLOUD-SEEDING

The American physicist Vincent Joseph Schaefer (b. 1906) had been working with Langmuir (see 1913) on the phenomenon of icing, particularly in reference to airplanes flying at high altitudes, which could develop layers of ice on their wings that could cause them to crash.

To study the production of ice crystals, Schaefer and Langmuir used a refrigerated box kept well below the freezing point of water. They hoped that in the box, water vapor would condense around dust particles to form ice crystals. It was important, however, to find just the proper types of dust particles to use as *seeds* for ice formation, and the experiments therefore continued for some time.

In July 1946, during a heat wave, Schaefer dropped some solid carbon dioxide into the box to cool it more effectively. Promptly, ice crystals formed and a miniature snowstorm whirled inside the box. Solid carbon dioxide might therefore help in *cloud-seeding*.

On November 13, 1946, Schaefer was flown by airplane over a cloud layer in western Massachusetts and dumped 6 pounds of pellets of frozen carbon dioxide. A snowstorm started.

In mild weather, such an artificial snowstorm started at high altitudes would, of course, turn to rain as it fell. Nevertheless, rain-making never became important. In the first place, it worked only if the proper kinds of clouds were present—in other words, if it was likely to rain in any case. In the second place, rain that might be helpful to some people would invariably be harmful to others, so that to make rain artificially would be to ask for an infinite amount of litigation.

IN ADDITION

With the end of World War II, there were reprisals against the Axis leaders, who were viewed as having deliberately started the war. Some nations had taken care of their traitors already, so that Vidkun Quisling (1887–1945) had been executed in Norway and Pierre Laval in France.

In 1946 a trial at Nuremberg sentenced twelve leading Nazis to death, including Hermann Göring and Joachim von Ribbentrop (1893–1946), who was Hitler's foreign minister. Göring cheated the hangman by committing suicide.

Europe was settling down to new rivalries. The Soviet Union established its domination over eastern Europe; on March 5, Churchill spoke of an *iron curtain* descending across Europe, separating the Soviet-dominated east from the democratic west. This marked the beginning of what came to be called the *cold war* between West and East.

The United Nations met for its first session on January 10, 1946. The League of Nations voted itself out of existence on April 18.

Victor Emmanuel III of Italy abdicated on May 9, 1946, and was succeeded by his son, Humbert II (b. 1904). Italy abolished the monarchy a month later, however, and became a republic. Behind the Iron Curtain, monarchies in Romania, Bulgaria, and Yugoslavia also vanished.

In Argentina, Juan Domingo Perón (1895–1974) became president, heading an authoritarian government that suppressed civil liberties.

In China, the end of the Japanese occupation did not affect the civil war that continued between Mao Tse-tung's Communists and Chiang Kai-shek. In southeast Asia, the Indo-Chinese, particularly the Vietnamese on the eastern shore, began a long struggle to drive out the French colonialists.

The Philippine Islands attained independence from the United States peacefully on July 4, 1946.

1947

PION

Yukawa had predicted the existence of a particle with mass between that of an electron and a proton. It would serve as an exchange particle, holding together the protons and neutrons within the nucleus despite the electromagnetic repulsion among the protons (see 1935, Strong Interaction). Anderson detected a particle of intermediate mass, the muon (see 1937), but it lacked the necessary properties to fulfill the role Yukawa had laid out for it.

Meanwhile, the English physicist Cecil Frank Powell (1903–1969) had worked out a new method of particle detection. Instead of having particles strike a cloud chamber and then photographing the results, he had the particles strike a photographic emulsion so that the result could be recorded directly.

In 1947 he exposed photographic plates in the Bolivian Andes and picked up evidence of a particle of intermediate mass that was *not* the one Anderson had discovered. Powell named it the *pi-meson,* which was eventually shortened to *pion.*

Whereas Anderson's muon had all the properties of an electron except for its greater mass, so that it was a *lepton,* the pion shared certain properties with the more massive particles and was lumped with them as a *hadron.* The pion interacted readily with protons and neutrons (as the muon did not) and had all the properties required for it to be Yukawa's predicted particle.

For this work, Powell was awarded the Nobel Prize for physics in 1950.

CARBON-14 DATING

Seven years before, Martin David Kamen had discovered carbon-14 and found it to have a surprisingly long half-life, some 5,700 years. In 1947 the American chemist Willard Frank Libby (1908–1980) put the isotope to important use.

It had turned out that cosmic ray bombardment converted some of the nitrogen-14 of the atmosphere into carbon-14. New carbon-14 was formed as old carbon-14 broke down radioactively, so that an equilibrium was achieved. A given very small quantity of it always remained in the Earth's atmosphere.

Libby reasoned that plants absorbed carbon dioxide in the course of photosynthesis, so that carbon atoms from the gas found their way into the molecules of plant tissues. This must include a very small quantity of carbon-14, since it was always present in atmospheric carbon dioxide. Despite the fact that this carbon-14 was present in terribly small amounts, its concentration could be determined because the beta particles it liberated could be detected with great precision.

Once a plant died, however, no further carbon-14 would be absorbed, so that what carbon-14 it contained would slowly break down without being replaced. By determining the concentration of carbon-14 in the remains of a once-living plant, the amount of time that had elapsed since the organism had died could be determined with surprising accuracy.

This meant that the age of old samples of wood, parchment, textiles, and so on could be determined, even if they were as old as 45,000 years or so.

This opened the door to judging the age of Egyptian mummies, of wood from structures built in prehistoric times, and of such historic and possibly ancient objects as the Dead Sea scrolls and the Shroud of Turin.

For this, Libby received the Nobel Prize for chemistry in 1960.

CRAB NEBULA AS RADIO SOURCE

It had been sixteen years since Jansky first detected radio waves from outer space (see 1932), but only since World War II had techniques been developed that permitted handling of the microwave radiation used in radar. These techniques could now be applied to astronomy.

In 1947 the Australian astronomer John C. Bolton had a radio telescope that was capable of locating a radio source with sufficient precision to associate it with an object that could be identified optically. Thus he found that the third strongest radio source in the sky was clearly the Crab Nebula, which was the remains of a great supernova explosion (see 1054).

The Crab Nebula was the first optical object discovered to be a radio source, and this was the first indication that radio astronomy might offer a technique for making discoveries that were not apparent from the simple study of visible light itself.

MARTIAN ATMOSPHERE

Since Schiaparelli had first detected the "canals" of Mars (see 1877), a feeling had persisted among nonastronomers, and even among some astronomers, that there might be life on Mars.

In 1947, however, the Dutch-born American astronomer Gerard Peter Kuiper (1905–1973) analyzed the reflected infrared light of Mars and reported that the Martian atmosphere seemed to be almost entirely carbon dioxide. There was no sign of nitrogen, oxygen, or water vapor. With that, the notion of an advanced civilization on Mars, or of any but possibly the simplest forms of life, went glimmering.

COENZYME A

The carbohydrates, fats, and proteins of food, in the course of their metabolism, all break down to an *acetyl group,* also called a *two-carbon fragment.* This is then built up again to form the substances characteristic of an organism's own tissues.

In 1947 Lipmann (see 1941) was able to isolate a substance that was essential to the transfer of the acetyl group from one compound to another. He called it *coenzyme A,* where the *A* stood for *acetyl group.* The structure of coenzyme A was found to include that of pantothenic acid, one of the B vitamins. Pantothenic acid is essential to life and must be present in our food, because the human body cannot form it within its own tissues, cannot form coenzyme A without it, and cannot run metabolic changes without coenzyme A.

CHLORAMPHENICOL

During World War II, penicillin (see 1939) and streptomycin (see 1940) had been isolated and the age of antibiotics had begun. In 1947 *chloramphenicol* was isolated from the same group of molds that had yielded streptomycin.

Chloramphenicol attacked many different microorganisms and was the first *broad-spectrum antibiotic* to be isolated. It was fairly toxic, however, and had to be used carefully.

HOLOGRAPHY

Photography had been in existence for over a century (see 1839). It worked by allowing a beam of light to be reflected from an object and fall upon a photographic film. The film recorded a two-dimensional pattern of the reflected beam, but the information was limited by the loss of the third dimension.

Suppose, instead, that a beam of light were split in two. One part would strike an object and be reflected with all the irregularities the object would impose on it. The second part would be reflected from a mirror, with no irregularities imposed on it. The two parts would meet at the photographic film, where the interference pattern would be recorded. The film, when developed, would then seem to be blank, but if light were made to pass through the film, it would take on the interference characteristics and produce a three-dimensional image, conveying far more information than an ordinary photograph. The three-dimensional image would be called a *holograph* (where the prefix means "the whole thing"—that is, all three dimensions).

This notion was first worked out in 1947 by a Hungarian-born British physicist, Dennis Gabor (1900–1979), but though the theory was sound, it could not be reduced to practice without further advances in optics. Once such advances took place, Gabor was awarded the Nobel Prize for physics, in 1971.

LAND CAMERA

In 1947 Land (see 1932) produced the *Land camera,* which produced not negatives but positive prints, completely developed soon after the photograph was taken.

The camera had a double roll of film consisting of ordinary negative film and a positive paper with sealed containers of chemicals between. The chemicals were released at the proper moment and developed the positive print automatically.

SUPERSONIC FLIGHT

Since the invention of the airplane (see 1903), planes had been going faster and faster, but propellers can only whirl so fast and it seemed that the speed limit for planes was bound to be less than the speed of sound, which is about 740 miles per hour. Once jet planes were devised during World War II (see 1941), however, the possibility of planes moving faster than sound *(supersonic flight)* arose.

The difficulty is that, at their fastest, air molecules move at the speed of sound. As long as a plane moves at a lesser speed, the air molecules ahead will move easily out of the way. At the speed of sound or beyond, the air molecules cannot move out of the way before being overtaken. They pile up ahead of the plane, which thus flies into compressed air.

This was viewed, dramatically, as flying

into a wall, and people began to speak of a *sound barrier*. Actually, there is no such thing. Bullets and shells can move faster than sound, as can the tip of a bullwhip. As air piles up, it eventually slips to one side and expands again, making a noise called a *sonic boom*. The crack of a bullwhip is a miniature sonic boom.

The compressing air does set up vibrations and stresses in a plane, however, and in order to attain and exceed the speed of sound, a plane must be carefully engineered to withstand those stresses.

On October 14, 1947, the American test pilot Charles Elwood Yeager, flying an X-1 rocket plane, was the first to fly faster than the speed of sound on level flight. He had *broken the sound barrier*.

TELEVISION IN THE HOME

Television had been a laboratory reality for twenty years, but methods for producing pictures on a screen were far too crude and far too expensive for everyday use by the general public.

By 1947, television sets could be produced that one could reasonably watch at home. The screens were small and fuzzy at first, and of course there were very few programs available for viewing. Improvements and elaborations came by leaps and bounds, however, and in a few years television was altering the concept of home entertainment, of advertisement, of show business, and even of politics.

IN ADDITION

On March 12, 1947, President Truman announced what came to be called the *Truman doctrine*, in which the United States vowed to aid countries threatened by Communism. On a less belligerent note, the United States offered the Marshall Plan, named after Secretary of State George Catlett Marshall (1880–1959). It extended economic aid to war-ravaged nations that needed it and proved enormously successful in helping to heal the wounds of western Europe.

The British Empire began a rapid breakup. India gained its independence from Great Britain, after two centuries of subservience, on August 15, 1947, whereupon Indian Hindus and Moslems began a bloody civil war.

The first report of what came to be known as *flying saucers*, or *unidentified flying objects* (*UFOs*), came on June 24, 1947. Since then, there have been innumerable sightings and circumstantial tales—and not one shred of real evidence for their existence.

1948

TRANSISTOR

When radio was in its infancy, crystals were used as rectifiers, allowing alternating current to pass in only one direction. These were unreliable, however, and such *crystal sets* were quickly replaced by sets containing radio tubes (see 1904), which had now been in use in various electronic instruments, including the new computers, for over forty years.

Radio tubes have to be large enough to enclose a vacuum, however. They are fragile, spring leaks, and frequently have to be replaced. They are also energy-hungry, and one must wait for the filament to heat up before the machine starts working.

In 1948 three physicists, William Bradford Shockley (b. 1910), Walter Houser Brattain (1902–1987), and John Bardeen (b. 1908)—all American, though Shockley was British-born—discovered a new kind of crystal. It consisted chiefly of germanium, which conducted an electric current less well than metals but better than such insulators as glass and rubber. Germanium—and silicon, which in a few years replaced germanium, since silicon was both cheaper and better—were considered examples of *semiconductors.*

If traces of certain impurities were added to the semiconductor, the crystal could serve as a rectifier or as an amplifier. It could, in short, perform any function that tubes could.

These semiconductors were solid (hence *solid-state devices*) and required no vacuum, so that they could be made quite small. They were rugged, so that they almost never needed replacing. They used very little energy, and they required no heating, so that they began working at once. A fellow worker, the American electrical engineer John Robinson Pierce (b. 1910), suggested that the device be called a *transistor,* because it transmitted current across a resistor.

With time, transistors replaced tubes completely. Transistors and the improvements that followed may well prove to be the most significant technological advance of the twentieth century.

LONG-PLAYING RECORDS

Until now, records had revolved at the rate of 78 times a minute. In 1948 the Hungarian-born American physicist Peter Carl Goldmark (1906–1977) developed a record that turned only 33⅓ times a minute. Between that and narrowing the track, or needle groove, six times more music could be placed on a single record than had been possible earlier. An entire symphony could now fit on one record.

CYBERNETICS

The American mathematician Norbert Weiner (1894–1964) spent World War II working on antiaircraft defense. To shoot down an attacking airplane, one must know the speed and direction of the plane's movements, the speed and direction of the wind, the speed of the projectile aimed at the airplane, and other factors as well. To do all this accurately and well, computers were needed that were far better than those available at the time.

Weiner grew interested in working out the mathematical basis of the communication of information and of the control of a system in the light of such communication. By 1948 he had summarized his work in a book entitled *Cybernetics.* This was the first important book devoted to computer control.

NUCLEAR STRUCTURE

The chemical properties of atoms depend on the arrangement of their electrons into *shells.* This is what makes the periodic table of the elements work (see 1916).

The nuclear properties of atoms must depend on the arrangement of the protons and neutrons in the nuclei. They, too, might be arranged in shells, but if so, this would be more difficult to determine than in the case of the outer electrons.

The German-born American physicist Maria Goeppert-Mayer (1906–1972) tried to work out the nature of the nuclear shells from the nuclear properties that had been observed for different atoms. She showed that nuclei that contained 2, 8, 20, 50, 82, or 126 protons or neutrons would be more stable than their neighbors in the periodic table. These were called *shell numbers,* or more dramatically but less scientifically, *magic numbers*. If 28 or 40 protons or neutrons were present, some stability resulted, and these were called *semimagic numbers*.

Goeppert-Mayer advanced her notions in 1948, at about the same time a German physicist, Johannes Hans Daniel Jensen (1907–1973), advanced the same views. As a result, Goeppert-Mayer and Jensen received shares of the Nobel Prize for physics in 1963.

QUANTUM ELECTRODYNAMICS

Making use of quantum theory, the American physicist Richard Phillips Feynman (1918–1988), in 1948, worked out equations governing the behavior of electrons and electromagnetic interactions generally; equations that allowed predictions of such phenomena to be made with far greater precision than had been possible until then.

This theory was called *quantum electrodynamics*, and it proved so successful that it has been used as a model for the attempted working out of equations governing the behavior of particles subjected to the weak and strong interactions.

For this work, Feynman received a share of the Nobel Prize for physics in 1965.

THE BIG BANG

LeMaitre had advanced the idea that the Universe originated as a condensed *cosmic egg*, which exploded and set its expansion into motion (see 1927).

In 1948 Gamow (see 1929) considered the consequences of this explosion (which he called the *big bang*) in far greater detail. He was the first to attempt to work out how the chemical elements were formed in the aftermath of the explosion.

He also predicted that the big bang had resulted in a vast surge of energy, and that the Universe had cooled off as it expanded until it was now, on the average, just a few degrees above absolute zero. As a result, there should be a background of microwave radiation coming equally from all parts of the sky, with a wavelength suitable to a universe at such a temperature.

MIRANDA

For almost a century, Uranus had been known to have four satellites. In 1948, however, Kuiper, who had described the atmosphere of Mars (see 1947), located a fifth Uranian satellite. It was smaller than the other four and closer to Uranus. Since three of the satellites had been named for spirits from Shakespeare's plays—Oberon and Titania from *A Midsummer Night's Dream* and Ariel from *The Tempest*—Kuiper named the new satellite *Miranda*, after the heroine of *The Tempest*.

Miranda was the first satellite to be named after a human being rather than a god, a goddess, or a spirit, although it was a fictional human being, to be sure.

NUCLEIC ACID-BASE BALANCE

It was now known that nucleic acids were large and very complex molecules, especially since Avery had shown that deoxyribonucleic acid (DNA) and not protein was the carrier of physical characteristics (see 1944). It was DNA, in short, that made up the genes of the chromosomes.

The question was: Just what was it about the structure of nucleic acids that made it possible for them to carry the vast amount of information that genes must carry in order to make human eggs develop into human beings and grasshopper eggs into grasshoppers, without the reverse ever happening?

Part of the structure of DNA was known to be four different *bases*. Two of them (adenine and guanine) were purines with a two-ring molecule, and two of them (cytosine and thymine) were pyrimidines with a one-ring molecule.

The Austrian-born American biochemist Erwin Chargaff (b. 1905) broke down nucleic acid molecules to their constituent bases and separated them by paper chromatography. He determined the quantity of each present and by 1948 was able to demonstrate that, in nucleic acids generally, the number of guanine units was equal to the number of cytosine units, and the number of adenine units was equal to the number of thymine units. This meant that the number of purine units was equal to the number of pyrimidine units.

This was a more important discovery than Chargaff apparently realized at the time, and he did not follow it up properly.

CYANOCOBALAMINE

Minot and Murphy had demonstrated that there was a dietary factor in liver that reversed the onslaught of pernicious anemia (see 1926). The nature of the factor was not easily determined, however, since its presence or absence could only be detected by following the slow changes in pernicious anemia patients.

In 1948, the American chemist Karl August Folkers (b. 1906) discovered that this factor, usually called vitamin B-12, was necessary for the growth of certain bacteria. The presence or absence of the vitamin in various fractions of liver could now easily be determined from the reaction of bacteria, and very soon, red crystals of the pure vitamin were isolated.

Eventually, analysis showed the vitamin to have perhaps the most complicated molecule known that was not simply a long chain of repeated single units. It was required by the body in exceedingly small doses, a thousandth that of other vitamins. The molecule possessed a cyanide group and a cobalt atom, neither of which was present in other known substances in living tissue. For that reason it was named *cyanocobalamine.*

Its existence made pernicious anemia easily treated without the necessity for eating excessive amounts of liver.

CORTISONE AND ARTHRITIS

The American physician Philip Showalter Hench (1896–1965) was interested in rheumatoid arthritis, a painful and crippling disease. Pregnancy and attacks of jaundice relieved its symptoms, so he conjectured that it was not a germ disease but a disorder of metabolism.

Hench tried various substances, including hormones, in the search for something that would relieve the symptoms. A decade earlier, the adrenocortical hormones had been isolated by Kendall (see 1935, Cortisone), and it occurred to Hench that these ought to be tried. His role as a colonel in the Army Medical Corps during World War II delayed him, but after the war he began working with the new hormones. He tried Compound E, also called *cortisone,* which had been isolated in 1946, and in 1948 found that it worked well. For this he received a share, along with Kendall, of the Nobel Prize for medicine and physiology in 1950.

Cortisone proved to be a tricky substance, however, to be used only with great care and judgment.

TETRACYCLINES

In 1948 a new antibiotic, chlortetracycline, discovered four years earlier by the American botanist Benjamin Minge Duggar (1872–1956), was placed on the market as *Aureomycin.* Its molecule was made up of four rings of atoms, and it was the first of a family of such antibiotics with the general name of *tetracyclines.* They were effective over a wide range of microorganisms and had low toxicity. They are now the most useful and least dangerous of the antibiotics.

TISSUE TRANSPLANTATION

It is important for the body to have weapons against foreign proteins, since these may well represent portions of parasites or their toxins that will cause sickness or death. The body needs an immune mechanism that can produce antibodies that will combine with the foreign proteins and render them harmless.

At times, however, foreign proteins help preserve life. A piece of tissue from one person may be transferred to another because the person receiving the tissue badly needs it. If the person receiving it then activates an immune mechanism to fight it off and reject it, it may mean death.

The American geneticist George Snell (b. 1903) studied this phenomenon. He believed the capacity of an organism to reject tissue from another organism involved genetic factors. Working with mice, by 1948 he had located the sites of specific genes *(histocompatibility genes)* that were concerned in the matter of acceptance or rejection. This was a step toward making tissue transplantation practical.

For this he received a share of the Nobel Prize for medicine and physiology in 1980.

VIRUS CULTURE

A great many of the advances in the fight against bacterial infection over the previous three-quarters of a century had resulted from the ability to grow pure bacterial cultures in the laboratory. This meant the bacteria could be studied easily and methods for slowing or stopping their growth could be developed.

Viruses, however, grow only within living cells, and working with organisms is much slower and less certain than working with Petri dishes. For this reason, viral diseases were far more difficult to fight than bacterial diseases.

Of course one needn't grow the viruses in adult organisms. They could be grown in the developing embryos in chicken eggs, or in those same embryo cells mixed with blood. The trouble was that although viruses would grow there, so would bacteria, and the bacteria would mask the viruses.

Once penicillin became available, however, it could be added to the chicken embryo broth. This prevented the growth of bacteria but did not affect the viruses. The American microbiologist John Franklin Enders (1897–1985) developed this technique in 1948, and it became useful in searching for ways to fight viral diseases, notably poliomyelitis *(infantile paralysis)*.

For this work, Enders, along with his colleagues Thomas Huckle Weller (b. 1915) and Frederick Chapman Robbins (b. 1916), shared the Nobel Prize for medicine and physiology in 1954.

STARCH CHROMATOGRAPHY

Chromatography of one sort or another had been in use since Tsvett's discovery of the technique (see 1906). In 1948 the American biochemists Stanford Moore (1913–1982) and William Howard Stein (1911–1980) refined the method by using starch as an absorptive medium. In this way they improved the manner in which amino acids and peptides could be separated, and the two shared the Nobel Prize for chemistry in 1972.

BATHYSCAPHE

Beebe explored the deeper layers of the ocean with his bathysphere (see 1934). The bathysphere, however, was a purely passive device suspended from a ship by a lifeline. It could not move or maneuver independently.

What was needed was a vessel that could move about in the ocean at any depth. Such a device was invented by the Swiss physicist Auguste Piccard, who had already explored the stratosphere by balloon (see 1931).

He called the new device a *bathyscaphe* (ship of the deep). It used a heavy ballast of iron pellets to take it down and a "balloon" containing gasoline to give it buoyancy. It jettisoned the iron pellets to rise.

In 1948 the bathyscaphe descended, with a man on board for the first time, to a depth of 4,500 feet. Over the next fifteen years, bathyscaphes penetrated the depths of the ocean, finding life at even the lowest levels.

IN ADDITION

In the United States, President Truman won reelection. Anti-Communist furor continued, with Alger Hiss (b. 1904) accused of being a Communist spy by the unsavory ex-Communist Whittaker Chambers (1901–1961).

The British mandate in Palestine came to an end, and a Jewish nation, Israel, came into existence on May 14, 1948, for the first time in nearly nineteen centuries.

Yugoslavia, under Josip Broz, universally known as *Tito* (1892–1980), broke away from Soviet influence and established a Communist regime hostile to the Soviet Union. The Soviet Union sent its army into Czechoslovakia on February 25, 1948, to make sure it remained firmly pro-Soviet.

West Berlin, which was inside the borders of East Germany but under Western rule, was blockaded on July 24, 1948, by the Soviets, who controlled East Berlin and East Germany. The United States began an airlift to supply West Berlin.

In India Gandhi was assassinated on January 30, 1948, by a Hindu extremist.

Korea became independent, but as two separate nations, a pro-Soviet North Korea and a pro-American South Korea under Synghman Rhee (1875–1965).

The Chinese civil war continued, with the Communists gaining the upper hand.

The first crossing of the Atlantic Ocean by jet planes took place in 1948, and air-conditioners were installed in automobiles for the first time.

Books in 1948 included *Sexual Behavior in the Human Male* by Alfred Charles Kinsey (1894–1956), a monumentally dull book of statistics that became a best-seller because the statistics dealt with sex.

1949

ICARUS

Witt had discovered Eros, an asteroid whose orbit carried it well within the orbit of Mars (see 1898). Since then, a number of asteroids had been discovered that could approach Earth more closely than any planet did. These were called *Earth-grazers*. Some had been discovered that approached the Sun more closely than even Venus did. These were called *Apollo-objects*, after the first asteroid of this type, *Apollo*, which had been discovered in 1937.

In 1949 Baade, who had discovered the asteroid Hidalgo (see 1920), discovered an asteroid that was an Earth-grazer, since it could come as close as 4,000,000 miles from Earth, and an Apollo-object too. Indeed, it approached the Sun more closely than Mercury did, skimming by at a distance of only 17,700,000 miles every 1.12 years. Baade named it Icarus, after the character in Greek mythology who lost his life when he flew on wax wings too close to the Sun.

NEREID

Kuiper, who had discovered Miranda (see 1948), reported another far distant satellite in 1949. For a century only one satellite had been known for Neptune, a large one named Triton. Now Kuiper detected a much smaller one, which he called *Nereid*. It circled Neptune in the most eccentric orbit of any known satellite.

ATOMIC CLOCK

Ever since Huygens had invented the pendulum clock (see 1654), scientists had depended on accurate time measurements in conducting their experiments and searched always for more and more accurate ways of measuring time.

Eventually the search for natural cyclic movements that were both precise and constant worked down to the molecular level. An ammonia molecule, for instance, vibrated back and forth, taking up its two possible tetrahedral positions alternately, about 24,000,000,000 times per second. At constant temperature, this vibration remained very constant.

In 1949 the American physicist Harold Lyons (b. 1913) was the first to harness this molecular vibration to time-keeping purposes. It was the first *atomic clock*. As atomic clocks of ever-greater precision were devised, physicists could eventually time events to a millionth of a trillionth of a second.

BERKELIUM AND CALIFORNIUM

For five years after being synthesized by Seaborg and his group (see 1944), curium (atomic number 96) had remained the most complex atom known.

In 1949 the more complex elements 97 and 98 were produced. Since this was done at the University of California in Berkeley, California, element 97 was named *berkelium* and 98 was named *californium*.

SOVIET FISSION BOMB

For four years the United States had had a monopoly on the nuclear fission, or atomic, bomb. There was a comfortable feeling among many Americans that the Soviets weren't smart enough to arrive at the necessary technology by themselves.

The Soviets had been working on fission bombs for years, however, and were clever enough to gain knowledge of some of the American techniques. On September 22, 1949, the Soviets exploded their first fission bomb, and Americans were jarred out of their complacency. In place of a world dominated by a benevolent America with a monopoly on a benevolent atomic bomb, there were now two superpowers that soon became fully capable of wiping each other out, along with the rest of the world.

The nations of the world descended into a nuclear nightmare, from which they have not yet awakened.

SICKLE-CELL ANEMIA

Sickle-cell anemia is a disease that produces distorted red cells, which are unable to transport oxygen properly. It particularly afflicts blacks, and those afflicted tend to die in childhood. It was first noted in 1910 by an

American physician, James Bryan Herrick (1861–1954).

It seemed clearly to be a genetic disease, and in 1949 Linus Pauling (see 1931) was able to show that a defective gene produced the abnormal hemoglobin. If this gene was present on only one of a chromosome pair, the carriers possessed *sickle-cell trait.* These carriers could lead reasonably normal lives and also be relatively resistant to malaria, since their blood was apparently less palatable to the malarial parasite.

With a defective gene on each of a chromosome pair, however, the carriers had sickle-cell anemia itself. The beneficial effect of a single dose kept the defective gene going, while the deleterious effect of a double dose tended to wipe it out. The total effect was to keep an equilibrium quantity in existence in some regions where malaria was particularly prevalent.

This was a case of a *molecular disease,* since an abnormal molecule, genetically produced, was the cause. Later a number of different abnormal hemoglobin molecules were found. *Hemoglobin A* was the normal molecule, and *hemoglobin S* the abnormal molecule that led to sickle-cell anemia.

Analysis of the amino acid content of these two varieties of hemoglobin eventually showed that they differed in a single amino acid out of a chain made up of three hundred or so. It was a remarkable display of the serious consequences of a tiny abnormality.

EMBRYONIC IMMUNOLOGICAL TOLERANCE

Snell had shown the genetic basis of the intolerance for foreign proteins that made tissue transplantation difficult (see 1948).

It occurred to an English anatomist, Peter Brian Medawar (1915–1987), that embryos might not yet have developed an immunological system capable of rejecting foreign proteins. And in fact, when he inoculated mice embryos with tissue cells from another strain, he found that rejection did not take place. Furthermore, when such embryos entered independent life and could form antibodies, they no longer treated cells from the other strain as foreign proteins.

By 1949 Medawar had shown how this technique might lead to reducing the difficulties of tissue transplantation. For this he received a share of the Nobel Prize for medicine and physiology in 1960.

ESSENTIAL AMINO ACIDS

Since the first amino acid had been discovered (see 1806), some twenty amino acids that occur commonly in protein molecules had been identified. The last of these was *threonine,* isolated in 1935 by the American biochemist William Cumming Rose (1887–1985).

Rose's dietary experiments showed that some amino acids could be manufactured in the human body from other amino acids, but some could not and had to be present as such in the diet. All the amino acids are essential to the body structure, but eight of them are essential *in the diet,* where human beings are concerned, and it is these eight that are considered the *essential amino acids.*

This was definitely established by Rose in 1949.

COMETARY STRUCTURE

The approach of a comet to the Sun results in the development of a hazy *coma* and a tail. The American astronomer Fred Lawrence Whipple (b. 1906) suggested in 1949 that this could be explained by supposing that comets were essentially icy in nature, made up of an admixture of silicate dust and gravel (and in some cases perhaps, a small rocky core). When heated by the Sun's closeness, cometary ice vaporized explosively, and the dust it contained formed the haze and the tail. In short, a comet is a "dirty snowball."

This suggestion was readily adopted by most astronomers, and few, if any, doubt it now.

IN ADDITION

The Chinese civil war ended fifteen years after Mao Tse-tung had led his people on the Long March, fleeing from the apparently victorious Chiang Kai-shek. Mao had won, and it was Chiang who now fled to the island of Taiwan.

In Europe the Soviets ended the Berlin blockade on May 12, 1949, in clear defeat. Two Germanies were established: the German Federal Republic (West Germany) on May 23, 1949, with its capital at Bonn; the German Democratic Republic (East Germany) on October 7, 1949, with its capital in East Berlin.

Transjordan broke away from Great Britain and eventually adopted the name of Jordan. The East Indies broke away from the Netherlands on December 27, 1949, and became the independent nation of Indonesia.

The United States and the nations of western Europe formed the *North Atlantic Treaty Organization* (NATO), pledging cooperation against the Soviet Union and its allies.

South Africa initiated a policy of *apartheid* (segregation of whites from other groups and the continued brutal subjection of nonwhites).

1950

COMETARY CLOUD

It had been clear for at least a century that comets were comparatively short-lived phenomena. At every approach to the Sun, a substantial quantity of a comet's mass must evaporate, never to return (assuming it to be a "dirty snowball"—see 1949). Even a large comet could only endure a few thousand returns to the neighborhood of the Sun, and some comets had actually been seen to fragment on their passage past the Sun.

Why then were not all comets long since gone? It seemed to the Dutch astronomer Jan Hendrik Oort (b. 1900) that a vast reservoir of them must exist. In 1950 he suggested that a huge spherical cloud of icy comets, perhaps a hundred billion of them altogether, must exist within an area 1 to 2 light-years from the Sun. They might represent the outermost material from the original nebula, left behind by the condensation of the inner material into the Sun and the planets 4.6 billion years ago and unchanged ever since.

Every once in a while, collisions or the gravitational pull of one of the nearer stars might alter rotational velocities and allow a comet to fall into the inner Solar System, where observers on Earth could see it. Oort estimated that perhaps 20 percent of the original cloud had thus been hurled inward (or outward into interstellar space), but an ample number remained to supply more comets for the indefinite future.

No direct evidence of this cometary cloud (sometimes called the *Oort cloud*) exists, but the notion is accepted by most astronomers.

PLUTO'S DIAMETER

Pluto had been discovered by considering the unexplained perturbation of Uranus's orbit, assuming that the gravitational pull of an undiscovered planet caused it, and calculating the place where that planet must be (see 1930). (Neptune had also been discovered in this way—see 1846—but its presence had not corrected all of the perturbation of Uranus, just most of it.)

For Pluto to have caused the perturbation, it would have to be several times as massive as Earth, and at first this was assumed to be the case, but it turned out to be considerably dimmer than would be expected of a massive planet, and that caused considerable consternation.

In 1950 Kuiper, the discoverer of Miranda (see 1948) and Nereid (see 1949), managed to see Pluto as a disk and to measure its apparent width. He showed that its diameter could not be more than 3,600 miles, which made it smaller than Mars and explained its dimness. If Pluto was as small as this, it could not be the source of the perturbation of Uranus, and it must have been located in the right spot by sheer coincidence.

This would seem to indicate that a planet massive enough to account for the perturbations of Uranus must still exist beyond Neptune's orbit, but if so, it has not yet been located.

TURING MACHINE

The excitement over computers after World War II roused the interest of an English mathematician, Alan Mathison Turing (1912–1954). (Turing had been instrumental in breaking the German secret code-machine during World War II, which had made it possible to anticipate German moves and thus been of major importance in helping to defeat the Nazis.)

In 1950 Turing showed that, in principle, a very simple machine capable of an astonishingly few moves could solve any problem capable of being formulated in mathematical terms. This *Turing machine* went a long way toward convincing workers in the field that something one might call *artificial intelligence* could exist.

Turing also pointed out how one could decide whether artificial intelligence had been achieved: if one could carry on a conversation with some hidden entity without being able to demonstrate from the conversation alone whether that entity was a human being or not, and if that entity was a machine, it displayed artificial intelligence.

GAME-PLAYING COMPUTERS

It was natural to think of computers at first simply as very fast calculating machines, different from Pascal's first adding device (see 1642) in degree but not in kind.

However, it quickly became apparent that computers could solve problems that human beings considered to require human thought. Thus in 1947 the American engineer Arthur L. Samuel (b. 1901) had worked out a checker-playing computer, refinements of which eventually proved capable of playing championship checkers.

Checkers, however, is a comparatively simple game. In 1950 the American mathematician Claude Elwood Shannon (b. 1916) suggested ways to design a chess-playing computer, and such machines have since been built and play excellent games of chess. Indeed, it is not beyond the limits of probability that such a machine might become the world's chess champion.

Such devices further exemplify the potential reality of artificial intelligence.

ENDOPLASMIC RETICULUM

The invention of the electron microscope (see 1932) made it possible to study the cell in far greater detail than ever before. An early expert in cellular electron microscopy was the Belgian cytologist Albert Claude (1898–1983).

In 1950 he discovered the *endoplasmic reticulum,* which serves as the structural background of the cell—a thin, fibrous skeleton, so to speak, that holds the various organelles in place. For this discovery, he received a share of the Nobel Prize for physiology and medicine in 1974.

CARBON-14 AS TRACER

Carbon-14 had first found a useful scientific purpose when applied as a dating technique by Libby (see 1947). By 1950 it had been obtained in sufficient quantity to be useful as a radioactive tracer.

In that year, the German-born American biochemist Konrad Emil Bloch (b. 1912) used both stable carbon-13 and radioactive carbon-14 as tracers and was able to show in detail the changes that occurred in building up the cholesterol molecule in the body from the acetyl group (see 1947).

As a result, Bloch received a share of the Nobel Prize for medicine and physiology in 1964.

IN ADDITION

In an attempt to unify the country, North Korea invaded South Korea on June 25, 1950, beginning the *Korean War.* The North Koreans made rapid progress, but United Nations forces (chiefly American) under General Douglas MacArthur (1880–1964) forced the North Koreans to retreat rapidly. By October 9, South Korea was free of the invaders and MacArthur drove northward into North Korea, approaching the Chinese border, despite Chinese warnings that they would intervene. When they did enter the war on November 26, MacArthur was caught completely by surprise and by the end of the year had been driven out of North Korea.

The Korean War exacerbated anti-Communist hysteria in the United States. Hiss was convicted of perjury, and a hitherto unknown senator from Wisconsin, Joseph Raymond McCarthy (1908–1957), began a four-year reign of terror, broadcasting unfounded accusations of Communism far and wide. This came to be called *McCarthyism.*

In Asia, Chinese forces invaded Tibet and supplied help to the Indochinese, who were fighting the French.

Among the books of 1950 was *Worlds in Collision* by Immanuel Velikovsky (1895–1979), which advanced astronomical and historical theories so nonsensical that it instantly became a cult classic.

The population of the world reached 2.5 billion, and the population of the United States passed the 150 million mark. London was still the largest city in the world with 8.1 million and New York was close behind.

1951

BREEDER REACTORS

The nuclear reactors built during the first decade of the atomic age depended on the fission of uranium-235. This is a relatively rare isotope of uranium, making up only 0.7 percent of the whole.

Then it was discovered that if such reactors were built with ordinary uranium-238 in a jacket around the core, enough neutrons could be made to stream into the uranium-238 to form plutonium atoms, which are fissionable. If ordinary thorium-232 is in the surrounding jacket, fissionable uranium-233 will be formed.

Such *breeder reactors* will actually breed more fissionable fuel in the jackets than is consumed in the core, and this means that all the uranium and thorium supply of the world can serve as potential fission fuel, rather than the rare uranium-235 alone. The vision of endless cheap energy reached its peak at this time, with few to predict that concerns about reactor safety and radioactive ash would destroy the American nuclear energy industry.

STELLARATOR

Nuclear fission was clearly not the ultimate nuclear energy source. It was well known that nuclear fusion, as in the conversion of hydrogen to helium taking place in the stars, would yield up to seven times as much energy, weight for weight, as nuclear fission.

Moreover, the ultimate fusion fuel, hydrogen, is available in virtually unlimited quantities on Earth and is easily obtained, as compared with the rarer and far more difficult to extract uranium and thorium needed as fission fuel. Still further, fusion produces less dangerous products than fission does.

However, whereas nuclear fission can be initiated at room temperature, through the use of slow neutrons, nuclear fusion requires the high temperatures and pressures found at the center of stars.

Uncontrolled fusion is comparatively simple. Some way can be found to include hydrogen and other light elements with an ordinary fission bomb in such a way that the fission explosion produces sufficient temperature and pressure to ignite the much more powerful and destructive fusion explosion. The result would be a *nuclear fusion bomb,* more popularly known as a *hydrogen bomb,* or an *H-bomb.* It can also be called a *thermonuclear device,* where *thermo-* means "heat," because that's what is needed.

The fact that the Soviet Union had developed its own fission bomb made some scientists, notably the violently anti-Soviet physicist Edward Teller (see 1939), feel that the United States should immediately develop a hydrogen bomb in order to maintain its preponderance of power. Other scientists, notably Robert Oppenheimer (see 1937), sickened at the thought of the damage such bombs could do and aware that the Soviets would then simply match it, were opposed. The political decision was to go ahead with the H-bomb, and Teller saw to it that Oppenheimer's opposition was used to put an end to his career.

Meanwhile, an attempt began to devise a way to produce *controlled* hydrogen fusion as a source of peaceful energy—a much harder task. It would be necessary to raise the temperature of hydrogen to tens or even hundreds of millions of degrees, while confining this ultrahot hydrogen long enough to allow fusion to begin.

At this temperature, however, hydrogen's tendency is to expand explosively and be gone unless it is tightly enclosed. No material container would do, for either the hot hydrogen would melt the container or the cold container would cool off the hydrogen.

The hydrogen might be confined by a magnetic field, however. It would then merely be necessary to devise some way of shaping a magnetic field that was strong enough to do the job.

In 1951 the American physicist Lyman Spitzer, Jr. (b. 1914) supervised the construction of a figure-8 device called a *stellarator* (from the Latin word for "star," because it was trying to duplicate what went on in stars) that might confine hot hydrogen gas efficiently. Later, a modified version developed in the Soviet Union and called a *Tokamak* (an abbreviation of a descriptive Russian phrase) was used.

In nearly forty years of trying, however, nuclear fusion ignition has not been achieved, though progress continues to be made.

HYDROGEN RADIATION

Van de Hulst had predicted, from theoretical considerations, that hydrogen atoms in space would emit microwave radiation with a wavelength of 21 centimeters (see 1944).

In 1951 Purcell, who had earlier helped work out the theory of nuclear magnetic resonance (see 1946), actually detected this radiation coming from outer space. This demonstrated the value of radio waves in detecting the presence of given atoms and molecules in interstellar space. The wavelengths of the radiation emitted were characteristic of different substances and acted as "fingerprints."

MILKY WAY STRUCTURE

The spiral structure of galaxies had first been described by Rosse (see 1845), but the structure of our own Milky Way Galaxy remained puzzling. Living within it as we do, we naturally lack the ability to view it from without, from which point of view its structure would become obvious.

By 1951, however, astronomers could detect radio wave emissions with great delicacy, and the American astronomer William Wilson Morgan (b. 1906) could make out the characteristic radio waves of ionized hydrogen coming from particularly hot, bright stars that were themselves characteristic of the spiral arms of galaxies. When several such lines of ionized hydrogen were found coming from our galaxy, it seemed convincing evidence that the Milky Way Galaxy had spiral arms. Our galaxy was thus revealed to be a spiral galaxy much like the Andromeda Galaxy. Our Sun is located in one of these spiral arms.

JUPITER XII

Seth Nicholson, who over a period of nearly four decades had discovered three of the small outer satellites of Jupiter, in 1951 discovered a fourth. This was the twelfth satellite of Jupiter to be detected and was therefore referred to as *Jupiter XII*.

Jupiter XII was only 20 miles in diameter and it eventually received the name *Ananke*.

SUPERCONDUCTIVITY THEORY

Four decades had elapsed since Kamerlingh Onnes had first discovered the phenomenon of superconductivity (see 1911). The reason that some metals and alloys lost all electrical resistance at temperatures near absolute zero remained obscure, however.

In 1951 John Bardeen, who had participated in the discovery of the transistor (see 1948), helped work out a theoretical consideration of the phenomenon, making use of quantum effects, that seemed to explain a good deal. It was good enough, in fact, for Bardeen to be awarded the Nobel Prize for physics in 1972. Since he had also shared in that prize in 1956 for the discovery of the transistor, he became the first person ever to receive two Nobel Prizes in physics.

UNIVAC

Mauchly and Eckert, who had designed ENIAC (see 1946), designed UNIVAC *(Universal Automatic Computer)* in 1951. It was the first computer to make use of magnetic tape and the first to be mass-produced rather than individually built by the designers for their own use. It marked the beginning, then, of computers in industry.

STEROID SYNTHESIS

Woodward, who had made it his business to synthesize the most complex of naturally occurring organic molecules (see 1944, Synthesis of Quinine), achieved the synthesis of cholesterol and cortisone in 1951. These were both *steroids,* with molecules possessing a characteristic four-ring structure.

ACETYLCOENZYME A

Lipmann had demonstrated the existence of coenzyme A (see 1947) as the carrier of the acetyl group, which represented a crucial crossroad in metabolism.

In 1951 the German biochemist Feodor Felix Konrad Lynen (1911–1979) studied the function of coenzyme A in the metabolism of fat molecules particularly. He was the first to isolate *acetylcoenzyme A,* the compound that acts as intermediary in the transference of the acetyl group from one compound to another.

This work paralleled the work done the year before by Bloch, who used carbon-14 as a tracer to work out the way coenzyme A was involved with cholesterol synthesis. As a result, Lynen shared with Bloch the Nobel Prize for physiology and medicine in 1964.

FLUORIDATION

The most common disease afflicting mankind is *caries,* more commonly known as tooth decay. The incidence until recently was almost 100 percent. Modern dentistry handles it by drilling away the affected area and substituting *fillings* of ceramic or metal.

Prevention, however, is better than cure, and dentists had noticed that people in certain areas in the United States rarely got caries. Their teeth also showed a mottling of the enamel, caused apparently by the fact that their drinking water had a higher-than-average content of fluoride ions.

The search began for a level of fluoride in the water that would protect against tooth decay without mottling or darkening tooth enamel. By 1951, projects for the careful fluoridation of water supplies were in progress, in the hope that this (and the use of fluoridated toothpaste) would substantially reduce the incidence of tooth decay.

IN ADDITION

In Korea the fighting simmered down and negotiations to end the war began.

1952

NUCLEAR FUSION BOMB

The American effort to produce a nuclear fusion bomb *(hydrogen bomb)* was soon successful. Hydrogen-2 fused at a lower temperature than hydrogen-1, and hydrogen-3 fused at a lower temperature still. Hydrogen-2 was a rare isotope of hydrogen, but there was enough present in the Earth's oceans to last humanity for billions of years. Hydrogen-3 was radioactive and had to be formed through nuclear reactions if enough was to be obtained for use. It was planned to fuse a mixture of hydrogen-2 and hydrogen-3 in liquid form by exposure to the temperatures and pressures produced by a fission bomb.

Such a fusion bomb was tested on a coral atoll in the Pacific Ocean on November 1, 1952, wiping out the atoll. The blast yielded energy equivalent to 10,000,000 tons (10 megatons) of TNT—five hundred times the 20-kiloton energy of the Hiroshima bomb.

Yet it did not give the United States security. Within a year, the Soviets had exploded a fusion bomb of their own. Both sides continually improved the efficiency and power of their fusion weapons, and Great Britain and China also acquired the technology. As Oppenheimer had foreseen (see 1951), the world descended further into the abyss of fear, from which it has not yet emerged.

EINSTEINIUM AND FERMIUM

Seaborg (see 1940 and 1944) and his team continued to make ever more complex atoms, and as the atoms were made, they were bombarded with small atomic nuclei. Some of these stuck to the complex nucleus of the atoms being bombarded so that still newer and more complex atoms were formed.

In 1952, however, complex atoms were formed in a different way. The ravening energies of the fusion bomb explosion in the Pacific (see above) had driven nuclei together and formed atoms even more complex than californium (element number 98), which was at the time the most complex known (see 1949). As a result, elements 99 and 100 were formed and detected. They were eventually named, respectively, *einsteinium* and *fermium* in honor of Albert Einstein and Enrico Fermi, who had died in the months preceding laboratory study of the elements.

KAONS AND HYPERONS

Pions had been discovered by Powell (see 1947), and since that was the intermediate-sized particle that Yukawa had predicted (see 1935, Strong Interaction), there seemed no reason to expect anything more.

In 1952, however, two Polish physicists, Marian Danysz and Jerzy Pniewski, discovered another intermediate-sized particle. It was about 3.5 times as massive as a pion but still only half as massive as a proton or neutron and was called a *K-meson,* which was sometimes abbreviated to *kaon*.

These same physicists went on that year to discover, among the products of cosmic ray collisions, a subatomic particle that was *more* massive than a proton or a neutron. This came to be called a *lambda particle.* It was 1.2 times as massive as a proton.

Other particles more massive than a proton were eventually discovered and called *hyperons.* Mesons, nucleons (protons and neutrons), and hyperons were all grouped together as *hadrons,* from a Greek word meaning "thick" or "strong," since they were all subject to the strong interaction. Electrons, muons, and neutrinos, subject to the weak interaction, were called *leptons,* from a Greek word meaning "weak."

While the number of leptons remained small, the number of hadrons rose steeply as the years passed, in the end reaching about a hundred. It was this that convinced physicists their notions of subatomic structure were incomplete; the large number of hadrons was too complicated a factor. There had to be some new view that would simplify the matter.

ORIGIN OF LIFE

Evidence of living things, in the form of traces of tiny bacterial cells, has been found in rocks that are 3.5 billion years old. Since the Earth is 4.6 billion years old, this means that in the first billion years of Earth's existence, living things must have evolved from nonliving chemicals. Presumably, they must have evolved from the molecules that made up the huge nebula of dust and gas out of which the Solar System was formed.

It is reasonably well established that the Universe is 99 percent hydrogen and helium in a 9-to-1 ratio, with oxygen, carbon, nitrogen, neon, sulfur, silicon, iron, and argon most important among the final 1 percent. Of these, helium, neon, and argon form no compounds. In the presence of a preponderance of hydrogen, the oxygen, carbon, nitrogen, and sulfur form water, methane, ammonia, and hydrogen sulfide respectively. Silicon combines with oxygen and various metals to form silicates, which are rocky substances, while iron mixes with other less common metals.

Earth therefore began with a core of nickel-iron, surrounded by a rocky mantle and crust. All of that was topped by an ocean of water and an atmosphere that may have consisted of ammonia, methane, and hydrogen sulfide (with some of each dissolved in the water). The simple compounds in the atmosphere and ocean would have been subjected to the energy of the Sun's radiation and, through the addition of such energy, might have built up very gradually into more complicated substances until objects complex enough to be considered alive were the result.

In 1952 the American chemist Stanley Lloyd Miller (b. 1930), working under Harold Urey (see 1931), attempted for the first time to check this possibility by experimentation.

Miller began with carefully purified and sterilized water and added an "atmosphere" of hydrogen, ammonia, and methane. He circulated this through his apparatus and past an electric discharge that would add energy. He kept this up for a week. He then analyzed the solution and found organic compounds and even a few of the simpler amino acids formed abiogenetically, that is, without the presence of life.

Others later continued the work, using other sources of energy and other mixtures of simple compounds. They also added to the mix more complex compounds that had been formed in other experiments of the sort. None of this served to ascertain the exact route by which life was formed, but it all helped make it seem likely that life *did* start by *some* route that involved the action of chemical and physical laws, so that no supernatural cause need be sought.

X-RAY DIFFRACTION OF DNA

The fine structure of DNA had to be worked out if indeed such molecules were the carriers of genetic information. Chargaff had made a start by showing that the number of purine groups was equal to the number of pyrimidine groups (see 1948), but there was much more to be done.

Since nucleic acids were long-chain polymers of nucleotides, there ought to be certain periodic regularities in the molecule (as in crystals—see 1914) that would show up in X-ray diffraction.

In 1952 the English biophysicist Rosalind Elsie Franklin (1920–1958) made careful X-ray diffraction studies of DNA, from which she deduced that the molecule ought to be helical (like a spiral staircase), with the phosphate groups that bound the units together located on the outside of the helix.

Franklin was a slow and careful worker, however, and did not wish to announce a result without carefully checking every step. This, combined with the fact that her coworkers tended to ignore her because she was a woman, resulted in the fruits of her labors going to benefit others.

INSULIN STRUCTURE

The use of paper chromatography (see 1944) had made it possible to determine the number of units of each amino acid in a particular protein molecule. If the molecules were broken down to fragments made up of two, three, or four amino acids connected together, these could also be isolated and identified.

Once the fragments were identified, one could reason out the structure of longer stretches of the molecules by arguing that a certain long stretch would give rise to shorter pieces, all of which would be found among the fragments, but would not give rise to shorter pieces that were not found there. In this way, little by little, working backward, one could work out the precise order of amino acids in the entire protein molecule.

A British biochemist, Frederick Sanger (b. 1918), used this painstaking method and, by 1952, could show that the molecule of the protein hormone insulin consisted of some fifty amino acids distributed among two interconnected chains. He was also able to demonstrate the exact order of the amino acids that made up each chain.

For this, he received the Nobel Prize for chemistry in 1958.

VIRUS NUCLEIC ACID

There was no doubt by this time that nucleic acid was the genetic carrier in plant and animal cells and even in bacteria. That left viruses.

Viruses consist of an internal core of nucleic acid, either DNA or RNA, or sometimes both, and an outer shell of protein. In 1952 Alfred Hershey (see 1945) showed that when bacteriophages attacked the bacterial cells that were its natural prey, it was the DNA that actually entered the cell and supervised the formation of both the DNA and protein components of the new viruses. The protein shell remained outside but presumably contained an enzyme that helped dissolve the cell wall and allow the entrance of the DNA.

Also in 1952, Joshua Lederberg (see 1946) showed that a bacteriophage could transfer genetic material from one infected cell to a new cell that it infected later. This meant that viruses could be used as tools for introducing genetic changes, or mutations. Lederberg termed this process *transduction;* it represented a technique with potential for genetic engineering.

NERVE GROWTH FACTOR

The Italian embryologist Rita Levi-Montalcini (b. 1909) worked on chick embryos during World War II under difficult conditions, since she was Jewish. After the war she continued her work and found that an implantation of certain tumors in chick embryos hastened nerve growth. By 1952 she could show that the *nerve growth factor* was a soluble substance released by the tumor. For this work she received the Nobel Prize for physiology and medicine in 1986.

RADIOIMMUNE ASSAY

By 1952 the American biophysicist Rosalyn Sussman Yalow (b. 1921) had worked out an extremely delicate method for locating antibodies and other biologically active substances present in the body in quantities so minute that they were detectable in no other way.

This is done by making use of a substance containing a radioactive atom that will combine with the biologically active material in question. The presence of the radioactive atom can be detected and the extent of combination determined with great delicacy. This makes possible numerous significant tests *(radioimmune assay)* for the presence of significant antibodies, greatly aiding medical diagnosis and allowing doctors to follow the course of events in medical treatment.

For this work, Yalow shared in the Nobel Prize for physiology and medicine in 1977.

REM SLEEP

Dreams were the subject of mystical theories in primitive times, being viewed as messages from some other realm not available to the waking senses. Freud (see 1900) initiated another form of dream analysis, but some think it has mystical components as well.

It was not till 1952 that straightforward observations were made concerning dreams, which didn't make use of the subjective reports of the dreamer. In that year the American psychologist William Charles Dement (b. 1928), studying sleeping subjects, noticed periods of rapid eye movement (REM) that sometimes persisted for minutes. During these periods of REM sleep, breathing, heartbeat, and blood pressure rose to waking levels. He noted that such REM sleep took up about a quarter of the sleeping time.

Sleepers who are awakened during these periods generally report that they were having a dream. Furthermore, a sleeper who is continually disturbed during these periods begins to suffer psychological distress. The periods of REM sleep are then multiplied during succeeding nights as though to make up for the lost dreaming.

It would seem, then, that dreaming has some important function in maintaining the working efficiency of the complex human brain. In this connection, REM sleep has been found to occur in infants to an even greater extent than in older individuals and to occur in mammals other than human beings, too. Exactly what the function of REM sleep and dreaming might be remains a matter of dispute, however.

TRANQUILIZERS

Sedatives are obviously useful drugs for quieting overexcited states and for inducing calm. The best-known sedatives prior to the 1950s were the barbiturates, but they diminished alertness and produced sleep.

In 1952 the American physician Robert Wallace Wilkins (b. 1906) studied a drug, *reserpine,* obtained from the root of an Indian shrub. He reported that it had a sedative effect *without* diminishing alertness or inducing sleep.

Reserpine and other drugs of its nature came to be called *tranquilizers.* They grew instantly popular with many people for use in the reduction (real or fancied) of tensions (real or fancied). They served, more seriously, as adjuncts to psychiatric treatment, for although they are in no sense a cure for any mental disease, they do calm violent patients without the use of harsh physical restraints, and calm patients may then more easily cooperate with the psychiatrist.

GAS CHROMATOGRAPHY

In 1952 A. J. P. Martin, who had helped evolve the technique of paper chromatography (see 1944), applied its principles to the separation of gases.

In his new technique, mixtures of gases or vapors were passed through a liquid solvent or over an adsorbing solid (that is, one to which gas molecules tend to cling) by means of a current of inert carrier gas, such as nitrogen or helium. The components of the gas mixture were pushed along with the carrier gas at different speeds, so that the carrier arrived at the other end with the components of the gas mixture separated. Such *gas chromatography* is particularly useful because of the speed of its separations and the great delicacy with which it can detect trace impurities.

ZONE REFINING

The American chemist William Gardner Pfann introduced the technique of *zone refining* in 1952. This calls for a rod, of germanium or silicon, for instance, placed within a circular heating element. The section of the rod that is enclosed softens and begins to melt, and as the rod is drawn through the hollow, the softened zone moves along its length. Any impurities in the rod tend to remain in the softened zone, so that they are washed to the end of the rod. After a few passes of this sort, the end can be cut off and what remains is unprecedentedly pure.

This is a most useful technique in preparing pure material that can then be "doped" with trace quantities of deliberately chosen impurities in concentrations suitable for solid-state electronic devices such as computers.

IN ADDITION

War hero Dwight David Eisenhower (1890–1969) became the thirty-fourth president of the United States.

George VI of Great Britain died on February 6, 1952, and was succeeded by his daughter, who reigned as Elizabeth II (b. 1926).

Farouk I of Egypt (1920–1965) was forced to abdicate on July 26, 1952, and the Egyptian monarchy came to an end.

1953

THE DOUBLE HELIX

The work of Chargaff and Franklin (see 1952) had supplied the information necessary to work out the structure of DNA. The key conceptual step was then taken by the English physicist Francis Harry Compton Crick (b. 1916) and the American biochemist James Dewey Watson (b. 1928). They made use of a key X-ray diffraction photograph taken by Franklin and made available to them by her boss, the New Zealand physicist Maurice Hugh Frederick Wilkins (b. 1916), apparently without Franklin's knowledge or permission.

In 1953 Watson and Crick suggested that DNA consisted of two chains of nucleotides arranged as a double helix, with the purine and pyrimidine bases facing each other and the phosphate links on the outside. Each two-ringed purine faced a one-ringed pyrimidine, so that the space between the two strands was constant. Adenines and thymines were paired, as were guanines and cytosines (thus accounting for Chargaff's results).

Each strand of the double helix was a model (or *template*) for the other. In cell division, each DNA double helix would separate into two strands, and each strand would build up its complementary strand on itself; an adenine fitting over every thymine on the strand, a thymine over every adenine, a guanine over every cytosine, and a cytosine over every guanine. In this way two double helixes would appear where there had been one before. Thus DNA underwent *replication* without changing its structure except for very occasional accidental errors, which represented mutations.

The Watson-Crick structure made so much sense that it was accepted at once. Watson, Crick, and Wilkins shared the Nobel Prize for physiology and medicine in 1962 as a result. By that time Franklin was dead, and her involvement did not have to be considered.

ISOTACTIC POLYMERS

For over forty years, chemists had been manufacturing polymers, long-chain molecules built up of simple units. Beginning with Bakelite (see 1909), these offered a variety of useful properties and came into steadily increasing use.

However, the technique of polymerization was a matter of random combination. The simple units were brought together and allowed to combine as they wished, so to speak. The result was that, instead of a single long chain, branches sometimes appeared. If there were atomic groups attached to the unit, these would jut outward in random directions along the chain. Such unpredictable aspects of the polymerization technique limited its usefulness.

In 1953 a German chemist, Karl Ziegler (1898–1973), discovered that he could use a resin, to which ions of metals such as aluminum or titanium were attached, as a catalyst in the production of polyethylene. Chains without branching were then formed. As a result, the new polyethylene was tougher and higher-melting than the old.

The Italian chemist Giulio Natta (1903–1979) carried on this sort of work and found that he could use such catalysts to insure that the side groups of a unit would all point in the same direction when a polymer formed. At Mrs. Natta's suggestion, these were called *isotactic polymers*, from Greek words meaning "ordered in the same way."

PLATE TECTONICS

For thirty years it had been known that there was a mountain range down the middle of the Atlantic Ocean. Eventually it was understood that this was part of a world-girdling range called the *Mid-Oceanic Ridge*.

In 1953 the American physicists Maurice Ewing (1906–1974) and Bruce Charles Heezen (1924–1977) discovered that a deep canyon ran the length of the ridge. It was called the *Great Global Rift*. There were places where the rift came quite close to land: it ran up the Red Sea between Africa and Arabia and skimmed the borders of the Pacific through the Gulf of California and up the coast of the state of California.

The rift seemed to break the Earth's crust into plates tightly joined as though fitted together by a skilled carpenter. They were therefore called *tectonic plates*, from a Greek word for "carpenter." The study of the evolution of the Earth's crust in terms of these plates is called *plate tectonics*, and it has totally revolutionized geology, explaining a great deal that had been mysterious before.

There are six large tectonic plates and a number of smaller ones, and it is along the boundaries of the plates that Earth's quakes and volcanoes seem to be concentrated. One plate, which includes most of the Pacific Ocean, and with boundaries of the eastern coast of Asia and the western coast of America, accounts for about 80 percent of the earthquake energy released on Earth.

BUBBLE CHAMBERS

At this time the most familiar device for detecting the paths of subatomic particles was the cloud chamber invented by Wilson (see 1911). The American physicist Donald Arthur Glaser (b. 1926) thought of reversing its principle.

In the cloud chamber, you have humid air

on the point of forming small droplets of liquid. What if you started, instead, with a liquid that was on the point of boiling and forming small bubbles of vapor? In the cloud chamber, a speeding charged particle would encourage the formation of droplets, and a line of droplets would mark out its path. In this new *bubble chamber,* a speeding charged particle would encourage the formation of bubbles, and a line of bubbles would mark out its path.

Since liquids are denser than gases, a speeding particle will slow more quickly in a bubble chamber than in a cloud chamber, curve more intensely, and reveal its properties more clearly. Then, too, there will be more collisions in the bubble chamber—more events will take place. Finally, if liquid hydrogen is used as the liquid, it will consist, for the most part, of electrons and single protons, and the simplicity of the background will make the results easier to interpret.

By 1953 Glaser had made his bubble chamber a practical reality and they have been an indispensable tool of subatomic investigations ever since. Glaser received the Nobel Prize for physics in 1960 as a result.

STRANGE PARTICLES

Kaons and hyperons are subject to the strong interaction and are formed in ways that involve the strong interaction. It would seem that they ought to break down by the strong interaction too, but they don't; they break down far more slowly by way of the weak interaction.

This is not to say that they break down slowly in any absolute sense—it takes just under a billionth of a second—but if they broke down by way of the strong reaction, it would only take a billionth of a billionth of a second. That they break down so slowly seemed strange, and they came to be called *strange particles.*

In 1953 the American physicist Murray Gell-Mann (b. 1929) tried to make sense out of this strangeness. He studied the properties of groups of two or three hadrons that differed only in the nature of their electric charge and assigned each group a kind of average electric charge.

In this way he was able to demonstrate that each group had a special property, which he called *strangeness,* that depended on the properties of the average electric charge. In the most familiar hadrons, such as the proton, neutron, and pion, the strangeness number was 0 and the property was negligible. For kaons and hyperons, however, the strangeness number was *not* 0 but +1, −1, +2, or −2.

For the strange particles to break down by the strong interaction, the strangeness number would have to remain unchanged, and since all the particles they might break down into had 0 strangeness, they couldn't break down in that way. Therefore, they were forced to break down by the weak interaction, in which the strangeness number didn't matter. Hence the comparatively long lifetime of the strange particles.

As a result of this and other work, Gell-Mann received the Nobel Prize for physics in 1969.

MASERS

Einstein had pointed out that if a photon of a certain size struck a molecule, the molecule would absorb the photon and rise to a higher energy level. If such a photon struck a molecule that was already in the higher energy level, the molecule would return to the lower energy level, emitting a photon of exactly the same wavelength and moving in exactly the same direction as the striking photon, which would itself continue to move on. Now there would be two photons, which would strike two other high-level molecules, so that you would end with four. In a very brief time, there would be a vast flood of photons, all of the same wavelength *(monochromatic)* and all moving in the same direction *(coherent radiation)*.

After World War II, when microwaves became extremely important in connection with radar and radio astronomy, the American physicist Charles Hard Townes (b. 1915) wondered if use could be made of this principle to produce a microwave beam of great intensity.

The ammonia molecule, for instance, vibrates 24 billion times a second under appropriate conditions. This could be converted into microwaves with a wavelength of 1¼ centimeters. Suppose ammonia molecules were then raised to a higher energy level through exposure to heat or electricity and exposed to a feeble beam of microwaves of the natural frequency of the ammonia molecule (1¼ centimeters). That should release a much stronger beam of the same wavelength.

By December 1953, Townes had a device that actually worked in this fashion. The process was described by the phrase *microwave amplification by stimulated emission of radiation*. Using the initials of the phrase, the instrument was referred to more briefly as a *maser*.

Two Soviet physicists, Aleksandr Mikhaylovich Prokhorov (b. 1916) and Nikolay Gennadiyevich Basov (b. 1922) worked out the theoretical basis of the maser independently of Townes at about the same time. As a result, all three shared in the Nobel Prize for physics in 1964.

HEART-LUNG MACHINE

A heart-lung machine (or *pump oxygenator*) is one that takes venous blood from the veins, oxygenates it by mixing it with air, and pumps it back into the arteries, thus bypassing lungs and heart. It makes it possible to stop the heart and perform open-heart surgery without endangering the patient's life.

The first successful heart-lung machine, devised by John G. Gibbon of the United States, was used in 1953. Since then it has been repeatedly improved and is now used in the coronary bypass operations that are routinely performed to relieve the life-threatening agony of angina pectoris.

TRANSISTORIZATION

The transistor had been invented by Shockley and his group (see 1948), but its performance was at first unreliable. However, reliability improved rapidly, and by 1953 the first significant transistorized instruments for use by the general public were introduced. These were hearing aids that were so small they could fit into the ear opening. They replaced, and worked better than, previous devices, which were heavy, bulky, and embarrassingly noticeable.

Meanwhile, Japan was working to produce transistorized radios, which would be much smaller than any till then in use and more reliable. The world was about to enter an age of *miniaturization*.

SPRAY CANS

In 1953 a new plastic valve mechanism was designed by the American inventor Robert H. Abplanalp (b. 1923), which made it possible to produce an aluminum spray can cheaply. The spray could be ejected by evaporating Freon (see 1930), which was a liquid that evaporated easily and did not produce high pressures.

As a result, there was a vast multiplication of spray cans and an enormous discharge of Freon vapors into the atmosphere. Freon is harmless in almost all respects, but it had an unexpected effect on the chemistry of the upper atmosphere. This had the potential for dire catastrophe—as was eventually discovered.

IN ADDITION

The Korean War came to an end with an armistice signed on July 17, 1953. There was no actual peace, but a demilitarized zone was set up along the border that had existed before the war began.

On March 5, 1953, Josef Stalin died. He was succeeded as head of the Soviet Union by Georgy Maksimilianovich Malenkov (b. 1902). The Soviet Union exploded its first fusion bomb on August 12, 1953, with physicist Andrey Dmitriyevich Sakharov (b. 1921) playing the role that Teller had played in the United States.

In Africa, a full-scale revolt against Great Britain began in Kenya. The rebels, who called themselves *Mau Mau* (the Hidden Ones), were led by Jomo Kenyatta (1894–1978).

Ibn Saūd of Saudi Arabia (1880–1953) died on November 9, 1953, and was succeeded by his son, Saud (1902–1969).

In the United States, Senator McCarthy was at the peak of his power.

1954

SALK VACCINE

Poliomyelitis *(infantile paralysis)* was a particularly frightening disease, because when it did not kill, it often paralyzed permanently, leaving people in wheelchairs or even iron lungs. What's more, it often hit young people.

Once the polio virus could be cultured in chick embryos, however, as shown by Enders and his group (see 1948), it was possible to experiment with it.

Thus the American microbiologist Jonas Edward Salk (b. 1914) tried to kill the virus, so that it would not give rise to the disease, but to leave it sufficiently intact that it would stimulate the growth of antibodies and lead to immunity should a living virus later invade.

First he tried his preparation *(Salk vaccine)* on children who had recovered from polio, to see if it raised the antibody content. Then in 1953 he dared to try it on children who had not had the disease, to see if antibodies would develop. They did, and within two years mass inoculation had begun. The dread disease became a thing of the past.

KIDNEY TRANSPLANT

When a vital human organ fails, death may be prevented if another organ can be transplanted. The transplanted organ may come from a living human being who can spare it, or from a human being dead in an accident so recent that the organ is still viable.

Unfortunately, human beings are allergic to one another, and the donated organ tends to be rejected, although scientists like Medawar (see 1949) were striving to find ways of reducing this tendency.

The first successful kidney transplant took place in December 1954 in Boston, from one identical twin to another. Since identical twins have the same genetic makeup, they have very little tendency to reject each other's organs. If one twin has two bad kidneys and the other two good ones, one of the good pair (which the donor can spare) can keep the dying twin alive. In this case, the twin that received the kidney lived on for eight years.

Many other kidney transplants have been carried through since, sometimes with considerable success, even among other than identical twins.

CONTROLLED FISSION REACTORS

Even before the first fission bomb had been exploded, a controlled nuclear reactor (although a very inefficient one) had been set up in Chicago in 1942. Its only function was to show that a fission bomb was possible.

Efforts were later made, however, to devise nuclear reactors efficient enough to serve as reasonable sources of controlled energy for peaceful uses. The fissioning uranium or plutonium would liberate heat at a moderate rate, and this heat would turn water into steam, which would turn a turbine and produce electricity.

Naturally, methods had to be devised to slow the fission reaction if it showed signs of proceeding too quickly and producing enough heat to result in a *meltdown*. A controlled nuclear reactor could not explode, since it was not enclosed strongly enough to build up the kind of heat and force that would lead to an explosion. It would, however, be capable of releasing a surge of nuclear radiation into the environment, so the pressure for safe operation was therefore strong.

The first nuclear reactor built to produce electric power for civilian use was put into action in the Soviet Union in June 1954. It was a very small one. Larger reactors were produced in Great Britain and in the United States soon after, and eventually they were distributed around the globe and began to contribute substantially to the world's energy supply, particularly in France and the Soviet Union.

Controlled reactors also came into use in another way. Submarines throughout both world wars had remained vulnerable because they had to surface periodically to recharge their batteries. Under the driving force of the Polish-born American naval officer Hyman George Rickover (1900–1986), a plan developed to equip American submarines with atomic reactors, which would require no recharging and could keep a submarine submerged for months at a time. The first nuclear-powered submarine, the *Nautilus*, was launched in January 1954.

Some nuclear-powered surface vessels were eventually built by the Soviet Union and the United States, but except for submarines, nuclear-powered forms of transportation did not catch on.

OXYTOCIN SYNTHESIS

At about the time that Sanger was working out the order of amino acids in the protein chains of the insulin molecule (see 1952), Vincent du Vigneaud (see 1942) was determining the exact makeup of the hormone *oxytocin*, produced by the posterior portion of the pituitary gland.

Oxytocin was a particularly simple protein hormone, its molecules made up of only eight amino acid residues arranged in a circle. Du Vigneaud went on in 1954 to synthesize oxytocin, combining just the right amino acids in the right order.

This was the first occasion on which a naturally occurring protein was synthesized and the synthetic protein shown to have just the same properties and abilities as the protein as it occurs in the body. For this, du Vigneaud was awarded the Nobel Prize for chemistry in 1955.

CHLOROPLAST ISOLATION

Since Pelletier and Caventou had isolated chlorophyll (see 1817), it had been well known that chlorophyll was essential to photosynthesis. However, no one had been able to make chlorophyll perform the task in the test-tube.

It had been nearly a century since von Sachs had discovered that chlorophyll was present in discrete organelles, called chloroplasts, within the plant cell (see 1862). It was natural to assume that in the cell, chlorophyll worked as a catalyst, not by itself but as part of an intricate system that was present intact in the chloroplast.

This could be shown if chloroplasts could be isolated intact from the cells, and made to show that they could then carry out photosynthesis in the test-tube. But chloroplasts are so flimsy that for a long time no procedure sufficed to extract them intact.

Finally in 1954 the Polish-American biochemist Daniel Israel Arnon (b. 1910) was able to obtain intact chloroplasts from disrupted spinach-leaf cells and demonstrate their ability to carry on photosynthesis outside the cell.

STRYCHNINE SYNTHESIS

In 1954 that master-synthesist Woodward (see 1944, Synthesis of Quinine) managed to synthesize the fearfully complicated (and poisonous) alkaloid strychnine, which had a molecule built up of seven intricately related rings of atoms.

GENETIC CODE

Granted that DNA contained the information that governed the inheritance of characteristics, it must do this by overseeing the manufacture of enyzmes, which in turn controlled the chemical reactions that went on inside cells. But how could DNA turn the trick? It was composed of chains of four different nucleotides, while enzymes, which were proteins, were composed of chains of twenty different amino acids.

In 1954 George Gamow (see 1929) suggested that it made no sense to try to line up an individual nucleotide with an individual amino acid, since there were too few of the former and too many of the latter. He pointed out that it must be necessary to deal with combinations of at least three nucleotides. If there were four different nucleotides, they could be built up into sixty-four different trinucleotides, or *codons*—more than enough to carry the information necessary to build up proteins.

Gamow got the details wrong in his scheme, but he had the right idea just the same. He was the first to conceive of a multinucleotide *genetic code.*

PHOTOVOLTAIC CELLS

Eighty years before, it had been discovered that the element selenium could conduct an electric current much more easily in the light than in the dark. It was eventually realized that the energy of sunlight knocked electrons loose from the selenium atoms, and it was those electrons that carried the current.

Selenium came to be used for small jobs. Thus a beam of light shone across a doorway into a selenium receiver and an electric current flowed that kept a door closed against the pull of a spring. If an approaching object interrupted the beam of light, the darkened selenium ceased conducting electricity and the door swung open. The device, popularly known as an *electric eye,* is more properly called a *photoelectric* or *photovoltaic cell.*

Selenium is not suitable for heavy jobs, as it is extremely inefficient, turning less than 1 percent of the energy of sunlight into electricity. In 1954, however, photovoltaic cells were devised that made use of semiconductors of the type used in transistors. Light kicked electrons out of place much more efficiently in their case; they turned about 4 percent of the sunlight into electricity. At that level, photovoltaic cells could also be referred to as *solar batteries*.

Photovoltaic cells continued to be improved, and eventually some with efficiencies of up to 30 percent appeared. As efficiency rose and the cost of manufacture fell, it seemed the time might come when electricity could be manufactured directly out of sunlight. If the Sun were used in that fashion, we would never run out of the supply and there would be no chemical pollution.

ROBOTS

The word *robot* (from a Czech word for "serf" or "slave") had been invented by the Czech playwright Karel Capek (1890–1938) in his play *R.U.R.*, first staged in Europe in 1920. Since Capek's time, the word has come to be applied to any manufactured device, usually envisaged as humanoid in shape (though it doesn't have to be) and made of metal (though again, it doesn't have to be), which is capable of doing work ordinarily done by human beings.

Although robots were much used in science fiction, the first patent wasn't taken out on a robotic device in real life until 1954. It was the work of the American inventor George C. Devol, Jr., who teamed up afterward with the American entrepreneur Joseph F. Engelberger (b. 1925), who had grown interested in robots as a result of reading *I, Robot* by Isaac Asimov (b. 1920).

For twenty years they continued to develop patents, but the manufacture of robots that were sufficiently cheap and compact to be used in industry had to await further advances in computers.

BEVATRON

Since the first particle accelerator was built by Cockcroft and Walton (see 1929), the energy produced by particle accelerators had increased enormously. By 1954, a particle accelerator capable of accelerating protons to an energy of 5 to 6 *billion electron volts (BeV)* had been built at the University of California. It was called the *Bevatron* because of the energy range of the particles it produced, and indeed it could produce particles in the energy range of fairly intense cosmic rays.

This meant it was no longer necessary to wait for cosmic rays to strike atoms in the atmosphere and produce interesting results, as Anderson had had to (see 1932). The necessary bombarding particles were always there, and in any reasonable quantities. The search could proceed much more surely in the laboratory, especially since particle accelerators continued to be made ever more powerful as time went on.

ORAL CONTRACEPTIVES

In a world that was overpopulated and growing more so, it seemed useful to find methods of reducing the birthrate. The most straightforward way of doing so was abstention from sex, but that wasn't really a practical solution. The trick would be to find a cheap and convenient way to lower the birthrate without interfering with sex.

It had been noted that there were natural ways of bringing this about, for during pregnancy and some parts of the menstrual cycle, women can safely indulge in sex without much chance, if any, of conception. It might be possible, then, to find some sort of hormone that, taken by mouth in pill form, as an *oral contraceptive,* would induce temporary sterility. The American biologist Gregory Goodwin Pincus (1903–1967) found such a hormone, and clinical tests in 1954 demonstrated its efficacy. Use of *the Pill* (as it was popularly termed) made possible sex without fear of pregnancy. This went a long way toward abolishing the double standard and encouraged the women's liberation movement, with its demand that women be treated on a par with men economically.

CONTACT LENSES

For some six centuries, people who were nearsighted, farsighted, or astigmatic had worn spectacles, or eyeglasses, to correct their vision (see 1249 and 1825). Eyeglasses, however, are a noticeable adjunct that call attention to a physical shortcoming. Furthermore, the myth arose that men who wore glasses were effeminate and that women who wore them were ugly. (The movies, in particular, helped propagate these mischievous ideas.) For that reason, it seemed useful to correct vision in a less conspicuous way.

As early as 1887, a German physician, Adolf Eugen Fick (1829–1901), had worked out the notion of *contact lenses,* small lenses that would just fit over the iris of the eye, correcting vision without anyone noticing it.

Glass in direct contact with the eye, however, would be irritating and dangerous. In 1954 plastic contact lenses were produced, which proved useful and popular, and contact lenses are now in common use.

IN ADDITION

Senator McCarthy attacked the United States army, accusing it of vague crimes and leftist leanings. The army challenged McCarthy on this, and the result was a congressional hearing on McCarthy's recent actions. On December 2, 1954, the Senate finally found the courage to condemn him for misconduct, and he declined rapidly to death by alcoholism less than three years later. It was the end of a nightmare.

In Indochina, the French were forced to leave northern Vietnam. Indochina was now divided into four independent nations: Laos, Cambodia, Communist-controlled North Vietnam, and South Vietnam, which remained under French protection.

In Africa Gamal Abdel Nasser (1918–1970) became prime minister of Egypt, while Algeria rose in revolt against the French colonialists.

1955

EXPLODING GALAXIES

Radio astronomy continued to show its practitioners that it could reveal information about the Universe not readily obtainable (if at all) by ordinary optical observations.

A radio source in Cygnus was unusually strong, and optical investigation of the region revealed a peculiarly shaped galaxy that looked rather like two galaxies undergoing a collision.

The Soviet astronomer Viktor Amazaspovich Ambartsumian (b. 1908) examined the nature of the radio source closely and suggested that it was really a galaxy in a state of vast explosion. That idea was borne out by later work.

This was another example of what are now known as *active galaxies,* in which events releasing enormous energies are taking place at the core. Whereas under optical observation, the Universe seemed to be serene and peaceful (except for the occasional novas and supernovas), radio astronomy began to show that it was a surprisingly violent place.

BIRTH OF STARS

The more massive a star, the brighter it is, the more rapidly it consumes its nuclear fuel, and the shorter its life on the main sequence (see 1914). The Sun came into being as a star only 4.5 billion years ago, some 10 billion years after the Universe itself came into being, and will remain on the main sequence only 5 or 6 billion years more.

Stars that are much more massive than the Sun can have lifetimes on the main sequence of less than a billion years, perhaps even only several million. Such stars that are still on the main sequence now must have been formed less than a billion years ago, or perhaps only a few million years ago. This leads to the thought that there may well be interstellar clouds out of which stars are forming right now.

For instance, there are reasons for thinking that the Orion nebula is an active star-former right now. In 1955 the American astronomer George Howard Herbig (b. 1920) detected two stars in the Orion nebula that had not been seen a few years earlier. This means we may have witnessed the actual birth of these stars.

JUPITER'S RADIO WAVES

Radio waves are not only emitted by stars and galaxies. In 1955 the American astronomer Kenneth Linn Franklin (b. 1923) detected radio waves emanating from the planet Jupiter. They were nonthermal; that is, they were not of the pattern that would be emitted simply because of the temperature of Jupiter's cloud layer. Speculation arose that they were the result of charged particles in motion in the neighborhood of Jupiter, and eventually this was found to be true.

PLUTO'S ROTATION

Little could be observed of Pluto because of its enormous distance from us, but in 1955 its light was seen to fluctuate slightly, with a period of 6.4 days. The obvious conclusion was that Pluto rotated on its axis once every 6.4 days and that one hemisphere reflected somewhat less light than the other.

ANTIPROTON

In the twenty-six years since Dirac had advanced his theory of antiparticles (see 1930), only the antielectron (positron) had been detected. Scientists were quite convinced that if the antielectron existed, the antiproton had to exist also. The antiproton, however, would have a mass 1837 times that of the antielectron and therefore require 1837 times the energy to be formed.

It was not practical to wait for one of the relatively few cosmic ray particles sufficiently energetic to form an antiproton. Once the bevatron was built, however (see 1954), energies capable of forming antiprotons were available in quantity.

In 1955 Segrè, who had first detected technetium (see 1937), and the American physicist Owen Chamberlain (b. 1920) bombarded copper for hours with protons possessing energies of 6.2 BeV. They worked out an elaborate system for detecting any antiprotons that might be formed, even amid large numbers of other particles of different charge and mass. The result was that, among 40,000 particles, they detected 60 antiprotons.

For this, Segrè and Chamberlain received the Nobel Prize for physics in 1959.

MENDELEVIUM

In 1955 Seaborg and his group (see 1940) bombarded einsteinium, element number 99, with protons and formed a few atoms of element 101. They named this new element *mendelevium*, after Mendeleyev, who had first worked out the periodic table (see 1869).

SYNTHETIC DIAMONDS

It had been known for nearly two centuries that diamonds were made up of carbon atoms, as graphite and coal were. It should be possible, in theory, then, to convert graphite into diamond. However, carbon atoms held each other so tightly that very high temperatures were required to shake them loose, and the high temperatures had to be combined with very high pressures in order to force the atoms into the more compact arrangement of diamond.

Moissan thought he had achieved the synthesis of diamond from graphite, but that proved to be a mistake. He might even have been the victim of a hoax, since he could not possibly have attained the temperatures and pressures required.

Bridgman's work on high pressure (see 1905) made the conversion possible at last, however, and in 1955 scientists managed to attain pressures of 100,000 atmospheres and temperatures of 2,500° C. In addition, they used chromium as a catalyst. As a result, they formed synthetic diamonds (indistinguishable from the "real thing") out of graphite. Eventually, with still higher temperatures and pressures, graphite was turned into diamond without the need for a catalyst.

FIELD ION MICROSCOPES

The art of magnifying the small reached a new plateau in 1955 when the German-born American physicist Erwin Wilhelm Mueller, who had developed the field-emission microscope (see 1937), devised the field ion microscope, which emitted beams of ions rather than electrons. This device strips positively charged helium ions off an extremely fine, curved needle tip, kept at liquid-hydrogen temperatures, and shoots them, in divergent paths, at a fluorescent screen. What appears on the screen then is a vastly magnified image (a million times or more) of the needle tip.

With such magnifications, individual atoms could be seen as dots and their arrangement studied.

NUCLEIC ACID FORMATION

Watson and Crick had worked out the structure of DNA and shown how the two strands of the double helix, when separated, could each form a second strand (see 1953).

The formation of the second strand, however, must surely require the catalytic services of an enzyme. In 1955 the Spanish-born American biochemist Severo Ochoa (b. 1905) isolated such an enzyme from the bacterium *Aztobacter vinelandii*. It was capable of catalyzing the formation of RNA-like substances from individual nucleotides.

Soon afterward the American biochemist Arthur Kornberg (b. 1918), who had been a student of Ochoa's, obtained another such enzyme from the bacterium *Escherichia coli*. It could catalyze the formation of DNA-like substances from individual nucleotides.

Now enzymes could be used to form nucleic acid chains made up of one, two, or three different nucleotides.

As a result, Ochoa and Kornberg received the Nobel Prize for medicine and physiology in 1959.

CYANOCOBALAMIN STRUCTURE

The British physicist Dorothy Mary Crowfoot Hodgkin (b. 1910) was interested in the structure of complex molecules in living tissue and, six years earlier, had completed the task of working out the atomic structure of penicillin. For the purpose, she used X-ray diffraction photographs and made use of a computer. This was the first direct use of a computer in solving a biochemical problem.

Hodgkin then went on to tackle the structure of cyanocobalamin (vitamin B-12). Its molecule was four times as massive as that of penicillin. Again Hodgkin made use of X-ray diffraction and a computer, but even so, the molecule was so complex that working out its structure took years.

Nevertheless, by 1955 Hodgkin had completed the job, and as a result was awarded the Nobel Prize for chemistry in 1964.

IN ADDITION

Winston Churchill resigned as prime minister of Great Britain and retired from public life. He was succeeded by his foreign minister, Anthony Eden (1897–1977).

Malenkov resigned as head of the Soviet government after less than two years on the job and was succeeded by Nikolay Aleksandrovich Bulganin (1895–1975).

Perón of Argentina was deposed in a military coup.

War began between North and South Vietnam, with China supporting the North and France the South.

In the United States, the Supreme Court began to outlaw various aspects of segregation. In Montgomery, Alabama, a black woman named Rosa Parks refused to give up her bus seat to a white man, a small act in itself but it led to increasing black demands to be treated like human beings.

1956

DETECTION OF THE NEUTRINO

Pauli had suggested that the neutrino existed (see 1931), but there had seemed no chance of ever detecting it. With no charge, and possibly no mass, and very little in the way of interaction with other particles, it had no handle, so to speak; there was nothing to seize it by.

Now, however, nuclear fission reactors existed, which could release floods of neutrinos. (Actually, the fission reaction involved the change of neutrons to protons, which liberated *antineutrinos* rather than neutrinos. However, if antineutrinos existed, no scientist alive would doubt for a minute that neutrinos existed also.)

Although antineutrinos scarcely interact with other particles, every once in a while one out of trillions will collide squarely with a proton, converting it to a neutron (the opposite of the change that produced the antineutrino in the first place) and liberating an antielectron (positron) as well. It was necessary to watch for the simultaneous appearance of a neutron and an antielectron, together with gamma rays of a certain energy arising after a certain interval.

In 1956 two American physicists, Frederick Reines (b. 1918) and Clyde Lorrain Cowan (b. 1919), set up the necessary detection system and snared a few antineutrinos.

Eventually, neutrinos themselves were detected in solar radiation.

CONSERVATION OF PARITY

Physicists had worked out conservation laws that dictated conservation of energy, momentum, angular momentum, and electric charge, among others. In every case, this meant that the total quantity of that property in a closed system (one that did not interact with objects outside the system) could not change, no matter what happened within the system. The assumption was that such conservation laws were universal.

The study of subatomic particles showed that these conservation laws held in the subatomic realm as well. In addition, new conservation laws were discovered, such as the *conservation of parity.* Parity was the quality of being either odd or even. Just as in numbers, odd parity plus odd parity equaled even parity; even parity plus even parity equaled even parity; but odd parity plus even parity equaled odd parity. Each particle was assigned a particular parity, either odd or even, so that the total of all the particles in a closed system was either odd or even. No matter what happened to the particles within the system, if it began even, it ended even, and if it began odd, it ended odd. At least, so it was assumed.

Then trouble arose with kaons. Sometimes kaons broke down to two pions, which together had even parity; and sometimes to three pions, which together had odd parity. It was concluded that there were two kinds of kaons, one with even parity and one with odd. However, no one could detect any difference between the two kinds of kaons or predict which one a particular kaon would be.

In 1956 two Chinese physicists, Yang Chen Ning (b. 1922) and Lee Tsung-dao (b. 1926), suggested that there was only one kind of kaon, but that since kaons broke down through the weak interaction, and since in the weak interaction parity was *not* necessarily conserved, then a kaon could break down into either two or three pions indiscriminately. They pointed out that if parity was conserved in the weak interaction, then in certain particle changes, electrons would come out in equal amounts, left and right. If parity was *not* conserved, then electrons would come out predominantly in one direction. The experiment was performed, and the electrons came out predominantly in one direction.

This meant that although parity seemed to be conserved in the strong interaction and the electromagnetic interaction, it was not conserved in the weak interaction. As a result, Yang and Lee received the Nobel Prize for physics in 1957.

This did not mean, by the way, that conservation of parity really broke down altogether. It might merely mean that parity had to be combined with another property for both to be conserved. For instance, if particles gave off electrons predominantly in one direction, antiparticles gave them off predominantly in the other direction. The combination was called C-P *(charge conjugation and parity),* so scientists decided there was a law of *C-P conservation.*

ANTINEUTRON

Once the antiproton was discovered (see 1955), its properties could be studied. If it encountered a proton, the two oppositely propertied particles annihilated each other and their masses were converted into energy in accordance with Einstein's equation (see

1905). If, however, a proton and antiproton did not actually collide but passed each other closely, they might not undergo annihilation, but the positive charge of the proton and the negative charge of the antiproton might neutralize each other across the small gap, leaving two uncharged particles. One would be the neutron, of course, but the other, as was shown in 1956, would have to be an *antineutron*.

This was a puzzle. When the notion of antiparticles had been raised, the only subatomic particles known were electrically charged. The electron was negative, so the antielectron had to be positive. The proton was positive, so the antiproton had to be negative. But since the neutron was uncharged, neither positive nor negative, what could the antineutron be to make it an antineutron?

As it turned out, the neutron, while neutral overall, must have equal amounts of positive and negative electric charge distributed within. This distribution must be not quite symmetrical, so that when the neutron spins (see 1925) it develops a magnetic field pointed in a particular direction. The antineutron must have its charge distributed asymmetrically in an opposite sense, so that when it spins in the same way as the neutron, the magnetic field points in the opposite direction.

A better understanding of charge distribution within neutral particles had to await further discoveries.

CONTINUOUS MASER

In the maser, as first developed by Townes (see 1953), the molecules were first raised to a high energy level and then allowed to drop down to a lower level, giving up the excess energy in a flash of coherent microwave radiation. There would then have to be a pause while the molecules were again raised to a higher level.

In 1956, however, the Dutch-born American physicist Nicolaas Bloembergen (b. 1920) devised a maser in which energy was on three levels rather than two, so that one of the upper levels could be storing while the other was emitting. For this *continuous maser*, he received a share of the Nobel Prize for physics in 1981.

TEMPERATURE OF VENUS

It had long been assumed that Venus, being closer to the Sun than Earth was, would be warmer than Earth, although its thick cloud layer would probably reflect much of the Sun's radiation. Furthermore, the cloud layer seemed to imply that Venus had a great deal of water, which might further moderate the weather, so that on the whole, Venus was thought to be a quite comfortable world.

However, every object gives off microwaves; the higher the temperature of the object, the shorter the wavelength of these microwaves. Once radio astronomy was developed, radio telescopes became delicate enough to detect microwaves being emitted by the planets.

In 1956 a team of American astronomers headed by Cornell H. Mayer studied the microwaves emitted by Venus's dark side. Their nature made it clear that something on Venus, either its surface or some layer in its atmosphere, was at a temperature far above the boiling point of water. The vision of Venus as a comfortable world was thus rudely shaken and was eventually destroyed.

RIBOSOMES

Once the electron microscope was turned upon the cell, far greater detail could be made out than had ever been seen before (see 1932). For instance, there were numerous small bodies, which were called *microsomes* (small bodies), distributed through the cytoplasm of the cell. The Romanian-born American physiologist George Emil Palade (b. 1912) studied them carefully by electron microscope and found they were not merely mitochondrial fragments, as some had thought, but independent bodies with a chemical composition quite different from that of mitochondria.

By 1956 Palade had shown that microsomes were rich in RNA (ribonucleic acid), and they were therefore renamed *ribosomes*. It was quickly realized that the ribosomes were the site of protein manufacture within the cell.

For this work, Palade was awarded a share of the Nobel Prize for physiology and medicine in 1974.

TRANSFER RNA

In 1956 the American biochemist Mahlon Bush Hoagland (b. 1921) discovered relatively small molecules of RNA in the cytoplasm. These came in different varieties, and Hoagland showed that each variety had the capacity to combine with a particular amino acid.

At the other end, the RNA molecule could combine with a particular spot on a ribosome. Once the RNA molecules were lined up properly, the amino acids at the other end of each would be lined up properly too and could easily combine to form particular proteins. Since these RNA molecules transferred the information from the ribosomes to the proteins, they were called *transfer RNA*.

But how did the transfer RNA manage to take up its positions to form the proper proteins? The DNA molecules in the chromosomes, which carried the information from parent to offspring, were buried deep in the cell nucleus, while the transfer RNA was in the cytoplasm.

As it happens, the nucleus contains RNA molecules also. Two French biologists, Jacques-Lucien Monod (1910–1976) and François Jacob (b. 1920), suggested that the information on the DNA molecule was transferred to an RNA molecule that had used one of the DNA strands as a model in its formation. These RNA molecules carried the information (message) out into the cytoplasm and were therefore called *messenger RNA*.

Each transfer-RNA molecule had a three-nucleotide combination at one end that fit three-nucleotide combinations at certain spots on the messenger RNA. So when the messenger RNA settled on the ribosome surface, various transfer-RNA molecules lined up, trinucleotide to appropriate trinucleotide, and the amino acids on the other end combined.

Thus information was transferred from the DNA in the chromosome to the messenger RNA, which traveled out from the nucleus to the ribosomes in the cytoplasm and gave the information to the transfer-RNA molecules, which transferred the information to the amino acids and formed the protein.

Three adjacent nucleotides along the DNA molecule plus the messenger-RNA molecule plus the transfer-RNA molecule made a particular amino acid (as Gamow—see 1929—had first suggested might be the case). What geneticists had to do was to figure out which trinucleotide translated into which amino acid. In other words, they had to work out the genetic code.

PITUITARY HORMONES

The structure of oxytocin had been worked out by du Vigneaud (see 1954) and that of insulin by Sanger (see 1952). The Chinese-born American biochemist Choh Hao Li (b. 1913) used the methods they had pioneered, to tackle the protein hormones manufactured by the pituitary gland.

In 1956 he showed that the molecule of ACTH (*adrenocorticotrophic hormone,* which stimulated the production of steroid hormones like cortisone by the adrenal cortex) was made up of 39 amino acids in a specific order. Furthermore, the entire chain of the natural hormone was not essential to its action. Fragments consisting of little more than half the chain demonstrated major activity.

Li also showed that the melanocyte-stimulating hormone (MSH) of the pituitary gland possessed an amino acid chain with the same order, in spots, as the ACTH chain.

Also in 1956, Li isolated human growth hormone from the pituitary gland and worked out the exact order of the 256 amino acids that made up its chain. Its structure was quite different from the analogous hormone in cattle or swine. Whereas some hormones can be useful to human beings even if obtained from other animals, this is not true of the growth hormone.

IN ADDITION

In the United States, Eisenhower won reelection easily. In Montgomery, Alabama, Martin Luther King, Jr. (1929–1968) emerged as a leader in the civil rights movement.

In Africa, Morocco became independent on January 1, 1956, Tunisia on March 20, Sudan on January 1, and the Gold Coast (as Ghana) on September 17.

In Asia, on February 29, 1956, Pakistan was recognized as a nation independent of India, with a western and eastern portion—West Pakistan and East Pakistan—separated by a wide stretch of Indian territory.

On July 26, 1956, Egypt nationalized the Suez Canal. On October 29, Israeli forces, with British and French encouragement, invaded the Sinai Peninsula and drove toward the Suez Canal. By the end of the month, the British and French were bombing the canal area. However, both the Soviet Union and the United States insisted that action against Egypt stop, and by the end of the year it was all over.

1957

SPUTNIK

Nearly three centuries before, Newton had pointed out how a rocket could put a vehicle into orbit around the Earth. After Germany had developed the V-2 rocket during World War II, both the United States and the Soviet Union began to think of placing a rocket in orbit. It was naturally taken for granted by all Americans that the United States, with its advanced technology, would be first in the field.

It came as an enormous shock to the United States then, when on October 4, 1957, the first satellite went into orbit—and was Soviet. It was called *Sputnik I* (the Russian word for "satellite"), and it began the *Space Age.*

JODRELL BANK

A quarter-century after Jansky had detected radio waves from space (see 1932), the first really large radio telescope was built. It was 250 feet across, and it was built at Jodrell Bank Experimental Station in Great Britain, under the leadership of the British astronomer Bernard Lovell (b. 1913).

It took six years to build and was sufficiently near completion in 1957 to allow it to track the Soviet satellite Sputnik as it revolved about the Earth.

DETAILS OF PHOTOSYNTHESIS

It was much harder to study photosynthesis in detail than many other reaction systems that proceed in living tissue. Photosynthesis will only work in intact chloroplasts, so that one must deal with reasonably intact living plant cells rather than with extracts or chopped-up preparations. Then, too, photosynthesis works so rapidly that it would seem almost beyond hope to see what was happening.

However, the biochemical techniques that had been developed over the past two decades made it possible. The American biochemist Melvin Calvin (b. 1911) subjected plant cells to carbon dioxide labeled with carbon-14, allowing exposure for no more than seconds before mashing and killing the cells. He then subjected the contents to paper chromatography. Substances containing carbon-14 were separated and studied.

Progress was slow, but little by little, Calvin and his group discovered and isolated the intermediate products, deduced how they must fit together, and built up a scheme of photosynthesis that made sense.

By 1957 the main strokes were filled out with detail, and Calvin was awarded the Nobel Prize for chemistry in 1961 as a result.

GIBBERELLINS

There are plant hormones that encourage growth, differentiation of tissue, budding, flowering, and so on. One group of these, the *gibberellins,* had been isolated from a fungus of the genus *Gibberella* (hence their name). They were studied first in Japan before World War II, but it was not until 1957 that an awareness of the compounds reached the west. Gibberellins are used to increase the size of plants, particularly in the cultivation of grapes.

INTERFERON

In 1957 a group headed by the British bacteriologist Alick Isaacs (1921–1967) showed that, under the stimulus of a virus invasion, cells would liberate a protein with antiviral properties. It would counter not only the virus that stimulated the production but other viruses as well. The protein, produced more quickly than antibodies, was named *interferon.*

Unfortunately, though the interferons are produced by many species, each species has its own. Only human interferon will work well in human beings, and humans produce it only in traces.

SABIN VACCINE

The Salk vaccine had been proved effective against poliomyelitis (see 1954). However, the Salk vaccine consisted of dead virus that had to be injected, and whose ability to stimulate antibody production might not be long-lasting. The Polish-American microbiologist Albert Bruce Sabin (b. 1906) thought it might be possible to find strains of polio that were too feeble to produce the disease even while alive but that would activate antibody formation and continue to do so as long as they remained in the body. Such live strains could be taken by mouth.

When Sabin thought he had the proper strains, judging by animal experiments, he tried them first on himself, then on prison volunteers. In 1957 the Sabin vaccine came into widespread use in the Soviet Union and eastern Europe. Three years afterward, it came into use in the United States as well.

PACEMAKER

The heart beats regularly, speeding up when exertion or emotion increases the oxygen requirements of the body and slowing down again in repose. For half a century, it had been known that a special patch of cells in the heart initiated the beat, and the patch was called, popularly, the pacemaker. When the pacemaker was diseased or damaged, the heartbeat could not be maintained properly and death might ensue.

Then an artificial pacemaker was devised that used a regular electrical pulse to initiate the heartbeat. At first such things were so bulky they had to be carried outside the body. The first pacemaker that was compact enough to be inserted under the skin in the patient's chest was devised in 1957 by the American physician Clarence Walton Lillehei. Pacemakers are now common among the elderly population.

TUNNEL DIODE

The Japanese physicist Leo Esaki (b. 1925) was working with tiny crystal rectifiers (*semiconductor diodes*) and found, in 1957, that on occasion their resistance decreased with current intensity rather than increasing as expected. This was caused by a *tunnel effect*, an ability on the part of electrons to penetrate barriers that were perhaps a hundred atoms thick, as though they were tunneling through. Electrons could do this because they had their wave aspects and could be, in their particle aspect, at any part of the wave. If the wave extended for a hundred-atom thickness, the electron could find itself, every once in a while, on the other side of an insulating barrier, even though this would have been thought impossible in classical physical theory.

The barrier-crossing electrons could be used for switching purposes, and *Esaki tunnel diodes* were ultrasmall and ultrafast. For his discovery, Esaki received a share of the Nobel Prize for physics in 1973.

BORAZON

A boron atom has one less electron than a carbon atom, while a nitrogen atom has one more electron than a carbon atom. If boron nitride, with a molecule consisting of one atom of boron and one of nitrogen, is heated and compressed under conditions that would form diamond from graphite (see 1955), the boron nitride takes on the diamond configuration too and is then known as *borazon*. This trick was accomplished in 1957.

The electron arrangement in borazon is just as in diamond. The slight asymmetry resulting from the alternation of boron and nitrogen nuclei keeps borazon from being quite as hard as diamond, but borazon is the more useful at high temperatures. At 900° C diamond will burn and turn into carbon dioxide, but borazon survives unchanged.

IN ADDITION

Prime Minister Eden of Great Britain, thoroughly discredited by the Suez crisis, resigned on January 9, 1957, and was replaced by Harold Macmillan (1894–1986).

President Eisenhower reacted to the Suez crisis by announcing the *Eisenhower doctrine,* which stated that the United States would give aid to any Mideast power trying to withstand Communist aggression.

1958

MÖSSBAUER EFFECT

Ordinarily, when an atom emits a gamma ray, it recoils. The wavelength of the gamma ray depends in part on the extent of this recoil. Since this varies somewhat from atom to atom, the gamma rays emitted show a spread of wavelength.

The German physicist Rudolf Ludwig Mössbauer (b. 1929) studied conditions under which atoms that were part of a crystal would emit a gamma ray in such a way that the recoil would be spread over all the atoms making up the crystal. The recoil is then vanishingly slight, and the gamma ray wavelength shows no spread due to that recoil. As a result, the crystal emits a sharply monochromatic beam of gamma rays, and this, discovered in 1958, is called the *Mössbauer effect*.

Gamma rays emitted in this way by one crystal will be easily absorbed by another crystal of the same type, but if the wavelength varies even slightly in either direction, absorption will not take place. For this work, Mössbauer received a share of the Nobel Prize for physics in 1961.

SOLAR X RAYS

The firing of rockets beyond the atmosphere made it possible to detect X rays coming from astronomical objects. Such X rays could not be detected from Earth's surface, because the Earth's atmosphere absorbed them.

In 1958 the American astronomer Herbert Friedman (b. 1916) observed the Sun during a total eclipse, by means of rocket-borne instruments, and detected X rays coming from the Sun's corona.

Two years before, he had shown that solar flares emitted X rays, but that was not surprising in view of the fact that flares were clearly very energetic solar explosions. X rays from the apparently quiet corona were more surprising, but this supported the claim of the Swedish physicist Bengt Edlen (b. 1906), who in 1940, as a result of studying the ultraviolet radiation from the Sun, had claimed that the corona must have a temperature of a million degrees or so.

This does not mean that the Solar corona is a great reservoir of heat. It is a volume of extraordinarily thin gas, where the individual atoms have a great deal of heat individually (hence the temperature), but where there are so few atoms that the total heat over the entire corona is not as great as the temperature indicates—by far.

MAGNETOSPHERE

In 1958 the United States entered the Space Age. The Soviet Union had placed two satellites in orbit in 1957, Sputnik I on October 4 and Sputnik II on November 3. The latter carried a dog, the first living animal to be placed into orbit.

The first successful American satellite was *Explorer I,* which was launched on January 31, 1958. It carried counters designed to estimate the number of charged particles in the upper atmosphere, and detected about the expected concentrations of particles at heights of up to several hundred miles, but at higher altitudes the number fell to zero.

Two other satellites launched soon afterward, one by the United States and one by the Soviet Union, recorded the same phenomenon.

The American physicist James Alfred Van Allen (b. 1914) did not believe the count could really fall to zero. He felt that what happened was that the count went so high it put the counter out of action.

When *Explorer IV* was launched by the United States on July 26, 1958, it carried special counters that were shielded with a thin layer of lead to keep out most of the radiation (rather like wearing sunglasses to protect the eyes). The radiation that penetrated the lead was not enough to overwhelm the counters, and now the count went up and up and up with increasing altitude—far higher than scientists had expected.

It appeared that surrounding the Earth, outside the atmosphere, there were belts containing high concentrations of charged particles that moved along the lines of force of Earth's magnetic field. These particles approached the Earth's surface in the neighborhood of the Earth's magnetic field. There they were responsible for the aurorae and, at times of unusually high concentration, for magnetic storms that affected the compass and electronic equipment. These belts were at first called the *Van Allen belts* but were eventually referred to as the *magnetosphere.*

This was the first important discovery—an entirely unexpected one—to be made as a result of the launching of artificial satellites.

NOBELIUM

The effort to form elements with higher and higher atomic numbers set a new record in 1958 with the formation of a few atoms with the atomic number of 102. There was some delay while the identity of the new atoms was confirmed, and when they were, the element was given the name of *nobelium* after Nobel (see 1866).

PHOTOCOPYING

An important aspect of office procedures is the copying of documents. Copying by hand is slow and cumbersome, and errors inevitably arise. Carbon paper and mimeograph machines were great improvements but usually messy.

An American physicist, Chester F. Carlson (1906–1968), strove to find a method of copying that would use dry powder, electric charge, and light. Because nothing moist is used, the procedure he found is called *xerography* (Greek for "dry writing"), and because light is used, it is called *photocopying.* It works by giving paper a positive electric charge and the powder a negative electric charge, so that the powder clings to those places where light does not penetrate and destroy the charge. In other words, the powder clings to the shadows cast by the opaque printing of the object being copied. The application of heat fixes the powder on the paper and the copy is produced. There is no mess and no moisture, and it can be done very quickly.

Carlson worked on the method for some twenty years and finally perfected it for office use in 1958. When he introduced the device, he called it *Xerox.*

IN ADDITION

The Middle East was growing increasingly turbulent. Faisal II of Iraq was assassinated on July 14, 1958, and the monarchy in that nation came to an end.

Civil war erupted in Lebanon, and on July 15, 1958, the United States landed marines there. They kept order, established a new government, and left.

The European empires continued to break up. France's hold over Tunisia, Guinea, and Madagascar slipped. The continuing revolt in Algeria destroyed the French Fourth Republic, which had been established after World War II. Charles de Gaulle returned to power, and on September 18, 1958, a *Fifth Republic* was established in which the president (de Gaulle) was granted much wider power.

In the Soviet Union, Nikita Khrushchev (1894–1971) replaced Bulganin as leader on March 27, 1958.

In 1958 rocket planes attained speeds equal to six times the speed of sound, and the nuclear submarine *Nautilus* crossed the Arctic Ocean under the ice, passing over the North Pole.

1959

MOON PROBES

On January 2, 1959, the Soviet Union launched *Lunik I.* This was the first rocket to surpass escape velocity (7 miles per second), so that it should recede from Earth indefinitely. It was aimed in the direction of the Moon, so that it was the first *Moon probe.* It missed the Moon by a considerable margin and took up an independent orbit about the Sun, so that it became the first *artificial planet.*

On September 12, 1959, the Soviet Union launched *Lunik II,* which was better aimed. It struck the Moon, and for the first time in history, an object made by human beings rested on the surface of another world.

Then, on October 4, 1959, two years to the day after the first satellite launch, the Soviet Union sent *Lunik III* behind the Moon. It sent back the first photographs of the far side, the side never before seen by human beings. Forty minutes of photos were taken from a height of 40,000 miles above the lunar surface. The photographs were fuzzy, but they showed that the far side was riddled with craters as the near side was although lacking the relatively crater-free maria, or *seas,* that existed on the near side. Why the two hemispheres are so different remains a mystery.

SHAPE OF THE EARTH

The previous year, the United States had launched a small satellite, *Vanguard I.* It revolved about the Earth in 2½ hours and its orbit could be studied in great detail. Its perigee (the point of its closest approach to the Earth) moved somewhat with each revolution, in part because of the gravitational pull of the Earth's equatorial bulge upon it.

By 1959, after *Vanguard I* had made thousands of revolutions, it was clear that the perigee was slightly more affected by the bulge south of the equator than by the bulge north of it. This meant that the Earth was a little bulgier (to the extent of about 25 feet) south of the equator than north.

The shape of the Earth was, in this way, more accurately determined than would have been possible by any reasonable Earth-bound observations. This was an indication of the way we could learn more about the Earth itself by going out into space.

SOLAR WIND

It had been recognized for some time that solar flares (see 1859) were highly energetic phenomena on the surface of the Sun. Occasionally, when a solar flare appeared on the solar surface, it was followed, after the lapse of some days, by a magnetic storm on Earth. Something was apparently emitted by the flares that eventually reached Earth.

The American physicist Eugene Newman Parker (b. 1927) had argued the year before that the Sun constantly emitted charged particles in every direction and that these drifted outward through the Solar System, passing Earth. He called them the *solar wind*.

A solar flare might therefore be the source of an unusually large gust of such particles, which on arriving at Earth would intensify the usual effects of the solar wind.

The existence of the solar wind was verified by Lunik II and Lunik III on their way toward the Moon in 1959, and later probes did the same.

SHAPE OF THE HEMOGLOBIN MOLECULE

Six years earlier, Sanger had determined the exact order of the amino acids in a long protein chain (see 1952, Insulin Structure), but even that was not sufficient to describe the structure of a protein totally. The amino acid chain bent and curved and was hooked together by a variety of chemical bonds.

Generally, an enzyme molecule did not work as a simple amino acid chain. It was the intricate three-dimensional folding of the amino acid chain that brought certain amino acids together and presented a surface on which chemical reactions could take place with great ease—though they could take place only with the greatest difficulty otherwise.

The Austrian-born British biochemist Max Ferdinand Perutz (b. 1914) undertook to work out the three-dimensional structure of hemoglobin. The logical way to do this was by the study of X-ray diffraction patterns, but the patterns obtained, while sufficient to be useful in working out the structure of small molecules such as vitamin B-12 or simple chains such as the DNA double helix (see 1953), were not helpful in working out the intricacies of the protein molecule.

Perutz found, however, that if he added a single atom of a heavy metal like gold or mercury to each molecule of protein, these atoms would strongly diffract X rays. The resultant pattern was sufficiently pronounced to give better hints to molecular structure.

By 1959 he had worked out the three-dimensional position of every atom in the hemoglobin molecule. His student, the British biochemist John Cowdery Kendrew (b. 1917), did the same for the similar but somewhat simpler molecule of myoglobin some time later.

As a result, Perutz and Kendrew shared the Nobel Prize for chemistry in 1962.

HOMO HABILIS

By now it was known that the two varieties of *Homo sapiens*, modern human beings and Neanderthal man (see 200,000 B.C.), had been preceded by the smaller-brained *Homo erectus* (see 500,000 B.C.), who may first have appeared on Earth as much as 1.5 million years ago. Before that there were various species of the genus *Australopithecus* (see 4,000,000 B.C.).

The australopithecenes, whose existence seems to have overlapped that of *Homo erectus*, were hominids in that they more closely resembled human beings than they resembled any ape, living or extinct. Nevertheless, they were sufficiently primitive to be kept out of genus *Homo*. Did they develop directly into *Homo erectus,* or was there an intermediate form?

The British anthropologist Louis Seymour Bazett Leakey (1903–1972), together with his wife, Mary, carefully explored the Olduvai Gorge in what is now Tanzania. There, on July 17, 1959, they came across the first fragments of a skull that, when pieced together, seemed to be a relic of the earliest known representative of genus *Homo,* a species that came into existence nearly 2 million years ago.

This species, intermediate between the australopithecines and *Homo erectus,* was eventually named *Homo habilis* (handy human). *Homo habilis* seemed to be the first hominid capable of shaping stone tools. Until then, all tools of earlier hominids, and of nonhominids for that matter, had been such things as leaves, twigs, shells, bones, and unmodified stones.

SPARK CHAMBER

Bubble chambers (see 1953) had turned out to be very useful, especially for the detection of ultra-short-lived particles. However, they cannot be triggered by desired events, but record everything. This means that myriads of tracks have to be searched through to find those of significance. Cloud chambers could be set for desired events only, but were insufficiently sensitive to show the newer particles. Something was needed that would be selective as cloud chambers and as sensitive as bubble chambers.

This need was met by the *spark chamber,* in which incoming particles ionized neon gas that was crossed by many metal plates. The ions then conducted an electric current that showed up as a visible line of sparks, marking the passage of the particles. The device could be adjusted to react only to those particles that were under study.

The first practical spark chamber was constructed in 1959 by two Japanese physicists, Saburo Fukui and Shotaro Miyamoto.

COLOR VISION

For a century it had been thought that three basic colors were sufficient to reproduce the entire color range, and to combine into white light. The three colors were red, green, and blue. These colors are, in fact, used in color television to produce all the other colors. Furthermore, in the retina of the eye there are three types of cells, one to react with each of the three basic colors.

In 1959, however, Land, who had invented Polaroid and the Land camera (see 1932), advanced a new theory of color vision. He maintained that only two different wavelengths of light are needed, and that they don't even have to be very sharp. One light could be ordinary white light with an average wavelength in the yellow-green. This would serve as the *short-wave light.* Red light would serve as the *long-wave light.* Thus, red and white in combination could present the full color range. Land produced a system of color photography based on this theory that reduced the cost of the process.

IN ADDITION

On January 3, Alaska joined the Union as the forty-ninth state, the first to be separated from the remaining states by a foreign power (Canada). Then, on August 21, Hawaii joined the Union as the fiftieth state, the first that was not part of the North American continent.

In Cuba, the corrupt dictator Fulgencio Batista y Zaldivar (1901–1973) fled the nation. Fidel Castro (b. 1927), who had been leading a rebellion against Batista for six years, became leader of the nation on February 16.

Japan introduced transistorized television sets,

and as these spread, the TV repairman, who had been almost a live-in adjunct of the sets during the first decade of their existence, gradually became an endangered species.

1960

LASER

The principle of the maser, which produced an intense, coherent, monochromatic beam of microwaves (see 1953), could be applied to any wavelength, including those of visible light. This had been pointed out by Townes (see 1953).

The first maserlike device capable of producing an intense, coherent, monochromatic beam of visible light was constructed in May 1960 by the American physicist Theodore Harold Maiman (b. 1927), making use of the three-level principle worked out by Bloembergen (see 1956).

Maiman designed a ruby cylinder with its ends carefully polished flat and parallel and covered with a thin silver film. Energy was fed into it from a flash lamp until it emitted a beam of red light.

The coherent light so produced had only a slight tendency to spread and could be concentrated into so tiny a point that, at that point, temperatures could be reached far higher than the surface of the Sun. The device was first called an *optical maser*, but since it could be described as *light amplification by stimulated emission of radiation,* the initials of that phrase were used and it came to be called a *laser*.

Lasers soon proliferated into many different types, with many different uses.

PROOF OF GENERAL RELATIVITY

The general theory of relativity, first advanced by Einstein (see 1916), had been confirmed in three ways, all of them astronomic and all of them very borderline: (1) the advance of Mercury's perihelion (see 1846), (2) the bending of light in a gravitational field (see 1919), and (3) the reddening of light in a gravitational field (see 1925).

Now, thanks to the Mössbauer effect (see 1958), it was possible to test general relativity in the laboratory. Suppose a beam of monochromatic gamma rays were shot down a shaft from the top of a building to the bottom. The Earth's gravitational field would be very slightly stronger at the bottom of the building than at the top, since the bottom is nearer the Earth's center. This strengthening of the gravitational field would serve to increase the wavelength of the gamma rays, according to general relativity. The strengthening of the field and consequent increase in wavelength would be almost immeasurably small but large enough to cause the absorption of the gamma rays by a crystal at the bottom of the shaft to decrease considerably.

This experiment, when conducted in 1960, proved to support general relativity (as have all experiments and observations since that time).

STANDARD METER

When the metric system was first established (see 1790), the basic standard of length (the *meter*) was set as 1/10,000,000 the distance from the equator to the North Pole. This distance could not be established with sufficient accuracy, however, so the standard meter was defined as the distance between two scratches on a rust-resistant platinum and iridium bar, which was kept in an air-conditioned vault in a Paris suburb, where it was immune to change with temperature alterations.

By 1960, however, scientific advance made it possible to turn to something far more constant. The General Conference of Weights and Measures set the standard meter equal to 1,650,763.73 wavelengths of one of the spectral lines of an isotope of the rare gas krypton. This gave the meter a thousand times the precision of the platinum-iridium bar.

INTEGRATED CIRCUITS

Transistors had been in existence for a dozen years (see 1948), constantly being made smaller and more reliable. By 1960 they could be made so small that it made no sense to try to handle them as separate units.

Instead, small pieces of thin silicon or some other semiconductor, about a quarter-inch square, were etched with tiny transistor circuits. These *chips* did the work of many transistors and were called *integrated circuits.*

The use of integrated circuits made computers smaller, cheaper, and more versatile. As time went on, more and more circuits—eventually thousands—could be etched into a single chip.

RESONANCE PARTICLES

The bubble chamber (see 1953) was excellent at detecting ultra-short-lived particles. The American physicist Luis Walter Alvarez (1911–1988) constructed huge bubble chambers and, beginning in 1960, detected particles that existed for only a few trillionths of a trillionth of a second before breaking down. The tracks they left in this incredibly short lifetime, even if they moved at the speed of light, were too small to detect directly. The existence of these so-called *resonance particles* could only be deduced from the nature of their longer-lived breakdown products.

The resonance particles, all hadrons (see 1952, Kaons and Hyperons), came to be discovered in great numbers until something like 150 had been found. It seemed impossible for so many different particles to exist separately, so the search was on for still simpler and far fewer particles that, in different combinations, might make up all these resonance particles.

For this work, Alvarez was awarded the Nobel Prize for physics in 1968.

SEA-FLOOR SPREADING

Once it was understood that the Earth's crust was divided up into a few large plates and some smaller ones, it seemed unlikely that the plates retained their position forever. Although Wegener's theory of continental drift (see 1912) seemed quite impossible, since the continents could not plow their way through the underlying rock, the similarity between the opposite coasts of the Atlantic Ocean might be accounted for in other ways.

In 1960 the American geophysicist Harry Hammond Hess (1906–1969) decided it was quite possible that molten magma from the mantle might ooze upward through the Great Global Rift (see 1953). That might force the North American and South American plates farther westward while the Eurasian and African plates were forced further eastward.

The Atlantic ocean would thus widen, in what came to be called *sea-floor spreading*, but its coasts would retain their shape from the time the continents were in contact.

Continental drift, then, was not the result of continents slowly floating on the underlying rock, as Wegener had thought. The continents were firmly fixed to the plates and could not move through them. But the plates themselves were forced apart in some places and forced together in others. Actual evidence for this new view was not long in coming.

WEATHER SATELLITES

Satellites intended primarily for the observation of the Earth began to be launched. The first was *Tiros I*, launched by the United States on April 1, 1960. In November *Tiros II* was launched, and in ten weeks it sent down over 20,000 photographs of vast stretches of Earth and its cloud cover. It took photos of a cyclone in New Zealand and of a patch of clouds in Oklahoma that was apparently spawning tornadoes.

These and other *weather satellites* that followed became indispensable in tracking hurricanes and other violent storms and in helping to predict weather. Innumerable lives must have been saved and enormous property damage avoided because of timely warnings of the approach of hurricanes (though they remain potentially dangerous and devastating even so). Such things must be considered when people argue that space exploration is only a matter of pyramid-building designed to appease national vanity and scientists' curiosity.

CYCLIC-AMP

Adenylic acid, also known as adenosine monophosphate (AMP), is one of the nucleotides that make up the molecular chains of nucleic acids. The American pharmacologist Earl Wilbur Sutherland, Jr. (1915–1974) had discovered it in tissue some years earlier, and in 1960 he worked out its structure. He found that the phosphate group was attached to the rest of the molecule in two different places rather than one. A ring of atoms was thus formed, and Sutherland called it *cyclic-AMP*.

Cyclic-AMP has a profound effect on the course of metabolism within a cell, since it apparently controls the way hormones can penetrate cells. For this work, Sutherland was awarded the Nobel Prize for physiology and medicine in 1971.

CHLOROPHYLL SYNTHESIS

In 1960 Woodward (see 1944), who specialized in synthesizing complex organic molecules, succeeded in synthesizing chlorophyll.

IN ADDITION

John Fitzgerald Kennedy (1917–1963) was elected the thirty-fifth president of the United States.

On May 1, 1960, a supersonic U-2 plane, spying on the Soviet Union for the United States, was shot down over Soviet territory by the Soviets. The United States said it was just off course, but the pilot was brought down alive and the American story was refuted.

Cuba nationalized its banks and industries, which hurt American industrial interests. The United States grew hostile, so Cuba turned to the Soviet Union for help.

The year saw most of the African colonies of France, Great Britain, and Belgium break away and become independent. In Europe, the island of Cyprus gained its independence from Great Britain on August 16, 1960.

World population passed 3 billion, and cities

outside Europe and the United States were beginning to increase rapidly in size. Tokyo passed both New York and London to become the largest city in the world with nearly 10 million people. The population of the United States reached 185 million; that of the Soviet Union 215 million.

1961

HUMAN BEINGS IN SPACE

The Soviet Union had put a dog into orbit four years before, where it was eventually put to death painlessly. A year before, the Soviet Union had put two dogs in space and brought them back alive. What followed was inevitable.

On April 21, 1961, Yury Alekseyevich Gagarin (1934–1968) was put into orbit by the Soviet Union in the spaceship *Vostok I.* He circled the Earth once in 89 minutes and was brought back alive. On August 6, 1961, a second Soviet cosmonaut, Gherman Stepanovich Titov (b. 1935), was put into orbit and circled the Earth seventeen times, remaining in space for a full day.

MICROWAVE REFLECTIONS FROM VENUS

Fifteen years before, microwaves had been reflected from the Moon. That had been comparatively easy. By 1961 the technique had advanced to the point where microwaves could be sent out to Venus, a hundred times as far away as the Moon. This was done, and reflections were received by five different groups, one Soviet, one British, and three American.

The microwaves traveled through space at the speed of light, and from the time measured between the emission of the original beam and the detection of the reflection, the distance of Venus and therefore the general scale of the Solar System could be determined much more accurately than had been possible from observations of the asteroid Eros (see 1941).

HELIOSPHERE

The previous year, the United States had put up *Echo I,* which expanded into a large sphere of aluminum foil from which microwave reflections could be obtained. By 1961 its orbit had been followed so carefully that the Belgian physicist Marcel Nicolet could calculate the drag of the thin wisps of atmosphere present at the heights through which the satellite was moving. The large volume of the satellite and its comparative lightness magnified that drag so that it was possible to deduce the density of the air at a height of from 200 to 600 miles above Earth's surface and to show that it consisted largely of helium. This region was known as the *heliosphere.*

Above it was a still more rarefied region consisting chiefly of hydrogen (the *protonosphere*). This was thought to continue thinning out for 40,000 miles before dwindling into the general density of the gases in interplanetary space.

QUARKS

In 1961 Gell-Mann, who had advanced an explanation of strange particles (see 1953), worked out a method for bringing order to the numerous hadrons that were being discovered.

He grouped them into families in which certain particle properties increased in value in a regular fashion and called it, whimsically, the *Eightfold Way*, with reference to certain Buddhist teachings. In doing so, he found that certain family groups had missing members, which he believed represented hadrons that had not yet been detected—much as Mendeleyev had predicted the existence of undiscovered elements from gaps in his periodic table (see 1869).

Independently, an Israeli physicist, Yuval Ne'emen (b. 1925), worked up similar groupings at about the same time.

In order to explain the existence of the families, Gell-Mann postulated the existence of unusual particles that he called *quarks* (from a phrase in *Finnegans Wake* by James Joyce). There were but a few of these quarks, each accompanied by an antiquark, and by grouping them in different combinations, either two or three at a time, the various hadrons could be accounted for.

The most startling thing about these quarks was that, if they were to produce hadrons properly, they had to have fractional electric charges. Some had charges of plus or minus ⅓ or ⅔. The notion of fractional charges was hard to take, but the quark theory explained so much that it was accepted perforce and earned for Gell-Mann the Nobel Prize for physics in 1969.

LAWRENCIUM

The attempt to produce more and more complex atoms did not cease. In 1961 a few atoms of element 103 were produced, and it was named *lawrencium* after Lawrence, the inventor of the cyclotron (see 1930), who had died three years earlier.

With lawrencium, the last of the actinides (see 1940) was discovered. There were now fifteen actinides as there were fifteen lanthanides.

THE GENETIC CODE

With the transition of genetic information from DNA to messenger RNA to transfer RNA to protein now well established (see 1956), the most important unanswered question was which trinucleotide corresponded to which amino acid. Without knowing that, one couldn't understand how information passed from DNA to proteins.

The American biochemist Marshall Warren Nirenberg (b. 1927) took the first step toward elucidating the problem. He made use of the enzyme discovered by Ochoa (see 1955) to form an RNA molecule out of a mixture that consisted of uridylic acid only. The RNA molecule that resulted was not one that existed naturally but a synthetic, *polyuridylic acid* (—U—U—U—U—U—), made up of a chain of uracil nucleotides only. The only possible trinucleotide in such an RNA molecule would be U—U—U.

When Nirenberg used polyuridylic acid as messenger RNA, it worked. Transfer-RNA molecules settled upon it and amino acids combined to form a protein. The protein was made up *only* of the amino acid phenylalanine.

This made it quite clear that the trinucleotide U—U—U was equivalent to the amino acid phenylalanine. Similar tactics revealed the equivalence of other trinucleotides to other amino acids until the entire genetic code was worked out.

As a result, a share of the Nobel Prize for medicine and physiology was awarded to Nirenberg in 1968. Of the two others who shared the prize, one was the Indian-born American chemist Har Gobind Khorana (b. 1922), who also did important work in elucidating the code.

GENE REGULATORS

Jacob and Monod, who had first suggested the existence of messenger RNA (see 1956), puzzled over the fact that different cells in a body had different chemistries even though they all had the same genetic makeup.

They decided that genes did not always work at the same rate, that there were devices that could slow down or speed up the workings of a particular gene, and that different cells had *gene regulators* that worked in different ways.

Eventually such gene regulators were isolated and found to be small protein molecules. As a result, Jacob and Monod, along with another coworker, André-Michael Lwoff (b. 1902), shared the Nobel Prize for physiology and medicine in 1965.

ELECTRONIC WATCH

In 1961 electronic watches were put on the market. Gone were the mainsprings and hair springs, gone the ticking sound, gone the daily necessity of winding. The watch's action was controlled by the vibration of a tiny tuning fork kept in steady action by a small battery. Replacing the battery once a year or so was all that was required.

IN ADDITION

On April 17, 1961, 1,600 Cuban exiles, trained by the C.I.A., landed at the Bahia de Cochinos *(Bay of Pigs)* on Cuba's southern coast. The C.I.A. was certain the Cuban people would rise against their oppressors, but they did not. Instead, Cuban armed forces mopped up the invaders quickly. Kennedy, unwilling to turn this into an American invasion, accepted it as an embarrassing fiasco.

Rafael Léonidas Trujillo Molina (1891–1961), right-wing dictator of the Dominican Republic, was assassinated on May 30, 1961.

Between August 15 and 17, 1961, East Germany built a wall around West Berlin, which it patrolled with armed guards. The flight of refugees from East Germany to West Berlin was reduced to a trickle.

The Soviet Union and China, which had been Communist allies since 1949, were coming to the parting of the ways. Nationalistic rivalries combined with Chinese suspicion of Soviet changes since the death of Stalin forced them apart.

In Africa the left-leaning leader of the Congo, Patrice Hemery Lumumba (1925–1961), was assassinated on January 17, 1961, by right-wing opponents in a civil war.

The Union of South Africa left the British Commonwealth on May 31, 1961, and declared itself a republic.

In the United States, President Kennedy established the Peace Corps, a means by which young volunteers could work in newly independent countries among people unused to self-rule. Biracial groups of *Freedom Riders* demonstrated for black civil rights in the South.

1962

AMERICAN IN SPACE

On February 20, 1962, the United States launched *Friendship 7*, which placed the first American in orbit. He was John Herschel Glenn, Jr. (b. 1921), who orbited the Earth three times and remained in space for 5 hours.

COMMUNICATIONS SATELLITES

Echo I (see 1961) could function as a passive communications satellite, since a radio beam could be aimed at it and the reflection would then reach Earth at a widely different spot. The reflection was too feeble to serve as anything more than a demonstration that it could be done, however.

On July 10, 1962, the United States launched *Telstar I*. It was a true communications satellite since it not only received radio waves but amplified them before sending them on. Thanks to *Telstar I* and to the numerous communications satellites that followed, it became possible to communicate easily, and in seconds, across continents and oceans. Communication capability is now truly global, and Earth is, in that respect, literally a global village.

VENUS PROBE

In the first five years of the Space Age, the Earth and the Moon had been the only targets. In 1962 the age of *planetary probes*—rockets designed to pass near other planets and send back information concerning them—was reached.

The first successful planetary probe was *Mariner 2*, launched by the United States on August 27, 1962. It was designed to approach the planet Venus, which comes closer to Earth than any other planet. *Mariner 2* was therefore a *Venus probe*, and it passed within 22,000 miles of Venus's cloud layer on December 14, 1962.

In the course of its trip, *Mariner 2* sent back irrefutable evidence of the existence of a solar wind (see 1959). It also detected the microwave radiation of the planet and showed definitely that its surface temperature was something like 475° C, hot enough to melt tin and lead and to boil mercury.

ROTATION OF VENUS

Although Venus approached Earth more closely than any other planet did, its period of rotation had remained mysterious. This was ironic, since the periods of rotation of other planets, even that of far distant Pluto (see 1955), were known. The reason for this was that Venus's thick and featureless cloud layer precluded any possibility of seeing its surface and detecting features that could be spotted as moving around the planet.

Yet although light waves could not penetrate the clouds, microwaves could. Furthermore, if a beam of microwaves is reflected from an object moving at right angles to the beam (as would be true of Venus if it were rotating), then the wavelength of the beam is broadened and distorted. From the extent of the distortion, the speed of rotational motion can be calculated.

In 1962 the American astronomers Roland L. Carpenter and Richard M. Goldstein were able to show that Venus had the astonishingly slow period of about 250 days. (The figure was later refined to 243.09 days.) What's more, Venus rotated in retrograde fashion, from east to west rather than from west to east as Earth and other bodies in the Solar System do. The reason why this is so is still not clear.

NOBLE GAS COMPOUNDS

Since the inert gases—helium, neon, argon, krypton, xenon, and radon—had been discovered, over half a century before, they had appeared to be truly inert, since their atoms would form bonds with no other atoms. This made increasing sense when their electronic structure was worked out, since in all cases their outermost electron shell was full so that they had no tendency to gain, lose, or share electrons.

This tendency was not absolute, however.

Linus Pauling (see 1931) had pointed out that the inert gas atoms grew less inert as their atomic number increased, so that those with higher atomic numbers might be induced to form a bond with fluorine, which was the most active of all elements and the most apt to snatch an electron from unlikely places.

In 1962 the British-born Canadian chemist Neil Bartlett (b. 1932) found that a compound, platinum fluoride, was almost as active as fluorine itself and easier to work with. He immersed it in xenon gas and the two substances combined to form xenon fluoroplatinate. This was the first known case of an inert gas atom forming a bond with any other atom or group of atoms.

Thereafter, other compounds involving fluorine or oxygen were formed, not only with xenon but with radon and krypton. The smaller inert gas molecules—argon, neon, and helium—remained inert, however.

As a result, chemists no longer liked to use the term *inert gases,* preferring *noble gases* as less likely to signify absolute inertness. Compounds such as xenon fluoroplatinate are therefore now referred to as *noble gas compounds.*

APPROACHING ABSOLUTE ZERO

After Giauque had introduced his technique for attaining very low temperatures (see 1925), temperatures as low as 1/50,000 of a degree above absolute zero had been reached. That was not yet the ultimate, of course. Absolute zero can never be truly attained but it can be approached ever more closely as new techniques are developed.

In 1962 the German-born British physicist Heinz London (1907–1970) used a mixture of the two helium isotopes helium-4 and helium-3 for the purpose. Ordinarily they mix perfectly, but at temperatures below 0.8°, they separate. Processes that allowed the two isotopes to mix and then separate repeatedly offered a powerful new way of cooling. Thanks to this and to other techniques involving helium-3 and nuclear magnetic properties, temperatures of 1/1,000,000 of a degree above absolute zero have been attained.

LIGHT-EMITTING DIODES

Light-emitting diodes are semiconductor devices (see 1948) that emit visible light as their electrons drop from a higher energy level to a lower one. The first practical device of this sort was produced in 1962.

Such diodes are now routinely used where light needs merely to be seen and not to illuminate. Thus, light-emitting diodes are used in digital clocks, in pocket computers, in elevator floor indicators, in taxi meters, and as signals in electronic equipment generally.

ENVIRONMENT

It isn't often that a book intended for the general public makes the world aware of a scientific problem, but it happened in 1962, when *Silent Spring,* by the American biologist Rachel Louise Carson (1907–1964), was published. Her account of the effect of indiscriminate use of pesticides on the environment was riveting. She described the possibility that pesticides would kill birds, for instance, to the point where spring would finally arrive without birdsong.

The book was largely responsible for a sudden increase in awareness of environmental dangers on the part of a large segment of the public.

IN ADDITION

Cuba, alarmed at the Bay of Pigs invasion of the previous year, was willing to accept the installation of Soviet missiles, and Khrushchev was willing to supply them. The United States, learning of this, placed an embargo about Cuba in October 1962, to prevent Soviet arms from reaching it, and for a while the two superpowers stood eye to eye. A nuclear war seemed closer than at any time before or since. Fortunately, a compromise was reached. The Soviet Union withdrew its missiles from Cuba, and the United States withdrew its missiles from Turkey.

The civil war in South Vietnam intensified, and the United States sent "observers," as well as money and arms, to the South Vietnam government. This was the beginning of the *Vietnam War.*

Algeria finally broke its ties with France and became an independent nation on July 3, 1962. Other nations to gain independence (all from the British Empire) were Burundi, Uganda, and Tanganyika in Africa; Jamaica and Trinidad in the West Indies; and West Samoa in the Pacific.

Color television was becoming popular.

1963

QUASARS

Among the radio sources located in the sky during the 1950s were a few that seemed confined to very small areas. These compact sources were known as 3C48, 3C147, 3C196, 3C273, and 3C286. The *3C* is short for *Third Cambridge Catalog of Radio Stars,* a listing compiled by the British astronomer Martin Ryle (1918–1984).

In 1960 these sources were pinpointed, by such men as the American astronomer Allan Rex Sandage (b. 1926) and the Australian astronomer Cyril Hazard, and found to originate in certain objects that looked like dim stars. It seemed strange that dim stars would be such strong radio sources, and the feeling arose that they might be something other than stars. They were referred to eventually as *quasistellar radio sources,* where *quasistellar* means "starlike," and this was eventually shortened to *quasars.*

The spectra of quasars proved puzzling; the lines could not be identified. In 1963, however, the Dutch-born American astronomer Maarten Schmidt (b. 1929) realized that the lines could be identified if they were viewed as lines that would ordinarily be in the ultraviolet but had been displaced by means of an enormous red-shift.

Such a red-shift would mean that the quasars were enormously distant, over a billion light-years away, and the fact that they could be seen at all at such a distance meant they were unusual indeed. It was finally decided that they were galaxies with extremely active centers (like particularly large Seyfert galaxies—see 1943). At their distances, only the centers could be seen, so that they had a starlike appearance. Some quasars have been detected that appear to be over 12 billion light-years away.

ARECIBO RADIO TELESCOPE

In 1963 the largest single radio telescope ever built was put into use. It was located about 8 miles south of Arecibo, Puerto Rico, and is about 1,000 feet across. It is not steerable, however, but is fixed in place.

X-RAY SOURCES

Through most of history, the only information reaching us from the stars had consisted of visible light waves. For the past thirty years, microwaves from astronomical objects had also been studied. Both light and microwaves can penetrate Earth's atmosphere, but there are other forms of radiation that cannot. They can only be studied by rockets, carrying appropriate instruments, that are sent up beyond the atmosphere.

Thus, X rays from astronomical objects are completely absorbed by the atmosphere and cannot be observed from Earth's surface. Rocketry, however, had made it possible to detect X rays in the solar radiation.

Still more interesting news came in 1963, when Bruno Rossi, who had done important work on cosmic rays, began to use rockets to observe whether solar X rays were reflected from the Moon. He did not succeed in detecting such radiation. However, he did detect it emerging from the Crab nebula and also from another supernova remnant in the constellation of Scorpio.

As rocket observation became more common and more sophisticated, many other X-ray sources beyond the Solar System were located.

HYDROXYLS IN SPACE

Trumpler had shown that there were thin wisps of matter in interstellar space (see 1930), and van de Hulst and Purcell had shown that hydrogen atoms were strewn across space (see 1944 and 1951).

It was reasonable to suppose that all this interstellar gas consisted of single atoms, since they would be spread out so thinly that the chances of collision would probably be too small to allow them to combine.

Yet suppose there were collisions. The three most common atoms are hydrogen, helium, and oxygen. Helium atoms don't combine with other atoms, but two hydrogen atoms might combine to form a hydrogen molecule, and a hydrogen and an oxygen atom might combine to form a *hydroxyl group.*

On Earth, a hydroxyl group is so active that it quickly combines with other atoms. The result is that such a group does not exist in the free state on Earth. In space, however, the chances of a hydroxyl group striking anything are so small that it might accumulate uncombined.

If so, hydroxyl groups ought to emit microwaves of characteristic wavelengths, and in 1963 two of those wavelengths were detected, indicating the presence of hydroxyl groups in interplanetary space.

WOMAN IN SPACE

On June 18, 1963, with the Space Age nearly six years old, the Soviet Union launched *Vostok 6,* bearing Valentina Vladimirovna Tereshkova (b. 1937), the first woman to be placed into orbit about the Earth.

MAGNETIC REVERSALS

As early as 1906, the French physicist Bernard Brunhes had noted that in some rocks, crystals magnetized in directions *opposite* to that of Earth's magnetic field. This was ignored at first, but slowly it came to be realized that the Earth's magnetic field strengthened and weakened. The weakening might go all the way to zero and then begin to strengthen in the opposite direction. It might be, then, that *magnetic reversals* occurred at intervals in Earth's history.

If the Atlantic Ocean's floor had spread with the upwelling of magma in the Global Rift (see 1960), then the sediments laid down on the sea bottom ought to exist in strips. If magnetic reversals had occurred, these strips ought to reflect them.

On investigation, this turned out to be the case. On either side of the rift were sediments with normal magnetization, farther away on either side were sediments with reversed magnetization, then normal again, then reversed, and so on, quite symmetrically on either side.

This offered the best evidence so far for the existence of sea-floor spreading, and also for the periodic reversal, at irregular intervals, of the magnetic field.

Of course, if two adjoining plates are pushed apart, two adjoining plates in some other part of the Earth must be pushed together. Thus the study of plate tectonics finally made sense out of mountain-building, volcanoes and earthquakes, the development of ocean deeps and island arcs. In short, plate tectonics became the central dogma of geology, as evolution is of biology, the atomic theory is of chemistry, and the conservation laws are of physics.

IN ADDITION

President Kennedy was shot to death in Dallas on November 22, 1963. He was succeeded by his vice president, Lyndon Baines Johnson (1908–1973), who became the thirty-sixth president of the United States.

Three weeks before the assassination, on November 1, the American-supported ruler of South Vietnam, Ngo Dinh Diem (1901–1963), was deposed by a military coup and killed.

On August 5, 1963, the United States, the Soviet Union, and Great Britain signed a pact prohibiting the testing of nuclear bombs in the atmosphere, in space, or under water, but permitting underground blasts.

The tranquilizer Valium was introduced and eventually became the most commonly prescribed drug.

1964

BACKGROUND MICROWAVE RADIATION

The German-born American physicist Arno Allan Penzias (b. 1933) and the American radio astronomer Robert Woodrow Wilson (b. 1936) were attempting to determine the characteristics of any radio-wave emission that might come from the outer regions of the Galaxy. They made use of a big horn-shaped antenna originally built to detect radio reflections from the Echo satellite (see 1961).

In May 1964 they found an excess of radio-wave emission that they could not explain. When they had accounted for all possible sources of error (including pigeon droppings inside the antenna), they found that there was a distinct *background microwave radiation*, coming from all directions with equal intensity.

They turned to the American physicist Robert Henry Dicke (b. 1916), who remembered that Gamow had predicted such background radiation would occur as a consequence of the big bang (see 1948).

The background radiation was characteristic of a universe with an average temperature of 3 degrees above absolute zero, and it could be assumed that the Universe had cooled to that average temperature from that prevailing at the moment of the big bang. The background radiation, as a fossil remnant (so to speak) of the big bang, finally established that event as the very likely mechanism whereby the Universe came into being.

For this discovery, Penzias and Wilson received a share of the Nobel Prize for physics in 1978.

OMEGA-MINUS PARTICLE

Gell-Mann had suggested the Eightfold Way as a means of ordering hadrons into groups (see 1961). In one of the groups he had set up was an empty spot that ought to hold a particle with a particular set of properties if his method of arrangement was valid. Included in those properties was a strangeness number of -2, something no known particle had.

In 1964 a particle was found with precisely the characteristics that Gell-Mann had predicted, including the strangeness number of -2. This had much the impact of the discovery of the missing elements whose characteristics Mendeleyev had predicted (see 1869). From that moment, the quark theory had to be taken very seriously indeed.

CPT SYMMETRY

After Lee and Yang had shown that parity was not conserved in the weak interaction (see 1956), parity was combined with a particle characteristic called *charge conjugation* (which told whether the particle in question was an ordinary particle or an antiparticle) with the idea that if parity was unbalanced in one direction in a particular particle, charge conjugation would be unbalanced in the other, and the two together would be conserved.

This was called *CP (charge-parity) conservation*. But in 1964 two American physicists, Val Logsden Fitch (b. 1923) and James Watson Cronin (b. 1931), found that CP was not always conserved either. Neutral kaons (see 1952) in their decay, on rare occasions violated CP conservation. In order to keep the symmetry, time (T) had to be added. Where CP was asymmetric in one direction, T was asymmetric in the other, and the combination retained symmetry.

Physicists now speak of *CPT symmetry*. Cronin and Fitch shared the Nobel Prize for physics in 1980.

TRANSFER-RNA STRUCTURE

Transfer-RNA molecules, which act as intermediates between the nucleotide chain in messenger RNA and the amino acid chain in proteins (see 1956), are relatively small.

In 1964 the molecule of alanine-transfer RNA (the particular transfer RNA that attaches itself to the amino acid alanine) was completely analyzed by a team headed by the American biochemist Robert William Holley (b. 1922). As Sanger had worked with insulin (see 1952), so Holley worked with transfer RNA, breaking it down to fragments, identifying the fragments, and working out how they fit together. The alanine-transfer-RNA molecule turned out to be made up of a chain of seventy-seven nucleotides.

The chains of this transfer-RNA molecule, and others that were analyzed later, seemed to form three lobes, rather like a three-leaf clover. For this work, Holley received a share of the Nobel Prize for medicine and physiology in 1968.

MULTIPERSON SPACEFLIGHT

On October 12, 1964, the Soviet Union launched *Voshkod 1*, which carried three cosmonauts into orbit. It was the first rocket to carry more than one human being.

RUTHERFORDIUM

In 1964 both Soviet and American researchers reported the formation of atoms of element number 104. There was some dispute as to priority and name. The Soviets named it *kurchatovium* after Igor Vasilyevich Kurchatov (1903–1960), who had led the team that developed the Soviet nuclear bomb. The Americans named it *rutherfordium* after Ernest Rutherford.

IN ADDITION

In the United States, President Lyndon Johnson ran for election in his own right and won. He pushed the *Tonkin Gulf Resolution* through Congress on August 7, 1964. This gave him a free hand in Vietnam and the Vietnam War went into high gear.

In the Soviet Union, Khrushchev lost power on September 13, 1964. He was succeeded by Aleksey Nikolayevich Kosygin (1904–1980) as premier and Leonid Ilyich Brezhnev (1906–1982) as party leader.

Additional African nations won their independence. Malawi, Zambia, and Tanzania were all formed out of former British colonies. Southern Rhodesia (known as Rhodesia after Northern Rhodesia became Zambia) remained under a white minority government ruled by Ian Douglas Smith (b. 1919).

India's prime minister, Jawaharlal Nehru (1889–1964), died on May 27, 1964.

1965

MARTIAN CRATERS

The Mars probe *Mariner 4* had been launched by the United States on November 28, 1964. On July 14, 1965, it passed within 6,000 miles of the Martian surface. As it did so, it took a series of twenty photographs, which were turned into microwave signals, beamed back to Earth, and there converted into photographs again.

For the first time in history, close-up views of the Martian surface could be seen by human beings. What those photographs showed were craters very much resembling those of the Moon. There was no sign of canals.

The photographs did not cover very much of the Martian surface, but the general impression was one of a Moonlike world, not an Earthlike one. The notion of artificial canals and an advanced civilization (see 1877), already shaky because of increased knowledge concerning the thin, oxygen-free atmosphere of Mars, now died a final death.

ROTATION OF MERCURY

Schiaparelli had suggested that Mercury turned one face to the Sun at all times (see 1889).

This had begun to seem doubtful, since if it were true, the side of Mercury facing away from the Sun should be extremely cold. Microwaves detected from the dark side of Mercury in 1962 had indicated that it was considerably warmer than would be expected if it were eternally dark.

In 1965 two American electrical engineers,

Rolf Buchanan Dyce (b. 1929) and Gordon H. Pettengill, working with microwave reflections from the Mercurian surface, were able to show that it turned on its axis in about 59 days, despite the fact that it revolved about the Sun in 88 days. This meant that every portion of the planet received sunlight at one time or another.

Eventually, the rotation was found to be 58.65 days, just two-thirds of the period of revolution, so that Mercury showed the same side to Earth every second revolution.

SPACEWALKS

In 1965 human beings were able to leave their orbiting rockets and, held by a tether, remain free in space—within their spacesuits, of course. This was referred to as a *spacewalk.*

The first to take a spacewalk was the Soviet cosmonaut Aleksei Leonov, who left the rocket ship *Voskhod II* on March 18, 1965. The American astronaut Edward Higgins White II (1930–1967) left his ship, *Gemini 4,* on June 3, 1965.

SPACE RENDEZVOUS

Maneuverability in space was increasing rapidly. On December 15, 1965, the American satellite *Gemini VII,* having been in space for fourteen days, approached within several feet of the previously launched *Gemini VI.* This was the first *space rendezvous.*

COMMUNICATIONS SATELLITES

On April 6, 1965, the United States launched *Early Bird,* the first communications satellite intended primarily for commercial use. It made available 240 voice circuits and one television channel. In this year the Soviet Union also began to send up communications signals.

VENUS PROBE

The Soviet Union was making repeated efforts to explore Venus. During the course of 1965, one of its Venus probes struck Venus and became the first human-made object to reach another planet.

HOLOGRAPHY

Gabor had worked out the theoretical basis of holography (see 1947), a system of photography that recorded the interference patterns of an ordinary and a reflected beam of light. Holography made it possible to set up a real image in space, something that amounted to a three-dimensional photograph.

Once the laser had been invented (see 1960) and come into use, it proved an ideal light source for the purpose. In 1965 Emmet N. Leith and Juris Upatnieks, at the University of Michigan, were able to produce the first holograms.

MICROFOSSILS

Until this time the earliest fossils had been found in rocks of the Cambrian era and were a little over 600 million years old. However, the Earth is 4,500 million years old. This meant that the first seven-eighths of Earth's history seemed to show no signs of life. This seemed unlikely, since the earliest fossils were already well developed and quite specialized forms of life that could not have come into existence without a long evolutionary history. The trouble was, though, that prior to the Cambrian era, organisms had not yet developed shells and other hard parts, and softer tissues do not fossilize easily.

In 1965, however, the American paleontologist Elso Sterrenberg Barghoorn (1913–1984) worked with tiny bits of carbonized material in very ancient rocks and showed that they might represent bacteria living in the early eons of Earth's history. When these bits of material were eventually studied by electron microscope, they showed themselves unmistakably to be *microfossils*, or remnants of simple cells. Some were found in rocks as old as 3,500 million years.

Consequently it appeared that life arose no later than a billion years after the formation of the Earth.

PROTEIN SYNTHESIS

With the structure of proteins known in detail in some cases, thanks to the work of scientists such as Sanger (see 1952) and Perutz (see 1959), it became possible to think of synthesizing them too.

In 1965 the American biochemist Robert Bruce Merrifield succeeded in synthesizing insulin. In that same year, a Welsh biochemist, David Phillips, synthesized lysozyme.

IN ADDITION

The United States began a program of bombing North Vietnam, to bring the war to a rapid end. It didn't work, but it increased discontent among the draft-liable college population, which saw the war enduring and the number of American troops in Vietnam rising.

In Indonesia, the left-leaning government was overthrown by an army revolt. Hundreds of thousands of "Communists," the name given to ethnic Chinese, were massacred.

Rhodesia declared its independence from Britain on November 11, 1965.

Black demonstrations continued in the southern United States. On February 1, 1965, a black march on Selma, Alabama, where blacks were not allowed to vote, was led by Martin Luther King.

1966

MOON AT CLOSE QUARTERS

On February 3, the Soviet Moon probe *Luna 9* made the first soft landing on the Moon (one in which the landing vessel was not destroyed in the process) and took some photographs of the approaching surface. On June 2, the American Moon probe, *Surveyor 1*, did the same, taking many more photographs of better quality.

On April 3, the Soviet probe *Luna 10* was the first to be placed into orbit about the Moon. The first of a series of American probes to go into orbit about the Moon (the *Luna Orbiters*) followed, and the United States was able to map the entire surface of the Moon, both the side facing us and the side away from us, in full detail.

SPACE-DOCKING

On March 16, the American satellite *Gemini VIII* linked up with another orbiting vessel. This was the first actual docking of one space vessel with another—a maneuver essential if human beings were to be sent to the Moon and brought safely back.

IN ADDITION

The Vietnam War continued to escalate, and so did protests in the United States and other nations.

In China, the *Cultural Revolution*, a period of enforced radical Marxism, began.

On January 19, 1966, Indira Gandhi (1917–1984), daughter of Jawaharlal Nehru, became prime minister of India.

The independence movement continued. In Africa, Botswana and Lesotho became independent, and in South America, Guyana. All had been British colonies.

1967

PULSARS

For some years there had been indications of radio sources in the sky that changed intensity after brief intervals, but radio telescopes at that time were not designed to catch such brief "twinkles."

Then, in Great Britain, the British astronomer Anthony Hewish (b. 1924) supervised the construction of a device with 2,048 separate receivers spread out in an array that covered an area of nearly 3 acres. This was designed to detect brief changes in microwave intensities.

In July 1967 the array was put to work, and within a month a graduate student, Jocelyn Bell, had detected bursts of microwaves from a place midway between Vega and Altair. The bursts were astonishingly brief, lasting only a thirtieth of a second. Even more astonishing, they followed one another with remarkable regularity—at intervals that were finally measured as 1.33730109 seconds.

This object was eventually called a *pulsating star*, a phrase quickly abbreviated to *pulsar*.

Eventually hundreds of pulsars were located. Hewish was awarded a share of the Nobel Prize for physics in 1974.

VENUS'S ATMOSPHERE

The Soviet Union continued to send out Venus probes that didn't last long because of the fearsome temperature and pressure of Venus's atmosphere.

By 1967 the probes had made it clear that Venus's atmosphere was about ninety times as dense as Earth's—far denser than anyone had expected. That atmosphere is about 96.6 percent carbon dioxide, and virtually all the rest is nitrogen. There is about as much nitrogen in Venus's atmosphere as in Earth's, but it is dwarfed by the overwhelming quantity of carbon dioxide. The carbon dioxide produces a *runaway greenhouse effect* that makes Venus the hottest planet in the Solar System, hotter even than Mercury.

SPACE CASUALTIES

The Space Age, now ten years old, saw its first human casualties. On January 27, 1967, three American astronauts died while an Apollo capsule was being tested on the ground. They were Virgil Ivan (Gus) Grissom (1926–1967), who had orbited in *Gemini 3* in 1965; Edward White, who had been the first American to take a spacewalk (see 1965), and Roger Bruce Chaffee (1935–1967).

On April 24, 1967, the Soviet spacecraft *Soyuz* made its first flight, but during the return to Earth, the ship became tangled in its parachute lines and the cosmonaut, Vladimir Mikhaylovich Komarov (1927–1967), who had piloted the first multiperson spacecraft (see 1964), died. He was the first person to die in the course of an actual spaceflight.

HEART TRANSPLANTS

On December 3, 1967, the South African surgeon Christiaan Neethling Barnard (b. 1922) performed the first successful heart transplant in history. The patient received another person's heart and went on to live with it for an additional year and a half.

A period followed when heart transplants were performed in some numbers, but their benefits proved dubious and the ethical problems enormous. Their popularity subsided.

CLONES

It is possible to produce a complete plant from a portion of one in a way that does not involve sexual reproduction. A plant twig, for instance, can be grafted to the branch of another tree, even when the other tree is of another species. The twig may well grow and flourish there, and such a twig is called a *clone,* from the Greek word for "twig."

Simple animals, not too specialized, can regenerate an entire organism from a relative scrap. Sponges, fresh-water hydras, flatworms, and starfish are all noted for this. The new organisms may also be called clones, by analogy.

Among vertebrates, cloning does not occur spontaneously. Suppose, though, that the nucleus of a living skin cell of one individual is placed into an ovum of another individual, the ovum's own nucleus having been removed. The chromosomes of the skin cell may then replicate and produce new cells with the genetic equipment of the introduced skin cell. The ovum may thus produce an organism, not of its own original species but of the species from which the skin cell was taken. This, too, would be a clone.

The technique of replacing one cell nucleus with another is tricky. It had first been successfully carried through fifteen years earlier by the American biologists Robert William Briggs (b. 1911) and Thomas J. King.

In 1967 the British biologist John B. Gurden applied this technique of a *nuclear transplantation* to transferring a cell from the intestine of a South African clawed frog to an egg cell of another individual of the same species. From that ovum, with its alien nucleus, a normal new individual developed—a clone of the one from which the nucleus was taken. This was the first clone produced of a vertebrate.

Amphibian ova are naked and unprotected, however. The ova of reptiles and birds are protected by shells, and the ova of mammals remain within the body. These require much more complicated techniques, and such cloning has not yet been achieved.

HAHNIUM

In 1967 the formation of element number 105 was reported in the United States, and it was named *hahnium,* after Otto Hahn (see 1917).

IN ADDITION

Egypt, Syria, and Jordan, all armed by the Soviet Union, attacked Israel on June 5, 1967. Israel, in six days (hence the *Six-Day War*), defeated them all.

The war in Vietnam continued as the United States vigorously bombed North Vietnam, which only increased protests within the United States. The general air of discontent also led to riots in the black ghettos of many cities.

The United States population topped 200 million in 1967. The Soviet Union had a population of 240 million.

1968

ELECTROWEAK INTERACTION

There are four known particle interactions: strong, weak, electromagnetic, and gravitational. These seem to account for all the events taking place in the observable universe, and they are completely different from one another. The electromagnetic and gravitational interactions are both long-range fields, with the electromagnetic enormously the stronger of the two. However, the electromagnetic interaction exhibits both attraction and repulsion, and these two properties tend to balance each other. The gravitational interaction, on the other hand, exhibits attraction only; it dominates the Universe as a whole, while the electromagnetic interaction dominates atomic and molecular structure.

The strong and the weak interactions are both short-range fields, making themselves felt only across nuclear distances, with the strong interaction much the stronger (as the name implies). The gravitational interaction affects all particles with mass, while the electromagnetic interaction affects only electrically charged particles, the strong interaction affects only hadrons, and the weak interaction is most prominent in connection with leptons.

The question arose among physicists of why there should be four different interactions. Was there not, perhaps, some way of showing that all four were but different aspects of one basic interaction, just as ice, water, and steam, despite their differences in properties, are nevertheless all aspects of the same "water substance"?

Einstein had tried to find a mathematical treatment that could cover both gravitational and electromagnetic interactions (at a time when the strong and weak interactions had not yet been discovered) and had failed.

In 1968, however, three men, the American physicists Steven Weinberg (b. 1933) and Sheldon Lee Glashow (b. 1932) and the Pakistani physicist Abdus Salam (b. 1926), independently worked out a mathematical treatment that included both the electromagnetic and weak interactions. Careful observations bore out the theory and showed that the two are, indeed, a single interaction at high enough temperatures. It is only as the temperature drops that the two aspects separate (as dropping temperature causes steam to liquefy to water, and then freeze to ice, with all three capable of existing simultaneously at appropriate temperatures and pressures).

As a result of this working out of what is called the *electroweak interaction,* Weinberg, Glashow, and Salam shared the Nobel Prize for physics in 1979.

SOLAR NEUTRINOS

Neutrinos had been detected but only in the form of antineutrinos produced by nuclear fission reactors (see 1956). The existence of antineutrinos made the existence of neutrinos themselves certain; still it would be useful to detect them directly.

The Sun produces its energy by the fusion of hydrogen to helium (see 1929). In the process, vast quantities of neutrinos are produced, some of which reach the Earth and a few of which would interact, under appropriate conditions, with detecting devices. This would prove their existence.

Frederick Reines, who with Cowan had first detected the antineutrino, now tried to detect neutrinos from the Sun. For the purpose, he set up a huge tank containing 100,000 gallons of tetrachlorethylene in a deep mine in South Dakota. There was enough rock and earth above the tank to absorb all radiation from the sky other than neutrinos. The tank was then exposed to neutrinos from the Sun for several months. Each neutrino that was absorbed by a chlorine atom in the tetracholorethylene would be converted to an argon atom, which could eventually be flushed out with helium.

By 1968 evidence of the existence of solar neutrinos was definitely obtained, but there were not enough of them. Calculations seemed to show that the Sun was actually producing, at most, only one-third of the neutrinos that it ought to be producing if current theories of nuclear activity at the Sun's core were correct. This *mystery of the missing neutrinos* has concerned astronomers ever since.

ASTROCHEMISTRY

When the hydroxyl group had been detected in interstellar gas clouds (see 1963), astronomers had been surprised. It seemed odd that enough individual atoms would strike each other and cling, forming two-atom combinations like the hydroxyl group, to be detectable at astronomic distances. It was thought there would be virtually no chance for combinations of three or more atoms.

The increasing ability to detect microwave radiation with great precision, however, led to further surprises. In 1968 microwave frequencies characteristic of water molecules (with three atoms each) and ammonia molecules (with four atoms each) were detected in interstellar gas clouds. This was the beginning of what came to be called *astrochemistry*. Since then, more and more complicated atom groupings have been detected, some involving as many as thirteen atoms.

All but the very simplest are composed of chains of carbon atoms, which once again points up the uniqueness of the carbon atom as a component of complex groupings, and therefore of life as we know it.

ROTATING NEUTRON STARS

Pulsars had been discovered (see 1967), and now the problem was to explain what produced pulsations in the range of seconds. Something had to be revolving, rotating, or pulsating at that rate, and nothing could be doing so that quickly on a cosmic scale unless it was simultaneously very small and very massive.

In 1968 Thomas Gold (b. 1920) suggested that pulsars were neutron stars (whose existence had been suggested by Zwicky—see 1934) and that they were rotating. Neutron stars, which could be as massive as ordinary stars but would be only some 14 kilometers or so across, would be composed of packed neutrons. They would have enormously intense magnetic fields, so that charged particles would be emitted only at the magnetic poles, and they would give off radiation as they followed curved paths. The beam of radiation would sweep past us once each revolution, and the revolution would take place in seconds.

If this were so, the pulsars would be losing substantial amounts of energy as they turned, and their periods of rotation (and therefore the frequency of their pulsations) should slowly be lengthening. Close observation showed that this was indeed the case, and the identification of pulsars as rotating neutron stars was accepted.

CIRCUMNAVIGATION OF THE MOON

On September 17, 1968, the Soviet probe *Zond 5,* with no crew aboard, circumnavigated the Moon. On December 24, 1968, the American probe *Apollo 8,* with three astronauts aboard—Frank Borman (b. 1928), James A. Lovell, Jr. (b. 1928), and William A. Anders (b. 1933)—circumnavigated the Moon ten times. The stage was finally set for a lunar landing.

IN ADDITION

On January 30, 1968, the day of the Vietnamese New Year celebration, or *Tet,* the Vietcong (those South Vietnamese fighting the American-backed government) and North Vietnamese launched an offensive against thirty South Vietnamese cities. While American forces managed to hold most of their ground and recapture what was briefly lost, the offensive made it clear that the American government had not been telling the truth about the progress of the war. Antiwar demonstrations reached a peak. On March 31 Johnson announced that he would not run for reelection, and the bombing of North Vietnam quickly stopped. On April 4 Martin Luther King, Jr., was assassinated, and on June 6 Robert Francis Kennedy (1925–1968), the younger brother of John F. Kennedy, was assassinated. That fall, Richard Milhous Nixon (b. 1913) was elected thirty-seventh president of the United States.

On March 16, 1968, American troops killed hundreds of civilian men, women, and children in the Vietnamese village of My Lai. News of the My Lai massacre was suppressed for nearly two years.

In Czechoslovakia, Soviet forces of occupation prevented the government's attempt to liberalize itself.

1969

HUMAN BEINGS ON THE MOON

At 4:18 P.M. eastern daylight savings time on July 20, 1969, Neil Alden Armstrong (b. 1930) and Edwin Eugene Aldrin, Jr. (b. 1930) brought the lunar module of *Apollo 11* to the surface of the Moon, while Michael Collins (b. 1930) remained in orbit about the Moon. Neil Armstrong stepped out, the first human being to set foot on any world other than the Earth, saying "That's one small step for a man, one giant leap for mankind." John Kennedy's goal of reaching the Moon by the end of the decade had been reached.

The two men remained on the Moon for 21 hours 37 minutes and returned to Earth safely at 12:51 P.M. eastern daylight savings time on July 24, eight days after takeoff. A second American ship landed on the Moon in November 1969, and astronauts remained on the Moon's surface for 15 hours.

COSMONAUT TRANSFER

On January 14, 1969, two Soviet spacecraft, each carrying cosmonauts, met in space, and cosmonauts passed from one ship to the other. It was the first time human beings had transferred from one spacecraft to another in flight and was another advance in human mobility in space.

OPTICAL PULSARS

Pulsars (see 1967), which had been shown to be rotating neutron stars (see 1968), ought, perhaps, to give off photons of all energies. There ought to be pulses not only of microwaves but of visible light, for instance. However, small light pulses are not as easily detected as microwave pulses are, and light, being more energetic, would not be given off in the intensities of microwaves.

It made sense, then, to look in places where neutron stars might have formed comparatively recently, since they would be particularly energetic. In particular, the Crab nebula was a good possibility. The light of its supernova explosion had reached Earth only a little over nine centuries previously (see 1054 and 1848), and the pulsar at its center had the shortest period of rotation (1/30 of a second) then known. That meant it was energy-rich.

In January 1969, sure enough, a star near the center of the Crab nebula was found to be blinking on and off thirty times a second, in time to the microwave pulses. It was the first *optical pulsar* to be detected. The Crab nebula pulsar was found to be emitting pulses of X rays as well.

ANTARCTIC METEORITES

One of the difficulties in studying meteorites is identifying them in the first place. A lump of nickel-iron that has clearly not been the result of metallurgical smelting can only be a meteorite. However, less than 10 percent of the meteorites that reach Earth are of nickel-iron, and many of those that fell in the past were collected by people who made use of iron, so that no iron meteorites are to be found at all in those areas where civilization has existed the longest.

The majority of meteorites, however, are stony in nature and not to be distinguished from ordinary rocks without careful analysis. Since one can't go about analyzing every rock, stony meteorites are not recognized unless they are actually seen to fall or unless they land in areas where surface rocks do not occur.

The ideal place for a stony meteorite to land and be at once detected is on an extensive ice cap. There, any rock sitting on top of an icy thickness of a mile or more *must* be a meteorite. Such objects are most likely to be found on the vast Antarctic ice cap, by far the largest in the world.

In 1969 a group of Japanese geologists came across nine closely spaced meteorites on the Antarctic ice cap. These roused the interest of scientists generally, and thousands of meteoric fragments have been found since, which have made it possible to study the subject in greater detail. There is some reason to believe, from delicate chemical analyses, that a few fragments may have reached us from the Moon and some perhaps even from Mars.

PROTEIN STRUCTURE

The techniques for elucidating protein structure had continued to advance since Sanger's work on the structure of insulin (see 1952). In 1969 the American biochemist Gerald Maurice Edelman (b. 1929) worked out the structure of a gamma globulin, a type of protein that exists in the blood and out of which various antibodies are formed. (Antibodies react with particular foreign proteins, so that they are essential to the body's immune mechanism.) For this, Edelman received a share of the Nobel Prize for physiology and medicine in 1972.

Also in 1969, D. C. Hodgkin (see 1955) completed our knowledge of the insulin molecule by working out its three-dimensional structure.

And still in 1969, the Chinese-born American biochemist Choh Hao Li (see 1943) synthesized the enzyme ribonuclease, putting every one of its 124 amino acids into a chain in the right order. Ribonuclease, which catalyzes ribonucleic acid's breakdown into smaller fragments, was the first enzyme to be synthesized.

ARTIFICIAL HEARTS

In comparison with most living organs, the heart is simple. It is primarily a pump designed to push the blood through the vessels of the circulatory system. It is not difficult to imagine that an artificial pump the size of the heart and similar in structure but powered from without might do the job.

The first attempt to place such an artificial heart inside a human being was made in 1969. The American surgeon Denton Cooley implanted a plastic heart designed by the Argentine-born American Domongo Liotta. The patient lived for nearly three days with the artificial heart before it was replaced with a transplanted natural heart.

CORONARY BYPASS

The heart receives blood from the coronary arteries, which branch off from the aorta near the point where the aorta leaves the heart. In other words, the very first share of the blood, as it emerges from the heart on its way to the body generally, is fed to the heart itself. There seems justice here, since the heart's labor on behalf of the body is both enormous and essential.

Unfortunately, the coronary arteries have a tendency to accumulate rough plaques on their inner surface, plaques that are rich in cholesterol. (This is particularly true when people eat too much of a cholesterol-rich diet.) These plaques narrow the bore of the arteries and allow less blood to reach the heart. The roughness also increases the likelihood of clot formation, which may stop the blood flow altogether.

Starving the heart of blood causes the severe pains of *angina pectoris*, and of course any serious stoppage causes a *heart attack* and death. A perfectly healthy heart can be immobilized because of such coronary mishaps.

In 1969 a surgical technique was developed of using veins or sometimes arteries from the patient's own body to lead the blood around the clogged portions of the coronaries, renewing the heart's blood supply. If more than one such portion is bypassed, we speak of a *double bypass*, a *triple bypass*, and so on.

Since 1969 *coronary bypass* operations have become extremely common, and while they do not necessarily lengthen life, they make what remains of it much more pain-free and also make it possible to indulge in exertion freely again. This is a great boon to many.

IN ADDITION

American troops in Vietnam reached a peak of nearly 550,000, yet victory seemed as far off as ever. Withdrawals began in the face of the ever-increasing disenchantment of the American public. The North Vietnamese president, Ho Chi Minh (1890–1969), died on September 3, 1969.

1970

BLACK HOLE EVAPORATION

Black holes (see 1916), it seemed, could only gain matter, never lose matter. If so, they were destined to grow indefinitely and would, in the end, consume all the matter of the Universe.

In 1970, however, the British physicist Stephen William Hawking (b. 1942) reasoned from quantum mechanical considerations that black holes might have a temperature. Therefore, if surrounded by an environment with a lower temperature, black holes would evaporate. Massive black holes with the mass of a star or of many stars would evaporate so slowly that they would endure for many, many times the present age of the Universe. As their mass decreased, however, their rate of evaporation would increase.

The view was developed, then, that the final status of the Universe would not be a collection of black holes but a thin expanding mélange of leptons and photons originating from evaporated black holes.

METEORITIC AMINO ACIDS

The Sri Lanka–born American biochemist Cyril Ponnamperuma (b. 1923) was continuing to attempt to produce molecules of biochemical interest from the primordial constituents of Earth's atmosphere, in the line of experimentation begun by Miller (see 1952).

This investigation into life's origins took an unusual turn in 1970 when Ponnamperuma studied a meteorite that had fallen in Australia the year before. It was of a rare kind called a *carbonaceous chondrite,* a fragile black material that contained measurable quantities of water and organic material. Ponnamperuma showed that five different amino acids of the kind that helped make up protein molecules were present in the meteorite.

These meteoritic amino acids did not originate in living tissue, for if they had (judging from the amino acids in living tissue on Earth), they would all have only one of two possible structural arrangements and would rotate the plane of polarized light (be *optically active*). The amino acids in the meteorite were optically inactive, meaning that they consisted of both structures in equal amounts, so that each canceled the optical activity of the other. This was to be expected if they had been formed by some process not involving life.

Combined with the findings of astrochemistry (see 1968), this made it seem ever more likely that chemical changes took place in nonliving systems, where conditions were favorable, that would inevitably lead in the direction of life.

GENE SYNTHESIS

Khorana, who had worked on the genetic code (see 1961), headed a research team that in 1970 succeeded in synthesizing a genelike molecule from scratch. That is, they did not use an already existing gene as a template but began with nucleotides and put them together in the right order.

Further exemplifying the strides made in the synthesis of complicated molecules, Li, who had synthesized the enzyme ribonuclease (see 1969), synthesized the still more complicated molecule of growth hormone in 1970.

RECOMBINANT DNA

In 1970 the American microbiologists Hamilton Othanel Smith (b. 1931) and Daniel Nathans (b. 1928) discovered an enzyme that could cut a molecule of DNA at certain specific sites. The resulting DNA fragments were still large enough to contain genetic information, and this work led to the formation of fragments that could recombine with each other to form new genes that did not exist in nature.

This technique of *recombinant DNA* became an important tool for geneticists and was a long step toward *genetic engineering,* in which genes could be modified, transferred, or designed.

For this work, Nathans and Smith shared the Nobel Prize for physiology and medicine in 1978.

REVERSE TRANSCRIPTASE

Ever since Watson and Crick had worked out the structure of DNA and shown how it could replicate itself (see 1953), it had been felt (and experimental work had supported the feeling) that genetic information flowed in a one-way fashion from DNA to RNA.

It usually turns out, however, that nature is more complicated than expected, and there are apparently loops in the flow that occasionally carry information from RNA back to DNA. In 1970 the American oncologist Howard Martin Temin (b. 1934), in his investigation of cancer cells, located an enzyme he called *reverse transcriptase,* which could affect the working of DNA in line with information received from RNA, thus making the DNA more responsive to the needs of the cell.

This same discovery was made independently by the American biochemist David Baltimore (b. 1938). Temin and Baltimore received shares of the Nobel Prize for physiology and medicine in 1975.

MEGAVITAMIN THERAPY

The necessity of vitamins in the diet had been recognized since the work of Eijkman (see 1896). The dosages required appeared to be small, however—in fact enzymatic in quantity.

Then it was suggested that the tiny recommended doses were merely those required to prevent the onset of serious disease and that much higher doses were common in primitive human diets that were heavy on fruits and vegetables. It was these much higher doses that might be needed for full health, and people began to speak of *megavitamin therapy.*

An outstanding proponent of this view was Linus Pauling (see 1931), who, beginning in 1970, recommended massive doses of vitamin C (ascorbic acid) for health. His views seem not to be accepted by most biochemists, but Pauling's voice is not one that can be lightly ignored.

FIBER OPTICS

Since current electricity had come into use (see 1800 and 1831), metallic wires, especially those of copper, had been used to conduct the current wherever it was needed.

By 1970 techniques had been developed to conduct light by means of fine, very clear glass fibers. The fibers were coated with plastic or with a second type of glass so chosen that any light that tended to travel out of the fiber into the coating would be totally reflected. In this way, light could follow the fiber around curves and corners. With the use of lasers, such light could be as easily modulated as electric currents, so that sound waves could be converted into light of varying amplitude and, at the other end, reconverted into sound waves.

Fiber optics, by replacing expensive copper with cheap glass and by using the tiny waves of light, which can carry enormous amounts of information, was instantly seen as having the potential of greatly extending communication by telephone.

SCANNING ELECTRON MICROSCOPE

In an ordinary electron microscope (see 1932), an electron beam, in a vacuum, passes through the sample being studied and leaves an imprint on the recording device beyond. The sample must be very thin if this is to work.

If a low-energy beam of electrons is used, however, it can scan the surface of a sample much as an electron beam scans the picture tube of a television set. The electrons will induce the surface to emit electrons of its own, and these induced electrons can be detected.

Such a *scanning electron microscope* produces a three-dimensional effect that gives more information about the surface and produces still greater magnifications than an ordinary electron microscope can do. In some cases it can even show the position of individual atoms.

The first practical scanning electron microscope was built in 1970 by the British-born American physicist Albert Victor Crewe.

PLANETARY SOFT LANDING

On August 17, 1970, the Soviet Union launched *Venera 7*, a Venus probe that reached the planet on December 15. An instrument package was dropped into the atmosphere and made a soft landing. This was the first soft landing of any human-made object on another planet. The instruments sent back information about the atmosphere and surface for 23 minutes before the extreme conditions of temperature and pressure destroyed it.

In 1970 the Soviet Moon probe *Lunik 17* landed without a crew on the Moon and returned safely to Earth. China and Japan each launched Moon rockets of their own in 1970.

APOLLO XIII

A near disaster in space took place in 1970 when *Apollo XIII*, en route to the Moon, underwent a loss in oxygen in the main chamber. The three astronauts crowded into the lunar module and maneuvered their way safely back to Earth while the world watched. Although the mission was a failure, the ingenious survival of the intrepid astronauts caught the admiration of all.

SUPERSONIC TRANSPORT

Once the *sound barrier* had been broken (see 1947) it became possible to build commercial jet planes that would routinely carry passengers at speeds greater than sound.

In 1970 such *supersonic transport* (usually abbreviated *SST*) came into use. The United States decided not to build them for reasons of noise and environmental damage, but Great Britain, France, and the Soviet Union did build them. Although they have worked well technologically, they have never proved a commercial success.

IN ADDITION

On April 30, 1970, President Nixon ordered troops into Cambodia to search for North Vietnamese "sanctuaries" (which were never found). This announcement caused college campuses in the

United States to explode in protest. On May 4, the National Guard fired into a crowd of protesting students at Kent State University in Ohio, killing four and wounding eight.

The Egyptian ruler Gamal Abdel Nasser (1918–1970) died on September 28, 1970, and was succeeded by Anwar Sadat (1918–1981).

In Syria a military coup on November 13, 1970, placed Hafiz al-Assad (b. 1928) in power. In Libya a military coup on January 16, 1970, placed Muammar Muhammad Al-Qaddafi (b. 1942) in power.

Charles de Gaulle died on November 9, 1970.

1971

MAPPING MARS

On May 30, 1971, the United States launched the Mars probe *Mariner 9,* and on November 13, 1971, it arrived at Mars and went into orbit, the first human-made object to be placed into orbit about another planet.

Mars was experiencing a planetwide dust storm as *Mariner 9* approached, but fortunately it was possible to have the probe study its small satellites. They were irregular potato-shaped bodies, with craters as "eyes." The longest diameter of Phobos was 17 miles, that of Deimos, 10 miles.

Eventually, when the dust storm died down, *Mariner 9* was able to take more than seven thousand photographs of Mars, which served to map it completely. There were no canals, althought there was a huge canyon stretching for thousands of miles. It was named *Valles Marineris.*

There were numerous craters, crowded mostly into one hemisphere, with volcanoes and jumbled terrain in the other. The largest volcano, *Olympus Mons,* reached a height of 15 miles above base level and had a base width of about 250 miles.

The atmosphere was only about one-hundredth the density of Earth's and consisted almost entirely of carbon dioxide. The temperature was too low for liquid water to exist at any time, and the ice caps of Mars may contain both frozen water and frozen carbon dioxide.

MOON ROCKS

Exploration of the Moon continued in 1971. *Apollo 14* reached the Moon on February 5, 1971, and its crew collected 98 pounds of Moon rocks that were brought back to Earth for analysis. They were the first samples of material collected by human beings on another world.

On July 30, 1971, *Apollo 15* landed on the Moon. It carried with it a *lunar rover,* a land vehicle designed to travel on the airless Moon. The astronauts traveled 17 miles using it and brought back more Moon rocks to Earth.

SPACE STATION

On April 19, 1971, the Soviet Union placed *Salyut 1* in orbit. It was the prototype of a space station intended for long-time habitation by relay teams of cosmonauts.

In July, however, three Soviet cosmonauts were found dead on board when the rocket *Soyuz 11* returned to Earth, because of loss of air from the cabin. It was the worst space disaster up to that point.

BLACK HOLE DETECTION

In 1971 an X-ray-detecting satellite found irregular changes in an X-ray source in the constellation of Cygnus, a source that had been named *Cygnus X-1.* Such irregular changes might be the result of matter circling a black hole in varying concentrations.

Cygnus X-1 was at once investigated with great care and found to exist in the immediate neighborhood of a large, hot, blue star about thirty times as massive as our Sun. The Canadian astronomer C. T. Bolt showed that this star and Cygnus X-1 were revolving about each other, and from the nature of the orbit, Cygnus X-1 had to be five to eight times as massive as our Sun. If Cygnus X-1 were a normal star, it would be easily visible. Since it was not, it must be a small, very dense object. It was too massive to be a neutron star, so it must be a black hole. Though this is not a clear-cut and absolute identification, astronomers are by and large satisfied that it is a black hole.

Since then, black holes have been observed, by similar indirect and not entirely reliable means, to exist in the centers of various galaxies, including perhaps our own.

MINI-BLACK HOLES

Hawking had proposed that black holes slowly evaporated and that the evaporation increased as mass decreased (see 1970). In 1971 he pointed out that at the time of the big bang, black holes of all sizes might have been created.

Some would be so small that the evaporation rate would be high enough to cause the last bit of explosive evaporation to take place right now, some 15 billion years after formation. Such mini-black holes might be common, and their existence could be proved from the specific characteristics of that final explosion.

The concept is an attractive one, but so far no astronomer has detected anything that could be interpreted as the final explosive evaporation of a mini-black hole.

POCKET CALCULATORS

In 1971 Texas Instruments placed on sale the first calculator that was easily portable. Making use of transistorized circuits, it weighed only 2½ pounds and cost merely $150. In subsequent years, both the weight and the cost decreased dramatically.

IN ADDITION

The *Pentagon Papers,* a classified discussion of the involvement of the United States in the Vietnam War, were leaked to the press by Daniel Ellsberg (b. 1932), who had formerly worked for the Defense Department. The documents showed how the government had deceived the American people, and this revelation further increased popular opposition to the war.

In March 1971 East Pakistan rebelled and, with Indian help, had established its independence from West Pakistan by the end of the year. It called itself Bangladesh.

On October 25, 1971, the United Nations voted to allow mainland China into the United Nations and to expel Taiwan.

1972

VITAMIN B-12 SYNTHESIS

Woodward (see 1944), who had spent his career synthesizing more and more complicated nonpolymeric molecules, finally spent ten years attempting the synthesis of perhaps the most complicated nonpolymeric molecule of them all—vitamin B-12. In 1972, he succeeded.

PUNCTUATED EVOLUTION

Since the development of Darwin's notion of evolution by natural selection (see 1858), the general feeling had been that evolution was a slow process but a steady one.

In 1972 this view was challenged by the American paleontologists Stephen Jay Gould and Niles Eldredge, who suggested what they called *punctuated evolution.* In this view, species are stable and persist virtually unchanged for a long time. Then, certain small groups of a particular species, subjected for some reason to a special environmental pressure, change comparatively rapidly and develop into a new species. Evolution, then, is a matter of stable situations punctuated by occasional periods of rapid change.

This view has not yet been generally accepted, but it is characteristic of the present-day turmoil in the matter of biological evolution. No biologist of any standing doubts that evolution has taken place, but some facets of the exact mechanism of evolution remain under dispute.

SPEED OF LIGHT

The first useful approximation of the speed of light had been made by Olaus Roemer (see 1675). Since then, measurements had become more and more precise, culminating for a while in Michelson's measurements (see 1927).

In October 1972, however, a research team headed by Kenneth M. Evenson, working with a chain of laser beams in Boulder, Colorado, obtained a figure for the speed of light that was far more precise than anything previously reached. He measured the speed as 186,282.3959 miles per second.

EARTH RESOURCES SATELLITES

In 1972 the United States launched *Landsat I,* the first satellite specifically designed to take large-scale photographs of Earth that would make it possible to study global resources. Not only did it give an overview of geological data, but it was capable of studying forest- and grain-growing areas, yielding data on normal and abnormal growths, plant disease, and so on.

Such Earth resource satellites are one of the many answers to those who ask why so much money and effort is expended on space when there are so many problems on Earth that are crying out for study. Space technology is a powerful tool for studying those problems.

The chief Soviet rocket achievement of the year was the Moon probe *Luna 20,* which, though without a crew, reached the Moon, made a soft landing, scooped up a sample of soil, and brought it safely back to Earth.

QUANTUM CHROMODYNAMICS

By now it was well established that quarks (see 1961) combine two at a time (a quark and an antiquark) to form mesons and three at a time to form protons, neutrons, and other hadrons.

Murray Gell-Mann (see 1953), who had originated the quark concept, labored to work out the rules governing the combination of quarks. He suggested that each quark came in three colors: red, blue, and green. (These colors are not to be taken literally but as an analogy.) Just as in the case of light, a red, a blue, and a green quark will combine to give a lack of color property (white). Only those combinations yielding white can exist.

In this way, Gell-Mann founded the study of *quantum chromodynamics,* on the model of quantum electrodynamics, which had proved to work so well (see 1948). Quarks, however, are much more complicated in their behavior in connection with the strong interaction than electrons are in connection with the electromagnetic interaction, and quantum chromodynamics is still being tinkered with.

CAT SCAN

X rays had been used in medical diagnosis for three-quarters of a century, but in all that time they had been used only to obtain a two-dimensional photograph of a three-dimensional body.

In 1972 a technique was introduced called *computerized axial tomographic scanning* (CAT scan), in which numerous X-ray "stills" were taken in such a way that they could be put together to form a three-dimensional image.

In another medical advance, the British surgeon John Charnley devised the first satisfactory plastic replacement for the fitting of the thighbone into the hip socket, in 1972, thus preventing crippling through joint degeneration.

LASER DISKS

Since the phonograph had been invented (see 1877), sound had been reproduced through the vibration of a needle running along a groove. Eventually, of course, both needle and groove wore out, so that sound reproduction became imperfect.

In 1972 *laser disks* (also called *compact disks*) became practical. Here the sound was picked up by a laser beam, which translated it into information recorded on flat disks in the form of microscopically small pits. These could then be picked up by other laser beams. There was no question of wear, more sound could be packed onto a given surface, and reproduction was nearer perfection than ever before.

IN ADDITION

In the United States, in what seemed to be an unimportant incident, five men were arrested in an attempt to burglarize the national headquarters of the Democratic party in the Watergate apartment complex. It was clear from the start that there were important political figures behind the five burglars. Nixon ran for reelection and won by a landslide.

Although the United States intensified its efforts in Vietnam, the South Vietnamese army was crumbling everywhere.

Elsewhere in Asia, Okinawa was returned to Japan after twenty-seven years of American occupation. Ceylon became a republic and changed its name to Sri Lanka.

In the Philippine Islands, Ferdinand Edralin Marcos (b. 1917) became a dictator, with enthusiastic American support.

A civil war had been simmering between Catholics and Protestants in Northern Ireland for three years. After a particularly bloody clash on January 30, 1972, Great Britain took over direct rule of the region, but the civil war continued.

1973

JUPITER PROBE

On March 2, 1972, a Jupiter probe, *Pioneer 10,* had been launched—the first probe intended to yield information concerning the outer Solar System. After passing safely through the asteroid belt, *Pioneer 10* reached the vicinity of Jupiter on December 3, 1973, and passed only 85,000 miles above Jupiter's surface, going right through the planet's magnetosphere.

Jupiter's magnetic field, forty times as energetic as Earth's, made itself felt at a distance of 4,300,000 miles from the planet.

From the data obtained by the probe, it was possible to build a picture of the planet's structure. It would seem that Jupiter is a ball of hot liquid hydrogen mixed with some helium (a constitution much like that of the Sun).

The temperature rises rapidly with distance beneath the visible cloud surface. At 600 miles below, it is already 3,600° C; at 1,800 miles below, it is 10,000° C; at 15,000 miles below, it is 20,000° C; and at the very center of Jupiter, it is 54,000° C. Below 15,000 miles, hydrogen takes on a metallic form.

Pioneer 10 carried a message from Earth etched into a 6- by 9-inch gold-plated aluminum slab. It showed a man and woman next to an outline of *Pioneer 10* drawn to scale. Also included were details of the Solar System and its location in the Universe relative to distant quasars.

SKYLAB

The first American orbiting object that might be considered a space station was *Skylab.* It was 118 feet long and was launched into orbit on May 14, 1973, about 270 miles above Earth's surface. On May 25, three astronauts were carried to Skylab and remained on it for twenty-eight days. A second crew remained for sixty days, and a third for eighty-four days. Surveys were taken of Earth's mineral resources and its crops and forests. Photographs of the Sun were also taken.

ORIGIN OF THE UNIVERSE

Scientists had come to accept the big bang as the manner in which the Universe had come into being, but that left one crucial question unanswered. Granted that all the matter in the Universe was originally compressed into a comparatively tiny body that expanded into the present Universe, where did that originally tiny body come from?

In 1973 the American physicist Edward P. Tryon pointed out that what we ordinarily think of as a vacuum is not truly a vacuum. It can give rise to subatomic particles that disappear before they can be detected, in accordance with quantum mechanics and the uncertainty principle.

He suggested that if we start with an infinite sea of nothingness, particles will appear and disappear. Every once in a while, a particle may appear that can devlop the mass of the Universe and begin to expand before it can disappear. The Universe may therefore be a random *quantum fluctuation* in a vacuum, so that it originated out of nothing.

The implications and the detailed development of such a universe have been argued over by astronomers ever since.

GENETIC ENGINEERING

It is one thing to understand the fundamental chemistry of the DNA molecules that make up the genes; it is another to be able to modify that chemistry. In 1973 two American biochemists, Stanley H. Cohen and Herbert W. Boyer, showed that when DNA was broken into fragments and these were combined into new genes (see 1970), the new genes could be inserted into bacterial cells, where they could be reproduced whenever the cells divided in two.

This was the beginning of *genetic engineering*. It offered a technique for something as simple and useful as modifying defective genes to make them normal, thus holding out the hope that genetic defects might someday be cured. It also offered the possibility of something as far-reaching as the ability to direct human evolution (with all the treacherous side effects that might involve).

PROTON DECAY

Since the successful unification of the electromagnetic and weak interactions (see 1968), physicists such as Glashow had been attempting to include the strong interaction as well under the umbrella of a single set of equations. This was being done, though it required several sets of modifications to make such a *Grand Unified Theory (GUT)* work, and it is still not entirely satisfactory today.

In 1973 Abdus Salam suggested that such a theory might imply that the proton was very slightly unstable.

Thus, in any group of protons, half might decay to positrons and neutrinos in perhaps 10^{33} years. That is a 1 followed by 33 zeroes, and that many years is a million trillion times the lifetime of our present universe to date. Still, given enough protons, a few might be found to decay in any given reasonably short period of time. Such proton decays have been searched for but have not yet been detected.

IN ADDITION

A cease-fire in Vietnam was agreed to by all parties on January 28, 1973, and the last American troops left Vietnam on March 29. After ten years of fighting and a death toll of about forty-six thousand, the United States had lost its first war.

Meanwhile, back in the United States, the Watergate burglary unraveled and implicated high government officials, including President Nixon, in a variety of unethical acts.

Chile's freely elected left-of-center president, Salvador Allende Gossens (1908–1973), was, with C.I.A. help, overthrown and killed by a military coup. General Augusto Pinochet Ugarte (b. 1916) became president, and at once clamped a repressive reactionary regime upon the nation.

Egypt and Syria attacked Israel on October 6, Yom Kippur, Israel's holiest day. Israel was caught by surprise, but after eighteen days, it was clearly winning the war, and a cease-fire was arranged.

As a side effect of the war, the Arab nations set up an oil embargo against the West.

1974

MAPPING OF MERCURY

Mariner 10 had been launched on November 3, 1973. On February 5, 1974, it passed by Venus just 3,600 miles above its cloud layer and then headed for Mercury. On March 19, 1974, it passed within 435 miles of Mercury's surface. It moved into an orbit about the Sun in such a way that it passed near Mercury a second and third time. On the third approach, it passed within 200 miles of Mercury's surface.

Mariner 10 confirmed Mercury's rotation rate and temperature and showed that it had no satellite and no significant atmosphere. It determined the diameter, mass, and density of Mercury with greater precision than had been possible before. In addition, it allowed about three-eighths of Mercury's surface to be mapped.

The photographs it took of Mercury showed a landscape that looked very much like that of the Moon. There were craters everywhere, the largest being 125 miles in diameter. Mercury is not as rich in "seas" as the Moon is. The largest one sighted is about 870 miles across and is called *Caloris* (heat). Mercury also has cliffs that are a couple of hundred miles long and about 1½ miles high.

In addition *Mariner 10* discovered that Mercury had a small magnetic field, about a hundredth as intense as the Earth's. This is puzzling, for Mercury does not rotate quickly enough to have a field, if current theories are correct.

FORMATION OF THE MOON

Over the course of the last century, three different types of suggestions had been made about the origin of the Moon. It was suggested first that the Moon was originally part of the Earth and pulled away as a result of centrifugal effect when the primordial Earth was spinning rapidly. However, Earth had never rotated rapidly enough to make such a pullaway possible.

Second was the thought that Earth and Moon had formed separately from the same swirl of planetesimals. But the Earth and Moon should then have much the same chemical composition, and they don't. Earth, for instance, has a large nickel-iron core while the Moon seems to have none at all.

Third was the thought that Earth and Moon were formed from different swirls of planetesimals, and at some time in the past the Earth captured the Moon. But the mechanics of such a Moon capture are difficult to work out.

These seemed the only three possibilities, and each one was so flawed that it looked as though the only way out of the mess was to decide that the Moon didn't really exist.

In 1974, however, the American astronomer William K. Hartmann suggested a fourth alternative. Suppose that in the early days of the Solar System, a planet the size of Mars (about one-tenth the mass of the Earth) had had a glancing collision with Earth. It would knock off part of the Earth's outer layers, which would coalesce into the Moon, while the colliding body coalesced with Earth. The nickel-iron cores of Earth and the colliding object would merge, while the Moon, formed from the outer layers of Earth, would lack such a core.

The suggestion was largely ignored, but eventually computer simulations of such a collision made the idea begin to look good.

At the moment, Hartmann's suggestion is preferred to any of the older ones.

LEDA

The twelfth satellite of Jupiter had been discovered among its four outermost satellites, circling at an average distance of 14,000,000 miles from Jupiter (see 1951). It was clear that if a thirteenth satellite existed, it would have to be smaller and dimmer than the others, for otherwise it would already have been discovered.

On September 10, 1974, the American astronomer Charles T. Kowall did detect a thirteenth Jovian satellite. It was also part of that outermost group, which now numbered five. This new satellite, eventually named Leda after one of Zeus's (Jupiter's) many loves in the Greek myths, seemed to be no more than 5 miles across.

FREON AND THE OZONE LAYER

Freon, which had been introduced by Midgley (see 1930), and similar compounds had first been used in air-conditioning and later in spray cans. The compounds contained chlorine and fluorine atoms attached to a carbon skeleton (chlorofluorocarbons) and seemed absolutely safe.

As such chemicals were released into the air by spray cans and eventually leaked out of air-conditioning units, they did not accumulate in such quantities as to present direct difficulties for any living organism.

On the other hand, some chlorofluorocarbons inevitably drifted upward into the atmosphere and there they encountered the ozone layer. Two American scientists, F. Sherwood Rowland and Mario Molina, pointed out that such chlorofluorocarbons had the potential for destroying the ozone layer, even if they were present in comparatively small amounts. And indeed, in recent years the ozone layer has been observed to be thinning.

As the ozone layer thins, more energetic ultraviolet light from the Sun will be able to reach Earth's surface, causing increases in the incidence of such problems as skin cancer and cataracts. Worse yet, the ultraviolet may be deadly to soil bacteria and ocean plankton, with incalculable effects on Earth's ecological balance.

TAUON

Up to this point, the leptons that were known included the electron, the electron neutrino, and their antiparticles and the muon, the muon neutrino, and their antiparticles. That amounted to eight leptons altogether.

In 1974 the American physicist Martin L. Perl found that when electrons and positrons (antielectrons) were smashed together at high energies, still a third variety of lepton was produced. This was named the *tau electron,* or in shorter version, the *tauon.* It presumably has a neutrino of its own, and there is surely an antiparticle for each, so that twelve leptons are now known.

The tauon is about 17 times as massive as the muon and about 3,500 times as massive as the electron. It is very unstable and lasts less than five-trillionths of a second before breaking down to a muon.

It may be that twelve leptons are all the leptons there are.

CHARMED QUARKS

While there may be only twelve leptons (see above), there are a large number of hadrons, beginning with the pion, which is the least massive hadron, through over a hundred more massive ones.

Leptons, however, are fundamental particles, which cannot be broken down into simpler ones (as far as we now know), while hadrons are composite particles and are made up of quarks. Quarks, like leptons, seem to be fundamental particles.

In 1974 three kinds of quarks were known: *u-quarks, d-quarks,* and *s-quarks.* (The letters stand for *up, down,* and *strange,* respectively, though sometimes the *s* is made to stand for *sideways* to match the other two.) However, theoretical considerations indicated that quarks ought to exist in pairs. Up-quarks and down-quarks were a pair; there ought to be a quark to serve as the pair of the strange-quark. Such a new quark was named a *c-quark* even before it was discovered, the *c* standing for *charmed.*

In 1974 the American physicist Burton Richter (b. 1931), using the enormous energies of the latest particle accelerators, produced a particle that, from its properties, had to include a c-quark in its makeup. Another American physicist, Samuel Chao Chung Ting (b. 1936), working independently, also produced a particle that had to contain a c-quark. The two shared the Nobel Prize for physics in 1976.

A third pair of quarks, the *t-quark* and the *b-quark* (which may stand for *top* and *bottom,* or for *truth* and *beauty,* depending on the level of whimsy), undoubtedly exist also. If so, there are twelve quarks (the ones I've mentioned and their antiparticles) to match the twelve leptons. This may be significant, though no one yet can explain why quarks and leptons should match each other in number, or why that number should be twelve.

IN ADDITION

The Watergate investigation continued, and the plight of the Nixon administration deepened. On March 1, 1974, seven important former White House officials were indicted. To escape impeachment and to preserve his pension, Nixon resigned on August 8, the first president ever to resign under fire. His vice president, Gerald Rudolph Ford (b. 1913) was sworn in as the thirty-eighth president of the United States. He gave Nixon a full pardon on September 8.

Turkey invaded the island of Cyprus on July 20, 1984, and divided it into Turkish and Greek portions.

Portugal agreed to give up its African colonies on July 27, 1974. Portugal was the first European nation to build up an overseas empire and the last to give it up.

1975

MICROCHIPS

Since transistors had first been developed (see 1948), they had been made steadily smaller, cheaper, and more reliable. By 1975 they had become so small, and the circuits upon them had been etched in so compact a manner, that they could be called *microchips.*

This meant that computers also could be made very small, very cheap, and very powerful. And this made possible personal computers, the first of which were introduced in

1975, foreshadowing the coming of word-processors and robots.

With the microchip, the computer was no longer suitable only for government and large industries. It began to invade the domain of the general public.

SURFACE OF VENUS

In 1975 the Soviet probes *Venera 9* and *Venera 10* managed to make successful soft landings on the planet Venus and to endure long enough to take photographs of a rocky surface. It was clear that sufficient light penetrated the cloud layer to make photographs possible.

ENDORPHINS

It was discovered in 1975 that the nervous system gives rise to compounds that alleviate pain. These consist of short chains of amino acids that seem to interact with the pain-receptors. Presumably, morphine and similar opiates work by mimicking the action of these substances, which are now called *endorphins.* The first part of the name indicates that they are "endogenously formed"; that is, formed within the human body. The second part of the name indicates their morphinelike action.

Endorphins may someday be used in pain control without the addictive qualities and other side effects of opiates.

IN ADDITION

On January 1, 1975, several important Watergate figures were convicted of obstructing justice.

The Middle East remained turbulent. There was civil war in Beirut, Lebanon, the Turks in Cyprus declared their portion of the island a separate nation, and King Faisal of Saudi Arabia (1906–1975) was assassinated on March 25. On the other hand, the Suez Canal was reopened to shipping.

Francisco Franco of Spain died on November 20, 1975, at the age of 82. On November 22, Spain accepted Juan Carlos (b. 1938) as king.

On October 9, 1975, Soviet scientist and dissenter Sakharov (see 1953) was the first Soviet citizen ever to win the Nobel Peace Prize.

1976

LIFE ON MARS

In 1975 two Mars probes had been launched by the United States, *Viking 1* on August 20 and *Viking 2* on September 9. Both went into orbit about Mars in mid-1976, and they took the best photographs of the Martian surface yet.

On July 20, 1976, *Viking 1* came down to the Martian surface at what would have been, on Earth, the edge of the tropical zone. Some weeks later *Viking 2* came down in a more northerly position. In coming down, they discovered that the Martian atmosphere, though chiefly carbon dioxide, was also 2.7 percent nitrogen and 1.6 percent argon.

The Martian surface was rocky, as Earth's surface is. The Martian surface, however, was richer in iron and sulfur and poorer in aluminum, sodium, and potassium. There was no sign of life on Mars on a scale visible to the eye.

The *Viking* probes were equipped to run experiments on Martian soil to see if any microscopic forms of life were present. The experiments were carried through with ambiguous results, but there was no trace of organic material in the soil, and this led astronomers to believe that certain lifelike responses to the experiments were the result of some odd chemical behavior of the soil.

There were signs of dry riverbeds, however, complete with tributaries. It may be that in ages past there was a reasonable supply of liquid water on Mars. If so, where did it go, and what led to such an extreme cooling of the planet?

PLUTO'S SURFACE

Pluto is a small planet, and it is ordinarily the most distant planet from the Sun by far. At perihelion, however, when it is closest to the Sun, it is a little closer to us than Neptune is. As luck would have it, Pluto had been approaching perihelion ever since it was discovered (see 1930), and by this time it could be seen from Earth about as well as it had ever been seen.

The spectrum of the light it reflected from the Sun was studied, and in 1976 it was decided that Pluto's surface was covered with frozen methane.

SYNTHETIC GENE

Khorana had synthesized a gene (see 1970). Now he went a step further and placed a synthetic gene inside a living cell. There it proved perfectly capable of functioning. This was the final proof (if any were needed) that the scientific conclusions about the structure of the gene were valid.

STRING THEORY

Suggestions were made in 1976 to the effect that, as the Universe cooled in the first instants after the big bang, flaws or creases would have appeared in the structure of space. These would form long one-dimensional *strings* containing huge masses, energies, and gravitational fields. Such thoughts persist, but observational evidence for the existence of such strings has not yet been obtained.

IN ADDITION

In the United States, Jimmy Carter (b. 1924) was elected the thirty-ninth president. This was the year that *Legionnaire's disease* attracted national attention. From July 21 to 24, an American Legion convention met in a Philadelphia hotel. By August 2, 180 of the attendees had come down with a pneumonialike disease, and 29 of them died. Apparently the germs for the disease had grown in the air-conditioning system of the hotel.

On September 9, 1976, Mao Tse-tung of China died.

1977

RINGS OF URANUS

On March 10, 1977, the planet Uranus moved in front of a ninth-magnitude star in the constellation of Libra. This occultation was observed by the American astronomer James L. Elliot from an airplane that took him high enough to minimize the distorting and obscuring effects of the lower atmosphere. The idea was to observe how the starlight dimmed as the atmosphere of Uranus approached the star and moved in front of it. That would yield information about the atmosphere of Uranus.

Some time before Uranus reached the star, the starlight suddenly dimmed and brightened several times. When Uranus had passed the star and it emerged, the same dimming-brightening pattern occurred in reverse. Apparently Uranus was surrounded by a series of thin concentric rings that were opaque enough to obscure starlight.

In this way, Saturn's position as the unique possessor of rings was lost, although nothing we know of can duplicate its rings, which are so large, bright, and beautiful.

CHIRON

On November 1, 1977, Kowall, the discoverer of Jupiter XIII (see 1974), was studying photographic plates, searching for asteroids in Jupiter's orbit. He detected something that might have been such an asteroid, but it was moving at only one-third the speed that would have associated it with Jupiter's orbit. It had to be somewhere in the neighborhood of Uranus.

Eventually Kowall worked out its orbit and found that it followed a markedly elliptical path that carried it out as far as Uranus at one end and as close as Saturn at the other, completing one turn about the Sun every 50.7 years. It seemed to be an asteroid, but it was the farthest one yet discovered, and by far. It was named Chiron, after the most famous centaur in Greek mythology.

INFLATIONARY UNIVERSE

The concept of the big bang did not answer all questions. Why should the Universe be divided into stars and galaxies if it started as a tiny, unimaginably hot blob of matter? Why didn't the gas expand into a vast cold blob of undifferentiated matter? Again, why should the Universe be so nearly flat? That is, why should it have a mass that put it so nearly at the borderline between the possibility of expanding forever and the possibility of someday starting to contract again?

In 1977 the American physicist Alan Guth suggested that, from the equations developed for the Grand Unified Theory (see 1973, Proton Decay), it could be argued that the Universe, in the instants after the big bang, underwent a sudden and exceedingly rapid inflation. (While objects within the Universe cannot move relative to one another more rapidly than the speed of light, the Universe as a whole could theoretically expand at any speed.)

This period of expansion can be used to explain some characteristics of the Universe that the classical big bang theory cannot, but the details are elusive, and the concept of an *inflationary universe* is still being worked on by cosmogonists.

VELA PULSAR

For nine years, the pulsar at the center of the Crab nebula had been the only optical pulsar known (see 1969). In 1977 a second optical pulsar was discovered, in the Vela nebula, which seems also to be the remnant of an old supernova explosion.

DEEP-SEA LIFE

In 1977 it was discovered that there were ocean vents, or "chimneys," that continually spewed hot water laden with minerals into the ocean. Bacteria could live in these surroundings by obtaining energy from the oxidation of sulfur compounds present in the spewings. Other life forms could live on the bacteria, so that at the other end of the food chain, large clams and tube-worms were supported.

Here was an entire society that lived on chemical change involving neither light nor photosynthesis. That such complex life forms could exist independently of photosynthesis had not been expected.

Independent research at this time discovered primitive forms of bacterial life that obtained energy by reducing carbon dioxide to methane. These *methanogens* live independently of oxygen.

Obviously bacterial life is tremendously versatile.

LUCY

In 1977 the American paleontologist Donald Johanson discovered a hominid fossil that was perhaps four million years old. Enough bones were dug up to make up about 40 percent of a complete skeleton. It was an australopithecine (see 1924), about 3½ feet tall. Its scientific name was *Australopithecus afarensis,* but since it was clearly the skeleton of a female, it was named, whimsically, Lucy.

The most interesting thing about Lucy is that she was obviously bipedal. From her hip and thighbones it is clear that she walked erect as easily as we do. It may be that this bipedal locomotion was the first adaptation that served to differentiate organisms that are more humanlike (hominids) from those that are more apelike (pongids).

NONBACTERIAL DNA

Bacterial DNA, which is easy to work with, is densely packed with meaningful genes. In other words, every part of bacterial DNA can be used as a blueprint for the synthesis of proteins. This is not surprising, since bacterial cells are much smaller than the cells of plants or animals, and there is no space to waste within them.

In 1977 it was discovered that the DNA in nonbacterial cells is *not* densely packed with genes. Much of the DNA molecule consists of stretches of nucleotides that appear to be meaningless and in any case are not used to manufacture proteins. Nonbacterial cells are large and can afford such waste, but it is hard to believe that they would. Nevertheless, if the "nonsense" portions of nonbacterial DNA have significance and purpose, what it is has yet to be discovered.

SMALLPOX AND AIDS

In 1977 the last case of smallpox was recorded, in Somalia. The smallpox virus is now thought to be extinct except for samples grown in laboratories for research purposes.

The ending of one scourge, however, seemed to have been balanced by the coming of another. In 1977 two male homosexuals in New York City were found to have a rare form of cancer, which was eventually recognized to be a symptom of a disease called *acquired immune deficiency syndrome,* usually abbreviated *AIDS.* This disease, usually fatal and so far incurable, spread rapidly and became as feared in the 1980s as smallpox was in the 1780s.

FIBER OPTICS

Fiber optics (see 1970) were used for the first time in experimental telephone setups and worked. Within a decade, they were being used in transatlantic cables.

BALLOON ANGIOPLASTY

Coronary atherosclerosis could now be treated by performing bypass operations (see 1969). An alternative, nonsurgical technique was developed in 1977. This called for tiny balloons being led into the affected arteries by means of catheters. The balloons then expanded and pressed back the plaques, widening the bore of the arteries. Such *balloon angioplasty* slowly became more popular as an alternative to bypass operations.

IN ADDITION

Menachem Begin (b. 1913), a conservative hardliner, became Israel's prime minister on June 21, 1977. Astonishingly enough, this heralded a rapprochement with Egypt.

1978

RADAR MAPPING OF VENUS

The United States launched *Pioneer Venus* on May 20, 1978, and it went into orbit about Venus on December 4, 1978. It sent several probes into Venus's atmosphere and found that the cloud layer contained droplets of sulfuric acid, that about 2.5 percent of the sunlight striking the cloud layer penetrated to the surface of the planet, and that the atmosphere was 96.6 percent carbon dioxide and 3.2 percent nitrogen. In view of its density, that meant Venus's atmosphere had more than three times as much nitrogen as Earth's.

Venus Pioneer also beamed radar waves at Venus, and from the reflections, details of the surface (otherwise invisible from outside the cloud layer) could be determined.

The surface of Venus does not appear to be broken into plates as Earth's is and is mostly of the type we associate with continents. It seems to have a huge supercontinent that covers about five-sixths of the total surface, with the remaining sixth a lowland that may once have contained water.

In the north is a large plateau named *Ishtar Terra,* about as large as the United States. On the eastern portion of the plateau is a mountain range. There is another and even larger plateau in the equatorial region called *Aphrodite Terra.* It too has mountains.

There are also canyons and what may be extinct volcanoes.

CHARON

On June 22, 1978, the American astronomer James W. Christy, examining photographs of Pluto, noted a distinct lump on one side. He checked other photographs and found that the lump shifted position. Finally he decided it was a satellite located some 12,500 miles from Pluto. (At Pluto's distance, this is not much of a separation as seen from Earth, which accounts for the long-delayed discovery.)

Christy named the satellite Charon, after the ferryman who took shades across the River Styx to Hades in Greek mythology.

Charon circles Pluto in 6.39 days, which is just the time it takes for Pluto to turn on its axis. The two bodies, Pluto and Charon, have slowed each other's rotational speed through tidal action until each perpetually presents one side to the other. They now revolve about a common center of gravity like two unequal parts of a dumbbell held together by a gravitational pull. (This is the only dumbbell-situation in the Universe that we have been able to observe so far.)

From the distance of separation and the time of revolution, it is possible to work out the total mass of the two bodies, which comes to about one-eighth the mass of the moon. Pluto is about 1,850 miles in diameter (far smaller than anyone had thought), and Charon is about 750 miles in diameter. Charon has 10 percent the mass of Pluto, so that the two are the nearest thing to a double planet we know. (The Earth-Moon is second in this respect, but the Moon has only 2 percent the mass of the Earth.)

ONCOGENES

In 1978 the American scientist Robert A. Weinberg and his colleagues successfully produced tumors in mice by the transfer of individual genes. The genes concerned were called *oncogenes,* the prefix being the one commonly used in medical terminology for "tumor."

The oncogene was found to be very similar to a normal gene. The two might differ, in fact, in a single amino acid along the chain. The picture therefore arises of a normal gene (a *protooncogene*), which is replicated with every cell division and, through some accidental change during one of the replications, becomes an oncogene.

VIRUS GENOME

Once the nature of genes had been determined (see 1944 and 1954), it became a dream of molecular biologists to work out the structure of all the genes of an organism (the *genome*).

In 1978 the first step was taken in this direction when the genome of a virus called SV40 was determined. Of course, viruses are the simplest living organisms and have the smallest genomes, but this points the way, ultimately, to the working out of the human genome.

TEST-TUBE BABY

On July 25, 1978, a normal baby was born in Great Britain of an egg that had been fertilized by a sperm in the glassware of a laboratory rather than within the mother's body. Such a *test-tube baby* offers a way for couples to have children when, for one reason or another, fertilization within the body cannot take place.

IN ADDITION

Israel and Egypt came to an agreement to foreswear war. This took place at Camp David, the presidential retreat in Maryland, under the eyes of President Carter.

Soviet cosmonauts set an endurance record of 96 days in space on March 16, 1978, and then set a new one of 139 days on September 2. Less happily, the fragments of Soviet satellite *Cosmos 1954,* which carried a load of uranium-235, fell to the Canadian Arctic on February 24, 1978, raising concerns about radioactive materials in orbit about the Earth.

1979

JUPITER'S SATELLITES

The probes *Voyager 1* and *Voyager 2* passed by Jupiter in March and July, respectively, of 1979. The most interesting result was that they gave humanity its first close look at the four Galilean satellites.

The satellites are subject to tidal heating from Jupiter, a heating that increases rapidly as distance from Jupiter decreases. Thus, Ganymede and Callisto, the outermost pair, are both covered with craters as might be expected and seem to be made up largely of icy material.

Europa, second-closest to Jupiter, is *not* cratered—it has, in fact, the smoothest solid surface yet seen in the Solar System—but cracked and criss-crossed with lines (rather like the old maps of Mars that showed the supposed canals). Apparently Europa is covered with a worldwide glacier that has liquid underneath. Any cosmic collision that might ordinarily cause a crater merely cracks the glacier, and the hole produced is then repaired by the formation of additional ice.

Io, the closest of the four satellites to Jupiter, has no water at all, the heating having driven it off. In fact, the inner heating produces active volcanoes, which the probes photographed in actual eruption. Sulfur dioxide exudes and breaks up into sulfur and oxygen. Io's surface is therefore covered by a yellow to red layer of sulfur, which fills in any but the most recent of craters. The eruptions also account for the thin gas that fills Io's orbit, forming a doughnut around Jupiter.

In addition to all of this, three more small satellites were discovered, all of them closer to Jupiter than any that had been discovered from Earth. This made sixteen satellites all told for Jupiter. Finally, very close to Jupiter, a thin ring of debris was found, so that Jupiter joined Saturn and Uranus in being a ringed planet.

EXTINCTION OF THE DINOSAURS

In 1979 the American scientist Walter Alvarez was trying to establish sedimentation rates in old sedimentary rocks in Italy. To do that, he made use of neutron-activation techniques that allowed him to determine with great precision the quantities of various rare elements present in the rocks.

He found to his surprise that a certain narrow layer of the rock contained some twenty-five times as much of the rare metal iridium as there was above and below it. The narrow layer that contained the iridium was sixty-five million years old and was therefore right at the boundary where the Mesozoic era yields to the Cenozoic era. At that boundary, dinosaurs and many other species of plants and animals had become extinct with surprising suddenness.

For years scientists had puzzled over that extinction and a variety of explanations had been offered, none of which had proved really satisfactory. Now it seemed to Alvarez that the excess of iridium couldn't be a coincidence. It had to have some connection with the "great dying," and it could only have come from some external body. Most of Earth's own iridium was in its iron core, so that the surface rocks were extremely short of it. A meteor, or even a comet, would be far richer in iridium than Earth's crust is.

Alvarez postulated, therefore, that a large asteroid or comet, several miles across, had struck Earth sixty-five million years ago, producing volcanic eruptions, tidal waves, fires, and so on. In addition, it may have splashed so much dust into the upper atmosphere that the radiation of the Sun was cut off for an extended period of time. All these varieties of disaster could have brought about enormous extinctions. (In fact, the litany of disaster makes it hard to see how any life at all could survive.)

This explanation for the extinction of the dinosaurs was met with considerable skepticism at first, but the evidence has been piling up since, and many now accept it.

GLUONS

Quarks are held together by the strong interaction (see 1935). This must involve an exchange particle (see 1935). Just as the electromagnetic interaction involves exchanges of photons, so quarks must undergo exchanges of a particle called, for obvious reasons, *gluon*.

Quarks are held in combination so tightly, however, that they have never been isolated as free particles. Gluons, too, are not easily isolated. In 1979, however, energetic subatomic-particle interactions produced some rudimentary indications of the production of gluons.

IN ADDITION

The Shah of Iran, Muhammad Rezi Pahlavi (1919–1980), went into exile on January 16, 1979. Ayatollah Ruholla Mussaui Khomeini (1900?–1989), the fanatical Shiite cleric who had been most instrumental in engineering the Shah's downfall from his exile in Paris, returned to Iran on February 1 and was in control of the government by February 11. Because the United States had been a strong supporter of the Shah, Iran turned anti-American, and when the dying Shah came to the United States for medical treatment, Iranian militants seized the American embassy in Tehran. On November 4, 1979, everyone in the embassy was taken hostage. The United States was helpless.

Anastasio Somoza, the corrupt dictator of Nicaragua (who had also been supported by the United States), was driven into exile on July 17, 1979. The new government resented American support of the previous government and became increasingly anti-American. In neighboring El Salvador, the right-wing government (supported by the United States) was having to fight increasingly militant rebels.

An increasingly intense civil war in Afghanistan between Moslem fundamentalists and groups supported by the Soviet Union led to an invasion by Soviet troops on December 27, 1979.

Egypt and Israel signed a formal peace treaty on March 26, 1979.

The United States and China established full diplomatic relations on January 1, 1979.

On May 3, 1979, Margaret Thatcher became the Conservative prime minister of Great Britain, the first woman to hold the post in British history.

1980

SATURNIAN SYSTEM

The probe *Voyager 1* passed by Saturn on November 12, 1980. *Voyager 2* followed not long after. A number of the satellites of Saturn were for the first time seen as more than points of light.

Titan, the largest satellite, was known to

have an atmosphere of methane, but methane turned out to be present in small quantities compared to nitrogen. (Nitrogen is a gas difficult to detect from Earth because its absorption characteristics are not easy to observe and study.) The Titanian atmosphere turned out to be 98 percent nitrogen and 2 percent methane and may be thicker than Earth's atmosphere. The atmospheric haze prevented any view of the surface, where there might be nitrogen lakes with dissolved polymers of methane and (a few speculate) possibly some form of life.

The other Saturnian satellites were, as might be expected, cratered. Mimas, the innermost of the nine sizable satellites, has a crater so large that the impact that produced it must have nearly shattered it.

Enceladus, the second of the nine, is comparatively smooth, while Hyperion is the least spherical and has a diameter that varies from 90 to 120 miles. Iapetus is a two-toned satellite, with one hemisphere much darker than the other, as though one side were icy and the other coated with dark dust. The reason for this is not yet clear.

The Saturn probes succeeded in finding eight satellites that were too small to be seen from Earth, bringing the total number to seventeen. Of the new satellites, five are closer to Saturn than Mimas is. Two satellites that are just inside Mimas's orbit are unusual in being *co-orbital*. That is, they share the same orbit, chasing each other around Saturn endlessly. This was the first known example of such co-orbital satellites.

The three new satellites beyond Mimas also represent unprecedented situations. The long-known satellite Dione was found to have a tiny co-orbital companion, *Dione B*, which circles Saturn at a point 60 degrees ahead of Dione. As a result, Saturn, Dione, and Dione B are always at the apices of an equilateral triangle. This is a comparatively stable gravitational position, called a *Trojan situation* because it is also the position of the Sun, Jupiter, and the Trojan asteroids (see 1906).

The satellite Tethys has two tiny companions, one 60 degrees ahead of it in orbit and one 60 degrees behind it. Clearly, the Saturnian satellite system is the richest and most complex in the Solar System.

The Saturnian rings were also found to be far more complex than had been thought. From a close view, they consist of hundreds, perhaps even thousands, of thin ringlets, which look like the grooves on a phonograph record. In places, dark streaks show up at right angles to the ringlets, like spokes on a wheel. Then too, a faint outermost ring seems to consist of three intertwined ringlets. None of this can be explained so far. It may be that a straightforward gravitational situation is being complicated by electromagnetic effects.

NEUTRINO MASS

In 1980 Frederick Reines (see 1956) reported on experiments that indicated neutrinos might have tiny quantities of mass (for years they had been thought to be massless).

Researchers in Moscow reported similar results in an entirely different experiment and thought the neutrino mass might be 1/13,000 of an electron.

If this were so, then the three neutrinos—the electron-neutrino, the muon-neutrino, and the tauon-neutrino—might well have slightly different masses and be capable of oscillating; that is, of continually changing from one to another.

If this were so, it would explain the fact that the Sun seems to emit only one-third the number of neutrinos it ought to. It should be emitting electron-neutrinos only, and detecting devices would only pick up electron-neutrinos. But if the neutrinos oscillated, the electron-neutrinos emitted by the Sun might change into muon-neutrinos and tauon-neutrinos en route to Earth and arrive here in equal quantities. The detecting devices would then pick up only the one-third that were electron-neutrinos.

Then, too, even if the neutrinos had only a tiny mass, there were so many neutrinos in the Universe that their total mass might make up a hundred times as much mass as everything else put together. The presence of this mass, ignored until now by astronomers, might explain how galaxies rotate, how clusters of galaxies hang together, and how the galaxies formed in the first place. All these things would make sense if the "mystery of the missing mass" were solved by attributing it to neutrinos.

Furthermore, the presence of missing mass in the form of neutrinos would be just sufficient to close the Universe; that is, to make certain that the Universe would someday begin to contract again.

The only trouble with all this is that the question of the neutrino's mass has not yet been confirmed, and it may well be a false alarm.

IN ADDITION

The invasion of Afghanistan by the Soviet Union met with general world disapproval.

The Iranians retained their American hostages all through the year, and an attempt by the United States to mount a rescue mission turned out to be a fiasco in which eight Americans died. The hostage crisis devastated the Carter presidency, and on November 4, 1980, Ronald Wilson Reagan (b. 1911) was elected fortieth president of the United States.

On September 22, 1980, war broke out between Iraq and Iran, and Iraq made large initial gains. The war quickly settled down to a long stalemate, however.

The population of the United States was now 226.5 million, having gone up 11.4 percent since 1970. World population was over 4 billion.

1981

SPACE SHUTTLE

Until now, all space vessels had been one-time operations, not reusable. It was clear that space exploration could be made more feasible if it were made less expensive by creating reusable vessels.

For that reason, the space shuttle was designed. Its purpose was to go into orbit and then return to Earth. It was not in itself designed to make spaceflight cheap; it was an expensive vessel. However, it would help engineers work out the techniques for developing a future generation of such vessels that would be cheaper.

The first shuttle flight took place on April 12, 1981, which happened, by coincidence, to be the twentieth anniversary of the first spaceflight, by Gagarin (see 1961). The shuttle left and returned safely. It was the first of over a score of such flights during the next four and a half years to be carried through safely.

NEPTUNE'S RINGS

The rings of Uranus had been discovered by the behavior of a star's light, which disappeared and reappeared as Uranus approached it and then left it (see 1977).

In 1981 similar phenomena were noted when Neptune passed in front of a star. But the pattern had been symmetrical as the star passed into and out of occultation with Uranus, and in the case of Neptune it was not.

The suggestion was therefore made that Neptune did not have symmetrical rings but merely arcs of ring material that did not stretch all the way around the planet. If so, this is another unprecedented situation.

IN ADDITION

Iran finally freed the United States hostages on January 20, 1980, after 444 days of imprisonment.

Anwar Sadat of Egypt was assassinated while watching a military parade on October 6, 1980.

1982

MILLISECOND PULSAR

The fastest pulsar (see 1967) known was in the Crab nebula and rotated some 30 times a second, emitting pulses at that rate. It was also the youngest pulsar known, and there didn't seem to be much chance of finding a pulsar still younger and more rapid.

In 1982, however, a pulsar was found that rotated more than twenty times as fast, at 642 times per second. Moreover, it did not show signs of being a particularly young pulsar. There was a suggestion that it was (or had been) part of a double star system and that, by absorbing matter from its companion, it had gained angular momentum and speeded up.

Other such *millisecond pulsars* (those that turn in the range of a thousandth of a second) have been discovered since.

MAGNETIC MONOPOLE

Maxwell's equations (see 1865) are not quite symmetrical with respect to electricity and magnetism. The asymmetry lies in this: Electricity exists as positive and negative charges, and these can be easily isolated—there are particles with positive charge only (positrons or protons) and particles with negative charge only (electrons or antiprotons). Magnetism, however, exists as north and south poles that do *not* seem to exist separately. Objects possessing magnetism always have both a north and a south pole. If an object could be found that was *only* a north pole or *only* a south pole *(magnetic monopoles)*, then Maxwell's equations could be made completely symmetrical.

By the Grand Unified Theory (see 1973, Proton Decay), it would seem that magnetic monopoles must exist but must be so monstrously massive that there could only have been enough energy to form them in the immediate instants following the big bang. Still, if they were formed then, they should still exist today, and scientists should be able to detect them.

Therefore, the physicist Blas Cabrera devised a setup that would produce an electric current if a magnetic monopole were to pass through it, and on February 14, 1982, it produced such a current.

That single detection has not yet been repeated, however, either by Cabrera or by anyone else, so the existence of the magnetic monopole remains in question.

JARVIK HEARTS

Artificial pumps to replace the human heart, at least temporarily, had been attempted (see 1969), but the best so far was devised by an American physician, Robert K. Jarvik (b. 1946). On December 1, 1982, a *Jarvik heart* was placed into the chest of a retired dentist, Barney Clark, who lived 112 days with it. Other patients were later similarly treated.

On the whole, though, the Jarvik heart was unsatisfactory. It required an outside energy source, so the patient was not freely mobile, and the quality of life was in general poor.

LASER PRINTERS

The first printers developed for word-processors were essentially automatic typewriters that typed about a line per second and were as noisy as typewriters.

In 1982 IBM put laser printers on the market. They printed silently and did some 30 lines per second.

IN ADDITION

On April 2, 1982, Argentina seized the Falkland Islands, one of the few scraps of the British Empire that yet remained. Reluctantly, the United States supported Great Britain, which landed forces on the islands on May 21, and forced an Argentinian surrender on June 15.

Israel handed back the Sinai Peninsula to Egypt on April 25, 1982, after occupying it for fifteen years. On the other hand, Israel invaded Lebanon, reaching the outskirts of Beirut before American pressure forced a halt.

In the Soviet Union, Leonid Brezhnev died and was succeeded by Yuri V. Andropov (1914–1984).

1983

W PARTICLES

The electroweak theory (see 1968) made it seem necessary for the weak interaction to involve three exchange particles, one positive (W^+), one negative (W^-), and one neutral (Z^0). They were massive particles, at least eighty times as massive as protons, which accounted for their being so elusive. It took a great deal of energy to form them.

By 1983, however, physicists had sufficient energy at their disposal to be able to detect all three particles. They did so and found that the particles had the predicted mass, too. That nailed the electroweak theory into place.

The experiments were planned and carried through by the Italian physicist Carlo Rubbia (b. 1934) and the Dutch physicist Simon van der Meer (b. 1925). They shared the Nobel Prize for physics in 1984.

EXTRASOLAR PLANETS

IRAS, a satellite designed to detect infrared radiation, picked up such radiation in the immediate neighborhood of the bright star Vega in 1983.

The best way of interpreting the radiation was to suppose that a ring of particles surrounded the star, rather like a rich asteroid belt. It was at once suggested that such a belt might be in the process of condensing into planets or be an indication that planets existed already.

In any case, this was the best evidence yet that stars other than the Sun might possess planets and that such *extrasolar planets* might even be common.

NUCLEAR WINTER

The speculative accounts of a comet strike on Earth sixty-five million years ago, which may have put an end to the dinosaurs, partly because a vast cloud of dust raised into the upper atmosphere by the strike had cut off sunlight (see 1979), inspired thoughts concerning nuclear war.

A group of people, of whom Carl Sagan (b. 1935) was most prominent, suggested that in the case of an all-out nuclear war, the explosion of thousands of nuclear bombs would raise enough dust to cut off sunlight for a considerable period and produce a *nuclear winter.* Mass starvation would then follow for the whole planet, not only for the loser of the nuclear exchange but for the "winner" as well, and for all the neutrals.

The predicted intensity of such a nuclear winter seems to have moderated a bit since the first pessimistic projections, but the side effects of a nuclear exchange in the form of fires, fallout, and disruption of the world's economy would surely be dreadful enough even if there were no nuclear winter at all.

IN ADDITION

On October 23, 1983, a car bomb on a suicide mission killed 241 United States Marines who had been sent to Lebanon to quiet the civil war.

On October 25, 1983, the United States Navy fell upon the microscopic island of Grenada and defeated six hundred Cubans.

On August 21, 1983, Benigno Aquino, a strong foe of the dictatorial Marcos regime in the Philippines, returned to his homeland and was shot to death the moment he got off the plane. Marcos appeared to be involved.

Menachem Begin resigned as Israeli premier and was succeeded by Yitzhak Shamir.

1984

DNA AND HUMAN EVOLUTION

DNA molecules change with time, forming mutations (see 1937). Presumably, if the DNA molecules of two different species are compared, then the more closely related the species are, the fewer the differences. And from the number of differences, one can tell, perhaps, the length of time it has taken the two species to differentiate from a common ancestor. Since the mutations are the result of chance changes, conclusions cannot be drawn with mathematical certainty, but they are suggestive.

In 1984 such DNA analysis was used to present reasons for supposing that human beings and chimpanzees were more closely related to each other, evolutionarily, than either was to gorillas or orangutans, and that human beings and chimpanzees diverged from a common ancestor some five to six million years ago.

BROWN DWARFS

In 1984 the red dwarf star Van Biesbroeck 8 was reported to have a still dimmer companion star. The companion was so small and dim that it seemed it could not be massive enough, or bright enough, to be shining by ordinary nuclear fusion. It was moderately heated to the point of just glowing, and it gave off radiation rich in infrared, by other types of nuclear reactions, perhaps.

If it were totally cold and didn't radiate in the visible region at all, it would be a *black dwarf*. Since it wasn't totally cold, it was called a *brown dwarf*. The existence of this particular brown dwarf was later disputed, but other examples were reported.

IN ADDITION

In the United States President Reagan ran for reelection and won.

In the USSR, Yuri Andropov (1914–1984) died on February 9 and was succeeded by Konstantin U. Chernenko (1911–1985).

Indira Gandhi of India was assassinated on October 31, 1984, and her son, Rajiv Gandhi (b. 1944) became prime minister in her stead.

1985

OZONE HOLE

A hole in the ozone layer over the Antarctic was detected by the British Antarctic survey, and the ozone concentration elsewhere was abnormally low. This was taken as disturbing confirmation of the deleterious effect of chlorofluorocarbons on ozone (see 1974).

PLUTO AND CHARON

The orbit of Charon about Pluto is so oriented that, at perihelion and aphelion, it alternately eclipses Pluto and is eclipsed by it for a period of some years. Naturally, this is much more important at perihelion, when Pluto is considerably nearer to us than at aphelion.

In 1985 Charon, which had been discovered near perihelion (see 1978), began to move in front of Pluto on the way to moving behind it. By studying the light of Pluto alone, when Charon was behind it, and the light from both bodies, when Charon was in front of it but not covering all of it, something could be deduced about the surfaces of the two bodies.

Thus, whereas Pluto's surface is frozen methane, the smaller Charon cannot hold on to methane. Its surface is frozen water.

IN ADDITION

In the USSR, Mikhail Gorbachev (b. 1931) was chosen to lead the country, which meant that for the first time in its history, it was headed by a young and vigorous person who was acquainted with Western ways and psychology.

1986

URANUS PROBE

On January 24, 1986, *Voyager 2* flew past Uranus and gave humanity its first close look at that distant planet discovered by Herschel (see 1781), as well as at its rings and satellites.

Uranus was found to rotate in 17.24 hours (previous estimates had been anywhere from 10 to 25 hours). It had, as expected, a magnetic field, but one that was tilted 60 degrees to the rotational axis.

Uranus's rings, detected from Earth nine years earlier, were confirmed. The five known satellites turned out to be somewhat larger than had been thought, and no fewer than ten small satellites were found to be circling at a smaller distance than that of Miranda, the already-known satellite nearest to Uranus.

Miranda was particularly astonishing. It was only 300 miles in diameter, so it could not have had enough internal heat to power geological change. Nevertheless, it had a veritable jumble of different surface features. It is thought that bombardment in its early days may have broken it into fragments that since then have coalesced every which way.

HALLEY'S COMET

Halley's comet returned for the third time since Halley had first worked out its orbit (see 1705). Its appearance in 1986 was unfortunately not a showy one, since it remained rather far from Earth even at its closest approach and could only be seen high in the sky from the southern hemisphere.

It was an unprecedented return, however, since it could now be studied by probes sent out by the Soviet Union and by the European Space Agency. The European probe, named Giotto (after the painter who had been the first to paint the comet realistically—see 1304), made the closest approach.

Comets had been pictured as "dirty snowballs" by Whipple (see 1949), and he was shown to be right, but Halley's comet turned out to be far dirtier than expected. While it lost ice each time it approached the Sun, it lost rocky particles to a much smaller extent and they accumulated on the surface, forming a kind of crust, through which the vapors formed by heated ice broke through at weak spots here and there. The result was that Halley's comet was dark black in color.

This meant it was larger than expected. Since it reflected so little light, a larger surface was required to produce its observed brightness.

IN ADDITION

On January 18, 1986, the United States space shuttle *Challenger* exploded in the first minute of its launch, killing the seven astronauts aboard. It was the first American loss of life in spaceflight, and since the United States was determined to do nothing more until the causes of the tragedy were worked out and measures were taken to prevent a repetition, the American space effort ground to a temporary halt.

In the Philippines, Marcos fled the country on February 22, and Corazon Aquino (b. 1933), widow of the assassinated Benigno Aquino, became president.

Olof Joachim Palme (1927–1986), the prime minister of Sweden, was assassinated on February 28, 1986.

In response to terrorist activity laid at the door of Libya, American planes bombed Tripoli, the Libyan capital, on April 14, 1986.

In the Soviet Union the worst nuclear accident to date took place at Chernobyl in the Ukraine when a nuclear reactor underwent a meltdown on April 28, 1986, spreading radioactive fallout over a broad area.

Toward the end of the year, it was revealed that the Reagan administration had sold arms to Iran in an effort to buy back hostages. It had then used the money to make illegal arms shipments to the Contras, who were fighting the Nicaraguan government at America's insistence. President Reagan pleaded ignorance.

1987

MAGELLANIC SUPERNOVA

The last supernova that had been visible in our own Milky Way Galaxy was one studied by Kepler in 1604. Since then, the nearest supernova had been one spied 2,300,000 light-years away in the Andromeda galaxy (see 1886), but at the time it was not known to be a supernova and was not carefully studied even with the instruments then available. Since then, the only supernovas to have been noted had been in still more distant galaxies.

In February 1987, however, a supernova was caught in its early explosive stage in the Large Magellanic Cloud. That was not in our own galaxy to be sure, but it was in the galaxy closest to our own, only 150,000 light-years away.

The explosion had been heralded by a spray of neutrinos, some of which were caught in recently devised *neutrino telescopes.* Undoubtedly, as more and better instruments of the sort are built, the sky will be regularly scanned for neutrinos that may herald supernova explosions.

The Magellanic supernova was carefully studied as the light built up and faded and as the nebulosity about it expanded and thinned. In 1988 the expected appearance of a pulsar at its center rotating two thousand times per second was reported.

WARM SUPERCONDUCTIVITY

Ever since superconductivity had been discovered by Kamerlingh Onnes (see 1911), scientists had tried to find materials that were superconductive at as high a temperature as possible in order to make super-efficient electrical conduction practical in everyday applications. All the metallic elements were studied, as were large numbers of alloys, but no superconductivity was found at temperatures higher than 23 degrees above absolute zero (23° K). This meant that superconductivity could only be experienced at liquid helium temperatures, and liquid helium is an expensive commodity.

Liquid hydrogen exists only at temperatures higher than 20° K, and the superconductivity transition point would have to be substantially higher than that to make it completely useful in a liquid hydrogen bath. Even if it could be done, liquid hydrogen, while less expensive than liquid helium, would surely give off hydrogen gas and create the danger of explosion.

Liquid nitrogen is much cheaper and easier to prepare and use than either liquid helium or liquid hydrogen. It is also quite safe to use. Unfortunately, it exists only at temperatures higher than 77° K, and that seemed hopelessly high.

In February 1987, however, a Swiss physicist, Karl Alex Mueller (b. 1927), and a German colleague, Johannes Georg Bednorz (b. 1950), studying ceramic substances (mixtures of metallic oxides) rather than metals, found they were obtaining superconductivity at temperatures of 30° K.

This broke with shattering impact on the scientific community. Everyone began studying ceramics, and superconductivity in the liquid nitrogen temperature range was obtained.

The difficulties that remain are that no good theoretical explanation of this astonishing behavior of ceramics exists, so that it is still a matter of "cookbook chemistry," as properties vary erratically from one ceramic to another and from one batch of a particular ceramic to another. Then too, ceramics are not easy to form into film or wire, so that much must be done before technological applications can be made.

Nevertheless, Mueller and Bednorz, without delay, shared the Nobel Prize for physics in 1987.

IN ADDITION

In the United States on October 19, 1987, the stock market lost 500 points. Thanks to lessons learned in 1929, the government was able to prevent the kind of disaster that followed the 1929 crash.

On December 8, 1987, Reagan of the United States and Gorbachev of the Soviet Union signed an agreement to eliminate nuclear weapons of intermediate range from Europe.

On May 17, 1987, an Iraqi missile mistakenly hit an American warship in the Persian Gulf. This led to an increased American presence in the gulf and to heightened American belligerence.

1988

DISTANT GALAXIES

New instruments and computerized techniques made it possible to detect galaxies with red shifts (see 1925) greater than any previously seen; greater even than those of quasars.

In 1988 some were detected that might be as much as seventeen billion light-years away. This was important with respect to the birth of the Universe. If the galaxies were seventeen billion light-years away, it had taken their light seventeen billion years to reach us, and we were seeing them as they existed seventeen billion years ago. This meant that even seventeen billion years ago the Universe was old enough to have formed galaxies.

The age of the Universe has not been determined precisely. It depends on knowledge concerning the distance of galaxies and the rate at which the Universe is expanding. These are uncertain quantities. The age of the Universe has been set at somewhere between ten and twenty billion years, with fifteen billion as the most probable.

If the distant galaxies are actually at the distance estimated, however, it would seem that the Universe must be older than we had thought. Again, any information we can get from those distant galaxies may tell us more about the formation and youth of galaxies, and that may change our ideas about how and when the Universe formed.

SHROUD OF TURIN

The Shroud of Turin is a linen cloth that seems to possess an image, front and back, of a bearded, long-haired man, which resembles the popular conception of Jesus. It was first displayed in France in the 1350s and in 1578 was taken to Turin, Italy.

Many people believed it to be the burial sheet of Jesus with the image produced by miraculous means. Others, more skeptical, assumed it to be a forgery produced not long before it was first shown.

In 1988 some of the linen was finally tested by the carbon-14 dating method worked out by Libby (see 1947). The results were clear. The linen had been part of living flax plants seven hundred years ago. The shroud, and presumably the image upon it, had been produced not long before it was first displayed and was thirteen centuries too late to have been the shroud of Jesus.

GREENHOUSE EFFECT

It had been known since Arrhenius (see 1884) had pointed it out that carbon dioxide in the atmosphere acted as a heat trap, making Earth's temperature warmer than it would otherwise be. (This was called the *greenhouse effect.*)

It was also known that the carbon dioxide content of the atmosphere had been rising steadily since 1900, partly because of the increasing use of coal and oil, which produce carbon dioxide when burned, and partly because of the cutting down of forests, which are the most efficient consumers of carbon dioxide.

As it happened, 1987 was the warmest year yet recorded by weather bureaus, and 1988 was warmer still. What's more, 1988 saw disastrous droughts in the United States and elsewhere. There was a feeling that the greenhouse effect was intensifying, and world concern began to intensify accordingly.

Rising temperatures would not only alter Earth's climate (probably for the worse) but could promote the melting of Earth's ice caps to produce a disastrous rise in sea level of up to 200 feet. The greenhouse effect, plus the thinning of the ozone layer, the steady rise in environmental pollution, and the inexorable increase in population, seemed to have placed the very habitability of our planet at risk, and a sense of crisis was beginning to pervade the world.

IN ADDITION

On May 14, 1988, the Soviet Union began to pull its troops out of Afghanistan.

On July 3, 1988, the United States mistakenly shot down a civilian Iranian airliner, killing 290 on board. When the world did not react with sufficient horror, Iran realized how isolated it had become. It agreed to a cease-fire, which at least temporarily ended the stalemated Gulf War on August 20.

George Herbert Walker Bush (b. 1924) was elected the forty-first president of the United States.

INDEX OF NAMES

A

B

C

D

E

F

G

H

I

J

K

N

Q

R

INDEX OF SUBJECTS

A

B

J

K

L

O

P

Q

R

U

X

Y

Z

ABOUT THE AUTHOR

ISAAC ASIMOV was born in the Soviet Union in 1920, was brought to the United States in 1923, and has been an American citizen since 1928. He was educated in the public school system of New York and obtained his B.S. in 1939, his M.A. in 1941, and his Ph.D. in 1948—all in chemistry, all from Columbia University.

He is professor of biochemistry at Boston University School of Medicine but hasn't worked at it since 1958 (nor is he paid).

He grew interested in science fiction in 1929 and sold his first story to a magazine in 1938. His first book appeared in 1950. Since his first sale he has written and sold 371 short pieces of fiction and some 3,000 nonfiction essays. He has published over 425 books of science fiction and nonfiction covering every branch of science and mathematics, history, literature, humor, and miscellaneous subjects.

He lives in New York City and is married to Janet Jeppson Asimov, a psychiatrist and also a science fiction writer. He has two children by an earlier marriage.